Pesticides 1995

Reference Book 500

Pesticides approved under The Control
of Pesticides Regulations 1986

London: HMSO

© Crown copyright 1995
Applications for reproduction should be made to HMSO
First published 1995

ISBN 0 11 242990 4

CONTENTS

	Page
Introduction	v
Chapter 1: How to use this book	vi
Chapter 2.1: The Legislation – The Food and Environment Protection Act 1985 and The Control of Pesticides Regulations 1986	xi
Chapter 2.2: The Legislation – Control of Substances Hazardous to Health Regulations 1988 (COSHH)	xv
Chapter 2.3: The EC Plant Protection Products Directive (91/414/EEC)	xvi
Chapter 2.4: The Biocides Directive	xviii
Chapter 3: The Approvals Process	xix
Annex A: Consents given by Ministers.	xxiv
Annex B: Products approved for use by aerial application.	xxx
Annex C: Off-label arrangements.	xlix
Annex D: Commodity substance approvals.	lvi
Annex E: Banned and severely restricted pesticides in the UK.	lxxxi
Annex F: Active Ingredients subject to the Poison's Law	lxxxiv
Annex G: List of ACP Published Evaluations.	lxxxv
Annex H: Products approved for use in or near water (other than for public hygiene and antifouling use).	lxxxvii
Annex I: List of authorised adjuvants	lxxxix

PART A: PSD Registered Products

Section 1: Professional Products	1
1.1 Herbicides	2
1.2 Fungicides	71
1.3 Insecticides	109
1.4 Vertebrate Control Products	125
1.5 Biological Agents	131
1.6 Miscellaneous	132

Section 2: Amateur Products 137
Section 3: Product Trade Name Index 175
Section 4: Active Ingredient Index 201

PART B: HSE Registered Products

Section 1: Wood Preservatives 215
Section 2: Surface biocides 299
Section 3: Insecticides 317
Section 4: Antifouling products 365
Section 5: Product Name Index 447
Section 6: Active Ingredient Index 473

INTRODUCTION

1. Pesticides 1995 contains lists of agricultural and non-agricultural pesticide products whose uses held either full or provisional approval in the UK as at **31 October 1994** for PSD products and HSE products. The exception is for those PSD products for which approvals were known on 1 November 1994 to be due to expire on or before 31 December 1994 and which therefore do not appear. Unless otherwise specified these products may be sold, supplied, stored, used and advertised *subject to the requirements of the pesticides legislation*, as explained in Chapter 2, and the conditions of their approvals, which are explained in Chapter 3.

2. Additional information on the products listed is available on product labels and literature; or from manufacturers and suppliers. You should read the product label before you buy a product or use it. *You have a duty to check that you know how to use a product safely in accordance with the statutory conditions of approval.* These approval conditions are set out on the product label and in the consents given by Ministers.

3. Further information on the approval status of pesticides may be obtained, in the case of PSD registered products listed in Part A, from the Pesticides Safety Directorate, External Relations Branch, Mallard House, Kings Pool, 3 Peasholme Green, York, YO1 2PX, or, in the case of HSE registered products listed in Part B, from the Health and Safety Executive, Pesticides Registration Section, Technology and Health Sciences Division, Magdalen House, Bootle, Merseyside, L20 3QZ.

1. HOW TO USE THIS BOOK

The introductory chapters give general guidance on the use of pesticides and the relevant legislative controls. The remainder of the text is divided into lists of approved products, which are indexed.

PART A

This contains products registered with the Approvals Group of the Pesticides Safety Directorate, an Executive Agency of the Ministry of Agriculture, Fisheries and Food (MAFF). It is divided into four sections:

Section 1: Contains products approved for professional use in agricultural, horticultural, forestry and other pest control areas. This is sub-divided into classification groups such as herbicides, fungicides, insecticides and vertebrate control products.

Within each of these sub-sections, approved products are listed by trade name under the appropriate alphabetically listed active ingredient along with details of the marketing company and the MAFF registration number.

For products with two or more active ingredients, the entry is repeated in respect of each, e.g.:
   2,4-D+Mecoprop and
   Mecoprop+2,4-D

Where there are three or more active ingredients, the secondary active ingredients in each entry are in alphabetical order, e.g.:
   Bentazone+Dichlorprop+Isoproturon
   Dichlorprop+Bentazone+Isoproturon
   Isoproturon+Bentazone+Dichlorprop

Products which have more than one area of use e.g. herbicide and fungicide, appear under all the appropriate activity headings.

Section 2: Products approved for amateur use feature in this section, although these products may also be used by professionals. The areas of use are detailed as part of the active ingredient heading.

Section 3: An alphabetical index of all active ingredient combinations with a cross-reference to the relevant part of each section by means of a code number.

Section 4: An alphabetical index of product trade names with a cross reference to the first entry of the product in each section by means of a code number.

Key to symbols

* * – some or all uses of this product are "approved for agricultural use" as defined by the Consents given by Ministers under Regulation 6 of the Control of Pesticides Regulations 1986. (For further information see Chapter 2.1, paragraph 9.)

* % – all approvals for advertisement, sale, supply and use of these products by the approval holder and his agents have been revoked; approvals for advertisement, sale, supply, storage and use by other persons, and for storage by the approval holder and his agents, remain for a limited period (up to two years).

* A – these products are approved for use in or near water as defined by the Consents given by Ministers under Regulation 6 of the Control of Pesticides Regulations 1986. (For further information see Chapter 3 paragraph 17.)

PART B

This covers products registered with the Health and Safety Executive (HSE) as wood preservatives, surface biocides, insecticides and antifouling products.

Within sections 1–4, relating to their area of use, approved products are listed below their active ingredients which are given a code number and listed alphabetically. In addition details of the marketing company, the definition of use code and HSE registration number are provided.

For products with two or more active ingredients, the entry is repeated for each active ingredient eg;

(1) 2-Phenylphenol and Cypermethrin
(2) Cypermethrin and 2-Phenylphenol

Products which have more than one area of use, e.g. wood preservation and surface biocide use, appear under each of the headings.

The area of use and user are indicated by the codes below:

Wood Preservatives
   WA – Amateur Use
   WP – Professional Use
   WI – Industrial Use

Surface Biocides
   SA – Amateur Use
   SP – Professional Use

Insecticides
   IA   – Amateur Use
   IP   – Professional Use
   IFSP – Food Storage Practice
   IAH – Animal Husbandry

Antifouling Products
   AQA – Aquaculture Use
   AVP – Professional Use (Deep Sea)
   AVA – Amateur Use (Yacht)

Amateur Use means that the product is freely available to the general public.

Professional Use means that the product can only be applied by operatives who are required to use pesticides during the course of their work, eg operatives employed by a remedial company.

Industrial Use means that the product can only be used in a factory situation.

Key to symbols

* these products have either been voluntarily withdrawn for commercial reasons or revoked and hold approval for storage, supply and use but not for advertisement and sale under The Control of Pesticides Regulations 1986.

\# these products no longer hold approval as they have been either revoked[1] for safety reasons or withdrawn for commercial reasons. As all conditions of approval have been revoked, there should be no products in this category on sale or being used.

Note[1] Excludes products revoked more than 5 years ago.

Active Ingredient Names

A number of active ingredients have several possible names. In order to compile the lists for sections 1–4, it has been necessary to standardise these names. Thus the active ingredient under which a product is listed may not be the same as that which appears on its label. These changes have been made for the purpose of Pesticides 1995 only and do not affect the name that should appear on the product label, which should be in accordance with the notice and schedule of approval.

The names that have been standardised are as follows:

Alkyldimethylbenzyl Ammonium Chloride	appears as **Benzalkonium Chloride**
Ammonium Alkyldimethylbenzyl Chloride	appears as **Benzalkonium Chloride**
Chromic Acetate	appears as **Chromium Acetate**
Copper (I)	appears as **Cuprous**
Copper (II)	appears as **Copper**
Didecyldimethylammonium Chloride	appears as **Dialkylmethyl Ammonium Chloride**
Diethoxy Cetostearyl Alcohol	appears as **Ceto-stearyl Diethoxylate**
Dioctyldimethylammonium Chloride	appears as **Dialkyldimethyl Ammonium Chloride**
Disodium Tetraborate	appears as **Sodium Tetraborate**

Lindane	appears as **Gamma-HCH**
Mono-ethoxy Oleic Acid	appears as **Oleyl Monoethoxylate**
Parachlorometacresol	appears as **4-Chloro-meta-cresol**
Paradichlorobenzene	appears as **1, 4-Dichlorobenzene**
Pyrethrum/Pyrethrins/ Pyrethrum extract	appears as **Pyrethrins**
Tributyltinmethacrylate Copolymer	appears as **Tributyltinmethacrylate**

Hydrates – Hydrated and anhydrous substances are listed as the anhydrous form.

Part B is divided into six sections;

SECTION 1: Wood Preservatives

SECTION 2: Surface Biocides

SECTION 3: Insecticides

SECTION 4: Antifouling Products

SECTION 5: An alphabetical index of product names cross-referenced to the relevant active ingredient code number(s).

SECTION 6: An alphabetical index of all active ingredient combinations cross-referenced to the relevant products by means of code numbers.

2. THE LEGISLATION

2.1 *Food and Environment Protection Act 1985 (FEPA)* and *The Control of Pesticides Regulations 1986 (COPR)*

Aims

1. Statutory powers to control pesticides are contained within Part III of FEPA. Section 16 of the Act describes the aims of the controls as being to:
   i. protect the health of human beings, creatures and plants;
   ii. safeguard the environment;
   iii. secure safe, efficient and humane methods of controlling pests; and
   iv. make information about pesticides available to the public.

2. The mechanism by which these aims are to be achieved is set out in regulations made under the Act. The Control of Pesticides Regulations 1986 (SI 1986/1510) define in detail those types of pesticides which are subject to control and those which are excluded; prescribe the *approvals* required before any pesticide may be sold, stored, supplied, used or advertised; and allow for general conditions on sale, supply, storage, advertisement, and use, including aerial application, of pesticides to be set out in *Consents*. The Consents have been updated since the Regulations came into force and the texts current at the time of printing are annexed to this Chapter (Annex A). New regulations are to be made in 1995 to implement the EC Plant Protection Products Directive (91/414/EEC). Further details are at Sections 2.3 and 2.4. Similar legislation exists in Northern Ireland and the majority of products approved for use in Great Britain are subsequently approved for use in Northern Ireland.

Definition of pesticides

3. Under the Food and Environment Protection Act 1985, a pesticide is any substance, preparation or organism prepared or used, among other uses, to protect plants or wood or other plant products from harmful organisms; to regulate the growth of plants; to give protection against harmful creatures; or

to render such creatures harmless. The term *pesticides* therefore has a very broad definition which embraces herbicides, fungicides, insecticides, rodenticides, soil-sterilants, wood preservatives and surface biocides among others. A more complete definition and details of pesticides which fall outside the scope of the legislation is given in Regulation 3 of the Control of Pesticides Regulations 1986.

What does the legislation require?

4. Firstly, the law requires that only pesticides *approved* by Ministers in the six responsible Departments (being MAFF, the Department of Employment (HSE), the Department of the Environment, the Department of Health, the Scottish Office Agriculture and Fisheries Department, the Welsh Office Agriculture Department, plus (for Northern Ireland approvals) the Department of Agriculture for Northern Ireland) shall be sold, supplied, used, stored, or advertised. There are three categories of approvals;

   i. an *experimental permit*, to enable testing and development to be carried out with a view to providing data needed to secure a higher level of approval;

   ii. a *provisional approval*, for a stipulated period, where there are some outstanding data. Where a date has been set for the submission of data this is referred to as the data submission deadline (see paragraph 6 below); and

   iii. a *full approval*, for an unstipulated period: although these will be superseded by time limited approvals as the Plant Protection Products Directive comes into effect.

5. Products granted only an *experimental permit* cannot be advertised or sold. Such products do *not* appear in this book.

6. A data submission deadline, as the name suggests, is the deadline by which the Registration Authorities must be in receipt of any requested data required to secure continued approval. Failure to meet this deadline will result in revocation of the approval for advertisement, sale, supply and use by the approval holder and his agents. Approval for advertisement, sale, supply, storage and use, by other parties, and storage by the approval holder and his agents will remain for a limited period of up to two years. PSD Products subject to this "phased revocation" are indicated by the symbol "%".

7. Products granted *provisional* or *full* approval are normally allowed to be advertised, sold, supplied, stored and used. The exception is when a provisional approval is subject to a restriction on the area to be treated, or the quantity to be supplied. In such cases it is necessary to withhold approval for advertisement.

8. New approvals are set out in 'The Pesticides Register' available monthly from HMSO book shops. For more details see inside the back cover.

9. Products listed in Part A and approved for agricultural use (as defined in the Consents) are indicated by the symbol "*". This formal classification is important for deciding whether the person storing, selling, supplying or using the pesticide needs to hold a recognised certificate of competence. The certification requirements are set out in Consents B and C(i), at Annex A at the end of this chapter. A leaflet covering certification requirements may be obtained from Pesticides Safety Directorate, Post Approvals Policy Branch, Mallard House, Kings Pool, 3 Peasholme Green, York YO1, 2PX.

10. Products approved for amateur use (as listed in Part A, Section 2) and those registered with the HSE (as listed in Part B) do not attract storekeeper, salesperson or operator certification under COPR.

11. Controls over the use of adjuvants and the tank mixing of pesticides are set out in the Consents, which should be consulted. The list of authorised adjuvants is at Annex I and is periodically updated and published in the 'Pesticides Register'.

12. Anyone who advertises, sells, supplies, stores or uses a pesticide is affected by the legislation, including people who use pesticides in their own homes, gardens and allotments.

Access to Information on Pesticides

13. A wide range of information about pesticides is published or made available by the Government. The independent Advisory Committee on Pesticides publishes an annual report on its activities, a summary of each meeting and detailed evaluations of new or reviewed pesticide active ingredients, (a full list of Published Evaluations as at 20 January 1995 is at ANNEX G). The data underlying published evaluations are available for inspection at York, and HSE Bootle, on application (for addresses see page v). Measures for wider access to data on older pesticides (those cleared

under PSPS) have also been announced and will be introduced once Parliamentary approval has been obtained.

Full results of pesticide residues monitoring in food stuffs and water are now published annually, as are usage surveys and reports of suspected incidents of human and wildlife poisoning. News of policy and procedural matters, product approvals and revocations are published monthly in "The Pesticides Register". For more details see inside the back cover.

Codes of Practice

14. Government Departments have produced guidance to help those working with pesticides meet the requirements of the legislation. These are:

i) The MAFF 'Code of Practice for Suppliers of Pesticides to Agriculture, Horticulture and Forestry, which gives practical guidance on the storage and transport of pesticides and the obligations on those who sell, supply and store for sale and supply. Copies are available from MAFF Publications, London SE99 7TP, priced £1.50. (PB0091); and

ii) the MAFF/HSC "Code of Practice for the Safe Use of Pesticides on Farms and Holdings" which promotes the safe use of pesticides. It covers the requirements of the Control of Pesticides Regulations 1986 (COPR) and the Control of Substances Hazardous to Health Regulations 1988 (COSHH). It is issued as a pack with posters and leaflets to make sure farmers and farm workers know about the safest possible practices. Copies are available from HMSO book shops and can be ordered from HMSO Publications Centre (0171 873 9090) priced £5.00 (ISBN 0-11-242892-4).

iii) the HSC "Approved Code of Practice for the Safe Use of Pesticides for Non-Agricultural Purposes" which provides advice to those in the non-agricultural sector on compliance with COSHH priced £4.00 (ISBN 0-11-885673-1).

iv) the HSC "Recommendations for Training Users of Non-Agricultural Pesticides". priced £2.00 (ISBN 0-11-885848-4).

v) the HSE/DOE "Remedial Timber Treatment in Buildings - A Guide to Good Practice and the Safe Use of Wood Preservatives". Copies are available from HMSO book shops and can be ordered from HMSO Publications Centre (0171 873 9090) priced £4.00 (ISBN 0-11-885-9870).

vi) the HSC "Approved Code of Practice for the Control of Substances Hazardous to Health in Fumigation Operations" provides practical guidance on the COSHH 1988 priced at £3.00 (ISBN 0–11–885469–0).

The above HSC and HSE publications are now available by mail order from:

HSE Books
P O Box 1999
SUDBURY
Suffolk
CO10 6FS

Tel. (01787) 881165

They can also be purchased from Dillons Bookstores and ordered from Ryman the Stationer and Ryman Computer Centres (0171–434–3000), but are no longer available from HMSO.

2.2 The Control of Substances Hazardous to Health Regulations 1988 (COSHH)

1. The COSHH Regulations apply to a wide range of pesticides used at work. They lay down essential requirements and a step-by-step approach for the control of exposure to hazardous substances. Failure to comply with COSHH, in addition to possibly exposing people to risk, constitutes an offence and is subject to penalties under the Health and Safety at Work etc. Act 1974.

What is a substance hazardous to health?

2. Substances (including pesticides and other chemicals) that are "hazardous to health" include substances labelled very toxic, toxic, harmful, irritant or corrosive, substances with occupational exposure limits, harmful micro-organisms, and substantial quantities of dust. They also include fumes, gases and other materials which could cause comparable harm to health.

What does COSHH require in respect of pesticides used at work?

3. Under COSHH employers and the self employed must make a suitable and sufficient COSHH assessment before work starts. This should include:

   a. An assessment of the risks to health arising from all operations involving the pesticide or its container – based principally on the

product label and using information supplied under S.6. of the Health and Safety at Work etc. Act;
b. the steps which need to be taken to prevent exposure or to achieve adequate control of exposure;
c. identification of other action necessary to comply with requirements for the use and maintenance of exposure control measures, monitoring, health surveillance, and the training and instruction of operators.

4. Pesticides should only be used when necessary, and if the benefit from using them significantly outweighs the risks to human health and the environment.

5. If a decision is reached to use a pesticide, the product selected should be one that poses the least risk to people, livestock and the environment yet is still effective in controlling the pest, disease or weed problem that has been identified.

6. Employers must ensure that the exposure of employees to substances hazardous to health by any route such as inhalation, ingestion, absorption through the skin or contact with the skin is either prevented or, where this is not reasonably practicable, adequately controlled by a range of measures. In order of preference, these should be:
   a. substitution (e.g. using a less hazardous pesticide);
   b. technical or engineering controls;
   c. operational controls;
   d. personal protective equipment (PPE).

2.3 The Plant Protection Products Directive (91/414/EEC)

1. Council Directive 91/414/EEC concerning the placing of plant protection products on the market is expected to be implemented in 1995 by means of the Plant Protection Products Regulations and will affect those products that fall within its scope. COPR will continue to apply to those products not covered by this directive.

2. The Directive is intended to harmonise arrangements for authorisation of plant protection products within the Community, although product authorisation will remain the responsibility of individual Member States. A list of pesticide active substances will be assembled in Annex I to the Directive. This will be built up over a period of time as existing active substances are

reviewed (under a collaborative EC Review Programme) and new ones authorised. Member States will only be able to authorise the marketing and use of plant protection products whose active substances are listed in Annex I (but see paragraph 7 below). The Pesticides Safety Directorate will be the authority in Great Britain for product authorisations under this Directive.

3. Commission Regulation 3600/92[1] and 933/94 lay down detailed rules for the review of existing active substances. Each active substance will be allocated to a Member State for review. A number of reviews will be carried out each year, with reviews of all active substances scheduled to have commenced by 2003. Eighty nine reviews were begun in 1993/94. Every producer will be obliged to notify the Commission whether or not the review of any given active substance is to be supported. Within, generally, twelve months data will have to be submitted to the Member State carrying out the review. A list setting out which Member State is to carry out which review was published on 5 April 1994.

4. Following the evaluation of the submitted data the Member State will send its report to the Commission with a recommendation as to whether the active substance should be listed on Annex I or not, with any conditions to be attached, or whether the active substance should be suspended from the market pending the provision of further data, or whether a decision on listing should be deferred pending the provision of further data.

5. The recommendation of the Member State will then be considered by the Standing Committee on Plant Health and a decision taken.

6. On completion of an EC review of an active substance Member States will review each of the products that contain it. If, on completion of the review, a decision is taken not to list the active substance on Annex I all product approvals must be revoked.

7. Under transitional arrangements Member States may:
   a. provisionally approve products containing new active substances not on the market by 25 July 1993 in advance of Annex I listing for periods of up to three years;
   b. maintain existing national approvals (in the case of Great Britain under the Control of Pesticides Regulations) until 25 July 2003, or completion of the review of the active substance, whichever is earlier; and

[1] Published in the Official Journal of the European Communities L366 of 15 December 1992, available from HMSO, 51 Nine Elms Lane, London SW8 5DR

c. authorise new products containing active substances on the Community market on 25 July 1993 under existing national rules.

8. The Directive provides for a system of "mutual recognition" whereby a pesticide product authorised by one Member State in accordance with the provisions of the Directive and where the active substance is in Annex I will, subject to certain conditions, be authorised by other Member States. Authorisation may be subject to conditions set to take account of differences such as climate and agricultural practice.

9. Uniform Principles (Annex VI of the Directive) establishing common criteria for evaluating products at a national level were agreed in June 1994. This will enable harmonisation of the evaluation of applications for pesticide products by Member States.

10. Detailed consideration of the impact of the Directive on approval procedures in the United Kingdom is still under way. A new handbook replacing the COPR Handbook will be published setting out the amended procedures.

2.4 Proposed Biocidal Products Directive

This Directive is expected to apply to all pesticides which are not plant protection products as well as disinfectants, water biocides and certain industrial preservatives. The Directive, as proposed, would introduce an authorisation scheme broadly similar to the one introduced by the Plant Protection Products Directive and described in 2.3 above. The Directive was published as a formal proposal in the European Commission's Official Journal dated 3 September 1993 reference number C239/3. Negotiations on the proposal began under the German presidency in July 1994. Adoption is not expected until mid 1995 at the earliest, with implementation in 1996.

3. THE APPROVALS PROCESS

1. All approvals are granted by Ministers in response to an application from a manufacturer, formulator, importer or distributor (or in certain circumstances a user) supported by the necessary data on safety, efficacy and, where relevant, humaneness. Approvals are normally granted only in relation to individual products and only for specified uses.

Statutory Conditions of Use

2. Statutory conditions of use are detailed on the product label (except in the case of 'off-label' approvals and commodity substances – see below) and will differ from product to product and use to use.

It is therefore important to *read the label (including any accompanying leaflet) carefully before purchase and use*. The label instructions which *must* be complied with (where they appear) are:

- (i) field of use;
- (ii) crop or situations for which treatment is permitted;
- (iii) maximum individual dose;
- (iv) maximum number of treatments;
- (v) maximum area or quantity which may be treated;
- (vi) latest time of application, or harvest interval;
- (vii) operator protection or training requirements;
- (viii) environmental protection requirements; and
- (ix) any other specific restrictions relating to particular pesticides.

3. In order to highlight the statutory conditions to the user the following steps have been taken:
  - i. Since 1 January 1992 any approved pesticides leaving the manufacturers premises must display the statutory conditions of use in a designated area on the label known as the 'statutory box'.

ii. Since 31 December 1994 all approved pesticides must include a statutory box on their label.

Approvals for Off-Label Uses

4. Users of pesticides may apply to have the approval of specific pesticides extended to cover uses additional to those approved and shown on the manufacturer's product label. Any such specific off-label approvals granted may have additional conditions of use attached to them. Use in these cases is undertaken at the user's choosing, and the commercial risk is entirely theirs. Users are required to be in possession of the relevant notice of approval. Approvals are published in "The Pesticides Register" and are available from local Agricultural Development and Advisory Services Offices (ADAS) and National Farmers Union Offices.

5. Since 1 January 1990 arrangements have been in place which permit many pesticide products to be used for additional specific minor uses, subject to adherence to various conditions. These arrangements were updated in December 1994 to current standards and additional guidance notes provided to improve clarity. Full details are given at Annex C at the end of this section. These arrangements are valid until 31 December 1999 with the extrapolations at Table 2 of the annex only valid until 31 December 1995.

Approval of Commodity Substances

6. Commodity substances are chemicals which have a variety of non-pesticidal uses and also have minor uses as pesticides. If such a substance is to be used as a pesticide, it requires approval under COPR, and is granted approval for use only.

7. *Sale, supply, storage and advertisement* of a commodity substance as a pesticide is an offence unless specific approval has been granted for the substance to be marketed as a pesticide under an approved label. Anyone wishing to sell, supply, store or advertise any of these substances specifically as pesticides therefore has to seek specific approval under COPR.

8. The commodity substances that can be used as a pesticide are listed below. For some substances separate approval is given for different fields of

use. The approvals are listed at Annex D and detail the conditions which a user must follow when the commodity substance is used as a pesticide;

4-chloro-*m*-cresol
Camphor
Carbon dioxide
Ethanol
Ethyl acetate
Formaldehyde
Isopropanol
Methyl bromide
Liquid Nitrogen
Strychnine hydrochloride
Sulphuric acid
Tetrachloroethylene
Thymol
Urea
White spirit

The approval for the use of ethylene dichloride as a commodity substance expired on 28 February 1994, as data to support its continued approval was not submitted

Note: A special derogation for the use of methyl bromide in food storage practice has been extended until further notice. The Registration Authorities maintain their aim of removing this derogation as soon as possible. Any changes in the approval of commodity substances will be announced in the Pesticides Register.

9. The supply of methyl bromide is restricted to those who can prove that they are competent and have the necessary experience to use it safely.

The British Pest Control Association (BPCA) provide training and their own certification which is available to both members and non members. Competence can be proved by the possession of the relevant BPCA certificate.

For further guidance on the use of methyl bromide and other fumigants please refer to COP 30 "The control of substances hazardous to health in fumigation operations" (see page xv).

List of Products Approved for Aerial Application

10. Products approved for use by aerial application, together with details of the crops on which they are approved for use are listed in Annex B at the end of this chapter. Only products approved for aerial application may be used for that purpose and regular returns have to be made to the Pesticides Usage Survey Group, MAFF, Central Science Laboratory, Hatching Green, Harpenden, Herts, AL5 2BD.

Amendment, Expiry, Suspension and Revocation of Approvals

11. Pesticide approvals may at any time be subject to review, amendment, suspension or revocation. Revocation of approval may occur for a number of reasons, for example the identification of safety concerns or an approval holder's failure to meet a data submission deadline. Where possible a "phased revocation" will be implemented, but when safety considerations make it necessary immediate revocation may take effect. On expiry or revocation of approvals it becomes unlawful to advertise, sell, supply, store or use the products for the uses concerned.

12. Suspension of approval may take place when it is anticipated that the action is temporary, e.g. where approval might be reinstated by the provision of required and satisfactory data.

13. A list of pesticide active ingredients banned or severely restricted in the UK can be found at Annex E.

Disposal of Pesticides

14. Farmers and growers will be aware of their legal liability in respect of the disposal of waste, and in particular their duty in law to ascertain that any waste is being disposed of properly. The waste disposal authorities (the County Councils in England, and District Councils in Wales and Scotland) will give advice on disposal matters and, when appropriate, the names of reputable specialist disposal contractors. The Government's "Code of Practice for the Safe Use of Pesticides on Farms and Holdings" (see page xiv) also provides guidance on the disposal of waste pesticides and their containers.

Authorised Adjuvants

15. An adjuvant is a substance other than water which is not in itself a pesticide but which enhances the effectiveness of the pesticide with which it is used. Consent C(i) permits the use of an adjuvant with a pesticide only in accordance with the conditions of approval of that pesticide or as varied subsequently in the lists of authorised adjuvants published by Ministers.

16. Adjuvants for use with agricultural pesticides have been categorised as surfactants, penetrants, extenders, wetting agents, sticking agents, spreading agents and fogging agents. A list of authorised adjuvants can be found at Annex I at the end of this chapter. Subsequent additions and amendments are published in "The Pesticides Register".

List of Products Approved for Use in or Near Water

17. Pesticides must not be used in or near water unless the product label specifically allows such use. "Guidelines for the use of herbicides on weeds in or near watercourses or lakes" (available from MAFF Publications, London SE99 7TP Tel 0645 556000) states "in or near water" includes drainage channels, streams, rivers, ponds, lakes, reservoirs, canals, dry ditches, areas designated for water storage and the banks or areas immediately adjacent to such water bodies.

18. Products approved for use in or near water (other than for public hygiene or antifouling use) are listed in Annex H at the end of this chapter.

ANNEX A
FOOD AND ENVIRONMENT PROTECTION ACT 1985 PART III

CONSENTS GIVEN BY MINISTERS

The advertisement, sale, supply, storage and use of pesticides is prohibited unless conditions of the appropriate Consent are met. The conditions at the time of going to print are set out below. Consents may be changed at any time. They are published in the official Gazettes and in 'The Pesticides Register'.

Consent A: Advertisement of Pesticides

Conditions subject to which consent to the advertisement of pesticides is given

1. An advertisement shall relate only to such uses of a pesticide as are permitted by the approval given in relation to that pesticide.

2. No advertisement shall contain any claim for safety in relation to the pesticide advertised which is not permitted on the label for that pesticide.

3. Any printed, pictorial, broadcast or recorded advertisement other than a notice at the point of sale intended to draw attention solely to product name and price, whether contained in a leaflet, poster, newspaper, magazine or other periodical, including an advertisement diffused through any broadcast or recorded medium, shall include:
   (a) a statement of the active ingredient of each pesticide mentioned in the advertisement, such statement being the name by which the active ingredient is identified in the approval given in relation to the pesticide in which it is contained, except that any price list consisting only of an indication of product availability and price need not state the active ingredients of each pesticide and that any advertisement of a range of pesticides need only state the active ingredients of those individual products which are identified by name;
   (b) a general warning as follows:
   "Read the label before you buy: Use pesticides safely"; and
   (c) where required by a condition of the approval given in relation to a pesticide mentioned in the advertisement, a statement of any special degree of risk to human beings, creatures, plants or the environment.

4. Any statement or warning given under Condition 3 shall be clearly presented:
   (a) in the case of a printed or pictorial advertisement, separately from any other text; and
   (b) in the case of a broadcast or recorded advertisement, shown or spoken separately.

Consent B: Sale, Supply and Storage of Pesticides

Conditions subject to which consent to the sale, supply and storage of pesticides is given

1. It shall be the duty of every employer to ensure that a person in his employment who may be required to sell, supply or store a pesticide during the course of that employment, is provided with such instruction and guidance as is necessary to enable that person to comply with the requirements in and under the regulations.

2. Any person who sells, supplies or stores a pesticide shall:
   (a) take all reasonable precautions, particularly with regard to storage and transport, to protect the health of human beings, creatures and plants, to safeguard the environment, and in particular avoid the pollution of water; and
   (b) be competent for the duties which he is called upon to perform.

3. No person shall sell, supply or otherwise market to the end-user an approved pesticide other than in the container supplied for that purpose by the holder of the approval of that pesticide, and under a label approved by the Ministers.

4. No person shall store for the purpose of sale or supply a pesticide approved for agricultural use in a quantity in excess of, at any one time, 200 kg or 200 litres or a similar mixed quantity unless he has obtained a certificate of competence recognised by the Ministers, or he stores that pesticide under the direct supervision of a person who holds such a certificate.

5. No person shall sell, supply or otherwise market to the end-user a pesticide approved for agricultural use unless he has obtained a certificate of competence recognised by the Ministers, or he sells or supplies that pesticide under the direct supervision of a person who holds such a certificate.

6. For the purposes of conditions 4 and 5, "a pesticide approved for agricultural use" means a pesticide (other than one with methyl bromide as an active ingredient) approved for use within one or more of the following fields of use:
   —agriculture and horticulture (including amenity horticulture)
   —forestry
   —in or near water (products for other than amateur, public hygiene or anti-fouling uses)
   —industrial herbicides (such as weed killers for use on land not intended for cropping).

Consent C(i): Use of Pesticides

Conditions subject to which consent to the use of pesticides is given

1. It shall be the duty of every employer to ensure that any person in his employment who may be required to use a pesticide during the course of that employment is provided with such instruction and guidance as is necessary to enable that person to comply with the requirements in and under the regulations.

2. Any person who uses a pesticide shall take all reasonable precautions to protect the health of human beings, creatures and plants, to safeguard the environment and in particular to avoid pollution of water.

3. No person shall use a pesticide in the course of his business or employment unless he has received adequate instructions and guidance in the safe efficient and humane use of pesticides and is competent for the duties which he is called upon to perform.

4. No person shall combine or mix for use two or more pesticides which are anticholinesterase compounds, unless such a mixture is in accordance with the conditions of an approval or the approved label of at least one of the pesticide product states that the mixture may be made; and no person shall combine or mix for use two or more pesticides if all the conditions of approval relating to this use cannot be complied with.

5. No person shall use a pesticide in conjunction with an adjuvant except in accordance with the conditions of the approval given originally in relation to that pesticide, or as varied subsequently by lists of authorised adjuvants published by the Ministers.

6. No person shall use in the course of a commercial service a pesticide approved for agricultural use unless:
   (i) he has obtained a certificate of competence recognised by the Ministers; or
   (ii) he uses the pesticide under the direct and personal supervision of a person who holds such a certificate; or
   (iii) he uses it in accordance with an approval, if any, for use in one of the following fields of use:
   home garden (amateur gardening)
   animal husbandry
   food storage practice
   vertebrate control (e.g. rodenticides and repellents)
   home kitchen and larder
   other domestic use
   wood preservative
   masonry biocides
   public hygiene/nuisance
   'other' industrial biocides
   anti-fouling paint
   'other' (as may be defined by the registration authority)

7. No person who was born later than 31 December 1964 shall use a pesticide approved for agricultural use, unless:
   (i) he has obtained a certificate of competence recognised by the Ministers; or
   (ii) he uses the pesticide under the direct and personal supervision of a person who holds such a certificate; or
   (iii) he uses it in accordance with an approval, if any, for one of the fields of use listed under paragraph 6(iii) above.

8. For the purposes of these conditions:
   (a) "adjuvant" has the meaning ascribed to it in Regulation 2(1) of the regulations.
   (b) "commercial service" means the application of a pesticide by a person:
      (i) to crops, land, produce, materials, buildings or the contents of buildings not in his or his employers' ownership or occupation; and
      (ii) to seed other than seed intended solely for use by that person or his employer.
   (c) "a pesticide approved for agricultural use" means a pesticide (other than one with methyl bromide as an active ingredient) approved for use within one or more of the following fields of use:
      —agriculture and horticulture (including amenity horticulture)
      —forestry
      —in or near water (products for other than amateur, public hygiene or anti-fouling uses)
      —industrial herbicides (such as weed killers for use on land not intended for cropping).

Consent C(ii): Aerial Application of Pesticides

Additional conditions subject to which consent to the use of pesticides applied from an aircraft in flight is given

1. No person shall undertake an aerial application unless he or his employer or main contractor holds an aerial application certificate granted under Article 42(2) of the Air Navigation Order 1985 (SI 1985/1643); and unless the pesticide to be used has been approved for the intended aerial application.

2. No person shall undertake an aerial application unless he, or a person on his behalf, has:
    (a) not less than 72 hours before the commencement of the aerial application, consulted the relevant authority if any part of land which is a Site of Special Scientific Interest, a National Nature Reserve, a Local Nature Reserve, or a Marine Nature Reserve, lies within a distance of three-quarters of one nautical mile from any part of the land to which the pesticide is to be applied;
    (b) not less than 72-hours before the commencement of the aerial application, consulted the water authority for the area in which he intends to apply the pesticide if the land to which he intends to apply the pesticide is adjacent to water;
    (c) obtained the consent of the water authority for the area in which the aerial application will take place if he intends to apply the pesticide for the purpose of controlling aquatic weeds or weeds on the banks of watercourses or lakes;
    (d) not less than 24 hours and (so far as is practicable) not more than 48 hours before the commencement of the aerial application, given notice of the intended operation to the Chief Environmental Health Officer for the district in which he intends to apply the pesticide;
    (e) not less than 24 hours and (so far as is practicable) not more than 48 hours before the commencement of the aerial application, given notice of the intended operation to the occupants of each building within 75 feet of any boundary of the land to which he intends to apply the pesticide, and to the owner, or his agent, of any livestock or crops within 75 feet of any boundary of the land on which he intends to apply the pesticide;
    (f) not less than 24 hours and (so far as is practicable) not more than 48 hours before the commencement of the aerial application, given notice of the intended operation to the person in charge of any hospital, school or other institution, any part of the curtilage of which lies within 500 feet of any flight path that he intends to use for the aerial application of the pesticide;
    (g) not less than 48 hours before the commencement of the aerial application, given notice of the intended operation to the appropriate reporting point of the local beekeepers' spray warning scheme operating within the district in which he intends to apply the pesticide.

A notice of an intended aerial application under paragraphs (e) or (f) of this condition shall be in writing and include details of the name, address and telephone number (if any), of the person intending to carry out the aerial application, the pesticide to be applied, the intended time and date of application and also an indication that similar details have been given to the Chief Environmental Health Officer for the district.

3. No person shall undertake an aerial application of a pesticide unless:
    (a) the wind velocity at the height of application at the place of intended application does not exceed 10 knots, unless the approval given in relation to that pesticide permits aerial application thereof when such wind velocity exceeds 10 knots;

(b) before the aerial application, he has provided and put in place within 200 feet of the land to which he intends to apply the pesticide, signs adequate to warn pedestrians and drivers of vehicles of the time and place of the intended application; and

(c) before the aerial application he has provided ground markers in all circumstances where a ground marker will assist the pilot to comply with the provisions of paragraph 5 below.

4. Any person who undertakes the aerial application of a pesticide shall:
   (a) keep and retain for not less than 3 years after each application, records of the nature, place and date of that application, the registration number of the aircraft used and the name and permanent address of the pilot of that aircraft, the name and quantity of pesticide applied, its rate of product and volume of application, type and specification of application system e.g. nozzle and its size, the method of application, the flight times of the aerial application, the speed and direction of the wind during that application and any unusual occurrences which affected that application;
   (b) provide by the end of January 1989, summaries of the records required by paragraph (a) of this condition relating to the 1988 calendar year, to Ministers, in a manner required by them under Section 16(11) of the Food and Environment Protection Act 1985;
   (c) provide within 30 days of the end of the calendar month to which the records required by paragraph (a) of this condition relate, summaries of those records to the Ministers, in a manner required by them under Section 16(11) of the Food and Environment Protection Act 1985.

5. The pilot of an aircraft engaged in an aerial application shall:
   (a) maintain the aircraft at a height of not less than 200 feet from ground level when flying over an occupied building or its curtilage;
   (b) maintain the aircraft at a horizontal distance from any occupied building and its curtilage, children's playground, sports ground or building containing livestock of:
      (i) not less than 100 feet, if he has the written consent of the occupier; and
      (ii) not less than 200 feet, in any other case.
   (c) maintain the aircraft at a height of not less than 250 feet from ground level over any motorway, or of not less than 100 feet from ground level over any other public highway, unless that public highway has been closed to traffic during the course of the application; and
   (d) confine the application of the pesticide to the land intended to be treated.

6. For the purposes of these conditions:
   (a) each of the following expressions has the meaning ascribed to it in Regulation 2(1) of the regulations – "aerial application", "approval", "curtilage", "local beekeepers' spray warning scheme";
   (b) each of the following expressions has the meaning ascribed to it in paragraph 7 of Schedule 4 to the regulations – "ground marker" and "water";
   (c) "Local Nature Reserve" means a nature reserve established by a local authority under Section 21 of the National Parks and Access to the Countryside Act 1949 and "the relevant authority" in regard thereto shall be the local authority which is providing or securing the provision of, the reserve;
   (d) "Marine Nature Reserve" means an area designated as such by the Secretary of State under Section 36 of the Wildlife and Countryside Act 1981, and the "relevant authority" in regard hereto shall be English Nature (previously the Nature Conservancy Council);

(e) "National Nature Reserve" means any land declared as such by English Nature (previously the Nature Conservancy Council) under Section 19 of the National Parks and Access to the Countryside Act 1949, or under Section 35 of the Wildlife and Countryside Act 1981, and "the relevant authority" in regard thereto shall be English Nature;

(f) "Site of Special Scientific Interest" means any area of land designated as such by English Nature (previously the Nature Conservancy Council) under Section 28 of the Wildlife and Countryside Act 1981, or in respect of which the Secretary of State has made an Order under Section 29 of the Wildlife and Countryside Act 1981, and "the relevant authority" in regard thereto shall be English Nature.

ANNEX B
PRODUCTS APPROVED FOR USE BY AERIAL APPLICATION

Products approved for use by aerial application, together with details of the crops on which they are approved for use are listed below. Only products approved for aerial application may be used for that purpose and regular returns have to be made to the Pesticides Usage Survey Group, MAFF, Central Science Laboratory, Hatching Green, Harpenden, Herts, AL5 2BD.

Note: Reference must be made to the label for conditions of use:

Active ingredient	Product (MAFF No.)	Crops/Uses
Alphacypermethrin	Fastac (02659)	Spring oilseed rape, winter oilseed rape
Asulam	Asulox (05235)	Rough upland intended for grazing, forestry plantation, non-crop land
	Asulox (06124)	Rough upland intended for grazing, forestry plantation, non-crop land
Benalaxyl + Mancozeb	Galben M (05091)	Potato (early, maincrop)
	Galben M (05092)	Potato (early, maincrop)
	Galben M (05904)	Potato
	Barclay Bezant (05914)	Potato (early, main crop)
	Clayton Benzeb (07081)	Potato (early, maincrop)
	Galben M (07220)	Early potato, maincrop potato
Benodanil	Calirus (00368)	Winter wheat, spring wheat, winter barley, spring barley
Benomyl	Benlate Fungicide (00229)	Winter barley, spring oilseed rape, winter oilseed rape, winter wheat spring barley, field bean, rye.
Benzoylprop-ethyl	Suffix (02037)	Wheat, winter field bean, spring field bean, oilseed rape, culinary mustard
Carbendazim	Ashlade Carbendazim Flowable (02662)	Winter wheat, winter barley, spring barley, oilseed rape
	Ashlade Carbendazim Flowable (06213)	Oilseed rape, spring barley, winter barley, winter wheat
	Battal FL (00215)	Green bean(dwarf), field bean, oilseed rape, winter wheat, winter & spring barley

Active ingredient	Product (MAFF No.)	Crops/Uses
Carbendazim—*continued*	Bavistin (00217)	French bean (dwarf), field bean, onion, oilseed rape, spring barley, winter barley, winter wheat, rye
	Bavistin DF (03848)	Oilseed rape, winter wheat, barley (winter, spring), onion, field bean, dwarf french bean, winter rye
	Bavistin FL (00218)	Spring wheat, winter wheat, rye, winter barley, spring barley, dwarf french bean, onion, oilseed rape
	Carbate Flowable (03341)	Field bean, oilseed rape, onion, rye, spring barley, winter barley, spring wheat, winter wheat.
	Derosal WDG (03404)	Field bean, spring barley, spring wheat, winter barley, winter wheat
	Derosal WDG (07316)	Spring barley, winter barley, field bean, spring wheat, winter wheat
	Hy-Carb (05933)	Barley, winter wheat, broad bean, field bean, dwarf bean
	Power Carbendazim 50 (03716)	Wheat, barley
	Quadrangle Hinge (04929)	Spring barley, winter barley, oilseed rape, winter wheat
	Stempor DG (02021)	Cereal, field bean, dwarf bean
	Stempor DG (06708)	Cereal, field bean, dwarf bean
	Supercarb (01560)	Spring barley, winter barley, winter wheat, field bean
	Top Farm Carbendazim-435 (05307)	Winter wheat, winter and spring barley, field bean, dwarf green bean, oilseed rape
	Tripart Defensor FL (02752)	Winter wheat, winter and spring barley, oilseed rape
Carbendazim + Chlorothalonil	Bravocarb (05119)	Winter wheat, winter barley, spring barley
Carbendazim + Mancozeb	Kombat WDG (04344)	Winter wheat, winter barley, winter oilseed rape
	Kombat WDG (05509)	Winter wheat, winter barley, winter oilseed rape
	Kombat WDG (05984)	Winter barley, winter wheat, winter oilseed rape
	Kombat WDG (07329)	Winter barley, winter oilseed rape, winter wheat
	Septal WDG (04279)	Winter wheat, winter barley, oilseed rape

Active ingredient	Product (MAFF No.)	Crops/Uses
Carbendazim + Mancozeb —continued	Septal WDG (05985)	Winter barley, winter wheat, winter oilseed rape
	Septal WDG (07281)	Winter barley, winter wheat, winter oilseed rape
Carbendazim + Maneb	Ashlade Mancarb FL (02928)	Cereal
	Ashlade Mancarb FL (06217)	Winter barley, winter wheat
	Campbells MC Flowable (03467)	Winter wheat, spring wheat, winter barley, spring barley
	Headland Dual (03782)	Winter wheat, spring wheat, winter barley, spring barley
	Tripart Legion (02997)	Winter wheat, winter barley
Carbendazim + Maneb + Sulphur	Bolda FL (03463)	Winter wheat, spring wheat, winter barley, spring barley, winter oilseed rape, spring oilseed rape
Carbendazim + Maneb + Tridemorph	Cosmic FL (03473)	Winter wheat, winter barley, spring barley
Carbendazim + Prochloraz	Sportak Alpha (03872)	Cereal
	Sportak Alpha (07222)	Cereal
Carbendazim + Triadimefon	200 Plus (05025)	Winter barley, winter rye
	Bayleton BM (00223)	Barley, wheat
Chlormequat	3C Chlormequat 460 (03916)	Wheat, barley, oats
	3C Chlormequat 600 (04079)	Wheat, barley, oats
	Arotex Extra (00117)	Spring barley, triticale
	Ashlade 4-60 CCC (06474)	Barley, oats, rye, wheat
	Ashlade 460 CCC (02894)	Wheat, barley, oats, rye
	Ashlade 700 5C (07046)	Spring oats, winter oats, spring wheat, winter wheat
	Ashlade 700 CCC (02912)	Wheat, oats, barley, rye
	Ashlade 700 CCC (06473)	Barley, oats, rye, wheat
	Atlas 3C: 645 Chlormequat (05710)	Spring wheat, winter wheat, oats
	Atlas 5C Chlormequat (03084)	Spring oats, winter oats, spring wheat, winter wheat
	Atlas Chlormequat 46 (05660)	Spring wheat, winter wheat, spring oats, winter oats
	Atlas Chlormequat 460:46 (06258)	Spring oats, spring wheat, winter oats, winter wheat
	Atlas Chlormequat 640 5C (05888)	Oats (spring, winter), wheat (spring, winter)
	Atlas Chlormequat 670 (06615)	Spring oats, winter oats, spring wheat, winter wheat
	Atlas Chlormequat 700 (03402)	Winter wheat, spring wheat, winter oats, spring oats
	Atlas Chlormequat 730 (06511)	Spring wheat, winter wheat, spring oats, winter oats

Active ingredient	Product (MAFF No.)	Crops/Uses
Chlormequat—*continued*	Atlas Quintacel (06524)	Spring oats, spring wheat, winter oats, winter wheat
	Atlas Terbine (06523)	Spring oats, spring wheat, winter oats, winter wheat
	Atlas Tricol (06522)	Spring oats, spring wheat, winter oats, winter wheat
	Atlas Tricol (07190)	Spring oats, spring wheat, winter oats, winter wheat
	Barclay Holdup (05132)	Spring wheat, winter wheat, spring oats, winter oats, winter barley
	Barclay Holdup (06799)	Winter barley, spring oats, winter oats, spring wheat, winter wheat
	Barleyquat B (07051)	Winter barley
	BASF 3C Chlormequat (03234)	Wheat, barley, oats, rye, triticale
	BASF 3C Chlormequat 600 (04077)	Wheat, barley, oats,
	BASF 3C Chlormequat 720 (06514)	Rye, spring oats, spring wheat, triticale, winter barley, winter oats, winter wheat
	BASF 3C Chlormequat 750 (06878)	Rye, spring oats, spring wheat, triticale, winter barley, winter oats, winter wheat
	Bettaquat B (07050)	Spring wheat, winter wheat
	CCC 700 (03366)	Wheat, barley, oats
	Headland Swift (04537)	Winter barley, winter wheat, spring barley, spring oats, spring wheat
	Hyquat (01094)	Wheat, oats
	Hyquat 70 (03364)	Wheat, oats, barley
	Hyquat 75 (03365)	Wheat, oats, barley
	Interlates Chlormequat 46 (00503)	Barley
	KW Chlormequat 600 (04078)	Wheat, barley, oats
	Mandops Barleyquat B (01256)	Barley, rye
	Mandops Barleyquat B (06001)	Barley, rye
	Mandops Bettaquat B (01257)	Wheat, oats
	Mandops Bettaquat B (06004)	Oats, wheat
	Mandops Chlormequat 700 (01264)	Wheat, oats, barley
	Mandops Chlormequat 700 (06002)	Barley, oats, wheat
	MSS Mircell (06939)	Spring oats, spring wheat, winter oats, winter wheat

Active ingredient	Product (MAFF No.)	Crops/Uses
Chlormequat—*continued*	New 5C Cycocel (01482)	Spring barley, winter barley, oats, rye, triticale, winter oilseed rape, spring wheat, winter wheat.
	New 5C Cycocel (01483)	Winter oilseed rape, spring wheat, winter wheat, oats, spring barley, (spring application), winter barley (autumn or spring application), triticale, rye
	PA Chlormequat 400 (01523)	Wheat, oats
	PA Chlormequat 460 (02549)	Wheat, oats
	Portman Chlormequat 700 (03465)	Cereal
	Stabilan 750 (03370)	Wheat, oats, barley
	Stabilan SC (07082)	Spring oats, winter oats, spring wheat, winter wheat
	Titan 670 (06944)	Spring oats, spring wheat, winter oats, winter wheat
	Titan 670 (07297)	Spring oats, winter oats, spring wheat, winter wheat
	Top Farm Chlormequat 640 (05323)	Winter wheat, spring wheat, winter oats, spring oats
	Tripart Brevis (03754)	Winter wheat, spring wheat, oats, barley, rye
2-Chloroethylphosphonic acid	Cerone (00462)	Winter barley
	Cerone (00463)	Winter barley
	Cerone (05443)	Winter barley
	Cerone (06185)	Winter barley
	Ibis (06816)	Winter barley
	Stantion (04784)	Winter barley
	Stantion (06205)	Winter barley
	Stefes Stance (06125)	Winter barley
	Unistar Ethephon 480 (06282)	Winter barley
Chlorothalonil	Barclay Corrib (05886)	Potato, winter wheat
	Barclay Corrib 500 (06392)	Field bean, potato, spring barley, spring wheat, winter barley, winter wheat
	BASF Bravo 500 (04939)	Potato, Winter Wheat
	BASF Bravo 500 (05637)	Potato, winter wheat
	BB Chlorothalonil (03320)	Wheat, potato.
	Bombardier (02675)	Spring wheat, winter wheat, potato.
	Bravo 500 (04945)	Winter wheat, potato
	Bravo 500 (05638)	Field bean, pea, potato, winter wheat
	Clortosip (06126)	Potato, wheat

Active ingredient	Product (MAFF No.)	Crops/Uses
Chlorothalonil—continued	Jupital (04946)	Potato, winter wheat
	Jupital (05554)	Potato, winter wheat
	Mainstay (05625)	Potato, winter wheat
	Sipcam UK Rover 500 (04165)	Winter wheat, spring wheat, winter barley, spring barley, field bean, potato
	Top Farm Chlorothalonil 500 (05926)	Winter wheat, potato
	Tripart Faber (04549)	Winter wheat, spring wheat, winter barley, spring barley, potato, field bean
	Tripart Faber (05505)	Wheat (winter), potato
Chlorothalonil + Carbendazim	Bravocarb (05119)	Winter wheat, winter barley, spring barley
Chlorothalonil + Cymoxanil	Ashlade Cyclops (04857)	Potato
	Cyclops (05661)	Potato
	Cyclops (06650)	Potato
	Guardian (05663)	Potato
	Guardian (06676)	Potato
	PP630 (05662)	Potato
	PP630 (06701)	Potato
Chlorotoluron	Ashlade Chlorotoluron 500 (06093)	Durum wheat, winter barley, triticale, winter wheat
	Dicurane 500 FW (00698)	Winter wheat, winter barley, durum wheat, triticale
	Dicurane 500 SC (05836)	Winter wheat, winter barley, triticale, durum wheat
	Dicurane 700 SC (04859)	Triticale, winter barley, winter wheat, durum wheat
	Lentipur CL 500 (05925)	Winter wheat, winter barley, durum wheat, triticale
	Ludorum (03059)	Winter wheat, winter barley, triticale, durum wheat
	Ludorum 700 (03999)	Winter barley, durum wheat, winter wheat, triticale
	Talisman (03109)	Winter barley, winter wheat, triticale, durum wheat
	Tol 7 (05084)	Winter barley, durum wheat, triticale, winter wheat
	TOL 7 (06484)	Winter barley, winter wheat, durum wheat, triticale
	Toro (04734)	Winter barley, winter wheat, durum wheat, triticale
	Toro (05927)	Winter barley, durum wheat, triticale, winter wheat
Chlorpropham + Linuron	Profalon (01640)	Ornamental (daffodil, narcissi, tulip)

Active ingredient	Product (MAFF No.)	Crops/Uses
Chlorpropham + Linuron —continued	Profalon (07331)	Ornamental (daffodil, narcissi, tulip)
Chlorpyrifos	Dursban 4 (00775)	Wheat, barley, oats
	Dursban 4 (05735)	Barley, oats, wheat
Copper oxychloride	Cuprokylt (00604)	Potato
Cymoxanil + Chlorothalonil	Cyclops (05661)	Potato
	Cyclops (06650)	Potato
	Guardian (05663)	Potato
	Guardian (06676)	Potato
	PP630 (05662)	Potato
	PP630 (06701)	Potato
Cymoxanil + Mancozeb	Ashlade Solace (05462)	Potato
	Ashlade Solace (06472)	Potato
	Besiege (05451)	Potato
	Clayton Krypton (06973)	Potato
	Curzate M (04343)	Potato
	Fytospore (00960)	Potato
	Fytospore (06517)	Potato
	Standon Cymoxanil Extra (06807)	Potato
	Stefes Blight Spray (05811)	Potato
	Systol M (03480)	Potato
	Systol M (07098)	Potato
Cymoxanil + Mancozeb + Oxadixyl	Ripost (04890)	Potato
	Ripost Pepite (06485)	Potato
	Riposte WDG (05053)	Potato
	Trustan (05022)	Potato
	Trustan WDG (05050)	Potato
Deltamethrin	Decis (00657)	Brussels sprout, cabbage, cauliflower, ornamental tree, pea, potato, spring oilseed rape, winter barley, winter oilseed rape, winter wheat
	Decis (06311)	Brussels sprout, cabbage, cauliflower, ornamental tree, pea, potato, spring oilseed rape, winter barley, winter oilseed rape, winter wheat
Demeton-S-methyl	Mifatox (01350)	Cereal, pea, potato, sugar beet, carrot, brassica, field bean
Desmetryne	Semeron 25 WP (01916)	Brassica
Diflubenzuron	Dimilin ODC-45 (04149)	Forestry (Corsican pine, Lodgepole pine, Scots pine)
	Dimilin ODC-45 (06655)	Forestry (Corsican pine, Lodgepole pine, Scots pine)

Active ingredient	Product (MAFF No.)	Crops/Uses
Dimethoate	Amos Dimethoate (05812)	Carrot, fodder beet, mangel, pea, potato, red beet, sugar beet
	Amos Dimethoate (06144)	Carrot, fodder beet, mangel, pea, potato, red beet, sugar beet
	Ashlade Dimethoate (04814)	Broccoli, Brussels sprout, cabbage, carrot, cauliflower, pea (combining and vining), potato, spring barley, sugar beet, wheat, winter barley
	Atlas Dimethoate 40 (03044)	Broccoli, Brussels sprout, cabbage, calabrese, carrot, cauliflower, pea, potato, sugar beet, swede, turnip
	BASF Dimethoate 40 (00199)	Pea, potato (ware), rye, spring barley, spring oats, spring wheat, sugar beet, triticale, winter barley, winter oats, winter wheat.
	Campbell's Dimethoate 40 (00398)	Cereal, pea, potato, sugar beet, fodder beet, mangel.
	CIA Dimethoate (05937)	Carrot, fodder beet, mangel, mangel steckling, potato, red beet, sugar beet, sugar beet steckling
	Cropsafe Dimethoate 40 (05153)	Wheat, barley, oats, pea, potato, fodder beet, sugar beet, mangel
	Danadim 400 (06488)	Barley, oats, pea, potato, sugar beet, triticale, wheat, winter rye
	Danadim Dimethoate 40 (07040)	Barley, oats, pea, potato, sugar beet, triticale, wheat, winter rye
	Danadim Dimethoate 40 (07351)	Winter barley, winter oats, pea, ware potato, sugar beet, triticale, wheat, winter rye
	Ipithoate (04694)	Wheat, barley, oats, winter rye, triticale, sugar beet, potato, pea
	Luxan Dimethoate (06558)	Carrot, fodder beet, mangel, pea, potato, red beet, sugar beet
	MTM Dimethoate 40 (05693)	Spring barley, spring oats, spring wheat, winter barley, winter oats, winter wheat, pea, potato, fodder beet, sugar beet, mangel
	Roxion (04813)	Sugar beet, potato, carrot, winter barley, spring barley, pea (combining and vining), wheat,

Active ingredient	Product (MAFF No.)	Crops/Uses
Dimethoate—*continued*		brassica (Brussels sprout, cabbage, cauliflower, broccoli, spring green)
	Roxion (07015)	Barley, broccoli, Brussels sprout, cabbage, carrot, cauliflower, pea (combining and vining), potato, sugar beet, spring green, wheat
	Top Farm Dimethoate (05936)	Carrot, fodder beet, mangel, mangel steckling, potato, red beet, sugar beet, sugar beet steckling
	Unicrop Dimethoate 40 (02265)	Cereal, pea, potato, sugar beet, herbage (seed crop)
	Unicrop Dimethoate 40 (06549)	Broccoli, Brussels sprout, cabbage, calabrese, carrot, cauliflower, pea, potato, sugar beet, swede, turnip, wheat
Disulfoton	Disyston P-10 (00715)	Brussels sprout, cabbage, carrot, cauliflower, sugar beet
Fenitrothion	Dicofen (00693)	Cereal, pea, wheat
	Unicrop Fenitrothion 50 (02267)	Wheat, pea
Fenpropimorph	BASF 421F (06127)	Spring barley, winter barley, field bean, spring oats, winter oats, rye, triticale, spring wheat, winter wheat
	Corbel (00578)	Spring barley, winter barley, field bean, spring oats, winter oats, spring wheat, winter wheat, rye, triticale.
	Keetak (06950)	Field bean, rye, spring barley, spring oats, spring wheat, triticale, winter barley, winter oats, winter wheat
	Mistral (04582)	Winter wheat, spring wheat, winter barley, spring barley, winter oats, spring oats, winter rye, triticale, field bean.
	Mistral (06199)	Field bean, spring barley, spring oats, spring rye, spring wheat, triticale, winter barley, winter oats, winter rye, winter wheat
Fentin acetate + Maneb	Brestan 60 (00325)	Potato
	Brestan 60 (07304)	Potato
Fentin hydroxide	Ashlade Flotin (03535)	Potato

Active ingredient	Product (MAFF No.)	Crops/Uses
Fentin hydroxide—*continued*	Ashlade Flotin (06223)	Potato
	Ashlade Flotin 2 (03783)	Potato
	Ashlade Flotin 2 (06224)	Potato
	CIA Fentin-475-WP (05550)	Potato
	Du-ter 50 (00778)	Potato
	Farmatin 560 (04595)	Potato
	Quadrangle Super-Tin 4L (03842)	Potato
	Super-Tin 4L (02995)	Potato
	Top Farm Fentin-475-WP (05504)	Potato
Fentin hydroxide + Metoxuron	Endspray (02799)	Potato
Flamprop-M-isopropyl	Commando (00564)	Barley, rye, triticale, wheat (durum, winter, spring)
	Commando (07005)	Barley, rye, triticale, wheat (durum, winter, spring)
	Cossack (05864)	Spring barley, winter barley, spring wheat, winter wheat
	Cossak (07006)	Spring barley, winter barley, spring wheat, winter wheat
	Quadrangle Gunner (04547)	Winter wheat, winter barley, spring barley, spring wheat, durum wheat
	Stefes Flamprop (05789)	Wheat, barley, durum wheat, rye, triticale
	Trooper (05863)	Spring barley, winter barley, spring wheat, winter wheat
	Trooper (07019)	Spring barley, winter barley, spring wheat, winter wheat
Heptenophos	Hostaquick (01079)	Cereal, pea, bean (broad, dwarf, field, runner), broccoli, Brussels sprout, cabbage, cauliflower.
	Hostaquick (07326)	Bean (broad, dwarf, field, runner), broccoli, Brussels sprout, cabbage, cauliflower, cereal, pea
Iprodione	Rovral Flo (04526)	Field bean
	Rovral Flo (06328)	Oilseed rape, pea (combining and vining), field bean, collard, cabbage
Iprodione + Thiophanate-methyl	Compass (06190)	Field bean, winter oilseed rape
Isoproturon	Barclay Guideline (06743)	Spring wheat, winter barley, winter rye, winter wheat, triticale

Active ingredient	Product (MAFF No.)	Crops/Uses
Isoproturon—*continued*	Hytane 500 SC (05848)	Cereal (Spring, autumn sown)
	Ki-Hara IPU 500 (06102)	Winter barley, spring wheat, winter wheat, winter rye, autumn sown triticale
	Landgold Isoproturon (06012)	Winter barley, winter wheat
	Landgold Isoproturon FL (06034)	Winter wheat, spring wheat, winter barley
	Phyto IPU (04380)	Winter wheat, spring wheat, winter barley, winter rye, triticale
	Sabre WDG (04719)	Winter wheat, winter barley, spring wheat, winter rye, triticale
	Stefes IPU (05776)	Winter wheat, winter barley, winter rye, spring wheat, triticale
	Top Farm IPU 500 (05978)	Spring wheat, winter barley, winter rye, winter triticale, winter wheat
	Tripart Pugil (06153)	Spring wheat, winter barley, winter rye, winter triticale, winter wheat
Linuron	Afalon (04665)	Potato, carrot, parsnip, parsley, celery
	Afalon (07299)	Carrot, celery, parsley, parsnip, potato
Linuron + Chlorpropham	Profalon (01640)	Ornamental (daffodil, narcissi, tulip)
	Profalon (07331)	Ornamental (daffodil, narcissi, tulip)
Mancozeb	Agrichem Mancozeb 80 (06354)	Potato
	Ashlade Mancozeb FL (03208)	Potato, spring wheat, winter wheat, spring barley, winter barley, spring oats, spring rye, winter rye, spring triticale, winter triticale
	Ashlade Mancozeb Flowable (06226)	Potato, spring barley, winter barley, oats, rye, triticale, spring wheat, winter wheat
	Barclay Manzeb 80 (05296)	Potato
	Barclay Manzeb 80 (05944)	Potato
	Barclay Manzeb 80W (05833)	Potato
	Dequiman MZ (03959)	Potato
	Dequiman MZ (06870)	Potato
	Dithane 945 (00719)	Potato
	Dithane 945 (04017)	Potato
	Dithane Dry Flowable (04251)	Potato

Active ingredient	Product (MAFF No.)	Crops/Uses
Mancozeb—*continued*	Dithane Dry Flowable (04255)	Potato
	Dithane Superflo (06290)	Potato
	Dithane Superflo (06593)	Potato
	Luxan Mancozeb Flowable (06812)	Potato
	Manazate 200 PI (07209)	Potato
	Manzate 200 (01281)	Potato
	Manzate 200 DF (06010)	Potato
	Nemispor (01473)	Potato, cereal
	Penncozeb (02716)	Potato
	Penncozeb (04609)	Potato
	Penncozeb (06873)	Potato
	Penncozeb (07089)	Potato
	Penncozeb WDG (06054)	Potato
	Penncozeb WDG (07095)	Potato
	Unicrop Flowable Mancozeb (04700)	Potato, wheat, barley, rye, oats, triticale
	Unicrop Mancozeb (05467)	Potato
Mancozeb + Benalaxyl	Barclay Bezant (05914)	Potato (early, maincrop)
	Clayton Benzeb (07081)	Potato (early, maincrop)
	Galben M (05091)	Potato (early, maincrop)
	Galben M (05092)	Potato (early, maincrop)
	Galben M (05904)	Potato
	Galben M (07220)	Potato (early, maincrop)
Mancozeb + Carbendazim	Kombat WDG (04344)	Winter wheat, winter barley, winter oilseed rape
	Kombat WDG (05509)	Winter wheat, winter barley, winter oilseed rape
	Kombat WDG (05984)	Winter barley, winter wheat, winter oilseed rape
	Kombat WDG (07329)	Winter barley, winter oilseed rape, winter wheat
	Septal WDG (04279)	Winter wheat, winter barley, oilseed rape
	Septal WDG (05985)	Winter barley, winter wheat, winter oilseed rape
	Septal WDG (07281)	Winter barley, winter wheat, winter oilseed rape
Mancozeb + Cymoxanil	Ashlade Solace (05462)	Potato
	Ashlade Solace (06472)	Potato
	Besiege (05451)	Potato
	Fytospore (00960)	Potato
	Fytospore (06517)	Potato
	Standon Cymoxanil Extra (06807)	Potato
	Stefes Blight Spray (05811)	Potato
	Systol M (03480)	Potato
	Systol M (07098)	Potato

Active ingredient	Product (MAFF No.)	Crops/Uses
Mancozeb + Cymoxanil + Oxadixyl	Ripost (04890)	Potato
	Trustan (05022)	Potato
	Trustan WDG (05050)	Potato
	Riposte WDG (05053)	Potato
	Ripost Pepite (06485)	Potato
Mancozeb + Metalaxyl	Fubol 58 WP (00927)	Potato
	Fubol 75WP (03462)	Potato
	Osprey 58 WP (05717)	Potato (early and maincrop)
	Osprey 58 WP (05931)	Potato (early and maincrop)
Mancozeb + Ofurace	Patafol Plus (02808)	Potato, oilseed rape
Mancozeb + Oxadixyl	Recoil (04039)	Potato
Maneb	Agrichem Maneb 80 (05474)	Potato
	Agrichem Maneb Dry Flowable (05515)	Potato
	Ashlade Maneb Flowable (02911)	Apple, blackcurrant, cereal, leaf brassica, root brassica, oilseed rape, pear, potato, sugar beet, tulip, spring barley, winter barley, spring wheat, winter wheat
	Ashlade Maneb Flowable (06477)	Potato
	BASF Maneb (00207)	Potato, winter wheat
	Campbell's X-Spor SC (03252)	Potato
	Headland Spirit (04548)	Potato
	Maneb 80 (01276)	Potato
	Manzate (02436)	Potato, cereal
	Mazin (01309)	Potato
	Mazin (06061)	Potato
	Quadrangle Maneb Flowable (06131)	Potato
	R H Maneb 80 (01796)	Potato
	Stefes Maneb DF (06418)	Potato
	Trimangol 80 (04294)	Flower bulb, potato, tomato
	Trimangol 80 (06070)	Potato
	Trimangol 80 (06871)	Potato
	Trimangol WDG (06992)	Potato
	Tripart Obex (06130)	Potato
	Unicrop Maneb (02274)	Wheat, barley, potato
	Unicrop Maneb 80 (05533)	Potato
	Unicrop Maneb 80 (06926)	Potato
	Unicrop Manguard DG (05516)	Potato
	X-Spor SC (06310)	Potato

Active ingredient	Product (MAFF No.)	Crops/Uses
Maneb + Carbendazim	Ashlade Mancarb FL (02928)	Cereal
	Ashlade Mancarb FL (06217)	Winter barley, Winter wheat
	Campbells MC Flowable (03467)	Winter wheat, spring wheat, winter barley, spring barley
	Headland Dual (03782)	Winter wheat, spring wheat, winter barley, spring barley
	Tripart Legion (02997)	Winter wheat, winter barley
Maneb + Carbendazim + Sulphur	Bolda FL (03463)	Winter wheat, spring wheat, winter barley, spring barley, winter oilseed rape, spring oilseed rape
Maneb + Carbendazim + Tridemorph	Cosmic FL (03473)	Winter wheat, winter barley, spring barley
Maneb + Fentin acetate	Brestan 60 (00325)	Potato
	Brestan 60 (07304)	Potato
Metalaxyl + Mancozeb	Fubol 58 WP (00927)	Potato
	Fubol 75WP (03462)	Potato
	Osprey 58 WP (05717)	Potato (early and maincrop)
	Osprey 58 WP (05931)	Potato (early and maincrop)
Metaldehyde	Doff Agricultural Slug Killer with Animal Repellent (06058)	Edible crop (around all outdoor and protected crop), Non-edible (around all outdoor and protected crop), Soil (bare)
	Doff Horticultural Slug Killer Blue Mini Pellets (05688)	Bare soil, edible crop, non-edible crop
	FP 107 (05861)	Bare soil, all outdoor edible and non-edible crops
	FP 107 (06666)	Bare soil, edible crop (outdoor), non-edible crop (outdoor)
	FP107 (06527)	Edible crop (protected and outdoor), non-edible crop (protected and outdoor), bare soil
	Gastrotox 6G Slug Pellets (04066)	Edible crop, Non-edible crop, Soil (bare)
	Mifaslug (01349)	Crop (all)
	Optimol (04903)	Bare soil, edible crop, non-edible crop (outdoor and protected)
	Optimol (05862)	Bare soil, edible and non-edible crops (outdoors)
	Optimol (06526)	Edible crop (protected and outdoor), non-edible crop (protected and outdoor), bare soil
	Optimol (06688)	Bare soil, edible and non-edible crop (outdoors)

Active ingredient	Product (MAFF No.)	Crops/Uses
Metaldehyde—*continued*	Quadrangle Mini Slug Pellets (01670)	Crop (all agricultural/horticultural)
	Slug Destroyer (03919)	Edible crop
	Slug Destroyer (07283)	Edible crop
	Slug Pellets (01558)	Crop (all)
	Super-flor 6 % Metaldehyde Slug Killer (05453)	Edible and non-edible crop (both protected and outdoors), bare soil
	Tripart Mini Slug Pellets (02207)	All agricultural and horticultural crops
	Unicrop 6% Mini Slug Pellets (02275)	Bare soil, edible crop and non-edible crop (both under protection and outdoors)
	Vassgro Mini Slug Pellets (03579)	Edible crop, non edible crop
Methabenzthiazuron	Tribunil (02169)	Barley (winter, spring), wheat (winter, spring (autumn sown), durum), winter rye, winter oats, triticale, grass (perennial ryegrass)
Methiocarb	Club (03800)	Bare soil, all outdoor edible and non edible crops
	Club (06645)	Bare soil, edible crop (outdoor), non-edible crop (outdoor)
	Club (07176)	Bare soil, edible (outdoor), non-edible (outdoor)
	Decoy (06535)	Bare soil, non-edible crop, outdoor edible crop
	Draza (00765)	Bare soil, all outdoor edible and non edible crops
	Draza 2 (04748)	Bare soil, outdoor crop (edible), outdoor crop (non-edible)
	Draza Plus (06553)	Bare soil, all outdoor edible crop, all outdoor non-edible crop
	Elvitox (06738)	Bare soil, outdoor non-edible crop, outdoor edible crop
	Epox (06737)	Bare soil, outdoor non-edible crop, outdoor edible crop
Metoxuron	Dosaflo (00753)	Winter wheat, winter barley
	Dosaflo (00754)	Winter wheat, winter barley
Metoxuron + Fentin hydroxide	Endspray (02799)	Potato
Monolinuron	Arresin (00118)	Potato (early and maincrop), dwarf french bean, leek
	Arresin (07303)	Bean (dwarf french), leek, potato (early and late crop)

Active ingredient	Product (MAFF No.)	Crops/Uses
Ofurace + Mancozeb	Patafol Plus (02808)	Potato, oilseed rape
Omethoate	Folimat (00912)	Winter wheat
Oxadixyl + Cymoxanil + Mancozeb	Ripost (04890)	Potato
	Trustan (05022)	Potato
	Trustan WDG (05050)	Potato
	Riposte WDG (05053)	Potato
	Ripost Pepite (06485)	Potato
Oxadixyl + Mancozeb	Recoil (04039)	Potato
Oxydemeton-methyl	Metasystox R (01333)	Cereal
	Metasystox R (05763)	Cereal
Permethrin	Permit (01577)	Cereal, leafy brassica, oilseed rape, pea
Phorate	BASF Phorate (00210)	Beet crop (sugar beet, fodder beet, red beet, mangel), field bean, broad bean, mangel
	Terrathion Granules (02106)	Field bean, broad bean
	MTM Phorate (05540)	Field bean, broad bean
	MTM Phorate (06609)	Broad bean, field bean
	BASF Phorate (06610)	Broad bean, field bean
Phosalone	Zolone Liquid (05231)	Cereal, brassica (seed crop), oilseed rape
	Zolone Liquid (06173)	Brassica (seed crop), cereal, oilseed rape
Pirimicarb	Aphox (00106)	Cereal (barley, durum wheat, oats, rye, triticale, wheat), pea, potato, sugar beet, sweetcorn, leafy brassica, maize, swede, oilseed rape, turnip, carrot, bean (field, dwarf french, broad, runner)
	Aphox (06633)	Barley, durum wheat, oats, rye, triticale, wheat.
	Barclay Pirimisect (06929)	Spring barley, spring oats, spring wheat, winter barley, winter oats, winter wheat
	Clayton Pirimicarb 50 SG (06972)	Barley, durum wheat, oats, rye, triticale, wheat
	Phantom (04519)	Cereal, pea, potato, sugar beet, bean (field, dwarf french, broad, runner), leaf brassica, maize, swede, sweetcorn, oilseed rape, turnip, carrot, barley, durum wheat, oats, rye, triticale, wheat
	Pirimicarb 50 DG (04063)	Barley, durum wheat, oats, rye, triticale, wheat.

Active ingredient	Product (MAFF No.)	Crops/Uses
Pirimicarb—*continued*	Portman Pirimicarb (06922)	Spring barley, spring oats, spring wheat, winter barley, winter oats, winter wheat
	Sapir (04518)	Cereal (barley, durum wheat, oats, rye, triticale, wheat),
	Sapir (06398)	Cereal (barley, durum wheat, oats, rye, triticale, wheat), pea, potato, sugar beet, bean (field, dwarf french, broad, runner), leafy brassica, swede, oilseed rape, turnip, sweetcorn, maize, carrot
	Stefes Pirimicarb (05758)	Wheat, durum wheat, barley, oats, rye, triticale, maize, sweetcorn, cauliflower, broccoli (including calabrese), cabbage, Brussels sprout, kale, chinese cabbage, collard, carrot, oilseed rape, swede, turnip, sugar beet, bean (field, dwarf french, broad, runner), potato, pea
	Unistar Pirimicarb 500 (06975)	Barley, durum wheat, oats, rye, triticale, wheat
Prochloraz	Sportak (03871)	Cereal, oilseed rape
	Sportak (07285)	Cereal, oilseed rape
Prochloraz + Carbendazim	Sportak Alpha (03872)	Cereal
	Sportak Alpha (07222)	Cereal
Prometryn	Gesagard 50 WP (00981)	Pea, potato, carrot, celery, leek
Propiconazole	Clayton Propiconazole (06415)	Spring barley, winter barley, spring wheat, winter wheat
	Mantis 250 EC (06240)	Winter barley, spring barley, spring oats, winter oats, spring rye, winter rye, spring wheat, winter wheat
	Radar (01683)	Winter wheat, winter barley, spring barley, spring wheat, rye, oats, triticale
	Radar (03000)	Winter wheat, winter barley, spring barley, spring wheat
	Radar (06747)	Spring barley, winter barley, spring oats, winter oats, spring rye, winter rye, spring wheat, winter wheat
	Standon Propiconazole (07037)	Oats, rye, spring barley, spring wheat, winter barley, winter wheat.

Active ingredient	Product (MAFF No.)	Crops/Uses
Propiconazole—continued	Stefes Restore (06267)	Spring barley, spring oats, spring rye, spring wheat, winter barley, winter oats, winter rye, winter wheat
	Tilt 250 EC (02138)	Oats, rye, winter wheat, winter barley, spring barley, spring wheat
Propiconazole + Tridemorph	Tilt Turbo 475 EC (03476)	Spring barley, spring wheat, winter barley, winter wheat.
Pyrazophos	Missile (03811)	Brussels sprout, barley (winter, spring)
Sulphur	Thiovit (02125)	Sugar beet
	Thiovit (05572)	Sugar Beet
Sulphur + Carbendazim + Maneb	Bolda FL (03463)	Winter wheat, spring wheat, winter barley, spring barley, winter oilseed rape, spring oilseed rape
Terbutryn	Prebane 500 SC (05851)	Cereal, rye, triticale
Thiometon	Ekatin (05281)	Wheat, durum wheat, barley, oats, triticale, rye
	WBC Systemic Aphicide (04989)	Wheat, durum wheat, barley, oats, triticale, rye
Tri-allate	Avadex BW Granular (00174)	Spring barley, winter barley, winter field bean, dried pea, vining pea, seed pea, winter wheat
	Avadex BW 15G (07117)	Spring barley, winter barley, winter field bean, dried pea, vining pea, seed pea, forage pea, winter wheat
Triadimefon	100-Plus (05112)	Winter wheat, spring wheat, winter rye, spring rye, winter oats, spring oats, winter barley, spring barley
	Bayleton (00221)	Winter barley, spring barley, winter oats, spring oats, winter rye, spring rye winter wheat, spring wheat.
	Bayleton 5 (00222)	Blackcurrant
	Standon Triadimefon 25 (05673)	Barley, oats, rye, wheat
Triadimefon + Carbendazim	200 Plus (05025)	Winter barley, winter rye
	Bayleton BM (00223)	Barley, wheat
Trichlorfon	Dipterex 80 (00711)	Beet crop, brassica, spinach

Active ingredient	Product (MAFF No.)	Crops/Uses
Tridemorph	Calixin (00369)	Spring barley, winter barley, spring oats, winter oats, winter wheat, swede, turnip
	Ringer (01557)	Winter wheat, barley, oats
	Standon Tridemorph 750 (05667)	Barley, oats, winter wheat, swede, turnip
Tridemorph + Carbendazim + Maneb	Cosmic FL (03473)	Winter wheat, winter barley, spring barley
Tridemorph + Propiconazole	Tilt Turbo 475 EC (03476)	Spring barley, spring wheat, winter barley, winter wheat.
Triforine	Saprol (01863)	Barley
Zineb-ethylene thiuram disulphide adduct	Polyram (02795)	Potato, spring barley, winter barley, winter wheat

ANNEX C
OFF-LABEL ARRANGEMENTS

THE REVISED LONG TERM ARRANGEMENTS FOR EXTENSION OF USE (1995)
SPECIFIC RESTRICTIONS FOR EXTENSION OF USE

Certain restrictions of use are necessary to ensure that the extension of use does not increase the risk to the operator, the consumer or the environment. For this reason **the following conditions must be followed** when applying pesticides under the terms of this scheme.

1. All safety precautions and statutory conditions relating to use (which are clearly identified in the statutory box on product labels) must be observed.

2. The method of application must be as stated on the pesticide label and in accordance with the relevant codes of practice and requirements under COSHH (Control of Substances Hazardous to Health).

3. Those planning to use hand held applicators to apply pesticides must ensure that this method of application is appropriate for the on-label uses (Note: unless otherwise stated spray applications to protected crops include hand held uses). Where this is not the case, users must only use pesticides which comply with the conditions at paragraph 11.

4. All reasonable precautions must be taken to safeguard wildlife and the environment.

5. Pesticides must only be used in the same situation as that on the product label i.e. outdoor or protected. Pesticides must not be used on protected crops, i.e. crops grown in glasshouses, poly tunnels, cloches or polythene covers or in any other building, unless the product label specifically allows use on that crop under protection. (N.B. unless specifically restricted to outdoor crops only, pesticides approved for use on tomatoes, cucumbers, lettuce, chrysanthemum and mushrooms include use under protection). Products approved for use only under protection must not be used outdoors.

6. Pesticides must not be used in or near water unless the product label specifically allows such use. Off-label use in or near water is not permitted under these arrangements. (In or near water includes drainage channels, streams, rivers, ponds, lakes, reservoirs, canals, dry ditches, areas designated for water storage. Use in or near coastal waters is also not permitted).

7. Off-label use by aerial application is not permitted under these arrangements.

8. Rodenticides and other vertebrate control agents are not included in these arrangements.

9. Use on land not intended for cropping, land not intended to bear vegetation, amenity grassland, managed amenity turf and amenity vegetation is not included in these arrangements (this includes areas such as paths, pavements, roads, ground around buildings, motorway verges, railway embankments, public parks, sports fields, upland areas, moorland areas, nature reserves etc.).

10. Pesticides classified as harmful, dangerous or extremely dangerous to bees must not be used during flowering on any crop (i.e. from first flower to complete petal fall) nor where bees are actively foraging (NB. If use during flowering is recommended for crops such as peas and oilseed rape this use relates *ONLY* to the label recommendations for these crops and must not be extrapolated to other crops under these arrangements). Applications of pesticides hazardous to bees must not be made when flowering weeds are present.

11. HAND HELD APPLICATIONS

When planning to use hand held equipment to apply a pesticide under these arrangements, users must ensure that hand held use is appropriate for the current on-label recommendations; spray applications to protected crops include hand held use unless stated otherwise. Where hand held use is not appropriate for the on-label recommendations hand held application should not be made if the pesticide label:

(a) prohibits hand held use;

(b) requires the use of personal protective clothing when using the pesticide diluted to the minimum volume rate recommended on the label for the dose required;

(c) is classified with one of the following hazard warnings:

"Corrosive", "very toxic", "toxic", or "risk of serious damage to eyes".

In other cases hand held application is permitted provided that:

(i) the concentration of the spray volume for the off-label use is no greater than the maximum concentration recommended on the pesticide label;

(ii) spray quality is at least as coarse as the British Crop Protection Council medium or coarse spray;

(iii) operators wear at least a coverall, gloves and rubber boots when applying pesticides below waist level. Use of a faceshield is also required for applications which are above waist height.

(iv) where there are label precautions with regard to buffer zone restrictions for vehicle mounted use, then users must observe a 2 m buffer zone when applying by hand held equipment.

12. BROADCAST AIR-ASSISTED USE

When planning to apply a pesticide under these arrangements by broadcast air-assisted sprayer (any equipment which broadcasts spray droplets produced by fan assistance and which carry outwards and upwards from the source of the spray), only pesticides with specific recommendations for such use or those currently approved for use on hops or any bush, cane and tree fruit can be used.

13. It is the responsibility of the user to ensure that the use does not result in any statutory UK Maximum Residue Levels (MRLs), as set out in statutory instrument No. 1985 of 1994: The Pesticides (Maximum Residue Levels in Crops, Food and Feeding Stuffs) Regulations 1994 (HMSO publication ISBN 0-11-044985-1) and any subsequent updates, being exceeded. These arrangements do not apply in either of the following situations:

(a) Where the MRL for the crop in column 1 of section IV "Crops used Wholly or Partly for Consumption by Humans or Livestock" is lower than the MRL for the crop in column 2.

(b) Where no MRL is set for the crop in Column 2 of section IV "Crops used Wholly or Partly for Consumption by Humans or Livestock" but a MRL at the limit of determination has been established for the crops in column 1.

In either of the above circumstances use on the crop in column 1 is *not* permitted.

14. These extensions of use only apply to label recommendations for use. Extrapolations must not be made from specific off-label approvals.

I. NON EDIBLE CROPS AND PLANTS

(a) Subject to the specific restrictions for extension of use set out above, pesticides provisionally or fully approved for use on any growing crop may be used on commercial agricultural and horticultural holdings and in forest nurseries on the following crops and plants

(i) hardy ornamental nursery stock, ornamental plants, ornamental bulbs and flowers and ornamental crops grown for seed where neither the seed nor any part of the plant is to be consumed by humans or animals;

(ii) forest nursery crops prior to final planting out.

(b) Subject to the specific restrictions for extension of use set out above, pesticides provisionally or fully approved for use on any growing *edible* crop may be used on commercial agricultural and horticultural holdings on non ornamental crops grown for seed where neither the seed nor any part of the plant is to be consumed by humans or animals. This extrapolation excludes use on potatoes, cereals, oilseeds, peas and beans grown for seed. Seed treatments themselves are not included in this extension of use.

(c) Subject to the specific restrictions for extension of use set out above, pesticides provisionally or fully approved for use on oilseed rape may be used on commercial agricultural and horticultural holdings on hemp grown for fibre. Seed treatments are not included in this extension of use.

Before making hand held applications see paragraph 11 of the specific restrictions for extension of use.

II. FARM FORESTRY AND ROTATIONAL COPPICING

Subject to the specific restrictions for extension of use set out above *herbicides* provisionally or fully approved for use on;

(i) cereals may be used in the first five years of establishment in farm forestry (including short rotation coppicing) on land previously under arable cultivation or improved grassland (as defined by the Farm Woodland Scheme);

(ii) cereals, oilseed rape, sugar beet, potatoes, peas and beans may be used in the first year of re-growth after cutting in coppices (short term, rotational, intensive wood production e.g. poplar or willow biofuel production) established on land previously under arable cultivation or improved grassland (as defined by the Farm Woodland Scheme).

III. NURSERY FRUIT CROPS

Subject to the specific restrictions for extension of use set out above, pesticides provisionally or fully approved for use on any crop for human or animal consumption may be used on commercial agricultural and horticultural holdings on nursery fruit trees, nursery vines prior to final planting out, bushes, canes and non-fruiting strawberry plants provided any fruit harvested for these crops within 1 year of treatment is destroyed. Applications must not be made where there are fruit present. If hand held or broadcast air assisted use is required see paragraphs 11 and 12, respectively, of the specific restrictions for extension of use.

IV. CROPS USED PARTLY OR WHOLLY FOR CONSUMPTION BY HUMANS OR LIVESTOCK

Subject to the specific restrictions for extension of use set out above pesticides may be used on commercial agricultural or horticultural holdings on the crops listed overleaf in the first column if they have been provisionally or fully approved for use on the crop(s) listed opposite them in the second column. These extrapolations do not include use in store which are subject to separate arrangements (at V)

TABLE ONE

Column 1: Minor use	Column 2: Crops on which use is provisionally or fully approved	Additional special conditions
A. ARABLE CROPS		
Poppy (grown for oilseed)	Sunflower	
Sesame	Sunflower	
Mustard	Oilseed rape	
Linseed	Oilseed rape	
Evening primrose	Oilseed rape	
Honesty	Oilseed rape	
Linola	Oilseed rape or linseed	
Flax (oilseed and fibre)	Oilseed rape or linseed	
Borage (grown for oilseed)	Oilseed rape	Seed treatments are not permitted
Grass seed crop	Wheat, barley, oats, rye, triticale	Treated crops must not be grazed or cut for fodder. Seed treatments are not permitted
Grass seed crop	Grass for grazing or fodder	
Rye	Wheat, barley	Treatments applied before second node detectable stage only
Triticale	Wheat, barley	
Durum wheat	Wheat	
Lupins	Combining peas or field beans	
B. FRUIT CROPS		
Almond } Application to the orchard floor ONLY	Apple or cherry or plum	For herbicides used on the orchard *floor* ONLY
Chestnut	Apple or cherry or plum	
Hazelnut	Apple or cherry or plum	
Walnut	Apple or cherry or plum	
For herbicides used on the orchard *floor* ONLY		
Quince	Apple or pear	
Crab apple	Apple or pear	
Almond	Products approved for use on two of the following: almond, chestnut, hazelnut and walnut	
Chestnut		
Hazelnut		
Walnut		
Nectarine	Peach	
Apricot	Peach	
Blackcurrant	Redcurrant	
Blackberry	Raspberry	
Rubus species (eg tayberry, loganberry)	Raspberry	
Dewberry	Raspberry	
Redcurrant	Blackcurrant	

Column 1: Minor use	Column 2: Crops on which use is provisionally or fully approved	Additional special conditions
Whitecurrant	Black or redcurrant	
Bilberry	Black or redcurrant	
Cranberry	Red or blackcurrant	
C. VEGETABLE CROPS		
Parsley root	Carrot or radish	
Fodder beet	Sugar beet	
Mangel	Sugar beet	
Horseradish	Carrot or radish	
Parsnip	Carrot	
Salsify	Carrot or celeriac	
Swede	Turnip	
Turnip	Swede	
Garlic	Bulb onion	
Shallot	Bulb onion	
Aubergine	Tomato	
Squash	Melon	
Pumpkin	Melon	
Marrow	Melon	
Watermelon	Melon	
Broccoli	Calabrese	
Calabrese	Broccoli	
Roscoff cauliflower	Cauliflower	
Collards	Kale	
Lamb's lettuce, frise, radicchio	Lettuce	
Beet leaves	Spinach	
Cress	Lettuce	
Scarole	Lettuce	
Leaf herbs and edible flowers*	Lettuce or spinach or parsley or sage or mint and tarragon	
Edible podded peas (e.g. mange-tout, sugar snap)	Edible podded beans	
Runner beans	Dwarf French beans	
Rhubarb	Celery	
Cardoon	Celery	
Edible fungi other than mushroon (e.g. oyster mushroom)	Mushroom	

* This extension of use applies to the following leaf herbs and edible flowers: angelica, balm, basil, bay, borage, burnet (salad), caraway, chamomile, chervil, chives, clary, coriander, dill, fennel, fenugreek, feverfew, hyssop, land cress, lovage, marjoram, marigold, mint, nasturtium, nettle, oregano, parsley, rocket, rosemary, rue, sage, savory, sorrel, tarragon, thyme, verbena (lemon), woodruff.

V. APPLICATION IN STORE ON CROPS USED PARTLY OR WHOLLY FOR HUMAN OR ANIMAL CONSUMPTION

Subject to the specific restrictions for extensions of use set out above pesticides may be used on commercial agricultural or horticultural holdings on the crops listed below in the first column if they have been provisionally or fully approved for use in store on the crops listed opposite them in the second column. *Seed treatments are not covered by this arrangement.*

Column 1: Minor Use *Column 2: Crops on which use is provisionally or fully approved*

Column 1: Minor Use	Column 2
Rye	Wheat
Barley	Wheat
Oats	Wheat
Buckwheat	Wheat
Millet	Wheat
Sorghum	Wheat
Triticale	Wheat
Dried peas	Dried beans
Dried beans	Dried peas
Mustard	Oilseed rape
Sunflower	Oilseed rape
Linola	Oilseed rape
Flax	Oilseed rape
Honesty	Oilseed rape
Poppy (grown for oilseed)	Oilseed rape
Borage (grown for oilseed)	Oilseed rape
Evening primrose	Oilseed rape
Sesame	Oilseed rape
Linseed	Oilseed rape

TABLE TWO

THE FOLLOWING EXTENSIONS OF USE WILL EXPIRE ON 31 DECEMBER 1995.

The following extensions of use are permitted for one year *only*, until 31 December 1995. After this date it will be a contravention of the Control of Pesticides Regulations 1986 to use these extrapolations.

Column 1: Minor use *Column 2: Crops on which use is provisionally or fully approved*

A: ARABLE CROPS

Column 1	Column 2
Poppy	Oilseed rape
Gold of Pleasure	Oilseed rape
Grass seed crops*	Wheat, barley, oats, rye, triticale
Oats	Wheat, barley – treatments applied before second node detectable stage ONLY
Durum wheat	Barley

Column 1: Minor use	Column 2: Crops on which use is provisionally or fully approved
B. FRUIT CROPS	
Redcurrant	Whitecurrant
Bilberry	Whitecurrant
C. VEGETABLE CROPS	
Beetroot	Carrot, radish
Celeriac	Carrot
Horseradish	Potato
Jerusalem artichoke	Potato, carrot, radish, turnip, swede
Salsify	Potato, radish
Garlic	Salad onion
Shallot	Salad onion
Pepper	Tomato
Squash	Cucumber, gherkin, courgette
Pumpkin	Cucumber, gherkin, courgette
Broccoli	Cauliflower
Calabrese	Cauliflower
Roscoff cauliflower	Broccoli, calabrese
Peas – non-edible podded (harvested green)	Vining peas
Broad beans	Vining peas, dwarf French beans
Runner beans	Edible podded peas
Beans – harvested dry	Peas, dwarf French, runner or broad beans
Peas – harvested dry	Vining peas or dwarf French, runner or broad beans
Fennel (as a vegetable)	Celery, leek
Kohlrabi	Celery, leek, cabbage

* This extension of use will be continuing in the revised Long Term Arrangements for Extension of Use at Table 1, with the proviso that treated grass seed crops are not consumed by livestock and seed treatments are not used.

ANNEX D
COMMODITY SUBSTANCE APPROVALS

HEALTH AND SAFETY EXECUTIVE

Food and Environment Protection Act 1985
Schedule: COMMODITY SUBSTANCE: **4-CHLORO-m-CRESOL**
Date of issue: 18 February 1993
Date of expiry: 28 February 2001

This approval is subject to the following conditions:

1 *FIELD OF USE*: ONLY AS A FUNGICIDE
2 *PEST AND USAGE AREA*: FOR THE CONTROL OF FUNGI ON INSECT SPECIMENS
3 *APPLICATION METHOD*: 4-CHLORO-m-CRESOL CRYSTALS IN A COLLECTING BOX

Operator protection:

(1) A written COSHH assessment must be made before using 4-chloro-m-cresol.

(2) Engineering control of operator exposure must be used where reasonably practicable in addition to the following items of personal protective equipment.

Operators must wear suitable protective clothing, including protective gloves and eye protection and a dust mask, when handling or applying the material.

(3) However engineering controls may replace personal protective equipment if a COSHH assessment shows they provide an equal or higher standard of protection.

Other specific restrictions:

(1) Operators should be provided with adequate information about the hazards of the substance and the precautions necessary for safe use. Sources of information include the supplier's Safety Data Sheet.

(2) Unprotected persons and animals must be excluded from any areas where treatment is taking place, and such areas should be ventilated after treatment.

(3) Must be used only by operators who are suitably trained and competent to carry out this work.

HEALTH AND SAFETY EXECUTIVE

Food and Environment Protection Act 1985
Schedule: COMMODITY SUBSTANCE: **CAMPHOR**
Date of issue: 18 February 1993
Date of expiry: 28 February 2001

This approval is subject to the following conditions:

1 *FIELD OF USE*: ONLY AS AN INSECT REPELLENT IN MUSEUMS AND BUILDINGS OF CULTURAL, ARTISTIC AND HISTORICAL INTEREST.
2 *PEST AND USAGE AREA*: FOR THE CONTROL OF FLYING AND CRAWLING INSECTS
3 *APPLICATION METHOD*: CRYSTALS OF CAMPHOR IN A SEALED SPECIMEN CASE

Operator protection:

(1) A written COSHH assessment must be made before using camphor. Operators should also observe the OES set out in HSE guidance note EH40/93 or subsequent issues.

(2) Engineering control of operator exposure must be used where reasonably practicable in addition to the following items of personal protective equipment.

Operators must wear suitable protective clothing, including protective gloves and eye protection, when handling or applying the material.

(3) However engineering controls may replace personal protective equipment if a COSHH assessment shows they provide an equal or higher standard of protection.

Other specific restrictions:

(1) Operators should be provided with adequate information about the hazards of the substance and the precautions necessary for safe use. Sources of information include the supplier's Safety Data Sheet.

(2) Unprotected persons and animals must be excluded from any areas where treatment is taking place, and such areas should be ventilated after treatment.

(3) Must be used only by operators who are suitably trained and competent to carry out this work.

Food and Environment Protection Act 1985
Control of Pesticides Regulations 1986 (SI 1986 No. 1510) : APPROVAL

In exercise of the powers conferred by Regulation 5 of the Control of Pesticides Regulations 1986 (SI 1986/1510) and of all other powers enabling them in that behalf, the Minister of Agriculture, Fisheries and Food and the Secretary of State hereby jointly give full approval for the use of:

Commodity substance: being 99.9% v/v **CARBON DIOXIDE** subject to the conditions set out below:

Date of issue: 8 October 1993

Use

Field of use: **Only as a rodenticide**

Situations: Trapped rodents.

Operator protection:

(1) Engineering control of operator exposure must be used where reasonably practicable in addition to the following personal protective equipment:

Operators must wear self-contained breathing apparatus when CO_2 levels are greater than 0.5% v/v.

(2) However, engineering controls may replace personal protective equipment if a COSHH assessment shows they provide an equal or higher standard of protection.

Other specific restrictions:

(1) Unprotected persons and non-target animals must be excluded from the treatment enclosures and from the area surrounding the treatment enclosures unless CO_2 levels are below 0.5% v/v.

(2) This substance must only be used by operators who are suitably trained and competent to carry out this work.

ADVISORY NOTE

This approval allows the use of CO_2 to destroy trapped rodent pests.

Food and Environment Protection Act 1985
Control of Pesticides Regulations 1986 (SI 1986 No. 1510) : APPROVAL

In exercise of the powers conferred by Regulation 5 of the Control of Pesticides Regulations 1986 (SI 1986/1510) and of all other powers enabling them in that behalf, the Minister of Agriculture, Fisheries and Food and the Secretary of State hereby jointly give full approval for the use of:

Commodity substance: being 99.9% v/v **CARBON DIOXIDE** subject to the conditions set out below:

Date of issue: 8 October 1993

Use

Field of use: **Only in vertebrate control**

Situations: Birds covered by general licences issued by the Agriculture and Environment Departments under Section 16(1) of the Wildlife and Countryside Act (1981) for the control of opportunistic bird species, where birds have been trapped or stupefied with alphachloralose/seconal.

Operator protection:

(1) Engineering control of operator exposure must be used where reasonably practicable in addition to the following personal protective equipment:

Operators must wear self-contained breathing apparatus when CO_2 levels are greater than 0.5% v/v.

(2) However, engineering controls may replace personal protective equipment if a COSHH assessment shows they provide an equal or higher standard of protection.

Other specific restrictions:

(1) Unprotected persons and non-target animals must be excluded from the treatment enclosures and from the area surrounding the treatment enclosures unless CO_2 levels are below 0.5% v/v.

(2) This substance must only be used by operators who are suitably trained and competent to carry out this work.

(3) Only to be used where a licence has been issued in accordance with Section 16(1) of the Wildlife and Countryside Act 1981 to permit the use of a substance otherwise prohibited under Section 5 of the Wildlife and Countryside Act 1981.

Food and Environment Protection Act 1985
Control of Pesticides Regulations 1986 (SI 1986 No. 1510) : APPROVAL

In exercise of the powers conferred by Regulation 5 of the Control of Pesticides Regulations 1986 (SI 1986/1510) and of all other powers enabling them in that behalf, the Minister of Agriculture, Fisheries and Food and the Secretary of State hereby jointly give full approval for the use of:

Commodity substance: being 99.9% v/v **CARBON DIOXIDE** subject to the conditions set out below:

Date of issue: 8 October 1993

Field of use: **Only as an insecticide, acaricide and rodenticide in food storage practice**

Situations: Raw and processed food commodities.

Operator protection:

(1) Engineering control of operator exposure must be used where reasonably practicable in addition to the following personal protective equipment:

Operators must wear self-contained breathing apparatus when CO_2 levels are greater than 0.5% v/v.

(2) However, engineering controls may replace personal protective equipment if a COSHH assessment shows they provide an equal or higher standard of protection.

Other specific restrictions:

(1) Unprotected persons and non-target animals must be excluded from the treatment enclosures and from the area surrounding the treatment enclosures unless CO_2 levels are below 0.5% v/v.

(2) This substance must only be used by operators who are suitably trained and competent to carry out this work.

ADVISORY NOTE

Ensure adequate ventilation of premises during all treatment and venting operations.

HEALTH AND SAFETY EXECUTIVE

Food and Environment Protection Act 1985
Schedule   COMMODITY SUBSTANCE: **CARBON DIOXIDE**
Date of issue: 18 February 1993
Date of Expiry: 28 February 2001

This approval is subject to the following conditions:

1 *FIELD OF USE*: ONLY AS A FUMIGANT IN WOOD PRESERVATION AND PUBLIC HYGIENE SITUATIONS.
2 *PEST AND USAGE AREA*: FOR THE CONTROL OF WOOD BORING AND OTHER FLYING AND CRAWLING INSECTS AND MITES
3 *APPLICATION METHOD*: INTRODUCTION INTO PURPOSE BUILT SEALED ENCLOSURES

Operator protection:

(1) A written COSHH assessment must be made before using carbon dioxide. Operators should also observe the OES set out in HSE guidance note EH40/93 or subsequent issues and COP30 "The control of substances hazardous to health in fumigation operations".

(2) Engineering control of operator exposure must be used where reasonably practicable in addition to the following items of personal protective equipment.

Operators must have access to suitable personal protective equipment including self contained breathing apparatus, when applying the gas, during the treatment process, and during venting.

(3) However engineering controls may replace personal protective equipment if a COSHH assessment shows they provide an equal or higher standard of protection.

Other specific restrictions:

(1) Operators should be provided with adequate information about the hazards of substance and the precautions necessary for safe use. Sources of information include the supplier's Safety Data Sheet.

(2) Unprotected persons and animals must be excluded from any areas where treatment is taking place, and such areas should be ventilated after treatment.

(3) Must be used only by operators who are suitably trained and competent to carry out this work.

HEALTH AND SAFETY EXECUTIVE

Food and Environment Protection Act 1985

Schedule: COMMODITY SUBSTANCE: **ETHANOL**

Date of issue: 18 February 1993

Date of expiry: 28 February 2001

This approval is subject to the following conditions:

1	*FIELD OF USE*:	i) AS AN INSECTICIDE IN MUSEUMS AND BUILDINGS OF CULTURAL, ARTISTIC AND HISTORICAL INTEREST. ii) AS A PRESERVATIVE IN MUSEUMS, AND BUILDINGS OF CULTURAL, ARTISTIC AND HISTORICAL INTEREST
2	*PEST AND USAGE AREA*:	FOR THE CONTROL OF FLYING AND CRAWLING INSECTS AND FUNGI
3	*APPLICATION METHOD*	i) IMMERSION IN A TANK ENCLOSED IN A FUME CUPBOARD ii) STORAGE OF SPECIMENS IN MATERIAL.

Operator protection

(1) A written COSHH assessment must be made before using ethanol. Operators should also observe the OES set out in HSE guidance note EH40/93 or subsequent issues.

(2) Engineering control of operator exposure must be used where reasonably practicable in addition to the following items of personal protective equipment.

Operators must wear suitable protective clothing, including protective gloves and eye protection, when handling or applying the material.

(3) However engineering controls may replace personal protective equipment if a COSHH assessment shows they provide an equal or higher standard of protection.

Other specific restrictions:

(1) Operators should be provided with adequate information about the hazards of the substance and the precautions necessary for safe use. Sources of information include the supplier's Safety Data Sheet.

(2) Unprotected persons and animals must be excluded from any areas where treatment is taking place, and such areas should be ventilated after treatment.

(3) Must be used only by operators who are suitably trained and competent to carry out this work.

HEALTH AND SAFETY EXECUTIVE

Food and Environment Protection Act 1985
Schedule: COMMODITY SUBSTANCE: **ETHYL ACETATE**
Date of issue: 18 February 1993
Date of expiry: 28 February 2001

This approval is subject to the following conditions;

1	*FIELD OF USE*:	ONLY AS AN INSECTICIDE IN MUSEUMS AND BUILDINGS OF CULTURAL, ARTISTIC AND HISTORICAL INTEREST.
2	*PEST AND USAGE AREA*:	FOR THE CONTROL OF FLYING AND CRAWLING INSECTS.
3	*APPLICATION METHOD;*	TREATMENT IN SEALED CONTAINERS.

Operation protection:

(1) A written COSHH assessment must be made before using ethyl acetate. Operators should also observe the OES set out in HSE guidance note EH40/93 or subsequent issues.

(2) Engineering control of operator exposure must be used where reasonably practicable in addition to the following items of personal protective equipment.

Operators must wear suitable protective clothing, including protective gloves and eye protection, when handling or applying the substance.

(3) However engineering controls may replace personal protective equipment if a COSHH assessment shows they provide an equal or higher standard of protection.

Other specific restrictions:

(1) Operators should be provided with adequate information about the hazards of the substance and the precautions necessary for safe use. Sources of information include the supplier's Safety Data Sheet.

(2) Unprotected persons and animals must be excluded from any areas where treatment is taking place, and such areas should be ventilated after treatment.

(3) Must be used only by operators who are suitably trained and competent to carry out this work.

CONTROL OF PESTICIDES REGULATIONS 1986 (SI 1986 NO 1510): APPROVAL

Level and Scope: Pursuant to Regulation 5 of the Control of Pesticides Regulations 1986 made under section 16 of the Food and Environment Protection Act 1985, the Minister of Agriculture, Fisheries and Food and the Secretary of State, acting jointly have granted provisional approval for the use of a

Commodity Substance: being **FORMALDEHYDE** (formalin 38-40% aqueous solution) subject to the conditions set out below:

Date of issue: 1 March 1991

Date of expiry: 28 February 2001 (see advisory note 3)

Use:

Field of Use: ONLY AS AN AGRICULTURAL/HORTICULTURAL FUNGICIDE AND STERILANT

Crops/Situations	*Maximum individual dose*	*Other specific restrictions* (1) Maximum concentration
Soil and compost sterilant, indoors and outdoors	As a drench: 0.5 litre formalin/m^2	1:4 parts water
Bulb dip	—	As a dip: 1:200 parts water
Mushroom houses	As a spray: 0.5 litre formalin/m^2	1:50 parts water
	As a fumigant: 400 ml formalin/$100m^3$ or 100 g of potassium permanganate added to 500 ml formalin/$100m^3$	
Greenhouse hygiene, boxes, pots etc.	As a fumigant: 400 ml formalin/$100m^3$ or 100 g of potassium permanganate added to 500 ml of formalin/$100m^3$	As a spray or dip: 1:50 parts water

Operator protection:

(1) A written COSHH assessment must be made before using formaldehyde. Operators should observe the Maximum Exposure Limit set out in HSE guidance note EH40/93 or subsequent issues and COP 30 "The control of substances hazardous to health in fumigation operations".

(2) Engineering control of operator exposure must be used where reasonably practicable in addition to the following personal protective equipment:

   (a) Operators must wear suitable respiratory equipment and other suitable protective equipment when handling or applying the fumigant.

   (b) Operators must wear suitable protective clothing and gloves when handling the concentrate.

(3) However, engineering controls may replace personal protective equipment if a COSHH assessment shows they provide an equal or higher standard of protection.

Other specific restrictions:

(1) Maximum concentration of formalin in water: see table.

(2) Operators must be supplied with a Section 6 (HSW) Safety Data Sheet before commencing work.

Advisory Note

1   Use as a disinfectant for equipment, greenhouse and public hygiene purposes are outside the Regulations.

2   Use of this substance for sterilising hatching eggs are outside the Regulations.

3   Data, as outlined in The Pesticides Register Issue No 8, 1990 (and subject to further amendment) must be submitted by interested parties by 28 February 1999 to ensure consideration of approval of this commodity substance beyond the stated expiry date.

CONTROL OF PESTICIDES REGULATIONS 1986 (SI 1986 NO 1510): APPROVAL

Level and Scope: Pursuant to Regulation 5 of the Control of Pesticides Regulations 1986 made under section 16 of the Food and Environment Protection Act 1985, the Minister of Agriculture, Fisheries and Food and the Secretary of State, acting jointly have granted provisional approval for the use of a

Commodity Substance: being **FORMALDEHYDE** (paraformaldehyde) subject to the conditions set out below:

Date of issue: 1 March 1991

Date of expiry: 28 February 2001 (see advisory note)

Use:

Field of Use: ONLY AS AN ANIMAL HUSBANDRY FUNGICIDE

Situations: Animal houses

Maximum individual dose: As a fumigant: 5g paraformaldehyde/m^3 or 20 g potassium permanganate added to 40 ml of formalin/m^3.

Operator protection:

(1) A written COSHH assessment must be made before using formaldehyde. Operators should observe the Maximum Exposure Limit set out in the HSE guidance note EH40/93 or subsequent issues and COP 30 "The control of substances hazardous to health in fumigation operations".

(2) Engineering control of operator exposure must be used where reasonably practicable in addition to the following personal protective equipment:

Operators must wear suitable respiratory equipment and other suitable protective equipment when handling or applying the fumigant.

(3) However, engineering controls may replace personal protective equipment if a COSHH assessment shows they provide an equal or higher standard of protection.

Other specific restrictions:

(1) Operators must remove livestock, feed, exposed milk and water and collect eggs before application.

(2) Operators must be supplied with a Section 6 (HSW) safety data sheet before commencing work.

Advisory Note

1   Data, as outlined in The Pesticides Register Issue No 8, 1990 (and subject to further amendment) must be submitted by interested parties by 28 February 1999 to ensure consideration of approval of this commodity substance beyond the stated expiry date.

HEALTH AND SAFETY EXECUTIVE

Food and Environment Protection Act 1985
Schedule: COMMODITY SUBSTANCE: **FORMALDEHYDE**
Date of issue: 18 February 1993
Date of expiry: 28 February 2001

This approval is subject to the following conditions:

1 *FIELD OF USE*:	i) AS AN INSECTICIDE IN MUSEUMS AND BUILDINGS OF CULTURAL, ARTISTIC AND HISTORICAL INTEREST ii) AS A PRESERVATIVE IN MUSEUMS AND BUILDINGS OF CULTURAL, ARTISTIC AND HISTORICAL INTEREST
2 *PEST AND USAGE AREA*	FOR THE CONTROL OF FLYING AND CRAWLING INSECTS AND FUNGI
3 *APPLICATION METHOD*:	i) TREATMENT IN SEALED CONTAINERS ii) STORAGE IN 5-l0% AQUEOUS SOLUTION

Operator protection:

(1) A written COSHH assessment must be made before using formaldehyde. Operators should also observe the MEL set out in HSE guidance note EH40/93 or subsequent issues.

(2) Engineering control of operator exposure must be used where reasonably practicable in addition to the following items of personal protective equipment.

Operators must wear suitable protective clothing, including protective gloves, eye protection and suitable respiratory protective equipment, when handling or applying the substance.

(3) However engineering controls may replace personal protective equipment if a COSHH assessment shows they provide an equal or higher standard of protection.

Other specific restrictions

(1) Operators should be provided with adequate information about the hazards of the substance and the precautions necessary for safe use. Sources of information include the supplier's Safety Data Sheet.

(2) Unprotected persons and animals must be excluded from any areas where treatment is taking place, and such areas should be ventilated after treatment.

(3) Must be used only by operators who are suitably trained and competent to carry out this work.

HEALTH AND SAFETY EXECUTIVE

Food and Environment Protection Act 1985

Schedule: COMMODITY SUBSTANCE: **ISOPROPANOL**

Date of issue: 18 February 1993

Date of expiry; 28 February 2001

This approval is subject to the following conditions;

1 *FIELD OF USE*:	ONLY AS A PRESERVATIVE IN MUSEUMS AND BUILDINGS OF CULTURAL, ARTISTIC AND HISTORICAL INTEREST.
2 *PEST AND USAGE AREA*:	FOR THE CONTROL OF FUNGI ON SPECIMENS
3 *APPLICATION METHOD*:	STORAGE OF SPECIMENS IN 50-60% AQUEOUS SOLUTION

Operator protection:

(1) A written COSHH assessment must be made before using isopropanol. Operators should also observe the OES set out in HSE guidance note EH40/93 or subsequent issues.

(2) Engineering control of operator exposure must be used where reasonably practicable in addition to the following items of personal protective equipment.

Operators must wear suitable protective clothing, including protective gloves and eye protection, when handling or applying the material.

(3) However engineering controls may replace personal protective equipment if a COSHH assessment shows they provide an equal or higher standard of protection.

Other specific restrictions:

(1) Operators should be provided with adequate information about the hazards of the substance and the precautions necessary for safe use. Sources of information include the supplier's Safety Data Sheet.

(2) Unprotected persons and animals must be excluded from any areas where treatment is taking place, and such areas should be ventilated after treatment.

(3) Must be used only by operators who are suitably trained and competent to carry out this work.

Food and Environment Protection Act 1985
Control of Pesticides Regulations 1986 (SI 1986 No. 1510) : APPROVAL

In exercise of the powers conferred by Regulation 5 of the Control of Pesticides Regulations 1986 (SI 1986/1510) and of all other powers enabling them in that behalf, the Minister of Agriculture, Fisheries and Food and the Secretary of State hereby jointly give full approval for the use of:

Commodity substance: being **METHYL BROMIDE** subject to the conditions set out below:

Date of issue: 22 November 1994
Field of use: **Only as a fumigant in vertebrate control**
Situations: Aircraft holds, ship holds.

Operator protection:

(1) A written COSHH assessment must be made before using methyl bromide.

(2) Engineering control of operator exposure must be used where reasonably practicable in addition to the following personal protective equipment:

Operators must wear suitable approved respiratory protective equipment and other suitable protective equipment when using the product.

Operators must *not* wear gloves or rubber boots when using the product.

(3) However, engineering controls may replace personal protective equipment if a COSHH assessment shows they provide an equal or higher standard of protection.

Other specific restrictions:

(1) Unprotected persons and animals must be kept out of aircraft holds or ship holds under fumigation or being aired following fumigation until any exposure levels are below the Occupational Exposure Standard.

(2) This product must only be used by professional operators of servicing companies, local authorities and Government departments who must be suitably trained and competent to carry out this work.

(3) Operators must refer to HSE guidance notes CS12, 'Fumigation with methyl bromide' and COP 30. 'The control of substances hazardous to health in fumigation operations' before using this product.

(4) Operators must be supplied with a Section 6 (HSWA) safety data sheet before commencing work.

(5) The container must not be re-used for any purpose.

ADVISORY NOTE

1. Approved products containing methyl bromide are available for agricultural, horticultural, food storage and space spray uses.

2. This field of use covers pesticides used for the control of rodents in aircraft holds and ship holds.

3. Data as outlined in the Pesticides Register issue No 8, 1990 (and subject to further amendment) must be submitted by interested parties by 28 February 1999 to ensure consideration of approval of this commodity substance beyond the stated expiry date.

HEALTH AND SAFETY EXECUTIVE

Food and Environment Protection Act 1985
Schedule: COMMODITY SUBSTANCE: **METHYL BROMIDE**
Date of Issue: 1 March 1991
Date of expiry: 28 February 2001 (see advisory note 3)

This approval is subject to the following conditions:

1. *FIELD OF USE:* ONLY AS A FUMIGANT IN PUBLIC HYGIENE
2. *PEST AND USAGE AREA:* FOR THE CONTROL OF FLYING AND CRAWLING INSECTS
3. *APPLICATION METHOD:* USERS SHOULD FOLLOW THE GUIDANCE GIVEN IN HSE GUIDANCE NOTE CS12 "FUMIGATION USING METHYL BROMIDE (BROMOETHANE)" (ISSUED 1986 AND REVISED IN 1991) AND COP 30 "THE CONTROL OF SUBSTANCES HAZARDOUS TO HEALTH IN FUMIGATION OPERATIONS"

Operator protection:

(1) A written COSHH assessment must be made before using methyl bromide.

(2) Engineering control of operator exposure must be used where reasonably practicable in addition to the following personal protective equipment:

Operators must wear suitable approved respiratory equipment and other suitable protective equipment when handling or applying the fumigant.

(3) However, engineering controls may replace personal protective equipment if a COSHH assessment shows they provide an equal or higher standard of protection.

Other specific restrictions:

(1) Must be used only by professional operators of servicing companies, local authorities and Government departments who must be suitably trained and competent to carry out this work.

(2) Operators must be supplied with a Section 6 (HSW) safety data sheet before commencing work.

(3) Unprotected persons and animals must be kept out of premises under fumigation or being aired following fumigation until any exposure levels are below the Occupational Exposure Standard.

Advisory Note

1 Approved products containing methyl bromide are available for agricultural, horticultural, food storage, rodenticide and space spray uses.

2 This field of use covers pesticides used for the control of harmful organism's, chiefly insects, detrimental to public health.

3 Data, as outlined in The Pesticides Register Issue no 8, 1990, (and subject to further amendment) must be submitted by interested parties by 28 February 1999 to ensure consideration of approval of this commodity substance beyond the stated expiry date.

HEALTH AND SAFETY EXECUTIVE

Food and Environment Protection Act 1985
Schedule: COMMODITY SUBSTANCE: **LIQUID NITROGEN**
Date of Issue: 18 February 1993
Date of expiry: 28 February 2001

This approval is subject to the following conditions:
1. *FIELD OF USE*: ONLY AS AN ACARICIDE IN DOMESTIC AND PUBLIC HYGIENE SITUATIONS.
2. *PEST AND USAGE AREA*: FOR THE CONTROL OF DUST MITES ON FURNISHINGS.
3. *APPLICATION METHOD*: SUITABLE LIQUID NITROGEN APPLICATOR.

Operator protection
(1) An assessment under The Management of Health and Safety at Work Regulation 1992 must be made before using liquid nitrogen.
(2) Operators must wear suitable protective clothing, including protective gloves, eye protection a dust mask and an oxygen monitor, when handling or applying the material.

Other specific restrictions:
(1) Operators should be provided with adequate information about the hazards of the substance and the precautions necessary for safe use. Sources of information include the supplier's Safety Data Sheet.
(2) Safe working procedures must be specified where it is possible that an oxygen deficient atmosphere could develop. Guidance on such procedures is given in HSE Guidance Note GS5 "Entry into confined spaces".
(3) Unprotected persons and animals must be excluded from any areas where treatment is taking place, and such areas should be ventilated after treatment.
(4) Must be used only by operators who are suitably trained and competent to carry out this work.

CONTROL OF PESTICIDES REGULATIONS 1986 (SI 1986 NO 1510): APPROVAL

Level and Scope: Pursuant to Regulation 5 of the Control of Pesticides Regulations 1986 made under section 16 of the Food and Environment Protection Act 1985, the Minister of Agriculture, Fisheries and Food and the Secretary of State, acting jointly have granted provisional approval for the use of a

Commodity Substance: being **STRYCHNINE HYDROCHLORIDE** subject to the conditions set out below:

Date of issue: 1 March 1991

Date of expiry: 28 February 1996 (see advisory note 3)

Sale or Supply

Label: Substance to be supplied with a label approved by the Registration authority.

Container: Substance to be supplied only in the original sealed packaging of the manufacturer and only in units of up to 2 g.

Storage: (1) Providers of a commercial service must not hold more strychnine than the amount specified by the authorising Agriculture Department.
(2) Substance must be stored in the original container under lock and key and only on the premises and under the control of the holder of an authority to purchase or a named individual.

Other: (1) Must only be supplied to the holders of an Authority to Purchase issued by the appropriate Agriculture Department (in England the Ministry of Agriculture, Fisheries and Food, in Wales the Welsh Office Agriculture Department, and in Scotland the Department of Agriculture and Fisheries for Scotland).
(2) Quantities of more than 8 g must only be supplied to providers of a commercial service.

Advertisement: The substance must not be advertised except that individual pharmacists may provide details of availability and price to a person authorised to purchase the substance.

Use:

Field of use: ONLY AS A VERTEBRATE CONTROL AGENT FOR THE DESTRUCTION OF MOLES UNDERGROUND

Situations: Commercial agricultural/horticultural land where public access is restricted. Grassland associated with aircraft landing strips, horse paddocks, gallops and race courses; golf courses; other areas specifically approved by Agricultural Departments.

Operator protection

(1) A written COSHH assessment must be made before using strychnine.

(2) Gloves must be worn when preparing and laying bait and when handling contaminated utensils.

Environmental protection:

(1) The substance must be prepared for applications with great care so that there is no contamination of the surface of the ground.

(2) Any prepared bait remaining at the end of the day must be buried.

Other specific restrictions:
(1) Authorities to purchase may only be issued to persons who satisfy the appropriate Agriculture Department that they are trained and competent in its use and can be entrusted with it.
(2) Access to strychnine must be restricted to those who hold the authority to purchase and to named individuals who satisfy the Agriculture Department.
(3) The substance must be used as and where directed by the appropriate Agriculture Department.
(4) Operators must be supplied with a Section 6 (HSW) Safety Data Sheet before commencing work.
(5) Providers of a commercial service must use suitable dedicated utensils capable of being washed clean. Such equipment must be cleaned after every treatment and washings disposed of in a mole run. The equipment must be stored securely between treatments.
(6) Non commercial users must use suitable disposable utensils which must be disposed of by burial on the land where the treatment takes place. They must not be retained for re-use.
(7) By 31 May, providers of a commercial service must advise the local office of the appropriate Agriculture Department of the treatments applied in the previous year, report the quantity of the substance held in store and the arrangements for secure storage.

Advisory Notes
1 Authorities to Purchase will be issued in addition to the Permits which are issued under the Poisons Rules. In deciding whether a person is fit to be entrusted with the substance, account will be taken of character and expertise.
2 Burial of contaminated materials should be in accordance with advice given in the Code of Practice for the Safe Use of Pesticides on Farms and Holdings (para 175).
3 Issue No 8 of the Pesticides Register 1990 detailed the data required for approval to continue beyond 28 February 1996. This data is currently under evaluation and full details will be published in the Pesticides Register once the outcome is known.

CONTROL OF PESTICIDES REGULATIONS 1986 (SI 1986 NO 1510): APPROVAL

Level and Scope: Pursuant to Regulation 5 of the Control of Pesticides Regulations 1986 made under section 16 of the Food and Environment Protection Act 1985, the Minister of Agriculture, Fisheries and Food and the Secretary of State, acting jointly have granted provisional approval for the use of a

Commodity Substance: being 77% clean **SULPHURIC ACID** subject to the conditions set out below:

Date of issue: 1 March 1991

Date of expiry: 28 February 1996 (see advisory note)

Use:

Field of Use: ONLY AS AN AGRICULTURAL DESICCANT

Crops	Maximum individual dose	Maximum number of treatments	Times of application
Potato (grown for canning)	800 litres substance/hectare	3 per crop*	1 May – 15 November
Potato (Other)	340 litres substance/hectare * See also other specific restrictions (9)	3 per crop*	1 May – 15 November
Linseed	280 litres substance/hectare	2 per crop	1 August – 15 November
Narcissus Bulbs	280 litres substance/hectare	1 per season	1 May – 15 November
Onions	280 litres substance/hectare	2 per crop	1 April – 30 August

Operator protection:

(1) A written COSHH assessment must be made before using sulphuric acid. Operators should observe the Occupational Exposure Standard set out in HSE guidance note EH40/90 or subsequent issues.

(2) Engineering control of operator exposure must be used where reasonably practicable in addition to the following personal protective equipment:

Operators must wear suitable protective clothing as listed below:

(a) When filling sprayer; spraying, carrying out adjustment to application equipment or cleaning equipment; re-entering the sprayed area within 24 hours.

Face shield of acid resistant type; acid resistant coveralls either single or combination garment (or for persons operating bulk installations, acid proof); gauntlet gloves either natural rubber or PVC material, rubber boots and acid proof apron.

(b) When re-entering the sprayed area between 24 and 96 hours.

Acid resistant coveralls, gloves and boots.

(3) However, engineering controls may replace personal protective equipment if a COSHH assessment shows they provide an equal or higher standard of protection.

(4) Operators must have liquid suitable for eye irrigation immediately available at all times throughout the operation.

Other specific restrictions:
(1) Must only be used by operators holding a relevant recognised certificate of competence in the use of the equipment for the application of sulphuric acid.
(2) Must not be applied using pedestrian controlled applicators or hand held equipment.
(3) All equipment must be constructed of materials suitable for use with or exposure to sulphuric acid.
(4) Application must be confined to the land intended to be treated.
(5) Spray must not be deposited within one metre of public footpaths.
(6) At least 24 hours written notice of the intended operation and the possibility of a hazard must be given to occupants of any premises and to the owner, or his agent, of any livestock or crops within 25 metres of any boundary of the land intended to be treated.
(7) Before the spraying takes place, readable notices must be posted on adjacent roads and paths warning passers by and drivers of vehicles of the time and place of the intended application and possibility of hazard. Notices to be kept in place for 96 hours following treatment
(8) Unprotected persons must be kept out of treated areas for at least 96 hours following treatment.
*(9) The maximum quantity to be applied to potatoes must not exceed 800 litres per hectare per crop.

Advisory Note
1 Issue No 8 of the Pesticides Register 1990 detailed the data required for approval to continue beyond 28 February 1996. This data is currently under evaluation and full details will be published in the Pesticides Register once the outcome is known.

HEALTH AND SAFETY EXECUTIVE

Food and Environment Protection Act 1985
Schedule: COMMODITY SUBSTANCE: **TETRACHLOROETHYLENE**
Date of issue: 18 February 1993
Date of expiry: 28 February 2001

This approval is subject to the following conditions:

1 *FIELD OF USE*: ONLY AS AN INSECTICIDE IN MUSEUMS AND BUILDINGS OF CULTURAL, ARTISTIC AND HISTORICAL INTEREST

2 *PEST AND USAGE AREA*: FOR THE CONTROL OF FLYING AND CRAWLING INSECTS ON TEXTILES.

3 *APPLICATION METHOD*: IMMERSION IN A TANK ENCLOSED IN A FUME CUPBOARD

Operator protection

(1) A written COSHH assessment must be made before using tetrachloroethylene. Operators should also observe the OES set out in HSE guidance note EH40/93 or subsequent issues.

(2) Engineering controls of operator exposure must be used where reasonably practicable in addition to the following items of personal protective equipment.

Operators must wear suitable protective clothing, including protective gloves and eye protection, when handling and using the material.

(3) However engineering controls may replace personal protective equipment if a COSHH assessment shows they provide an equal or higher standard of protection.

Other specific restrictions

(1) Operators should be provided with adequate information about the hazards of the substance and the precautions necessary for safe use. Sources of information include the supplier's Safety Data Sheet.

(2) Unprotected persons and animals must be excluded from any areas where treatment is taking place, and such areas should be ventilated after treatment.

(3) Must be used only by operators who are suitably trained and competent to carry out this work.

HEALTH AND SAFETY EXECUTIVE

Food and Environment Protection Act 1985
Schedule: COMMODITY SUBSTANCE: **THYMOL**
Date of issue: 18 February 1993
Date of expiry: 28 February 2001

This approval is subject to the following conditions

1 *FIELD OF USE*: ONLY AS A FUNGICIDE IN MUSEUMS AND BUILDINGS OF CULTURAL, ARTISTIC AND HISTORICAL INTEREST.
2 *PEST AND USAGE AREA*: FOR THE CONTROL OF FUNGI.
3 *APPLICATION METHOD*: TREATMENT IN SEALED TREATMENT CABINETS

Operator protection:

(1) A written COSHH assessment must be made before using thymol.
(2) Engineering control of operator exposure must be used where reasonably practicable in addition to the following items of personal protective equipment.

   Operators must wear suitable protective clothing, including protective gloves and eye protection, when handling or applying the material.
(3) However engineering controls may replace personal protective equipment if a COSHH assessment shows they provide an equal or higher standard of protection.

Other specific restrictions:

(1) Operators should be provided with adequate information about the hazards of the substance and the precautions necessary for safe use. Sources of information include the supplier's Safety Data Sheet.
(2) Unprotected persons and animals must be excluded from any areas where treatment is taking place, and such areas should be ventilated after treatment.
(3) Must be used only by operators who are suitably trained and competent to carry out this work.

CONTROL OF PESTICIDES REGULATIONS 1986 (SI 1986 NO 1510): APPROVAL

Level and Scope: Pursuant to Regulation 5 of the Control of Pesticides Regulations 1986 made under section 16 of the Food and Environment Protection Act 1985, the Minister of Agriculture, Fisheries and Food and the Secretary of State, acting jointly have granted provisional approval for the use of a

Commodity Substance: being **UREA** subject to the conditions set out below:

Date of issue: 1 March 1991

Date of expiry: 28 February 2001 (see advisory note 4)

Use:

Field of Use: ONLY AS A HOME GARDEN FUNGICIDE

Crop/Situation	*Maximum individual dose*	*Maximum number of treatments*	*Latest time of application*
Cut stumps of trees	1 litre of 37% w/v aqueous solution/m^2 of cut stump	One per cut stump	At felling
Apple and pear trees	0.07 litres of 7% w/v solution/m^2	One per tree per year	Post harvest, pre leaf-fall

Operator protection:

When used at work, the following must be observed:

(1) Engineering control of operator exposure must be used where reasonably practicable in addition to the following personal protective equipment:

Operators must wear suitable protective gloves when mixing urea solution.

(2) However, engineering controls may replace personal protective equipment if a COSHH assessment shows they provide an equal or higher standard of protection.

Advisory Notes

1 Pesticides approved for amateur use may be used by professional operators without user certification.

2 For the treatment of cut stumps of trees the following dyes may be used at 0.04% w/v concentration when mixed with solution of urea:
   (i) Kenacid Turquoise V5898 (CI Acid Blue 42045)
   (ii) Denacid Turquoise AN 200 } (CI Acid Blue 9 (42090)
        Duasyn Acid Blue AE-20)

3 Operators should wear rubber or other chemical-proof gloves, a nuisance dust mask and a cotton/terylene overall when handling dye powder.

4 Data, as outlined in The Pesticides Register Issue No 8, 1990 (and subject to further amendment) must be submitted by interested parties by 28 February 1999 to ensure consideration of approval of this commodity substance beyond the stated expiry date.

HEALTH AND SAFETY EXECUTIVE

Food and Environment Protection Act 1985
Schedule: COMMODITY SUBSTANCE: **WHITE SPIRIT**
Date of issue: 18 February 1993
Date of expiry: 28 February 2001

This approval is subject to the following conditions;

1. *FIELD OF USE*: ONLY AS AN INSECTICIDE IN MUSEUMS AND BUILDINGS OF CULTURAL, ARTISTIC AND HISTORICAL INTEREST
2. *PEST AND USAGE AREA*: FOR THE CONTROL OF FLYING AND CRAWLING INSECTS ON TEXTILES
3. *APPLICATION METHOD*: IMMERSION IN A TANK ENCLOSED IN A FUME CUPBOARD

Operator protection;

(1) A written COSHH assessment must be made before using white spirit. Operators should also observe the OES set out in HSE guidance note EH40/93 or subsequent issues.

(2) Engineering controls of operator exposure must be used where reasonably practicable in addition to the following items of personal protective equipment.

Operators must wear suitable protective clothing, including protective gloves and eye protection, when handling and using the material.

(3) However engineering controls may replace personal protective equipment if a COSHH assessment shows they provide an equal or higher standard of protection.

Other specific restrictions

(1) Operators must be provided with adequate information on hazards and precautions. Sources of information may include the supplier's Safety Data Sheet.

(2) Unprotected persons and animals must be excluded from any areas where treatment is taking place, and such areas should be ventilated after treatment.

(3) Must be used only by operators who are suitably trained and competent to carry out this work.

HEALTH AND SAFETY EXECUTIVE

Food and Environment Protection Act 1985

COMMODITY SUBSTANCE: **ETHYLENE DICHLORIDE**

The approval for use of ethylene dichloride as a commodity substance expired on 28 February 1994, as data to support its continued approval was not submitted.

ANNEX E
BANNED AND SEVERELY RESTRICTED PESTICIDES IN THE UK

Substances banned as pesticides in the UK as at 30 November 1994

Pesticide	Date	Reason for Ban
Aldrin	1989	Environmental hazard (persistent organochlorine). Banned under the 'Prohibition Directive'
Antu (thiourea)	1966	Evidence of carcinogenicity
Azobenzene	1975	Evidence of carcinogenicity
Binapacryl	1987	Evidence of carcinogenicity
Bitertanol	1985	Products voluntarily withdrawn due to evidence of teratogenicity
Cadmium compounds	1965	Evidence of carcinogenicity
Calcium arsenate	1968	High acute toxicity. Persistence in soil. Evidence of carcinogenicity
Campheclor	1984	Environmental hazard. Banned under the 'Prohibition Directive'.
Captafol	1989	Evidence of carcinogenicity
Chlordane	1992	Environmental hazard (persistent organochlorine). Banned under the 'Prohibition Directive'.
Chlordecone	1977	Evidence of carcinogenicity
Choline, potassium & sodium salts of maleic hydrazide containing more than 1 mg/kg of free hydrazine expressed on the basis of the acid equivalent	1991	Banned under the EC "Prohibition Directive"
Cyhexatin	1988	Evidence of teratogenicity
DDT	1964–1984	Environmental hazard (persistent organochlorine). High acute toxicity
Dicofol containing less than 78% of PP dicofol or more than 1g/kg DDT & DDT-related compounds	1991	Banned under the EC "Prohibition Directive"
Dieldrin	1989	Environmental hazard (persistent organochlorine)
Dinoseb (dinoseb acetate, dinoseb amine)	1988	Evidence of teratogenicity (potential danger to operators)

Dinoterb	as Dinoseb	
DNOC	1989	Evidence of teratogenicity in related dinitro compounds
Ethylene dibromide	1985	Evidence of carcinogenicity
Ethylene oxide	1990	Evidence of carcinogenicity
Endrin	as DDT	
HCH (containing less than 99% of the gamma isomer)	1979	Banned under the EC "Prohibition Directive"
Hexachlorobenzene	1975	Environmental hazard (persistent organochlorine)
Heptachlor	1981	Environmental hazard (persistent organochlorine). Banned under the 'Prohibition Directive'.
Maleic hydrazide & its salts other than its choline, potassium salts	1991	Banned under the EC "Prohibition Directive"
Mercuric chloride (mercury II chloride)	1965	High mammalian toxicity
Mercuric oxide (mercury oxide)	1992	Banned under the EC "Prohibition Directive"
Mercurous chloride	1992	Banned under the EC "Prohibition Directive"
Methyl mercury	1971	Environmental hazard (bird deaths associated with accumulation in the food chain)
Nitrofen	1981	Possible mutagenic, carcinogenic & teratogenic
Organomercury compounds	1992	Banned under the EC "Prohibition Directive"
Phenylmercury salicylate	1972	Acute toxicity accumulation in the environment
Potassium arsenite	1961	Acute toxicity to livestock & wildlife
Quintozene containing more than 1 g/kg of HCB or more than 10 g/kg pentachlorobenzene	1991	Banned under the EC "Prohibition Directive"
Selenium compounds eg sodium selenate	1962	Acute toxicity to humans & livestock
Sodium arsenite	1961	As potassium arsenite
1,1,2,2-tetrachloroethane	1969	Acute (narcotic) and chronic (liver damage) toxicity to humans

Substances severely restricted as pesticides in the UK as at 30 November 1994

Pesticide	Date last reviewed	Restriction
2 Aminobutane	1990	Permitted only on seed potatoes.
Atrazine	1992	Approval for use on non-crop land (excluding home garden use) revoked. Approvals for aerial use revoked. Restrictions on number of applications to crops.
Bromoxynil	1987	Approval for home garden use and application via hand-held applicators revoked other than via hand-held lances and pedestrian controlled sprayers (use via knapsack sprayers is not permitted). Timing restriction on grassland, leeks and onions.
	1990	Timing restrictions on some crops.
Dithiocarbamates (mancozeb, maneb, thiram, zineb)	1990	Restriction on number and time of applications to protected lettuce.
Inorganic fluorides eg sodium fluoride	1966	Not permitted for agricultural/horticultural food storage or amateur use.
Ioxynil	(as bromoxynil)	
Pentachlorophenol (PCP), Sodium Pentachlorophenoxide & Pentachlorophenyl Laurate	1992	Use as a wood preservative restricted to professional and industrial operators. Use as a surface biocide restricted to professional operators. Professional use permitted only in buildings of artistic, cultural and historic interest or in emergencies to treat against dry rot fungus or cubic rot fungus. Use by professionals must be notified prior to treatment, in accordance with agreed forms.
Simazine	(as atrazine)	
Tributyltin oxide (TBTO)	1987	Use in antifouling products restricted to boats over 25 metres in overall length.
	1990	Use as a wood preservative and surface biocide restricted to industrial process and in paste formulations applied by professional operators.
	1992	A maximum concentration of 1% only, within tributyltin copolymer antifouling products for professional (ie deep sea) use

ANNEX F
CHEMICALS SUBJECT TO THE POISONS LAW

Certain products in this book are subject to the provisions of the Poisons Act 1972, the Poisons List order 1982 and the Poisons Rules 1982 (copies of all these are obtainable from HMSO). These Rules include general and specific provisions for the storage and sale and supply of listed non-medicine poisons.

The chemicals approved for use in the UK and included in this book are specified under Parts I and II of the Poisons List as follows:

Part I Poisons (sale restricted to registered retail pharmacists)

aluminium phosphide
chloropicrin
fluoroacetamide
magnesium phosphide

methyl bromide
Sodium cyanide
Strychnine

Part II Poisons (sale restricted to registered retail pharmacists and listed sellers registered with local authority)

aldicarb
alphachloralose
ammonium bifluoride
azinphos-methyl
carbofuran (a)
chlorfenvinphos (a,b)
demeton-S-methyl
demeton-S-methyl sulphone
dichlorvos (a,e)
disulfoton (a)
DNOC*
drazoxolon (b)
endosulfan
fentin acetate
fentin hydroxide
fonofos (a)
formaldehyde
mephosfolan

methomyl (f)
mevinphos
nicotine (c)
omethoate
oxamyl (a)
oxydemeton-methyl
paraquat (d)
phorate (a)
pirimiphos-ethyl (b)
quinalphos
sodium fluoride
sulphuric acid
thiofanox (a)
thiometon
triazophos (a)
vamidothion
zinc phosphide

(a) Granular formulations which do not contain more than 12% w/w of this, or a combination of similarly flagged poisons, are exempt.
(b) Treatments on seeds are exempt
(c) Formulations containing not more than 7.5% of nicotine are exempt
(d) Pellets containing not more than 5% paraquat ion are exempt
(e) Preparations in aerosol dispensers containing not more than 1% w/w ai are exempt. Materials impregnated with dichlorvos for slow release are exempt.
(f) Solid substances containing not more than 1% w/w ai are exempt.

* Approvals for use of DNOC products revoked 31 December 1989

ANNEX G
List of ACP Published Evaluations

PUBLISHED EVALUATIONS, JANUARY 1995.

1. Flocoumafen	£3.00	
2. Quizalofop-ethyl	£3.50	
3. Cyfluthrin	£4.00	
4. Ethoprophos	£5.25	
5. Benfuracarb	£3.75	
6. RH 3866	£3.75	
7. DPX M6316	£3.50	
8. Azaconazole	£2.75	
9. Oxine copper	£3.25	
10. Fluasifop-p-butyl	£5.50	
11. Flusilazole	£10.50	
12. Bifenthrin	£3.50	
13. IPBC	£4.25	
14. Daminozide	£4.75	
15. Tributyltin napthenate	£2.75	
16. Ethylene bisdithiocarbamates (1)	£10.00	
17. Fenoxaprop-p-ethyl	£8.00	
18. Fenoxaprop ethyl	£10.25	
19. HOE 070542 Triazole Coformulant	£7.50	
20. PP321 (Lambda-cyhalothrin)	£4.75	
21. Cyhalothrin	£3.75	
22. Alachlor (1)	£8.50	
23. Fenpropathin	£9.50	
24. Tributyltin oxide (1)	£6.75	
25. Fentin hydroxide	£3.75	
26. Fenbutatin oxide	£2.75	
27. Fentine acetate	£3.00	
28. Iprodione	£3.75	
29. 2-Aminobutane	£4.00	
30. Dimethoate (1)	£3.50	
31. Cycloxydim	£12.25	
32. Dinocap	£8.90	
33. Glufosinate-ammonium	£21.50	
34. Vinclozolin	£11.75	
35. Diazinon (1)	£25.00	
36. Ethylene bisdithiocarbamates (2)	£25.00	
37. Chlorsulfuron	£4.20	
38. Metsulfuron methyl	£7.60	
39. Thifensulfuron methyl	£5.00	
40. Teflubenzuron	£1.50	
41. Alachlor (2)	£7.00	
42. Tefluthrin	£13.00	
43. Diazinon (2)	£2.75	
44. Hydroprene	£5.50	
45. SAN 619F (Cyproconazole)	£25.00	
46. Diclofop-methyl	£25.00	
47. Gamma-HCH (Lindane 1)	£4.00	
48. Triasulfuron	£25.00	
49. Thiodicarb	£25.00	
50. Fluoroglycofen-ethyl	£25.00	
51. Atrazine (1)	£9.50	
52. Simazine (1)	£9.50	
53. Guazatine	£9.50	
54. Thiabendazole	£10.00	
55. Esfenvalerate	£25.00	
56. Thiophanate-methyl	£11.00	
57. Benomyl	£14.00	
58. Carbendazim	£18.00	
59. Review of the use of Grain Protectants in the UK	£4.75	
60. Abamectin	£13.50	
61. Tribenuron methyl	£15.00	
62. Propamocarb hydrochloride	£21.00	
63. Methyl bromide	£12.00	
64. Gamma-HCH (Lindane 2)	£18.50	
65. Tebuconazole (1)	£25.00	
66. Imazaquin	£17.00	
67. Fenpropidin	£17.00	
68. 2,4-D	£25.00	
69. Tolclofos-methyl	£16.00	
70. Tralkoxydim	£25.00	
71. Atrazine (2)	£20.00	
72. Simazine (2)	£18.50	
73. Imidacloprid	£23.00	
74. Tributyltin oxide (2)	£4.50	
75. Desmedipham	£17.50	
76. Oxydemeton-methyl	£11.00	
77. Demeton-s-methyl	£4.00	
78. Fenpiclonil	£18.50	
79. Dimefuron	£11.50	
80. Propiconazole	£8.00	
81. Buprofezin	£19.00	
82. 2-Phenyl Phenol	£8.00	

83. Omethoate (1)	£13.50	102. Hydramethylnon	£6.50
84. Triazophos	£24.00	103. Commodity Substances	£6.00
85. S-Methoprene	£5.50	104. Lambda-Cyhalothrin	£3.00
86. Dimethoate (2)	£21.50	105. Bti (2)	£4.00
87. Chlorpropham	£5.50	106. Difenoconazole	£14.00
88. Tebuconazole (2)	£7.00	107. Chlorfenvinphos	£28.50
89. Cyromazine	£11.00	108. Epoxiconazole	£25.00
90. Kathon 886	£9.00	109. Aldicarb	£9.50
91. Amidosulfuron	£21.00	110. Tributyltin	£4.50
92. Bitertanol	£11.50	111. Triorganotin	£16.50
93. Anilazine	£19.50	112. Chlorothalonil	£22.50
94. Propaquizafop	£23.00	113. Vinclozolin	£12.50
95. Mecoprop	£17.00	114. Pentachlorophenol	£17.50
96. Mecoprop-P	£17.00	115. 3-Iodo-2-Propynyl-n-butyl Carbamate	£8.50
97. Triazoxide	£15.00	116. Cyromazine	£6.00
98. Phorate	£14.00	117. Diclofop-methyl (2)	£4.50
99. Dimethomorph	£21.50	118. Fomesafen	£15.00
100. Fluazinam	£17.00		
101. Bti (1)	£3.50		

Total cost of all evaluations (£)   £1389.45

Evaluations are available from the ACP Secretariat at Pesticides Safety Directorate, Room 308, Mallard House, Kings Pool, 3 Peasholme Green, York, YO1 2PX Tel. 01904 455705

ANNEX H
PRODUCTS APPROVED FOR USE IN OR NEAR WATER
(other than for Public Hygiene and Antifouling use)

Products approved for use in or near water as at 31 October 1994, are listed below. Before you use any product in or near water you should first consult the appropriate water regulatory body (National Rivers Authority/Local Rivers Purification Authority or in Scotland, the Local River Purification Board). Always read the label before use.

Product name	Marketing company	MAFF No.
Algae Kit	Ciba Agriculture	04545
Asulox	Rhone-Poulenc Agriculture	06124
Asulox	Rhone-Poulenc Agriculture Ltd	05235
Atlas 2,4-D	Atlas Interlates Ltd	03052
Barclay Gallup Amenity	Barclay Chemicals (UK) Ltd	06753
Blanc-kit	Intercel (UK)	04546
Bos MH 180	Bos Chemicals Ltd	04327
Casoron G	Imperial Chemical Industries Plc	00448
Casoron G	Zeneca Professional Products	06854
Casoron GSR	Imperial Chemical Industries Plc	00451
Casoron GSR	Zeneca Professional Products	06856
Clarosan 1FG	Ciba Agriculture	03859
Clayton Swath	Clayton Plant Protection (UK) Ltd	06715
Dormone	Rhone-Poulenc Agriculture Ltd	05412
Fydulan	Imperial Chemical Industries Plc	00958
Fydulan	Zeneca Professional Products	06823
Glyfonex	Danagri APS	06955
Glyphogan	Makhteshim-Agan (UK) Ltd	05784
Helosate	Helm AG	06499
Krenite	Du Pont (UK) Ltd	01165
Midstream	Imperial Chemical Industries Plc	01348
Midstream	Zeneca Professional Products	06824
MON 44068 Pro	Monsanto Plc	06815
MON 52276	Schering Agriculture	06949
MSS 2,4-D Amine	Mirfield Sales Services Ltd	01391
Reglone	Imperial Chemical Industries Plc	04444
Reglone	Zeneca Crop Protection	06703
Regulox K	Rhone-Poulenc Environmental Products	05405
Roundup	Monsanto Plc	01828
Roundup	Protectacrop Ltd	05532
Roundup 03947	Schering Agrochemicals Ltd	03947
Roundup A	Complete Weed Control Ltd	05463
Roundup Biactive	Monsanto Plc	06941
Roundup Biactive Dry	Monsanto Plc	06942
Roundup Pro	Monsanto Plc	04146
Roundup Pro Biactive	Monsanto Plc	06954

Product name	Marketing company	MAFF No.
Spasor	Rhone-Poulenc Environmental Products	03436
Spasor	Rhone-Poulenc Environmental Products	07211
Stetson	Monsanto Plc	06956

ANNEX I
LIST OF AUTHORISED ADJUVANTS

The following adjuvants are authorised for use with pesticides.

Adjuvant No	Product	Marketing Company	Use/Category
0001	High Trees Galion	Service Chemicals Ltd	Agricultural/ Horticultural Wetter and Spreader
0002	Tradename ADJ	Farm Protection Ltd	Agricultural Wetter and Sticker
0003	Farmon Blue	Farm Protection	Agricultural/ Horticultural Wetter and Spreader
0004	High Trees Mixture B	Service Chemicals Ltd	Agricultural Wetter and Spreader
0005	Cropspray 11E	Chiltern Farm Chemicals Ltd	Agricultural/ Horticultural Adjuvant Oil
0006	Tripart Cropspray 11E	Tripart Farm Chemicals Ltd	Agricultural/ Horticultural Adjuvant Oil
0009	High Trees Non-Ionic Wetter	Service Chemicals Ltd	Agricultural/ Horticultural Wetter and Spreader
0010	Quadrangle Cropspray 11E	Quadrangle Ltd	Agricultural/ Horticultural Adjuvant Oil
0011	Codacide Oil	Microcide Ltd	Agricultural/ Horticultural/ Forestry and Industrial Wetter, Sticker and Extender
0013	Actipron	Bayer UK Ltd	Agricultural Wetter
0015	Spreadite Liquid	Dow Agriculture	Agricultural Wetter and Spreader
0016	Tripart Minax	Tripart Farm Chemicals Ltd	Agricultural/ Horticultural Wetter and Spreader
0017	Exell	Truchem Ltd	Agricultural Surfactant
0018	Event	Industrial Detergents (UK) Ltd	Agricultural Surfactant
0019	Adder	Rhone Poulenc Agriculture	Agricultural Wetter
0020	Hyspray	Fine Agrochemicals Ltd	Agricultural Wetter and Spreader

Adjuvant No	Product	Marketing Company	Use/Category
0021	Atlas Adjuvant Oil	Atlas Interlates Ltd	Agricultural Adjuvant Oil
0022	Atlas Adze	Atlas Interlates Ltd	Agricultural/Horticultural Wetter and Spreader
0023	Atlas Adherbe	Atlas Interlates Ltd	Agricultural Adjuvant Oil
0024	Clifton Glyphosate Additive	Clifton Chemicals Ltd	Agricultural Wetter and Sticker
0025	Planet	Industrial Detergents (UK) Ltd	Agricultural Surfactant
0026	Genamin T-200 CS	Monsanto Plc	Agricultural Surfactant
0027	Sprayprover	Fine Agrochemicals Ltd	Agricultural Wetter
0028	Clifton Wetter	Clifton Chemicals Ltd	Agricultural Wetter
0029	Team Four 80	Monsanto Plc	Agricultural Surfactant
0030	Genamin T-200NF	Monsanto Plc	Agricultural Surfactant
0031	Emerald	Intracrop Ltd	Agricultural/Horticultural Extender
0033	Agral	ICI Agrochemicals	Agricultural/Horticultural Wetter
0034	Spreader	Pan Britannica Industries Ltd	Agricultural/Horticultural Wetter
0035	Vassgro Spreader	L W Vass (Agricultural) Ltd	Agricultural/Horticultural Wetter
0037	Spraymate Bond	Newman Agrochemicals Ltd	Agricultural Surfactant
0038	Spraymate LI 700	Newman Agrochemicals Ltd	Agricultural Wetter
0039	Nu-Film-P	Intracrop Ltd	Agricultural/Horticultural Sticker and Wetter
0042	Ethokem T/25	Midkem Ltd	Agricultural Wetter
0043	Spray Save	Antrad (UK) Ltd	Agricultural Wetter
0044	Frigate	ISK Biotech Europe Ltd	Agricultural Wetter and Sticker
0045	High Trees Wayfarer	Service Chemicals Ltd	Agricultural Surfactant
0047	Citowett	BASF (UK) Ltd	Agricultural Wetter
0049	Ethokem C/12	Midkem Ltd	Agricultural Wetter and Spreader
0050	Libsorb	Atlas Interlates Ltd	Agricultural/Horticultural Wetter and Spreader

Adjuvant No	Product	Marketing Company	Use/Category
0052	Ethokem	Midkem Ltd	Agricultural Wetter and Spreader
0053	Sprayfast	Mandops (Agricultural Specialists) Ltd	Agricultural/ Horticultural Wetter, Sticker and Extender
0054	Quadrangle Quad-Fast	Quadrangle Ltd	Agricultural/ Horticultural Sticker
0055	Nu-Film-17	Intracrop Ltd	Agricultural/ Horticultural Extender
0056	Agstock Addwett	Agstock Chemicals Ltd	Agricultural Wetter, Spreader and Sticker
0057	Agriwet	ABM Chemicals Ltd	Agricultural Wetter
0058	Agrisorb	ABM Chemicals Ltd	Agricultural Adjuvant Oil
0059	Lo Dose	Quadrangle Ltd	Agricultural Wetter and Sticker
0060	Enhance	Midkem Ltd	Agricultural/ Horticultural Wetter and Spreader
0061	Power Spray Save	Power Agrochemicals	Agricultural Wetter
0062	Activator 90	Newman Agrochemicals Ltd	Agricultural Wetter
0066	Cutinol	Midkem Ltd	Agricultural Adjuvant Oil
0067	Ashlade Adjuvant Oil	Ashlade Formulations Ltd	Agricultural Extender
0068	Enhance Low Foam	Midkem Ltd	Agricultural Wetter and Spreader
0069	Sterox NJ	Monsanto Plc	Agricultural Surfactant
0071	Stick-It	Quadrangle Ltd	Agricultural/ Horticultural Wetter
0072	Fyzol 11E	Schering Agrochemicals Ltd	Agricultural Adjuvant Oil
0073	Headland Guard	Headland Agrochemicals Ltd	Agricultural Sticker and Extender
0074	Headland Intake	Headland Agrochemicals Ltd	Agricultural Wetter
0075	Mangard	Mandops (Agrochemical Specialists) Ltd	Agricultural/ Horticultural Sticker and Extender
0079	Team Surfactant	Monsanto Plc	Agricultural Wetter
0080	Top Up Surfactant	Farmers Crop Chemicals Ltd	Agricultural Wetter

Adjuvant No	Product	Marketing Company	Use/Category
0081	Power Non-Ionic Wetter	Power Agrichemicals Ltd	Agricultural/Horticultural Wetter
0085	Swirl	Shell Chemicals UK Ltd	Agricultural Adjuvant Oil
0086	Concorde	ICI Plc	Agricultural Adjuvant Oil
0087	Minder	Stoller Chemical Ltd	Agricultural/Horticultural Wetter
0094	Tonic	Brown Butlin Group	Agricultural Sticker and Extender
0096	Tripart Tenax	Tripart Farm Chemicals	Agricultural Sticker and Extender
0097	Tripart Acer	Tripart Farm Chemicals	Agricultural Wetter
0098	Forestry Bee	Top Farm Formulations Ltd	Agricultural Wetter
0099	Pro-Mix	Service Chemicals Ltd	Agricultural/Horticultural Wetter
0100	Non-Ionic 90	Top Farm Formulations Ltd	Agricultural/Horticultural Wetter
0102	Ethywet	Top Farm Formulations Ltd	Agricultural/Horticultural Wetter
0103	Anphix	ANP Developments Ltd	Agricultural Wetter
0106	GS 800 Adjuvant	Midkem Ltd	Agricultural Wetter
0107	Keystone	Farm Protection Ltd	Agricultural/Horticultural Wetter
0108	Tripart Minax	Tripart Farm Chemicals Ltd	Agricultural/Horticultural Wetter and Spreader
0109	Jogral	Industrial Detergents (UK) Ltd	Agricultural Spreader
0110	Agropen	Industrial Detergents (UK) Ltd	Agricultural Wetter and Spreader
0111	Solar	Industrial Detergents (UK) Ltd	Agricultural Wetter
0112	Quadrangle Q900 Non-Ionic Wetter	Quadrangle Ltd	Agricultural/Horticultural Wetter
0113	Polycote Prime Polymer	Ciba Geigy Agrochemicals	Agricultural Sticker and Extender
0114	Polycote Polymer	Ciba Geigy Agrochemicals	Agricultural Sticker and Extender
0115	Polycote Pedigree Polymer	Ciba Geigy Agrochemicals	Agricultural Sticker and Extender
0116	Rapide	Intracrop Ltd	Agricultural Wetter
0117	Tripart Lentus	Tripart Farm Chemicals Ltd	Agricultural Sticker and Extender
0118	Clifton Alkyl 90	Clifton Chemicals Ltd	Agricultural/Horticultural Wetter and Spreader

Adjuvant No	Product	Marketing Company	Use/Category
0119	Du Pont Adjuvant	Du Pont (UK) Ltd	Agricultural/ Horticultural Wetter
0120	Farmon Wetter	Farm Protection Ltd	Agricultural/ Horticultural Wetter and Spreader
0121	Barclay Dryfast XL	Barclay Chemicals (UK) Ltd	Agricultural/ Horticultural Wetter, Sticker and Extender
0122	Stefes Spread and Seal	Stefes Plant Protection Ltd	Agricultural/ Horticultural Wetter, Sticker and Extender
0123	Barclay Dryfast	Barclay Chemicals (UK) Ltd	Agricultural/ Horticultural Wetter, Sticker and Extender
0124	Amos Non-Ionic Wetter	Kommer-Brookwick	Agricultural/ Horticultural Wetter and Spreader
0125	Intracrop BLA	Intracrop	Agricultural Sticker and Extender
0126	Barclay Actol	Barclay Chemicals (UK) Ltd	Agricultural/ Horticultural Adjuvant Oil
0127	SM99	Newman Agrochemicals Ltd	Agricultural/ Horticultural Adjuvant Oil
0128	Frigate	ISK Biotech Europe Ltd	Agricultural Wetter and Sticker
0129	Lo Dose	Quadrangle Agrochemicals	Agricultural Wetter and Sticker
0130	Cropspray 11E	Atlantis Oil & Chemical Co Ltd	Agricultural/ Horticultural Adjuvant Oil
0131	Sprayfast	Mandops (UK) Ltd	Agricultural/ Horticultural Wetter, Sticker and Extender
0132	Mangard	Mandops (UK) Ltd	Agricultural/ Horticultural Sticker and Extender
0134	SM99	Newman Agrochemicals Ltd	Agricultural/ Horticultural Adjuvant Oil
0135	Ashlade Adjuvant Oil	Ashlade Formulations Ltd	Agricultural Extender

Adjuvant No	Product	Marketing Company	Use/Category
0136	Axiom	Bayer UK Ltd	Agricultural/Horticultural Adjuvant Oil
0137	Cropoil	Chiltern Farm Chemicals Ltd	Agricultural/Horticultural Adjuvant Oil
0138	Cropoil	Quadrangle Agrochemicals Ltd	Agricultural/Horticultural Adjuvant Oil
0139	Luxan Non-Ionic Wetter	Luxan (UK) Ltd	Agricultural/Horticultural Wetter and Spreader
0140	Planet	Ideal Manufacturing Ltd	Agricultural Wetter
0141	Solar	Ideal Manufacturing Ltd	Agricultural Wetter
0142	Jogral	Ideal Manufacturing Ltd	Agricultural Wetter
0143	Agropen	Ideal Manufacturing Ltd	Agricultural Wetter
0144	Spraymate Spraymac	Newman Agrochemicals Ltd	Agricultural Wetter
0145	Spraymate Spray-fix	Newman Agrochemicals Ltd	Agricultural Sticker and Extender
0146	Ethokem	Techsol Manufacturing	Agricultural Wetter
0147	Enhance	Techsol Manufacturing	Agricultural/Horticultural Wetter and Spreader
0148	Enhance Low Foam	Techsol Manufacturing	Agricultural Wetter and Spreader
0149	Ethokem C12	Techsol Manufacturing	Agricultural Wetter and Spreader
0150	GS 800 Adjuvant	Techsol Manufacturing	Agricultural Wetter
0151	Cutinol	Techsol Manufacturing	Agricultural Adjuvant Oil
0153	Atplus 463	Zeneca Crop Protection	Agricultural Adjuvant Oil
0154	Agral	Zeneca Crop Protection	Agricultural/Horticultural Wetter
0155	Concorde	Zeneca Crop Protection	Agricultural Adjuvant Oil
0156	Nickerson Seed Film Coat SF 06578DB	Nickerson Seeds Ltd	Agricultural Sticker
0157	Nickerson Seed Film Coat SF 06578	Nickerson Seeds Ltd	Agricultural Sticker
0158	Nickerson Seed Film Coat SF 06711-40	Nickerson Seeds Ltd	Agricultural Sticker
0159	Nickerson Seed Film Coat SF 40268-30	Nickerson Seeds Ltd	Agricultural Sticker
0160	Nickerson Seed Film Coat SF 06578-R	Nickerson Seeds Ltd	Agricultural Sticker

Adjuvant No	Product	Marketing Company	Use/Category
0161	Mixture B	Service Chemicals Ltd	Agricultural/Horticultural/Forestry Wetter
0162	Galion	Intracrop	Agricultural/Horticultural Wetter and Spreader
0163	Output	Zeneca Crop Protection	Agricultural Adjuvant Oil
0164	Comulin	Joseph Batson & Co Ltd	Agricultural/Horticultural Adjuvant Oil
0165	Lyrol	Joseph Batson & Co Ltd	Agricultural/Horticultural Adjuvant Oil
0166	Atlas Companion	Atlas Interlates Ltd	Agricultural/Horticultural Wetter and Spreader
0167	Swirl	Cyanamid of GB Ltd	Agricultural Adjuvant Oil
0168	Ryda	Interagro (UK) Ltd	Agricultural Wetter
0169	Conka	Interagro (UK) Ltd	Agricultural Adjuvant Oil
0170	Sylgard 309	Dow Corning Ltd	Agricultural Wetter and Spreader
0172	Fyzol 11E	AgrEvo UK Crop Protection Ltd	Agricultural Adjuvant Oil
0173	Intracrop Bla-Tex	Intracrop Ltd	Agricultural Sticker and Extender
0174	Top Up	Farmers Crop Chemicals Ltd	Agricultural Wetter
0175	Intracrop Rapide Beta	Intracrop Ltd	Agricultural Wetter

PART A 1

PSD Registered Products

1
PROFESSIONAL PRODUCTS

PROFESSIONAL PRODUCTS: HERBICIDES

Product Name	Marketing Company	Reg. No.

1.1 Herbicides
including growth regulators, defoliants, rooting agents and desiccants

1 Alloxydim-sodium

*	Clout	Embetec Crop Protection	MAFF 04752

2 Amidosulfuron

*	Eagle	AgrEvo UK Crop Protection Ltd	MAFF 07318
*	Eagle	Hoechst UK Ltd	MAFF 06980
*	Pursuit	AgrEvo UK Crop Protection Ltd	MAFF 07333
*	Pursuit	Hoechst UK Ltd	MAFF 06981

3 Amitrole

*	Loft	A H Marks & Co Ltd	MAFF 06030
*	MSS Aminotriazole 80% WP	Mirfield Sales Services Ltd	MAFF 04374
*	MSS Aminotriazole Technical	Mirfield Sales Services Ltd	MAFF 04645
*	Weedazol - TL	A H Marks & Co Ltd	MAFF 02349
*	Weedazol-TL	Bayer Plc	MAFF 02979

4 Amitrole + *Atrazine*

*%	New Atraflow Plus	Rhone-Poulenc Agriculture Ltd	MAFF 05402

5 Amitrole + *Bromacil* + *Diuron*

*	BR Destral	Rhone-Poulenc Environmental Products	MAFF 05184

6 Amitrole + *2,4-D* + *Diuron*

*	Trik	Smyth-Morris Chemicals	MAFF 02182
*	Weedazol Total	A H Marks & Co Ltd	MAFF 02351

7 Amitrole + *Diuron*

*%	Amizol-D	A H Marks & Co Ltd	MAFF 00093
*	Orchard Herbicide	Hoechst UK Ltd	MAFF 03379

8 Amitrole + *Simazine*

*	Alpha Simazol	Makhteshim-Agan (UK) Ltd	MAFF 04799
*	Alpha Simazol T	Makhteshim-Agan (UK) Ltd	MAFF 04874
*%	CDA Simflow Plus	Rhone-Poulenc Agriculture Ltd	MAFF 05401
*%	Clearway	Rhone-Poulenc Agriculture Ltd	MAFF 05328
*	Mascot Highway WP	Complete Weed Control Ltd	MAFF 05600
*	MSS Simazine/Aminotriazole 43% FL	Mirfield Sales Services Ltd	MAFF 04361

* These products are "approved for agricultural use". For further details refer to page vii.
% These products are subject to the staged revocation procedure. For further details refer to page vii.
A These products are approved for use in or near water. For further details refer to page vii.

PROFESSIONAL PRODUCTS: HERBICIDES

Product Name	Marketing Company	Reg. No.

8 Amitrole + *Simazine*—continued

*%	Primatol SE 500 FW	Ciba-Geigy Agrochemicals	MAFF 01638
*%	Primatol SE 500 SC	Ciba-Geigy Agrochemicals	MAFF 05852
*	Ritefeed Simazine/Aminotriazole 43% FL	Ritefeed Ltd	MAFF 05087
*%	Simflow Plus	Rhone-Poulenc Agriculture Ltd	MAFF 05406

9 Ammonium sulphamate

*	Amcide	Battle Hayward & Bower Ltd	MAFF 04246

10 Anthracene oil

*	Sterilite Hop Defoliant	Coventry Chemicals Ltd	MAFF 05060

11 Asulam

*	Asulox	Embetec Crop Protection	MAFF 04413
A*	Asulox	Rhone-Poulenc Agriculture	MAFF 06124
A*	Asulox	Rhone-Poulenc Agriculture Ltd	MAFF 05235

12 Atrazine

*	Alpha Atrazine 50 SC	Makhteshim-Agan (UK) Ltd	MAFF 04877
*	Alpha Atrazine 50 WP	Alpha (GB) Ltd	MAFF 04793
*	Atlas Atrazine	Atlas Interlates Ltd	MAFF 03097
*	Atraflow	Burts & Harvey	MAFF 00160
*%	Atraflow	Rhone-Poulenc Agriculture Ltd	MAFF 05388
*%	Gesaprim 500 FW	Ciba-Geigy Agrochemicals	MAFF 00982
*	Gesaprim 500 SC	Ciba Agriculture	MAFF 05845
*	Ipitrax	I Pi Ci	MAFF 04693
*	MSS Atrazine 50 FL	Mirfield Sales Services Ltd	MAFF 01398
*	MSS Atrazine 80 WP	Mirfield Sales Services Ltd	MAFF 04360
*	Unicrop Atrazine 50	Universal Crop Protection Ltd	MAFF 02645
*	Unicrop Flowable Atrazine	Universal Crop Protection Ltd	MAFF 02268
*	Unicrop Flowable Atrazine	Universal Crop Protection Ltd	MAFF 05446

13 Atrazine + *Amitrole*

*%	New Atraflow Plus	Rhone-Poulenc Agriculture Ltd	MAFF 05402

14 Atrazine + *Sodium chlorate*

*	Atlacide Extra Dusting Powder	Chipman Ltd	MAFF 00124

15 Aziprotryne

*	Brasoran 50 WP	Ciba Agriculture	MAFF 00316

---

* These products are "approved for agricultural use". For further details refer to page vii.
% These products are subject to the staged revocation procedure. For further details refer to page vii.
A These products are approved for use in or near water. For further details refer to page vii.

PROFESSIONAL PRODUCTS: HERBICIDES

Product Name	Marketing Company	Reg. No.

16 Benazolin
* Galtak 50 SC	AgrEvo UK Crop Protection Ltd	MAFF 07258
* Galtak 50 SC	Schering Agrochemicals Ltd	MAFF 04928

17 Benazolin + *Bromoxynil* + *Ioxynil*
* Asset	AgrEvo UK Crop Protection Ltd	MAFF 07243
* Asset	Schering Agrochemicals Ltd	MAFF 03824

18 Benazolin + *Clopyralid*
* Benazalox	AgrEvo UK Crop Protection Ltd	MAFF 07246
* Benazalox	Schering Agrochemicals Ltd	MAFF 03858
* Landgold Benazolin Plus	Landgold & Co Ltd	MAFF 06048

19 Benazolin + *Clopyralid* + *Dimefuron*
* Scorpio	AgrEvo UK Crop Protection Ltd	MAFF 07276
* Scorpio	Schering Agriculture	MAFF 05859

20 Benazolin + *2,4-D* + *MCPA*
* Legumex Extra	AgrEvo UK Crop Protection Ltd	MAFF 07262

21 Benazolin + *2,4-Db* + *MCPA*
* Legumex Extra	Schering Agrochemicals Ltd	MAFF 03869
* Setter 33	AgrEvo UK Crop Protection Ltd	MAFF 07282
*% Setter 33	Dow Chemical Co Ltd	MAFF 04377
* Setter 33	DowElanco Ltd	MAFF 05623
* Setter 33	Schering Agrochemicals Ltd	MAFF 04376

22 Benazolin + *Dicamba* + *Dichlorprop*
*% Celt	Schering Agrochemicals Ltd	MAFF 04464

23 Benazolin + *Dimefuron*
* Scorpio 400	AgrEvo UK Crop Protection Ltd	MAFF 07277
* Scorpio 400	Schering Agriculture	MAFF 05858

24 Bentazone
* Basagran	BASF Plc	MAFF 00188

25 Bentazone + *Cyanazine* + *2,4-Db*
* Topshot	Cyanamid Of GB Ltd	MAFF 07178
* Topshot	Shell Chemicals UK Ltd	MAFF 03576

26 Bentazone + *MCPA* + *MCPB*
* Acumen	BASF Plc	MAFF 00028

* These products are "approved for agricultural use". For further details refer to page vii.
% These products are subject to the staged revocation procedure. For further details refer to page vii.
A These products are approved for use in or near water. For further details refer to page vii.

PROFESSIONAL PRODUCTS: HERBICIDES

Product Name	Marketing Company	Reg. No.
27 Bentazone + *MCPB*		
* Pulsar	BASF Plc	MAFF 04002
28 Benzoylprop-ethyl		
* Suffix	Shell Chemicals UK Ltd	MAFF 02037
29 Bifenox + *Chlorotoluron*		
*% Dicurane Duo 446 FW	Ciba-Geigy Agrochemicals	MAFF 04839
* Dicurane Duo 446 SC	Ciba Agriculture	MAFF 05839
*% Dicurane Duo 495 FW	Ciba-Geigy Agrochemicals	MAFF 03860
*% Dicurane Duo 495 SC	Ciba Agriculture	MAFF 05838
30 Bifenox + *Cyanazine*		
* FR 1442	Rhone-Poulenc Agriculture Ltd	MAFF 04926
* Vantage	Cyanamid of GB Ltd.	MAFF 07188
* Vantage	Shell Chemicals UK Ltd	MAFF 04927
31 Bifenox + *Dicamba*		
*% Quickstep	Farm Protection Ltd	MAFF 04749
* RP 283	Rhone-Poulenc Agriculture Ltd	MAFF 06178
*% RP 283	Rhone-Poulenc Agriculture Ltd	MAFF 04767
* SP 283	Sandoz Agro Ltd	MAFF 04863
32 Bifenox + *Isoproturon*		
* Invicta	Farm Protection	MAFF 04825
* RP 4169	Rhone-Poulenc Agriculture Ltd	MAFF 05801
33 Bifenox + *Mecoprop*		
*% EXP 8506	Rhone-Poulenc Crop Protection	MAFF 04886
34 Bromacil		
* Alpha Bromacil 80 WP	Makhteshim-Agan (UK) Ltd	MAFF 04802
*% Borocil 1.5	Rhone-Poulenc Agriculture Ltd	MAFF 05063
* Hyvar X	Du Pont (UK) Ltd	MAFF 01105
35 Bromacil + *Amitrole* + *Diuron*		
* BR Destral	Rhone-Poulenc Environmental Products	MAFF 05184
36 Bromacil + *Diuron*		
* Borocil K	Rhone-Poulenc Environmental Products	MAFF 05183
*% Krovar 1	Du Pont (UK) Ltd	MAFF 01166

* These products are "approved for agricultural use". For further details refer to page vii.
% These products are subject to the staged revocation procedure. For further details refer to page vii.
A These products are approved for use in or near water. For further details refer to page vii.

PROFESSIONAL PRODUCTS: HERBICIDES

Product Name	Marketing Company	Reg. No.

37 Bromacil + Picloram
* Hydon	Chipman Ltd	MAFF 01088

38 Bromoxynil + Benazolin + Ioxynil
* Asset	AgrEvo UK Crop Protection Ltd	MAFF 07243
* Asset	Schering Agrochemicals Ltd	MAFF 03824

39 Bromoxynil + Clopyralid
* Vindex	DowElanco Ltd	MAFF 05470
*% Vindex	Quadrangle Ltd	MAFF 04049

40 Bromoxynil + Clopyralid + Fluroxypyr + Ioxynil
* Crusader S	DowElanco Ltd	MAFF 05174

41 Bromoxynil + Dichlorprop + Ioxynil + MCPA
* Atlas Minerva	Atlas Interlates Ltd	MAFF 03046

42 Bromoxynil + Diflufenican + Ioxynil
* Capture	Rhone-Poulenc Agriculture Ltd	MAFF 06881
* Quartz BL	Rhone-Poulenc Agriculture Ltd	MAFF 06123

43 Bromoxynil + Ethofumesate + Ioxynil
* Leyclene	AgrEvo UK Crop Protection Ltd	MAFF 07263
* Leyclene	Schering Agrochemicals Ltd	MAFF 05285
* Trapper	DowElanco Ltd	MAFF 06325

44 Bromoxynil + Fluroxypyr
* Sickle	DowElanco Ltd	MAFF 05187

45 Bromoxynil + Fluroxypyr + Ioxynil
* Advance	DowElanco Ltd	MAFF 05173

46 Bromoxynil + Ioxynil
* Agri HBN 400	Agri-Export Ltd	MAFF 04257
* Alpha Biotril Plus	Makhteshim-Agan (UK) Ltd	MAFF 04878
* Alpha Briotril	Makhteshim-Agan (UK) Ltd	MAFF 04876
* Alpha Briotril Plus 19/19	Makhteshim-Agan (UK) Ltd	MAFF 04740
* Deloxil	A H Marks & Co Ltd	MAFF 00663
* Deloxil	AgrEvo UK Crop Protection Ltd	MAFF 07313
* Deloxil	Hoechst UK Ltd	MAFF 00664
* Deloxil 400	AgrEvo UK Crop Protection Ltd	MAFF 07314
* Deloxil 400	Hoechst UK Ltd	MAFF 06613
*% FR 1001	Rhone-Poulenc Agriculture Ltd	MAFF 05233

---

* These products are "approved for agricultural use". For further details refer to page vii.
% These products are subject to the staged revocation procedure. For further details refer to page vii.
A These products are approved for use in or near water. For further details refer to page vii.

PROFESSIONAL PRODUCTS: HERBICIDES

Product Name	Marketing Company	Reg. No.

46 Bromoxynil + *Ioxynil*—continued

*	Oxytril CM	Rhone-Poulenc Agriculture	MAFF 06201
*	Oxytril CM	Rhone-Poulenc Crop Protection	MAFF 04605
*	Percept	MTM Agrochemicals Ltd	MAFF 05481
*%	RP 1001	Rhone-Poulenc Crop Protection	MAFF 04835
*	Status	Rhone-Poulenc Agriculture Ltd	MAFF 06906
*	Stellox 380 EC	Ciba Agriculture	MAFF 04169
*	Stellox 60 WG	Ciba Agriculture	MAFF 04626

47 Bromoxynil + *Ioxynil* + *Isoproturon*

*%	Astrol	Rhone-Poulenc Agriculture Ltd	MAFF 05441
*	Doublet	Rhone-Poulenc Agriculture	MAFF 06198
*	Doublet	Rhone-Poulenc Agriculture Ltd	MAFF 05356
*%	Twin-Tak	Rhone-Poulenc Agriculture Ltd	MAFF 05379

48 Bromoxynil + *Ioxynil* + *Isoproturon* + *Mecoprop*

*	Terset	Rhone-Poulenc Agriculture Ltd	MAFF 06171
*%	Terset	Rhone-Poulenc Agriculture Ltd	MAFF 05227

49 Bromoxynil + *Ioxynil* + *Mecoprop*

*%	Brittox	May & Baker Ltd	MAFF 00332
*	RP 83/8	Rhone-Poulenc Agriculture	MAFF 06169
*%	RP 83/8	Rhone-Poulenc Crop Protection	MAFF 05681
*	Swipe 560 EC	Ciba Agriculture	MAFF 02057

50 Bromoxynil + *Ioxynil* + *Triasulfuron*

*	Teal	Ciba Agriculture	MAFF 06117
*	Teal G	Ciba Agriculture	MAFF 06118

51 Bromoxynil + *Ioxynil* + *Trifluralin*

*	Masterspray	Pan Britannica Industries Ltd	MAFF 02971

52 Carbendazim + *Tecnazene*

*	Hickstor 6 Plus MBC	Hickson & Welch Ltd	MAFF 04176
%	Hortag Carbotec	Hortag Chemicals Ltd	MAFF 04135
*	Hortag Tecnacarb	Avon Packers Ltd	MAFF 02929
*	New Arena Plus	Hickson & Welch Ltd	MAFF 04598
*	New Hickstor 6 Plus MBC	Hickson & Welch Ltd	MAFF 04599
	Tripart Arena Plus	Hickson & Welch Ltd	MAFF 05602

53 Carbetamide

*	Carbetamex	Embetec Crop Protection	MAFF 04415
*	Carbetamex	Rhone-Poulenc Agriculture	MAFF 06186
*	Carbetamex	Rhone-Poulenc Crop Protection	MAFF 05236

---

* These products are "approved for agricultural use". For further details refer to page vii.
% These products are subject to the staged revocation procedure. For further details refer to page vii.
A These products are approved for use in or near water. For further details refer to page vii.

PROFESSIONAL PRODUCTS: HERBICIDES

Product Name	Marketing Company	Reg. No.
54 Carbetamide + *Dimefuron*		
*% Pradone Plus	Rhone-Poulenc Agriculture Ltd	MAFF 05857
55 Chlorbufam + *Chloridazon*		
* Alicep	BASF Plc	MAFF 00077
56 Chloridazon		
* Atlas Silver	Atlas Interlates Ltd	MAFF 03547
* Barclay Champion	Barclay Chemicals (UK) Ltd	MAFF 06903
* Barclay Claddagh	Barclay Chemicals (UK) Ltd	MAFF 06330
* Better DF	Sipcam UK Ltd	MAFF 06250
* Better Flowable	Sipcam UK Ltd	MAFF 04924
* Gladiator	Tripart Farm Chemicals Ltd	MAFF 00986
* Gladiator DF	Tripart Farm Chemicals Ltd	MAFF 06342
* Hyzon	Agrichem Ltd	MAFF 01106
* Luxan Chloridazon	Luxan (UK) Ltd	MAFF 06279
* Luxan Chloridazon	Luxan (UK) Ltd	MAFF 06304
* Luxan Chloridazon	Luxan (UK) Ltd	MAFF 06555
* New Murbetex Fl	DowElanco Ltd	MAFF 04235
* PA Weedmaster	Portman Agrochemicals Ltd	MAFF 02550
*% Phyto-Chloridazon	Phyto Research Pflanzenschutz GmbH	MAFF 04652
* Portman Weedmaster	Portman Agrochemicals Ltd	MAFF 06018
*% Power Chloridazon	Kommer-Brookwick	MAFF 05754
* Power Chloridazon	Power Agrichemicals Ltd	MAFF 02717
* Pyramin DF	BASF Plc	MAFF 03438
* Pyramin FL	BASF Plc	MAFF 01661
* Starter Flowable	Truchem Ltd	MAFF 03421
* Stefes Chloridazon	Stefes Plant Protection Ltd	MAFF 05777
* Takron	BASF Plc	MAFF 06237
* Tripart Gladiator 2	Tripart Farm Chemicals Ltd	MAFF 06618
* Trojan SC	AgrEvo UK Crop Protection Ltd	MAFF 07298
* Trojan SC	Schering Agrochemicals Ltd	MAFF 03926
57 Chloridazon + *Chlorbufam*		
* Alicep	BASF Plc	MAFF 00077
58 Chloridazon + *Chlorpropham + Fenuron + Propham*		
* Atlas Electrum	Atlas Interlates Ltd	MAFF 03548
59 Chloridazon + *Ethofumesate*		
* Magnum	BASF Plc	MAFF 01237
* Spectron	AgrEvo UK Crop Protection Ltd	MAFF 07284
* Spectron	Schering Agrochemicals Ltd	MAFF 03828

*   These products are "approved for agricultural use". For further details refer to page vii.
% These products are subject to the staged revocation procedure. For further details refer to page vii.
A These products are approved for use in or near water. For further details refer to page vii.

PROFESSIONAL PRODUCTS: HERBICIDES

Product Name	Marketing Company	Reg. No.

60 Chloridazon + *Lenacil*

* Advizor	Du Pont (UK) Ltd	MAFF 06571
* Advizor	Farm Protection Ltd	MAFF 00032
* Advizor S	Du Pont (UK) Ltd	MAFF 06960
* Varmint	Pan Britannica Industries Ltd	MAFF 01561

61 Chloridazon + *Propachlor*

* Ashlade CP	Ashlade Formulations Ltd	MAFF 06481
*% Ashlade CP	Ashlade Formulations Ltd	MAFF 02852

62 Chlormequat

* 3C Chlormequat 460	Pennine Chemical Services Ltd	MAFF 03916
* 3C Chlormequat 600	Pennine Chemical Services Ltd	MAFF 04079
*% 5 Star Chlormequat	Rhone-Poulenc Crop Protection	MAFF 05208
*% 5C Chlormequat Plus	Rhone-Poulenc Agriculture Ltd	MAFF 05205
*% ABM Chlormequat 40	Rhone-Poulenc Crop Protection	MAFF 05206
*% ABM Chlormequat 67.5	Rhone-Poulenc Crop Protection	MAFF 05207
* Adjust	Mandops (UK) Ltd	MAFF 05589
*% Alpha Chlormequat 460	Alpha (GB) Ltd	MAFF 00506
* Alpha Chlormequat 460	Makhteshim-Agan (UK) Ltd	MAFF 04804
* Alpha Pentagan	Makhteshim-Agan (UK) Ltd	MAFF 04794
*% Alpha Pentagan Extra	Alpha (GB) Ltd	MAFF 02987
* Alpha Pentagan Extra	Makhteshim-Agan (UK) Ltd	MAFF 04796
*% Arotex Extra	Imperial Chemical Industries Plc	MAFF 00117
*% Ashlade 460 CCC	Ashlade Formulations Ltd	MAFF 02894
* Ashlade 460 CCC	Ashlade Formulations Ltd	MAFF 06474
* Ashlade 5C	Ashlade Formulations Ltd	MAFF 06227
*% Ashlade 5C	Ashlade Formulations Ltd	MAFF 04727
* Ashlade 700 5C	Ashlade Formulations Ltd	MAFF 07046
* Ashlade 700 CCC	Ashlade Formulations Ltd	MAFF 06473
*% Ashlade 700 CCC	Ashlade Formulations Ltd	MAFF 02912
* Atlas 3C:645 Chlormequat	Atlas Interlates Ltd	MAFF 05710
* Atlas 5C Chlormequat	Atlas Interlates Ltd	MAFF 03084
* Atlas Chlormequat 46	Atlas Interlates Ltd	MAFF 05660
* Atlas Chlormequat 460:46	Atlas Interlates Ltd	MAFF 06258
*% Atlas Chlormequat 640 5C	Atlas Interlates Ltd	MAFF 05888
*% Atlas Chlormequat 670	Atlas Interlates Ltd	MAFF 06615
* Atlas Chlormequat 700	Atlas Interlates Ltd	MAFF 03402
*% Atlas Chlormequat 730	Atlas Interlates Ltd	MAFF 06511
* Atlas Quintacel	Atlas Interlates Ltd	MAFF 06524
* Atlas Terbine	Atlas Interlates Ltd	MAFF 06523
* Atlas Tricol	Atlas Interlates Ltd	MAFF 06522
* Atlas Tricol	Atlas Interlates Ltd	MAFF 07190
* Barclay Chlormequat 72.5	Barclay Chemicals (UK) Ltd	MAFF 05118
* Barclay Holdup	Barclay Chemicals (UK) Ltd	MAFF 05132

* These products are "approved for agricultural use". For further details refer to page vii.
% These products are subject to the staged revocation procedure. For further details refer to page vii.
A These products are approved for use in or near water. For further details refer to page vii.

PROFESSIONAL PRODUCTS: HERBICIDES

Product Name	Marketing Company	Reg. No.
62 Chlormequat—*continued*		
* Barclay Holdup	Barclay Chemicals (UK) Ltd	MAFF 06799
* Barclay Holdup 460	Barclay Chemicals (UK) Ltd	MAFF 05458
* Barleyquat B	Mandops (UK) Ltd	MAFF 07051
* BASF 3C Chlormequat	BASF Plc	MAFF 03234
* BASF 3C Chlormequat 600	BASF Plc	MAFF 04077
* BASF 3C Chlormequat 720	BASF Plc	MAFF 06514
* BASF 3C Chlormequat 750	BASF Plc	MAFF 06878
* Bettaquat B	Mandops (UK) Ltd	MAFF 07050
* CCC 700	Farmers Crop Chemicals Ltd	MAFF 03366
*% Cleanacres PDR 675	Cleanacres Ltd	MAFF 04037
* Dimanquat	Rhone-Poulenc Agriculture	MAFF 06163
*% Dimanquat	Rhone-Poulenc Agriculture Ltd	MAFF 05211
* Fargro Chlormequat	Fargro Ltd	MAFF 02600
* Headland Swift	WBC Technology Ltd	MAFF 04537
* Helmsman	Rhone-Poulenc Agriculture	MAFF 06189
*% Helmsman	Rhone-Poulenc Crop Protection	MAFF 05459
* Hyquat	Agrichem Ltd	MAFF 01094
* Hyquat 70	Agrichem Ltd	MAFF 03364
* Hyquat 75	Agrichem Ltd	MAFF 03365
*% Interlates Chlormequat 46	Interlates Ltd	MAFF 00503
*% KW Chlormequat 600	Kenneth Wilson (Holdings) Ltd	MAFF 04078
* Mandops Barleyquat B	Mandops (UK) Ltd	MAFF 06001
* Mandops Barleyquat B	Mandops Ltd	MAFF 01256
* Mandops Bettaquat B	Mandops (UK) Ltd	MAFF 06004
* Mandops Bettaquat B	Mandops Ltd	MAFF 01257
* Mandops Chlormequat 460	Mandops (UK) Ltd	MAFF 06090
* Mandops Chlormequat 460	Mandops Ltd	MAFF 01263
* Mandops Chlormequat 700	Mandops (Agrochemical Specialists) Ltd	MAFF 06002
* Mandops Chlormequat 700	Mandops Ltd	MAFF 01264
*% Mandops Helestone	Mandops Ltd	MAFF 01269
* Manipulator	Mandops (UK) Ltd	MAFF 05871
* MSS Chlormequat 40	Mirfield Sales Services Ltd	MAFF 01401
* MSS Chlormequat 460	Mirfield Sales Services Ltd	MAFF 03935
* MSS Chlormequat 60	Mirfield Sales Services Ltd	MAFF 03936
* MSS Chlormequat 70	Mirfield Sales Services Ltd	MAFF 03937
* MSS Mircell	Mirfield Sales Services Ltd	MAFF 06939
* New 5C Cycocel	BASF Plc	MAFF 01482
* New 5C Cycocel	Cyanamid of GB Ltd	MAFF 01483
* PA Chlormequat 400	Portman Agrochemicals Ltd	MAFF 01523
* PA Chlormequat 460	Portman Agrochemicals Ltd	MAFF 02549
* PDR 675	Rhone-Poulenc Agriculture	MAFF 06180
*% PDR 675	Rhone-Poulenc Crop Protection	MAFF 05721

* These products are "approved for agricultural use". For further details refer to page vii.
% These products are subject to the staged revocation procedure. For further details refer to page vii.
A These products are approved for use in or near water. For further details refer to page vii.

PROFESSIONAL PRODUCTS: HERBICIDES

Product Name	Marketing Company	Reg. No.
62 Chlormequat—*continued*		
*% Phyto-CCC	Phyto Research Pflanzenschutz GmbH	MAFF 04357
* Podquat	Mandops Ltd	MAFF 03003
* Portman Chlormequat 700	Portman Agrochemicals Ltd	MAFF 03465
* Portman Supaquat	Portman Agrochemicals Ltd	MAFF 03466
* Quadrangle Chlormequat 700	Quadrangle Ltd	MAFF 03401
* RP Chlormequat 40	Rhone-Poulenc Crop Protection	MAFF 05997
* RP Chlormequat 67.5	Rhone-Poulenc Crop Protection	MAFF 05998
* Stabilan 5C	Chemie Linz UK Ltd	MAFF 07082
* Stabilan 750	Chemie Linz UK Ltd	MAFF 03370
* Standup 700	L W Vass (Agricultural) Ltd	MAFF 03522
* Stefes CCC	Stefes Crop Protection Ltd	MAFF 05959
* Stefes CCC 640	Stefes Plant Protection Ltd	MAFF 06993
* Stefes CCC 700	Stefes Plant Protection Ltd	MAFF 07116
* Stefes CCC 720	Stefes Plant Protection Ltd	MAFF 05834
* Stefes K2	Stefes Plant Protection Ltd	MAFF 07054
* Titan	AgrEvo UK Crop Protection Ltd	MAFF 07296
* Titan	Schering Agrochemicals Ltd	MAFF 03925
* Titan 670	AgrEvo UK Crop Protection Ltd	MAFF 07297
* Titan 670	Schering Agrochemicals Ltd	MAFF 06944
* Top Farm Chlormequat 640	Top Farm Formulations Ltd	MAFF 05323
* Tripart 5C	Tripart Farm Chemicals Ltd	MAFF 04726
* Tripart Brevis	Tripart Farm Chemicals Ltd	MAFF 03754
* Tripart Brevis 2	Tripart Farm Chemicals Ltd	MAFF 06612
* Tripart Chlormequat 460	Tripart Farm Chemicals Ltd	MAFF 03685
63 Chlormequat + *2-Chloroethylphosphonic acid*		
* Strate	Quadrangle Ltd	MAFF 05103
* Sypex	BASF Plc	MAFF 04650
* Terpal C	BASF Plc	MAFF 07062
* Upgrade	Rhone-Poulenc Agriculture	MAFF 06177
* Upgrade	Rhone-Poulenc Crop Protection	MAFF 04624
64 Chlormequat + *Imazaquin*		
* Meteor	Cyanamid of GB Ltd	MAFF 06505
* Talgard	Cyanamid Of GB Ltd	MAFF 07187
65 2-Chloroethylphosphonic acid		
* Cerone	A H Marks & Co Ltd	MAFF 00462
*% Cerone	Imperial Chemical Industries Plc	MAFF 00463
* Cerone	Rhone-Poulenc Agriculture Ltd	MAFF 05443
* Cerone	Rhone-Poulenc Agriculture	MAFF 06185
*% Cerone	Zeneca Crop Protection	MAFF 06643

* These products are "approved for agricultural use". For further details refer to page vii.
% These products are subject to the staged revocation procedure. For further details refer to page vii.
A These products are approved for use in or near water. For further details refer to page vii.

PROFESSIONAL PRODUCTS: HERBICIDES

Product Name	Marketing Company	Reg. No.
65 2-Chloroethylphosphonic acid—*continued*		
* Ethrel C	Hortichem Ltd	MAFF 06995
*% Ethrel C	Imperial Chemical Industries Plc	MAFF 00810
* Ibis	Rhone-Poulenc Agriculture Ltd	MAFF 06816
*% Stantion	Rhone-Poulenc Agriculture Ltd	MAFF 04784
* Stantion	Rhone-Poulenc Agriculture	MAFF 06205
* Stefes Stance	Stefes Plant Protection Ltd	MAFF 06125
* Terpal	Clifton Chemicals Ltd	MAFF 05026
* Unistar Ethephon 480	Unistar Ltd	MAFF 06282
66 2-Chloroethylphosphonic acid + *Chlormequat*		
* Strate	Quadrangle Ltd	MAFF 05103
* Sypex	BASF Plc	MAFF 04650
* Terpal C	BASF Plc	MAFF 07062
* Upgrade	Rhone-Poulenc Agriculture	MAFF 06177
* Upgrade	Rhone-Poulenc Crop Protection	MAFF 04624
67 2-Chloroethylphosphonic acid + *Mepiquat*		
* Sypex M	BASF Plc	MAFF 06810
* Terpal	BASF Plc	MAFF 02103
68 2-Chloroethylphosphonic acid + *Mepiquat chloride*		
* Stefes Mepiquat	Stefes Plant Protection Ltd	MAFF 06970
69 Chlorotoluron		
* Alpha Chlortoluron 500	Makhteshim-Agan (UK) Ltd	MAFF 04848
* Ashlade Chlortoluron 500	Ashlade Formulations Ltd	MAFF 06093
* Chlortoluron 500	Pan Britannica Industries Ltd	MAFF 03686
*% Dicurane 500 FW	Ciba-Geigy Agrochemicals	MAFF 00698
* Dicurane 500 SC	Ciba Agriculture	MAFF 05836
* Dicurane 700 SC	Ciba Agriculture	MAFF 04859
* Lentipur CL 500	Chemie Linz UK Ltd	MAFF 05925
*% Ludorum	Ashlade Formulations Ltd	MAFF 03059
* Ludorum 700	Ashlade Formulations Ltd	MAFF 03999
*% Phyto-Toluron	Phyto Research Pflanzenschutz GmbH	MAFF 04491
* Portman Chlortoluron	Portman Agrochemicals Ltd	MAFF 03068
* Power Chlortoluron 500 FC	Power Agrichemicals Ltd	MAFF 03499
* Stefes Toluron	Stefes Plant Protection Ltd	MAFF 05779
* Talisman	Farmers Crop Chemicals Ltd	MAFF 03109
* Tol 7	Ashlade Formulations Ltd	MAFF 06484
*% Tol 7	Ashlade Formulations Ltd	MAFF 05084
* Toro	Sipcam UK Ltd	MAFF 05927
*% Toro	Sipcam UK Ltd	MAFF 04734
* Tripart Culmus	Tripart Farm Chemicals Ltd	MAFF 06619

* These products are "approved for agricultural use". For further details refer to page vii.
% These products are subject to the staged revocation procedure. For further details refer to page vii.
A These products are approved for use in or near water. For further details refer to page vii.

PROFESSIONAL PRODUCTS: HERBICIDES

Product Name	Marketing Company	Reg. No.

70 Chlorotoluron + *Bifenox*

*%	Dicurane Duo 446 FW	Ciba-Geigy Agrochemicals	MAFF 04839
*	Dicurane Duo 446 SC	Ciba Agriculture	MAFF 05839
*%	Dicurane Duo 495 FW	Ciba-Geigy Agrochemicals	MAFF 03860
*%	Dicurane Duo 495 SC	Ciba Agriculture	MAFF 05838

71 Chlorotoluron + *Isoxaben*

*%	Crimson 514 FW	Ciba-Geigy Agrochemicals	MAFF 04830
*%	Crimson 514 SC	Ciba Agriculture	MAFF 05835
*%	Crimson 514 SC	DowElanco Ltd	MAFF 05760

72 Chlorotoluron + *Pendimethalin*

*	Totem	Cyanamid of GB Ltd	MAFF 04670

73 Chlorotoluron + *Trifluralin*

*%	Dicurane Combi 540 FW	Ciba-Geigy Agrochemicals	MAFF 04628
*%	Dicurane Combi 540 SC	Ciba Agriculture	MAFF 05837

74 Chlorphonium

*	Phosfleur 1.5	Perifleur Products Ltd	MAFF 05750
*	Phosfleur 10% Liquid	Perifleur Products Ltd	MAFF 01587

75 Chlorpropham

*	Atlas CIPC 40	Atlas Interlates Ltd	MAFF 03049
*	Atlas Herbon Pabrac	Atlas Interlates Ltd	MAFF 03997
	BL500	Wheatley Chemical Co Ltd	MAFF 00279
*	Campbell's CIPC 40%	J D Campbell & Sons Ltd	MAFF 00389
*	Croptex Pewter	Hortichem Ltd	MAFF 02507
*	MTM CIPC 40	MTM Agrochemicals Ltd	MAFF 05895
%	Mirvale 500 HN	Ciba-Geigy Agrochemicals	MAFF 01360
*	MSS CIPC 40 EC	Mirfield Sales Services Ltd	MAFF 01403
*	MSS CIPC 5 G	Mirfield Sales Services Ltd	MAFF 01402
*	MSS CIPC 50 LF	Mirfield Sales Services Ltd	MAFF 03285
*	MSS CIPC 50 M	Mirfield Sales Services Ltd	MAFF 01404
*	Triherbicide CIPC	Atochem Agri BV	MAFF 06426
*	Triherbicide CIPC	Elf Atochem Agri BV	MAFF 06874
*%	Triherbicide CIPC	Pennwalt Chemicals Ltd	MAFF 04293
	Warefog 25	Mirfield Sales Services Ltd	MAFF 06776
%	Warefog 25	Wheatley Chemical Co Ltd	MAFF 02323

76 Chlorpropham + *Chloridazon* + *Fenuron* + *Propham*

*	Atlas Electrum	Atlas Interlates Ltd	MAFF 03548

---

* These products are "approved for agricultural use". For further details refer to page vii.
% These products are subject to the staged revocation procedure. For further details refer to page vii.
A These products are approved for use in or near water. For further details refer to page vii.

PROFESSIONAL PRODUCTS: HERBICIDES

Product Name	Marketing Company	Reg. No.

77 Chlorpropham + *Cresylic acid* + *Fenuron* + *Propham*

* Marks PCF Beet Herbicide	A H Marks & Co Ltd	MAFF 02538
* MSS Sugar Beet Herbicide	Mirfield Sales Services Ltd	MAFF 02447

78 Chlorpropham + *Diuron* + *Propham*

* Atlas Pink C	Atlas Interlates Ltd	MAFF 03095

79 Chlorpropham + *Fenuron*

* Atlas Red	Atlas Interlates Ltd	MAFF 03091
* Croptex Chrome	Hortichem Ltd	MAFF 02415

80 Chlorpropham + *Fenuron* + *Propham*

* Atlas Gold	Atlas Interlates	MAFF 03086
* MTM Sugar Beet Herbicide	MTM Agrochemicals Ltd	MAFF 05044

81 Chlorpropham + *Linuron*

* Profalon	AgrEvo UK Crop Protection Ltd	MAFF 07331
* Profalon	Hoechst UK Ltd	MAFF 01640

82 Chlorpropham + *Pentanochlor*

* Atlas Brown	Atlas Interlates Ltd	MAFF 03835

83 Chlorpropham + *Propham*

*% Amos Gro-Stop	Kommer-Brookwick	MAFF 06145
Atlas Indigo	Atlas Interlates Ltd	MAFF 03087
Luxan Gro-Stop	Luxan (UK) Ltd	MAFF 06559
Pommetrol M	Sam Fletcher Ltd	MAFF 01615
% Power Gro-Stop	Kommer-Brookwick	MAFF 06142
% Power Gro-Stop	Power Agrichemicals Ltd	MAFF 02719

84 Chlorthal dimethyl

* Dacthal W 75	ISK Biosciences Ltd	MAFF 05556
*% Dacthal W-75	SDS Biotech UK Ltd	MAFF 02629
* Dacthal W75	Hortichem Ltd	MAFF 05500

85 Chlorthal dimethyl + *Propachlor*

*% Decimate	Fermenta ASC Europe Ltd	MAFF 04858
* Decimate	ISK Biosciences Ltd	MAFF 05626

86 Chlortoluron

* Top Farm Toluron 500	Top Farm Formulations Ltd	MAFF 05986

---

* These products are "approved for agricultural use". For further details refer to page vii.
% These products are subject to the staged revocation procedure. For further details refer to page vii.
A These products are approved for use in or near water. For further details refer to page vii.

PROFESSIONAL PRODUCTS: HERBICIDES

Product Name	Marketing Company	Reg. No.

87 Clopyralid
*	Dow Shield	DowElanco Ltd	MAFF 05578
*	Format	DowElanco Ltd	MAFF 05592
*%	Lontrel 100	Dow Chemical Co Ltd	MAFF 03940
*	Lontrel 100	DowElanco Ltd	MAFF 05737

88 Clopyralid + *Benazolin*
*	Benazalox	AgrEvo UK Crop Protection Ltd	MAFF 07246
*	Benazalox	Schering Agrochemicals Ltd	MAFF 03858
*	Landgold Benazolin Plus	Landgold & Co Ltd	MAFF 06048

89 Clopyralid + *Benazolin* + *Dimefuron*
*	Scorpio	AgrEvo UK Crop Protection Ltd	MAFF 07276
*	Scorpio	Schering Agriculture	MAFF 05859

90 Clopyralid + *Bromoxynil*
*	Vindex	DowElanco Ltd	MAFF 05470
*%	Vindex	Quadrangle Ltd	MAFF 04049

91 Clopyralid + *Bromoxynil* + *Fluroxypyr* + *Ioxynil*
*	Crusader S	DowElanco Ltd	MAFF 05174

92 Clopyralid + *Cyanazine*
*	Coupler SC	Cyanamid Of GB Ltd	MAFF 07022
*	Coupler SC	Shell Chemicals UK Ltd	MAFF 03393

93 Clopyralid + *Dichlorprop* + *MCPA*
*	Lontrel Plus	DowElanco Ltd	MAFF 05269
*%	Lontrel Plus	ICI Agrochemicals	MAFF 01226
*%	Lontrel Plus	Zeneca Crop Protection	MAFF 06683

94 Clopyralid + *Diflufenican* + *MCPA*
*	Spearhead	Rhone-Poulenc Environmental Products	MAFF 07342

95 Clopyralid + *Fluroxypyr* + *Ioxynil*
*	Hotspur	DowElanco Ltd	MAFF 05185
*%	Hotspur	Farm Protection Ltd	MAFF 05301

96 Clopyralid + *Fluroxypyr* + *MCPA*
*	Bofix	DowElanco Ltd	MAFF 06377

---

* These products are "approved for agricultural use". For further details refer to page vii.
% These products are subject to the staged revocation procedure. For further details refer to page vii.
A These products are approved for use in or near water. For further details refer to page vii.

PROFESSIONAL PRODUCTS: HERBICIDES

Product Name	Marketing Company	Reg. No.

97 Clopyralid + Ioxynil
* Escort — DowElanco Ltd — MAFF 05466

98 Clopyralid + Mecoprop
*% Seloxone — Dow Chemical Co Ltd — MAFF 01914
*% Seloxone — DowElanco Ltd — MAFF 05743
*% Seloxone — Imperial Chemical Industries Plc — MAFF 01915

99 Clopyralid + Propyzamide
* Matrikerb — Pan Britannica Industries Ltd — MAFF 01308
* Matrikerb — Rohm & Haas (UK) Ltd — MAFF 02443

100 Clopyralid + Triclopyr
* Grazon 90 — DowElanco Ltd — MAFF 05456

101 Cresylic acid + Chlorpropham + Fenuron + Propham
* Marks PCF Beet Herbicide — A H Marks & Co Ltd — MAFF 02538
* MSS Sugar Beet Herbicide — Mirfield Sales Services Ltd — MAFF 02447

102 Cyanazine
* Fortrol — Cyanamid Of GB Ltd — MAFF 07009
* Fortrol — Shell Chemicals UK Ltd — MAFF 00924
* Match — Cyanamid Of GB Ltd — MAFF 07010
* Match — Shell Chemicals UK Ltd — MAFF 04769
* Reply — Cyanamid Of GB Ltd — MAFF 07084

103 Cyanazine + Bentazone + 2,4-Db
* Topshot — Cyanamid Of GB Ltd — MAFF 07178
* Topshot — Shell Chemicals UK Ltd — MAFF 03576

104 Cyanazine + Bifenox
* FR 1442 — Rhone-Poulenc Agriculture Ltd — MAFF 04926
* Vantage — Cyanamid of GB Ltd. — MAFF 07188
* Vantage — Shell Chemicals UK Ltd — MAFF 04927

105 Cyanazine + Clopyralid
* Coupler SC — Cyanamid Of GB Ltd — MAFF 07022
* Coupler SC — Shell Chemicals UK Ltd — MAFF 03393

106 Cyanazine + Fluroxypyr
*% Spitfire — Dow Chemical Co Ltd — MAFF 04324
* Spitfire — DowElanco Ltd — MAFF 05747

* These products are "approved for agricultural use". For further details refer to page vii.
% These products are subject to the staged revocation procedure. For further details refer to page vii.
A These products are approved for use in or near water. For further details refer to page vii.

PROFESSIONAL PRODUCTS: HERBICIDES

Product Name	Marketing Company	Reg. No.
107 Cyanazine + *Linuron*		
* Stay-Kleen	Shell Chemicals UK Ltd + Chemical Spraying Co Ltd	MAFF 02018
108 Cyanazine + *Mecoprop*		
* Cleaval	Cyanamid Of GB Ltd	MAFF 07142
* Cleaval	Shell Chemicals UK Ltd	MAFF 00541
* FCC Topcorn Extra	Farmers Crop Chemicals Ltd	MAFF 05514
* FCC Topcorn Extra	Farmers Crop Chemicals Ltd	MAFF 07143
* MTM Eminent	MTM Agrochemicals Ltd	MAFF 05513
* MTM Eminent	MTM Agrochemicals Ltd	MAFF 07144
109 Cyanazine + *Terbuthylazine*		
* Angle 567 SC	Ciba Agriculture	MAFF 06254
110 Cycloxydim		
* Landgold Cycloxydim	Landgold & Co Ltd	MAFF 06269
* Laser	BASF Plc	MAFF 05251
* Stratos	BASF Plc	MAFF 06891
111 2,4-D		
* 2,4-D 50%	Chemie Linz UK Ltd	MAFF 02762
* Agrichem 2,4-D	Agrichem Ltd	MAFF 04098
* Agricorn D	Farmers Crop Chemicals Ltd	MAFF 00056
A* Atlas 2,4-D	Atlas Interlates Ltd	MAFF 03052
*% BASF 2,4-D Ester 480	BASF Plc	MAFF 02980
* BH 2,4-D Ester 50	Rhone-Poulenc Environmental Products	MAFF 05390
* Campbell's Destox	J D Campbell & Sons Ltd	MAFF 00397
* Campbell's Dioweed 50	J D Campbell & Sons Ltd	MAFF 00401
*% CDA Dicotox Extra	Rhone-Poulenc Environmental Products	MAFF 04757
* Cyanamid D50	Cyanamid Of GB Ltd	MAFF 07024
* Dicotox Extra	Rhone-Poulenc Environmental Products	MAFF 05330
A* Dormone	Rhone-Poulenc Agriculture Ltd	MAFF 05412
*% Farmon 2,4-D	Farm Protection	MAFF 05703
*% Farmon 2,4-D	Farm Protection Ltd	MAFF 00826
* For-ester	Synchemicals Ltd	MAFF 00914
* HY-D	Agrichem Ltd	MAFF 06278
* Headland Staff	Headland Agrochemicals Ltd	MAFF 07189
* Marks 2,4-D-A	A H Marks & Co Ltd	MAFF 01282
* Mega D	AKZO Chemicals Ltd	MAFF 01319
A* MSS 2,4-D Amine	Mirfield Sales Services Ltd	MAFF 01391

* These products are "approved for agricultural use". For further details refer to page vii.
% These products are subject to the staged revocation procedure. For further details refer to page vii.
A These products are approved for use in or near water. For further details refer to page vii.

PROFESSIONAL PRODUCTS: HERBICIDES

Product Name	Marketing Company	Reg. No.
111 2,4-D—*continued*		
* MSS 2,4-D Ester	Mirfield Sales Services Ltd	MAFF 01393
* Nomix 2,4-D Herbicide	Nomix-Chipman Ltd	MAFF 06394
* Palormone D	Universal Crop Protection Ltd	MAFF 01534
* Ritefeed 2,4-D Amine	Ritefeed Ltd	MAFF 05309
* Shell D50	Shell Chemicals UK Ltd	MAFF 01931
* Silvapron D	BP Chemicals Ltd	MAFF 01935
* Syford	Synchemicals Ltd	MAFF 02062
*% Weedone LV4	A H Marks & Co Ltd	MAFF 02358
112 2,4-D + *Amitrole* + *Diuron*		
* Trik	Smyth-Morris Chemicals	MAFF 02182
* Weedazol Total	A H Marks & Co Ltd	MAFF 02351
113 2,4-D + *Benazolin* + *MCPA*		
* Legumex Extra	AgrEvo UK Crop Protection Ltd	MAFF 07262
114 2,4-D + *Dicamba*		
* Longlife plus	ICI Agrochemicals	MAFF 04735
*% New Estermone	Synchemicals Ltd	MAFF 02939
* New Estermone	Vitax Ltd	MAFF 06336
115 2,4-D + *Dicamba* + *Ferrous sulphate*		
* Renovator	Zeneca Professional Products	MAFF 06860
116 2,4-D + *Dicamba* + *Ioxynil*		
* Super Verdone	Imperial Chemical Industries Plc	MAFF 02051
* Super Verdone	Zeneca Professional Products	MAFF 06862
117 2,4-D + *Dicamba* + *Mecoprop*		
* New Formulation Weed and Brush Killer	Synchemicals Ltd	MAFF 02961
* New Formulation Weed and Brushkiller	Vitax Ltd	MAFF 07072
118 2,4-D + *Dicamba* + *Triclopyr*		
* Broadshot	Cyanamid Of GB Ltd	MAFF 07141
* Broadshot	Shell Chemicals UK Ltd	MAFF 03056
119 2,4-D + *Dichlorprop*		
* Marks Polytox-M	A H Marks & Co Ltd	MAFF 01301
* Polymone X	Universal Crop Protection Ltd	MAFF 01613

* These products are "approved for agricultural use". For further details refer to page vii.
% These products are subject to the staged revocation procedure. For further details refer to page vii.
A These products are approved for use in or near water. For further details refer to page vii.

PROFESSIONAL PRODUCTS: HERBICIDES

Product Name	Marketing Company	Reg. No.
120 2,4-D + *Dichlorprop* + *MCPA* + *Mecoprop*		
* Campbell's New Camppex	J D Campbell & Sons Ltd	MAFF 00414
121 2,4-D + *Ferrous sulphate* + *Mecoprop*		
*% Supergreen Feed, Weed and Mosskiller	Rhone-Poulenc Agriculture Ltd	MAFF 05445
122 2,4-D + *Mecoprop*		
*% BH CMPP/2,4-D	Burts & Harvey	MAFF 00249
* BH CMPP/2,4-D	Rhone-Poulenc Environmental Products	MAFF 05393
*% CDA CMPP/2,4-D	Rhone-Poulenc Agriculture Ltd	MAFF 05411
* CDA Supertox 30	Rhone-Poulenc Environmental Products	MAFF 04664
* Cleanrun	Richardsons Fertilisers Ltd	MAFF 04759
*% Cleanrun	Zeneca Professional Products	MAFF 06857
* Com-Trol	Certified Laboratories Ltd	MAFF 03343
* Longlife Cleanrun	Zeneca Professional Products	MAFF 06914
* Mascot Selective Weedkiller	Rigby Taylor Ltd	MAFF 03423
* Nomix Turf Selective Herbicide	Nomix-Chipman Ltd	MAFF 06777
* Nomix Turf Selective LC Herbicide	Nomix-Chipman Ltd	MAFF 05973
* Select-Trol	Chemsearch	MAFF 03342
*% Selective Weedkiller	Anteco Ltd	MAFF 05601
* Selective Weedkiller	Yule Catto Consumer Chemicals Ltd	MAFF 06579
* Supertox 30	Rhone-Poulenc Agriculture Ltd	MAFF 05340
*% Sydex	Synchemicals Ltd	MAFF 02061
* Sydex	Vitax Ltd	MAFF 06412
*% Verdone CDA	Imperial Chemical Industries Plc	MAFF 03410
* Zennapron	BP Oil Ltd	MAFF 02377
123 2,4-D + *Mecoprop-P*		
* Mascot Selective-P	Rigby Taylor Ltd	MAFF 06105
124 2,4-D + *Picloram*		
* Atladox HI	Chipman Ltd	MAFF 00126
* Atladox HI	Chipman Ltd	MAFF 05559
*% Tordon 101	Dow Chemical Co Ltd	MAFF 02154
* Tordon 101	DowElanco Ltd	MAFF 05816
125 2,4-DB		
* Campbell's DB Straight	J D Campbell & Sons Ltd	MAFF 00394
* MTM DB Straight	MTM Agrochemicals Ltd	MAFF 06814
* Marks 2,4-DB	A H Marks & Co Ltd	MAFF 01283

* These products are "approved for agricultural use". For further details refer to page vii.
% These products are subject to the staged revocation procedure. For further details refer to page vii.
A These products are approved for use in or near water. For further details refer to page vii.

PROFESSIONAL PRODUCTS: HERBICIDES

Product Name	Marketing Company	Reg. No.

126 2,4-DB + Benazolin + MCPA

* Legumex Extra	Schering Agrochemicals Ltd	MAFF 03869
* Setter 33	AgrEvo UK Crop Protection Ltd	MAFF 07282
*% Setter 33	Dow Chemical Co Ltd	MAFF 04377
* Setter 33	DowElanco Ltd	MAFF 05623
* Setter 33	Schering Agrochemicals Ltd	MAFF 04376

127 2,4-DB + Bentazone + Cyanazine

* Topshot	Cyanamid Of GB Ltd	MAFF 07178
* Topshot	Shell Chemicals UK Ltd	MAFF 03576

128 2,4-DB + Linuron + MCPA

* Alistell	Farm Protection Ltd	MAFF 05726
*% Alistell	Farm Protection Ltd	MAFF 00080
* Alistell	Zeneca Crop Protection	MAFF 06515
* Clovacorn Extra	Farmers Crop Chemicals Ltd	MAFF 02596

129 2,4-DB + MCPA

* Agrichem DB Plus	Agrichem Ltd	MAFF 00044
* Campbell's Redlegor	J D Campbell & Sons Ltd	MAFF 00421
* Marks 2,4-DB Extra	A H Marks & Co Ltd	MAFF 01284
* MSS 2,4-DB + MCPA	Mirfield Sales Services Ltd	MAFF 01392

130 Dalapon + Dichlobenil

A* Fydulan	Imperial Chemical Industries Plc	MAFF 00958
A* Fydulan	Zeneca Professional Products	MAFF 06823

131 Daminozide

* B-Nine	Uniroyal Chemical Ltd	MAFF 04468
* Dazide	Fine Agrochemicals Ltd	MAFF 02691

132 Desmedipham + Ethofumesate + Phenmedipham

* Betanal Congress	AgrEvo UK Crop Protection Ltd	MAFF 07110
* Betanal Progress	AgrEvo UK Crop Protection Ltd	MAFF 07111
* Betanal Ultima	Quadrangle Agrochemicals Ltd	MAFF 07039

133 Desmedipham + Phenmedipham

* Betanal Compact	AgrEvo UK Crop Protection Ltd	MAFF 07247
* Betanal Compact	Schering Agrochemicals Ltd	MAFF 06780
* Betanal Quorum	AgrEvo UK Crop Protection Ltd	MAFF 07252
* Betanal Quorum	Schering Agrochemicals Ltd	MAFF 06782
* Betanal Rostrum	AgrEvo UK Crop Protection Ltd	MAFF 07253
* Betanal Rostrum	Schering Agrochemicals Ltd	MAFF 06781

---

* These products are "approved for agricultural use". For further details refer to page vii.
% These products are subject to the staged revocation procedure. For further details refer to page vii.
A These products are approved for use in or near water. For further details refer to page vii.

PROFESSIONAL PRODUCTS: HERBICIDES

Product Name	Marketing Company	Reg. No.
134 Desmetryne		
* Semeron 25 WP	Ciba Agriculture	MAFF 01916
135 Dicamba		
*% Banvel 4S	Sandoz Crop Protection Ltd	MAFF 05511
* Tracker	Shell Chemicals UK Ltd	MAFF 03847
* Tracker	Shell Chemicals UK Ltd	MAFF 07149
136 Dicamba + *Benazolin* + *Dichlorprop*		
*% Celt	Schering Agrochemicals Ltd	MAFF 04464
137 Dicamba + *Bifenox*		
*% Quickstep	Farm Protection Ltd	MAFF 04749
* RP 283	Rhone-Poulenc Agriculture Ltd	MAFF 06178
*% RP 283	Rhone-Poulenc Agriculture Ltd	MAFF 04767
* SP 283	Sandoz Agro Ltd	MAFF 04863
138 Dicamba + *2,4-D*		
* Longlife plus	ICI Agrochemicals	MAFF 04735
*% New Estermone	Synchemicals Ltd	MAFF 02939
* New Estermone	Vitax Ltd	MAFF 06336
139 Dicamba + *2,4-D* + *Ferrous sulphate*		
* Renovator	Zeneca Professional Products	MAFF 06860
140 Dicamba + *2,4-D* + *Ioxynil*		
* Super Verdone	Imperial Chemical Industries Plc	MAFF 02051
* Super Verdone	Zeneca Professional Products	MAFF 06862
141 Dicamba + *2,4-D* + *Mecoprop*		
* New Formulation Weed and Brush Killer	Synchemicals Ltd	MAFF 02961
* New Formulation Weed and Brushkiller	Vitax Ltd	MAFF 07072
142 Dicamba + *2,4-D* + *Triclopyr*		
* Broadshot	Cyanamid Of GB Ltd	MAFF 07141
* Broadshot	Shell Chemicals UK Ltd	MAFF 03056
143 Dicamba + *Maleic hydrazide*		
* Dockmaster	Rhone-Poulenc Environmental Products	MAFF 05932

* These products are "approved for agricultural use". For further details refer to page vii.
% These products are subject to the staged revocation procedure. For further details refer to page vii.
A These products are approved for use in or near water. For further details refer to page vii.

PROFESSIONAL PRODUCTS: HERBICIDES

Product Name	Marketing Company	Reg. No.

144 Dicamba + *Maleic hydrazide* + *MCPA*

* Mazide Selective	Vitax Ltd	MAFF 05753
* Synchemicals Mazide Selective	Synchemicals Ltd	MAFF 02070

145 Dicamba + *MCPA*

* Banvel M	Sandoz Agro Ltd	MAFF 05794

146 Dicamba + *MCPA* + *Mecoprop*

* Banlene Plus	AgrEvo UK Crop Protection Ltd	MAFF 07245
* Banlene Plus	Schering Agrochemicals Ltd	MAFF 03851
*% Banvel MP	Sandoz Crop Protection Ltd	MAFF 05792
* Campbell's Field Marshall	J D Campbell & Sons Ltd	MAFF 00406
*% Campbell's Grassland Herbicide	J D Campbell & Sons Ltd	MAFF 00407
* Campbell's Grassland Herbicide	MTM Agrochemicals Ltd	MAFF 06157
* Docklene	AgrEvo UK Crop Protection Ltd	MAFF 07257
* Docklene	Schering Agriculture	MAFF 03863
* Fisons Tritox	Fisons Plc	MAFF 03013
* Headland Relay	WBC Technology Ltd	MAFF 03778
* Herrisol	Bayer Plc	MAFF 01048
* Hyprone	Agrichem Ltd	MAFF 01093
* Hysward	Agrichem Ltd	MAFF 01096
* Mascot Super Selective	Rigby Taylor Ltd	MAFF 03621
* New Bandock	Shell Chemicals UK Ltd	MAFF 05497
* Pasturol	Farmers Crop Chemicals Ltd	MAFF 01545
* Springcorn Extra	Farmers Crop Chemicals Ltd	MAFF 02004
Tetralex Plus	Cyanamid Of GB Ltd	MAFF 07146
* Tetralex Plus	Shell Chemicals UK Ltd	MAFF 02115
* Tribute	Chipman Ltd	MAFF 03470
* Tribute	Nomix-Chipman Ltd	MAFF 06921

147 Dicamba + *MCPA* + *Mecoprop-P*

* Field Marshal	MTM Agrochemicals Ltd	MAFF 06077
* MTM Grassland Herbicide	MTM Agrochemicals Ltd	MAFF 06089
* Mascot Super Selective-P	Rigby Taylor Ltd	MAFF 06106
* MSS Mircam Plus	Mirfield Sales Services Ltd	MAFF 01416
* Pasturol D	Farmers Crop Chemicals Ltd	MAFF 07033
* Quadrangle Quadban	Quadrangle Agrochemicals	MAFF 07090

148 Dicamba + *Mecoprop*

* Banvel P	Sandoz Agro Ltd	MAFF 05793
* Condox	Zeneca Crop Protection	MAFF 06519
* Di Farmon	Farm Protection Ltd	MAFF 03882
* Endox	Farmers Crop Chemicals Ltd	MAFF 00798
* Farmon Condox	Farm Protection Ltd	MAFF 03883

* These products are "approved for agricultural use". For further details refer to page vii.
% These products are subject to the staged revocation procedure. For further details refer to page vii.
A These products are approved for use in or near water. For further details refer to page vii.

PROFESSIONAL PRODUCTS: HERBICIDES

Product Name	Marketing Company	Reg. No.

148 Dicamba + Mecoprop—continued
* Hyban	Agrichem Ltd	MAFF 01084
* Hygrass	Agrichem (International) Ltd	MAFF 01090

149 Dicamba + Mecoprop + Triclopyr
* Fettel	Farm Protection Ltd	MAFF 02516
* Fettel	Zeneca Crop Protection	MAFF 06399

150 Dicamba + Mecoprop-P
* MSS Mircam	Mirfield Sales Services Ltd	MAFF 01415

151 Dicamba + Paclobutrazol
* Holdfast D	Imperial Chemical Industries Plc	MAFF 05056
* Holdfast D	Zeneca Professional Products	MAFF 06858

152 Dicamba + Triasulfuron
* Accord	AgrEvo UK Crop Protection Ltd	MAFF 07240
* Accord	Schering Agrochemicals Ltd	MAFF 06496
* Banvel T	Sandoz Agro Ltd	MAFF 06497
* Framolene	Ciba Agriculture	MAFF 06495

153 Dichlorophen + 4-Indol-3-ylbutyric acid + 1-Naphthylacetic acid
* Synergol	Silvaperl Products Ltd	MAFF 04594

154 Dichlobenil
*% BH Prefix D	Burts & Harvey	MAFF 03688
A* Casoron G	Imperial Chemical Industries Plc	MAFF 00448
A* Casoron G	Zeneca Professional Products	MAFF 06854
* Casoron G4	Imperial Chemical Industries Plc	MAFF 02406
* Casoron G4	Zeneca Crop Protection	MAFF 06855
A* Casoron GSR	Imperial Chemical Industries Plc	MAFF 00451
A* Casoron GSR	Zeneca Professional Products	MAFF 06856
* Prefix D	Cyanamid Of GB Ltd	MAFF 07013
* Prefix D	Shell Chemicals UK Ltd	MAFF 01631

155 Dichlobenil + Dalapon
* Fydulan	Imperial Chemical Industries Plc	MAFF 00958
* Fydulan	Zeneca Professional Products	MAFF 06823

156 Dichlorophen
* Enforcer	Zeneca Professional Products	MAFF 07079
* Mascot Mosskiller	Rigby Taylor Ltd	MAFF 02439
* Nomix-Chipman Mosskiller	Nomix-Chipman Ltd	MAFF 06271

---

* These products are "approved for agricultural use". For further details refer to page vii.
% These products are subject to the staged revocation procedure. For further details refer to page vii.
A These products are approved for use in or near water. For further details refer to page vii.

PROFESSIONAL PRODUCTS: HERBICIDES

Product Name	Marketing Company	Reg. No.

156 Dichlorophen—*continued*

*% Panacide M	BDH Ltd	MAFF 01536
* Panacide M	Coalite Chemicals	MAFF 05611
*% Panacide TS	BDH Ltd	MAFF 01537
* Panacide Technical Solution	Coalite Chemicals	MAFF 05612
* Ritefeed Dichlorophen	Ritefeed Ltd	MAFF 05265
* Super Mosstox	Rhone-Poulenc Agriculture Ltd	MAFF 05339

157 Dichlorophen + *Ferrous sulphate*

* Aitken's Lawn Sand Plus	R Aitken Ltd	MAFF 04542
* SHL Lawn Sand Plus	Sinclair Horticulture & Leisure Ltd	MAFF 04439

158 Dichlorprop

* Campbell's Redipon	J D Campbell & Sons Ltd	MAFF 00419
* Marks Polytox-K	A H Marks & Co Ltd	MAFF 01300
* MSS 2,4-DP	Mirfield Sales Services Ltd	MAFF 01394
* Redipon	MTM Agrochemicals Ltd	MAFF 07192

159 Dichlorprop + *Benazolin* + *Dicamba*

*% Celt	Schering Agrochemicals Ltd	MAFF 04464

160 Dichlorprop + *Bromoxynil* + *Ioxynil* + *MCPA*

* Atlas Minerva	Atlas Interlates Ltd	MAFF 03046

161 Dichlorprop + *Clopyralid* + *MCPA*

* Lontrel Plus	DowElanco Ltd	MAFF 05269
*% Lontrel Plus	ICI Agrochemicals	MAFF 01226
*% Lontrel Plus	Zeneca Crop Protection	MAFF 06683

162 Dichlorprop + *2,4-D*

* Marks Polytox-M	A H Marks & Co Ltd	MAFF 01301
* Polymone X	Universal Crop Protection Ltd	MAFF 01613

163 Dichlorprop + *2,4-D* + *MCPA* + *Mecoprop*

* Campbell's New Camppex	J D Campbell & Sons Ltd	MAFF 00414

164 Dichlorprop + *Ferrous sulphate* + *MCPA*

* SHL Turf and Weed and Mosskiller	William Sinclair Horticulture Ltd	MAFF 04438

165 Dichlorprop + *Ioxynil* + *MCPA*

* Certrol-PA	J W Chafer Ltd	MAFF 00470

---

* These products are "approved for agricultural use". For further details refer to page vii.
% These products are subject to the staged revocation procedure. For further details refer to page vii.
A These products are approved for use in or near water. For further details refer to page vii.

PROFESSIONAL PRODUCTS: HERBICIDES

Product Name	Marketing Company	Reg. No.
166 Dichlorprop + *MCPA*		
* Campbells Redipon Extra	J D Campbell & Sons Ltd	MAFF 00420
*% Hemoxone	Imperial Chemical Industries Plc	MAFF 01043
* MSS 2,4-DP + MCPA	Mirfield Sales Services Ltd	MAFF 01396
* SHL Turf Feed and Weed	William Sinclair Horticulture Ltd	MAFF 04437
*% Seritox 50	Rhone-Poulenc Agriculture Ltd	MAFF 05225
* Seritox 50	Rhone-Poulenc Agriculture Ltd	MAFF 06170
* Seritox Turf	Rhone-Poulenc Environmental Products	MAFF 06802
*% Seritox Turf	Rhone-Poulenc Environmental Products	MAFF 05337
167 Diclofop-methyl		
* Hoegrass	AgrEvo UK Crop Protection Ltd	MAFF 07323
* Hoegrass	Hoechst UK Ltd	MAFF 01063
* Hoegrass 280	AgrEvo UK Crop Protection Ltd	MAFF 07324
* Hoegrass 280	Hoechst UK Ltd	MAFF 04718
* Hoegrass EW	AgrEvo UK Crop Protection Ltd	MAFF 07325
* Hoegrass EW	Hoechst UK Ltd	MAFF 06425
168 Diclofop-methyl + *Fenoxaprop-P-ethyl*		
* Tigress	AgrEvo UK Crop Protection Ltd	MAFF 07337
* Tigress	Hoechst UK Ltd	MAFF 05976
169 Difenzoquat		
* Avenge 2	AgrEvo UK Crop Protection Ltd	MAFF 07244
* Avenge 2	Cyanamid Of GB Ltd	MAFF 03241
* Avenge 2	Schering Agrochemicals Ltd	MAFF 04470
170 Diflufenican		
* EXP 4005	Rhone-Poulenc Agriculture Ltd	MAFF 05690
171 Diflufenican + *Bromoxynil* + *Ioxynil*		
* Capture	Rhone-Poulenc Agriculture Ltd	MAFF 06881
* Quartz BL	Rhone-Poulenc Agriculture Ltd	MAFF 06123
172 Diflufenican + *Clopyralid* + *MCPA*		
* Spearhead	Rhone-Poulenc Environmental Products	MAFF 07342
173 Diflufenican + *Isoproturon*		
* Adition	Rhone-Poulenc Agriculture	MAFF 06184
*% Adition	Rhone-Poulenc Agriculture	MAFF 05146
* Clayton Fenican IPU	Clayton Plant Protection (UK) Ltd	MAFF 06759

* These products are "approved for agricultural use". For further details refer to page vii.
% These products are subject to the staged revocation procedure. For further details refer to page vii.
A These products are approved for use in or near water. For further details refer to page vii.

PROFESSIONAL PRODUCTS: HERBICIDES

Product Name	Marketing Company	Reg. No.

173 Diflufenican + *Isoproturon*—continued

* Cougar	May & Baker Ireland Ltd	MAFF 04675
* Cougar	Rhone-Poulenc Ireland Ltd	MAFF 06611
* Javelin	Rhone-Poulenc Agriculture	MAFF 06192
*% Javelin	Rhone-Poulenc Agriculture Ltd	MAFF 05095
* Javelin Gold	Rhone-Poulenc Agriculture	MAFF 06200
* Javelin Gold	Rhone-Poulenc Crop Protection	MAFF 05555
* Landgold DFF 625	Landgold & Co Ltd	MAFF 06274
* Landgold FF550	Landgold & Co Ltd	MAFF 06053
* Oyster	Rhone-Poulenc Agriculture	MAFF 07345
* Panther	Rhone-Poulenc Agriculture	MAFF 06491
*% Panther	Rhone-Poulenc Crop Protection	MAFF 05189
*% Stalker	Rhone-Poulenc Agriculture Ltd	MAFF 05557
* Tolkan Turbo	Rhone-Poulenc Agriculture	MAFF 06795
* Zodiac TX	Rhone-Poulenc Agriculture Ltd	MAFF 06775

174 Diflufenican + *Trifluralin*

* Ardent	Rhone-Poulenc Agriculture	MAFF 06203
* Ardent	Rhone-Poulenc Agriculture Ltd	MAFF 04248
*% Hawk 440 SC	Ciba-Geigy Agriculture	MAFF 05803

175 Dikegulac

* Atrinal	Burts & Harvey	MAFF 00170
* Atrinal	Rhone-Poulenc Agriculture Ltd	MAFF 05389

176 Dimefuron + *Benazolin*

* Scorpio 400	AgrEvo UK Crop Protection Ltd	MAFF 07277
* Scorpio 400	Schering Agriculture	MAFF 05858

177 Dimefuron + *Benazolin* + *Clopyralid*

* Scorpio	AgrEvo UK Crop Protection Ltd	MAFF 07276
* Scorpio	Schering Agriculture	MAFF 05859

178 Dimefuron + *Carbetamide*

*% Pradone Plus	Rhone-Poulenc Agriculture Ltd	MAFF 05857

179 Diphenamid

* Enide 50 W	Imperial Chemical Industries Plc	MAFF 00800
* Enide 50W	Zeneca Crop Protection	MAFF 06660

180 Diquat

* Barclay Desiquat	Barclay Chemicals (UK) Ltd	MAFF 04969
* CIA Diquat-200	Combined Independent Agronomists Ltd	MAFF 05551

---

* These products are "approved for agricultural use". For further details refer to page vii.
% These products are subject to the staged revocation procedure. For further details refer to page vii.
A These products are approved for use in or near water. For further details refer to page vii.

PROFESSIONAL PRODUCTS: HERBICIDES

Product Name	Marketing Company	Reg. No.

180 Diquat—continued

* Clayton Quatrow	Clayton Plant Protection (UK) Ltd	MAFF 07175
* Diquat 200	Top Farm Formulations Ltd	MAFF 05141
* Landgold Diquat	Landgold & Co Ltd	MAFF 05974
A* Midstream	Imperial Chemical Industries Plc	MAFF 01348
A* Midstream	Zeneca Professional Products	MAFF 06824
A* Reglone	Imperial Chemical Industries Plc	MAFF 04444
*% Reglone	Imperial Chemical Industries Plc	MAFF 01713
A* Reglone	Zeneca Crop Protection	MAFF 06703
* Standon Diquat	Standon Chemicals Ltd	MAFF 05587
* Stefes Diquat	Stefes Plant Protection Ltd	MAFF 05493
* Top Farm Diquat-200	Top Farm Formulations Ltd	MAFF 05552

181 Diquat + *Paraquat*

* Farmon PDQ	Farm Protection	MAFF 02886
* PDQ	Zeneca Crop Protection	MAFF 06518
* Parable	Imperial Chemical Industries Plc	MAFF 03805
* Parable	Zeneca Crop Protection	MAFF 06692
* Precede	Imperial Chemical Industries Plc	MAFF 05732
* Precede	Zeneca Crop Protection	MAFF 06702

182 Diquat + *Paraquat* + *Simazine*

* Soltair	Imperial Chemical Industries Plc	MAFF 03601
* Soltair	Zeneca Crop Protection	MAFF 06772

183 Diuron

* Chipko Diuron 80	Chipman Ltd	MAFF 00497
* Chipman Diuron Flowable	Nomix-Chipman Ltd	MAFF 05701
* Diuron 50 FL	Rhone-Poulenc Agriculture Ltd	MAFF 05198
* Diuron 50 FL	Staveley Chemicals Ltd	MAFF 02814
* Diuron 80 FL	Rhone-Poulenc Environmental Products	MAFF 05003
* Diuron 80% WP	Rhone-Poulenc Agriculture Ltd	MAFF 05199
* Diuron 80% WP	Staveley Chemicals Ltd	MAFF 00730
* Freeway	Rhone-Poulenc Environmental Products	MAFF 06047
* Hoechst Diuron	Hoechst UK Ltd	MAFF 01056
* Karmex	Du Pont (UK) Ltd	MAFF 01128
* MTM Diuron 80	MTM Agrochemicals Ltd	MAFF 04958
*% MSS Diuron 50 FL	Mirfield Sales Services Ltd	MAFF 04786
* MSS Diuron 50FL	Mirfield Sales Services Ltd	MAFF 07160
* Promark Diuron	Hoechst UK Ltd	MAFF 03906
* Unicrop Flowable Diuron	Universal Crop Protection Ltd	MAFF 02270

---

* These products are "approved for agricultural use". For further details refer to page vii.
% These products are subject to the staged revocation procedure. For further details refer to page vii.
A These products are approved for use in or near water. For further details refer to page vii.

PROFESSIONAL PRODUCTS: HERBICIDES

Product Name	Marketing Company	Reg. No.
184 Diuron + *Amitrole*		
*% Amizol-D	A H Marks & Co Ltd	MAFF 00093
* Orchard Herbicide	Hoechst UK Ltd	MAFF 03379
185 Diuron + *Amitrole* + *Bromacil*		
* BR Destral	Rhone-Poulenc Environmental Products	MAFF 05184
186 Diuron + *Amitrole* + *2,4-D*		
* Trik	Smyth-Morris Chemicals	MAFF 02182
* Weedazol Total	A H Marks & Co Ltd	MAFF 02351
187 Diuron + *Bromacil*		
* Borocil K	Rhone-Poulenc Environmental Products	MAFF 05183
*% Krovar 1	Du Pont (UK) Ltd	MAFF 01166
188 Diuron + *Chlorpropham* + *Propham*		
* Atlas Pink C	Atlas Interlates Ltd	MAFF 03095
189 Diuron + *Glyphosate*		
* Touche	Nomix-Chipman Ltd	MAFF 06797
190 Diuron + *Paraquat*		
* Dexuron	Chipman Ltd	MAFF 00689
* Dexuron	Nomix-Chipman Ltd	MAFF 07169
191 Dodecylbenzyl trimethyl ammonium chloride		
* Country Fresh Disinfectant	Dimex Ltd	MAFF 04443
*% Resistone PC	Rhone-Poulenc Agriculture Ltd	MAFF 05196
192 EPTC		
*% Eptam 6E	Farm Protection Ltd	MAFF 03944
193 Ethofumesate		
* Atlas Thor	Atlas Interlates Ltd	MAFF 06966
* Barclay Keeper 200	Barclay Chemicals (UK) Ltd	MAFF 05266
* Kemiron	Pan Britannica Industries Ltd	MAFF 05883
* Kemiron Flow	Kemira Agro Benelux SA	MAFF 06740
* Landgold Ethofumesate 200	Landgold & Co Ltd	MAFF 06257
* Nortron	AgrEvo UK Crop Protection Ltd	MAFF 07266
* Nortron	Schering Agrochemicals Ltd	MAFF 03853
* Standon Ethofumesate 200	Standon Chemicals Ltd	MAFF 05668
* Stefes Fumat	Stefes Plant Protection Ltd	MAFF 05525
* Top Farm Ethofumesate 200	Top Farm Formulations Ltd	MAFF 05987

* These products are "approved for agricultural use". For further details refer to page vii.
% These products are subject to the staged revocation procedure. For further details refer to page vii.
A These products are approved for use in or near water. For further details refer to page vii.

PROFESSIONAL PRODUCTS: HERBICIDES

Product Name	Marketing Company	Reg. No.

194 Ethofumesate + *Bromoxynil* + *Ioxynil*

* Leyclene	AgrEvo UK Crop Protection Ltd	MAFF 07263
* Leyclene	Schering Agrochemicals Ltd	MAFF 05285
* Trapper	DowElanco Ltd	MAFF 06325

195 Ethofumesate + *Chloridazon*

* Magnum	BASF Plc	MAFF 01237
* Spectron	AgrEvo UK Crop Protection Ltd	MAFF 07284
* Spectron	Schering Agrochemicals Ltd	MAFF 03828

196 Ethofumesate + *Desmedipham* + *Phenmedipham*

* Betanal Congress	AgrEvo UK Crop Protection Ltd	MAFF 07110
* Betanal Progress	AgrEvo UK Crop Protection Ltd	MAFF 07111
* Betanal Ultima	Quadrangle Agrochemicals Ltd	MAFF 07039

197 Ethofumesate + *Metamitron* + *Phenmedipham*

* Bayer UK 407 WG	Bayer UK Ltd	MAFF 06313
* CX 171	AgrEvo UK Crop Protection Ltd	MAFF 06803
* Goltix Triple	Bayer UK Ltd	MAFF 06314

198 Ethofumesate + *Phenmedipham*

* Betanal Maestro	AgrEvo UK Crop Protection Ltd	MAFF 07249
* Betanal Maestro	Schering Agrochemicals Ltd	MAFF 06022
* Betanal Montage	AgrEvo UK Crop Protection Ltd	MAFF 07250
* Betanal Montage	Schering Agrochemicals Ltd	MAFF 06024
* Betanal Quadrant	AgrEvo UK Crop Protection Ltd	MAFF 07251
* Betanal Quadrant	Schering Agrochemicals Ltd	MAFF 06023
* Betanal Tandem	AgrEvo UK Crop Protection Ltd	MAFF 07254
* Betanal Tandem	Schering Agrochemicals Ltd	MAFF 03857

199 Fatty acids

* Koppert De Moss	Koppert (UK) Ltd	MAFF 03568

200 Fenoxaprop-ethyl

* Cheetah R	AgrEvo UK Crop Protection Ltd	MAFF 07308
* Cheetah R	Hoechst UK Ltd	MAFF 04932
* Clayton Fencer	Clayton Plant Protection (UK) Ltd	MAFF 06525
* Landgold Fenoxaprop	Landgold & Co Ltd	MAFF 06352

201 Fenoxaprop-P-ethyl

* Cheetah Super	AgrEvo UK Crop Protection Ltd	MAFF 07309
* Cheetah Super	Hoechst UK Ltd	MAFF 05825
* Wildcat	AgrEvo UK Crop Protection Ltd	MAFF 07340
* Wildcat	Hoechst UK Ltd	MAFF 06363

* These products are "approved for agricultural use". For further details refer to page vii.
% These products are subject to the staged revocation procedure. For further details refer to page vii.
A These products are approved for use in or near water. For further details refer to page vii.

PROFESSIONAL PRODUCTS: HERBICIDES

Product Name	Marketing Company	Reg. No.

202 Fenoxaprop-P-ethyl + *Diclofop-methyl*
* Tigress — AgrEvo UK Crop Protection Ltd — MAFF 07337
* Tigress — Hoechst UK Ltd — MAFF 05976

203 Fenoxaprop-P-ethyl + *Isoproturon*
* Puma X — AgrEvo UK Crop Protection Ltd — MAFF 07332
* Puma X — Hoechst UK Ltd — MAFF 06768
* Whip X — AgrEvo UK Crop Protection Ltd — MAFF 07339
* Whip X — Hoechst UK Ltd — MAFF 06771

205 Fentin hydroxide + *Metoxuron*
*% Endspray — Pan Britannica Industries Ltd — MAFF 02799

206 Fenuron + *Chloridazon* + *Chlorpropham* + *Propham*
* Atlas Electrum — Atlas Interlates Ltd — MAFF 03548

207 Fenuron + *Chlorpropham*
* Atlas Red — Atlas Interlates Ltd — MAFF 03091
* Croptex Chrome — Hortichem Ltd — MAFF 02415

208 Fenuron + *Chlorpropham* + *Cresylic acid* + *Propham*
* Marks PCF Beet Herbicide — A H Marks & Co Ltd — MAFF 02538
* MSS Sugar Beet Herbicide — Mirfield Sales Services Ltd — MAFF 02447

209 Fenuron + *Chlorpropham* + *Propham*
* Atlas Gold — Atlas Interlates — MAFF 03086
* MTM Sugar Beet Herbicide — MTM Agrochemicals Ltd — MAFF 05044

210 Ferrous sulphate
* Elliot's Lawn Sand — Thomas Elliott Ltd — MAFF 04860
* Elliot's Mosskiller — Thomas Elliott Ltd — MAFF 04909
* Fisons Greenmaster Autumn — Fisons Plc — MAFF 03211
* Fisons Greenmaster Mosskiller — Fisons Plc — MAFF 00881
* Maxicrop Mosskiller and Conditioner — Maxicrop International Ltd — MAFF 04635
* SHL Lawn Sand — Sinclair Horticulture & Leisure Ltd — MAFF 05254
* Taylor's Lawn Sand — Rigby Taylor Ltd — MAFF 04451
*% Verdant No 12 Mosskiller — T Parker & Sons (Turf Management) Ltd — MAFF 04931
*% Verdant No.7 Lawn Sand — T Parker & Sons (Turf Management) Ltd — MAFF 04862
* Vitagrow Lawn Sand — Vitagrow (Fertilisers) Ltd — MAFF 05097
* Vitax Micro Gran 2 — Vitax Ltd — MAFF 04541
* Vitax Turf Tonic — Vitax Ltd — MAFF 04354

---

* These products are "approved for agricultural use". For further details refer to page vii.
% These products are subject to the staged revocation procedure. For further details refer to page vii.
A These products are approved for use in or near water. For further details refer to page vii.

PROFESSIONAL PRODUCTS: HERBICIDES

Product Name	Marketing Company	Reg. No.

211 Ferrous sulphate + 2,4-D + Dicamba
* Renovator	Zeneca Professional Products	MAFF 06860

212 Ferrous sulphate + 2,4-D + Mecoprop
*% Supergreen Feed, Weed and Mosskiller	Not applicable	MAFF 05445

213 Ferrous sulphate + Dichlorophen
* Aitken's Lawn Sand Plus	R Aitken Ltd	MAFF 04542
* SHL Lawn Sand Plus	Sinclair Horticulture & Leisure Ltd	MAFF 04439

214 Ferrous sulphate + Dichlorprop + MCPA
* SHL Turf and Weed and Mosskiller	William Sinclair Horticulture Ltd	MAFF 04438

215 Flamprop-M-isopropyl
* Commando	Cyanamid Of GB Ltd	MAFF 07005
* Commando	Shell Chemicals UK Ltd	MAFF 00564
* Cossack	Cyanamid Of GB Ltd	MAFF 07006
* Cossack	Shell Chemicals UK Ltd	MAFF 05864
* Quadrangle Gunner	Quadrangle Ltd	MAFF 04547
* Stefes Flamprop	Stefes Plant Protection Ltd	MAFF 05789
* Trooper	Cyanamid Of GB Ltd	MAFF 07019
* Trooper	Shell Chemicals UK Ltd	MAFF 05863

216 Fluazifop-P-butyl
* Barclay Winner	Barclay Chemicals (UK) Ltd	MAFF 06986
* Chorus	Zeneca Crop Protection	MAFF 06761
* Citadel	Zeneca Crop Protection	MAFF 06762
* Corral	Quadrangle Agrochemicals Ltd	MAFF 05686
* Corral	Zeneca Crop Protection	MAFF 06647
* Fusilade 250 EW	Zeneca Crop Protection	MAFF 06531
* Fusilade 5	Imperial Chemical Industries Plc	MAFF 02833
* Fusilade 5	Zeneca Crop Protection	MAFF 06669
* Grapple	Zeneca Crop Protection	MAFF 06532
* Landgold Fluazifop-P	Landgold & Co Ltd	MAFF 06020
* PP005	Zeneca Crop Protection	MAFF 06530
* PP007	Farm Protection Ltd	MAFF 06533
* Power Prime	Power Agrichemicals Ltd	MAFF 04747
* Standon Fluazifop-P	Standon Chemicals Ltd	MAFF 06060
* Stefes Slayer	Stefes Plant Protection Ltd	MAFF 06277
* Wizzard	Farm Protection Ltd	MAFF 06534
* Wizzard	Zeneca Crop Protection	MAFF 06521

* These products are "approved for agricultural use". For further details refer to page vii.
% These products are subject to the staged revocation procedure. For further details refer to page vii.
A These products are approved for use in or near water. For further details refer to page vii.

PROFESSIONAL PRODUCTS: HERBICIDES

Product Name	Marketing Company	Reg. No.
217 Fluoroglycofen-ethyl		
* Compete 20	BASF Plc	MAFF 06589
* Compete 5	BASF Plc	MAFF 06550
218 Fluoroglycofen-ethyl + *Isoproturon*		
* Compete Forte	Hoechst UK Ltd	MAFF 06606
* Competitor	AgrEvo UK Crop Protection Ltd	MAFF 07310
* Competitor	Hoechst UK Ltd	MAFF 06605
* Effect	AgrEvo UK Crop Protection Ltd	MAFF 07319
* Effect	Hoechst UK Ltd	MAFF 06607
219 Fluoroglycofen-ethyl + *Mecoprop-P*		
* Compete Mix 20 PVA	BASF Plc	MAFF 07223
* Estrad	BASF Plc	MAFF 07196
* Estrad Duplo	BASF Plc	MAFF 06793
220 Fluoroglycofen-ethyl + *Triasulfuron*		
* Compete Mix A	Rohm & Haas (UK) Ltd	MAFF 05992
* Satis 15 WP	Ciba Agriculture	MAFF 05991
221 Fluroxypyr		
* Barclay Hurler	Barclay Chemicals (UK) Ltd	MAFF 06952
* Clayton Fluroxypyr	Clayton Plant Protection (UK) Ltd	MAFF 06356
* Landgold Fluroxypyr	Landgold & Co Ltd	MAFF 06080
* Standon Fluroxypyr	Standon Chemicals Ltd	MAFF 07055
* Starane 2	DowElanco Ltd	MAFF 05496
* Stefes Fluroxypyr	Stefes Plant Protection Ltd	MAFF 05977
222 Fluroxypyr + *Bromoxynil*		
* Sickle	DowElanco Ltd	MAFF 05187
223 Fluroxypyr + *Bromoxynil* + *Clopyralid* + *Ioxynil*		
* Crusader S	DowElanco Ltd	MAFF 05174
224 Fluroxypyr + *Bromoxynil* + *Ioxynil*		
* Advance	DowElanco Ltd	MAFF 05173
225 Fluroxypyr + *Clopyralid* + *Ioxynil*		
* Hotspur	DowElanco Ltd	MAFF 05185
*% Hotspur	Farm Protection Ltd	MAFF 05301
226 Fluroxypyr + *Clopyralid* + *MCPA*		
* Bofix	DowElanco Ltd	MAFF 06377

* These products are "approved for agricultural use". For further details refer to page vii.
% These products are subject to the staged revocation procedure. For further details refer to page vii.
A These products are approved for use in or near water. For further details refer to page vii.

PROFESSIONAL PRODUCTS: HERBICIDES

Product Name	Marketing Company	Reg. No.
227 Fluroxypyr + *Cyanazine*		
*% Spitfire	Dow Chemical Co Ltd	MAFF 04324
* Spitfire	DowElanco Ltd	MAFF 05747
228 Fluroxypyr + *Ioxynil*		
* Stexal	DowElanco Ltd	MAFF 05188
229 Fluroxypyr + *Mecoprop-P*		
* Bastion T	DowElanco Ltd	MAFF 06011
230 Fluroxypyr + *Thifensulfuron-methyl* + *Tribenuron-methyl*		
* DP 353	Du Pont (UK) Ltd	MAFF 07036
* Starane Super	DowElanco Ltd	MAFF 07035
231 Fluroxypyr + *Triclopyr*		
* Doxstar	DowElanco Ltd	MAFF 06050
232 Fosamine-ammonium		
A* Krenite	Du Pont (UK) Ltd	MAFF 01165
233 Gibberellins		
* Berelex	Imperial Chemical Industries Plc	MAFF 00231
* Berelex	Zeneca Crop Protection	MAFF 06637
* Regulex	Imperial Chemical Industries Plc	MAFF 03010
*% Regulex	Zeneca Crop Protection	MAFF 06704
234 Glufosinate ammonium		
* Challenge	AgrEvo UK Crop Protection Ltd	MAFF 07306
* Challenge	Hoechst UK Ltd	MAFF 05176
* Challenge 2	AgrEvo UK Crop Protection Ltd	MAFF 07307
* Challenge 2	Hoechst UK Ltd	MAFF 05175
* Dash	Hoechst UK Ltd	MAFF 05177
* Hoe 39866 SL06	AgrEvo UK Crop Protection Ltd	MAFF 07322
* HOE 39866 SL06	Hoechst UK Ltd	MAFF 05178
* Harvest	AgrEvo UK Crop Protection Ltd	MAFF 07321
* Harvest	Hoechst UK Ltd	MAFF 06337
235 Glufosinate ammonium + *Monolinuron*		
* Conquest	AgrEvo UK Crop Protection Ltd	MAFF 07311
* Conquest	Hoechst UK Ltd	MAFF 05679

* These products are "approved for agricultural use". For further details refer to page vii.
% These products are subject to the staged revocation procedure. For further details refer to page vii.
A These products are approved for use in or near water. For further details refer to page vii.

PROFESSIONAL PRODUCTS: HERBICIDES

Product Name	Marketing Company	Reg. No.
236 Glyphosate		
* Apache	Zeneca Crop Protection	MAFF 06748
* Barclay Dart	Barclay Chemicals (UK) Ltd	MAFF 05129
* Barclay Gallup	Barclay Chemicals (UK) Ltd	MAFF 05020
* Barclay Gallup	Barclay Chemicals (UK) Ltd	MAFF 05161
* Barclay Gallup Amenity	Barclay Chemicals (UK) Ltd	MAFF 06732
A* Barclay Gallup Amenity	Barclay Chemicals (UK) Ltd	MAFF 06753
* CDA Spasor	Rhone-Poulenc Agriculture Ltd	MAFF 06414
* Clarion	Du Pont (UK) Ltd	MAFF 06749
* Clayton Glyphosate	Clayton Plant Protection (UK) Ltd	MAFF 06608
* Clayton Glyphosate 360	Clayton Plant Protection (UK) Ltd	MAFF 05316
A* Clayton Swath	Clayton Plant Protection (UK) Ltd	MAFF 06715
* Clean-Up-360	Top Farm Formulations Ltd	MAFF 05076
* Clear-Up	Agpack Ltd	MAFF 05164
A* Glyfonex	Danagri APS	MAFF 06955
* Glyfosate - 360	Top Farm Formulations Ltd	MAFF 05319
A* Glyphogan	Makhteshim-Agan (UK) Ltd	MAFF 05784
A* Helosate	Helm AG	MAFF 06499
* Hilite	Chipman Ltd	MAFF 06261
*% Holdup	Dalmant Ltd	MAFF 04721
*% Holdup	Dalmant Ltd	MAFF 04722
*% Holdup	Dalmant Ltd	MAFF 04723
*% Holdup	Dalmant Ltd	MAFF 04724
* I T Glyphosate	I T Agro Ltd	MAFF 07212
* Landgold Glyphosate 360	Landgold & Co Ltd	MAFF 05929
*% MON 35010	Monsanto Plc	MAFF 06375
A* MON 44068 Pro	Monsanto Plc	MAFF 06815
A* MON 52276	Schering Agriculture	MAFF 06949
* Mogul	Monsanto Plc	MAFF 07076
* Muster	Imperial Chemical Industries Plc	MAFF 03549
* Muster	Zeneca Crop Protection	MAFF 06685
* Muster LA	Zeneca Professional Products	MAFF 05762
* Outlaw	Barclay Chemicals (UK) Ltd	MAFF 07357
* Portland Glyphosate 480	Portman Agrochemicals Ltd	MAFF 07194
* Portman Glider	Portman Agrochemicals Ltd	MAFF 04695
* Portman Glyphosate	Portman Agrochemicals Ltd	MAFF 05891
* Portman Glyphosate 360	Portman Agrochemicals Ltd	MAFF 04699
* Power Glyphosate 360	Power Agrichemicals Ltd	MAFF 04714
A* Roundup	Monsanto Plc	MAFF 01828
A*% Roundup	Protectacrop Ltd	MAFF 05532
A* Roundup	Schering Agrochemicals Ltd	MAFF 03947
A*% Roundup A	Complete Weed Control Ltd	MAFF 05463
A* Roundup Biactive	Monsanto Plc	MAFF 06941
A* Roundup Biactive Dry	Monsanto Plc	MAFF 06942
* Roundup Four 80	Monsanto Plc	MAFF 03176

* These products are "approved for agricultural use". For further details refer to page vii.
% These products are subject to the staged revocation procedure. For further details refer to page vii.
A These products are approved for use in or near water. For further details refer to page vii.

PROFESSIONAL PRODUCTS: HERBICIDES

Product Name	Marketing Company	Reg. No.

236 Glyphosate—*continued*

A*	Roundup Pro	Monsanto Plc	MAFF 04146
A*	Roundup Pro Biactive	Monsanto Plc	MAFF 06954
*	Roundup Two 40	Monsanto Agricultural Co	MAFF 04538
A*	Spasor	Rhone-Poulenc Environmental Products	MAFF 03436
A*	Spasor	Rhone-Poulenc Environmental Products	MAFF 07211
*	Stacato	Sipcam UK Ltd	MAFF 05892
*	Stampede	Zeneca Crop Protection	MAFF 06327
*	Standon Glyphosate 360	Standon Chemicals Ltd	MAFF 05582
*	Stefes Complete	Stefes Plant Protection Ltd	MAFF 06084
*	Stefes Glyphosate	Stefes Plant Protection Ltd	MAFF 05819
*	Stefes Kickdown	Stefes Plant Protection Ltd	MAFF 06329
*	Stefes Kickdown 2	Stefes Plant Protection Ltd	MAFF 06548
A*	Stetson	Monsanto Plc	MAFF 06956
*	Sting	Monsanto Plc	MAFF 02789
*	Sting CT	Monsanto Plc	MAFF 04754
*	Stirrup	Chipman Ltd	MAFF 04174
*	Stirrup	Nomix-Chipman Ltd	MAFF 06132
*	Sulfosate	Zeneca Crop Protection	MAFF 06750
*	Touchdown	Zeneca Crop Protection	MAFF 06326
*	Touchdown LA	Zeneca Professional Products	MAFF 06444
*	Unistar Glyfosate 360	Unistar Ltd	MAFF 05928
*	Unistar Glyphosate 360	Unistar Ltd	MAFF 06332

237 Glyphosate + *Diuron*

*	Touche	Nomix-Chipman Ltd	MAFF 06797

238 Glyphosate + *Simazine*

*%	Rival	Monsanto Plc	MAFF 03377
*%	Ultra-sonic	Rigby Taylor Ltd	MAFF 03546

239 Hexazinone

*%	Velpar Liquid	Du Pont (UK) Ltd	MAFF 02288

240 Imazamethabenz-methyl

*	Dagger	Cyanamid of GB Ltd	MAFF 03737

241 Imazamethabenz-methyl + *Isoproturon*

*	Pinnacle 400	Cyanamid of GB Ltd	MAFF 04687

---

* These products are "approved for agricultural use". For further details refer to page vii.
% These products are subject to the staged revocation procedure. For further details refer to page vii.
A These products are approved for use in or near water. For further details refer to page vii.

PROFESSIONAL PRODUCTS: HERBICIDES

Product Name	Marketing Company	Reg. No.
242 Imazapyr		
* Arsenal	Chipman Ltd	MAFF 02904
* Arsenal	Cyanamid of GB Ltd	MAFF 04064
* Arsenal	Nomix-Chipman Ltd	MAFF 05537
* Arsenal 50	Chipman Ltd	MAFF 03774
* Arsenal 50	Cyanamid of GB Ltd	MAFF 04070
* Arsenal 50	Nomix-Chipman Ltd	MAFF 05567
* Arsenal 50 F	Cyanamid of GB Ltd	MAFF 05856
* Claymore	Cyanamid of GB Ltd	MAFF 05534
243 Imazaquin + *Chlormequat*		
* Meteor	Cyanamid of GB Ltd	MAFF 06505
* Talgard	Cyanamid Of GB Ltd	MAFF 07187
244 Indol-3-ylacetic acid		
* Rhizopon A Powder	Fargro Ltd	MAFF 07131
* Rhizopon A Tablets	Fargro Ltd	MAFF 07132
245 4-Indol-3-ylbutyric acid		
* Chrysotek Beige	Fargro Ltd	MAFF 07125
* Chryzotop Green	Fargro Ltd	MAFF 07129
* Rhizopon AA Powder (1%)	Fargro Ltd	MAFF 07127
* Rhizopon AA Powder (0.5%)	Fargro Ltd	MAFF 07126
* Rhizopon AA Powder (2%)	Fargro Ltd	MAFF 07128
* Rhizopon AA Tablets	Fargro Ltd	MAFF 07130
* Seradix 1	Embetec Crop Protection	MAFF 04680
* Seradix 1	Rhone-Poulenc Agriculture	MAFF 06191
* Seradix 1	Rhone-Poulenc Agriculture Ltd	MAFF 05240
* Seradix 2	Embetec Crop Protection	MAFF 04681
* Seradix 2	Rhone-Poulenc Agriculture	MAFF 06193
* Seradix 3	Embetec Crop Protection	MAFF 04682
* Seradix 3	Rhone-Poulenc Agriculture	MAFF 06194
* Seradix No 2	Rhone-Poulenc Agriculture Ltd	MAFF 05241
* Seradix No 3	Rhone-Poulenc Agriculture Ltd	MAFF 05242
246 4-Indol-3-ylbutyric acid + *Dichlorophen* + *1-Naphthylacetic acid*		
* Synergol	Silvaperl Products Ltd	MAFF 04594
247 Ioxynil		
* Actrilawn 10	Rhone-Poulenc Environmental Products	MAFF 05247
*% Actrilawn 10	Rhone-Poulenc Environmental Products	MAFF 04431

* These products are "approved for agricultural use". For further details refer to page vii.
% These products are subject to the staged revocation procedure. For further details refer to page vii.
A These products are approved for use in or near water. For further details refer to page vii.

PROFESSIONAL PRODUCTS: HERBICIDES

Product Name	Marketing Company	Reg. No.

247 Ioxynil—*continued*

*	Totril	Rhone-Poulenc Agriculture Ltd	MAFF 06116
*%	Totril	Rhone-Poulenc Agriculture Ltd	MAFF 05434

248 Ioxynil + *Benazolin* + *Bromoxynil*

*	Asset	AgrEvo UK Crop Protection Ltd	MAFF 07243
*	Asset	Schering Agrochemicals Ltd	MAFF 03824

249 Ioxynil + *Bromoxynil*

*	Agri HBN 400	Agri-Export Ltd	MAFF 04257
*	Alpha Biotril Plus	Makhteshim-Agan (UK) Ltd	MAFF 04878
*	Alpha Briotril	Makhteshim-Agan (UK) Ltd	MAFF 04876
*	Alpha Briotril Plus 19/19	Makhteshim-Agan (UK) Ltd	MAFF 04740
*	Deloxil	A H Marks & Co Ltd	MAFF 00663
*	Deloxil	AgrEvo UK Crop Protection Ltd	MAFF 07313
*	Deloxil	Hoechst UK Ltd	MAFF 00664
*	Deloxil 400	AgrEvo UK Crop Protection Ltd	MAFF 07314
*	Deloxil 400	Hoechst UK Ltd	MAFF 06613
*%	FR 1001	Rhone-Poulenc Agriculture Ltd	MAFF 05233
*	Oxytril CM	Rhone-Poulenc Agriculture	MAFF 06201
*	Oxytril CM	Rhone-Poulenc Crop Protection	MAFF 04605
*	Percept	MTM Agrochemicals Ltd	MAFF 05481
*%	RP 1001	Rhone-Poulenc Crop Protection	MAFF 04835
*	Status	Rhone-Poulenc Agriculture Ltd	MAFF 06906
*	Stellox 380 EC	Ciba Agriculture	MAFF 04169
*	Stellox 60 WG	Ciba Agriculture	MAFF 04626

250 Ioxynil + *Bromoxynil* + *Clopyralid* + *Fluroxypyr*

*	Crusader S	DowElanco Ltd	MAFF 05174

251 Ioxynil + *Bromoxynil* + *Dichlorprop* + *MCPA*

*	Atlas Minerva	Atlas Interlates Ltd	MAFF 03046

252 Ioxynil + *Bromoxynil* + *Diflufenican*

*	Capture	Rhone-Poulenc Agriculture Ltd	MAFF 06881
*	Quartz BL	Rhone-Poulenc Agriculture Ltd	MAFF 06123

253 Ioxynil + *Bromoxynil* + *Ethofumesate*

*	Leyclene	AgrEvo UK Crop Protection Ltd	MAFF 07263
*	Leyclene	Schering Agrochemicals Ltd	MAFF 05285
*	Trapper	DowElanco Ltd	MAFF 06325

254 Ioxynil + *Bromoxynil* + *Fluroxypyr*

*	Advance	DowElanco Ltd	MAFF 05173

* These products are "approved for agricultural use". For further details refer to page vii.
% These products are subject to the staged revocation procedure. For further details refer to page vii.
A These products are approved for use in or near water. For further details refer to page vii.

PROFESSIONAL PRODUCTS: HERBICIDES

Product Name	Marketing Company	Reg. No.

255 Ioxynil + Bromoxynil + Isoproturon

*%	Astrol	Rhone-Poulenc Agriculture Ltd	MAFF 05441
*	Doublet	Rhone-Poulenc Agriculture	MAFF 06198
*	Doublet	Rhone-Poulenc Agriculture Ltd	MAFF 05356
*%	Twin-Tak	Rhone-Poulenc Agriculture Ltd	MAFF 05379

256 Ioxynil + Bromoxynil + Isoproturon + Mecoprop

*	Terset	Rhone-Poulenc Agriculture Ltd	MAFF 06171
*%	Terset	Rhone-Poulenc Agriculture Ltd	MAFF 05227

257 Ioxynil + Bromoxynil + Mecoprop

*%	Brittox	May & Baker Ltd	MAFF 00332
*	RP 83/8	Rhone-Poulenc Agriculture	MAFF 06169
*%	RP 83/8	Rhone-Poulenc Crop Protection	MAFF 05681
*	Swipe 560 EC	Ciba Agriculture	MAFF 02057

258 Ioxynil + Bromoxynil + Triasulfuron

*	Teal	Ciba Agriculture	MAFF 06117
*	Teal G	Ciba Agriculture	MAFF 06118

259 Ioxynil + Bromoxynil + Trifluralin

*	Masterspray	Pan Britannica Industries Ltd	MAFF 02971

260 Ioxynil + Clopyralid

*	Escort	DowElanco Ltd	MAFF 05466

261 Ioxynil + Clopyralid + Fluroxypyr

*	Hotspur	DowElanco Ltd	MAFF 05185
*%	Hotspur	Farm Protection Ltd	MAFF 05301

262 Ioxynil + 2,4-D + Dicamba

*	Super Verdone	Imperial Chemical Industries Plc	MAFF 02051
*	Super Verdone	Zeneca Professional Products	MAFF 06862

263 Ioxynil + Dichlorprop + MCPA

*	Certrol-PA	J W Chafer Ltd	MAFF 00470

264 Ioxynil + Fluroxypyr

*	Stexal	DowElanco Ltd	MAFF 05188

265 Ioxynil + Isoproturon + Mecoprop

*%	Musketeer	Hoechst UK Ltd	MAFF 01461
*	Post-Kite	Schering Agrochemicals Ltd	MAFF 03832

* These products are "approved for agricultural use". For further details refer to page vii.
% These products are subject to the staged revocation procedure. For further details refer to page vii.
A These products are approved for use in or near water. For further details refer to page vii.

PROFESSIONAL PRODUCTS: HERBICIDES

Product Name	Marketing Company	Reg. No.
266 Ioxynil + *Mecoprop*		
*% Synox	Vitax Ltd	MAFF 05707
267 Isoproturon		
* Alpha Isoproturon 500	Makhteshim-Agan (UK) Ltd	MAFF 05882
* Alpha Isoproturon 650	Makhteshim-Agan (UK) Ltd	MAFF 07034
* Arelon	AgrEvo UK Crop Protection Ltd	MAFF 06716
* Arelon	Hoechst UK Ltd	MAFF 04544
* Arelon WDG	AgrEvo UK Crop Protection Ltd	MAFF 07302
* Arelon WDG	Hoechst UK Ltd	MAFF 04494
* Atlas Protall	Atlas Interlates Ltd	MAFF 06340
* Auger	Rhone-Poulenc Agriculture	MAFF 06581
* Barclay Guideline	Barclay Chemicals (UK) Ltd	MAFF 06743
* Barclay Proton	Barclay Chemicals (UK) Ltd	MAFF 05262
* Chiltern Isoproturon	Chiltern Farm Chemicals Ltd	MAFF 07157
* Graminon 500 SC	Ciba Agriculture	MAFF 06032
* Graminon 500SC	Ciba Agriculture	MAFF 06055
* Hytane 500 SC	Ciba Agriculture	MAFF 05848
*% Hytane 500FW	Ciba-Geigy Agrochemicals	MAFF 01098
*% IPU 500	Rhone-Poulenc Agriculture Ltd	MAFF 05696
*% IPU 500 Herbicide	Rhone-Poulenc Agriculture Ltd	MAFF 05232
* Isoguard	Gharda Chemicals Ltd	MAFF 06778
* Isoguard	Portman Agrochemicals Ltd	MAFF 06072
* Isoproturon 500	AgrEvo UK Crop Protection Ltd	MAFF 06718
* Isoproturon 500	Hoechst UK Ltd	MAFF 06065
* Isoproturon 553	AgrEvo UK Crop Protection Ltd	MAFF 06719
* Isoproturon 553	Hoechst UK Ltd	MAFF 06071
* Ki-Hara IPU 500	Ki-Hara Chemicals Ltd	MAFF 06102
* Landgold Isoproturon	Landgold & Co Ltd	MAFF 06012
* Landgold Isoproturon FC	Landgold & Co Ltd	MAFF 06021
* Landgold Isoproturon FL	Landgold & Co Ltd	MAFF 06034
* Landgold Isoproturon SC	Landgold & Co Ltd	MAFF 05989
* MSS Iprofile	Mirfield Sales Services Ltd	MAFF 06341
*% Phyto IPU	Phyto Research Pflanzenschutz GmbH	MAFF 04380
* Portman Isotop	Portman Agrochemicals Ltd	MAFF 03434
*% Power Swing	Power Agrichemicals Ltd	MAFF 04434
* Quintil 500	Phytorus	MAFF 06800
* Sabina 500 SC	Ciba Agriculture	MAFF 06033
* Sabina 500 SC	Ciba Agriculture	MAFF 06056
* Sabre	AgrEvo UK Crop Protection Ltd	MAFF 06717
* Sabre	Schering Agrochemicals Ltd	MAFF 04148
*% Sabre WDG	Schering Agrochemicals Ltd	MAFF 04719
* Stefes IPU	Stefes Plant Protection Ltd	MAFF 05776

* These products are "approved for agricultural use". For further details refer to page vii.
% These products are subject to the staged revocation procedure. For further details refer to page vii.
A These products are approved for use in or near water. For further details refer to page vii.

PROFESSIONAL PRODUCTS: HERBICIDES

Product Name	Marketing Company	Reg. No.

267 Isoproturon—*continued*

* Tolkan Liquid	Rhone-Poulenc Agriculture	MAFF 06172
* Tolkan Liquid	Rhone-Poulenc Agriculture Ltd	MAFF 04562
* Top Farm IPU 500	Top Farm Formulations Ltd	MAFF 05978
* Tripart Pugil	Tripart Farm Chemicals Ltd	MAFF 06153

268 Isoproturon + *Bifenox*

* Invicta	Farm Protection	MAFF 04825
* RP 4169	Rhone-Poulenc Agriculture Ltd	MAFF 05801

269 Isoproturon + *Bromoxynil* + *Ioxynil*

*% Astrol	Rhone-Poulenc Agriculture Ltd	MAFF 05441
* Doublet	Rhone-Poulenc Agriculture	MAFF 06198
* Doublet	Rhone-Poulenc Agriculture Ltd	MAFF 05356
*% Twin-Tak	Rhone-Poulenc Agriculture Ltd	MAFF 05379

270 Isoproturon + *Bromoxynil* + *Ioxynil* + *Mecoprop*

* Terset	Rhone-Poulenc Agriculture Ltd	MAFF 06171
*% Terset	Rhone-Poulenc Agriculture Ltd	MAFF 05227

271 Isoproturon + *Diflufenican*

* Adition	Rhone-Poulenc Agriculture	MAFF 06184
*% Adition	Rhone-Poulenc Agriculture	MAFF 05146
* Clayton Fenican IPU	Clayton Plant Protection (UK) Ltd	MAFF 06759
* Cougar	May & Baker Ireland Ltd	MAFF 04675
* Cougar	Rhone-Poulenc Ireland Ltd	MAFF 06611
* Javelin	Rhone-Poulenc Agriculture	MAFF 06192
*% Javelin	Rhone-Poulenc Agriculture Ltd	MAFF 05095
* Javelin Gold	Rhone-Poulenc Agriculture	MAFF 06200
* Javelin Gold	Rhone-Poulenc Crop Protection	MAFF 05555
* Landgold DFF 625	Landgold & Co Ltd	MAFF 06274
* Landgold FF550	Landgold & Co Ltd	MAFF 06053
* Oyster	Rhone-Poulenc Agriculture	MAFF 07345
* Panther	Rhone-Poulenc Agriculture	MAFF 06491
*% Panther	Rhone-Poulenc Crop Protection	MAFF 05189
*% Stalker	Rhone-Poulenc Agriculture Ltd	MAFF 05557
* Tolkan Turbo	Rhone-Poulenc Agriculture	MAFF 06795
* Zodiac TX	Rhone-Poulenc Agriculture Ltd	MAFF 06775

272 Isoproturon + *Fenoxaprop-P-ethyl*

* Puma X	AgrEvo UK Crop Protection Ltd	MAFF 07332
* Puma X	Hoechst UK Ltd	MAFF 06768
* Whip X	AgrEvo UK Crop Protection Ltd	MAFF 07339
* Whip X	Hoechst UK Ltd	MAFF 06771

* These products are "approved for agricultural use". For further details refer to page vii.
% These products are subject to the staged revocation procedure. For further details refer to page vii.
A These products are approved for use in or near water. For further details refer to page vii.

PROFESSIONAL PRODUCTS: HERBICIDES

	Product Name	Marketing Company	Reg. No.

273 Isoproturon + *Fluoroglycofen-ethyl*

*	Compete Forte	Hoechst UK Ltd	MAFF 06606
*	Competitor	AgrEvo UK Crop Protection Ltd	MAFF 07310
*	Competitor	Hoechst UK Ltd	MAFF 06605
*	Effect	AgrEvo UK Crop Protection Ltd	MAFF 07319
*	Effect	Hoechst UK Ltd	MAFF 06607

274 Isoproturon + *Imazamethabenz-methyl*

*	Pinnacle 400	Cyanamid of GB Ltd	MAFF 04687

275 Isoproturon + *Ioxynil* + *Mecoprop*

*%	Musketeer	Hoechst UK Ltd	MAFF 01461
*	Post-Kite	Schering Agrochemicals Ltd	MAFF 03832

276 Isoproturon + *Isoxaben*

*%	Fanfare 469 FW	Ciba-Geigy Agrochemicals	MAFF 04020
*	Fanfare 469 SC	Ciba Agriculture	MAFF 05841
*	IPSO	DowElanco Ltd	MAFF 05736
*%	IPSO	Elanco Products Ltd	MAFF 04089

277 Isoproturon + *Metsulfuron-methyl*

*%	Oracle	Du Pont (UK) Ltd	MAFF 04027

278 Isoproturon + *Pendimethalin*

*	Encore	Cyanamid of GB Ltd	MAFF 04737
*	Jolt	Cyanamid of GB Ltd	MAFF 05488
*	Trump	Cyanamid of GB Ltd	MAFF 03687

279 Isoproturon + *Simazine*

*%	Harlequin 500 FW	Ciba-Geigy Agrochemicals	MAFF 04625
*	Harlequin 500 SC	Ciba Agriculture	MAFF 05847

280 Isoproturon + *Trifluralin*

*	Autumn Kite	AgrEvo UK Crop Protection Ltd	MAFF 07119
*	Autumn Kite	Schering Agrochemicals Ltd	MAFF 03830
*	Cantor	Portman Agrochemicals Ltd	MAFF 07217
*	Debut	Portman Agrochemicals Ltd	MAFF 06882

281 Isoxaben

*	Flexidor	DowElanco Ltd	MAFF 05121
*%	Flexidor	Elanco Products Ltd	MAFF 03481
*	Flexidor 125	DowElanco Ltd	MAFF 05104
*%	Gallery	DowElanco Ltd	MAFF 05983

* These products are "approved for agricultural use". For further details refer to page vii.
% These products are subject to the staged revocation procedure. For further details refer to page vii.
A These products are approved for use in or near water. For further details refer to page vii.

PROFESSIONAL PRODUCTS: HERBICIDES

Product Name	Marketing Company	Reg. No.

281 Isoxaben—*continued*

* Gallery 125	DowElanco Ltd	MAFF 06889
* Knot Out	Vitax Ltd	MAFF 05163
* Tripart Ratio	Tripart Farm Chemicals Ltd	MAFF 04659

282 Isoxaben + *Chlorotoluron*

*% Crimson 514 FW	Ciba-Geigy Agrochemicals	MAFF 04830
*% Crimson 514 SC	Ciba Agriculture	MAFF 05835
*% Crimson 514 SC	DowElanco Ltd	MAFF 05760

283 Isoxaben + *Isoproturon*

*% Fanfare 469 FW	Ciba-Geigy Agrochemicals	MAFF 04020
* Fanfare 469 SC	Ciba Agriculture	MAFF 05841
* IPSO	DowElanco Ltd	MAFF 05736
*% IPSO	Elanco Products Ltd	MAFF 04089

284 Isoxaben + *Methabenzthiazuron*

* Glytex	Bayer Plc	MAFF 04230

285 Isoxaben + *Terbuthylazine*

* Skirmish 495 SC	Ciba Agriculture	MAFF 05692

286 Lenacil

* Clayton Lenacil 80W	Clayton Plant Protection (UK) Ltd	MAFF 07074
* Stefes Lenacil	Stefes Plant Protection Ltd	MAFF 07103
* Venzar Flowable	Du Pont (UK) Ltd	MAFF 06907
* Venzar Weedkiller	Du Pont (UK) Ltd	MAFF 02293
* Vizor	Du Pont (UK) Ltd	MAFF 06572
* Vizor	Farm Protection Ltd	MAFF 05072
*% Vizor	Farm Protection Ltd	MAFF 02315

287 Lenacil + *Chloridazon*

* Advizor	Du Pont (UK) Ltd	MAFF 06571
* Advizor	Farm Protection Ltd	MAFF 00032
* Advizor S	Du Pont (UK) Ltd	MAFF 06960
* Varmint	Pan Britannica Industries Ltd	MAFF 01561

288 Lenacil + *Linuron*

* Lanslide	Pan Britannica Industries Ltd	MAFF 01184

289 Lenacil + *Phenmedipham*

* DUK-880	Du Pont (UK) Ltd	MAFF 04121

---

* These products are "approved for agricultural use". For further details refer to page vii.
% These products are subject to the staged revocation procedure. For further details refer to page vii.
A These products are approved for use in or near water. For further details refer to page vii.

PROFESSIONAL PRODUCTS: HERBICIDES

Product Name	Marketing Company	Reg. No.
290 Linuron		
* Afalon	AgrEvo UK Crop Protection Ltd	MAFF 07299
* Afalon	Hoechst UK Ltd	MAFF 04665
* Afalon EC	AgrEvo UK Crop Protection Ltd	MAFF 07300
* Afalon EC	Hoechst UK Ltd	MAFF 00034
* Alpha Linuron 50 SC	Makhteshim-Agan (UK) Ltd	MAFF 06967
* Alpha Linuron 50 WP	Makhteshim-Agan (UK) Ltd	MAFF 04870
* Ashlade Linuron FL	Ashlade Formulations Ltd	MAFF 06221
*% Ashlade Linuron FL	Ashlade Formulations Ltd	MAFF 03439
* Atlas Linuron	Atlas Interlates Ltd	MAFF 03054
* Campbell's Linuron 45% Flowable	J D Campbell & Sons Ltd	MAFF 00408
*% Linuron 50	Du Pont (UK) Ltd	MAFF 01209
* Linuron Flowable	Pan Britannica Industries Ltd	MAFF 02965
* Liquid Linuron	Pan Britannica Industries Ltd	MAFF 01556
* Rotalin	Farm Protection	MAFF 06253
* Rotalin	Farm Protection Ltd	MAFF 01827
291 Linuron + *Chlorpropham*		
* Profalon	AgrEvo UK Crop Protection Ltd	MAFF 07331
* Profalon	Hoechst UK Ltd	MAFF 01640
292 Linuron + *Cyanazine*		
* Stay-Kleen	Shell Chemicals UK Ltd + Chemical Spraying Co Ltd	MAFF 02018
293 Linuron + *2,4-Db* + *MCPA*		
* Alistell	Farm Protection Ltd	MAFF 05726
*% Alistell	Farm Protection Ltd	MAFF 00080
* Alistell	Zeneca Crop Protection	MAFF 06515
* Clovacorn Extra	Farmers Crop Chemicals Ltd	MAFF 02596
294 Linuron + *Lenacil*		
* Lanslide	Pan Britannica Industries Ltd	MAFF 01184
295 Linuron + *Terbutryn*		
* Tempo	Farm Protection	MAFF 02736
296 Linuron + *Trietazine*		
* Bronox	AgrEvo UK Crop Protection Ltd	MAFF 07255
* Bronox	Schering Agrochemicals Ltd	MAFF 03864
297 Linuron + *Trifluralin*		
* Ashlade Flint	Ashlade Formulations Ltd	MAFF 04471
*% Atlas Janus	Atlas Interlates Ltd	MAFF 03085

* These products are "approved for agricultural use". For further details refer to page vii.
% These products are subject to the staged revocation procedure. For further details refer to page vii.
A These products are approved for use in or near water. For further details refer to page vii.

PROFESSIONAL PRODUCTS: HERBICIDES

Product Name	Marketing Company	Reg. No.

297 Linuron + *Trifluralin*—continued

* Campbell's Trifluron	J D Campbell & Sons Ltd	MAFF 02682
* Chandor	DowElanco Ltd	MAFF 05631
*% Chandor	Elanco Products Ltd	MAFF 00483
* Ipicombi TL	I Pi Ci	MAFF 04608
* Janus	Montedison UK Ltd	MAFF 02805
* Linnet	Pan Britannica Industries Ltd	MAFF 01555
* Neminfest	Enichem UK Ltd	MAFF 05822
* Neminfest	Montedison UK Ltd	MAFF 02546
* Neminfest	Sipcam UK Ltd	MAFF 07219
*% Quadrangle Onslaught	Quadrangle Ltd	MAFF 02548
*% Tri-farmon FL	Farm Protection Ltd	MAFF 02870
* Triplen Combi	Sipcam UK Ltd	MAFF 05939

298 Maleic hydrazide

* Antergon MH 180	Uniroyal Chemical Ltd	MAFF 06233
*% Antergon MH180	KVK Agro A/S	MAFF 05873
A*% Bos MH 180	Bos Chemicals Ltd	MAFF 04327
* Bos MH 180	Uniroyal Chemical Ltd	MAFF 06502
*% Burtolin	Burts & Harvey	MAFF 00355
* Burtolin	Rhone-Poulenc Agriculture Ltd	MAFF 05399
* CDA Regulox K	Rhone-Poulenc Ltd	MAFF 04756
* Fazor	DowElanco Ltd	MAFF 05558
* Fazor	Uniroyal Chemical Ltd	MAFF 05461
* MSS MH 18	Mirfield Sales Services Ltd	MAFF 03065
*% Regulox K	Burts & Harvey	MAFF 01716
A* Regulox K	Rhone-Poulenc Environmental Products	MAFF 05405
* Royal MH 180	Uniroyal Chemical Ltd	MAFF 07043
* Synchemicals Mazide 25	Synchemicals Ltd	MAFF 02067

299 Maleic hydrazide + *Dicamba*

* Dockmaster	Rhone-Poulenc Environmental Products	MAFF 05932

300 Maleic hydrazide + *Dicamba* + *MCPA*

* Mazide Selective	Vitax Ltd	MAFF 05753
* Synchemicals Mazide Selective	Synchemicals Ltd	MAFF 02070

301 MCPA

* Agrichem MCPA 25	Agrichem Ltd	MAFF 00050
* Agrichem MCPA 50	Agrichem Ltd	MAFF 04097
* Agricorn 500	Farmers Crop Chemicals Ltd	MAFF 00055
* Agritox 50	Nufarm UK Ltd	MAFF 07400

---

* These products are "approved for agricultural use". For further details refer to page vii.
% These products are subject to the staged revocation procedure. For further details refer to page vii.
A These products are approved for use in or near water. For further details refer to page vii.

PROFESSIONAL PRODUCTS: HERBICIDES

Product Name	Marketing Company	Reg. No.

301 MCPA—continued

*	Agritox 50	Rhone-Poulenc Agriculture Ltd	MAFF 05219
*	Agritox 50	Rhone-Poulenc Agriculture Ltd	MAFF 06161
*%	Agroxone 50	Imperial Chemical Industries Plc	MAFF 03629
*%	Agroxone 50	Zeneca Crop Protection	MAFF 06654
*	Atlas MCPA	Atlas Interlates Ltd	MAFF 03055
*	BASF MCPA Amine 50	BASF Plc	MAFF 00209
*%	BH MCPA 75	Burts & Harvey	MAFF 03363
*	BH MCPA 75	Rhone-Poulenc Agriculture Ltd	MAFF 05395
*	CIA MCPA 500	Combined Independent Agronomists Ltd	MAFF 05874
*	Campbell's MCPA 25	J D Campbell & Sons Ltd	MAFF 00409
*	Campbells MCPA 50	J D Campbell & Sons Ltd	MAFF 00381
*	Empal	Universal Crop Protection Ltd	MAFF 00795
*	HY-MCPA	Agrichem Ltd	MAFF 06293
*	Headland Spear	Headland Agrochemicals Ltd	MAFF 07115
*	Headland Spear	WBC Technology Ltd	MAFF 05455
*	Headland Spear	WBC Technology Ltd	MAFF 06733
*	Luxan MCPA 50	Luxan (UK) Ltd	MAFF 06912
*	Marks MCPA 50A	A H Marks & Co Ltd	MAFF 02710
*	Marks MCPA P30	A H Marks & Co Ltd	MAFF 01292
*	Marks MCPA S.25	A H Marks & Co Ltd	MAFF 01294
*	Marks MCPA SP	A H Marks & Co Ltd	MAFF 01293
*	MCPA 25%	Chemie Linz UK Ltd	MAFF 02766
*	Mega M	AKZO Chemicals Ltd	MAFF 01320
*	MSS MCPA 50	Mirfield Sales Services Ltd	MAFF 01412
*	Phenoxylene 50	AgrEvo UK Crop Protection Ltd	MAFF 07163
*	Phenoxylene 50	Schering Agrochemicals Ltd	MAFF 03854
*	Quadrangle MCPA 50	Quadrangle Agrochemicals Ltd	MAFF 06935
*	Top Farm MCPA 500	Top Farm Formulations Ltd	MAFF 05930
*	Tripart MCPA 50	Tripart Farm Chemicals Ltd	MAFF 02206

302 MCPA + *Benazolin* + *2,4-D*

*	Legumex Extra	AgrEvo UK Crop Protection Ltd	MAFF 07262

303 MCPA + *Benazolin* + *2,4-Db*

*	Legumex Extra	Schering Agrochemicals Ltd	MAFF 03869
*	Setter 33	AgrEvo UK Crop Protection Ltd	MAFF 07282
*%	Setter 33	Dow Chemical Co Ltd	MAFF 04377
*	Setter 33	DowElanco Ltd	MAFF 05623
*	Setter 33	Schering Agrochemicals Ltd	MAFF 04376

304 MCPA + *Bentazone* + *MCPB*

*	Acumen	BASF Plc	MAFF 00028

---

* These products are "approved for agricultural use". For further details refer to page vii.
% These products are subject to the staged revocation procedure. For further details refer to page vii.
A These products are approved for use in or near water. For further details refer to page vii.

PROFESSIONAL PRODUCTS: HERBICIDES

Product Name	Marketing Company	Reg. No.
305 MCPA + *Bromoxynil* + *Dichlorprop* + *Ioxynil*		
* Atlas Minerva	Atlas Interlates Ltd	MAFF 03046
306 MCPA + *Clopyralid* + *Dichlorprop*		
* Lontrel Plus	DowElanco Ltd	MAFF 05269
*% Lontrel Plus	ICI Agrochemicals	MAFF 01226
*% Lontrel Plus	Zeneca Crop Protection	MAFF 06683
307 MCPA + *Clopyralid* + *Diflufenican*		
* Spearhead	Rhone-Poulenc Environmental Products	MAFF 07342
308 MCPA + *Clopyralid* + *Fluroxypyr*		
* Bofix	DowElanco Ltd	MAFF 06377
309 MCPA + *2,4-D* + *Dichlorprop* + *Mecoprop*		
* Campbell's New Camppex	J D Campbell & Sons Ltd	MAFF 00414
310 MCPA + *2,4-Db*		
* Agrichem DB Plus	Agrichem Ltd	MAFF 00044
* Campbell's Redlegor	J D Campbell & Sons Ltd	MAFF 00421
* Marks 2,4-DB Extra	A H Marks & Co Ltd	MAFF 01284
* MSS 2,4-DB + *MCPA*	Mirfield Sales Services Ltd	MAFF 01392
311 MCPA + *2,4-Db* + *Linuron*		
* Alistell	Farm Protection Ltd	MAFF 05726
*% Alistell	Farm Protection Ltd	MAFF 00080
* Alistell	Zeneca Crop Protection	MAFF 06515
* Clovacorn Extra	Farmers Crop Chemicals Ltd	MAFF 02596
312 MCPA + *Dicamba*		
* Banvel M	Sandoz Agro Ltd	MAFF 05794
313 MCPA + *Dicamba* + *Maleic hydrazide*		
* Mazide Selective	Vitax Ltd	MAFF 05753
* Synchemicals Mazide Selective	Synchemicals Ltd	MAFF 02070
314 MCPA + *Dicamba* + *Mecoprop*		
* Banlene Plus	AgrEvo UK Crop Protection Ltd	MAFF 07245
* Banlene Plus	Schering Agrochemicals Ltd	MAFF 03851
*% Banvel MP	Sandoz Crop Protection Ltd	MAFF 05792
* Campbell's Field Marshall	J D Campbell & Sons Ltd	MAFF 00406
*% Campbell's Grassland Herbicide	J D Campbell & Sons Ltd	MAFF 00407

* These products are "approved for agricultural use". For further details refer to page vii.
% These products are subject to the staged revocation procedure. For further details refer to page vii.
A These products are approved for use in or near water. For further details refer to page vii.

PROFESSIONAL PRODUCTS: HERBICIDES

Product Name	Marketing Company	Reg. No.

314 MCPA + *Dicamba* + *Mecoprop*—continued

*	Campbell's Grassland Herbicide	MTM Agrochemicals Ltd	MAFF 06157
*	Docklene	AgrEvo UK Crop Protection Ltd	MAFF 07257
*	Docklene	Schering Agriculture	MAFF 03863
*	Fisons Tritox	Fisons Plc	MAFF 03013
*	Headland Relay	WBC Technology Ltd	MAFF 03778
*	Herrisol	Bayer Plc	MAFF 01048
*	Hyprone	Agrichem Ltd	MAFF 01093
*	Hysward	Agrichem Ltd	MAFF 01096
*	Mascot Super Selective	Rigby Taylor Ltd	MAFF 03621
*	New Bandock	Shell Chemicals UK Ltd	MAFF 05497
*	Pasturol	Farmers Crop Chemicals Ltd	MAFF 01545
*	Springcorn Extra	Farmers Crop Chemicals Ltd	MAFF 02004
	Tetralex Plus	Cyanamid Of GB Ltd	MAFF 07146
*	Tetralex Plus	Shell Chemicals UK Ltd	MAFF 02115
*	Tribute	Chipman Ltd	MAFF 03470
*	Tribute	Nomix-Chipman Ltd	MAFF 06921

315 MCPA + *Dicamba* + *Mecoprop-P*

*	Field Marshal	MTM Agrochemicals Ltd	MAFF 06077
*	MTM Grassland Herbicide	MTM Agrochemicals Ltd	MAFF 06089
*	Mascot Super Selective-P	Rigby Taylor Ltd	MAFF 06106
*	MSS Mircam Plus	Mirfield Sales Services Ltd	MAFF 01416
*	Pasturol D	Farmers Crop Chemicals Ltd	MAFF 07033
*	Quadrangle Quadban	Quadrangle Agrochemicals	MAFF 07090

316 MCPA + *Dichlorprop*

*	Campbells Redipon Extra	J D Campbell & Sons Ltd	MAFF 00420
*%	Hemoxone	Imperial Chemical Industries Plc	MAFF 01043
*	MSS 2,4-DP + MCPA	Mirfield Sales Services Ltd	MAFF 01396
F*	SHL Turf Feed and Weed	William Sinclair Horticulture Ltd	MAFF 04437
*%	Seritox 50	Rhone-Poulenc Agriculture Ltd	MAFF 05225
*	Seritox 50	Rhone-Poulenc Agriculture Ltd	MAFF 06170
*	Seritox Turf	Rhone-Poulenc Environmental Products	MAFF 06802
*%	Seritox Turf	Rhone-Poulenc Environmental Products	MAFF 05337

317 MCPA + *Dichlorprop* + *Ferrous sulphate*

*	SHL Turf and Weed and Mosskiller	William Sinclair Horticulture Ltd	MAFF 04438

318 MCPA + *Dichlorprop* + *Ioxynil*

*	Certrol-PA	J W Chafer Ltd	MAFF 00470

---

* These products are "approved for agricultural use". For further details refer to page vii.
% These products are subject to the staged revocation procedure. For further details refer to page vii.
A These products are approved for use in or near water. For further details refer to page vii.

PROFESSIONAL PRODUCTS: HERBICIDES

Product Name	Marketing Company	Reg. No.
319 MCPA + MCPB		
* Bellmac Plus	MTM Agrochemicals Ltd	MAFF 06794
*% Campbell's Bellmac Plus	J D Campbell & Sons Ltd	MAFF 00385
* MSS MCPB + MCPA	Mirfield Sales Services Ltd	MAFF 01413
* Trifolex-Tra	Cyanamid Of GB Ltd	MAFF 07147
* Trifolex-Tra	Shell Chemicals UK Ltd	MAFF 02175
* Tropotox Plus	Rhone-Poulenc Agriculture Ltd	MAFF 05228
* Tropotox Plus	Rhone-Poulenc Crop Protection	MAFF 06156
320 MCPA + Mecoprop		
* Fisons Greenmaster Extra	Fisons Plc	MAFF 03130
321 MCPB		
* Bellmac Straight	J D Campbell & Sons Ltd	MAFF 06428
*% Campbell's Bellmac Straight	J D Campbell & Sons Ltd	MAFF 00386
* Cropsafe MCPB	Hortichem Ltd	MAFF 04383
* Marks MCPB	A H Marks & Co Ltd	MAFF 03497
* Tropotox	Rhone-Poulenc Agriculture	MAFF 06179
* Tropotox	Rhone-Poulenc Crop Protection	MAFF 05963
322 MCPB + Bentazone		
* Pulsar	BASF Plc	MAFF 04002
323 MCPB + Bentazone + MCPA		
* Acumen	BASF Plc	MAFF 00028
324 MCPB + MCPA		
* Bellmac Plus	MTM Agrochemicals Ltd	MAFF 06794
*% Campbell's Bellmac Plus	J D Campbell & Sons Ltd	MAFF 00385
* MSS MCPB + MCPA	Mirfield Sales Services Ltd	MAFF 01413
* Trifolex-Tra	Cyanamid Of GB Ltd	MAFF 07147
* Trifolex-Tra	Shell Chemicals UK Ltd	MAFF 02175
* Tropotox Plus	Rhone-Poulenc Agriculture Ltd	MAFF 05228
* Tropotox Plus	Rhone-Poulenc Crop Protection	MAFF 06156
325 Mecoprop		
*% Amos CMPP	Kommer-Brookwick	MAFF 05962
* Amos CMPP	Luxan (UK) Ltd	MAFF 06556
* Atlas CMPP	Atlas Interlates Ltd	MAFF 03050
* BH CMPP Extra	Rhone-Poulenc Agriculture Ltd	MAFF 05392
* Campbell's CMPP	J D Campbell & Sons Ltd	MAFF 02918
* Clenecorn	Farmers Crop Chemicals Ltd	MAFF 00542
* Clovotox	Rhone-Poulenc Environmental Products	MAFF 05354

---

* These products are "approved for agricultural use". For further details refer to page vii.
% These products are subject to the staged revocation procedure. For further details refer to page vii.
A These products are approved for use in or near water. For further details refer to page vii.

PROFESSIONAL PRODUCTS: HERBICIDES

Product Name	Marketing Company	Reg. No.

325 Mecoprop—*continued*

*	CMPP 60	Chemie Linz UK Ltd	MAFF 02764
*	Compitox Extra	Nufarm (UK) Ltd	MAFF 07398
*	Compitox Extra	Rhone-Poulenc Agriculture Ltd	MAFF 06162
*%	Compitox Extra	Rhone-Poulenc Agriculture Ltd	MAFF 05221
*	Headland Charge	Headland Agrochemicals Ltd	MAFF 04495
*	Hymec	Agrichem Ltd	MAFF 01091
*	Iso Cornox 57	AgrEvo UK Crop Protection Ltd	MAFF 07261
*	Iso Cornox 57	Schering Agrochemicals Ltd	MAFF 03868
*	Landgold CMPP F	Landgold & Co Ltd	MAFF 06276
*	Luxan CMPP	Luxan (UK) Ltd	MAFF 06037
*	Luxan CMPP 600	Luxan (UK) Ltd	MAFF 05909
*	Marks Mecoprop E	A H Marks & Co Ltd	MAFF 01296
*	Marks Mecoprop K	A H Marks & Co Ltd	MAFF 01297
*	Mascot Cloverkiller	Rigby Taylor Ltd	MAFF 02438
*%	Methoxone	Imperial Chemical Industries Plc	MAFF 01334
*	MSS CMPP	Mirfield Sales Services Ltd	MAFF 01405
*%	Power CMPP	Power Agrichemicals Ltd	MAFF 02616
*%	Quad CMPP 600	Quadrangle Ltd	MAFF 04497
*%	Standon CMPP F	Standon Chemicals Ltd	MAFF 05659
*	Top Farm CMPP	Top Farm Formulations Ltd	MAFF 05468
*	Unistar CMPP	Unistar Ltd	MAFF 06353

326 Mecoprop + *Bifenox*

*%	EXP 8506	Rhone-Poulenc Crop Protection	MAFF 04886

327 Mecoprop + *Bromoxynil* + *Ioxynil*

*%	Brittox	May & Baker Ltd	MAFF 00332
*	RP 83/8	Rhone-Poulenc Agriculture	MAFF 06169
*%	RP 83/8	Rhone-Poulenc Crop Protection	MAFF 05681
*	Swipe 560 EC	Ciba Agriculture	MAFF 02057

328 Mecoprop + *Bromoxynil* + *Ioxynil* + *Isoproturon*

*	Terset	Rhone-Poulenc Agriculture Ltd	MAFF 06171
*%	Terset	Rhone-Poulenc Agriculture Ltd	MAFF 05227

329 Mecoprop + *Clopyralid*

*%	Seloxone	Dow Chemical Co Ltd	MAFF 01914
*%	Seloxone	DowElanco Ltd	MAFF 05743
*%	Seloxone	Imperial Chemical Industries Plc	MAFF 01915

330 Mecoprop + *Cyanazine*

*	Cleaval	Cyanamid Of GB Ltd	MAFF 07142
*	Cleaval	Shell Chemicals UK Ltd	MAFF 00541

* These products are "approved for agricultural use". For further details refer to page vii.
% These products are subject to the staged revocation procedure. For further details refer to page vii.
A These products are approved for use in or near water. For further details refer to page vii.

PROFESSIONAL PRODUCTS: HERBICIDES

Product Name	Marketing Company	Reg. No.

330 Mecoprop + Cyanazine—*continued*

* FCC Topcorn Extra	Farmers Crop Chemicals Ltd	MAFF 05514
* FCC Topcorn Extra	Farmers Crop Chemicals Ltd	MAFF 07143
* MTM Eminent	MTM Agrochemicals Ltd	MAFF 05513
* MTM Eminent	MTM Agrochemicals Ltd	MAFF 07144

331 Mecoprop + 2,4-D

*% BH CMPP/2,4-D	Burts & Harvey	MAFF 00249
* BH CMPP/2,4-D	Rhone-Poulenc Environmental Products	MAFF 05393
*% CDA CMPP/2,4-D	Rhone-Poulenc Agriculture Ltd	MAFF 05411
* CDA Supertox 30	Rhone-Poulenc Environmental Products	MAFF 04664
* Cleanrun	Richardsons Fertilisers Ltd	MAFF 04759
*% Cleanrun	Zeneca Professional Products	MAFF 06857
* Com-Trol	Certified Laboratories Ltd	MAFF 03343
* Longlife Cleanrun	Zeneca Professional Products	MAFF 06914
* Mascot Selective Weedkiller	Rigby Taylor Ltd	MAFF 03423
* Nomix Turf Selective Herbicide	Nomix-Chipman Ltd	MAFF 06777
* Nomix Turf Selective LC Herbicide	Nomix-Chipman Ltd	MAFF 05973
* Select-Trol	Chemsearch	MAFF 03342
*% Selective Weedkiller	Anteco Ltd	MAFF 05601
* Selective Weedkiller	Yule Catto Consumer Chemicals Ltd	MAFF 06579
* Supertox 30	Rhone-Poulenc Agriculture Ltd	MAFF 05340
*% Sydex	Synchemicals Ltd	MAFF 02061
* Sydex	Vitax Ltd	MAFF 06412
*% Verdone CDA	Imperial Chemical Industries Plc	MAFF 03410
* Zennapron	BP Oil Ltd	MAFF 02377

332 Mecoprop + 2,4-D + Dicamba

* New Formulation Weed and Brush Killer	Synchemicals Ltd	MAFF 02961
* New Formulation Weed and Brushkiller	Vitax Ltd	MAFF 07072

333 Mecoprop + 2,4-D + Dichlorprop + MCPA

* Campbell's New Camppex	J D Campbell & Sons Ltd	MAFF 00414

334 Mecoprop + 2,4-D + Ferrous sulphate

*% Supergreen Feed, Weed and Mosskiller	Not applicable	MAFF 05445

---

* These products are "approved for agricultural use". For further details refer to page vii.
% These products are subject to the staged revocation procedure. For further details refer to page vii.
A These products are approved for use in or near water. For further details refer to page vii.

PROFESSIONAL PRODUCTS: HERBICIDES

Product Name	Marketing Company	Reg. No.

335 Mecoprop + *Dicamba*

*	Banvel P	Sandoz Agro Ltd	MAFF 05793
*	Condox	Zeneca Crop Protection	MAFF 06519
*	Di Farmon	Farm Protection Ltd	MAFF 03882
*	Endox	Farmers Crop Chemicals Ltd	MAFF 00798
*	Farmon Condox	Farm Protection Ltd	MAFF 03883
*	Hyban	Agrichem Ltd	MAFF 01084
*	Hygrass	Agrichem (International) Ltd	MAFF 01090

336 Mecoprop + *Dicamba* + *MCPA*

*	Banlene Plus	AgrEvo UK Crop Protection Ltd	MAFF 07245
*	Banlene Plus	Schering Agrochemicals Ltd	MAFF 03851
*%	Banvel MP	Sandoz Crop Protection Ltd	MAFF 05792
*	Campbell's Field Marshall	J D Campbell & Sons Ltd	MAFF 00406
*%	Campbell's Grassland Herbicide	J D Campbell & Sons Ltd	MAFF 00407
*	Campbell's Grassland Herbicide	MTM Agrochemicals Ltd	MAFF 06157
*	Docklene	AgrEvo UK Crop Protection Ltd	MAFF 07257
*	Docklene	Schering Agriculture	MAFF 03863
*	Fisons Tritox	Fisons Plc	MAFF 03013
*	Headland Relay	WBC Technology Ltd	MAFF 03778
*	Herrisol	Bayer Plc	MAFF 01048
*	Hyprone	Agrichem Ltd	MAFF 01093
*	Hysward	Agrichem Ltd	MAFF 01096
*	Mascot Super Selective	Rigby Taylor Ltd	MAFF 03621
*	New Bandock	Shell Chemicals UK Ltd	MAFF 05497
*	Pasturol	Farmers Crop Chemicals Ltd	MAFF 01545
*	Springcorn Extra	Farmers Crop Chemicals Ltd	MAFF 02004
	Tetralex Plus	Cyanamid Of GB Ltd	MAFF 07146
*	Tetralex Plus	Shell Chemicals UK Ltd	MAFF 02115
*	Tribute	Chipman Ltd	MAFF 03470
*	Tribute	Nomix-Chipman Ltd	MAFF 06921

337 Mecoprop + *Dicamba* + *Triclopyr*

*	Fettel	Farm Protection Ltd	MAFF 02516
*	Fettel	Zeneca Crop Protection	MAFF 06399

338 Mecoprop + *Ioxynil*

*%	Synox	Vitax Ltd	MAFF 05707

339 Mecoprop + *Ioxynil* + *Isoproturon*

*%	Musketeer	Hoechst UK Ltd	MAFF 01461
*	Post-Kite	Schering Agrochemicals Ltd	MAFF 03832

---

* These products are "approved for agricultural use". For further details refer to page vii.
% These products are subject to the staged revocation procedure. For further details refer to page vii.
A These products are approved for use in or near water. For further details refer to page vii.

PROFESSIONAL PRODUCTS: HERBICIDES

Product Name	Marketing Company	Reg. No.
340 Mecoprop + *MCPA*		
* Fisons Greenmaster Extra	Fisons Plc	MAFF 03130
341 Mecoprop-P		
* Astix	Rhone-Poulenc Agriculture Ltd	MAFF 06174
*% Astix	Rhone-Poulenc Agriculture Ltd	MAFF 04472
* Astix K	Rhone-Poulenc Agriculture Ltd	MAFF 06904
* BAS 03729H	Mirfield Sales Services Ltd	MAFF 05912
* Duplosan	BASF Plc	MAFF 05889
* Duplosan New System CMPP	BASF Plc	MAFF 04481
* Landgold Mecoprop-P	Landgold & Co Ltd	MAFF 06052
* Landgold Mecoprop-P 600	Landgold & Co Ltd	MAFF 06461
*% Marks Mecoprop BAI	A H Marks & Co Ltd	MAFF 04509
* Mascot Cloverkiller-P	Rigby Taylor Ltd	MAFF 06099
* MSS Mirprop	Mirfield Sales Services Ltd	MAFF 05911
* MSS Optica	Mirfield Sales Services Ltd	MAFF 04973
* Optica	A H Marks & Co Ltd	MAFF 05814
* Optica	Bayer UK Ltd	MAFF 04922
* Standon Mecoprop-P	Standon Chemicals Ltd	MAFF 05651
* Stefes Mecoprop-P	Stefes Plant Protection Ltd	MAFF 05780
* Stefes Mecoprop-P2	Stefes Plant Protection Ltd	MAFF 06239
* Unistar CMPP	Unistar Ltd	MAFF 06306
342 Mecoprop-P + *2,4-D*		
* Mascot Selective-P	Rigby Taylor Ltd	MAFF 06105
343 Mecoprop-P + *Dicamba*		
* MSS Mircam	Mirfield Sales Services Ltd	MAFF 01415
344 Mecoprop-P + *Dicamba* + *MCPA*		
* Field Marshal	MTM Agrochemicals Ltd	MAFF 06077
* MTM Grassland Herbicide	MTM Agrochemicals Ltd	MAFF 06089
* Mascot Super Selective-P	Rigby Taylor Ltd	MAFF 06106
* MSS Mircam Plus	Mirfield Sales Services Ltd	MAFF 01416
* Pasturol D	Farmers Crop Chemicals Ltd	MAFF 07033
* Quadrangle Quadban	Quadrangle Agrochemicals	MAFF 07090
345 Mecoprop-P + *Fluoroglycofen-ethyl*		
* Compete Mix 20 PVA	BASF Plc	MAFF 07223
* Estrad	BASF Plc	MAFF 07196
* Estrad Duplo	BASF Plc	MAFF 06793
346 Mecoprop-P + *Fluroxypyr*		
* Bastion T	DowElanco Ltd	MAFF 06011

*   These products are "approved for agricultural use". For further details refer to page vii.
%   These products are subject to the staged revocation procedure. For further details refer to page vii.
A   These products are approved for use in or near water. For further details refer to page vii.

PROFESSIONAL PRODUCTS: HERBICIDES

Product Name	Marketing Company	Reg. No.
347 Mecoprop-P + *Thifensulfuron-methyl*		
* Duet	Du Pont (UK) Ltd	MAFF 05169
348 Mecoprop-P + *Triasulfuron*		
* Raven	Ciba Agriculture	MAFF 06119
349 Mefluidide		
* Check Turf II	Certified Laboratories Ltd	MAFF 04463
* Echo	Imperial Chemical Industries Plc	MAFF 04744
* Echo	Imperial Chemical Industries Plc	MAFF 05618
* Embark	Gordon International Corporation	MAFF 04810
* Gro-Tard II	Chemsearch	MAFF 04462
* Mowchem	Rhone-Poulenc Agriculture Ltd	MAFF 05333
*% Mowchem	Rhone-Poulenc Environmental Products	MAFF 04811
350 Mepiquat + *2-Chloroethylphosphonic acid*		
* Sypex M	BASF Plc	MAFF 06810
* Terpal	BASF Plc	MAFF 02103
351 Mepiquat chloride + *2-Chloroethylphosphonic acid*		
* Stefes Mepiquat	Stefes Plant Protection Ltd	MAFF 06970
352 Metamitron		
* Goltix WG	Bayer Plc	MAFF 02430
* Landgold Metamitron	Landgold & Co Ltd	MAFF 06287
* Lektan	Bayer Plc	MAFF 06111
* Power Countdown	Power Agrichemicals Ltd	MAFF 04760
* Power Countdown	Power Agrichemicals Ltd	MAFF 04770
* Quartz	Bayer Plc	MAFF 05167
* Stefes 7G	Stefes Plant Protection Ltd	MAFF 06350
* Stefes Metamitron	Stefes Plant Protection Ltd	MAFF 05821
* Tripart Accendo	Tripart Farm Chemicals Ltd	MAFF 06110
353 Metamitron + *Ethofumesate* + *Phenmedipham*		
* Bayer UK 407 WG	Bayer UK Ltd	MAFF 06313
* CX 171	AgrEvo UK Crop Protection Ltd	MAFF 06803
* Goltix Triple	Bayer UK Ltd	MAFF 06314
354 Metazachlor		
* Barclay Metaza	Barclay Chemicals (UK) Ltd	MAFF 07354
* Butisan S	BASF Plc	MAFF 00357
* Comet	BASF Plc	MAFF 06817
* Standon Metazachlor 50	Standon Chemicals Ltd	MAFF 05581

* These products are "approved for agricultural use". For further details refer to page vii.
% These products are subject to the staged revocation procedure. For further details refer to page vii.
A These products are approved for use in or near water. For further details refer to page vii.

PROFESSIONAL PRODUCTS: HERBICIDES

Product Name	Marketing Company	Reg. No.
355 Methabenzthiazuron		
* Tribunil	Bayer Plc	MAFF 02169
* Tribunil WG	Bayer Plc	MAFF 03260
356 Methabenzthiazuron + *Isoxaben*		
* Glytex	Bayer Plc	MAFF 04230
357 Metoxuron		
* Deftor	Sandoz Agro Ltd	MAFF 05545
*% Dosaflo	Imperial Chemical Industries Plc	MAFF 00753
* Dosaflo	Sandoz Agro Ltd	MAFF 00754
*% Dosaflo	Zeneca Crop Protection	MAFF 06677
* Endspray II	Pan Britannica Industries Ltd	MAFF 05079
358 Metoxuron + *Fentin hydroxide*		
*% Endspray	Pan Britannica Industries Ltd	MAFF 02799
359 Metribuzin		
* Lexone 70 DF	Du Pont (UK) Ltd	MAFF 04991
* Power Metribuzin 70	Power Agrichemicals Ltd	MAFF 04674
* Sencorex WG	Bayer Plc	MAFF 03755
360 Metsulfuron-methyl		
* Ally	Du Pont (UK) Ltd	MAFF 02977
* Ally WSB	Du Pont (UK) Ltd	MAFF 06588
* Clayton Metsulfuron	Clayton Plant Protection (UK) Ltd	MAFF 06734
*% FCL Metsulfuron-Methyl-20	Farm Chemicals Ltd	MAFF 05820
* Jubilee	Du Pont (UK) Ltd	MAFF 06082
* Jubilee 20DF	Du Pont (UK) Ltd	MAFF 06136
* Landgold Metsulfuron	Landgold & Co Ltd	MAFF 06280
* Lorate 20DF	Du Pont (UK) Ltd	MAFF 06135
* Standon Metsulfuron	Standon Chemicals Ltd	MAFF 05670
361 Metsulfuron-methyl + *Isoproturon*		
*% Oracle	Du Pont (UK) Ltd	MAFF 04027
362 Metsulfuron-methyl + *Thifensulfuron-methyl*		
* Harmony M	Du Pont (UK) Ltd	MAFF 03990
363 Monolinuron		
* Arresin	AgrEvo UK Crop Protection Ltd	MAFF 07303
* Arresin	Hoechst UK Ltd	MAFF 00118

* These products are "approved for agricultural use". For further details refer to page vii.
% These products are subject to the staged revocation procedure. For further details refer to page vii.
A These products are approved for use in or near water. For further details refer to page vii.

PROFESSIONAL PRODUCTS: HERBICIDES

Product Name	Marketing Company	Reg. No.

364 Monolinuron + *Glufosinate ammonium*

*	Conquest	AgrEvo UK Crop Protection Ltd	MAFF 07311
*	Conquest	Hoechst UK Ltd	MAFF 05679

365 Monolinuron + *Paraquat*

*%	Gramonol 5	Hoechst UK Ltd	MAFF 00994
*	Gramonol Five	Imperial Chemical Industries Plc	MAFF 00995
*	Gramonol Five	Zeneca Crop Protection	MAFF 06673

366 1-Naphthylacetic acid

*	Rhizopon B Powder (0.1%)	Fargro Ltd	MAFF 07133
*	Rhizopon B Powder (0.2%)	Fargro Ltd	MAFF 07134
*	Rhizopon B Tablets	Fargro Ltd	MAFF 07135
*%	Tipoff	Imperial Chemical Industries Plc	MAFF 02139
*	Tipoff	Universal Crop Protection Ltd	MAFF 05878

367 1-Naphthylacetic acid + *Dichlorophen* + *4-Indol-3-ylbutyric acid*

*	Synergol	Silvaperl Products Ltd	MAFF 04594

368 2-Naphthyloxyacetic acid

*	Betapal Concentrate	Synchemicals Ltd	MAFF 00234

369 Napropamide

*	Banweed	Hortichem Ltd	MAFF 05280
*	Banweed	Zeneca Crop Protection	MAFF 06672
*	Devrinol	Embetec Crop Protection	MAFF 05170
*	Devrinol	Imperial Chemical Industries Plc	MAFF 05171
*	Devrinol	Rhone-Poulenc Agriculture	MAFF 06195
*	Devrinol	Zeneca Crop Protection	MAFF 06653

370 Oxadiazon

*	Ronstar 2G	Rhone-Poulenc Agriculture	MAFF 06492
*	Ronstar 2G	Rhone-Poulenc Agriculture Ltd	MAFF 05433
*	Ronstar Liquid	Rhone-Poulenc Agriculture	MAFF 06493
*	Ronstar Liquid	Rhone-Poulenc Agriculture	MAFF 06766
*	Ronstar Liquid	Rhone-Poulenc Agriculture Ltd	MAFF 05431

371 Paclobutrazol

*	Bonzi	ICI Agrochemicals	MAFF 05058
*	Bonzi	Zeneca Crop Protection	MAFF 06640
*	Cultar	ICI Agrochemicals	MAFF 05055
*	Cultar	Zeneca Crop Protection	MAFF 06649

---

* These products are "approved for agricultural use". For further details refer to page vii.
% These products are subject to the staged revocation procedure. For further details refer to page vii.
A These products are approved for use in or near water. For further details refer to page vii.

PROFESSIONAL PRODUCTS: HERBICIDES

Product Name	Marketing Company	Reg. No.
372 Paclobutrazol + *Dicamba*		
* Holdfast D	Imperial Chemical Industries Plc	MAFF 05056
* Holdfast D	Zeneca Professional Products	MAFF 06858
373 Paraquat		
* Barclay Total	Barclay Chemicals (UK) Ltd	MAFF 05260
* CIA Paraquat 200	Top Farm Formulations Ltd	MAFF 05520
* Dextrone X	Nomix-Chipman Ltd	MAFF 00687
* Gramoxone 100	AgrEvo UK Crop Protection Ltd	MAFF 07260
* Gramoxone 100	Imperial Chemical Industries Plc	MAFF 00997
* Gramoxone 100	Schering Agrochemicals Ltd	MAFF 03867
* Gramoxone 100	Zeneca Crop Protection	MAFF 06674
* Landgold Paraquat	Landgold & Co Ltd	MAFF 06025
* Power Paraquat	Power Agrichemicals Ltd	MAFF 04539
* Power Paraquat	Power Agrichemicals Ltd	MAFF 04725
* Scythe	Cyanamid of GB Ltd	MAFF 02455
* Scythe LC	Cyanamid of GB Ltd	MAFF 05877
* Speedway	Garden and Professional Products	MAFF 02732
* Speedway	Zeneca Professional Products	MAFF 06861
* Speedway Liquid	Imperial Chemical Industries Plc	MAFF 04365
* Speedway Liquid	Zeneca Professional Products	MAFF 06825
* Standon Paraquat	Standon Chemicals Ltd	MAFF 05621
* Stefes Paraquat	Stefes Plant Protection Ltd	MAFF 05134
* Top Farm Paraquat	Top Farm Formulations Ltd	MAFF 05075
* Top Farm Paraquat 200	Top Farm Formulations Ltd	MAFF 05519
374 Paraquat + *Diquat*		
* Farmon PDQ	Farm Protection	MAFF 02886
* PDQ	Zeneca Crop Protection	MAFF 06518
* Parable	Imperial Chemical Industries Plc	MAFF 03805
* Parable	Zeneca Crop Protection	MAFF 06692
* Precede	Imperial Chemical Industries Plc	MAFF 05732
* Precede	Zeneca Crop Protection	MAFF 06702
375 Paraquat + *Diquat* + *Simazine*		
* Soltair	Imperial Chemical Industries Plc	MAFF 03601
* Soltair	Zeneca Crop Protection	MAFF 06772
376 Paraquat + *Diuron*		
* Dexuron	Chipman Ltd	MAFF 00689
* Dexuron	Nomix-Chipman Ltd	MAFF 07169

* These products are "approved for agricultural use". For further details refer to page vii.
% These products are subject to the staged revocation procedure. For further details refer to page vii.
A These products are approved for use in or near water. For further details refer to page vii.

PROFESSIONAL PRODUCTS: HERBICIDES

Product Name	Marketing Company	Reg. No.

377 Paraquat + *Monolinuron*

*%	Gramonol 5	Hoechst UK Ltd	MAFF 00994
*	Gramonol Five	Imperial Chemical Industries Plc	MAFF 00995
*	Gramonol Five	Zeneca Crop Protection	MAFF 06673

378 Pendimethalin

*	Aspire	Cyanamid of GB Ltd	MAFF 05919
*	Sovereign 330 EC	Ciba Agriculture	MAFF 04614
*	Sovereign 330 EC	Cyanamid of GB Ltd	MAFF 04513
*	Stomp 400 SC	Cyanamid of GB Ltd	MAFF 04183

379 Pendimethalin + *Chlorotoluron*

*	Totem	Cyanamid of GB Ltd	MAFF 04670

380 Pendimethalin + *Isoproturon*

*	Encore	Cyanamid of GB Ltd	MAFF 04737
*	Jolt	Cyanamid of GB Ltd	MAFF 05488
*	Trump	Cyanamid of GB Ltd	MAFF 03687

381 Pendimethalin + *Prometryn*

*	Monarch	Cyanamid of GB Ltd	MAFF 05160

382 Pendimethalin + *Simazine*

*	Deuce	Cyanamid of GB Ltd	MAFF 06746
*	Merit	Cyanamid of GB Ltd	MAFF 04976

383 Penmedipham

*	Campbell's Beetup	J D Campbell & Sons Ltd	MAFF 02680

384 Pentanochlor

*	Atlas Solan 40	Atlas Interlates Ltd	MAFF 03834
*	Croptex Bronze	Hortichem Ltd	MAFF 04087

385 Pentanochlor + *Chlorpropham*

*	Atlas Brown	Atlas Interlates Ltd	MAFF 03835

386 Phenmedipham

*	Atlas Protrum K	Atlas Interlates Ltd	MAFF 03089
*	Beetomax	Fine Agrochemicals Ltd	MAFF 03129
*	Beetup	MTM Agrochemicals Ltd	MAFF 05299
*	Betanal E	AgrEvo UK Crop Protection Ltd	MAFF 07248
*	Betanal E	Schering Agrochemicals Ltd	MAFF 03862
*	Betaren 120	Pan Britannica Industries Ltd	MAFF 05718

* These products are "approved for agricultural use". For further details refer to page vii.
% These products are subject to the staged revocation procedure. For further details refer to page vii.
A These products are approved for use in or near water. For further details refer to page vii.

PROFESSIONAL PRODUCTS: HERBICIDES

Product Name	Marketing Company	Reg. No.

386 Phenmedipham—continued

	Product Name	Marketing Company	Reg. No.
*	Betosip	Sipcam UK Ltd	MAFF 06787
*	Betosip 114	Sipcam UK Ltd	MAFF 05910
*	Cirrus	AgrEvo UK Crop Protection Ltd	MAFF 07256
*	Cirrus	Quadrangle Ltd	MAFF 06367
*	Cirrus	Schering Agriculture	MAFF 05879
*	Crop Phenmedipham	Crop Consultants	MAFF 04263
*%	Goliath	Rhone-Poulenc Agriculture Ltd	MAFF 05214
*	Headland Dephend	Headland Agrochemicals Ltd	MAFF 04925
*	Herbasan	Pan Britannica Industries Ltd	MAFF 07161
*	Hickson Phenmedipham	Hickson & Welch Ltd	MAFF 02825
*	Kemifam E	Kemira Cropcare Ltd	MAFF 06104
*	Kemifam E	Pan Britannica Industries Ltd	MAFF 07150
*	Luxan Phenmedipham	Luxan (UK) Ltd	MAFF 06933
*%	Phyto-Medipham	Phyto Research Pflanzenschutz GmbH	MAFF 04493
*%	Pistol 25	Rhone-Poulenc Agriculture Ltd	MAFF 05215
*	Portman Betalion	Portman Agrochemicals Ltd	MAFF 04677
*%	Quadrangle Pistol 400	Quadrangle Ltd	MAFF 03444
*	Stefes Forte	Stefes Plant Protection Ltd	MAFF 06427
*	Stefes Medipham	Stefes Plant Protection Ltd	MAFF 05778
*	Tripart Beta	Tripart Farm Chemicals Ltd	MAFF 03111
*	Tripart Beta 2	Tripart Farm Chemicals Ltd	MAFF 06510
*%	Tripart Beta 2	Tripart Farm Chemicals Ltd	MAFF 04831
*	Vangard	Farmers Crop Chemicals Ltd	MAFF 02743

387 Phenmedipham + *Desmedipham*

	Product Name	Marketing Company	Reg. No.
*	Betanal Compact	AgrEvo UK Crop Protection Ltd	MAFF 07247
*	Betanal Compact	Schering Agrochemicals Ltd	MAFF 06780
*	Betanal Quorum	AgrEvo UK Crop Protection Ltd	MAFF 07252
*	Betanal Quorum	Schering Agrochemicals Ltd	MAFF 06782
*	Betanal Rostrum	AgrEvo UK Crop Protection Ltd	MAFF 07253
*	Betanal Rostrum	Schering Agrochemicals Ltd	MAFF 06781

388 Phenmedipham + *Desmedipham* + *Ethofumesate*

	Product Name	Marketing Company	Reg. No.
*	Betanal Congress	AgrEvo UK Crop Protection Ltd	MAFF 07110
*	Betanal Progress	AgrEvo UK Crop Protection Ltd	MAFF 07111
*	Betanal Ultima	Quadrangle Agrochemicals Ltd	MAFF 07039

389 Phenmedipham + *Ethofumesate*

	Product Name	Marketing Company	Reg. No.
*	Betanal Maestro	AgrEvo UK Crop Protection Ltd	MAFF 07249
*	Betanal Maestro	Schering Agrochemicals Ltd	MAFF 06022
*	Betanal Montage	AgrEvo UK Crop Protection Ltd	MAFF 07250
*	Betanal Montage	Schering Agrochemicals Ltd	MAFF 06024

---

* These products are "approved for agricultural use". For further details refer to page vii.
% These products are subject to the staged revocation procedure. For further details refer to page vii.
A These products are approved for use in or near water. For further details refer to page vii.

PROFESSIONAL PRODUCTS: HERBICIDES

Product Name	Marketing Company	Reg. No.

389 Phenmedipham + *Ethofumesate*—continued

* Betanal Quadrant	AgrEvo UK Crop Protection Ltd	MAFF 07251
* Betanal Quadrant	Schering Agrochemicals Ltd	MAFF 06023
* Betanal Tandem	AgrEvo UK Crop Protection Ltd	MAFF 07254
* Betanal Tandem	Schering Agrochemicals Ltd	MAFF 03857

390 Phenmedipham + *Ethofumesate* + *Metamitron*

* Bayer UK 407 WG	Bayer UK Ltd	MAFF 06313
* CX 171	AgrEvo UK Crop Protection Ltd	MAFF 06803
* Goltix Triple	Bayer UK Ltd	MAFF 06314

391 Phenmedipham + *Lenacil*

* DUK-880	Du Pont (UK) Ltd	MAFF 04121

392 Picloram

* Tordon 22K	Chipman Ltd	MAFF 02152
* Tordon 22K	Chipman Ltd	MAFF 05790
* Tordon 22K	DowElanco Ltd	MAFF 05083

393 Picloram + *Bromacil*

* Hydon	Chipman Ltd	MAFF 01088

394 Picloram + *2,4-D*

* Atladox HI	Chipman Ltd	MAFF 00126
* Atladox HI	Chipman Ltd	MAFF 05559
*% Tordon 101	Dow Chemical Co Ltd	MAFF 02154
* Tordon 101	DowElanco Ltd	MAFF 05816

395 Prometryn

* Alpha Prometryne 50 WP	Makhteshim-Agan (UK) Ltd	MAFF 04871
* Alpha Prometryne 80 WP	Makhteshim-Agan (UK) Ltd	MAFF 04795
* Atlas Prometryne 50 WP	Atlas Interlates Ltd	MAFF 03502
* Gesagard 50 WP	Ciba Agriculture	MAFF 00981

396 Prometryn + *Pendimethalin*

* Monarch	Cyanamid of GB Ltd	MAFF 05160

397 Prometryn + *Terbutryn*

* Peaweed	Pan Britannica Industries Ltd	MAFF 03248
* Spudweed	Pan Britannica Industries Ltd	MAFF 04965

---

* These products are "approved for agricultural use". For further details refer to page vii.
% These products are subject to the staged revocation procedure. For further details refer to page vii.
A These products are approved for use in or near water. For further details refer to page vii.

PROFESSIONAL PRODUCTS: HERBICIDES

Product Name	Marketing Company	Reg. No.
398 Propachlor		
*% Albrass	Imperial Chemical Industries Plc	MAFF 00069
*% Albrass	Zeneca Crop Protection	MAFF 06630
* Albrass SC	ICI Agrochemicals	MAFF 05759
*% Albrass SC	Zeneca Crop Protection	MAFF 06631
* Alpha Propachlor 50 SC	Makhteshim-Agan (UK) Ltd	MAFF 04873
* Alpha Propachlor 65 WP	Makhteshim-Agan (UK) Ltd	MAFF 04807
* Atlas Orange	Atlas Interlates Ltd	MAFF 03096
* Atlas Propachlor	Atlas Interlates Ltd	MAFF 06462
* Croptex Amber	Hortichem Ltd	MAFF 03078
* PA Propachlor Flowable	Portman Agrochemicals Ltd	MAFF 02784
* Portman Propachlor 50 FL	Portman Agrochemicals Ltd	MAFF 06892
* Ramrod 20 Granular	Hortichem Ltd	MAFF 05806
* Ramrod 20 Granular	Monsanto Plc	MAFF 01687
* Ramrod Flowable	Monsanto Plc	MAFF 01688
* Sentinel 2	Tripart Farm Chemicals Ltd	MAFF 05140
* Tripart Sentinel	Tripart Farm Chemicals Ltd	MAFF 03250
399 Propachlor + *Chloridazon*		
* Ashlade CP	Ashlade Formulations Ltd	MAFF 06481
*% Ashlade CP	Ashlade Formulations Ltd	MAFF 02852
400 Propachlor + *Chlorthal dimethyl*		
*% Decimate	Fermenta ASC Europe Ltd	MAFF 04858
* Decimate	ISK Biosciences Ltd	MAFF 05626
401 Propaquizafop		
* Falcon	Cyanamid Of GB Ltd	MAFF 07025
* Shogun 100 EC	Ciba Agriculture	MAFF 07026
402 Propham + *Chloridazon* + *Chlorpropham* + *Fenuron*		
* Atlas Electrum	Atlas Interlates Ltd	MAFF 03548
403 Propham + *Chlorpropham*		
*% Amos Gro-Stop	Kommer-Brookwick	MAFF 06145
Atlas Indigo	Atlas Interlates Ltd	MAFF 03087
Luxan Gro-Stop	Luxan (UK) Ltd	MAFF 06559
Pommetrol M	Sam Fletcher Ltd	MAFF 01615
% Power Gro-Stop	Kommer-Brookwick	MAFF 06142
% Power Gro-Stop	Power Agrichemicals Ltd	MAFF 02719
404 Propham + *Chlorpropham* + *Cresylic acid* + *Fenuron*		
* Marks PCF Beet Herbicide	A H Marks & Co Ltd	MAFF 02538
* MSS Sugar Beet Herbicide	Mirfield Sales Services Ltd	MAFF 02447

* These products are "approved for agricultural use". For further details refer to page vii.
% These products are subject to the staged revocation procedure. For further details refer to page vii.
A These products are approved for use in or near water. For further details refer to page vii.

PROFESSIONAL PRODUCTS: HERBICIDES

Product Name	Marketing Company	Reg. No.

405 Propham + *Chlorpropham* + *Diuron*
* Atlas Pink C	Atlas Interlates Ltd	MAFF 03095

406 Propham + *Chlorpropham* + *Fenuron*
* Atlas Gold	Atlas Interlates	MAFF 03086
* MTM Sugar Beet Herbicide	MTM Agrochemicals Ltd	MAFF 05044

407 Propyzamide
* Barclay Piza 500	Barclay Chemicals (UK) Ltd	MAFF 05283
* CIA Propyzamide 500	Top Farm Formulations Ltd	MAFF 05483
* Campbell's Rapier	J D Campbell & Sons Ltd	MAFF 03985
* Clayton Propel	Clayton Plant Protection (UK) Ltd	MAFF 06073
* FCC Rapier	Farmers Crop Chemicals Ltd	MAFF 04192
* Kerb 50 W	Pan Britannica Industries Ltd	MAFF 01133
* Kerb 50 W	Rohm & Haas (UK) Ltd	MAFF 02986
* Kerb Flo	Pan Britannica Industries Ltd	MAFF 04521
* Kerb Flo	Rohm & Haas (UK) Ltd	MAFF 02759
* Kerb Granules	Pan Britannica Industries Ltd	MAFF 01135
* Kerb Granules	Rohm & Haas (UK) Ltd	MAFF 01136
* Landgold Propyzamide 50	Landgold & Co Ltd	MAFF 05916
* Rapier	MTM Agrochemicals Ltd	MAFF 05314
* Standon Propyzamide 50	Standon Chemicals Ltd	MAFF 05569
* Stefes Pride	Stefes Plant Protection Ltd	MAFF 05616
* Top Farm Propyzamide 500	Top Farm Formulations Ltd	MAFF 05484

408 Propyzamide + *Clopyralid*
* Matrikerb	Pan Britannica Industries Ltd	MAFF 01308
* Matrikerb	Rohm & Haas (UK) Ltd	MAFF 02443

409 Pyridate
* Barclay Pirate	Barclay Chemicals (UK) Ltd	MAFF 07104
* Lentagran 45 WP	Ciba Agriculture	MAFF 05074
*% Lentagran 50% WP	Chemie Linz UK Ltd	MAFF 02767
* Lentagran WP	Ciba Agriculture	MAFF 06331

410 Quizalofop-ethyl
* Everest	Schering Agrochemicals Ltd	MAFF 07073
* Mission	AgrEvo UK Crop Protection Ltd	MAFF 07264
* Mission	Schering Agrochemicals Ltd	MAFF 06439
* Pilot	AgrEvo UK Crop Protection Ltd	MAFF 07268
* Pilot	Schering Agrochemicals Ltd	MAFF 03837
* Stefes Biggles	Stefes Plant Protection Ltd	MAFF 07083

---

* These products are "approved for agricultural use". For further details refer to page vii.
% These products are subject to the staged revocation procedure. For further details refer to page vii.
A These products are approved for use in or near water. For further details refer to page vii.

PROFESSIONAL PRODUCTS: HERBICIDES

Product Name	Marketing Company	Reg. No.
411 Sethoxydim		
* Checkmate	Rhone-Poulenc Agriculture	MAFF 06129
412 Simazine		
* Alpha Simazine 50 SC	Alpha (GB) Ltd	MAFF 04801
* Alpha Simazine 50 WP	Makhteshim-Agan (UK) Ltd	MAFF 04879
* Alpha Simazine 80 WP	Alpha (GB) Ltd	MAFF 04800
* Ashlade Simazine 50 FL	Ashlade Formulations Ltd	MAFF 06482
*% Ashlade Simazine 50 FL	Ashlade Formulations Ltd	MAFF 02885
* Atlas Simazine	Atlas Interlates Ltd	MAFF 05610
* Gesatop 50 WP	Ciba Agriculture	MAFF 00983
*% Gesatop 500 FW	Ciba-Geigy Agrochemicals	MAFF 00984
* Gesatop 500 SC	Ciba Agriculture	MAFF 05846
* Mascot Simazine 2 % Granular	Complete Weed Control Ltd	MAFF 05597
* MSS Simazine 50 FL	Mirfield Sales Services Ltd	MAFF 01418
* MSS Simazine 80 WP	Mirfield Sales Services Ltd	MAFF 04362
*% New Simflow	Rhone-Poulenc Agriculture Ltd	MAFF 05403
* Simazine SC	DowElanco Ltd	MAFF 05745
*% Simflow	Burts & Harvey	MAFF 01954
*% Simflow	Rhone-Poulenc Agriculture Ltd	MAFF 05417
* Syngran	Synchemicals Ltd	MAFF 02079
* Unicrop Flowable Simazine	Universal Crop Protection Ltd	MAFF 02271
* Unicrop Flowable Simazine	Universal Crop Protection Ltd	MAFF 05447
* Unicrop Simazine 50	Universal Crop Protection Ltd	MAFF 02646
*% Weedex S 2 FG	Ciba-Geigy Agrochemicals	MAFF 02355
* Weedex S 2 FG	Hortichem Ltd	MAFF 04223
413 Simazine + *Amitrole*		
* Alpha Simazol	Makhteshim-Agan (UK) Ltd	MAFF 04799
* Alpha Simazol T	Makhteshim-Agan (UK) Ltd	MAFF 04874
*% CDA Simflow Plus	Rhone-Poulenc Agriculture Ltd	MAFF 05401
*% Clearway	Rhone-Poulenc Agriculture Ltd	MAFF 05328
* Mascot Highway WP	Complete Weed Control Ltd	MAFF 05600
* MSS Simazine/Aminotriazole 43% FL	Mirfield Sales Services Ltd	MAFF 04361
*% Primatol SE 500 FW	Ciba-Geigy Agrochemicals	MAFF 01638
*% Primatol SE 500 SC	Ciba-Geigy Agrochemicals	MAFF 05852
* Ritefeed Simazine/Aminotriazole 43% FL	Ritefeed Ltd	MAFF 05087
*% Simflow Plus	Rhone-Poulenc Agriculture Ltd	MAFF 05406
414 Simazine + *Diquat* + *Paraquat*		
* Soltair	Imperial Chemical Industries Plc	MAFF 03601
* Soltair	Zeneca Crop Protection	MAFF 06772

* These products are "approved for agricultural use". For further details refer to page vii.
% These products are subject to the staged revocation procedure. For further details refer to page vii.
A These products are approved for use in or near water. For further details refer to page vii.

PROFESSIONAL PRODUCTS: HERBICIDES

Product Name	Marketing Company	Reg. No.

415 Simazine + *Glyphosate*

*%	Rival	Monsanto Plc	MAFF 03377
*%	Ultra-sonic	Rigby Taylor Ltd	MAFF 03546

416 Simazine + *Isoproturon*

*%	Harlequin 500 FW	Ciba-Geigy Agrochemicals	MAFF 04625
*	Harlequin 500 SC	Ciba Agriculture	MAFF 05847

417 Simazine + *Pendimethalin*

*	Deuce	Cyanamid of GB Ltd	MAFF 06746
*	Merit	Cyanamid of GB Ltd	MAFF 04976

418 Simazine + *Trietazine*

*%	Aventox SC	Dow Chemical Co Ltd	MAFF 04229
*	Aventox SC	DowElanco Ltd	MAFF 05629
*	Remtal SC	AgrEvo UK Crop Protection Ltd	MAFF 07270
*	Remtal SC	Schering Agrochemicals Ltd	MAFF 03827

419 Sodium chlorate

*	Ace-Sodium Chlorate [Fire Suppressed Weedkiller]	Ace Chemicals Ltd	MAFF 06413
*	Arpal Non Selex Powder	R P Adams Ltd	MAFF 04565
*	Atlacide Soluble Powder Weedkiller	Chipman Ltd	MAFF 00125
*	Centex	Chemsearch	MAFF 00456
*	Cooke's Professional Sodium Chlorate Weedkiller with Fire Depressant	Cooke's Chemicals (Sales) Ltd	MAFF 06796
*	Cooke's Weedclear	Cooke's Chemicals (Sales) Ltd	MAFF 06512
*%	Delsanex Chlorate Weedkiller	Diversey Ltd	MAFF 00666
*	Deosan Chlorate Weedkiller	Diversey Ltd	MAFF 05636
*%	Dimex TWK	Dimex Ltd	MAFF 04867
*	Doff Sodium Chlorate Weedkiller	Doff Portland Ltd	MAFF 06049
*	Gem Sodium Chlorate Weedkiller	Joseph Metcalf Ltd	MAFF 04276
*	Lever Industrial Sodium Chlorate Weedkiller with Fire Depressant	Lever Industrial Ltd	MAFF 06355
*	Sodium Chlorate (Fire suppressed) Weedkiller	Ace Chemicals Ltd	MAFF 04449
*	Sodium chlorate	Marlow Chemical Co Ltd	MAFF 06294
*	TWK Total Weedkiller	Yule Catto Consumer Chemicals Ltd	MAFF 06393

420 Sodium chlorate + *Atrazine*

*	Atlacide Extra Dusting Powder	Chipman Ltd	MAFF 00124

---

* These products are "approved for agricultural use". For further details refer to page vii.
% These products are subject to the staged revocation procedure. For further details refer to page vii.
A These products are approved for use in or near water. For further details refer to page vii.

PROFESSIONAL PRODUCTS: HERBICIDES

Product Name	Marketing Company	Reg. No.
421 Sodium monochloroacetate		
* Atlas Somon	Atlas Interlates Ltd	MAFF 03045
* Croptex Steel	Hortichem Ltd	MAFF 02418
422 Sodium silver thiosulphate		
* Argylene	Fargro Ltd	MAFF 03386
423 TCA		
* NaTA	AgrEvo UK Crop Protection Ltd	MAFF 07330
* NATA	Hoechst UK Ltd	MAFF 01467
424 Tebutam		
* Comodor 600	Agrichem (International) Ltd	MAFF 06808
* Comodor 600	Ciba Agriculture	MAFF 06792
* Comodor 600	Farm Protection	MAFF 04453
* Comodor 600	ICI Agrochemicals	MAFF 05010
* Comodor 600	Zeneca Crop Protection	MAFF 06646
425 Tebuthiuron		
* Bushwacker Spray	Rhone-Poulenc Environmental Products	MAFF 05249
*% Perflan 80W	DowElanco Ltd	MAFF 06286
426 Tecnazene		
* Atlas Tecgran 100	Atlas Interlates Ltd	MAFF 05574
* Atlas Tecnazene 6% Dust	Atlas Interlates Ltd	MAFF 06351
* Bygran F	Wheatley Chemical Co Ltd	MAFF 00365
Hickstor 3	Hickson & Welch Ltd	MAFF 03105
* Hickstor 5	Hickson & Welch Ltd	MAFF 03180
Hickstor 6	Hickson & Welch Ltd	MAFF 03106
* Hystore 10	Agrichem (International) Ltd	MAFF 03581
Hytec	Agrichem Ltd	MAFF 01099
* Hytec 6	Agrichem Ltd	MAFF 03580
* New Hickstor 6	Hickson & Welch Ltd	MAFF 04221
Tripart Arena 10 G	Tripart Farm Chemicals Ltd	MAFF 05603
Tripart Arena 3	Tripart Farm Chemicals Ltd	MAFF 05605
* Tripart Arena 5G	Tripart Farm Chemicals Ltd	MAFF 05604
*% Tripart Arena 5G	Tripart Farm Chemicals Ltd	MAFF 04441
% Tripart Arena 6%	Tripart Farm Chemicals Ltd	MAFF 02574
* Tripart New Arena 6	Tripart Farm Chemicals Ltd	MAFF 05813
427 Tecnazene + *Carbendazim*		
* Hickstor 6 Plus MBC	Hickson & Welch Ltd	MAFF 04176
% Hortag Carbotec	Hortag Chemicals Ltd	MAFF 04135

* These products are "approved for agricultural use". For further details refer to page vii.
% These products are subject to the staged revocation procedure. For further details refer to page vii.
A These products are approved for use in or near water. For further details refer to page vii.

PROFESSIONAL PRODUCTS: HERBICIDES

Product Name	Marketing Company	Reg. No.
427 Tecnazene + *Carbendazim*—continued		
* Hortag Tecnacarb	Avon Packers Ltd	MAFF 02929
* New Arena Plus	Hickson & Welch Ltd	MAFF 04598
* New Hickstor 6 Plus MBC	Hickson & Welch Ltd	MAFF 04599
Tripart Arena Plus	Hickson & Welch Ltd	MAFF 05602
428 Tecnazene + *Thiabendazole*		
Hytec Super	Agrichem (International) Ltd	MAFF 01100
New Arena TBZ 6	Tripart Farm Chemicals Ltd	MAFF 05606
% New Arena TBZ 6	Tripart Farm Chemicals Ltd	MAFF 04906
% New Hickstor TBZ 6	Hickson & Welch Ltd	MAFF 04905
429 Terbacil		
* Sinbar	Du Pont (UK) Ltd	MAFF 01956
430 Terbuthylazine + *Cyanazine*		
* Angle 567 SC	Ciba Agriculture	MAFF 06254
431 Terbuthylazine + *Isoxaben*		
* Skirmish 495 SC	Ciba Agriculture	MAFF 05692
432 Terbuthylazine + *Terbutryn*		
* Opogard 500 SC	Ciba Agriculture	MAFF 05850
*% Opogard 500FW	Ciba-Geigy Agrochemicals	MAFF 01514
433 Terbutryn		
* Alpha Terbutryne 50 SC	Makhteshim-Agan (UK) Ltd	MAFF 04809
* Alpha Terbutryne 50 WP	Makhteshim-Agan (UK) Ltd	MAFF 04875
A* Clarosan 1FG	Ciba Agriculture	MAFF 03859
*% Prebane 500 FW	Ciba-Geigy Agrochemicals	MAFF 01627
*% Prebane 500 SC	Ciba Agriculture	MAFF 05851
434 Terbutryn + *Linuron*		
* Tempo	Farm Protection	MAFF 02736
435 Terbutryn + *Prometryn*		
* Peaweed	Pan Britannica Industries Ltd	MAFF 03248
* Spudweed	Pan Britannica Industries Ltd	MAFF 04965
436 Terbutryn + *Terbuthylazine*		
* Opogard 500 SC	Ciba Agriculture	MAFF 05850
*% Opogard 500FW	Ciba-Geigy Agrochemicals	MAFF 01514

* These products are "approved for agricultural use". For further details refer to page vii.
% These products are subject to the staged revocation procedure. For further details refer to page vii.
A These products are approved for use in or near water. For further details refer to page vii.

PROFESSIONAL PRODUCTS: HERBICIDES

Product Name	Marketing Company	Reg. No.
437 Terbutryn + *Trietazine*		
* Senate	AgrEvo UK Crop Protection Ltd	MAFF 07279
* Senate	Schering Agrochemicals Ltd	MAFF 04275
438 Terbutryn + *Trifluralin*		
* Alpha Terbalin 35 SC	Makhteshim-Agan (UK) Ltd	MAFF 04792
* Alpha Terbalin 350 CS	Makhteshim-Agan (UK) Ltd	MAFF 07118
* Ashlade Summit	Ashlade Formulations Ltd	MAFF 06214
*% Ashlade Summit	Ashlade Formulations Ltd	MAFF 04496
439 Thiabendazole + *Tecnazene*		
Hytec Super	Agrichem (International) Ltd	MAFF 01100
New Arena TBZ 6	Tripart Farm Chemicals Ltd	MAFF 05606
% New Arena TBZ 6	Tripart Farm Chemicals Ltd	MAFF 04906
% New Hickstor TBZ 6	Hickson & Welch Ltd	MAFF 04905
440 Thifensulfuron-methyl		
* Crackshot	Shell Chemicals UK Ltd	MAFF 06540
* DUK 118	Du Pont (UK) Ltd	MAFF 04596
* Prospect	Du Pont (UK) Ltd	MAFF 06541
441 Thifensulfuron-methyl + *Fluroxypyr* + *Tribenuron-methyl*		
* DP 353	Du Pont (UK) Ltd	MAFF 07036
* Starane Super	DowElanco Ltd	MAFF 07035
442 Thifensulfuron-methyl + *Mecoprop-P*		
* Duet	Du Pont (UK) Ltd	MAFF 05169
443 Thifensulfuron-methyl + *Metsulfuron-methyl*		
* Harmony M	Du Pont (UK) Ltd	MAFF 03990
444 Thifensulfuron-methyl + *Tribenuron-methyl*		
* DUK 110	Du Pont (UK) Ltd	MAFF 06266
445 Tralkoxydim		
* Grasp	Imperial Chemical Industries Plc	MAFF 06545
* Grasp	Zeneca Crop Protection	MAFF 06675
* Splendor	Imperial Chemical Industries Plc	MAFF 06546
* Splendor	Zeneca Crop Protection	MAFF 06707
446 Tri-allate		
* Avadex BW	Monsanto Plc	MAFF 00173
* Avadex BW 15G	Monsanto Plc	MAFF 07117

* These products are "approved for agricultural use". For further details refer to page vii.
% These products are subject to the staged revocation procedure. For further details refer to page vii.
A These products are approved for use in or near water. For further details refer to page vii.

PROFESSIONAL PRODUCTS: HERBICIDES

Product Name	Marketing Company	Reg. No.
446 Tri-allate—*continued*		
* Avadex BW 480	Monsanto Plc	MAFF 04742
* Avadex BW Granular	Monsanto Plc	MAFF 00174
447 Triasulfuron		
* Lo-Gran 20WG	Ciba Agriculture	MAFF 05993
* Viper	Ciba Agriculture	MAFF 06765
448 Triasulfuron + *Bromoxynil* **+** *Ioxynil*		
* Teal	Ciba Agriculture	MAFF 06117
* Teal G	Ciba Agriculture	MAFF 06118
449 Triasulfuron + *Dicamba*		
* Accord	AgrEvo UK Crop Protection Ltd	MAFF 07240
* Accord	Schering Agrochemicals Ltd	MAFF 06496
* Banvel T	Sandoz Agro Ltd	MAFF 06497
* Framolene	Ciba Agriculture	MAFF 06495
450 Triasulfuron + *Fluoroglycofen-ethyl*		
* Compete Mix A	Rohm & Haas (UK) Ltd	MAFF 05992
* Satis 15 WP	Ciba Agriculture	MAFF 05991
451 Triasulfuron + *Mecoprop-P*		
* Raven	Ciba Agriculture	MAFF 06119
452 Tribenuron-methyl		
* Quantum	Du Pont (UK) Ltd	MAFF 06270
453 Tribenuron-methyl + *Fluroxypyr* **+** *Thifensulfuron-methyl*		
* DP 353	Du Pont (UK) Ltd	MAFF 07036
* Starane Super	DowElanco Ltd	MAFF 07035
454 Tribenuron-methyl + *Thifensulfuron-methyl*		
* DUK 110	Du Pont (UK) Ltd	MAFF 06266
455 Triclopyr		
* Chipman Garlon 4	Chipman Ltd	MAFF 06016
*% Garlon 2	Dow Chemical Co Ltd	MAFF 00975
* Garlon 2	DowElanco Ltd	MAFF 05682
* Garlon 2	Imperial Chemical Industries Plc	MAFF 03767
* Garlon 2	Zeneca Crop Protection	MAFF 06616
* Garlon 4	DowElanco Ltd	MAFF 05090

* These products are "approved for agricultural use". For further details refer to page vii.
% These products are subject to the staged revocation procedure. For further details refer to page vii.
A These products are approved for use in or near water. For further details refer to page vii.

PROFESSIONAL PRODUCTS: HERBICIDES

Product Name	Marketing Company	Reg. No.
455 Triclopyr—*continued*		
*% Timbrel	Dow Chemical Co Ltd	MAFF 04108
* Timbrel	DowElanco Ltd	MAFF 05815
456 Triclopyr + *Clopyralid*		
* Grazon 90	DowElanco Ltd	MAFF 05456
457 Triclopyr + *2,4-D* + *Dicamba*		
* Broadshot	Cyanamid Of GB Ltd	MAFF 07141
* Broadshot	Shell Chemicals UK Ltd	MAFF 03056
458 Triclopyr + *Dicamba* + *Mecoprop*		
* Fettel	Farm Protection Ltd	MAFF 02516
* Fettel	Zeneca Crop Protection	MAFF 06399
459 Triclopyr + *Fluroxypyr*		
* Doxstar	DowElanco Ltd	MAFF 06050
460 Trietazine + *Linuron*		
* Bronox	AgrEvo UK Crop Protection Ltd	MAFF 07255
* Bronox	Schering Agrochemicals Ltd	MAFF 03864
461 Trietazine + *Simazine*		
*% Aventox SC	Dow Chemical Co Ltd	MAFF 04229
* Aventox SC	DowElanco Ltd	MAFF 05629
* Remtal SC	AgrEvo UK Crop Protection Ltd	MAFF 07270
* Remtal SC	Schering Agrochemicals Ltd	MAFF 03827
462 Trietazine + *Terbutryn*		
* Senate	AgrEvo UK Crop Protection Ltd	MAFF 07279
* Senate	Schering Agrochemicals Ltd	MAFF 04275
463 Trifluralin		
* Alpha Trifluralin 48 EC	Makhteshim-Agan (UK) Ltd	MAFF 04798
* Ashlade Trimaran	Ashlade Formulations Ltd	MAFF 06228
*% Ashlade Trimaran	Ashlade Formulations Ltd	MAFF 04485
* Atlas Trifluralin	Atlas Interlates Ltd	MAFF 03051
* Campbell's Trifluralin	J D Campbell & Sons Ltd	MAFF 00425
* Digermin	Montedison UK Ltd	MAFF 00701
* Digermin	Sipcam UK Ltd	MAFF 07221
* Ipifluor	I Pi Ci	MAFF 04692
* MTM Trifluralin	J D Campbell & Sons Ltd	MAFF 05313
* MSS Trifluralin 48 EC	Mirfield Sales Services Ltd	MAFF 05144

* These products are "approved for agricultural use". For further details refer to page vii.
% These products are subject to the staged revocation procedure. For further details refer to page vii.
A These products are approved for use in or near water. For further details refer to page vii.

PROFESSIONAL PRODUCTS: HERBICIDES

Product Name	Marketing Company	Reg. No.

463 Trifluralin—*continued*

* Portman Trifluralin	Portman Agrochemicals Ltd	MAFF 05751
* Portman Trifluralin 480	Portman Agrochemicals Ltd	MAFF 02957
* Treflan	DowElanco Ltd	MAFF 05817
*% Treflan	Elanco Products Ltd	MAFF 03174
* Trigard	Farmers Crop Chemicals Ltd	MAFF 02178
* Tripart Trifluralin 48 EC	Tripart Farm Chemicals Ltd	MAFF 02215
* Triplen	Sipcam UK Ltd	MAFF 05897
* Tristar	Pan Britannica Industries Ltd	MAFF 02219

464 Trifluralin + *Bromoxynil* + *Ioxynil*

* Masterspray	Pan Britannica Industries Ltd	MAFF 02971

465 Trifluralin + *Chlorotoluron*

*% Dicurane Combi 540 FW	Ciba-Geigy Agrochemicals	MAFF 04628
*% Dicurane Combi 540 SC	Ciba Agriculture	MAFF 05837

466 Trifluralin + *Diflufenican*

* Ardent	Rhone-Poulenc Agriculture	MAFF 06203
* Ardent	Rhone-Poulenc Agriculture Ltd	MAFF 04248
*% Hawk 440 SC	Ciba-Geigy Agriculture	MAFF 05803

467 Trifluralin + *Isoproturon*

* Autumn Kite	AgrEvo UK Crop Protection Ltd	MAFF 07119
* Autumn Kite	Schering Agrochemicals Ltd	MAFF 03830
* Cantor	Portman Agrochemicals Ltd	MAFF 07217
* Debut	Portman Agrochemicals Ltd	MAFF 06882

468 Trifluralin + *Linuron*

* Ashlade Flint	Ashlade Formulations Ltd	MAFF 04471
*% Atlas Janus	Atlas Interlates Ltd	MAFF 03085
* Campbell's Trifluron	J D Campbell & Sons Ltd	MAFF 02682
* Chandor	DowElanco Ltd	MAFF 05631
*% Chandor	Elanco Products Ltd	MAFF 00483
* Ipicombi TL	I Pi Ci	MAFF 04608
* Janus	Montedison UK Ltd	MAFF 02805
* Linnet	Pan Britannica Industries Ltd	MAFF 01555
* Neminfest	Enichem UK Ltd	MAFF 05822
* Neminfest	Montedison UK Ltd	MAFF 02546
* Neminfest	Sipcam UK Ltd	MAFF 07219
*% Quadrangle Onslaught	Quadrangle Ltd	MAFF 02548
*% Tri-farmon FL	Farm Protection Ltd	MAFF 02870
* Triplen Combi	Sipcam UK Ltd	MAFF 05939

* These products are "approved for agricultural use". For further details refer to page vii.
% These products are subject to the staged revocation procedure. For further details refer to page vii.
A These products are approved for use in or near water. For further details refer to page vii.

PROFESSIONAL PRODUCTS: HERBICIDES

Product Name	Marketing Company	Reg. No.
469 Trifluralin + *Terbutryn*		
* Alpha Terbalin 35 SC	Makhteshim-Agan (UK) Ltd	MAFF 04792
* Alpha Terbalin 350 CS	Makhteshim-Agan (UK) Ltd	MAFF 07118
* Ashlade Summit	Ashlade Formulations Ltd	MAFF 06214
*% Ashlade Summit	Ashlade Formulations Ltd	MAFF 04496

---

* These products are "approved for agricultural use". For further details refer to page vii.
% These products are subject to the staged revocation procedure. For further details refer to page vii.
A These products are approved for use in or near water. For further details refer to page vii.

PROFESSIONAL PRODUCTS: FUNGICIDES

Product Name   *Marketing Company*   *Reg. No.*

1.2 Fungicides
including bactericides

470 2-Aminobutane
*	CSC 2-Aminobutane	Chemical Spraying Co Ltd	MAFF 03224
*	CSC 2-Aminobutane	Chemical Spraying Co Ltd	MAFF 05036
*	Hortichem 2-Aminobutane	Hortichem Ltd	MAFF 06147

471 Anilazine
*	Dyrene	Bayer Plc	MAFF 06982
*	Dyrene 720	Bayer Plc	MAFF 07235

472 Benalaxyl + *Mancozeb*
*	Barclay Bezant	Barclay Chemicals (UK) Ltd	MAFF 05914
*	Clayton Benzeb	Clayton Plant Protection (UK) Ltd	MAFF 07081
*%	Galben M	Dow Chemical Co Ltd	MAFF 03809
*	Galben M	DowElanco Ltd	MAFF 05092
*	Galben M	EniMont UK Ltd	MAFF 05091
*	Galben M	Enichem UK Ltd	MAFF 05904
*	Galben M	Montedison UK Ltd	MAFF 02753
*	Galben M	Sipcam UK Ltd	MAFF 07220

473 Benodanil
*	Calirus	BASF Plc	MAFF 00368
*%	Compo Benodanil	BASF Plc	MAFF 03459
*%	Mascot Clearing	Rigby Taylor Ltd	MAFF 03422

474 Benomyl
*	Benlate Fungicide	Du Pont (UK) Ltd	MAFF 00229

475 Bitertanol + *Fuberidazole*
*	Sibutol	Bayer Plc	MAFF 07238
*	Sibutol Flowable	Bayer Plc	MAFF 06983

476 Bitertanol + *Fuberidazole* + *Triadimenol*
*	Cereline	Bayer Plc	MAFF 07239
*	Cereline Flowable	Bayer Plc	MAFF 06984

477 Bupirimate
*	Nimrod	Imperial Chemical Industries Plc	MAFF 01498
*	Nimrod	Zeneca Crop Protection	MAFF 06686

---

*   These products are "approved for agricultural use". For further details refer to page vii.
%   These products are subject to the staged revocation procedure. For further details refer to page vii.
A   These products are approved for use in or near water. For further details refer to page vii.

PROFESSIONAL PRODUCTS: FUNGICIDES

Product Name	Marketing Company	Reg. No.
478 Bupirimate + *Triforine*		
* Nimrod T	Imperial Chemical Industries Plc	MAFF 01499
* Nimrod T	Zeneca Professional Products	MAFF 06859
479 Captan		
* Alpha Captan 49 Flowable	Makhteshim-Agan (UK) Ltd	MAFF 04818
* Alpha Captan 50 WP	Makhteshim-Agan (UK) Ltd	MAFF 04797
* Alpha Captan 83 WP	Makhteshim-Agan (UK) Ltd	MAFF 04806
*% Captan 83	Dow Chemical Co Ltd	MAFF 04231
* Captan 83	DowElanco Ltd	MAFF 05630
* PP Captan 80WG	ICI Agrochemicals	MAFF 05826
* PP Captan 80WG	Zeneca Crop Protection	MAFF 06696
* PP Captan 83	ICI Agrochemicals	MAFF 01619
* PP Captan 83	Zeneca Crop Protection	MAFF 06697
480 Captan + *Lindane*		
* Gammalex	ICI Agrochemicals	MAFF 00965
* Gammalex	Zeneca Crop Protection	MAFF 06396
481 Captan + *Nuarimol*		
*% Kapitol	DowElanco Ltd	MAFF 05051
*% Kapitol WDG	DowElanco Ltd	MAFF 05085
482 Captan + *Penconazole*		
* Topas C 50WP	Ciba Agriculture	MAFF 03232
483 Carbendazim + *Mancozeb*		
* Kombat WDG	AgrEvo UK Crop Protection Ltd	MAFF 07329
484 Carbendazim		
* Ashlade Carbendazim Flowable	Ashlade Formulations Ltd	MAFF 06213
*% Ashlade Carbendazim Flowable	Ashlade Formulations Ltd	MAFF 02662
* BASF Turf Systemic Fungicide	BASF Plc	MAFF 05774
*% Battal FL	Farm Protection Ltd	MAFF 00215
* Bavistin	BASF Plc	MAFF 00217
* Bavistin DF	BASF Plc	MAFF 03848
* Bavistin FL	BASF Plc	MAFF 00218
* Campbell's Carbendazim 50% Flowable	J D Campbell & Sons Ltd	MAFF 02681
* Carbate Flowable	Pan Britannica Industries Ltd	MAFF 03341
* Delsene 50 DF	Du Pont (UK) Ltd	MAFF 02692
* Derosal Liquid	AgrEvo UK Crop Protection Ltd	MAFF 07315
* Derosal Liquid	Hoechst UK Ltd	MAFF 00671

* These products are "approved for agricultural use". For further details refer to page vii.
% These products are subject to the staged revocation procedure. For further details refer to page vii.
A These products are approved for use in or near water. For further details refer to page vii.

PROFESSIONAL PRODUCTS: FUNGICIDES

Product Name	Marketing Company	Reg. No.

484 Carbendazim—continued

*	Derosal WDG	AgrEvo UK Crop Protection Ltd	MAFF 07316
*	Derosal WDG	Hoechst UK Ltd	MAFF 03404
*	Fisons Turfclear	Fisons Plc	MAFF 06275
*	Fisons Turfclear	Fisons Plc	MAFF 02253
*	HY-CARB	Agrichem Ltd	MAFF 05933
*	Headland Addstem	Headland Agrochemicals Ltd	MAFF 06755
*	Mascot Systemic Turf Fungicide	Rigby Taylor Ltd	MAFF 02839
*	Maxim	Farmers Crop Chemicals Ltd	MAFF 02712
*	Power Carbendazim 50	Power Agrichemicals Ltd	MAFF 03716
*	Quadrangle Hinge	Quadrangle Ltd	MAFF 04929
*%	Rite Weed Carbendazim	Rite Feed Ltd	MAFF 04193
*	Stefes C-Flo	Stefes Plant Protection Ltd	MAFF 07052
*	Stefes Carbendazim Flo	Stefes Plant Protection Ltd	MAFF 05677
*	Stempor DG	Imperial Chemical Industries Plc	MAFF 02021
*	Stempor DG	Zeneca Crop Protection	MAFF 06708
*	Stempor WG	ICI Agrochemicals	MAFF 05709
*	Stempor WG	Zeneca Crop Protection	MAFF 06709
*	Supercarb	Pan Britannica Industries Ltd	MAFF 01560
*	Top Farm Carbendazim - 435	Top Farm Formulations Ltd	MAFF 05307
*	Tripart Defensor FL	Tripart Farm Chemicals Ltd	MAFF 02752

485 Carbendazim + *Chlorothalonil*

*	Bravocarb	ISK Biosciences Ltd	MAFF 05119
*	Greenshield	Zeneca Professional Products	MAFF 06763

486 Carbendazim + *Chlorothalonil* + *Maneb*

*	Ashlade Mancarb Plus	Ashlade Formulations Ltd	MAFF 06222
*%	Ashlade Mancarb Plus	Ashlade Formulations Ltd	MAFF 04440
*	Tripart Victor	Tripart Farm Chemicals Ltd	MAFF 04359

487 Carbendazim + *Cyproconazole*

*	Alto Combi	Cyanamid of GB Ltd	MAFF 05067
*	Alto Combi	Sandoz Agro Ltd	MAFF 05066

488 Carbendazim + *Flusilazole*

*	Contrast	Du Pont (UK) Ltd	MAFF 06150
*	Finish	Du Pont (UK) Ltd	MAFF 06742
*	Punch C	Du Pont (UK) Ltd	MAFF 04690
*	Punch C	Du Pont (UK) Ltd	MAFF 06801
*	Standon Flusilazole Plus	Standon Chemicals Ltd	MAFF 06728

---

* These products are "approved for agricultural use". For further details refer to page vii.
% These products are subject to the staged revocation procedure. For further details refer to page vii.
A These products are approved for use in or near water. For further details refer to page vii.

PROFESSIONAL PRODUCTS: FUNGICIDES

Product Name	Marketing Company	Reg. No.
489 Carbendazim + *Flutriafol*		
* Early Impact	Imperial Chemical Industries Plc	MAFF 02915
* Early Impact	Zeneca Crop Protection	MAFF 06659
* Pacer	Imperial Chemical Industries Plc	MAFF 06242
* Pacer	Zeneca Crop Protection	MAFF 06690
* Palette	Imperial Chemical Industries Plc	MAFF 06241
* Palette	Zeneca Crop Protection	MAFF 06691
490 Carbendazim + *Iprodione*		
* Calidan	Rhone-Poulenc Agriculture Ltd	MAFF 06536
* Vitesse	Rhone-Poulenc Environmental Products	MAFF 06537
491 Carbendazim + *Mancozeb*		
* Kombat WDG	AgrEvo UK Crop Protection Ltd	MAFF 07201
* Kombat WDG	Hoechst UK Ltd	MAFF 04344
* Kombat WDG	Hoechst UK Ltd	MAFF 05984
* Kombat WDG	Rohm & Haas (UK) Ltd	MAFF 05509
* Septal WDG	AgrEvo UK Crop Protection Ltd	MAFF 07200
* Septal WDG	AgrEvo UK Crop Protection Ltd	MAFF 07281
* Septal WDG	Schering Agrochemicals Ltd	MAFF 04279
* Septal WDG	Schering Agrochemicals Ltd	MAFF 05985
492 Carbendazim + *Maneb*		
*% Ashlade Mancarb FL	Ashlade Formulations Ltd	MAFF 02928
*% Ashlade Mancarb FL	Ashlade Formulations Ltd	MAFF 06217
*% Campbell's MC Flowable	J D Campbell & Sons Ltd	MAFF 03467
* Headland Dual	Headland Agrochemicals Ltd	MAFF 03782
* MC Flowable	MTM Agrochemicals Ltd	MAFF 06295
* Multi-W FL	Pan Britannica Industries Ltd	MAFF 04131
* New Squadron	Quadrangle Agrochemicals Ltd	MAFF 06756
* Protector	Procam Agriculture Ltd	MAFF 07075
* Squadron	Quadrangle Agrochemicals Ltd	MAFF 05981
* Tripart Legion	Tripart Farm Chemicals Ltd	MAFF 06113
*% Tripart Legion	Tripart Farm Chemicals Ltd	MAFF 02997
493 Carbendazim + *Maneb* + *Sulphur*		
*% Bolda FL	Farm Protection Ltd	MAFF 03463
494 Carbendazim + *Maneb* + *Tridemorph*		
* Cosmic FL	BASF Plc	MAFF 03473
495 Carbendazim + *Metalaxyl*		
* Ridomil MBC 60 WP	Ciba Agriculture	MAFF 01804

* These products are "approved for agricultural use". For further details refer to page vii.
% These products are subject to the staged revocation procedure. For further details refer to page vii.
A These products are approved for use in or near water. For further details refer to page vii.

PROFESSIONAL PRODUCTS: FUNGICIDES

Product Name	Marketing Company	Reg. No.
496 Carbendazim + *Prochloraz*		
* Hobby	Ciba Agriculture	MAFF 06309
* Hobby	Ciba Agriculture	MAFF 06437
* Sportak Alpha	AgrEvo UK Crop Protection Ltd	MAFF 07222
* Sportak Alpha	Schering Agrochemicals Ltd	MAFF 03872
497 Carbendazim + *Propiconazole*		
* Hispor 45WP	Ciba Agriculture	MAFF 01050
* STR 301	Ciba-Geigy Agriculture	MAFF 06288
* Sparkle 45 WP	Ciba Agriculture	MAFF 04968
498 Carbendazim + *Tecnazene*		
% Arena Plus	Tripart Farm Chemicals Ltd	MAFF 04258
* Hickstor 6 Plus MBC	Hickson & Welch Ltd	MAFF 04176
% Hortag Carbotec	Hortag Chemicals Ltd	MAFF 04135
* Hortag Tecnacarb	Avon Packers Ltd	MAFF 02929
* New Arena Plus	Hickson & Welch Ltd	MAFF 04598
* New Hickstor 6 Plus MBC	Hickson & Welch Ltd	MAFF 04599
Tripart Arena Plus	Hickson & Welch Ltd	MAFF 05602
499 Carbendazim + *Thiram*		
* DAL-CT	Agrichem (International) Ltd	MAFF 06248
*% DAL CT	Agrichem Ltd	MAFF 04029
500 Carbendazim + *Triadimefon*		
*% 200 Plus	Dalgety Agriculture Ltd	MAFF 05025
* 200-Plus	Bayer Plc	MAFF 05070
* Bayleton BM	Bayer Plc	MAFF 00223
501 Carbendazim + *Triadimenol*		
* Tiara	Bayer Plc	MAFF 05278
502 Carboxin + *Imazalil* + *Thiabendazole*		
* Cerevax Extra	Imperial Chemical Industries Plc	MAFF 02501
* Cerevax Extra	Zeneca Crop Protection	MAFF 06642
* Vitaflo Extra	Uniroyal Chemical Ltd	MAFF 07394
* Vitavax Extra	Uniroyal Chemical Ltd	MAFF 07334
503 Carboxin + *Lindane* + *Thiram*		
*% Vitavax RS	DowElanco Ltd	MAFF 02310
* Vitavax RS	Uniroyal Chemical Ltd	MAFF 06029

* These products are "approved for agricultural use". For further details refer to page vii.
% These products are subject to the staged revocation procedure. For further details refer to page vii.
A These products are approved for use in or near water. For further details refer to page vii.

PROFESSIONAL PRODUCTS: FUNGICIDES

Product Name	Marketing Company	Reg. No.
504 Carboxin + *Thiabendazole*		
* Cerevax	Imperial Chemical Industries Plc	MAFF 02500
* Cerevax	Zeneca Crop Protection	MAFF 06641
* Vitavax	Uniroyal Chemical Ltd	MAFF 07205
505 Chlorothalonil		
* ASCE 3748	ISK Biosciences Ltd	MAFF 07352
* Barclay Corrib	Barclay Chemicals (UK) Ltd	MAFF 05886
* Barclay Corrib 500	Barclay Chemicals (UK) Ltd	MAFF 06392
*% BASF Bravo 500	BASF Plc	MAFF 04939
* BB Chlorothalonil	Brown Butlin Group	MAFF 03320
* Bombardier	Universal Crop Protection Ltd	MAFF 02675
* Bravo 500	BASF Plc	MAFF 05637
*% Bravo 500	Fermenta ASC Europe Ltd	MAFF 04945
* Bravo 500	ISK Biosciences Ltd	MAFF 05638
* Bravo 720	ISK Biosciences Ltd	MAFF 05544
*% Bravo 720	SDS Biotech UK Ltd	MAFF 04044
* Chiltern Chlorothalonil 500	Chiltern Farm Chemicals Ltd	MAFF 04961
* Clortosip	Sipcam UK Ltd	MAFF 06126
* Contact 75	ISK Biosciences Ltd	MAFF 05563
*% Contact 75	SDS Biotech UK Ltd	MAFF 04772
* Daconil Turf	Imperial Chemical Industries Plc	MAFF 03658
* Daconil Turf	Zeneca Professional Products	MAFF 06867
* Duomo	Sipcam UK Ltd	MAFF 06152
*% Jupital	Fermenta ASC Europe Ltd	MAFF 04946
* Jupital	ISK Biosciences Ltd	MAFF 05554
* Landgold Chlorothalonil 50	Landgold & Co Ltd	MAFF 06265
* Mainstay	Quadrangle Ltd	MAFF 05625
* Miros DF	Sipcam UK Ltd	MAFF 04966
* Repulse	Imperial Chemical Industries Plc	MAFF 02724
* Repulse	Zeneca Crop Protection	MAFF 06705
* Rover DF	Sipcam UK Ltd	MAFF 06151
* Sipcam UK Rover 500	Sipcam UK Ltd	MAFF 04165
* Standon Chorothalonil 50	Standon Chemicals Ltd	MAFF 05922
* Top Farm Chlorothalonil 500	Top Farm Formulations Ltd	MAFF 05926
* Tripart Faber	Tripart Farm Chemicals Ltd	MAFF 04549
* Tripart Faber	Tripart Farm Chemicals Ltd	MAFF 05505
* Ultrafaber	Tripart Farm Chemicals Ltd	MAFF 05627
*% Ultrafaber	Tripart Farm Chemicals Ltd	MAFF 05155
506 Chlorothalonil + *Carbendazim*		
* Bravocarb	ISK Biosciences Ltd	MAFF 05119
* Greenshield	Zeneca Professional Products	MAFF 06763

* These products are "approved for agricultural use". For further details refer to page vii.
% These products are subject to the staged revocation procedure. For further details refer to page vii.
A These products are approved for use in or near water. For further details refer to page vii.

PROFESSIONAL PRODUCTS: FUNGICIDES

Product Name	Marketing Company	Reg. No.

507 Chlorothalonil + *Carbendazim* + *Maneb*

*	Ashlade Mancarb Plus	Ashlade Formulations Ltd	MAFF 06222
*%	Ashlade Mancarb Plus	Ashlade Formulations Ltd	MAFF 04440
*	Tripart Victor	Tripart Farm Chemicals Ltd	MAFF 04359

508 Chlorothalonil + *Cymoxanil*

*	Ashlade Cyclops	ICI Agrochemicals	MAFF 04857
*%	Cyclops	ICI Agrochemicals	MAFF 05661
*	Cyclops	Zeneca Crop Protection	MAFF 06650
*	Guardian	ICI Agrochemicals	MAFF 05663
*	Guardian	Zeneca Crop Protection	MAFF 06676
*%	PP630	Imperial Chemical Industries Plc	MAFF 05662
*%	PP630	Zeneca Crop Protection	MAFF 06701

509 Chlorothalonil + *Cyproconazole*

*	Alto Elite	Sandoz Agro Ltd	MAFF 05069
*	Octolan	Sandoz Agro Ltd	MAFF 06256
*	SAN 703	Sandoz Agro Ltd	MAFF 06255

510 Chlorothalonil + *Fenpropimorph*

*	BAS 438	BASF (UK) Ltd	MAFF 03451
*	Corbel CL	BASF (UK) Ltd	MAFF 04196
*%	Mistral CT	Rhone-Poulenc Agriculture	MAFF 06176
*%	Mistral CT	Rhone-Poulenc Crop Protection	MAFF 04543

511 Chlorothalonil + *Flutriafol*

*	Halo	Farm Protection Ltd	MAFF 05573
*	Halo	Zeneca Crop Protection	MAFF 06520
*	Halo 300	Zeneca Crop Protection	MAFF 06951
*	Impact Excel	Imperial Chemical Industries Plc	MAFF 03758
*	Impact Excel	Zeneca Crop Protection	MAFF 06680
*%	Impact Excel 375	Imperial Chemical Industries Plc	MAFF 03757
*	Impact Excel 375	Zeneca Crop Protection	MAFF 06681

512 Chlorothalonil + *Metalaxyl*

*%	Folio 575 FW	Ciba-Geigy Agrochemicals	MAFF 04122
*	Folio 575 SC	Ciba Agriculture	MAFF 05843

513 Chlorothalonil + *Propiconazole*

*%	Sambarin 312.5 FW	Ciba-Geigy Agrochemicals	MAFF 04685
*	Sambarin 312.5 SC	Ciba Agriculture	MAFF 05809
*	Sambarin TP	Ciba Agriculture	MAFF 04094

---

* These products are "approved for agricultural use". For further details refer to page vii.
% These products are subject to the staged revocation procedure. For further details refer to page vii.
A These products are approved for use in or near water. For further details refer to page vii.

PROFESSIONAL PRODUCTS: FUNGICIDES

Product Name	Marketing Company	Reg. No.
514 Copper ammonium carbonate		
* Croptex Fungex	Hortichem Ltd	MAFF 02888
515 Copper complex - bordeaux		
* Wetcol 3 Copper Fungicide	Ford Smith & Co Ltd	MAFF 02360
516 Copper oxychloride		
* Cuprokylt	Universal Crop Protection Ltd	MAFF 00604
* Cuprokylt L	Universal Crop Protection Ltd	MAFF 02769
* Cuprosana H	Universal Crop Protection Ltd	MAFF 00605
*% Headland Inorganic Liquid Copper	WBC Technology Ltd	MAFF 04919
*% Sprayon Inorganic Liquid Copper	Sprayon Agricultural Products Ltd	MAFF 04918
* Vitigran	AgrEvo UK Crop Protection Ltd	MAFF 07338
* Vitigran	Hoechst UK Ltd	MAFF 02313
517 Copper oxychloride + *Maneb* + *Sulphur*		
* Ashlade SMC Flowable	Ashlade Formulations Ltd	MAFF 06494
*% Ashlade SMC Flowable	Ashlade Formulations Ltd	MAFF 04560
*% Rearguard	Universal Crop Protection Ltd	MAFF 04938
518 Copper oxychloride + *Metalaxyl*		
* Ridomil Plus 50 WP	Ciba Agriculture	MAFF 01803
519 Copper sulphate + *Sulphur*		
* Top Cop	Stoller Chemical Ltd	MAFF 04553
520 Cymoxanil + *Chlorothalonil*		
* Ashlade Cyclops	ICI Agrochemicals	MAFF 04857
*% Cyclops	ICI Agrochemicals	MAFF 05661
* Cyclops	Zeneca Crop Protection	MAFF 06650
* Guardian	ICI Agrochemicals	MAFF 05663
* Guardian	Zeneca Crop Protection	MAFF 06676
*% PP630	Imperial Chemical Industries Plc	MAFF 05662
*% PP630	Zeneca Crop Protection	MAFF 06701
521 Cymoxanil + *Mancozeb*		
* Ashlade Solace	Ashlade Formulations Ltd	MAFF 06472
*% Ashlade Solace	Ashlade Formulations Ltd	MAFF 05462
* Besiege	Du Pont (UK) Ltd	MAFF 05451
* Clayton Krypton	Clayton Plant Protection (UK) Ltd	MAFF 06973
* Curzate M	Du Pont (UK) Ltd	MAFF 04343
* Fytospore	Farm Protection	MAFF 00960
* Fytospore	Zeneca Crop Protection	MAFF 06517

* These products are "approved for agricultural use". For further details refer to page vii.
% These products are subject to the staged revocation procedure. For further details refer to page vii.
A These products are approved for use in or near water. For further details refer to page vii.

PROFESSIONAL PRODUCTS: FUNGICIDES

Product Name	Marketing Company	Reg. No.
521 Cymoxanil + *Mancozeb*—continued*		
* Standon Cymoxanil Extra	Standon Chemicals Ltd	MAFF 06807
* Stefes Blight Spray	Stefes Plant Protection Ltd	MAFF 05811
* Systol M	Quadrangle Agrochemicals Ltd	MAFF 07098
* Systol M	Quadrangle Ltd	MAFF 03480
522 Cymoxanil + *Mancozeb* + *Oxadixyl*		
* Ripost	Sandoz Agro Ltd	MAFF 04890
* Ripost Pepite	Sandoz Agro Ltd	MAFF 06485
*% Ripost WDG	Sandoz Crop Protection Ltd	MAFF 05053
* Trustan	Du Pont (UK) Ltd	MAFF 05022
* Trustan WDG	Du Pont (UK) Ltd	MAFF 05050
523 Cyproconazole		
* Alto 100 SL	Sandoz Agro Ltd	MAFF 05065
* Aplan	Sandoz Agro Ltd	MAFF 06121
* Barclay Shandon	Barclay Chemicals (UK) Ltd	MAFF 06464
* Divora	Sandoz Agro Ltd	MAFF 06122
* SAN 619	Sandoz Agro Ltd	MAFF 06120
524 Cyproconazole + *Carbendazim*		
* Alto Combi	Cyanamid of GB Ltd	MAFF 05067
* Alto Combi	Sandoz Agro Ltd	MAFF 05066
525 Cyproconazole + *Chlorothalonil*		
* Alto Elite	Sandoz Agro Ltd	MAFF 05069
* Octolan	Sandoz Agro Ltd	MAFF 06256
* SAN 703	Sandoz Agro Ltd	MAFF 06255
526 Cyproconazole + *Mancozeb*		
* Alto Eco	Sandoz Agro Ltd	MAFF 05068
527 Cyproconazole + *Prochloraz*		
* Sportak Delta 460	AgrEvo UK Crop Protection Ltd	MAFF 07224
* Sportak Delta 460	Schering Agrochemicals Ltd	MAFF 05107
* Tiptor	AgrEvo UK Crop Protection Ltd	MAFF 07294
* Tiptor	AgrEvo UK Crop Protection Ltd	MAFF 07295
* Tiptor	Cyanamid of GB Ltd	MAFF 05106
* Tiptor	Sandoz Agro Ltd	MAFF 05105
* Tiptor	Schering Agrochemicals Ltd	MAFF 05831
* Tiptor	Schering Agrochemicals Ltd	MAFF 06971

* These products are "approved for agricultural use". For further details refer to page vii.
% These products are subject to the staged revocation procedure. For further details refer to page vii.
A These products are approved for use in or near water. For further details refer to page vii.

PROFESSIONAL PRODUCTS: FUNGICIDES

Product Name	Marketing Company	Reg. No.
528 Cyproconazole + *Tridemorph*		
* Alto Major	Sandoz Agro Ltd	MAFF 06979
* Moot	Sandoz Agro Ltd	MAFF 06990
* San 735	Sandoz Agro Ltd	MAFF 06991
529 Dichlofluanid		
* Elvaron	Bayer Plc	MAFF 00789
* Elvaron WG	Bayer Plc	MAFF 04855
530 Dichlorophen		
* 50-50 Liquid Mosskiller	Vitax Ltd	MAFF 07191
* Halophen RE 49	McMillan Technical Services Ltd	MAFF 04636
* Super Mosstox	Rhone-Poulenc Agriculture Ltd	MAFF 05339
531 Dicloran		
* Fumite Dicloran Smoke	Octavius Hunt Ltd	MAFF 00930
532 Difenoconazole		
* Plover 250EC	Ciba Agriculture	MAFF 07232
533 Dimethomorph		
* Acrobat	Cyanamid UK	MAFF 06987
534 Dimethomorph + *Mancozeb*		
* Acrobat MZ	Cyanamid UK	MAFF 06988
* Invader	Cyanamid UK	MAFF 06989
535 Dinocap		
* Karathane Liquid	Rohm & Haas (UK) Ltd	MAFF 01126
* Karathane WP	Rohm & Haas (UK) Ltd	MAFF 01127
536 Dithianon		
* Dithianon Flowable	Cyanamid Of GB Ltd	MAFF 07007
* Dithianon Flowable	Shell Chemicals UK Ltd	MAFF 04065
537 Dithianon + *Penconazole*		
*% Topas D275 FW	Ciba-Geigy Agrochemicals	MAFF 04915
* Topas D275 SC	Ciba Agriculture	MAFF 05855
538 Dodemorph		
* F 238	BASF Plc	MAFF 00206

---

* These products are "approved for agricultural use". For further details refer to page vii.
% These products are subject to the staged revocation procedure. For further details refer to page vii.
A These products are approved for use in or near water. For further details refer to page vii.

PROFESSIONAL PRODUCTS: FUNGICIDES

Product Name	Marketing Company	Reg. No.
539 Dodine		
* Barclay Dodex	Barclay Chemicals (UK) Ltd	MAFF 05655
* Radspor FL	Truchem Ltd	MAFF 01685
540 Epoxiconazole		
* Epic	BASF Plc	MAFF 07237
* Opus	BASF Plc	MAFF 07236
541 Epoxiconazole + *Fenpropimorph*		
* Eclipse	BASF Plc	MAFF 07361
* Opus Team	BASF Plc	MAFF 07362
542 Epoxiconazole + *Tridemorph*		
* Ensign	BASF Plc	MAFF 07364
* Opus Plus	BASF Plc	MAFF 07363
543 Ethirimol		
* Milgo E	Imperial Chemical Industries Plc	MAFF 01357
* Milgo E	Zeneca Crop Protection	MAFF 06684
* Milstem	Imperial Chemical Industries Plc	MAFF 01358
544 Ethirimol + *Flutriafol* + *Imazalil* + *Thiabendazole*		
* Ferrax IM	Imperial Chemical Industries Plc	MAFF 05142
* Ferrax IM	Zeneca Crop Protection	MAFF 06663
545 Ethirimol + *Flutriafol* + *Thiabendazole*		
* Ferrax	Bayer UK Ltd	MAFF 05284
* Ferrax	DowElanco Ltd	MAFF 04917
* Ferrax	Imperial Chemical Industries Plc	MAFF 02827
* Ferrax	Zeneca Crop Protection	MAFF 06662
546 Ethirimol + *Fuberidazole* + *Triadimenol*		
* Bay UK 292	Bayer Plc	MAFF 04335
* Militan	ICI Agrochemicals	MAFF 06095
* Militan	Zeneca Crop Protection	MAFF 06711
547 Etridiazole		
* AAterra WP	Imperial Chemical Industries Plc	MAFF 03795
* AAterra WP	Zeneca Crop Protection	MAFF 06625
548 Fenarimol		
* Rimidin	Rigby Taylor Ltd	MAFF 05907
* Rubigan	DowElanco Ltd	MAFF 05489
*% Rubigan	Elanco Products Ltd	MAFF 02926

* These products are "approved for agricultural use". For further details refer to page vii.
% These products are subject to the staged revocation procedure. For further details refer to page vii.
A These products are approved for use in or near water. For further details refer to page vii.

PROFESSIONAL PRODUCTS: FUNGICIDES

Product Name	Marketing Company	Reg. No.
549 Fenpiclonil		
* Beret 050 FS	Ciba Agriculture	MAFF 06805
550 Fenpiclonil + *Imazalil*		
* Beret Extra 060 FS	Ciba Agriculture	MAFF 06806
551 Fenpropidin		
* Mallard 750 EC	Ciba Agriculture	MAFF 06315
* Patrol	Imperial Chemical Industries Plc	MAFF 03317
* Patrol	Zeneca Crop Protection	MAFF 06693
* Tern 750 EC	Ciba Agriculture	MAFF 05643
*% Tern 750 EC	Ciba-Geigy Agrochemicals	MAFF 05256
552 Fenpropidin + *Propiconazole*		
* Legend	ICI Agrochemicals	MAFF 04978
* Legend	Zeneca Crop Protection	MAFF 06682
* Opal	Zeneca Crop Protection	MAFF 06938
* Sheen 550 EC	Ciba Agriculture	MAFF 05980
* Zulu 550 EC	Ciba Agriculture	MAFF 06594
* Zulu 550 EC	Ciba Agriculture	MAFF 06937
553 Fenpropimorph		
* Aura 750 EC	Ciba Agriculture	MAFF 05705
* Aura 750 EC	Ciba-Geigy Agrochemicals	MAFF 05300
* BAS 421F	BASF Plc	MAFF 06127
* Corbel	BASF Plc	MAFF 00578
* Keetak	BASF Plc	MAFF 06950
* Landgold Fenpropimorph 750	Landgold & Co Ltd	MAFF 06319
* Mistral	Ciba Agriculture	MAFF 06943
* Mistral	Rhone-Poulenc Agriculture	MAFF 06199
* Mistral	Rhone-Poulenc Agriculture Ltd	MAFF 04582
* Standon Fenpropimorph 750	Standon Chemicals Ltd	MAFF 05654
* Widgeon 750 EC	Ciba Agriculture	MAFF 06101
554 Fenpropimorph + *Chlorothalonil*		
* BAS 438	BASF (UK) Ltd	MAFF 03451
* Corbel CL	BASF (UK) Ltd	MAFF 04196
*% Mistral CT	Rhone-Poulenc Agriculture	MAFF 06176
*% Mistral CT	Rhone-Poulenc Crop Protection	MAFF 04543
555 Fenpropimorph + *Epoxiconazole*		
* Eclipse	BASF Plc	MAFF 07361
* Opus Team	BASF Plc	MAFF 07362

* These products are "approved for agricultural use". For further details refer to page vii.
% These products are subject to the staged revocation procedure. For further details refer to page vii.
A These products are approved for use in or near water. For further details refer to page vii.

PROFESSIONAL PRODUCTS: FUNGICIDES

Product Name	Marketing Company	Reg. No.
556 Fenpropimorph + *Flusilazole*		
* BAS 48500F	BASF Plc	MAFF 06784
* Colstar	Du Pont (UK) Ltd	MAFF 06783
557 Fenpropimorph + *Flusilazole* + *Tridemorph*		
* Bingo	BASF Plc	MAFF 06920
* DUK 51	Du Pont (UK) Ltd	MAFF 06764
558 Fenpropimorph + *Lindane* + *Thiram*		
* Lindex Plus FS Seed Treatment	DowElanco Ltd	MAFF 03934
559 Fenpropimorph + *Prochloraz*		
* Jester	Ciba Agriculture	MAFF 06586
* Sprint	AgrEvo UK Crop Protection Ltd	MAFF 07291
* Sprint	Schering Agrochemicals Ltd	MAFF 03986
* Sprint HF	AgrEvo UK Crop Protection Ltd	MAFF 07292
* Sprint HF	Schering Agrochemicals Ltd	MAFF 06490
560 Fenpropimorph + *Propiconazole*		
* Decade 500 EC	Ciba Agriculture	MAFF 05757
*% Decade 500 EC	Ciba-Geigy Agrochemicals	MAFF 05711
* Glint 500 EC	Ciba Agriculture	MAFF 04126
* Mantle 425 EC	Ciba Agriculture	MAFF 05715
561 Fenpropimorph + *Tridemorph*		
* BAS 46402F	BASF Plc	MAFF 03313
* Gemini	BASF Plc	MAFF 05684
562 Fentin Hydroxide		
* Farmatin 560	AgrEvo UK Crop Protection Ltd	MAFF 07320
563 Fentin acetate + *Maneb*		
* Brestan 60	AgrEvo UK Crop Protection Ltd	MAFF 07304
* Brestan 60	Hoechst UK Ltd	MAFF 00325
* Brestan 60 SP	AgrEvo UK Crop Protection Ltd	MAFF 07305
* Brestan 60 SP	Hoechst UK Ltd	MAFF 06042
564 Fentin hydroxide		
*% Ashlade Flotin	Ashlade Formulations Ltd	MAFF 03535
* Ashlade Flotin	Ashlade Formulations Ltd	MAFF 06223
* Ashlade Flotin 2	Ashlade Formulations Ltd	MAFF 06224
*% Ashlade Flotin 2	Ashlade Formulations Ltd	MAFF 03783

* These products are "approved for agricultural use". For further details refer to page vii.
% These products are subject to the staged revocation procedure. For further details refer to page vii.
A These products are approved for use in or near water. For further details refer to page vii.

PROFESSIONAL PRODUCTS: FUNGICIDES

Product Name	Marketing Company	Reg. No.
564 Fentin hydroxide—*continued*		
* CIA Fentin-475-WP	Combined Independent Agronomists Ltd	MAFF 05550
* Du-Ter 50	AgrEvo UK Crop Protection Ltd	MAFF 07317
* Du-Ter 50	Hoechst UK Ltd	MAFF 06416
*% Du-ter 50	Imperial Chemical Industries Plc	MAFF 00778
*% Farmatin 560	Farm Protection Ltd	MAFF 04595
* Farmatin 560	Hoechst UK Ltd	MAFF 06417
*% Quadrangle Super-Tin 4L	Quadrangle Ltd	MAFF 03842
* Super-Tin 4L	Chiltern Farm Chemicals Ltd	MAFF 02995
*% Top Farm Fentin-475-WP	Top Farm Formulations Ltd	MAFF 05504
565 Fentin hydroxide + *Metoxuron*		
*% Endspray	Pan Britannica Industries Ltd	MAFF 02799
566 Ferbam + *Maneb* + *Zineb*		
*% Trimanzone PVA	Brian Lewis Agriculture Ltd	MAFF 05289
* Trimanzone PVA	Elf Atochem Agri BV	MAFF 06877
* Trimanzone WP	Brian Lewis Agriculture Ltd	MAFF 05860
* Trimanzone WP	Elf Atochem Agri BV	MAFF 06876
567 Fluazinam		
* FD 4058	Zeneca Crop Protection	MAFF 07093
* Frowncide	ISK Biotech Europe Ltd	MAFF 07094
* Legacy	ISK Biosciences Ltd	MAFF 07401
* Salvo	Zeneca Crop Protection	MAFF 07092
* Shirlan	Zeneca Crop Protection	MAFF 07091
568 Flusilazole		
* DUK 747	Du Pont (UK) Ltd	MAFF 05775
* Genie	Du Pont (UK) Ltd	MAFF 05272
* Landgold Flusilazole 400	Landgold & Co Ltd	MAFF 05908
* Lyric	Du Pont (UK) Ltd	MAFF 06543
* Punch	Du Pont (UK) Ltd	MAFF 05273
* Sanction	Du Pont (UK) Ltd	MAFF 04773
569 Flusilazole + *Carbendazim*		
* Contrast	Du Pont (UK) Ltd	MAFF 06150
* Finish	Du Pont (UK) Ltd	MAFF 06742
* Punch C	Du Pont (UK) Ltd	MAFF 04690
* Punch C	Du Pont (UK) Ltd	MAFF 06801
* Standon Flusilazole Plus	Standon Chemicals Ltd	MAFF 06728

* These products are "approved for agricultural use". For further details refer to page vii.
% These products are subject to the staged revocation procedure. For further details refer to page vii.
A These products are approved for use in or near water. For further details refer to page vii.

PROFESSIONAL PRODUCTS: FUNGICIDES

Product Name	Marketing Company	Reg. No.
570 Flusilazole + *Fenpropimorph*		
* BAS 48500F	BASF Plc	MAFF 06784
* Colstar	Du Pont (UK) Ltd	MAFF 06783
571 Flusilazole + *Fenpropimorph* + *Tridemorph*		
* Bingo	BASF Plc	MAFF 06920
* DUK 51	Du Pont (UK) Ltd	MAFF 06764
572 Flusilazole + *Tridemorph*		
* Fusion	Du Pont (UK) Ltd	MAFF 04908
* Meld	BASF Plc	MAFF 04914
573 Flutriafol		
* Clayton Flutriafol	Clayton Plant Protection (UK) Ltd	MAFF 05972
* Impact	Imperial Chemical Industries Plc	MAFF 03028
* Impact	Zeneca Crop Protection	MAFF 06679
* Landgold Flutriafol	Landgold & Co Ltd	MAFF 06244
* Pointer	Imperial Chemical Industries Plc	MAFF 06243
* Pointer	Zeneca Crop Protection	MAFF 06695
* PP 450	Imperial Chemical Industries Plc	MAFF 02748
* PP 450	Zeneca Crop Protection	MAFF 06700
* Standon Flutriafol	Standon Chemicals Ltd	MAFF 05940
574 Flutriafol + *Carbendazim*		
* Early Impact	Imperial Chemical Industries Plc	MAFF 02915
* Early Impact	Zeneca Crop Protection	MAFF 06659
* Pacer	Imperial Chemical Industries Plc	MAFF 06242
* Pacer	Zeneca Crop Protection	MAFF 06690
* Palette	Imperial Chemical Industries Plc	MAFF 06241
* Palette	Zeneca Crop Protection	MAFF 06691
575 Flutriafol + *Chlorothalonil*		
* Halo	Farm Protection Ltd	MAFF 05573
* Halo	Zeneca Crop Protection	MAFF 06520
* Halo 300	Zeneca Crop Protection	MAFF 06951
* Impact Excel	Imperial Chemical Industries Plc	MAFF 03758
* Impact Excel	Zeneca Crop Protection	MAFF 06680
*% Impact Excel 375	Imperial Chemical Industries Plc	MAFF 03757
* Impact Excel 375	Zeneca Crop Protection	MAFF 06681
576 Flutriafol + *Ethirimol* + *Imazalil* + *Thiabendazole*		
* Ferrax IM	Imperial Chemical Industries Plc	MAFF 05142
* Ferrax IM	Zeneca Crop Protection	MAFF 06663

* These products are "approved for agricultural use". For further details refer to page vii.
% These products are subject to the staged revocation procedure. For further details refer to page vii.
A These products are approved for use in or near water. For further details refer to page vii.

PROFESSIONAL PRODUCTS: FUNGICIDES

Product Name	Marketing Company	Reg. No.
577 Flutriafol + *Ethirimol* + *Thiabendazole*		
* Ferrax	Bayer UK Ltd	MAFF 05284
* Ferrax	DowElanco Ltd	MAFF 04917
* Ferrax	Imperial Chemical Industries Plc	MAFF 02827
* Ferrax	Zeneca Crop Protection	MAFF 06662
578 Flutriafol + *Imazalil* + *Thiabendazole*		
* Vincit IM	Zeneca Crop Protection	MAFF 06380
579 Flutriafol + *Iprodione*		
*% Cyclone	Rhone-Poulenc Agriculture	MAFF 06175
* Cyclone	Rhone-Poulenc Crop Protection	MAFF 05000
580 Flutriafol + *Thiabendazole*		
* Vincit L	Imperial Chemical Industries Plc	MAFF 04322
581 Fosetyl-aluminium		
* Aliette	Hortichem Ltd	MAFF 02484
* Aliette	Rhone-Poulenc Agriculture	MAFF 05648
* Aliette	Rhone-Poulenc Agriculture Ltd	MAFF 05439
582 Fuberidazole + *Bitertanol*		
* Sibutol	Bayer Plc	MAFF 07238
* Sibutol Flowable	Bayer Plc	MAFF 06983
583 Fuberidazole + *Bitertanol* + *Triadimenol*		
* Cereline	Bayer Plc	MAFF 07239
* Cereline Flowable	Bayer Plc	MAFF 06984
584 Fuberidazole + *Ethirimol* + *Triadimenol*		
* Bay UK 292	Bayer Plc	MAFF 04335
* Militan	ICI Agrochemicals	MAFF 06095
* Militan	Zeneca Crop Protection	MAFF 06711
585 Fuberidazole + *Imazalil* + *Triadimenol*		
* Baytan IM	Bayer Plc	MAFF 00226
586 Fuberidazole + *Triadimenol*		
* Baytan	Bayer Plc	MAFF 00225
* Baytan	Imperial Chemical Industries Plc	MAFF 03946
* Baytan	Zeneca Crop Protection	MAFF 06636
* Baytan Flowable	Bayer Plc	MAFF 02593

* These products are "approved for agricultural use". For further details refer to page vii.
% These products are subject to the staged revocation procedure. For further details refer to page vii.
A These products are approved for use in or near water. For further details refer to page vii.

PROFESSIONAL PRODUCTS: FUNGICIDES

Product Name	Marketing Company	Reg. No.

586 Fuberidazole + *Triadimenol*—continued

* Baytan Flowable	Imperial Chemical Industries Plc	MAFF 03845
* Baytan Flowable	Zeneca Crop Protection	MAFF 06635

587 Furalaxyl

* Fongarid 25 WP	Ciba Agriculture	MAFF 03595

588 Guazatine

* Murbenine	Uniroyal Chemical Ltd	MAFF 07199
* Panoctine	Rhone-Poulenc Agriculture	MAFF 06207
* Panoctine	Rhone-Poulenc Agriculture Ltd	MAFF 05012
* Rappor Liquid Seed Treatment	DowElanco Ltd	MAFF 04900
* Ravine	Rhone-Poulenc Agriculture Ltd	MAFF 07193

589 Guazatine + *Imazalil*

* Murbenine Plus	Uniroyal Chemical Ltd	MAFF 07202
* Panoctine Plus	Rhone-Poulenc Agriculture	MAFF 06208
* Panoctine Plus	Rhone-Poulenc Agriculture Ltd	MAFF 05013
* Rappor Plus Liquid Seed Treatment	Rhone-Poulenc Agriculture Ltd	MAFF 04899
* Ravine Plus	Rhone-Poulenc Agriculture	MAFF 07343

590 Hymexazol

* Tachigaren 70 WP	Sumitomo Corporation	MAFF 02649

591 Imazalil

Clinafarm Spray	Janssen Pharmaceutica NV	MAFF 06436
* Fungaflor	Brinkman UK Ltd	MAFF 05968
* Fungaflor	Hortichem Ltd	MAFF 03891
* Fungaflor	Hortichem Ltd	MAFF 05967
*% Fungaflor	Janssen Pharmaceutica NV	MAFF 03452
*% Fungaflor C	Rhone-Poulenc Crop Protection	MAFF 05181
* Fungaflor Smoke	Brinkman UK Ltd	MAFF 06009
* Fungaflor Smoke	Hortichem Ltd	MAFF 05969
*% Fungaflor Smoke	Janssen Pharmaceutica NV	MAFF 03599
* Fungazil 100 SL	Embetec Crop Protection	MAFF 05652
* Fungazil 100 SL	Rhone-Poulenc Agriculture	MAFF 06202

592 Imazalil + *Carboxin* + *Thiabendazole*

* Cerevax Extra	Imperial Chemical Industries Plc	MAFF 02501
* Cerevax Extra	Zeneca Crop Protection	MAFF 06642
* Vitaflo Extra	Uniroyal Chemical Ltd	MAFF 07394
* Vitavax Extra	Uniroyal Chemical Ltd	MAFF 07334

---

* These products are "approved for agricultural use". For further details refer to page vii.
% These products are subject to the staged revocation procedure. For further details refer to page vii.
A These products are approved for use in or near water. For further details refer to page vii.

PROFESSIONAL PRODUCTS: FUNGICIDES

Product Name	Marketing Company	Reg. No.

593 Imazalil + *Ethirimol* + *Flutriafol* + *Thiabendazole*

* Ferrax IM	Imperial Chemical Industries Plc	MAFF 05142
* Ferrax IM	Zeneca Crop Protection	MAFF 06663

594 Imazalil + *Fenpiclonil*

* Beret Extra 060 FS	Ciba Agriculture	MAFF 06806

595 Imazalil + *Flutriafol* + *Thiabendazole*

* Vincit IM	Zeneca Crop Protection	MAFF 06380

596 Imazalil + *Fuberidazole* + *Triadimenol*

* Baytan IM	Bayer Plc	MAFF 00226

597 Imazalil + *Guazatine*

* Murbenine Plus	Uniroyal Chemical Ltd	MAFF 07202
* Panoctine Plus	Rhone-Poulenc Agriculture	MAFF 06208
* Panoctine Plus	Rhone-Poulenc Agriculture Ltd	MAFF 05013
* Rappor Plus Liquid Seed Treatment	Rhone-Poulenc Agriculture Ltd	MAFF 04899
* Ravine Plus	Rhone-Poulenc Agriculture	MAFF 07343

598 Imazalil + *Pencycuron*

* Monceren IM	Bayer Plc	MAFF 06259
* Monceren IM Flowable	Bayer Plc	MAFF 06731

599 Imazalil + *Thiabendazole*

* Extratect Flowable	MSD Agvet	MAFF 05507

600 Iprodione

* CDA Rovral	Rhone-Poulenc Environmental Products	MAFF 04679
* Rovral Flo	Rhone-Poulenc Agriculture	MAFF 06328
* Rovral Flo	Rhone-Poulenc Crop Protection	MAFF 04526
* Rovral Green	Rhone-Poulenc Environmental Products	MAFF 05702
* Rovral Liquid FS	Rhone-Poulenc Agriculture	MAFF 06366
* Rovral WP	Rhone-Poulenc Agriculture	MAFF 06091
* Rovral WP	Rhone-Poulenc Agriculture Ltd	MAFF 04647
* Turbair Rovral	Pan Britannica Industries Ltd	MAFF 02248

601 Iprodione + *Carbendazim*

* Calidan	Rhone-Poulenc Agriculture Ltd	MAFF 06536
* Vitesse	Rhone-Poulenc Environmental Products	MAFF 06537

---

* These products are "approved for agricultural use". For further details refer to page vii.
% These products are subject to the staged revocation procedure. For further details refer to page vii.
A These products are approved for use in or near water. For further details refer to page vii.

PROFESSIONAL PRODUCTS: FUNGICIDES

Product Name	Marketing Company	Reg. No.
602 Iprodione + *Flutriafol*		
*% Cyclone	Rhone-Poulenc Agriculture	MAFF 06175
* Cyclone	Rhone-Poulenc Crop Protection	MAFF 05000
603 Iprodione + *Thiophanate-methyl*		
* Compass	Rhone-Poulenc Agriculture	MAFF 06190
* Compass	Rhone-Poulenc Agriculture Ltd	MAFF 04580
* FR 1433	Rhone-Poulenc Crop Protection	MAFF 04994
604 Lindane + *Captan*		
* Gammalex	ICI Agrochemicals	MAFF 00965
* Gammalex	Zeneca Crop Protection	MAFF 06396
605 Lindane + *Carboxin* + *Thiram*		
*% Vitavax RS	DowElanco Ltd	MAFF 02310
* Vitavax RS	Uniroyal Chemical Ltd	MAFF 06029
606 Lindane + *Fenpropimorph* + *Thiram*		
* Lindex Plus FS Seed Treatment	DowElanco Ltd	MAFF 03934
607 Lindane + *Thiabendazole* + *Thiram*		
* Hysede FL	Agrichem Ltd	MAFF 02863
608 Lindane + *Thiram*		
* Hydraguard	Agrichem Ltd	MAFF 03278
609 Mancozeb		
* Agrichem Mancozeb 80	Agrichem (International) Ltd	MAFF 06354
*% Ashlade Mancozeb FL	Ashlade Formulations Ltd	MAFF 03208
* Ashlade Mancozeb Flowable	Ashlade Formulations Ltd	MAFF 06226
* Barclay Manzeb 80	Barclay Chemicals (UK) Ltd	MAFF 05944
*% Barclay Manzeb 80	Barclay Chemicals (UK) Ltd	MAFF 05296
* Barclay Manzeb 80W	Barclay Chemicals (UK) Ltd	MAFF 05833
* Dequiman MZ	Elf Atochem Agri BV	MAFF 06870
* Dequiman MZ	Pennwalt Chemicals Ltd	MAFF 03959
* Dithane 945	Pan Britannica Industries Ltd	MAFF 00719
* Dithane 945	Pan Britannica Industries Ltd	MAFF 04017
* Dithane Dry Flowable	Pan Britannica Industries Ltd	MAFF 04251
* Dithane Dry Flowable	Pan Britannica Industries Ltd	MAFF 04255
* Dithane Superflo	Pan Britannica Industries Ltd	MAFF 06290
* Dithane Superflo	Pan Britannica Industries Ltd	MAFF 06593
* Karamate Dry Flo	Rohm & Haas (UK) Ltd	MAFF 04250
* Karamate N	Rohm & Haas (UK) Ltd	MAFF 01125

* These products are "approved for agricultural use". For further details refer to page vii.
% These products are subject to the staged revocation procedure. For further details refer to page vii.
A These products are approved for use in or near water. For further details refer to page vii.

PROFESSIONAL PRODUCTS: FUNGICIDES

Product Name	Marketing Company	Reg. No.
609 Mancozeb—*continued*		
* Luxan Mancozeb Flowable	Luxan (UK) Ltd.	MAFF 06812
* Manzate 200	Du Pont (UK) Ltd	MAFF 01281
* Manzate 200 DF	Du Pont (UK) Ltd	MAFF 06010
* Manzate 200 PI	Du Pont (UK) Ltd	MAFF 07209
* Nemispor	EniMont UK Ltd	MAFF 05506
* Nemispor	Montedison UK Ltd	MAFF 01473
* Nemispor	Sipcam UK Ltd	MAFF 07348
* Penncozeb	Cyanamid of GB Ltd	MAFF 07089
* Penncozeb	Elf Atochem Agri BV	MAFF 06873
* Penncozeb	Shell Chemicals UK Ltd	MAFF 02716
* Penncozeb	Shell Chemicals UK Ltd	MAFF 04609
* Penncozeb Flowable	Pennwalt Chemicals Ltd	MAFF 04395
* Penncozeb WDG	Cyanamid of GB Ltd	MAFF 07095
* Penncozeb WDG	Shell Chemicals UK Ltd	MAFF 06054
* Portman Mandate 80	Portman Agrochemicals Ltd	MAFF 06320
* Unicrop Flowable Mancozeb	Universal Crop Protection Ltd	MAFF 04700
* Unicrop Mancozeb	Universal Crop Protection Ltd	MAFF 05467
610 Mancozeb + *Benalaxyl*		
* Barclay Bezant	Barclay Chemicals (UK) Ltd	MAFF 05914
* Clayton Benzeb	Clayton Plant Protection (UK) Ltd	MAFF 07081
*% Galben M	Dow Chemical Co Ltd	MAFF 03809
* Galben M	DowElanco Ltd	MAFF 05092
* Galben M	EniMont UK Ltd	MAFF 05091
* Galben M	Enichem UK Ltd	MAFF 05904
* Galben M	Montedison UK Ltd	MAFF 02753
* Galben M	Sipcam UK Ltd	MAFF 07220
611 Mancozeb + *Carbendazim*		
* Kombat WDG	AgrEvo UK Crop Protection Ltd	MAFF 07329
612 Mancozeb + *Carbendazim*		
* Kombat WDG	AgrEvo UK Crop Protection Ltd	MAFF 07201
* Kombat WDG	Hoechst UK Ltd	MAFF 04344
* Kombat WDG	Hoechst UK Ltd	MAFF 05984
* Kombat WDG	Rohm & Haas (UK) Ltd	MAFF 05509
* Septal WDG	AgrEvo UK Crop Protection Ltd	MAFF 07200
* Septal WDG	AgrEvo UK Crop Protection Ltd	MAFF 07281
* Septal WDG	Schering Agrochemicals Ltd	MAFF 04279
* Septal WDG	Schering Agrochemicals Ltd	MAFF 05985

* These products are "approved for agricultural use". For further details refer to page vii.
% These products are subject to the staged revocation procedure. For further details refer to page vii.
A These products are approved for use in or near water. For further details refer to page vii.

PROFESSIONAL PRODUCTS: FUNGICIDES

Product Name	Marketing Company	Reg. No.
613 Mancozeb + *Cymoxanil*		
* Ashlade Solace	Ashlade Formulations Ltd	MAFF 06472
*% Ashlade Solace	Ashlade Formulations Ltd	MAFF 05462
* Besiege	Du Pont (UK) Ltd	MAFF 05451
* Clayton Krypton	Clayton Plant Protection (UK) Ltd	MAFF 06973
* Curzate M	Du Pont (UK) Ltd	MAFF 04343
* Fytospore	Farm Protection	MAFF 00960
* Fytospore	Zeneca Crop Protection	MAFF 06517
* Standon Cymoxanil Extra	Standon Chemicals Ltd	MAFF 06807
* Stefes Blight Spray	Stefes Plant Protection Ltd	MAFF 05811
* Systol M	Quadrangle Agrochemicals Ltd	MAFF 07098
* Systol M	Quadrangle Ltd	MAFF 03480
614 Mancozeb + *Cymoxanil* + *Oxadixyl*		
* Ripost	Sandoz Agro Ltd	MAFF 04890
* Ripost Pepite	Sandoz Agro Ltd	MAFF 06485
*% Ripost WDG	Sandoz Crop Protection Ltd	MAFF 05053
* Trustan	Du Pont (UK) Ltd	MAFF 05022
* Trustan WDG	Du Pont (UK) Ltd	MAFF 05050
615 Mancozeb + *Cyproconazole*		
* Alto Eco	Sandoz Agro Ltd	MAFF 05068
616 Mancozeb + *Dimethomorph*		
* Acrobat MZ	Cyanamid UK	MAFF 06988
* Invader	Cyanamid UK	MAFF 06989
617 Mancozeb + *Metalaxyl*		
* Fubol 58 WP	Ciba Agriculture	MAFF 00927
* Fubol 75 WP	Ciba Agriculture	MAFF 03462
*% Osprey 58 WP	Ciba Agriculture	MAFF 05931
* Osprey 58WP	Ciba Agriculture	MAFF 05717
618 Mancozeb + *Ofurace*		
*% Patafol Plus	Imperial Chemical Industries Plc	MAFF 02808
619 Mancozeb + *Oxadixyl*		
* Recoil	Sandoz Agro Ltd	MAFF 04039
*% Recoil	Schering Agrochemicals Ltd	MAFF 04038
620 Mancozeb + *Propamocarb hydrochloride*		
* Tattoo	AgrEvo UK Crop Protection Ltd	MAFF 07293
* Tattoo	Schering Agrochemicals Ltd	MAFF 06745

* These products are "approved for agricultural use". For further details refer to page vii.
% These products are subject to the staged revocation procedure. For further details refer to page vii.
A These products are approved for use in or near water. For further details refer to page vii.

PROFESSIONAL PRODUCTS: FUNGICIDES

Product Name	Marketing Company	Reg. No.

621 Maneb

	Product Name	Marketing Company	Reg. No.
*	Agrichem Maneb 80	Agrichem (International) Ltd	MAFF 05474
*	Agrichem Maneb Dry Flowable	Agrichem (International) Ltd	MAFF 05515
*%	Amos Maneb 80	Kommer-Brookwick	MAFF 05568
*	Amos Maneb 80	Luxan (UK) Ltd	MAFF 06560
*%	Amos Maneb 800 WP	Kommer-Brookwick	MAFF 05964
*	Ashlade Maneb FL	Ashlade Formulations Ltd	MAFF 06477
*%	BASF Maneb	BASF Plc	MAFF 00207
*%	Campbell's X-Spor SC	J D Campbell & Sons Ltd	MAFF 03252
*	Headland Spirit	Headland Agrochemicals Ltd	MAFF 04548
*	Luxan Maneb 80	Luxan (UK) Ltd	MAFF 06570
*	Luxan Maneb 800 WP	Luxan (UK) Ltd	MAFF 06561
*	Maneb 80	Rohm & Haas (UK) Ltd	MAFF 01276
*%	Manzate	Du Pont (UK) Ltd	MAFF 02436
*	Mazin	Universal Crop Protection Ltd	MAFF 06061
*%	Mazin	Universal Crop Protection Ltd	MAFF 01309
*%	Power Maneb 80	Power Agrichemicals Ltd	MAFF 03236
*	Quadrangle Maneb Flowable	Quadrangle Agrochemicals Ltd	MAFF 06131
*	R H Maneb 80	Rohm & Haas (UK) Ltd	MAFF 01796
*	Stefes Maneb DF	Stefes Plant Protection Ltd	MAFF 06418
*	Trimangol 80	Atochem Agri BV	MAFF 06070
*	Trimangol 80	Elf Atochem Agri BV	MAFF 06871
*%	Trimangol 80	Pennwalt Chemicals Ltd	MAFF 04294
*	Trimangol WDG	Elf Atochem Agri BV	MAFF 06992
*	Tripart Obex	Tripart Farm Chemicals Ltd	MAFF 06130
.*	Unicrop Flowable Maneb	Universal Crop Protection Ltd	MAFF 05546
*%	Unicrop Maneb	Universal Crop Protection Ltd	MAFF 02274
*	Unicrop Maneb 80	Universal Crop Protection Ltd	MAFF 05533
*	Unicrop Maneb 80	Universal Crop Protection Ltd	MAFF 06926
*	Unicrop Manguard DG	Universal Crop Protection Ltd	MAFF 05516
*	X-Spor SC	MTM Agrochemicals Ltd	MAFF 06310

622 Maneb + *Carbendazim*

	Product Name	Marketing Company	Reg. No.
*%	Ashlade Mancarb FL	Ashlade Formulations Ltd	MAFF 02928
*%	Ashlade Mancarb FL	Ashlade Formulations Ltd	MAFF 06217
*%	Campbell's MC Flowable	J D Campbell & Sons Ltd	MAFF 03467
*	Headland Dual	Headland Agrochemicals Ltd	MAFF 03782
*	MC Flowable	MTM Agrochemicals Ltd	MAFF 06295
*	Multi-W FL	Pan Britannica Industries Ltd	MAFF 04131
*	New Squadron	Quadrangle Agrochemicals Ltd	MAFF 06756
*	Protector	Procam Agriculture Ltd	MAFF 07075
*	Squadron	Quadrangle Agrochemicals Ltd	MAFF 05981
*	Tripart Legion	Tripart Farm Chemicals Ltd	MAFF 06113
*%	Tripart Legion	Tripart Farm Chemicals Ltd	MAFF 02997

---

* These products are "approved for agricultural use". For further details refer to page vii.
% These products are subject to the staged revocation procedure. For further details refer to page vii.
A These products are approved for use in or near water. For further details refer to page vii.

PROFESSIONAL PRODUCTS: FUNGICIDES

Product Name	Marketing Company	Reg. No.

623 Maneb + *Carbendazim* + *Chlorothalonil*

*	Ashlade Mancarb Plus	Ashlade Formulations Ltd	MAFF 06222
*%	Ashlade Mancarb Plus	Ashlade Formulations Ltd	MAFF 04440
*	Tripart Victor	Tripart Farm Chemicals Ltd	MAFF 04359

624 Maneb + *Carbendazim* + *Sulphur*

*%	Bolda FL	Farm Protection Ltd	MAFF 03463

625 Maneb + *Carbendazim* + *Tridemorph*

*	Cosmic FL	BASF Plc	MAFF 03473

626 Maneb + *Copper oxychloride* + *Sulphur*

*	Ashlade SMC Flowable	Ashlade Formulations Ltd	MAFF 06494
*%	Ashlade SMC Flowable	Ashlade Formulations Ltd	MAFF 04560
*%	Rearguard	Universal Crop Protection Ltd	MAFF 04938

627 Maneb + *Fentin acetate*

*	Brestan 60	AgrEvo UK Crop Protection Ltd	MAFF 07304
*	Brestan 60	Hoechst UK Ltd	MAFF 00325
*	Brestan 60 SP	AgrEvo UK Crop Protection Ltd	MAFF 07305
*	Brestan 60 SP	Hoechst UK Ltd	MAFF 06042

628 Maneb + *Ferbam* + *Zineb*

*%	Trimanzone PVA	Brian Lewis Agriculture Ltd	MAFF 05289
*	Trimanzone PVA	Elf Atochem Agri BV	MAFF 06877
*	Trimanzone WP	Brian Lewis Agriculture Ltd	MAFF 05860
*	Trimanzone WP	Elf Atochem Agri BV	MAFF 06876

629 Maneb + *Zinc*

*%	Ashlade Maneb FL	Ashlade Formulations Ltd	MAFF 02911
*	Barclay Manzeb Flow	Barclay Chemicals (UK) Ltd	MAFF 05872
*	Manex	Chiltern Farm Chemicals Ltd	MAFF 05731
*	Quadrangle Manex	Quadrangle Ltd	MAFF 03406

630 Maneb + *Zineb*

*%	Trithac	Dow Chemical Co Ltd	MAFF 04241

631 Metalaxyl

*	Apron 350 FS	Ciba Agriculture	MAFF 06726
*%	Apron 350 FS (Without Stickers)	Ciba Agriculture	MAFF 04245
*	Polycote Universal	Seedcote Systems Ltd	MAFF 05942

---

* These products are "approved for agricultural use". For further details refer to page vii.
% These products are subject to the staged revocation procedure. For further details refer to page vii.
A These products are approved for use in or near water. For further details refer to page vii.

PROFESSIONAL PRODUCTS: FUNGICIDES

Product Name	Marketing Company	Reg. No.
632 Metalaxyl + *Carbendazim*		
* Ridomil MBC 60 WP	Ciba Agriculture	MAFF 01804
633 Metalaxyl + *Chlorothalonil*		
*% Folio 575 FW	Ciba-Geigy Agrochemicals	MAFF 04122
* Folio 575 SC	Ciba Agriculture	MAFF 05843
634 Metalaxyl + *Copper oxychloride*		
* Ridomil Plus 50 WP	Ciba Agriculture	MAFF 01803
635 Metalaxyl + *Mancozeb*		
* Fubol 58 WP	Ciba Agriculture	MAFF 00927
* Fubol 75 WP	Ciba Agriculture	MAFF 03462
*% Osprey 58 WP	Ciba Agriculture	MAFF 05931
* Osprey 58WP	Ciba Agriculture	MAFF 05717
636 Metalaxyl + *Thiabendazole*		
* Apron T69 WS	Ciba Agriculture	MAFF 06725
*% Apron T69 WS (Without Stickers)	Ciba Agriculture	MAFF 03123
*% Polycote Select	Ciba Agriculture	MAFF 05943
* Polycote Select	Seedcote Systems Ltd	MAFF 06727
637 Metalaxyl + *Thiabendazole* + *Thiram*		
* Apron Combi 453 FS	Ciba Agriculture	MAFF 04110
* Apron Combi FS	Ciba Agriculture	MAFF 07203
638 Metalaxyl + *Thiram*		
*% Favour 600 FW	Ciba-Geigy Agrochemicals	MAFF 04000
* Favour 600 SC	Ciba Agriculture	MAFF 05842
639 Metoxuron + *Fentin hydroxide*		
*% Endspray	Pan Britannica Industries Ltd	MAFF 02799
640 Myclobutanil		
* RH-3866 ST	DowElanco Ltd	MAFF 06547
* Systhane 40W	Rohm & Haas (UK) Ltd	MAFF 04267
* Systhane 6 Flo	AgrEvo UK Crop Protection Ltd	MAFF 07334
* Systhane 6 Flo	Hoechst UK Ltd	MAFF 03921
* Systhane 6 Flo	T P Whelehan Son & Co Ltd	MAFF 06551
* Systhane 6W	Pan Britannica Industries Ltd	MAFF 04570
* Systhane 6W	Pan Britannica Industries Ltd	MAFF 04571

* These products are "approved for agricultural use". For further details refer to page vii.
% These products are subject to the staged revocation procedure. For further details refer to page vii.
A These products are approved for use in or near water. For further details refer to page vii.

PROFESSIONAL PRODUCTS: FUNGICIDES

Product Name	Marketing Company	Reg. No.

641 Nitrothal-isopropyl + *Zineb-ethylene thiuram disulphide adduct*
*% Pallitop — BASF Plc — MAFF 01533

642 Nuarimol
* Chemtech Nuarimol — Chemtech (Crop Protection) Ltd — MAFF 04517
* Flarepath — DowElanco Ltd — MAFF 05502
* Triminol — DowElanco Ltd — MAFF 05818
*% Triminol — Elanco Products Ltd — MAFF 02467

643 Nuarimol + *Captan*
*% Kapitol — DowElanco Ltd — MAFF 05051
*% Kapitol WDG — DowElanco Ltd — MAFF 05085

644 Octhilinone
* Pancil T — Rohm & Haas (UK) Ltd — MAFF 01540

645 Ofurace + *Mancozeb*
*% Patafol Plus — Imperial Chemical Industries Plc — MAFF 02808

646 Oxadixyl + *Cymoxanil* + *Mancozeb*
* Ripost — Sandoz Agro Ltd — MAFF 04890
* Ripost Pepite — Sandoz Agro Ltd — MAFF 06485
*% Ripost WDG — Sandoz Crop Protection Ltd — MAFF 05053
* Trustan — Du Pont (UK) Ltd — MAFF 05022
* Trustan WDG — Du Pont (UK) Ltd — MAFF 05050

647 Oxadixyl + *Mancozeb*
* Recoil — Sandoz Agro Ltd — MAFF 04039
*% Recoil — Schering Agrochemicals Ltd — MAFF 04038

648 Oxadixyl + *Thiabendazole* + *Thiram*
*% Leap — DowElanco Ltd — MAFF 05157
* Leap — ICI Agrochemicals — MAFF 05158
* Leap — Sandoz Agro Ltd — MAFF 05156
* Leap — Zeneca Crop Protection — MAFF 06397

649 Oxycarboxin
* Plantvax 20 — Uniroyal Chemical Ltd — MAFF 01600
* Plantvax 75 — Applied Horticulture Ltd — MAFF 01601
* Ringmaster — Rhone-Poulenc Agriculture Ltd — MAFF 05334

---

* These products are "approved for agricultural use". For further details refer to page vii.
% These products are subject to the staged revocation procedure. For further details refer to page vii.
A These products are approved for use in or near water. For further details refer to page vii.

PROFESSIONAL PRODUCTS: FUNGICIDES

Product Name	Marketing Company	Reg. No.
650 Parahydroxyphenylsalicylamide		
*% Fumispore	Laboratoire de Chimie et de Biologie	MAFF 04745
651 Penconazole		
* Topas 100EC	Ciba Agriculture	MAFF 03231
652 Penconazole + *Captan*		
* Topas C 50WP	Ciba Agriculture	MAFF 03232
653 Penconazole + *Dithianon*		
*% Topas D275 FW	Ciba-Geigy Agrochemicals	MAFF 04915
* Topas D275 SC	Ciba Agriculture	MAFF 05855
654 Pencycuron		
* Monceren	Bayer UK Ltd	MAFF 04882
* Monceren DS	Bayer Plc	MAFF 04160
* Monceren FS	Bayer Plc	MAFF 04907
655 Pencycuron + *Imazalil*		
* Monceren IM	Bayer Plc	MAFF 06259
* Monceren IM Flowable	Bayer Plc	MAFF 06731
656 Permethrin + *Thiram*		
* Combinex	Pan Britannica Industries Ltd	MAFF 00562
657 Potassium sorbate + *Sodium metabisulphite* + *Sodium propionate*		
* Brimstone Plus	Mandops Ltd	MAFF 03695
658 Prochloraz		
* Barclay Eyetak	Barclay Chemicals (UK) Ltd	MAFF 06813
* Fisons Octave	Fisons Plc	MAFF 03416
* Octave	AgrEvo UK Crop Protection Ltd	MAFF 07267
* Octave	Schering Agrochemicals Ltd	MAFF 03989
* Power Prochloraz 400	Power Agrichemicals Ltd	MAFF 04732
* Prelude 20 LF	Ciba-Geigy Agriculture	MAFF 06083
* Prelude 20LF	AgrEvo UK Crop Protection Ltd	MAFF 07269
* Prelude 20LF	Agrichem Ltd	MAFF 04371
* Prelude 20LF	Schering Agrochemicals Ltd	MAFF 04370
* Sporgon 50 WP	Darmycel (UK)	MAFF 03829
* Sportak	AgrEvo UK Crop Protection Ltd	MAFF 07285
* Sportak	Schering Agrochemicals Ltd	MAFF 03871
* Sportak 45	AgrEvo UK Crop Protection Ltd	MAFF 07286

* These products are "approved for agricultural use". For further details refer to page vii.
% These products are subject to the staged revocation procedure. For further details refer to page vii.
A These products are approved for use in or near water. For further details refer to page vii.

PROFESSIONAL PRODUCTS: FUNGICIDES

Product Name	Marketing Company	Reg. No.
658 Prochloraz—*continued*		
* Sportak 45	Schering Agrochemicals Ltd	MAFF 03815
* Sportak 45 HF	AgrEvo UK Crop Protection Ltd	MAFF 07287
* Sportak 45 HF	Schering Agriculture	MAFF 05724
* Sportak Focus HF	AgrEvo UK Crop Protection Ltd	MAFF 07288
* Sportak Focus HF	Schering Agrochemicals Ltd	MAFF 05961
* Sportak HF	AgrEvo UK Crop Protection Ltd	MAFF 07289
* Sportak HF	Schering Agrochemicals Ltd	MAFF 05960
* Sportak Sierra HF	AgrEvo UK Crop Protection Ltd	MAFF 07290
* Sportak Sierra HF	Schering Agrochemicals Ltd	MAFF 05913
659 Prochloraz + *Carbendazim*		
* Hobby	Ciba Agriculture	MAFF 06309
* Hobby	Ciba Agriculture	MAFF 06437
* Sportak Alpha	AgrEvo UK Crop Protection Ltd	MAFF 07222
* Sportak Alpha	Schering Agrochemicals Ltd	MAFF 03872
660 Prochloraz + *Cyproconazole*		
* Sportak Delta 460	AgrEvo UK Crop Protection Ltd	MAFF 07224
* Sportak Delta 460	Schering Agrochemicals Ltd	MAFF 05107
* Tiptor	AgrEvo UK Crop Protection Ltd	MAFF 07294
* Tiptor	AgrEvo UK Crop Protection Ltd	MAFF 07295
* Tiptor	Cyanamid of GB Ltd	MAFF 05106
* Tiptor	Sandoz Agro Ltd	MAFF 05105
* Tiptor	Schering Agrochemicals Ltd	MAFF 05831
* Tiptor	Schering Agrochemicals Ltd	MAFF 06971
661 Prochloraz + *Fenpropimorph*		
* Jester	Ciba Agriculture	MAFF 06586
* Sprint	AgrEvo UK Crop Protection Ltd	MAFF 07291
* Sprint	Schering Agrochemicals Ltd	MAFF 03986
* Sprint HF	AgrEvo UK Crop Protection Ltd	MAFF 07292
* Sprint HF	Schering Agrochemicals Ltd	MAFF 06490
662 Propamocarb hydrochloride		
* Advocate	AgrEvo UK Crop Protection Ltd	MAFF 07241
* Advocate	Schering Agrochemicals Ltd	MAFF 06539
* Fisons Filex	Fisons Plc	MAFF 00869
663 Propamocarb hydrochloride + *Mancozeb*		
* Tattoo	AgrEvo UK Crop Protection Ltd	MAFF 07293
* Tattoo	Schering Agrochemicals Ltd	MAFF 06745

* These products are "approved for agricultural use". For further details refer to page vii.
% These products are subject to the staged revocation procedure. For further details refer to page vii.
A These products are approved for use in or near water. For further details refer to page vii.

PROFESSIONAL PRODUCTS: FUNGICIDES

Product Name	Marketing Company	Reg. No.
664 Propiconazole		
* Clayton Propiconazole	Clayton Plant Protection (UK) Ltd	MAFF 06415
* Landgold Propiconazole	Landgold & Co Ltd	MAFF 06291
* Mantis 250 EC	Ciba Agriculture	MAFF 06240
* Radar	Farm Protection	MAFF 03000
*% Radar	Imperial Chemical Industries Plc	MAFF 01683
* Radar	Zeneca Crop Protection	MAFF 06678
* Radar	Zeneca Crop Protection	MAFF 06747
* Standon Propiconazole	Standon Chemicals Ltd	MAFF 07037
* Stefes Restore	Stefes Plant Protection Ltd	MAFF 06267
* Tilt 250 EC	Ciba Agriculture	MAFF 02138
* Top Farm Propiconazole 250	Top Farm Formulations Ltd	MAFF 05938
665 Propiconazole + *Carbendazim*		
* Hispor 45WP	Ciba Agriculture	MAFF 01050
* STR 301	Ciba-Geigy Agriculture	MAFF 06288
* Sparkle 45 WP	Ciba Agriculture	MAFF 04968
666 Propiconazole + *Chlorothalonil*		
*% Sambarin 312.5 FW	Ciba-Geigy Agrochemicals	MAFF 04685
* Sambarin 312.5 SC	Ciba Agriculture	MAFF 05809
* Sambarin TP	Ciba Agriculture	MAFF 04094
667 Propiconazole + *Fenpropidin*		
* Legend	ICI Agrochemicals	MAFF 04978
* Legend	Zeneca Crop Protection	MAFF 06682
* Opal	Zeneca Crop Protection	MAFF 06938
* Sheen 550 EC	Ciba Agriculture	MAFF 05980
* Zulu 550 EC	Ciba Agriculture	MAFF 06594
* Zulu 550 EC	Ciba Agriculture	MAFF 06937
668 Propiconazole + *Fenpropimorph*		
* Decade 500 EC	Ciba Agriculture	MAFF 05757
*% Decade 500 EC	Ciba-Geigy Agrochemicals	MAFF 05711
* Glint 500 EC	Ciba Agriculture	MAFF 04126
* Mantle 425 EC	Ciba Agriculture	MAFF 05715
669 Propiconazole + *Tridemorph*		
* Tilt Turbo 475 EC	Ciba Agriculture	MAFF 03476
670 Propineb		
* Antracol	Bayer Plc	MAFF 00104

* These products are "approved for agricultural use". For further details refer to page vii.
% These products are subject to the staged revocation procedure. For further details refer to page vii.
A These products are approved for use in or near water. For further details refer to page vii.

PROFESSIONAL PRODUCTS: FUNGICIDES

Product Name	Marketing Company	Reg. No.
671 Pyrazophos		
* Afugan	AgrEvo UK Crop Protection Ltd	MAFF 07301
* Afugan	Hoechst UK Ltd	MAFF 00037
* Missile	AgrEvo UK Crop Protection Ltd	MAFF 07153
* Missile	Hoechst UK Ltd	MAFF 03811
672 Pyrifenox		
* Dorado	Imperial Chemical Industries Plc	MAFF 05700
* Dorado	Zeneca Crop Protection	MAFF 06657
* SL 471 200 EC	Ciba Agriculture	MAFF 06035
673 Quinomethionate		
* Morestan	Bayer Plc	MAFF 01376
674 Quintozene		
* Botrilex	Imperial Chemical Industries Plc	MAFF 00308
*% Brabant PCNB 20 % Dust	Bos Chemicals Ltd	MAFF 00313
*% Brabant PCNB 20 % Dust	Uniroyal Chemical Ltd	MAFF 05475
* Quintozene Wettable Powder	Rhone-Poulenc Environmental Products	MAFF 05404
* Terraclor 20 D	Uniroyal Chemical Ltd	MAFF 06578
675 Sodium metabisulphite + *Potassium sorbate* + *Sodium propionate*		
* Brimstone Plus	Mandops Ltd	MAFF 03695
676 Sodium propionate + *Potassium sorbate* + *Sodium metabisulphite*		
* Brimstone Plus	Mandops Ltd	MAFF 03695
677 Sulphur		
*% Amos Micro-Sulphur	Kommer-Brookwick	MAFF 05971
*% Ashlade Sulphur FL	Ashlade Formulations Ltd	MAFF 02812
* Ashlade Sulphur FL	Ashlade Formulations Ltd	MAFF 06478
* Atlas Sulphur 80 FL	Atlas Interlates Ltd	MAFF 03802
* Headland Sulphur	Headland Agrochemicals Ltd	MAFF 03714
* Kumulus DF	BASF Plc	MAFF 04707
* Kumulus S	BASF Plc	MAFF 01170
* Luxan Micro-Sulphur	Luxan (UK) Ltd	MAFF 06565
* MTM Sulphur Flowable	MTM Agrochemicals Ltd	MAFF 05312
* Microsul Flowable Sulphur	Stoller Chemical Ltd	MAFF 03907
* Microthiol Special	Elf Atochem Agri SA	MAFF 06268
* MSS Sulphur 80	Mirfield Sales Services Ltd	MAFF 05752
* MSS Sulphur 80 WP	Mirfield Sales Services Ltd	MAFF 03225
* Power Micro-Sulphur	Power Agrichemicals Ltd	MAFF 02721

* These products are "approved for agricultural use". For further details refer to page vii.
% These products are subject to the staged revocation procedure. For further details refer to page vii.
A These products are approved for use in or near water. For further details refer to page vii.

PROFESSIONAL PRODUCTS: FUNGICIDES

Product Name	Marketing Company	Reg. No.
677 Sulphur—*continued*		
* Solfa	Atlas Interlates Ltd	MAFF 06959
*% Solfa	Farm Protection Ltd	MAFF 03529
* Stoller Flowable Sulphur	Stoller Chemical Ltd	MAFF 03760
* Thiovit	Pan Britannica Industries Ltd	MAFF 02125
* Thiovit	Sandoz Agro Ltd	MAFF 05572
* Tripart Imber	Tripart Farm Chemicals Ltd	MAFF 04050
678 Sulphur + *Carbendazim* + *Maneb*		
*% Bolda FL	Farm Protection Ltd	MAFF 03463
679 Sulphur + *Copper oxychloride* + *Maneb*		
* Ashlade SMC Flowable	Ashlade Formulations Ltd	MAFF 06494
*% Ashlade SMC Flowable	Ashlade Formulations Ltd	MAFF 04560
*% Rearguard	Universal Crop Protection Ltd	MAFF 04938
680 Sulphur + *Copper sulphate*		
* Top Cop	Stoller Chemical Ltd	MAFF 04553
682 Tar oils		
* Sterilite Tar Oil Winter Wash 60 % Stock Emulsion	Coventry Chemicals Ltd	MAFF 05061
* Sterilite Tar Oil Winter Wash 80 % Miscible Quality	Coventry Chemicals Ltd	MAFF 05062
683 Tebuconazole		
* Clayton Tebucon	Clayton Plant Protection (UK) Ltd	MAFF 07045
* Folicur	Bayer Plc	MAFF 06386
* Folicur EC	Bayer Plc	MAFF 06385
* Raxil	Bayer Plc	MAFF 06460
684 Tebuconazole + *Triadimenol*		
* Garnet	Bayer Plc	MAFF 06391
* Ruby	Bayer Plc	MAFF 06389
* Silvacur	Bayer Plc	MAFF 06387
685 Tebuconazole + *Triazoxide*		
* Raxil S	Bayer Plc	MAFF 06974
686 Tebuconazole + *Tridemorph*		
* Allicur	Bayer Plc	MAFF 06468
* BAS 91580F	BASF Plc	MAFF 06469
* Bayer UK 300	Bayer Plc	MAFF 06788
* BUK 92300F	BASF Plc	MAFF 06789

*   These products are "approved for agricultural use". For further details refer to page vii.
%   These products are subject to the staged revocation procedure. For further details refer to page vii.
A   These products are approved for use in or near water. For further details refer to page vii.

PROFESSIONAL PRODUCTS: FUNGICIDES

	Product Name	Marketing Company	Reg. No.

687 Tecnazene

	Product Name	Marketing Company	Reg. No.
*	Atlas Tecgran 100	Atlas Interlates Ltd	MAFF 05574
*	Atlas Tecnazene 6% Dust	Atlas Interlates Ltd	MAFF 06351
*	Bygran F	Wheatley Chemical Co Ltd	MAFF 00365
	Bygran S	Wheatley Chemical Co Ltd	MAFF 00366
	Fusarex 10% Granules	Imperial Chemical Industries Plc	MAFF 00955
	Fusarex 10% Granules	Zeneca Crop Protection	MAFF 06668
	Fusarex 6% Dust	Imperial Chemical Industries Plc	MAFF 00954
	Fusarex 6% Dust	Zeneca Crop Protection	MAFF 06667
	Hickstor 3	Hickson & Welch Ltd	MAFF 03105
*	Hickstor 5	Hickson & Welch Ltd	MAFF 03180
	Hickstor 6	Hickson & Welch Ltd	MAFF 03106
*	Hickstor 10	Hickson & Welch Ltd	MAFF 03121
	Hortag Tecnazene 10% Granules	Hortag Chemicals Ltd	MAFF 03966
	Hortag Tecnazene Double Dust	Hortag Chemicals Ltd	MAFF 01072
	Hortag Tecnazene Potato Dust	Hortag Chemicals Ltd	MAFF 01074
	Hortag Tecnazene Potato Granules	Hortag Chemicals Ltd	MAFF 01075
*	Hystore 10	Agrichem (International) Ltd	MAFF 03581
	Hytec	Agrichem Ltd	MAFF 01099
*	Hytec 6	Agrichem Ltd	MAFF 03580
*	New Hickstor 6	Hickson & Welch Ltd	MAFF 04221
	Nebulin	Wheatley Chemical Co Ltd	MAFF 01469
	New Hystore	Agrichem Ltd	MAFF 01485
%	Tripart Arena 10G	Tripart Farm Chemicals Ltd	MAFF 03122
%	Tripart Arena 3	Tripart Farm Chemicals Ltd	MAFF 03818
	Tripart Arena 10 G	Tripart Farm Chemicals Ltd	MAFF 05603
	Tripart Arena 3	Tripart Farm Chemicals Ltd	MAFF 05605
*	Tripart Arena 5G	Tripart Farm Chemicals Ltd	MAFF 05604
*%	Tripart Arena 5G	Tripart Farm Chemicals Ltd	MAFF 04441
%	Tripart Arena 6%	Tripart Farm Chemicals Ltd	MAFF 02574
*	Tripart New Arena 6	Tripart Farm Chemicals Ltd	MAFF 05813

688 Tecnazene + *Carbendazim*

	Product Name	Marketing Company	Reg. No.
%	Arena Plus	Tripart Farm Chemicals Ltd	MAFF 04258
*	Hickstor 6 Plus MBC	Hickson & Welch Ltd	MAFF 04176
%	Hortag Carbotec	Hortag Chemicals Ltd	MAFF 04135
*	Hortag Tecnacarb	Avon Packers Ltd	MAFF 02929
*	New Arena Plus	Hickson & Welch Ltd	MAFF 04598
*	New Hickstor 6 Plus MBC	Hickson & Welch Ltd	MAFF 04599
	Tripart Arena Plus	Hickson & Welch Ltd	MAFF 05602

689 Tecnazene + *Thiabendazole*

Product Name	Marketing Company	Reg. No.
Hortag Tecnazene Plus	Hortag Chemicals Ltd	MAFF 01073
Hytec Super	Agrichem (International) Ltd	MAFF 01100

* These products are "approved for agricultural use". For further details refer to page vii.
% These products are subject to the staged revocation procedure. For further details refer to page vii.
A These products are approved for use in or near water. For further details refer to page vii.

PROFESSIONAL PRODUCTS: FUNGICIDES

	Product Name	Marketing Company	Reg. No.

689 Tecnazene + *Thiabendazole*—continued

	New Arena TBZ 6	Tripart Farm Chemicals Ltd	MAFF 05606
%	New Arena TBZ 6	Tripart Farm Chemicals Ltd	MAFF 04906
%	New Hickstor TBZ 6	Hickson & Welch Ltd	MAFF 04905
	Storite SS	MSD Agvet	MAFF 02034

690 Thiabendazole

*	Ceratotect	MSD Agvet	MAFF 03554
	Hykeep	Agrichem (International) Ltd	MAFF 06744
*	Hymush	Agrichem Ltd	MAFF 01092
*	Storite Clear Liquid	MSD Agvet	MAFF 02032
*	Storite Flowable	MSD Agvet	MAFF 02033
*%	Tecto Dust	MSD Agvet	MAFF 04985
*	Tecto Flowable Turf Fungicide	Synchemicals Ltd	MAFF 02094
*	Tecto Flowable Turf Fungicide	Vitax Ltd	MAFF 06273
*%	Tecto Wettable Powder	MSD Agvet	MAFF 02095

691 Thiabendazole + *Carboxin*

*	Cerevax	Imperial Chemical Industries Plc	MAFF 02500
*	Cerevax	Zeneca Crop Protection	MAFF 06641
*	Vitavax	Uniroyal Chemical Ltd	MAFF 07205

692 Thiabendazole + *Carboxin* + *Imazalil*

*	Cerevax Extra	Imperial Chemical Industries Plc	MAFF 02501
*	Cerevax Extra	Zeneca Crop Protection	MAFF 06642
*	Vitaflo Extra	Uniroyal Chemical Ltd	MAFF 07394
*	Vitavax Extra	Uniroyal Chemical Ltd	MAFF 07334

693 Thiabendazole + *Ethirimol* + *Flutriafol*

*	Ferrax	Bayer UK Ltd	MAFF 05284
*	Ferrax	DowElanco Ltd	MAFF 04917
*	Ferrax	Imperial Chemical Industries Plc	MAFF 02827
*	Ferrax	Zeneca Crop Protection	MAFF 06662

694 Thiabendazole + *Ethirimol* + *Flutriafol* + *Imazalil*

*	Ferrax IM	Imperial Chemical Industries Plc	MAFF 05142
*	Ferrax IM	Zeneca Crop Protection	MAFF 06663

695 Thiabendazole + *Flutriafol*

*	Vincit L	Imperial Chemical Industries Plc	MAFF 04322

696 Thiabendazole + *Flutriafol* + *Imazalil*

*	Vincit IM	Zeneca Crop Protection	MAFF 06380

---

* These products are "approved for agricultural use". For further details refer to page vii.
% These products are subject to the staged revocation procedure. For further details refer to page vii.
A These products are approved for use in or near water. For further details refer to page vii.

PROFESSIONAL PRODUCTS: FUNGICIDES

Product Name	Marketing Company	Reg. No.
697 Thiabendazole + *Imazalil*		
* Extratect Flowable	MSD Agvet	MAFF 05507
698 Thiabendazole + *Lindane* + *Thiram*		
* Hysede FL	Agrichem Ltd	MAFF 02863
699 Thiabendazole + *Metalaxyl*		
* Apron T69 WS	Ciba Agriculture	MAFF 06725
*% Apron T69 WS (Without Stickers)	Ciba Agriculture	MAFF 03123
*% Polycote Select	Ciba Agriculture	MAFF 05943
* Polycote Select	Seedcote Systems Ltd	MAFF 06727
700 Thiabendazole + *Metalaxyl* + *Thiram*		
* Apron Combi 453 FS	Ciba Agriculture	MAFF 04110
* Apron Combi FS	Ciba Agriculture	MAFF 07203
701 Thiabendazole + *Oxadixyl* + *Thiram*		
*% Leap	DowElanco Ltd	MAFF 05157
* Leap	ICI Agrochemicals	MAFF 05158
* Leap	Sandoz Agro Ltd	MAFF 05156
* Leap	Zeneca Crop Protection	MAFF 06397
702 Thiabendazole + *Tecnazene*		
Hortag Tecnazene Plus	Hortag Chemicals Ltd	MAFF 01073
Hytec Super	Agrichem (International) Ltd	MAFF 01100
New Arena TBZ 6	Tripart Farm Chemicals Ltd	MAFF 05606
% New Arena TBZ 6	Tripart Farm Chemicals Ltd	MAFF 04906
% New Hickstor TBZ 6	Hickson & Welch Ltd	MAFF 04905
Storite SS	MSD Agvet	MAFF 02034
703 Thiabendazole + *Thiram*		
*% Ascot 480 FS	Ciba Agriculture	MAFF 04047
* HY-TL	Agrichem (International) Ltd	MAFF 06246
* HY-VIC	Agrichem (International) Ltd	MAFF 06247
*% HY-VIC	Agrichem Ltd	MAFF 02858
*% Hy-TL	Agrichem Ltd	MAFF 03352
704 Thiophanate-methyl		
*% CDA Mildothane	Rhone-Poulenc Environmental Products	MAFF 04154
* Cercobin Liquid	Rhone-Poulenc Agriculture	MAFF 06188
*% Cercobin Liquid	Rhone-Poulenc Crop Protection	MAFF 05139
* Mildothane Liquid	Dow Chemical Co Ltd	MAFF 04345

* These products are "approved for agricultural use". For further details refer to page vii.
% These products are subject to the staged revocation procedure. For further details refer to page vii.
A These products are approved for use in or near water. For further details refer to page vii.

PROFESSIONAL PRODUCTS: FUNGICIDES

Product Name	Marketing Company	Reg. No.

704 Thiophanate-methyl—*continued*

* Mildothane Liquid	DowElanco Ltd	MAFF 05244
* Mildothane Liquid	Embetec Crop Protection	MAFF 05722
* Mildothane Liquid	Hortichem Ltd	MAFF 04552
* Mildothane Liquid	Rhone-Poulenc Agriculture	MAFF 06211
* Mildothane Turf Liquid	Rhone-Poulenc Environmental Products	MAFF 05331

705 Thiophanate-methyl + *Iprodione*

* Compass	Rhone-Poulenc Agriculture	MAFF 06190
* Compass	Rhone-Poulenc Agriculture Ltd	MAFF 04580
* FR 1433	Rhone-Poulenc Crop Protection	MAFF 04994

706 Thiram

* Agrichem Flowable Thiram	Agrichem (International) Ltd	MAFF 06245
* Hy-Flo	Agrichem (International) Ltd	MAFF 04637
* Robinson's Thiram 60	Agrichem (International) Ltd	MAFF 04638
* Unicrop Thianosan DG	Universal Crop Protection Ltd	MAFF 05454

707 Thiram + *Carbendazim*

* DAL-CT	Agrichem (International) Ltd	MAFF 06248
*% DAL CT	Agrichem Ltd	MAFF 04029

708 Thiram + *Carboxin* + *Lindane*

*% Vitavax RS	DowElanco Ltd	MAFF 02310
* Vitavax RS	Uniroyal Chemical Ltd	MAFF 06029

709 Thiram + *Fenpropimorph* + *Lindane*

* Lindex Plus FS Seed Treatment	DowElanco Ltd	MAFF 03934

710 Thiram + *Lindane*

* Hydraguard	Agrichem Ltd	MAFF 03278

711 Thiram + *Lindane* + *Thiabendazole*

* Hysede FL	Agrichem Ltd	MAFF 02863

712 Thiram + *Metalaxyl*

*% Favour 600 FW	Ciba-Geigy Agrochemicals	MAFF 04000
* Favour 600 SC	Ciba Agriculture	MAFF 05842

713 Thiram + *Metalaxyl* + *Thiabendazole*

* Apron Combi 453 FS	Ciba Agriculture	MAFF 04110
* Apron Combi FS	Ciba Agriculture	MAFF 07203

* These products are "approved for agricultural use". For further details refer to page vii.
% These products are subject to the staged revocation procedure. For further details refer to page vii.
A These products are approved for use in or near water. For further details refer to page vii.

PROFESSIONAL PRODUCTS: FUNGICIDES

Product Name	Marketing Company	Reg. No.

714 Thiram + *Oxadixyl* + *Thiabendazole*

*%	Leap	DowElanco Ltd	MAFF 05157
*	Leap	ICI Agrochemicals	MAFF 05158
*	Leap	Sandoz Agro Ltd	MAFF 05156
*	Leap	Zeneca Crop Protection	MAFF 06397

715 Thiram + *Permethrin*

*	Combinex	Pan Britannica Industries Ltd	MAFF 00562

716 Thiram + *Thiabendazole*

*%	Ascot 480 FS	Ciba Agriculture	MAFF 04047
*	HY-TL	Agrichem (International) Ltd	MAFF 06246
*	HY-VIC	Agrichem (International) Ltd	MAFF 06247
*%	HY-VIC	Agrichem Ltd	MAFF 02858
*%	Hy-TL	Agrichem Ltd	MAFF 03352

717 Tolclofos-methyl

*	Fisons Basilex	Fisons Plc	MAFF 02847
*	Rizolex	AgrEvo UK Crop Protection Ltd	MAFF 07271
*	Rizolex	Schering Agrochemicals Ltd	MAFF 03826
*	Rizolex 50 WP	AgrEvo UK Crop Protection Ltd	MAFF 07272
*	Rizolex 50 WP	Schering Agrochemicals Ltd	MAFF 04452
*	Rizolex Flowable	AgrEvo UK Crop Protection Ltd	MAFF 07273
*	Rizolex Flowable	Schering Agrochemicals Ltd	MAFF 03719

718 Triadimefon

*	100 Plus	Dalgety Agriculture Ltd	MAFF 05112
*	Bayleton	Bayer Plc	MAFF 00221
*	Bayleton 5	Bayer UK Ltd	MAFF 00222
*	Landgold Triadimefon 25	Landgold & Co Ltd	MAFF 06139
*	Standon Triadimefon 25	Standon Chemicals Ltd	MAFF 05673

719 Triadimefon + *Carbendazim*

*%	200 Plus	Dalgety Agriculture Ltd	MAFF 05025
*	200-Plus	Bayer Plc	MAFF 05070
*	Bayleton BM	Bayer Plc	MAFF 00223

720 Triadimenol

*	Bayfidan	Bayer Plc	MAFF 02672
*	Spinnaker	Cyanamid Of GB Ltd	MAFF 07023
*	Spinnaker	Shell Chemicals UK Ltd	MAFF 03941

---

* These products are "approved for agricultural use". For further details refer to page vii.
% These products are subject to the staged revocation procedure. For further details refer to page vii.
A These products are approved for use in or near water. For further details refer to page vii.

PROFESSIONAL PRODUCTS: FUNGICIDES

Product Name	Marketing Company	Reg. No.
721 Triadimenol + *Bitertanol* + *Fuberidazole*		
* Cereline	Bayer Plc	MAFF 07239
* Cereline Flowable	Bayer Plc	MAFF 06984
722 Triadimenol + *Carbendazim*		
* Tiara	Bayer Plc	MAFF 05278
723 Triadimenol + *Ethirimol* + *Fuberidazole*		
* Bay UK 292	Bayer Plc	MAFF 04335
* Militan	ICI Agrochemicals	MAFF 06095
* Militan	Zeneca Crop Protection	MAFF 06711
724 Triadimenol + *Fuberidazole*		
* Baytan	Bayer Plc	MAFF 00225
* Baytan	Imperial Chemical Industries Plc	MAFF 03946
* Baytan	Zeneca Crop Protection	MAFF 06636
* Baytan Flowable	Bayer Plc	MAFF 02593
* Baytan Flowable	Imperial Chemical Industries Plc	MAFF 03845
* Baytan Flowable	Zeneca Crop Protection	MAFF 06635
725 Triadimenol + *Fuberidazole* + *Imazalil*		
* Baytan IM	Bayer Plc	MAFF 00226
726 Triadimenol + *Tebuconazole*		
* Garnet	Bayer Plc	MAFF 06391
* Ruby	Bayer Plc	MAFF 06389
* Silvacur	Bayer Plc	MAFF 06387
727 Triadimenol + *Tridemorph*		
* Dorin	Bayer Plc	MAFF 03292
* Jasper	BASF Plc	MAFF 06044
728 Triazoxide + *Tebuconazole*		
* Raxil S	Bayer Plc	MAFF 06974
729 Tridemorph		
* Calixin	BASF Plc	MAFF 00369
*% Ringer	Pan Britannica Industries Ltd	MAFF 01557
* Standon Tridemorph 750	Standon Chemicals Ltd	MAFF 05667
730 Tridemorph + *Carbendazim* + *Maneb*		
* Cosmic FL	BASF Plc	MAFF 03473

---

* These products are "approved for agricultural use". For further details refer to page vii.
% These products are subject to the staged revocation procedure. For further details refer to page vii.
A These products are approved for use in or near water. For further details refer to page vii.

PROFESSIONAL PRODUCTS: FUNGICIDES

Product Name	Marketing Company	Reg. No.
731 Tridemorph + *Cyproconazole*		
* Alto Major	Sandoz Agro Ltd	MAFF 06979
* Moot	Sandoz Agro Ltd	MAFF 06990
* San 735	Sandoz Agro Ltd	MAFF 06991
732 Tridemorph + *Epoxiconazole*		
* Ensign	BASF Plc	MAFF 07364
* Opus Plus	BASF Plc	MAFF 07363
733 Tridemorph + *Fenpropimorph*		
* BAS 46402F	BASF Plc	MAFF 03313
* Gemini	BASF Plc	MAFF 05684
734 Tridemorph + *Fenpropimorph* + *Flusilazole*		
* Bingo	BASF Plc	MAFF 06920
* DUK 51	Du Pont (UK) Ltd	MAFF 06764
735 Tridemorph + *Flusilazole*		
* Fusion	Du Pont (UK) Ltd	MAFF 04908
* Meld	BASF Plc	MAFF 04914
736 Tridemorph + *Propiconazole*		
* Tilt Turbo 475 EC	Ciba Agriculture	MAFF 03476
737 Tridemorph + *Tebuconazole*		
* Allicur	Bayer Plc	MAFF 06468
* BAS 91580F	BASF Plc	MAFF 06469
* Bayer UK 300	Bayer Plc	MAFF 06788
* BUK 92300F	BASF Plc	MAFF 06789
738 Tridemorph + *Triadimenol*		
* Dorin	Bayer Plc	MAFF 03292
* Jasper	BASF Plc	MAFF 06044
739 Triforine		
* Fairy Ring Destroyer	Vitax Ltd	MAFF 05541
* Funginex	Vitax Ltd	MAFF 05824
* Saprol	Cyanamid Of GB Ltd	MAFF 07016
* Saprol	Shell Chemicals UK Ltd	MAFF 01863
* Triforine Liquid Seed Treatment	Cyanamid Of GB Ltd	MAFF 07018
* Triforine Liquid Seed Treatment	Shell Chemicals UK Ltd	MAFF 02177

* These products are "approved for agricultural use". For further details refer to page vii.
% These products are subject to the staged revocation procedure. For further details refer to page vii.
A These products are approved for use in or near water. For further details refer to page vii.

PROFESSIONAL PRODUCTS: FUNGICIDES

Product Name	Marketing Company	Reg. No.
740 Triforine + *Bupirimate*		
* Nimrod T	Imperial Chemical Industries Plc	MAFF 01499
* Nimrod T	Zeneca Professional Products	MAFF 06859
741 Vinclozolin		
* BASF Turf Protectant Fungicide	BASF Plc	MAFF 05457
* Mascot Contact Turf Fungicide	Rigby Taylor Ltd	MAFF 02711
*% Ronilan	BASF Plc	MAFF 01821
* Ronilan DF	BASF Plc	MAFF 04456
* Ronilan FL	BASF Plc	MAFF 02960
742 Zinc + *Maneb*		
*% Ashlade Maneb FL	Ashlade Formulations Ltd	MAFF 02911
* Barclay Manzeb Flow	Barclay Chemicals (UK) Ltd	MAFF 05872
* Manex	Chiltern Farm Chemicals Ltd	MAFF 05731
* Quadrangle Manex	Quadrangle Ltd	MAFF 03406
743 Zineb		
* Hortag Zineb Wettable	Hortag Chemicals Ltd	MAFF 03752
* Tritoftorol	Elf Atochem Agri BV	MAFF 06872
*% Tritoftorol	Pennwalt Chemicals Ltd	MAFF 04290
*% Turbair Zineb	Pan Britannica Industries Ltd	MAFF 02251
* Unicrop Zineb	Universal Crop Protection Ltd	MAFF 02279
744 Zineb + *Ferbam* + *Maneb*		
*% Trimanzone PVA	Brian Lewis Agriculture Ltd	MAFF 05289
* Trimanzone PVA	Elf Atochem Agri BV	MAFF 06877
* Trimanzone WP	Brian Lewis Agriculture Ltd	MAFF 05860
* Trimanzone WP	Elf Atochem Agri BV	MAFF 06876
745 Zineb + *Maneb*		
*% Trithac	Dow Chemical Co Ltd	MAFF 04241
746 Zineb-ethylene thiuram disulphide adduct		
* Polyram	BASF Plc	MAFF 02795
747 Zineb-ethylene thiuram disulphide adduct + *Nitrothal-isopropyl*		
*% Pallitop	BASF Plc	MAFF 01533

---

* These products are "approved for agricultural use". For further details refer to page vii.
% These products are subject to the staged revocation procedure. For further details refer to page vii.
A These products are approved for use in or near water. For further details refer to page vii.

PROFESSIONAL PRODUCTS: INSECTICIDES

Product Name   *Marketing Company*   *Reg. No.*

1.3 Insecticides
including acaricides and nematicides

748 Abamectin
*	Dynamec	MSD Agvet	MAFF 06804
*%	Dynamec	MSD Agvet	MAFF 06078

749 Aldicarb
*	Landgold Aldicarb 10G	Landgold & Co Ltd	MAFF 06036
*	Standon Aldicarb 10 G	Standon Chemicals Ltd	MAFF 05915
*	Temik 10G	Rhone-Poulenc Agriculture	MAFF 04339
*	Temik 10G	Rhone-Poulenc Agriculture	MAFF 06210
*	Temik 10G	Rhone-Poulenc Agriculture Ltd	MAFF 05422

750 Aldicarb + *Lindane*
*	Sentry	Embetec Crop Protection	MAFF 06209
*%	Sentry	Rhone-Poulenc Agriculture Ltd	MAFF 05421

751 Alphacypermethrin
*	Acquit	Du Pont (UK) Ltd	MAFF 07000
*	Acquit	Shell Chemicals UK Ltd	MAFF 05553
*	Apex	Shell Chemicals UK Ltd	MAFF 06154
*	Clayton Alpha-Cyper	Clayton Plant Protection (UK) Ltd	MAFF 07065
*	Contest	Cyanamid of GB Ltd	MAFF 07001
*	Fastac	Cyanamid Of GB Ltd	MAFF 07008
*	Fastac	Shell Chemicals UK Ltd	MAFF 02659
*	Stefes Alphacypermethrin	Stefes Plant Protection Ltd	MAFF 05800

752 Aluminium phosphide
	Fumitoxin	Igrox Chemicals Ltd	MAFF 04207
	Phostek	Killgerm Chemicals Ltd	MAFF 05115
	Phostoxin I	Rentokil Ltd	MAFF 05694
	Rentokil Phostoxin	Rentokil Ltd	MAFF 01775

753 Amitraz
*	Mitac 20	AgrEvo UK Crop Protection Ltd	MAFF 07265
*	Mitac 20	Schering Agrochemicals Ltd	MAFF 03870
%	Taktic	Smith Kline Animal Health Ltd	MAFF 04026
	Taktic Building Sprays	Animal Health Business Unit, Hoechst UK Ltd	MAFF 06504

---

* These products are "approved for agricultural use". For further details refer to page vii.
% These products are subject to the staged revocation procedure. For further details refer to page vii.
A These products are approved for use in or near water. For further details refer to page vii.

PROFESSIONAL PRODUCTS: INSECTICIDES

Product Name	Marketing Company	Reg. No.
754 Azamethiphos		
Alfacron 10 WP	Ciba Agriculture	MAFF 02832
Alfracon 50 WP	Ciba Agriculture	MAFF 06552
Farm Fly Spray 10 WP	Rentokil Ltd	MAFF 05274
755 Bendiocarb		
* Ficam ULV	Cambridge Animal and Public Health Ltd	MAFF 06905
* Garvox 3G	AgrEvo UK Crop Protection Ltd	MAFF 07259
* Garvox 3G	Schering Agrochemicals Ltd	MAFF 03866
* Seedox SC	AgrEvo UK Crop Protection Ltd	MAFF 07278
* Seedox SC	DowElanco Ltd	MAFF 05935
* Seedox SC	Schering Agrochemicals Ltd	MAFF 03856
756 Benfuracarb		
* Oncol 10 G	Zeneca Crop Protection	MAFF 07063
* Oncol 10G	Farm Protection Ltd	MAFF 04265
* Oncol 10G	Zeneca Crop Protection	MAFF 07020
757 Bifenthrin		
* Talstar	DowElanco Ltd	MAFF 04916
* Talstar	FMC Corporation NV	MAFF 06913
758 Bromophos + *Resmethrin*		
Turbair Kilsect Super	Pan Britannica Industries Ltd	MAFF 02241
759 Buprofezin		
* Applaud	Zeneca Crop Protection	MAFF 06900
760 Captan + *Lindane*		
* Gammalex	ICI Agrochemicals	MAFF 00965
* Gammalex	Zeneca Crop Protection	MAFF 06396
761 Carbaryl		
* Agri Carbaryl 85 WP	Agri-Export Ltd	MAFF 04256
Microcarb Suspendable Powder	Micro-Biologicals Ltd	MAFF 01339
*% Murvin 85	Dow Chemical Co Ltd	MAFF 04234
* Murvin 85	DowElanco Ltd	MAFF 05740
Scattercarb	Micro-Biologicals Ltd	MAFF 01886
*% Sevin 85 Sprayable	Rhone-Poulenc Agriculture Ltd	MAFF 04957
* Thinsec	Imperial Chemical Industries Plc	MAFF 02463
* Thinsec	Zeneca Crop Protection	MAFF 06710

* These products are "approved for agricultural use". For further details refer to page vii.
% These products are subject to the staged revocation procedure. For further details refer to page vii.
A These products are approved for use in or near water. For further details refer to page vii.

PROFESSIONAL PRODUCTS: INSECTICIDES

Product Name	Marketing Company	Reg. No.
762 Carbofuran		
* Barclay Carbosect	Barclay Chemicals (UK) Ltd	MAFF 05512
* Nex	Tripart Farm Chemicals Ltd	MAFF 05165
* Rampart	Sipcam UK Ltd	MAFF 05166
* Stefes Carbofuran	Stefes Plant Protection Ltd	MAFF 06969
* Throttle	Bayer Plc	MAFF 05204
*% Throttle	Quadrangle Ltd	MAFF 05018
* Yaltox	Bayer Plc	MAFF 02371
763 Carbosulfan		
* Landgold Carbosulfan 10G	Landgold & Co Ltd	MAFF 06046
* Marshal 10 G	Rhone-Poulenc Agriculture	MAFF 06165
* Marshal 10 G	Rhone-Poulenc Crop Protection	MAFF 04646
* Power Chase	Power Agrichemicals Ltd	MAFF 04787
* Standon Carbosulfan 10G	Standon Chemicals Ltd	MAFF 05671
764 Carboxin + *Lindane* + *Thiram*		
*% Vitavax RS	DowElanco Ltd	MAFF 02310
* Vitavax RS	Uniroyal Chemical Ltd	MAFF 06029
765 Chlorfenvinphos		
* Birlane 24	Cyanamid Of GB Ltd	MAFF 07002
* Birlane 24	Shell Chemicals UK Ltd	MAFF 00275
* Birlane Granules	Cyanamid Of GB Ltd	MAFF 07003
* Birlane Granules	Shell Chemicals UK Ltd	MAFF 00276
* Birlane Liquid Seed Treatment	Cyanamid Of GB Ltd	MAFF 07004
* Birlane Liquid Seed Treatment	Rhone-Poulenc Agriculture Ltd	MAFF 05436
* Birlane Liquid Seed Treatment	Rhone-Poulenc Agriculture Ltd	MAFF 06376
* Birlane Liquid Seed Treatment	Shell Chemicals UK Ltd	MAFF 06019
*% Sapecron 10 FG	Ciba-Geigy Agrochemicals	MAFF 01860
* Sapecron 240 EC	Ciba Agriculture	MAFF 01861
* Sedanox Granules	Bayer UK Ltd	MAFF 05033
766 Chlorpyrifos		
* Alpha Chlorpyrifos 48 EC	Makhteshim-Agan (UK) Ltd	MAFF 04821
* Barclay Clinch	Barclay Chemicals (UK) Ltd	MAFF 06148
*% Crossfire	Dow Chemical Co Ltd	MAFF 03718
* Crossfire	DowElanco Ltd	MAFF 05706
* Crossfire	Rhone-Poulenc Agriculture Ltd	MAFF 05329
*% Dursban 4	Dow Chemical Co Ltd	MAFF 00775
* Dursban 4	DowElanco Ltd	MAFF 05735
*% Dursban 5G	DowElanco Ltd	MAFF 05306
* Lorsban T	DowElanco Ltd	MAFF 05970
* Spannit	Pan Britannica Industries Ltd	MAFF 01992

* These products are "approved for agricultural use". For further details refer to page vii.
% These products are subject to the staged revocation procedure. For further details refer to page vii.
A These products are approved for use in or near water. For further details refer to page vii.

PROFESSIONAL PRODUCTS: INSECTICIDES

Product Name	Marketing Company	Reg. No.
766 Chlorpyrifos—*continued*		
* Spannit Granules	Pan Britannica Industries Ltd	MAFF 04048
* Suscon Green Soil Insecticide	Fargro Ltd	MAFF 06312
* Talon	Farmers Crop Chemicals Ltd	MAFF 06017
*% Talon	Farmers Crop Chemicals Ltd	MAFF 03375
* Tripart Audax	Tripart Farm Chemicals Ltd	MAFF 05523
767 Chlorpyrifos + *Dimethoate*		
* Atlas Sheriff	Atlas Interlates Ltd	MAFF 04114
*% Sheriff	Dow Chemical Co Ltd	MAFF 04112
* Sheriff	DowElanco Ltd	MAFF 05744
768 Chlorpyrifos + *Disulfoton*		
* Twinspan	Pan Britannica Industries Ltd	MAFF 02255
769 Chlorpyrifos-methyl		
% Reldan 50	Dow Chemical Co Ltd	MAFF 02556
* Reldan 50	DowElanco Ltd	MAFF 05742
770 Clofentezine		
* Apollo 50 SC	AgrEvo UK Crop Protection Ltd	MAFF 07242
* Apollo 50 SC	Schering Agrochemicals Ltd	MAFF 03996
771 Cyfluthrin		
* Baythroid	Bayer Plc	MAFF 04273
772 Cypermethrin		
* Ambush C	Imperial Chemical Industries Plc	MAFF 00087
* Ambush C	Zeneca Crop Protection	MAFF 06632
*% Amos Cypermethrin 10	Kommer-Brookwick	MAFF 05619
* Ashlade Cypermethrin 10 EC	Ashlade Formulations Ltd	MAFF 06229
*% Ashlade Cypermethrin 10 EC	Ashlade Formulations Ltd	MAFF 03609
* Barclay Cypersect XL	Barclay Chemicals (UK) Ltd	MAFF 06509
* Chemtech Cypermethrin 10 EC	Chemtech (Crop Protection) Ltd	MAFF 04827
* Clayton Cyperten	Clayton Plant Protection (UK) Ltd	MAFF 07041
* Clayton Meteor	Clayton Plant Protection (UK) Ltd	MAFF 06074
* Cymbush	Imperial Chemical Industries Plc	MAFF 05887
* Cymbush	Zeneca Crop Protection	MAFF 06652
* Cyperkill 10	Mitchell Cotts Chemicals Ltd	MAFF 04119
* Cyperkill 25	Mitchell Cotts Chemicals Ltd	MAFF 03741
* Cyperkill 5	Mitchell Cotts Chemicals Ltd	MAFF 00625
* Cypertox	Farmers Crop Chemicals Ltd	MAFF 05122
*% Hortag Cypermethrin 10 EC	Hortag Chemicals Ltd	MAFF 04147

* These products are "approved for agricultural use". For further details refer to page vii.
% These products are subject to the staged revocation procedure. For further details refer to page vii.
A These products are approved for use in or near water. For further details refer to page vii.

PROFESSIONAL PRODUCTS: INSECTICIDES

Product Name	Marketing Company	Reg. No.
772 Cypermethrin—*continued*		
* Luxan Cypermethrin	Luxan (UK) Ltd	MAFF 06283
* Quadrangle Cyper 10	Quadrangle Ltd	MAFF 03242
* Ripcord	Cyanamid Of GB Ltd	MAFF 07014
* Ripcord	Shell Chemicals UK Ltd	MAFF 01808
* Standon Cypermethrin	Standon Chemicals Ltd	MAFF 06818
* Stefes Cypermethrin	Stefes Plant Protection Ltd	MAFF 05635
* Stefes Cypermethrin 2	Stefes Plant Protection Ltd	MAFF 05719
* Toppel 10	Farm Protection Ltd	MAFF 02990
* Toppel 10	Zeneca Crop Protection	MAFF 06516
* Vassgro Cypermethrin Insecticide	L W Vass (Agricultural) Ltd	MAFF 03240
773 Cyromazine		
Neporex 2 SG	Ciba Agriculture	MAFF 06985
774 Deltamethrin		
*% Barclay Calypso	Barclay Chemicals (UK) Ltd	MAFF 05565
% Crackdown	Wellcome Foundation Ltd	MAFF 02412
* Decis	AgrEvo UK Crop Protection Ltd	MAFF 07172
* Decis	Hoechst UK Ltd	MAFF 06311
*% Decis	Hoechst UK Ltd	MAFF 00657
* Standon Deltamethrin	Standon Chemicals Ltd	MAFF 07053
*% Stefes Deltamethrin	Stefes Plant Protection Ltd	MAFF 05678
* Thripstick	Aquaspersions Ltd	MAFF 02134
775 Deltamethrin + *Heptenophos*		
* Decis Quick	AgrEvo UK Crop Protection Ltd	MAFF 07312
* Decis Quick	Hoechst UK Ltd	MAFF 03117
776 Deltamethrin + *Pirimicarb*		
* Evidence	AgrEvo UK Crop Protection Ltd	MAFF 06934
777 Demeton-S-methyl		
* Campbell's DSM	J D Campbell & Sons Ltd	MAFF 00405
* Mifatox	Farmers Crop Chemicals Ltd	MAFF 01350
778 Diazinon		
*% Basudin 40WP	Ciba-Geigy Agrochemicals	MAFF 00214
* Basudin 5FG	Ciba Agriculture	MAFF 00213
* Darlingtons Diazinon Granules	Darmycel (UK)	MAFF 05674

* These products are "approved for agricultural use". For further details refer to page vii.
% These products are subject to the staged revocation procedure. For further details refer to page vii.
A These products are approved for use in or near water. For further details refer to page vii.

PROFESSIONAL PRODUCTS: INSECTICIDES

Product Name	Marketing Company	Reg. No.
779 Dichlorvos		
* Darmycel Dichlorvos	Darlingtons Mushroom Laboratories	MAFF 05699
Nuvan 500 EC	Ciba Agriculture	MAFF 03861
780 Dicofol		
* Alpha Dicofol 18.5 EC	Makhteshim-Agan (UK) Ltd	MAFF 04824
*% Fumite Dicofol Smoke	Octavius Hunt Ltd	MAFF 00931
* Kelthane	Rohm & Haas (UK) Ltd	MAFF 01131
781 Dicofol + *Tetradifon*		
* Childion	Hortichem Ltd	MAFF 03821
* Childion	Imperial Chemical Industries Plc	MAFF 00486
* Childion	Zeneca Crop Protection	MAFF 06644
* Turbair Acaricide	Pan Britannica Industries Ltd	MAFF 02231
782 Dienochlor		
*% Pentac Aquaflow	Dow Chemical Co Ltd	MAFF 04238
* Pentac Aquaflow	DowElanco Ltd	MAFF 05741
* Pentac Aquaflow	Sandoz Agro Ltd	MAFF 05697
783 Diflubenzuron		
* Dimilin ODC-45	Imperial Chemical Industries Plc	MAFF 04149
* Dimilin ODC-45	Zeneca Crop Protection	MAFF 06655
* Dimilin WP	Imperial Chemical Industries Plc	MAFF 03810
* Dimilin WP	Zeneca Crop Protection	MAFF 06656
784 Dimethoate		
* Amos Dimethoate	Kommer-Brookwick	MAFF 05812
*% Amos Dimethoate	Kommer-Brookwick	MAFF 06144
* Ashlade Dimethoate	Ashlade Formulations Ltd	MAFF 04814
* Atlas Dimethoate	Atlas Interlates Ltd	MAFF 03044
* BASF Dimethoate 40	BASF Plc	MAFF 00199
* CIA Dimethoate	Combined Independent Agronomists Ltd	MAFF 05937
* Campbell's Dimethoate 40	J D Campbell & Sons Ltd	MAFF 00398
*% Cropsafe Dimethoate 40	Hortichem Ltd	MAFF 05153
*% Danadim 400	Cheminova Agro A/S	MAFF 06488
* Danadim Dimethoate 40	Cheminova Agro A/S	MAFF 07040
* Danadim Dimethoate 40	Cheminova UK Ltd	MAFF 07351
*% Ipithoate	I Pi Ci	MAFF 04694
* Luxan Dimethoate	Luxan (UK) Ltd	MAFF 06558
* MTM Dimethoate 40	MTM Agrochemicals Ltd	MAFF 05693

---

* These products are "approved for agricultural use". For further details refer to page vii.
% These products are subject to the staged revocation procedure. For further details refer to page vii.
A These products are approved for use in or near water. For further details refer to page vii.

PROFESSIONAL PRODUCTS: INSECTICIDES

Product Name	Marketing Company	Reg. No.
784 Dimethoate—*continued*		
* P A Dimethoate 40	Portman Agrochemicals Ltd	MAFF 01527
* Portman Dimethoate EC	Portman Agrochemicals Ltd	MAFF 03190
* Roxion	Cyanamid Of GB Ltd	MAFF 07015
* Roxion	Shell Chemicals UK Ltd	MAFF 04813
* Top Farm Dimethoate	Top Farm Formulations Ltd	MAFF 05936
* Turbair Systemic Insecticide	Pan Britannica Industries Ltd	MAFF 02250
* Unicrop Dimethoate 40	Universal Crop Protection Ltd	MAFF 06549
*% Unicrop Dimethoate 40	Universal Crop Protection Ltd	MAFF 02265
785 Dimethoate + *Chlorpyrifos*		
* Atlas Sheriff	Atlas Interlates Ltd	MAFF 04114
*% Sheriff	Dow Chemical Co Ltd	MAFF 04112
* Sheriff	DowElanco Ltd	MAFF 05744
786 Disulfoton		
* Campbell's Disulfoton FE 10	J D Campbell & Sons Ltd	MAFF 00402
* Disyston FE-10	Bayer Plc	MAFF 00714
* Disyston P-10	Bayer Plc	MAFF 00715
* MTM Disulfoton P10	MTM Agrochemicals Ltd	MAFF 05094
* Solvigran 10	Sandoz Products Ltd	MAFF 02585
787 Disulfoton + *Chlorpyrifos*		
* Twinspan	Pan Britannica Industries Ltd	MAFF 02255
788 Disulfoton + *Fonofos*		
* Doubledown	Imperial Chemical Industries Plc	MAFF 03943
* Doubledown	Zeneca Crop Protection	MAFF 06658
789 Disulfoton + *Quinalphos*		
* Knave	Hortichem Ltd	MAFF 02534
790 Endosulfan		
* Thiodan 20 EC	AgrEvo UK Crop Protection Ltd	MAFF 07335
* Thiodan 20 EC	Hoechst UK Ltd	MAFF 02122
* Thiodan 35 EC	AgrEvo UK Crop Protection Ltd	MAFF 07336
* Thiodan 35 EC	Hoechst UK Ltd	MAFF 02123
* Thiodan 35 EC	Hoechst UK Ltd	MAFF 07182
791 Esfenvalerate		
* Sumi-Alpha	Cyanamid Of GB Ltd	MAFF 07207
* Sumi-Alpha	Shell Chemicals UK Ltd	MAFF 05945

* These products are "approved for agricultural use". For further details refer to page vii.
% These products are subject to the staged revocation procedure. For further details refer to page vii.
A These products are approved for use in or near water. For further details refer to page vii.

PROFESSIONAL PRODUCTS: INSECTICIDES

Product Name	Marketing Company	Reg. No.
792 Ethiofencarb		
* Croneton	Bayer Plc	MAFF 00593
793 Ethoprophos		
* Mocap 10G	Rhone-Poulenc Agriculture	MAFF 06773
*% Mocap 10G	Rhone-Poulenc Agriculture Ltd	MAFF 04935
794 Etrimfos		
Satisfar	Nickersons Seed Specialists Ltd	MAFF 04180
Satisfar	Sandoz Agro Ltd	MAFF 03694
Satisfar Dust	Nickersons Seed Specialists Ltd	MAFF 04085
Satisfar Dust	Sandoz Agro Ltd	MAFF 04142
795 Fatty acids		
* Safers Insecticidal Soap	Koppert (UK) Ltd	MAFF 06064
* Safers Insecticidal Soap	Safer Ltd	MAFF 07197
* Savona	Koppert (UK) Ltd	MAFF 03137
* Savona	Koppert (UK) Ltd	MAFF 06057
796 Fenbutatin oxide		
* Torque	Cyanamid Of GB Ltd	MAFF 07148
* Torque	Imperial Chemical Industries Plc	MAFF 03303
* Torque	Shell Chemicals UK Ltd	MAFF 02466
* Torque	Zeneca Crop Protection	MAFF 07021
797 Fenitrothion		
Antec Durakil	Antec International Ltd	MAFF 05147
* Dicofen	Pan Britannica Industries Ltd	MAFF 00693
* Unicrop Fenitrothion 50	Universal Crop Protection Ltd	MAFF 02267
798 Fenitrothion + *Permethrin* + *Resmethrin*		
Turbair Grain Store Insecticide	Pan Britannica Industries Ltd	MAFF 02238
799 Fenpropathrin		
* Meothrin	Cyanamid Of GB Ltd	MAFF 07206
* Meothrin	Shell Chemicals UK Ltd	MAFF 04993
800 Fenpropimorph + *Lindane* + *Thiram*		
* Lindex Plus FS Seed Treatment	DowElanco Ltd	MAFF 03934
801 Fenvalerate		
* Sumicidin	Cyanamid Of GB Ltd	MAFF 07208
* Sumicidin	Shell Chemicals UK Ltd	MAFF 02568

* These products are "approved for agricultural use". For further details refer to page vii.
% These products are subject to the staged revocation procedure. For further details refer to page vii.
A These products are approved for use in or near water. For further details refer to page vii.

PROFESSIONAL PRODUCTS: INSECTICIDES

Product Name	Marketing Company	Reg. No.
802 Fonofos		
* Cudgel	Imperial Chemical Industries Plc	MAFF 04349
* Cudgel	Zeneca Crop Protection	MAFF 06648
* Dyfonate 10G	Farm Protection Ltd	MAFF 03945
* Fonofos Seed Treatment	Imperial Chemical Industries Plc	MAFF 04051
* Sonar	ICI Agrochemicals	MAFF 05865
* Sonar	Zeneca Crop Protection	MAFF 06706
803 Fonofos + *Disulfoton*		
* Doubledown	Imperial Chemical Industries Plc	MAFF 03943
* Doubledown	Zeneca Crop Protection	MAFF 06658
804 Heptenophos		
* Hostaquick	AgrEvo UK Crop Protection Ltd	MAFF 07326
* Hostaquick	Hoechst UK Ltd	MAFF 01079
805 Heptenophos + *Deltamethrin*		
* Decis Quick	AgrEvo UK Crop Protection Ltd	MAFF 07312
* Decis Quick	Hoechst UK Ltd	MAFF 03117
806 Imidacloprid		
Gaucho	Bayer UK Ltd	MAFF 06590
807 Iodofenphos		
* Elocril 50 WP	Ciba Agriculture	MAFF 00787
808 Lambda cyhalothrin		
*% Hallmark	ICI Agrochemicals	MAFF 04466
* Hallmark	Zeneca Crop Protection	MAFF 06434
* Landgold Lambda-C	Landgold & Co Ltd	MAFF 06097
*% PP 321	ICI Agrochemicals	MAFF 04467
* PP 321	Zeneca Crop Protection	MAFF 06698
*% PP 321 Oilseed Rape Summer Pests	ICI Agrochemicals	MAFF 04648
* PP 321 Oilseed Rape Summer Pests	Zeneca Crop Protection	MAFF 06699
* Standon Lambda-C	Standon Chemicals Ltd	MAFF 05672
809 Lindane ($\geq$ 99% Gamma-HCH)		
* Atlas Steward	Atlas Interlates Ltd	MAFF 03062
* Fumite Lindane 10	Octavius Hunt Ltd	MAFF 00933
* Fumite Lindane 40	Octavius Hunt Ltd	MAFF 00934
* Fumite Lindane Pellets	Octavius Hunt Ltd	MAFF 00937
* Gamma Col	ICI Agrochemicals	MAFF 00964
* Gamma Col	Zeneca Crop Protection	MAFF 06670

* These products are "approved for agricultural use". For further details refer to page vii.
% These products are subject to the staged revocation procedure. For further details refer to page vii.
A These products are approved for use in or near water. For further details refer to page vii.

PROFESSIONAL PRODUCTS: INSECTICIDES

Product Name	Marketing Company	Reg. No.

809 Lindane—*continued*

*%	Gamma HCH Dust	Hortag Chemicals Ltd	MAFF 04392
*	Gamma-Col Turf	ICI Agrochemicals	MAFF 05450
*	Gamma-Col Turf	Zeneca Professional Products	MAFF 06826
*	Gammasan 30	ICI Agrochemicals	MAFF 00969
*	Gammasan 30	Zeneca Crop Protection	MAFF 06671
*	Kotol FS	Rhone-Poulenc Agriculture Ltd	MAFF 06968
*	Lindane 20	Pan Britannica Industries Ltd	MAFF 01553
*	Lindane Flowable	Pan Britannica Industries Ltd	MAFF 02610
%	Murfume Grain Store Smoke	DowElanco Ltd	MAFF 05575
*%	Murfume Lindane Smoke	DowElanco Ltd	MAFF 05576
*	New Kotol	Embetec Crop Protection	MAFF 05081
*	New Kotol	Rhone-Poulenc Agriculture Ltd	MAFF 06212
*	Unicrop Leatherjacket Pellets	Universal Crop Protection Ltd	MAFF 02272
*	Wireworm FS Seed Treatment	Dow Agriculture	MAFF 05640
*	Wireworm Liquid Seed Treatment	DowElanco Ltd	MAFF 05641

810 Lindane + Aldicarb

*%	Sentry	Rhone-Poulenc Agriculture Ltd	MAFF 05421
*	Sentry	Embetec Crop Protection	MAFF 06209

811 Lindane + Captan

*	Gammalex	ICI Agrochemicals	MAFF 00965
*	Gammalex	Zeneca Crop Protection	MAFF 06396

812 Lindane + Carboxin + Thiram

*%	Vitavax RS	DowElanco Ltd	MAFF 02310
*	Vitavax RS	Uniroyal Chemical Ltd	MAFF 06029

813 Lindane + Fenpropimorph + Thiram

*	Lindex Plus FS Seed Treatment	DowElanco Ltd	MAFF 03934

814 Lindane + Thiabendazole + Thiram

*	Hysede FL	Agrichem Ltd	MAFF 02863

815 Lindane + Thiophanate-methyl

*	Castaway Plus	Rhone-Poulenc Environmental Products	MAFF 05327
*	CDA Castaway Plus	Rhone-Poulenc Environmental Products	MAFF 04758

816 Lindane + Thiram

*	Hydraguard	Agrichem Ltd	MAFF 03278

---

* These products are "approved for agricultural use". For further details refer to page vii.
% These products are subject to the staged revocation procedure. For further details refer to page vii.
A These products are approved for use in or near water. For further details refer to page vii.

PROFESSIONAL PRODUCTS: INSECTICIDES

Product Name	Marketing Company	Reg. No.
817 Malathion		
Ban-Mite	Johnsons Veterinary Products Ltd	MAFF 06039
* MTM Malathion 60	MTM Agrochemicals Ltd	MAFF 05714
818 Mephosfolan		
* Cytro-Lane	Cyanamid Of GB Ltd	MAFF 00626
819 Metaldehyde		
*% Optimol	ICI Agrochemicals	MAFF 05862
820 Methiocarb		
* Club	Bayer Plc	MAFF 07176
*% Club	Imperial Chemical Industries Plc	MAFF 03800
*% Club	Zeneca Crop Protection	MAFF 06645
* Decoy	Bayer Plc	MAFF 06535
* Draza	Bayer UK Ltd	MAFF 00765
* Draza 2	Bayer Plc	MAFF 04748
* Draza Plus	Bayer Plc	MAFF 06553
* Elvitox	Bayer Plc	MAFF 06738
* Epox	Bayer Plc	MAFF 06737
821 Methomyl		
*% Lannate 20L	Du Pont (UK) Ltd	MAFF 01183
% Sorex Golden Fly Bait	Sorex Ltd	MAFF 02731
822 Methomyl + *(Z)-9-Tricosene*		
Golden Malrin Fly Bait	Sanofi Animal Health Ltd	MAFF 05579
% Improved Golden Malrin Fly Bait	Zoecon Corporation	MAFF 01111
823 Methoprene		
* Apex 5E	Sandoz Speciality Pest Control Ltd	MAFF 05730
824 Mineral oil		
* Hortichem Spraying Oil	Hortichem Ltd	MAFF 03816
825 Nicotine		
* Campbell's Nico Soap	J D Campbell & Sons Ltd	MAFF 00416
* Nicotine 40% Shreds	DowElanco Ltd	MAFF 05725
* XL All Insecticide	Synchemicals Ltd	MAFF 02369
* XL All Nicotine 95%	Synchemicals Ltd	MAFF 02370
826 Omethoate		
*% Folimat	Bayer Plc	MAFF 00912

* These products are "approved for agricultural use". For further details refer to page vii.
% These products are subject to the staged revocation procedure. For further details refer to page vii.
A These products are approved for use in or near water. For further details refer to page vii.

PROFESSIONAL PRODUCTS: INSECTICIDES

Product Name	Marketing Company	Reg. No.
827 Oxamyl		
* Fielder	Du Pont (UK) Ltd	MAFF 05279
* Power Blade	Power Agrichemicals Ltd	MAFF 04743
* Vydate 10G	Du Pont (UK) Ltd	MAFF 02322
828 Oxydemeton-methyl		
*% Metasystox R	Bayer Plc	MAFF 01333
*% Metasystox R	Bayer Plc	MAFF 05763
829 Permethrin		
* Ambush	Imperial Chemical Industries Plc	MAFF 00086
* Fumite Permethrin Smoke Generators	Octavius Hunt Ltd	MAFF 00940
Geest Fumite 300 MK2	Octavius Hunt Ltd	MAFF 06249
*% Murfume Permethrin Smoke	DowElanco Ltd	MAFF 05577
* Permasect 10 EC	Mitchell Cotts Chemicals Ltd	MAFF 03920
* Permasect 25 EC	Mitchell Cotts Chemicals Ltd	MAFF 01576
* Permit	Pan Britannica Industries Ltd	MAFF 01577
* Turbair Permethrin	Pan Britannica Industries Ltd	MAFF 02246
830 Permethrin + *Fenitrothion* + *Resmethrin*		
Turbair Grain Store Insecticide	Pan Britannica Industries Ltd	MAFF 02238
831 Permethrin + *Thiram*		
* Combinex	Pan Britannica Industries Ltd	MAFF 00562
832 Phorate		
* BASF Phorate	BASF Plc	MAFF 00210
* BASF Phorate	BASF Plc	MAFF 06610
*% Campbell's Phorate	J D Campbell & Sons Ltd	MAFF 00418
* MTM Phorate	MTM Agrochemicals Ltd	MAFF 06609
*% MTM Phorate	MTM Agrochemicals Ltd	MAFF 05540
*% Terrathion Granules	Farmers Crop Chemicals Ltd	MAFF 02106
833 Phosalone		
* Zolone Liquid	Rhone-Poulenc Agriculture	MAFF 06173
* Zolone Liquid	Rhone-Poulenc Agriculture Ltd	MAFF 05430
*% Zolone Liquid	Rhone-Poulenc Crop Protection	MAFF 05231
834 Pirimicarb		
* Aphox	Imperial Chemical Industries Plc	MAFF 00106
* Aphox	Zeneca Crop Protection	MAFF 06633
* Barclay Pirimisect	Barclay Chemicals (UK) Ltd	MAFF 06929

* These products are "approved for agricultural use". For further details refer to page vii.
% These products are subject to the staged revocation procedure. For further details refer to page vii.
A These products are approved for use in or near water. For further details refer to page vii.

PROFESSIONAL PRODUCTS: INSECTICIDES

Product Name	Marketing Company	Reg. No.

834 Pirimicarb—continued

*% Barclay Pirimisect	Barclay Chemicals (UK) Ltd	MAFF 05305
* Clayton Pirimicarb 50 SG	Clayton Plant Protection (UK) Ltd	MAFF 06972
* Landgold Pirimicarb 50	Landgold & Co Ltd	MAFF 06238
* Phantom	Bayer Plc	MAFF 04519
*% Pirimicarb 50 WG	Schering Agrochemicals Ltd	MAFF 04063
* Pirimor	Imperial Chemical Industries Plc	MAFF 01594
* Pirimor	Zeneca Crop Protection	MAFF 06694
* Portman Pirimicarb	Portman Agrochemicals Ltd	MAFF 06922
* Power Demo H	Power Agrichemicals Ltd	MAFF 05042
* Sapir	Farm Protection	MAFF 04518
* Sapir	Zeneca Crop Protection	MAFF 06398
* Standon Pirimicarb 50	Standon Chemicals Ltd	MAFF 05622
* Standon Pirimicarb H	Standon Chemicals Ltd	MAFF 05669
* Stefes Pirimicarb	Stefes Plant Protection Ltd	MAFF 05758
* Unistar Pirimicarb 500	Unistar Ltd	MAFF 06975

835 Pirimicarb + *Deltamethrin*

* Evidence	AgrEvo UK Crop Protection Ltd	MAFF 06934

836 Pirimiphos-ethyl

*% Fernex Granules	Applied Horticulture Ltd	MAFF 00862

837 Pirimiphos-methyl

Actellic Dust	Imperial Chemical Industries Plc	MAFF 00011
Actellic Dust	Zeneca Crop Protection	MAFF 06626
Actellic Smoke Generator 10	Imperial Chemical Industries Plc	MAFF 00017
Actellic Smoke Generator 10	Zeneca Public Health	MAFF 06864
* Actellic Smoke Generator No 20	Zeneca Crop Protection	MAFF 06627
* Actellic Smoke Generator No. 20	Imperial Chemical Industries Plc	MAFF 00018
* Actellifog	Imperial Chemical Industries Plc	MAFF 00019
* Actellifog	Zeneca Crop Protection	MAFF 06628
*% Blex	Imperial Chemical Industries Plc	MAFF 00284
* Blex	Zeneca Crop Protection	MAFF 06639
* Fumite Pirimiphos Methyl Smoke	Octavius Hunt Ltd	MAFF 00941

838 Propoxur

* Fumite Propoxur Smoke	Octavius Hunt Ltd	MAFF 00942
*% Murfume Propoxur Smoke	DowElanco Ltd	MAFF 05739
Pigeon Loft Spray	Natural Granen	MAFF 06542

---

* These products are "approved for agricultural use". For further details refer to page vii.
% These products are subject to the staged revocation procedure. For further details refer to page vii.
A These products are approved for use in or near water. For further details refer to page vii.

PROFESSIONAL PRODUCTS: INSECTICIDES

Product Name	Marketing Company	Reg. No.
839 Pyrethrins		
Alfadex	Ciba Agriculture	MAFF 00074
Turbair Flydown	Pan Britannica Industries Ltd	MAFF 05482
Turbair Kilsect Short Life Grade	Pan Britannica Industries Ltd	MAFF 02240
Turbair Super Flydown	Pan Britannica Industries Ltd	MAFF 02249
840 Pyrethrins + *Resmethrin*		
* Pynosect 30 Fogging Solution	Mitchell Cotts Chemicals Ltd	MAFF 01650
* Pynosect 30 Water Miscible	Mitchell Cotts Chemicals Ltd	MAFF 01653
841 Quinalphos		
* Savall	Sandoz Agro Ltd	MAFF 01864
842 Quinalphos + *Disulfoton*		
* Knave	Hortichem Ltd	MAFF 02534
843 Quinalphos + *Thiometon*		
* Tombel	Hortichem Ltd	MAFF 04124
* Tombel	Sandoz Agro Ltd	MAFF 05288
844 Resmethrin		
* Turbair Resmethrin Extra	Pan Britannica Industries Ltd	MAFF 02247
845 Resmethrin + *Bromophos*		
Turbair Kilsect Super	Pan Britannica Industries Ltd	MAFF 02241
846 Resmethrin + *Fenitrothion* + *Permethrin*		
Turbair Grain Store Insecticide	Pan Britannica Industries Ltd	MAFF 02238
847 Resmethrin + *Pyrethrins*		
* Pynosect 30 Fogging Solution	Mitchell Cotts Chemicals Ltd	MAFF 01650
* Pynosect 30 Water Miscible	Mitchell Cotts Chemicals Ltd	MAFF 01653
848 Rotenone		
* Devcol Liquid Derris	Devcol Ltd	MAFF 06063
* Liquid Derris	Ford Smith & Co Ltd	MAFF 01213
849 Sulphur		
* Ashlade Sulphur FL	Ashlade Formulations Ltd	MAFF 06478
* Kumulus DF	BASF Plc	MAFF 04707
* MSS Sulphur 80	Mirfield Sales Services Ltd	MAFF 05752

* These products are "approved for agricultural use". For further details refer to page vii.
% These products are subject to the staged revocation procedure. For further details refer to page vii.
A These products are approved for use in or near water. For further details refer to page vii.

PROFESSIONAL PRODUCTS: INSECTICIDES

	Product Name	Marketing Company	Reg. No.
	850 Tar oils		
*%	Mortegg Emulsion	Dow Chemical Co Ltd	MAFF 04233
*	Mortegg Emulsion	DowElanco Ltd	MAFF 05738
*	Sterilite Tar Oil Winter Wash 60 % Stock Emulsion	Coventry Chemicals Ltd	MAFF 05061
*	Sterilite Tar Oil Winter Wash 80 % Miscible Quality	Coventry Chemicals Ltd	MAFF 05062
	851 Teflubenzuron		
*	Nemolt	Cyanamid Of GB Ltd	MAFF 07012
*	Nemolt	Shell Chemicals UK Ltd	MAFF 04274
	852 Tefluthrin		
*	Force ST	ICI Agrochemicals	MAFF 05109
*	Force ST	Zeneca Crop Protection	MAFF 06665
	853 Tetradifon		
*	Tedion V-18 EC	Hortichem Ltd	MAFF 03820
	854 Tetradifon + *Dicofol*		
*	Childion	Hortichem Ltd	MAFF 03821
*	Childion	Imperial Chemical Industries Plc	MAFF 00486
*	Childion	Zeneca Crop Protection	MAFF 06644
*	Turbair Acaricide	Pan Britannica Industries Ltd	MAFF 02231
	855 Thiabendazole + *Lindane* + *Thiram*		
*	Hysede FL	Agrichem Ltd	MAFF 02863
	856 Thiofanox		
*%	Dacamox 5 G	Rhone-Poulenc Agriculture Ltd	MAFF 05444
	857 Thiometon		
*	Ekatin	Sandoz Agro Ltd	MAFF 05281
*%	WBC Systemic Aphicide	WBC Technology Ltd	MAFF 04989
	858 Thiometon + *Quinalphos*		
*	Tombel	Hortichem Ltd	MAFF 04124
*	Tombel	Sandoz Agro Ltd	MAFF 05288
	859 Thiophanate-methyl + *Lindane*		
*	Castaway Plus	Rhone-Poulenc Environmental Products	MAFF 05327
*	CDA Castaway Plus	Rhone-Poulenc Environmental Products	MAFF 04758

* These products are "approved for agricultural use". For further details refer to page vii.
% These products are subject to the staged revocation procedure. For further details refer to page vii.
A These products are approved for use in or near water. For further details refer to page vii.

PROFESSIONAL PRODUCTS: INSECTICIDES

Product Name	Marketing Company	Reg. No.

860 Thiram + *Carboxin* + *Lindane*

*% Vitavax RS	DowElanco Ltd	MAFF 02310
* Vitavax RS	Uniroyal Chemical Ltd	MAFF 06029

861 Thiram + *Fenpropimorph* + *Lindane*

* Lindex Plus FS Seed Treatment	DowElanco Ltd	MAFF 03934

862 Thiram + *Lindane*

* Hydraguard	Agrichem Ltd	MAFF 03278

863 Thiram + *Lindane* + *Thiabendazole*

* Hysede FL	Agrichem Ltd	MAFF 02863

864 Thiram + *Permethrin*

* Combinex	Pan Britannica Industries Ltd	MAFF 00562

865 Triazophos

* Hostathion	AgrEvo UK Crop Protection Ltd	MAFF 07327
* Hostathion	Hoechst UK Ltd	MAFF 01080

866 Trichlorfon

* Dipterex 80	Bayer Plc	MAFF 00711

867 (Z)-9-Tricosene + *Methomyl*

Golden Malrin Fly Bait	Sanofi Animal Health Ltd	MAFF 05579
% Improved Golden Malrin Fly Bait	Zoecon Corporation	MAFF 01111

---

* These products are "approved for agricultural use". For further details refer to page vii.
% These products are subject to the staged revocation procedure. For further details refer to page vii.
A These products are approved for use in or near water. For further details refer to page vii.

1.4 Vertebrate Control Products
Including rodenticide, mole killers and bird repellents

868 Alphachloralose

	Product Name	Marketing Company	Reg. No.
	Alpha Chloralose (Pure)	Rodent Control Ltd	MAFF 00082
	Alpha Chloralose 4% Ready Mixed	Rodent Control Ltd	MAFF 00083
	Klearwell Alphachloralose Bait	Layson Ltd	MAFF 02604
	Rentokil Alphachloralose	Rentokil Ltd	MAFF 01720
	Rentokil Alphachloralose Concentrate	Rentokil Ltd	MAFF 01721
%	Rentokil Alphalard	Rentokil Ltd	MAFF 02451
	Rodextra Mouse Bait	Rodentex Ltd	MAFF 01809
	Townex Mouse Poison	Town & County Pest Services Ltd	MAFF 02163

869 Aluminium ammonium sulphate

	Product Name	Marketing Company	Reg. No.
	Curb	Sphere Laboratories (London) Ltd	MAFF 02480
	Guardsman B	Chiltern Farm Chemicals Ltd	MAFF 05494
	Guardsman L Crop Spray	Sphere Laboratories (London) Ltd	MAFF 03605
	Guardsman M	Chiltern Farm Chemicals Ltd	MAFF 05495
	Guardsman STP	Sphere Laboratories (London) Ltd	MAFF 03606
	Liquid Curb Crop Spray	Sphere Laboratories (London) Ltd	MAFF 03164
	Narsty	Mandops (UK) Ltd	MAFF 06003
	Narsty	Mandops Ltd	MAFF 02860

870 Aluminium phosphide

	Product Name	Marketing Company	Reg. No.
%	Alutal	J A Kent Services (East Midlands) Ltd	MAFF 04104
	Amos Talunex	Kommer-Brookwick	MAFF 05571
	Amos Talunex	Kommer-Brookwick	MAFF 05615
%	Amos Talunex	Kommer-Brookwick	MAFF 06143
	Amos Talunex	Luxan (UK) Ltd	MAFF 06567
	Luxan Talunex	Luxan (UK) Ltd	MAFF 06563
	Phostek	Killgerm Chemicals Ltd	MAFF 05116
	Phostek	Killgerm Chemicals Ltd	MAFF 05115
	Rentokil Phostoxin	Rentokil Ltd	MAFF 01775
%	Talunex	Kommer-Brookwick	MAFF 06141
%	Talunex	Power Agrichemicals Ltd	MAFF 04739

871 Bone oil

	Product Name	Marketing Company	Reg. No.
%	Rabbit Smear Liquid	Craven Chemical Co Ltd	MAFF 01682

---

* These products are "approved for agricultural use". For further details refer to page vii.
% These products are subject to the staged revocation procedure. For further details refer to page vii.
A These products are approved for use in or near water. For further details refer to page vii.

PROFESSIONAL PRODUCTS: VERTEBRATE CONTROL PRODUCTS

Product Name	Marketing Company	Reg. No.
872 Brodifacoum		
Brodifacoum	Sorex Ltd	MAFF 04660
Brodifacoum Bait	Sorex Ltd	MAFF 00336
Brodifacoum Bait Blocks	Sorex Ltd	MAFF 04590
Brodifacoum Sewer Bait	Sorex Ltd	MAFF 00337
Klerat	ICI Agrochemicals	MAFF 04698
Klerat	Zeneca Public Health	MAFF 06869
Klerat Mouse Tube	Imperial Chemical Industries Plc	MAFF 04243
Klerat Mouse Tube	Zeneca Public Health	MAFF 06830
Klerat Wax Blocks	ICI Agrochemicals	MAFF 04746
Klerat Wax Blocks	Zeneca Public Health	MAFF 06827
873 Bromadiolone		
Deadline Place Packs	Rentokil Ltd	MAFF 06624
Rentokil Biotrol Plus Outdoor Rat Killer (for service use only)	Rentokil Ltd	MAFF 03707
Rentokil Bromard	Rentokil Ltd	MAFF 01727
Rentokil Bromatrol	Rentokil Ltd	MAFF 05077
Rentokil Bromatrol Contact Dust	Rentokil Ltd	MAFF 01729
Rentokil Deadline	Rentokil Ltd	MAFF 05078
Rentokil Deadline Contact Dust	Rentokil Ltd	MAFF 01736
Rentokil Deadline Liquid Concentrate	Rentokil Ltd	MAFF 01737
Rentokil Liquid Bromatrol	Rentokil Ltd	MAFF 03645
Rentokil Rodine C	Rentokil Ltd	MAFF 04777
Slaymor	Ciba Agriculture	MAFF 01958
Slaymor Bait Bags	Ciba Agriculture	MAFF 03183
% Slaymor Liquid Concentrate	Ciba-Geigy Agrochemicals	MAFF 04826
874 Calciferol		
Hyperkil	Antec International Ltd	MAFF 04310
Rentokil Deerat Concentrate	Rentokil Ltd	MAFF 01738
875 Calciferol + *Difenacoum*		
Sorexa CD	Sorex Ltd	MAFF 03514
Sorexa CD Concentrate	Sorex Ltd	MAFF 03513
876 Chlorophacinone		
% Drat	Embetec Crop Protection	MAFF 04418
Drat	Rhone-Poulenc Agriculture Ltd	MAFF 05238
Drat Bait	Rhone-Poulenc Agriculture Ltd	MAFF 05239
Drat Rat & Mouse Bait	Battle Hayward & Bower Ltd	MAFF 00764
Endorats	American Products	MAFF 06503
Karate Ready To Use Rat and Mouse Bait	Lever Industrial Ltd	MAFF 05321

* These products are "approved for agricultural use". For further details refer to page vii.
% These products are subject to the staged revocation procedure. For further details refer to page vii.
A These products are approved for use in or near water. For further details refer to page vii.

PROFESSIONAL PRODUCTS: VERTEBRATE CONTROL PRODUCTS

Product Name	Marketing Company	Reg. No.
876 Chlorophacinone—*continued*		
Karate Ready To Use Rodenticide Sachets	Lever Industrial Ltd	MAFF 05890
Killgerm Sakarat Special	Killgerm Chemicals Ltd	MAFF 02844
Klearwell Drat Rodent Poison	Layson Ltd	MAFF 01157
Rat-Eyre Rat and Mouse Bait	Vermin Eradication Supplies	MAFF 03952
Ridento Ready to use Rat Bait	Ace Chemicals Ltd	MAFF 03804
* Rout	Samuel McCausland Ltd	MAFF 06729
Rout Rat Packs	Samuel McCausland Ltd	MAFF 06730
Ruby Rat	J V Heatherington	MAFF 06059
877 Coumatetralyl		
Racumin Master Mix	Bayer Plc	MAFF 01677
Racumin Rat Bait	Bayer Plc	MAFF 01679
Racumin Tracking Powder	Bayer Plc	MAFF 01681
Townex Sachets	Town & County Pest Services Ltd	MAFF 02164
878 Difenacoum		
% Killgercide Ready To Use Cut Wheat Bait	Killgerm Chemicals Ltd	MAFF 03879
Killgerm Rat Rod	Killgerm Chemicals Ltd	MAFF 05154
% Killgerm Rat Rods	Killgerm Chemicals Ltd	MAFF 04071
% Killgerm Ratak Cut Wheat Rat Bait	Killgerm Chemicals Ltd	MAFF 04448
* Killgerm Ratak Cut Wheat Rat Bait	Killgerm Chemicals Ltd	MAFF 06919
% Killgerm Ratak Rat Pellets	Killgerm Chemicals Ltd	MAFF 04446
* Killgerm Ratak Rat Pellets	Killgerm Chemicals Ltd	MAFF 06918
% Killgerm Ratak Whole Wheat Rat Bait	Killgerm Chemicals Ltd	MAFF 04447
* Killgerm Ratak Whole Wheat Rat Bait	Killgerm Chemicals Ltd	MAFF 06917
Killgerm Wax Bait	Killgerm Chemicals Ltd	MAFF 04096
Neokil	Sorex Ltd	MAFF 05564
Neosorexa Bait	Sorex Ltd	MAFF 01474
Neosorexa Concentrate	Sorex Ltd	MAFF 01475
Neosorexa Liquid Concentrate	Sorex Ltd	MAFF 04640
Neosorexa Ratpacks	Sorex Ltd	MAFF 04653
% Pest-Pel	Chemsearch	MAFF 03743
Ratak	Imperial Chemical Industries Plc	MAFF 02586
Ratak	Zeneca Public Health	MAFF 06832
Ratak Wax Blocks	Imperial Chemical Industries Plc	MAFF 04217
Ratak Wax Blocks	Zeneca Public Health	MAFF 06829
Rataway	Deosan Ltd	MAFF 05560
Rataway Bait Bags	Deosan Ltd	MAFF 05562
Rataway W	Deosan Ltd	MAFF 05561
Rentokil Fentrol	Rentokil Ltd	MAFF 01747
% Rentokil Fentrol Contact Dust	Rentokil Ltd	MAFF 01748

* These products are "approved for agricultural use". For further details refer to page vii.
% These products are subject to the staged revocation procedure. For further details refer to page vii.
A These products are approved for use in or near water. For further details refer to page vii.

PROFESSIONAL PRODUCTS: VERTEBRATE CONTROL PRODUCTS

Product Name	Marketing Company	Reg. No.

878 Difenacoum—*continued*

	Rentokil Fentrol Gel	Rentokil Ltd	MAFF 01749
*	Sorexa D	Sorex Ltd	MAFF 06879
%	Triple XXX	Certified Laboratories Ltd	MAFF 03742

879 Difenacoum + *Calciferol*

Sorexa CD	Sorex Ltd	MAFF 03514
Sorexa CD Concentrate	Sorex Ltd	MAFF 03513

880 Diphacinone

Ditrac All-Weather Bait Bar	Bell Laboratories Inc	MAFF 07227
Ditrac All-Weather Blox	Bell Laboratories Inc	MAFF 07228
Ditrac Rat and Mouse Bait	Bell Laboratories Inc	MAFF 07170
Ditrac Super Blox All-Weather	Bell Laboratories Inc	MAFF 07229
PCQ Rat and Mouse Bait	Bell Laboratories Inc	MAFF 02823
Rodent Cake	Bell Laboratories Inc	MAFF 04299
Tomcat All-Weather Blox	Bell Laboratories Inc	MAFF 07230
Tomcat All-Weather Rodent Bar	Bell Laboratories Inc	MAFF 07231
Tomcat Rat and Mouse Bait	Antec International Ltd	MAFF 07171

881 Flocoumafen

Storm	Sorex Ltd	MAFF 03710
Storm Mouse Bait Blocks	Sorex Ltd	MAFF 04850

882 Lindane (≥ 99% Gamma-HCH)

	HCH 50% Tracking Dust	Rodent Control Ltd	MAFF 01042
%	Rentokil Lindane Contact Dust	Rentokil Ltd	MAFF 01759
%	Rentokil Mouse Dust	Rentokil Ltd	MAFF 01772

883 Quassia

	Dog Off	Fieldspray Ltd	MAFF 04397
%	Hoppit	Fieldspray Division, Nilco Chemical Co Ltd	MAFF 05477
%	Hoppit	Fieldspray Ltd	MAFF 04396

884 Seconal

Seconal	Killgerm Chemicals Ltd	MAFF 04715

885 Sodium cyanide

Cymag	Imperial Chemical Industries Plc	MAFF 00623
Cymag	Zeneca Crop Protection	MAFF 06651

---

* These products are "approved for agricultural use". For further details refer to page vii.
% These products are subject to the staged revocation procedure. For further details refer to page vii.
A These products are approved for use in or near water. For further details refer to page vii.

PROFESSIONAL PRODUCTS: VERTEBRATE CONTROL PRODUCTS

Product Name	Marketing Company	Reg. No.
886 Sulphonated cod liver oil		
Scuttle	Fine Agrochemicals Ltd	MAFF 04559
Scuttle	Fine Agrochemicals Ltd	MAFF 06232
887 Warfarin		
Dethmor Plusbait	Gerhardt Pharmaceuticals Ltd	MAFF 00684
Dethmor Warfarin 5	Gerhardt Pharmaceuticals Ltd	MAFF 00685
Grey Squirrel Liquid Conc	Rodent Control Ltd	MAFF 01009
Grey Squirrel Liquid Concentrate	Killgerm Chemicals Ltd	MAFF 06455
Grovex R15 Liquid Warfarin	W H Groves & Family Ltd	MAFF 01021
Grovex R18 Sewer Bait	W H Groves & Family Ltd	MAFF 01023
Grovex Warfarin 5	W H Groves & Family Ltd	MAFF 01019
Grovex Warfarin 5 (Inert Base)	W H Groves & Family Ltd	MAFF 01020
Kemkill Sewer Bait 0.025%	Kemkill Pest Control Services Ltd	MAFF 02855
Kemkill Sewer Bait 0.05%	Kemkill Pest Control Services Ltd	MAFF 04120
Killgerm Sewercide Cut Wheat Rat Bait	Killgerm Chemicals Ltd	MAFF 03761
Killgerm Sewercide Whole Wheat Rat Bait	Killgerm Chemicals Ltd	MAFF 03759
Kilmol	Callisto Aviaries Ltd	MAFF 01151
Klearwell Blue Warfarin 5	Layson Ltd	MAFF 01156
Klearwell Oiled Warfarin Sewer Bait	Layson Ltd	MAFF 01158
Klearwell Ready Mixed Warfarin	Layson Ltd	MAFF 01159
Rentokil Biotrol	Rentokil Ltd	MAFF 01723
% Rentokil Biotrol Concentrate	Rentokil Ltd	MAFF 01724
Rentokil Grey Squirrel Bait	Rentokil Ltd	MAFF 04164
% Rentokil Liquid Warfarin Concentrate	Rentokil Ltd	MAFF 04226
Rodentex Rat Killer	Rodentex Ltd	MAFF 01810
Rodentex Universal Rat Bait	Rodentex Ltd	MAFF 01811
Sakarat Concentrate	Killgerm Chemicals Ltd	MAFF 01849
Sakarat Ready To Use (Cut Wheat Base)	Killgerm Chemicals Ltd	MAFF 04340
Sakarat Ready To Use (Whole Wheat Base)	Killgerm Chemicals Ltd	MAFF 01850
Sakarat X Ready To Use Warfarin Rat Bait	Killgerm Chemicals Ltd	MAFF 01851
Sakarat X Warfarin Concentrate	Killgerm Chemicals Ltd	MAFF 01852
Sewarin Extra	Killgerm Chemicals Ltd	MAFF 03426
Sewarin Extra Ready For Use Bait	Killgerm Chemicals Ltd	MAFF 01929
Sewarin P	Killgerm Chemicals Ltd	MAFF 01930
Sorex Warfarin 250 ppm Rat Bait	Sorex Ltd	MAFF 07371
Sorex Warfarin 500 ppm Rat Bait	Sorex Ltd	MAFF 07372
Sorex Warfarin Rat Bait	Sorex Ltd	MAFF 03101
Sorex Warfarin Sewer Bait	Sorex Ltd	MAFF 07373

* These products are "approved for agricultural use". For further details refer to page vii.
% These products are subject to the staged revocation procedure. For further details refer to page vii.
A These products are approved for use in or near water. For further details refer to page vii.

PROFESSIONAL PRODUCTS: VERTEBRATE CONTROL PRODUCTS

Product Name	Marketing Company	Reg. No.
887 Warfarin—*continued*		
Sorexa Plus	Sorex Ltd	MAFF 01986
Warfarin 0.025% Ready For Use (Cereal Base)	Kemkill Pest Control Services Ltd	MAFF 02580
Warfarin 0.05% Ready For Use (Cereal Base)	Kemkill Pest Control Services Ltd	MAFF 02579
Warfarin 0.5% Master Mix (Cereal Base)	Kemkill Pest Control Services Ltd	MAFF 02575
Warfarin 0.5% Master Mix (Talc Base)	Kemkill Pest Control Services Ltd	MAFF 02576
Warfarin 1% Master Mix (Cereal Base)	Kemkill Pest Control Services Ltd	MAFF 02577
Warfarin 1% Master Mix (Talc Base)	Kemkill Pest Control Services Ltd	MAFF 02578
Warfarin 1% Tracking Dust (Talc Base)	Kemkill Pest Control Services Ltd	MAFF 02581
Warfarin Concentrate	Battle Hayward & Bower Ltd	MAFF 02325
Warfarin Concentrate	Sorex Ltd	MAFF 02326
Warfarin Master Mix 0.5% (Cereal Base)	Rodent Control Ltd	MAFF 02327
Warfarin Master Mix 1.0% (Cereal Base)	Rodent Control Ltd	MAFF 02329
Warfarin RTU	Sorex Ltd	MAFF 02335
Warfarin Ready For Use 0.025%	Rodent Control Ltd	MAFF 02332
Warfarin Ready For Use 0.05%	Rodent Control Ltd	MAFF 02331
Warfarin Ready Mixed Bait	Battle Hayward & Bower Ltd	MAFF 02333
Warfarin Sewer Bait	Sorex Ltd	MAFF 02336
Warfarin Sewer Treatment	Rodent Control Ltd	MAFF 02337
Warfarin Soluble 1%	Rodent Control Ltd	MAFF 02338
888 Zinc phosphide		
Grovex Zinc Phosphide	Killgerm Chemicals Ltd	MAFF 06230
% Grovex Zinc Phosphide	W H Groves & Family Ltd	MAFF 01030
RCR Zinc Phosphide	Killgerm Chemicals Ltd	MAFF 06231
% Zinc Phosphide	Rodent Control Ltd	MAFF 02378
ZP Rodent Bait	Bell Laboratories Inc	MAFF 02822
889 Ziram		
AAprotect	Universal Crop Protection Ltd	MAFF 03784

---

* These products are "approved for agricultural use". For further details refer to page vii.
% These products are subject to the staged revocation procedure. For further details refer to page vii.
A These products are approved for use in or near water. For further details refer to page vii.

PROFESSIONAL PRODUCTS: BIOLOGICAL AGENTS

Product Name   *Marketing Company*   *Reg. No.*

1.5 Biological Pesticides

890 Bacillus thuringiensis

	Product Name	Marketing Company	Reg. No.
*	Bactospeine WP	Koppert (UK) Ltd	MAFF 02913
*%	Biobit Flowable Concentrate	Novo Enzyme Products Ltd	MAFF 04834
*	Certan	Sandoz Agro Ltd	MAFF 00465
*	Dipel	English Woodlands Ltd	MAFF 03214
*	Dipel	Pan Britannica Industries Ltd	MAFF 06308
*	Dipel	Pan Britannica Industries Ltd	MAFF 06577
*	Fargro Thuricide HP	Fargro Ltd	MAFF 05276
*	Novosol Flowable Concentrate	Ashlade Formulations Ltd	MAFF 06566
*	Thuricide HP	Sandoz Agro Ltd	MAFF 02136

891 Trichoderma viride

*	Binab T	HDRA Sales Ltd	MAFF 00264

892 Verticillium lecanii

*	Mycotal	Koppert (UK) Ltd	MAFF 04782
*	Vertalec	Koppert (UK) Ltd	MAFF 04781

---

* These products are "approved for agricultural use". For further details refer to page vii.
% These products are subject to the staged revocation procedure. For further details refer to page vii.
A These products are approved for use in or near water. For further details refer to page vii.

PROFESSIONAL PRODUCTS: MISCELLANEOUS

Product Name	Marketing Company	Reg. No.

1.6 Miscellaneous
Including molluscicides, lumbricides, soil sterilants and fumigants

893 Aldicarb
* Landgold Aldicarb 10G	Landgold & Co Ltd	MAFF 06036
* Standon Aldicarb 10 G	Standon Chemicals Ltd	MAFF 05915

894 Aldicarb + *Lindane*
* Sentry	Embetec Crop Protection	MAFF 06209

895 Aluminium phosphide
Detia Gas EX-B	Northern Fumigation Services Ltd	MAFF 03791
Detia Gas EX-T	Northern Fumigation Services Ltd	MAFF 03792

896 Carbaryl
* Twister	Rhone-Poulenc Agriculture Ltd	MAFF 05162
* Twister Flow	Rhone-Poulenc Environmental Products	MAFF 05712

897 Carbendazim
* Fisons Turfclear	Fisons Plc	MAFF 06275

898 Carbofuran
* Barclay Carbosect	Barclay Chemicals (UK) Ltd	MAFF 05512
* Nex	Tripart Farm Chemicals Ltd	MAFF 05165
* Rampart	Sipcam UK Ltd	MAFF 05166
* Throttle	Bayer Plc	MAFF 05204
*% Throttle	Quadrangle Ltd	MAFF 05018

899 Carbosulfan
* Landgold Carbosulfan 10G	Landgold & Co Ltd	MAFF 06046
* Marshal 10 G	Rhone-Poulenc Agriculture	MAFF 06165
* Marshal 10 G	Rhone-Poulenc Crop Protection	MAFF 04646
* Standon Carbosulfan 10G	Standon Chemicals Ltd	MAFF 05671

900 Chloropicrin
* Chloropicrin Fumigant	Dewco-Lloyd Ltd	MAFF 04216

---

* These products are "approved for agricultural use". For further details refer to page vii.
% These products are subject to the staged revocation procedure. For further details refer to page vii.
A These products are approved for use in or near water. For further details refer to page vii.

PROFESSIONAL PRODUCTS: MISCELLANEOUS

Product Name	Marketing Company	Reg. No.

901 Chloropicrin + *Methyl bromide*

Brom-O-Gas	Great Lakes Chemical (Europe) Ltd	MAFF 04508
Methyl Bromide 98%	Bromine & Chemicals Ltd	MAFF 01335

902 Chlorpropham + *Cresylic acid* + *Fenuron* + *Propham*

*	Marks PCF Beet Herbicide	A H Marks & Co Ltd	MAFF 02538
*	MSS Sugar Beet Herbicide	Mirfield Sales Services Ltd	MAFF 02447

903 Cresylic acid

*	Brays Emulsion	Garden and Professional Products	MAFF 00323

904 Cresylic acid + *Chlorpropham* + *Fenuron* + *Propham*

*	Marks PCF Beet Herbicide	A H Marks & Co Ltd	MAFF 02538
*	MSS Sugar Beet Herbicide	Mirfield Sales Services Ltd	MAFF 02447

905 Dazomet

*	Basamid	BASF Plc	MAFF 00192
*	Basamid	Hortichem Ltd	MAFF 07204
*	Dazomet 98	DowElanco Ltd	MAFF 05113
*%	Salvo	Stauffer Chemical Ltd	MAFF 01853

906 1,2-Dichloropropane + *1,3-Dichloropropene*

*	Cyanamid DD Soil Fumigant	Cyanamid Of GB Ltd	MAFF 07017
*	Shell DD Soil Fumigant	Shell Chemicals UK Ltd	MAFF 01932

907 1,3-Dichloropropene

*%	Telone 2000	Dow Agriculture	MAFF 04500
*	Telone 2000	DowElanco Ltd	MAFF 05748
*%	Telone II	Dow Chemical Co Ltd	MAFF 02097
*	Telone II	DowElanco Ltd	MAFF 05749

908 1,3-Dichloropropene + *1,2-Dichloropropane*

*	Cyanamid DD Soil Fumigant	Cyanamid Of GB Ltd	MAFF 07017
*	Shell DD Soil Fumigant	Shell Chemicals UK Ltd	MAFF 01932

909 Ethoprophos

*	Mocap 10G	Rhone-Poulenc Agriculture	MAFF 06773
*%	Mocap 10G	Rhone-Poulenc Agriculture Ltd	MAFF 04935

910 Fenuron + *Chlorpropham* + *Cresylic acid* + *Propham*

*	Marks PCF Beet Herbicide	A H Marks & Co Ltd	MAFF 02538
*	MSS Sugar Beet Herbicide	Mirfield Sales Services Ltd	MAFF 02447

* These products are "approved for agricultural use". For further details refer to page vii.
% These products are subject to the staged revocation procedure. For further details refer to page vii.
A These products are approved for use in or near water. For further details refer to page vii.

PROFESSIONAL PRODUCTS: MISCELLANEOUS

Product Name	Marketing Company	Reg. No.
911 Formaldehyde		
*% Cubisan	Diversey Ltd	MAFF 00601
912 Lindane + *Aldicarb*		
* Sentry	Embetec Crop Protection	MAFF 06209
913 Lindane + *Thiophanate-methyl*		
* Castaway Plus	Rhone-Poulenc Environmental Products	MAFF 05327
* CDA Castaway Plus	Rhone-Poulenc Environmental Products	MAFF 04758
914 Magnesium phosphide		
Rentokil Degesch Phostoxin Plates	Rentokil Ltd	MAFF 01739
915 Metaldehyde		
* Amos Metaldehyde	Kommer-Brookwick	MAFF 05756
*% Amos Metaldehyde	Kommer-Brookwick	MAFF 06140
* Chiltern Hardy	Chiltern Farm Chemicals Ltd	MAFF 06948
* Doff Agricultural Slug Killer with Animal Repellent	Doff Portland Ltd	MAFF 06058
* Doff Horticultural Slug Killer Blue Mini Pellets	Doff Portland Ltd	MAFF 05688
* Doff Metaldehyde Slug Killer Mini Pellets	Doff Portland Ltd	MAFF 00741
* Escar-go 3	Chiltern Farm Chemicals Ltd	MAFF 06075
* Escar-go 6	Chiltern Farm Chemicals Ltd	MAFF 06076
*% FP107	ICI Agrochemicals	MAFF 05861
* FP107	Imperial Chemical Industries Plc	MAFF 06527
* FP107	Zeneca Crop Protection	MAFF 06666
* Fisons Helarion	Fisons Plc	MAFF 02520
* Gastrotox 6G Pellets	Truchem Ltd	MAFF 04066
* Luxan Metaldehyde	Luxan (UK) Ltd	MAFF 06564
* Metarex	B De Sangosse SA	MAFF 04910
* Metarex RG	De Sangosse UK	MAFF 06754
*% Metarex RG	Phosyn Chemical Ltd	MAFF 05255
* Mifaslug	Farmers Crop Chemicals Ltd	MAFF 01349
* Morgan's Blue Mini Slug Killer Pellets	David Morgan (Nottingham) Ltd	MAFF 06431
* Optimol	Farm Protection	MAFF 04903
* Optimol	Imperial Chemical Industries Plc	MAFF 06526
* Optimol	Zeneca Crop Protection	MAFF 06688
* Quadrangle Mini Slug Pellets	Quadrangle Agrochemicals Ltd	MAFF 01670
* Slug Destroyer	AgrEvo UK Crop Protection Ltd	MAFF 07283
* Slug Destroyer	Schering Agrochemicals Ltd	MAFF 03919

* These products are "approved for agricultural use". For further details refer to page vii.
% These products are subject to the staged revocation procedure. For further details refer to page vii.
A These products are approved for use in or near water. For further details refer to page vii.

PROFESSIONAL PRODUCTS: MISCELLANEOUS

Product Name	Marketing Company	Reg. No.

915 **Metaldehyde**—*continued*

* Slug Pellets	Pan Britannica Industries Ltd	MAFF 01558
* Super-flor 6 % Metaldehyde Slug Killer Mini Pellets	Collingham Marketing Ltd	MAFF 05453
* Tripart Mini Slug Pellets	Tripart Farm Chemicals Ltd	MAFF 02207
* Unicrop 6 % Mini Slug Pellets	Universal Crop Protection Ltd	MAFF 02275
*% Vassgro Mini Slug Pellets	L W Vass (Agricultural) Ltd	MAFF 03579

916 **Metam-sodium**

* Campbell's Metham Sodium	J D Campbell & Sons Ltd	MAFF 00412
* Sistan	Universal Crop Protection Ltd	MAFF 01957

917 **Methiocarb**

* Club	Bayer Plc	MAFF 07176
*% Club	Imperial Chemical Industries Plc	MAFF 03800
*% Club	Zeneca Crop Protection	MAFF 06645
* Decoy	Bayer Plc	MAFF 06535
* Draza	Bayer UK Ltd	MAFF 00765
* Draza 2	Bayer Plc	MAFF 04748
* Draza Plus	Bayer Plc	MAFF 06553
* Draza ST	Bayer Plc	MAFF 05315
* Elvitox	Bayer Plc	MAFF 06738
* Epox	Bayer Plc	MAFF 06737

918 **Methyl bromide**

Fumyl-O-Gas	Brian Jones & Associates Ltd	MAFF 04833
Mebrom 100	Mebrom NV	MAFF 04869
Mebrom 98	Mebrom NV	MAFF 04779
Methyl Bromide 100%	Bromine & Chemicals Ltd	MAFF 01336
Rentokil Methyl Bromide	Rentokil Ltd	MAFF 05646
Sobrom BM 100	Brian Jones & Associates Ltd	MAFF 04381
Sobrom BM 98	Brian Jones & Associates Ltd	MAFF 04189

919 **Methyl bromide** + *Chloropicrin*

Brom-O-Gas	Great Lakes Chemical (Europe) Ltd	MAFF 04508
Methyl Bromide 98%	Bromine & Chemicals Ltd	MAFF 01335

920 **Oxamyl**

* Fielder	Du Pont (UK) Ltd	MAFF 05279

921 **Propham** + *Chlorpropham* + *Cresylic acid* + *Fenuron*

* Marks PCF Beet Herbicide	A H Marks & Co Ltd	MAFF 02538
* MSS Sugar Beet Herbicide	Mirfield Sales Services Ltd	MAFF 02447

* These products are "approved for agricultural use". For further details refer to page vii.
% These products are subject to the staged revocation procedure. For further details refer to page vii.
A These products are approved for use in or near water. For further details refer to page vii.

PROFESSIONAL PRODUCTS: MISCELLANEOUS

Product Name	Marketing Company	Reg. No.

922 Thiodicarb

*% Genesis	Embetec Crop Protection	MAFF 05990
* Genesis	Rhone-Poulenc Agriculture	MAFF 06168

923 Thiophanate-methyl + *Lindane*

* Castaway Plus	Rhone-Poulenc Environmental Products	MAFF 05327
* CDA Castaway Plus	Rhone-Poulenc Environmental Products	MAFF 04758

---

* These products are "approved for agricultural use". For further details refer to page vii.
% These products are subject to the staged revocation procedure. For further details refer to page vii.
A These products are approved for use in or near water. For further details refer to page vii.

2
AMATEUR PRODUCTS

Product Name	Marketing Company	Reg. No.

Amateur Products

924 Alloxydim-sodium *(Herbicide)*

% Weed Out	Rhone-Poulenc Agriculture Ltd	MAFF 04531
Weed-out Couchgrass Killer	Pan Britannica Industries Ltd	MAFF 05947

925 Alphachloralose *(Vertebrate control)*

Rentokil Alphakil	Rentokil Ltd	MAFF 01722

926 Aluminium ammonium sulphate *(Vertebrate control)*

Bio Catapult	Pan Britannica Industries Ltd	MAFF 07195
Curb (Garden Pack)	Sphere Laboratories (London) Ltd	MAFF 03983
Scoot	Garotta Products Ltd	MAFF 03706
Stay-Off	Synchemicals Ltd	MAFF 02019

927 Aluminium sulphate *(Miscellaneous)*

6 X Slug Killer	Organic Concentrates Ltd	MAFF 04702
% Chempak Slug Killer	Chempak Ltd	MAFF 04921
Fertosan Slug and Snail Powder	Fertosan Products Ltd	MAFF 00864
Growing Success Slug Killer	Growing Success Organics Ltd	MAFF 04386
% Septico Slug Killer	Septico Organic Products	MAFF 02562

928 Amitrole + Atrazine *(Herbicide)*

Deeweed	Arable & Bulb Chemicals Ltd	MAFF 00659
Doff Total Path Weedkiller	Doff Portland Ltd	MAFF 04632
Murphy Path Weed Killer	Fisons Plc	MAFF 03630
Wilko Path Weedkiller	Wilkinson Home & Garden Stores	MAFF 04633
Wilko Path Weedkiller	Wilkinson Home & Garden Stores	MAFF 06976
% Wilko Path Weedkiller	Wilkinson Home & Garden Stores	MAFF 05773

929 Amitrole + Atrazine + 2,4-D *(Herbicide)*

% Rentokil Path & Patio Weedkiller	Rentokil Ltd	MAFF 02873

930 Amitrole + 2,4-D + Diuron + Simazine *(Herbicide)*

Hytrol	Agrichem (International) Ltd	MAFF 04540

931 Amitrole + Diquat + Paraquat + Simazine *(Herbicide)*

Pathclear	ICI Garden and Professional Products	MAFF 01546
Pathclear	Zeneca Garden Care	MAFF 06845

* These products are "approved for agricultural use". For further details refer to page vii.
% These products are subject to the staged revocation procedure. For further details refer to page vii.
A These products are approved for use in or near water. For further details refer to page vii.

AMATEUR PRODUCTS

Product Name	Marketing Company	Reg. No.

932 Amitrole + MCPA *(Herbicide)*
% Murphy Problem Weeds Killer — Fisons Plc — MAFF 05295

933 Amitrole + MCPA + Simazine *(Herbicide)*
% Fisons Path Weeds Killer — Fisons Plc — MAFF 02700

934 Amitrole + Simazine *(Herbicide)*

Path and Drive Weedkiller	May & Baker Garden Care	MAFF 04762
Path and Drive Weedkiller	Pan Britannica Industries Ltd	MAFF 05958
% Path and Drive Weedkiller	Rhone-Poulenc Agriculture Ltd	MAFF 05351
% Super Weedex	Fisons Plc	MAFF 02054

935 Ammonium sulphamate *(Herbicide)*

Amcide	Battle Hayward & Bower Ltd	MAFF 00089
Root-Out	Dax Products Ltd	MAFF 03510

936 Atrazine + Amitrole *(Herbicide)*

Deeweed	Arable & Bulb Chemicals Ltd	MAFF 00659
Doff Total Path Weedkiller	Doff Portland Ltd	MAFF 04632
Murphy Path Weed Killer	Fisons Plc	MAFF 03630
Wilko Path Weedkiller	Wilkinson Home & Garden Stores	MAFF 04633
Wilko Path Weedkiller	Wilkinson Home & Garden Stores	MAFF 06976
% Wilko Path Weedkiller	Wilkinson Home & Garden Stores	MAFF 05773

937 Atrazine + Amitrole + 2,4-D *(Herbicide)*
% Rentokil Path & Patio Weedkiller — Rentokil Ltd — MAFF 02873

938 Bacillus thuringiensis *(Biological)*

Bactospeine WP	Koppert (UK) Ltd	MAFF 05675
Bio 'BT' Caterpillar Killer	Pan Britannica Industries Ltd	MAFF 06307
Dipel	English Woodlands Ltd	MAFF 04210
Dipel	Pan Britannica Industries Ltd	MAFF 06576
Nature's Answer	Fisons Plc	MAFF 05275
Natures Friends Bactospeine	Zeneca Garden Care	MAFF 07070

939 Benazolin + 2,4-D + Dicamba + Dichlorophen + Dichlorprop + Mecoprop *(Herbicide)*

Boots Total Lawn Treatment	The Boots Company Plc	MAFF 00293
Vitax Green Up Lawn Feed 'N' Weed Plus Mosskiller	Vitax Ltd	MAFF 05570

---

* These products are "approved for agricultural use". For further details refer to page vii.
% These products are subject to the staged revocation procedure. For further details refer to page vii.
A These products are approved for use in or near water. For further details refer to page vii.

AMATEUR PRODUCTS

Product Name	Marketing Company	Reg. No.
940 Bendiocarb *(Insecticide)*		
B & Q Ant Killer	B & Q Plc	MAFF 04880
B & Q Woodlice killer	B & Q Plc	MAFF 07155
% Camco Ant Powder	Camco Ltd	MAFF 04024
Camco Insect Powder	Cambridge Animal and Public Health Ltd	MAFF 05111
Co-op Garden Maker Ant Killer	Co-Operative Wholesale Society Ltd	MAFF 04883
Delta Insect Powder	Cambridge Animal and Public Health Ltd	MAFF 04025
Devcol Ant Killer	Devcol Ltd	MAFF 05867
Do It All Ant Killer	Do-It-All Ltd	MAFF 04854
Doff Ant Control Powder	Doff Portland Ltd	MAFF 04881
Doff Wasp Nest Killer	Doff Portland Ltd	MAFF 06114
Doff Woodlice Killer	Doff Portland Ltd	MAFF 06081
Ficam Insect Powder	Cambridge Animal and Public Health Ltd	MAFF 05286
Focus Ant Killer	Focus DIY Ltd	MAFF 06235
Great Mills Ant Killer	Great Mills (Retail) Ltd	MAFF 04852
Great Mills Woodlice Killer	Great Mills (Retail) Ltd	MAFF 07066
Homebase Ant Killer	Sainsbury's Homebase House and Garden Centres	MAFF 06236
Homebase Woodlice Killer	Sainsbury's Homebase House and Garden Centres	MAFF 06994
Portland Brand Ant Killer	Doff Portland Ltd	MAFF 07068
Secto Ant and Crawling Insect Powder	Secto Co Ltd	MAFF 05009
Secto Wasp Killer Powder	Secto Co Ltd	MAFF 05517
Texas Ant killer	Texas Home Care	MAFF 07069
Vapona Woodlice Killer	Sara Lee Household & Personal Care	MAFF 06592
Wilko Ant Destroyer	Wilkinson Home & Garden Stores	MAFF 04853
Wilko Woodlice Killer	Wilkinson Home & Garden Stores	MAFF 07156
Woolworths Ant Killer	Woolworths Plc	MAFF 07067
941 Benomyl *(Fungicide)*		
Benlate	Garden and Professional Products	MAFF 02491
942 Bifenthrin *(Insecticide)*		
Polysect Insecticide	Monsanto Agricultural Co	MAFF 06908
Polysect Insecticide	Monsanto Agricultural Co	MAFF 06909
Polysect Insecticide Ready To Use	Monsanto Agricultural Co	MAFF 06910
Polysect Insecticide Ready To Use	Monsanto Agricultural Co	MAFF 06911

*   These products are "approved for agricultural use". For further details refer to page vii.
%   These products are subject to the staged revocation procedure. For further details refer to page vii.
A   These products are approved for use in or near water. For further details refer to page vii.

AMATEUR PRODUCTS

Product Name	Marketing Company	Reg. No.
943 Bioallethrin + *Permethrin* (Insecticide)		
Floracid	Perycut Insectengun Ltd	MAFF 04577
Floracid	Perycut Insectengun Ltd	MAFF 06798
Longer Lasting Bug Gun	ICI Garden and Professional Products	MAFF 06137
Longer Lasting Bug Gun	Zeneca Garden Care	MAFF 06916
New Spraydex Greenfly Killer	Spraydex Ltd	MAFF 04847
Spraydex Greenfly Killer	Spraydex Ltd	MAFF 06587
944 Bitumen (Insecticide)		
Corry's Fruit Tree Grease	Vitax Ltd	MAFF 05676
945 Bitumen (Miscellaneous)		
PBI Arbrex Pruning Compound	Pan Britannica Industries Ltd	MAFF 00111
946 Bone oil (Vertebrate control)		
Renardine	Gilbertson & Page Ltd	MAFF 04402
Renardine	Roebuck-Eyot Ltd	MAFF 06378
Renardine 72/2	Roebuck-Eyot Ltd	MAFF 06769
947 Borax (Insecticide)		
Nippon Ant Killer Liquid	Vitax Ltd	MAFF 05270
948 Borax + *Carbaryl* (Insecticide)		
% Boots Ant Killer	The Boots Company Plc	MAFF 04088
949 Brodifacoum (Vertebrate control)		
Mouser	Imperial Chemical Industries Plc	MAFF 03213
Mouser	Zeneca Public Health	MAFF 06831
950 Bromadiolone (Vertebrate control)		
Bromadeth Rat and Mouse Killer	Gerhardt Pharmaceuticals Ltd	MAFF 03515
Rentokil Biotrol Plus Outdoor Rat Killer	Rentokil Ltd	MAFF 02936
Rentokil Rodine C (For Home Garden Use)	Rentokil Ltd	MAFF 03318
Rentokil Total Mouse Killer System	Rentokil Ltd	MAFF 02994
951 Bromophos (Insecticide)		
% PBI Bromophos	Pan Britannica Industries Ltd	MAFF 00340
952 Bupirimate + *Pirimicarb* + *Triforine* (Fungicide)(Insecticide)		
Roseclear	Imperial Chemical Industries Plc	MAFF 01826
Roseclear	Zeneca Garden Care	MAFF 06849

* These products are "approved for agricultural use". For further details refer to page vii.
% These products are subject to the staged revocation procedure. For further details refer to page vii.
A These products are approved for use in or near water. For further details refer to page vii.

AMATEUR PRODUCTS

Product Name	Marketing Company	Reg. No.

953 Bupirimate + Triforine (Fungicide)
Nimrod T	Imperial Chemical Industries Plc	MAFF 03982
Nimrod T	Zeneca Garden Products	MAFF 06843

954 Butoxycarboxim (Insecticide)
Plant Pin	Phostrogen Ltd	MAFF 01599

955 Calciferol + Difenacoum (Vertebrate control)
Karate Mouse Bait	Lever Industrial Ltd	MAFF 05866

956 Captan + 1-Naphthylacetic acid (Fungicide)(Herbicide)
	Doff Hormone Rooting Powder	Doff Portland Ltd	MAFF 01065
	Murphy Hormone Rooting Powder	Fisons Plc	MAFF 03618
%	New Strike	May & Baker Garden Care	MAFF 04430
	New Strike	Pan Britannica Industries Ltd	MAFF 05956
	Rooting Powder	Vitax Ltd	MAFF 06334

957 Carbaryl (Insecticide)
%	Murphy Lawn Pest Killer	Murphy Home & Garden Products	MAFF 04433

958 Carbaryl (Miscellaneous)
Autumn and Winter Toplawn	Pan Britannica Industries Ltd	MAFF 04138
Fisons Water On Lawn Pest Killer	Fisons Plc	MAFF 05896
Murphy Lawn Pest Killer	Fisons Plc	MAFF 05924

959 Carbaryl + Borax (Insecticide)
%	Boots Ant Killer	The Boots Company Plc	MAFF 04088

960 Carbaryl + Rotenone (Insecticide)
Boots Garden Insect Powder with Derris	The Boots Company Plc	MAFF 02849

961 Carbendazim (Fungicide)
Boots Garden Fungicide	The Boots Company Plc	MAFF 02401
Doff Plant Disease Control	Doff Portland Ltd	MAFF 07159
Murphy Systemic Action Fungicide	Fisons Plc	MAFF 04992
Murphy Zap Cap Systemic Fungicide	Fisons Plc	MAFF 04678
PBI Supercarb	Pan Britannica Industries Ltd	MAFF 03981

962 Carbendazim + Copper oxychloride + Permethrin + Sulphur
(Fungicide)(Insecticide)
%	Bio Multiveg	Pan Britannica Industries Ltd	MAFF 03116

---

* These products are "approved for agricultural use". For further details refer to page vii.
% These products are subject to the staged revocation procedure. For further details refer to page vii.
A These products are approved for use in or near water. For further details refer to page vii.

AMATEUR PRODUCTS

Product Name	Marketing Company	Reg. No.

963 Chlorpyrifos + *Diazinon* *(Insecticide)*
    Chlorophos   Pan Britannica Industries Ltd   MAFF 03604

964 Citronella oil + *Methyl nonyl ketone* *(Vertebrate control)*
    Secto Keep Off   Secto Co Ltd   MAFF 04960

965 Copper *(Fungicide)*
    Bordeaux Mixture   Vitax Ltd   MAFF 07162

966 Copper complex – bordeaux *(Fungicide)*
    Bordeaux Mixture   Synchemicals Ltd   MAFF 00297

967 Copper oxychloride *(Fungicide)*
    Murphy Traditional Copper Fungicide   Fisons Plc   MAFF 04585

968 Copper oxychloride + *Carbendazim* + *Permethrin* + *Sulphur*
(Fungicide)(Insecticide)
%   Bio Multiveg   Pan Britannica Industries Ltd   MAFF 03116

969 Copper sulphate *(Fungicide)*
    B & Q Garden Fungicide Spray   B & Q Plc   MAFF 06285
    PBI Cheshunt Compound   Pan Britannica Industries Ltd   MAFF 00485
    Spraydex General Purpose Fungicide   Spraydex Ltd   MAFF 02865

970 Copper sulphate + *Ferrous sulphate* *(Miscellaneous)*
    Snailaway   Interpet Ltd   MAFF 02457

971 Coumatetralyl *(Vertebrate control)*
    PBI Racumin Mouse Bait   Pan Britannica Industries Ltd   MAFF 01678
    PBI Racumin Rat Bait   Pan Britannica Industries Ltd   MAFF 01680

972 Cresylic acid *(Miscellaneous)*
    Armillatox   Armillatox Ltd   MAFF 00115
    Armillatox   Armillatox Ltd   MAFF 06234
    Clean Up   Garden and Professional Products   MAFF 00539
    Medo   Vitax Ltd   MAFF 05787

973 2,4-D *(Herbicide)*
%   Double H Lawn Feed and Weed   TriTrade Ltd   MAFF 04300
    Good Life Lawn Feed and Weed   Humber Fertilizers Ltd   MAFF 04242

974 2,4-D + *2-3-6 TBA* *(Herbicide)*
    Touchweeder   Thomas Elliott Ltd   MAFF 02864

---

*   These products are "approved for agricultural use". For further details refer to page vii.
%   These products are subject to the staged revocation procedure. For further details refer to page vii.
A   These products are approved for use in or near water. For further details refer to page vii.

AMATEUR PRODUCTS

Product Name	Marketing Company	Reg. No.

975 2,4-D + *Amitrole* + *Atrazine* (Herbicide)

% Rentokil Path & Patio Weedkiller	Rentokil Ltd	MAFF 02873

976 2,4-D + *Amitrole* + *Diuron* + *Simazine* (Herbicide)

Hytrol	Agrichem (International) Ltd	MAFF 04540

977 2,4-D + *Benazolin* + *Dicamba* + *Dichlorophen* + *Dichlorprop* + *Mecoprop* (Herbicide)

Boots Total Lawn Treatment	The Boots Company Plc	MAFF 00293
Vitax Green Up Lawn Feed 'N' Weed Plus Mosskiller	Vitax Ltd	MAFF 05570

978 2,4-D + *Dicamba* (Herbicide)

B & Q Granular Weed and Feed	B & Q Plc	MAFF 05294
B & Q Lawnweed Spray	B & Q Plc	MAFF 05804
Bio Lawn Weed Killer	Pan Britannica Industries Ltd	MAFF 00268
Elliot's Easy-Weed	Thomas Elliott Ltd	MAFF 06251
Fisons Lawn Spot Weeder	Fisons Plc	MAFF 00886
% Green Up Lawn Feed and Weed	Synchemicals Ltd	MAFF 02897
Green Up Lawn Feed and Weed	Vitax Ltd	MAFF 06419
Green Up Weedfree Lawn Weedkiller	Vitax Ltd	MAFF 05322
% Green Up Weedfree Spot Weedkiller for Lawns	Synchemicals Ltd	MAFF 03253
Green Up Weedfree Spot Weedkiller for Lawns	Vitax Ltd	MAFF 06321
Lawn Builder Plus Weed Control	O M Scott & Sons Ltd	MAFF 06343
Lawn Weed Gun	ICI Garden and Professional Products	MAFF 04407
Lawn Weed Gun	Zeneca Garden Care	MAFF 06838
Lawnsman Weed and Feed	Imperial Chemical Industries Plc	MAFF 02535
Lawnsman Weed and Feed	Zeneca Garden Care	MAFF 06842
PBI Toplawn	Pan Britannica Industries Ltd	MAFF 02145
Tesco Lawn Feed 'n' Weed	Tesco Stores Ltd	MAFF 05617

979 2,4-D + *Dicamba* + *Ferrous sulphate* (Herbicide)

B & Q Triple Action Lawn Care	B & Q Plc	MAFF 05282
Greensward	Imperial Chemical Industries Plc	MAFF 03931
Greensward	Zeneca Garden Care	MAFF 06841
Triple Action Grass Hopper	Imperial Chemical Industries Plc	MAFF 03932
Triple Action Grasshopper	Zeneca Garden Care	MAFF 06852

980 2,4-D + *Dicamba* + *Mecoprop* (Herbicide)

New Formulation SBK Brushwood Killer	Vitax Ltd	MAFF 05043

---

* These products are "approved for agricultural use". For further details refer to page vii.
% These products are subject to the staged revocation procedure. For further details refer to page vii.
A These products are approved for use in or near water. For further details refer to page vii.

AMATEUR PRODUCTS

Product Name	Marketing Company	Reg. No.
981 2,4-D + *Dichlorprop* (Herbicide)		
B & Q Lawn Feed and Weed	B & Q Plc	MAFF 05487
B & Q Lawn Spot Weeder	B & Q Plc	MAFF 05486
B & Q Lawn Weedkiller	B & Q Plc	MAFF 05324
Doff Lawn Feed and Weed	Doff Portland Ltd	MAFF 05117
Doff Lawn Spot Weeder	Doff Portland Ltd	MAFF 03995
Doff Nettle Gun	Doff Portland Ltd	MAFF 07158
Doff New Formula Lawn Weedkiller	Doff Portland Ltd	MAFF 05666
Fisons Clover-kil	Fisons Plc	MAFF 05808
Fisons Ready-to-Use Clover-kil	Fisons Plc	MAFF 05585
Fisons Ready-to-Use Lawn Weedkiller	Fisons Plc	MAFF 05868
Fisons Water-on Lawn Weedkiller	Fisons Plc	MAFF 05807
Focus Lawn Weedkiller	Focus DIY Ltd	MAFF 06096
Great Mills Lawn Spot Weeder	Great Mills (Retail) Ltd	MAFF 05014
Murphy Clover-Kil	Fisons Plc	MAFF 05271
Murphy Lawn Weedkiller	Fisons Plc	MAFF 05586
Murphy Lawn Weedkiller and Lawn Tonic	Murphy Home & Garden Products	MAFF 03619
Wilko Lawn Spot Weeder	Wilkinson Home & Garden Stores	MAFF 05130
Wilko Lawn Weedkiller	Wilkinson Home & Garden Stores	MAFF 03749
Woolworth Lawn Weedkiller	Woolworths Plc	MAFF 07064
Woolworths Lawn Spot Weeder	Doff Portland Ltd	MAFF 07105
982 2,4-D + *Dichlorprop* + *Mecoprop* (Herbicide)		
Boots Kill-A-Weed	The Boots Company Plc	MAFF 03523
Boots Lawn Weedkiller	The Boots Company Plc	MAFF 02677
983 2,4-D + *Ferrous sulphate* + *Mecoprop* (Herbicide)		
Asda Lawn Feed and Weed Moss Killer	Asda Stores Ltd	MAFF 03819
Gem Lawn Weed and Feed Plus Mosskiller	Gem Gardening	MAFF 04488
Green Up Feed and Weed Plus Mosskiller	Vitax Ltd	MAFF 05491
Green Up Feed and Weed Plus Mosskiller	Vitax Ltd	MAFF 06513
Notcutts Feed and Weed with Mosskiller	Notcutts Garden Centres Ltd	MAFF 05124
% Supergreen Feed, Weed and Mosskiller	May & Baker Garden Care	MAFF 05349
Supergreen Triple (Feed, Weed and Mosskiller)	Pan Britannica Industries Ltd	MAFF 05955
Vitagrow Feed and Weed with Mosskiller	Vitagrow (Fertilisers) Ltd	MAFF 05098

* These products are "approved for agricultural use". For further details refer to page vii.
% These products are subject to the staged revocation procedure. For further details refer to page vii.
A These products are approved for use in or near water. For further details refer to page vii.

AMATEUR PRODUCTS

Product Name	Marketing Company	Reg. No.
983 2,4-D + Ferrous sulphate + Mecoprop (Herbicide)—continued		
Weed 'N' Feed Extra	Steetley Chemicals Ltd	MAFF 03450
Weed 'N' Feed Extra	Vitax Ltd	MAFF 06506
984 2,4-D + Mecoprop (Herbicide)		
Asda Lawn Weed and Feed	Asda Stores Ltd	MAFF 04761
Elliott's Turf Feed and Weed	Thomas Elliott Ltd	MAFF 05046
Gem Lawn Weed and Feed	Gem Gardening	MAFF 04486
Gem Lawn Weed and Feed 4	Gem Gardening	MAFF 04487
% Green Up Lawn Spot Weedkiller	Vitax Ltd	MAFF 05086
J. Arthur Bower's Lawn Food with Weedkiller	Sinclair Horticulture & Leisure Ltd	MAFF 04301
% Lawn Feed and Weed Granules	May & Baker Garden Care	MAFF 05263
Lawn Feed and Weed Granules	Pan Britannica Industries Ltd	MAFF 05902
% Lawn Feed and Weed Granules	Rhone-Poulenc Agriculture Ltd	MAFF 05343
% Lawn Spot Weed Granules	May & Baker Garden Care	MAFF 05291
Lawn Spot Weed Granules	Pan Britannica Industries Ltd	MAFF 05903
% Spot Weeder	Rhone-Poulenc Agriculture Ltd	MAFF 05347
Spraydex Lawn Spot Weeder	Spraydex Ltd	MAFF 03141
Supergreen Double (Feed and Weed)	Pan Britannica Industries Ltd	MAFF 05949
Supergreen Double (Feed and Weed)	Rhone-Poulenc Agriculture Ltd	MAFF 05348
Supertox	Rhone-Poulenc Agriculture Ltd	MAFF 05350
Supertox Lawn Weedkiller	Pan Britannica Industries Ltd	MAFF 05948
Supertox Spot	Pan Britannica Industries Ltd	MAFF 05951
% Supertox Spot Weeder	May & Baker Garden Care	MAFF 05168
% Verdant F6 Combined Fertilizer and Weedkiller	T Parker & Sons (Turf Management) Ltd	MAFF 05200
Verdant No. 6 Feed and Weed	T Parker & Sons (Turf Management) Ltd	MAFF 05047
Verdone 2	Imperial Chemical Industries Plc	MAFF 03271
Verdone 2	Zeneca Garden Care	MAFF 06853
Wilko Lawn Feed 'n' Weed	Wilkinson Home & Garden Stores	MAFF 04403
Wilko Lawn Food and Weedkiller	Wilkinson Home & Garden Stores	MAFF 04302
985 2,4-D + Mecoprop-P (Herbicide)		
Green Up Lawn Spot Weedkiller	Vitax Ltd	MAFF 06028
Proctors Lawn Weed and Feed	H & T Proctor	MAFF 04669
Vitax 'Green Up' Granular Lawn Feed and Weed	Vitax Ltd	MAFF 06158
986 Dalapon (Herbicide)		
Synchemicals Couch and Grass Killer	Synchemicals Ltd	MAFF 02735

---

* These products are "approved for agricultural use". For further details refer to page vii.
% These products are subject to the staged revocation procedure. For further details refer to page vii.
A These products are approved for use in or near water. For further details refer to page vii.

AMATEUR PRODUCTS

Product Name	Marketing Company	Reg. No.
987 Diazinon + Chlorpyrifos *(Insecticide)*		
Chlorophos	Pan Britannica Industries Ltd	MAFF 03604
988 Dicamba + Benazolin + 2,4-D + Dichlorophen + Dichlorprop + Mecoprop *(Herbicide)*		
Boots Total Lawn Treatment	The Boots Company Plc	MAFF 00293
Vitax Green Up Lawn Feed 'N' Weed Plus Mosskiller	Vitax Ltd	MAFF 05570
989 Dicamba + 2,4-D *(Herbicide)*		
B & Q Granular Weed and Feed	B & Q Plc	MAFF 05294
B & Q Lawnweed Spray	B & Q Plc	MAFF 05804
Bio Lawn Weed Killer	Pan Britannica Industries Ltd	MAFF 00268
Elliot's Easy-Weed	Thomas Elliott Ltd	MAFF 06251
Fisons Lawn Spot Weeder	Fisons Plc	MAFF 00886
% Green Up Lawn Feed and Weed	Synchemicals Ltd	MAFF 02897
Green Up Lawn Feed and Weed	Vitax Ltd	MAFF 06419
Green Up Weedfree Lawn Weedkiller	Vitax Ltd	MAFF 05322
% Green Up Weedfree Spot Weedkiller for Lawns	Synchemicals Ltd	MAFF 03253
Green Up Weedfree Spot Weedkiller for Lawns	Vitax Ltd	MAFF 06321
Lawn Builder Plus Weed Control	O M Scott & Sons Ltd	MAFF 06343
Lawn Weed Gun	ICI Garden and Professional Products	MAFF 04407
Lawn Weed Gun	Zeneca Garden Care	MAFF 06838
Lawnsman Weed and Feed	Imperial Chemical Industries Plc	MAFF 02535
Lawnsman Weed and Feed	Zeneca Garden Care	MAFF 06842
PBI Toplawn	Pan Britannica Industries Ltd	MAFF 02145
Tesco Lawn Feed 'n' Weed	Tesco Stores Ltd	MAFF 05617
990 Dicamba + 2,4-D + Ferrous sulphate *(Herbicide)*		
B & Q Triple Action Lawn Care	B & Q Plc	MAFF 05282
Greensward	Imperial Chemical Industries Plc	MAFF 03931
Greensward	Zeneca Garden Care	MAFF 06841
Triple Action Grass Hopper	Imperial Chemical Industries Plc	MAFF 03932
Triple Action Grasshopper	Zeneca Garden Care	MAFF 06852
991 Dicamba + 2,4-D + Mecoprop *(Herbicide)*		
New Formulation SBK Brushwood Killer	Vitax Ltd	MAFF 05043

* These products are "approved for agricultural use". For further details refer to page vii.
% These products are subject to the staged revocation procedure. For further details refer to page vii.
A These products are approved for use in or near water. For further details refer to page vii.

AMATEUR PRODUCTS

Product Name	Marketing Company	Reg. No.
992 Dicamba + *Dichlorprop* + *MCPA* (*Herbicide*)		
B & Q Liquid Weed and Feed	B & Q Plc	MAFF 05293
Boots Nettle and Bramble Weedkiller	The Boots Company Plc	MAFF 03455
Groundclear	Pan Britannica Industries Ltd	MAFF 05953
Groundclear Spot	Pan Britannica Industries Ltd	MAFF 05950
Lawnsman Liquid Weed and Feed	Imperial Chemical Industries Plc	MAFF 03610
Liquid Weed and Feed	ICI Garden and Professional Products	MAFF 05869
Liquid Weed and Feed	Zeneca Garden Care	MAFF 06887
% Ready To Use Groundclear	May & Baker Garden Care	MAFF 05704
Woolworth Liquid Lawn Feed and Weed	Woolworths Plc	MAFF 07060
993 Dicamba + *Dichlorprop* + *Mecoprop* (*Herbicide*)		
New Supertox	Pan Britannica Industries Ltd	MAFF 06128
994 Dicamba + *MCPA* + *Mecoprop* (*Herbicide*)		
Bio Weed Pencil	Pan Britannica Industries Ltd	MAFF 04054
Biospot	Pan Britannica Industries Ltd	MAFF 05071
Fisons Evergreen Feed and Weed Liquid	Fisons Plc	MAFF 05664
Fisons Lawncare Liquid	Fisons Plc	MAFF 03625
Homebase Lawn Feed and Weed liquid	Homebase Ltd	MAFF 06086
995 Dicamba + *Mecoprop* (*Herbicide*)		
Elliott's Touchweeder	Thomas Elliott Ltd	MAFF 06252
996 Dichlobenil (*Herbicide*)		
Casoron G-4	Vitax Ltd	MAFF 05734
Casoron G4	Vitax Ltd	MAFF 06866
Path and Shrub Guard	ICI Garden Products	MAFF 05733
Path and Shrub Guard	Zeneca Garden Care	MAFF 06844
997 Dichlorophen (*Herbicide*)		
Bio Moss Killer	Pan Britannica Industries Ltd	MAFF 00270
Mossgun	Imperial Chemical Industries Plc	MAFF 03326
Mossgun	Zeneca Garden Care	MAFF 06821
Murphy Super Mosskiller Ready To Use	Fisons Plc	MAFF 05120

* These products are "approved for agricultural use". For further details refer to page vii.
% These products are subject to the staged revocation procedure. For further details refer to page vii.
A These products are approved for use in or near water. For further details refer to page vii.

AMATEUR PRODUCTS

Product Name	Marketing Company	Reg. No.

998 Dichlorophen + *Benazolin* + *2,4-D* + *Dicamba* + *Dichlorprop* + *Mecoprop* *(Herbicide)*

Boots Total Lawn Treatment	The Boots Company Plc	MAFF 00293
Vitax Green Up Lawn Feed 'N' Weed Plus Mosskiller	Vitax Ltd	MAFF 05570

999 Dichlorophen + *1-Naphthylacetic acid* *(Herbicide)*

Bio Roota	Pan Britannica Industries Ltd	MAFF 00271

1000 Dichlorprop + *Benazolin* + *2,4-D* + *Dicamba* + *Dichlorophen* + *Mecoprop* *(Herbicide)*

Boots Total Lawn Treatment	The Boots Company Plc	MAFF 00293
Vitax Green Up Lawn Feed 'N' Weed Plus Mosskiller	Vitax Ltd	MAFF 05570

1001 Dichlorprop + *2,4-D* *(Herbicide)*

B & Q Lawn Feed and Weed	B & Q Plc	MAFF 05487
B & Q Lawn Spot Weeder	B & Q Plc	MAFF 05486
B & Q Lawn Weedkiller	B & Q Plc	MAFF 05324
Doff Lawn Feed and Weed	Doff Portland Ltd	MAFF 05117
Doff Lawn Spot Weeder	Doff Portland Ltd	MAFF 03995
Doff Nettle Gun	Doff Portland Ltd	MAFF 07158
Doff New Formula Lawn Weedkiller	Doff Portland Ltd	MAFF 05666
Fisons Clover-kil	Fisons Plc	MAFF 05808
Fisons Ready-to-Use Clover-kil	Fisons Plc	MAFF 05585
Fisons Ready-to-Use Lawn Weedkiller	Fisons Plc	MAFF 05868
Fisons Water-on Lawn Weedkiller	Fisons Plc	MAFF 05807
Focus Lawn Weedkiller	Focus DIY Ltd	MAFF 06096
Great Mills Lawn Spot Weeder	Great Mills (Retail) Ltd	MAFF 05014
Murphy Clover-Kil	Fisons Plc	MAFF 05271
Murphy Lawn Weedkiller	Fisons Plc	MAFF 05586
Murphy Lawn Weedkiller and Lawn Tonic	Murphy Home & Garden Products	MAFF 03619
Wilko Lawn Spot Weeder	Wilkinson Home & Garden Stores	MAFF 05130
Wilko Lawn Weedkiller	Wilkinson Home & Garden Stores	MAFF 03749
Woolworth Lawn Weedkiller	Woolworths Plc	MAFF 07064
Woolworths Lawn Spot Weeder	Doff Portland Ltd	MAFF 07105

1002 Dichlorprop + *2,4-D* + *Mecoprop* *(Herbicide)*

Boots Kill-A-Weed	The Boots Company Plc	MAFF 03523
Boots Lawn Weedkiller	The Boots Company Plc	MAFF 02677

---

* These products are "approved for agricultural use". For further details refer to page vii.
% These products are subject to the staged revocation procedure. For further details refer to page vii.
A These products are approved for use in or near water. For further details refer to page vii.

AMATEUR PRODUCTS

	Product Name	Marketing Company	Reg. No.
	1003 Dichlorprop + *Dicamba* + *MCPA* (Herbicide)		
	B & Q Liquid Weed and Feed	B & Q Plc	MAFF 05293
	Boots Nettle and Bramble Weedkiller	The Boots Company Plc	MAFF 03455
	Groundclear	Pan Britannica Industries Ltd	MAFF 05953
	Groundclear Spot	Pan Britannica Industries Ltd	MAFF 05950
	Lawnsman Liquid Weed and Feed	Imperial Chemical Industries Plc	MAFF 03610
	Liquid Weed and Feed	ICI Garden and Professional Products	MAFF 05869
	Liquid Weed and Feed	Zeneca Garden Care	MAFF 06887
%	Ready To Use Groundclear	May & Baker Garden Care	MAFF 05704
	Woolworth Liquid Lawn Feed and Weed	Woolworths Plc	MAFF 07060
	1004 Dichlorprop + *Dicamba* + *Mecoprop* (Herbicide)		
	New Supertox	Pan Britannica Industries Ltd	MAFF 06128
	1005 Dichlorprop + *Ferrous sulphate* + *MCPA* (Herbicide)		
%	J Arthur Bower's Feed and Weed Mosskiller	Sinclair Horticulture & Leisure Ltd	MAFF 04188
	J Arthur Bower's Granular Feed, Weed and Mosskiller	Sinclair Horticulture & Leisure Ltd	MAFF 06041
%	J Arthur Bower's Granular Feed, Weed and Mosskiller	Sinclair Horticulture & Leisure Ltd	MAFF 04473
	J Arthur Bower's Granular Feed, Weed and Mosskiller	William Sinclair Horticulture Ltd	MAFF 07042
	Wilko Lawn Feed, Weed and Mosskiller	Wilkinson Home & Garden Stores	MAFF 04602
	1006 Dichlorprop + *MCPA* (Herbicide)		
%	Boots Lawn Weed and Feed Soluble Powder	The Boots Company Plc	MAFF 02403
	Doff Lawn Weed and Feed Soluble Powder	Doff Portland Ltd	MAFF 05708
	J Arthur Bower's Granular Feed and Weed	Sinclair Horticulture & Leisure Ltd	MAFF 06040
%	J Arthur Bower's Granular Lawn Feed and Weed	Sinclair Horticulture & Leisure Ltd	MAFF 04480
%	J Arthur Bowers Lawn Feed and Weed	Sinclair Horticulture & Leisure Ltd	MAFF 04186
	Wilko Soluble Lawn Food and Weedkiller	Wilkinson Home & Garden Stores	MAFF 04391
	1007 Dichlorvos (*Insecticide*)		
	Fix Up Slow Release Fly Killer	Secto Co Ltd	MAFF 00897
	Sectovap Greenhouse Pest Killer	Secto Co Ltd	MAFF 03650

* These products are "approved for agricultural use". For further details refer to page vii.
% These products are subject to the staged revocation procedure. For further details refer to page vii.
A These products are approved for use in or near water. For further details refer to page vii.

AMATEUR PRODUCTS

Product Name	Marketing Company	Reg. No.

1008 Difenacoum *(Vertebrate control)*
    Ratak — Imperial Chemical Industries Plc — MAFF 05152
    Ratak — Zeneca Public Health — MAFF 06828
    Sorexa D Mouse Killer — Sorex Ltd — MAFF 06901

1009 Difenacoum + Calciferol *(Vertebrate control)*
    Karate Mouse Bait — Lever Industrial Ltd — MAFF 05866

1010 Dikegulac *(Herbicide)*
    Cutlass — Imperial Chemical Industries Plc — MAFF 00617
    Cutlass — Zeneca Garden Care — MAFF 06839

1011 Dimethoate *(Insecticide)*
    Boots Greenfly and Blackfly Killer — The Boots Company Plc — MAFF 02402
    Doff Systemic Insecticide — Doff Portland Ltd — MAFF 02658

1012 Dimethoate + Permethrin *(Insecticide)*
    Bio Long Last — Pan Britannica Industries Ltd — MAFF 00269

1013 Diquat + Amitrole + Paraquat + Simazine *(Herbicide)*
    Pathclear — ICI Garden and Professional Products — MAFF 01546
    Pathclear — Zeneca Garden Care — MAFF 06845

1014 Diquat + Paraquat *(Herbicide)*
    Weedol — Imperial Chemical Industries Plc — MAFF 02357
    Weedol — Zeneca Garden Care — MAFF 06863

1015 Diuron + Amitrole + 2,4-D + Simazine *(Herbicide)*
    Hytrol — Agrichem (International) Ltd — MAFF 04540

1016 Diuron + Simazine *(Herbicide)*
    Total Weedkiller Granules — Pan Britannica Industries Ltd — MAFF 05946
  % Total Weedkiller Granules — Rhone-Poulenc Agriculture Ltd — MAFF 05352

1017 Fatty acids *(Herbicide)*
    Bio Speedweed — Pan Britannica Industries Ltd — MAFF 06134

1018 Fatty acids *(Insecticide)*
  % B & Q Houseplant Insecticide Ready-to-Use — B & Q Plc — MAFF 06349
  % B & Q Natural Pest Killer — B & Q Plc — MAFF 06345
    Bio Friendly Pest Pistol — Pan Britannica Industries Ltd — MAFF 04911

* These products are "approved for agricultural use". For further details refer to page vii.
% These products are subject to the staged revocation procedure. For further details refer to page vii.
A These products are approved for use in or near water. For further details refer to page vii.

AMATEUR PRODUCTS

Product Name	Marketing Company	Reg. No.

1018 Fatty acids *(Insecticide)—continued*

	Product Name	Marketing Company	Reg. No.
	Bio Pest Pistol	Pan Britannica Industries Ltd	MAFF 07233
	Devcol 928 Insecticide Concentrate	Devcol Morgan Ltd	MAFF 07140
	Devcol Green/Blackfly Killer	Devcol Morgan Ltd	MAFF 07216
	Devcol Houseplant Insecticide Ready For Use	Devcol Morgan Ltd	MAFF 06999
	Devcol Natural Pest Gun	Devcol Morgan Ltd	MAFF 06997
	Devcol Natural Pest Killer	Devcol Morgan Ltd	MAFF 06996
	Devcol Pest Killer	Devcol Morgan Ltd	MAFF 06998
	Devcol Whitefly Insecticide	Devcol Morgan Ltd	MAFF 07213
	Fisons Nature's Answer Organic Insecticide RTU	Fisons Plc	MAFF 06528
	Greenco GR1	Cumulas Organics	MAFF 06006
	Greenco GR1	David Morgan (Nottingham) Ltd	MAFF 06005
	Greenco GR3	Cumulas Organics	MAFF 06008
	Greenco GR3	David Morgan (Nottingham) Ltd	MAFF 06007
	Greenco GR3 Pest Jet	Cumulas Organics	MAFF 07215
	Greenco GR3 Pest Jet	David Morgan (Nottingham) Ltd	MAFF 07214
%	Homebase Natural Pest Killer	Septico Organic Products	MAFF 06347
%	ICI Insecticidal Soap	Zeneca Garden Products	MAFF 06348
%	Natural Bug Gun	Zeneca Garden Products	MAFF 06346
	Phostrogen House Plant Insecticide	Phostrogen Ltd	MAFF 06538
	Phostrogen Safer's Garden Insecticide Concentrate	Phostrogen Ltd	MAFF 05499
	Phostrogen Safer's Ready-To-Use Fruit and Vegetable Insecticide	Phostrogen Ltd	MAFF 04329
	Phostrogen Safer's Ready-To-Use House Plant Insecticide	Phostrogen Ltd	MAFF 04328
	Phostrogen Safer's Ready-To-Use Rose and Flower Insecticide and Flower Insecticide	Phostrogen Ltd	MAFF 04341
	Safer's Insecticidal Soap Ready to Use	Safer Ltd	MAFF 06573
%	Savona Ready to Use	Koppert (UK) Ltd	MAFF 06296
%	Savona Ready-For Use	Koppert (UK) Ltd	MAFF 03138
	Savona Rose Spray	Koppert (UK) Ltd	MAFF 06297
%	Savona Rose Spray	Koppert (UK) Ltd	MAFF 03517

1019 Fatty acids + *Sulphur* *(Fungicide)(Insecticide)*

	Product Name	Marketing Company	Reg. No.
	Fisons Nature's Answer Fungicide and Insect Killer	Fisons Plc	MAFF 06767

1020 Fenarimol + *Permethrin* *(Fungicide)(Insecticide)*

	Product Name	Marketing Company	Reg. No.
	Murphy Zap Cap Combined Insecticide and Fungicide	Fisons Plc	MAFF 03924

---

* These products are "approved for agricultural use". For further details refer to page vii.
% These products are subject to the staged revocation procedure. For further details refer to page vii.
A These products are approved for use in or near water. For further details refer to page vii.

AMATEUR PRODUCTS

Product Name	Marketing Company	Reg. No.
1021 Fenitrothion *(Insecticide)*		
Doff Garden Insect Powder	Doff Portland Ltd	MAFF 00972
PBI Fenitrothion	Pan Britannica Industries Ltd	MAFF 01552
1022 Ferrous sulphate *(Herbicide)*		
Asda Lawn Sand	Asda Stores Ltd	MAFF 04520
Asda Lawn Sand	Asda Stores Ltd	MAFF 06062
B & Q Mosskiller for Lawns	B & Q Plc	MAFF 05827
Boots Lawn Moss Killer and Fertiliser	The Boots Company Plc	MAFF 02494
Chempak Lawn Sand	Chempak Ltd	MAFF 05723
Doff Lawn Mosskiller and Fertilizer	Doff Portland Ltd	MAFF 05689
Fisons Autumn Extra	Fisons Plc	MAFF 03525
Fisons Lawn Sand	Fisons Plc	MAFF 00885
Fisons Mosskil Extra	Fisons Plc	MAFF 03267
Gem Lawn Sand	Gem Gardening	MAFF 04555
Green Up Mossfree	Synchemicals Ltd	MAFF 03270
Green Up Mossfree	Vitax Ltd	MAFF 05639
Homebase Lawn Feed and Mosskiller	Fisons Plc	MAFF 06107
ICI Mosskiller For Lawns	Imperial Chemical Industries Plc	MAFF 01389
J Arthur Bower's Lawn Sand	Sinclair Horticulture & Leisure Ltd	MAFF 04083
J Arthur Bowers Lawn Sand	William Sinclair Horticulture Ltd	MAFF 07028
Maxicrop Mosskiller and Lawn Tonic	Maxicrop International Ltd	MAFF 04661
Moss Control Plus Lawn Fertilizer	O M Scott & Sons Ltd	MAFF 05613
Mosskiller For Lawns	Zeneca Garden Care	MAFF 06820
Murphy Lawn Feed and Moss Killer	Murphy Home & Garden Products	MAFF 07100
PBI Velvas	Pan Britannica Industries Ltd	MAFF 02291
Phostrogen Soluble Lawn Moss Killer and Lawn Tonic	Phostrogen Ltd	MAFF 07112
Premier Autumn Lawn Feed with Mosskiller	Premier Way Ltd	MAFF 07049
Vitax Lawn Sand	Vitax Ltd	MAFF 04352
Wilko Lawn Sand	Wilkinson Home & Garden Stores	MAFF 04084
1023 Ferrous sulphate + *2,4-D* + *Dicamba* *(Herbicide)*		
B & Q Triple Action Lawn Care	B & Q Plc	MAFF 05282
Greensward	Imperial Chemical Industries Plc	MAFF 03931
Greensward	Zeneca Garden Care	MAFF 06841
Triple Action Grass Hopper	Imperial Chemical Industries Plc	MAFF 03932
Triple Action Grasshopper	Zeneca Garden Care	MAFF 06852
1024 Ferrous sulphate + *2,4-D* + *Mecoprop* *(Herbicide)*		
Asda Lawn Feed and Weed Moss Killer	Asda Stores Ltd	MAFF 03819
Gem Lawn Weed and Feed Plus Mosskiller	Gem Gardening	MAFF 04488

* These products are "approved for agricultural use". For further details refer to page vii.
% These products are subject to the staged revocation procedure. For further details refer to page vii.
A These products are approved for use in or near water. For further details refer to page vii.

AMATEUR PRODUCTS

Product Name	Marketing Company	Reg. No.

1024 Ferrous sulphate + 2,4-D + Mecoprop (Herbicide)—continued

	Product Name	Marketing Company	Reg. No.
	Green Up Feed and Weed Plus Mosskiller	Vitax Ltd	MAFF 05491
	Green Up Feed and Weed Plus Mosskiller	Vitax Ltd	MAFF 06513
	Notcutts Feed and Weed with Mosskiller	Notcutts Garden Centres Ltd	MAFF 05124
%	Supergreen Feed, Weed and Mosskiller	May & Baker Garden Care	MAFF 05349
	Supergreen Triple (Feed, Weed and Mosskiller)	Pan Britannica Industries Ltd	MAFF 05955
	Vitagrow Feed and Weed with Mosskiller	Vitagrow (Fertilisers) Ltd	MAFF 05098
	Weed 'N' Feed Extra	Steetley Chemicals Ltd	MAFF 03450
	Weed 'N' Feed Extra	Vitax Ltd	MAFF 06506

1025 Ferrous sulphate + *Dichlorprop* + *MCPA* (Herbicide)

	Product Name	Marketing Company	Reg. No.
%	J Arthur Bower's Feed and Weed Mosskiller	Sinclair Horticulture & Leisure Ltd	MAFF 04188
	J Arthur Bower's Granular Feed, Weed and Mosskiller	Sinclair Horticulture & Leisure Ltd	MAFF 06041
%	J Arthur Bower's Granular Feed, Weed and Mosskiller	Sinclair Horticulture & Leisure Ltd	MAFF 04473
	J Arthur Bower's Granular Feed, Weed and Mosskiller	William Sinclair Horticulture Ltd	MAFF 07042
	Wilko Lawn Feed, Weed and Mosskiller	Wilkinson Home & Garden Stores	MAFF 04602

1026 Ferrous sulphate + *MCPA* + *Mecoprop* (Herbicide)

Product Name	Marketing Company	Reg. No.
Fisons Evergreen Extra	Fisons Plc	MAFF 03890
Great Mills Lawn Feed, Weed and Mosskiller	Great Mills (Retail) Ltd	MAFF 06109
Homebase Lawn Feed Weed and Mosskiller	Homebase Ltd	MAFF 06108
Murphy Lawn Feed, Weed and Mosskiller	Murphy Home & Garden Products	MAFF 07058
Premier Lawn Feed, Weed and Mosskiller	Premier Way Ltd	MAFF 07154
Woolworths Lawn Feed, Weed and Mosskiller	Woolworths Plc	MAFF 06724

1027 Ferrous sulphate + *Copper sulphate* (Miscellaneous)

Product Name	Marketing Company	Reg. No.
Snailaway	Interpet Ltd	MAFF 02457

---

* These products are "approved for agricultural use". For further details refer to page vii.
% These products are subject to the staged revocation procedure. For further details refer to page vii.
A These products are approved for use in or near water. For further details refer to page vii.

AMATEUR PRODUCTS

Product Name	Marketing Company	Reg. No.
1028 Garlic oil + *Orange peel oil* + *Orange pith oil* (Vertebrate control)		
Growing Success Cat Repellent	Growing Success Organics Ltd	MAFF 06262
1029 Glufosinate ammonium *(Herbicide)*		
Great Mills Fast Action Weedkiller Ready to Use	Great Mills (Retail) Ltd.	MAFF 07114
Homebase Contact Action Weedkiller	Homebase Ltd	MAFF 06735
Homebase Contact Action Weedkiller Ready-to-Use Sprayer	Homebase Ltd	MAFF 06736
Murphy Ultra-Weed	Fisons Plc	MAFF 05180
Murphy Weedmaster	Fisons Plc	MAFF 05179
Murphy Weedmaster Ready-to-use Sprayer	Fisons Plc	MAFF 05531
1030 Glyphosate *(Herbicide)*		
B & Q Complete Weedkiller	B & Q Plc	MAFF 05290
B & Q Complete Weedkiller Ready To Use	B & Q Plc	MAFF 05292
B & Q Complete Weedkiller Ready To Use	B & Q Plc	MAFF 06722
% Boots Systemic Weed and Grass Killer	The Boots Company Plc	MAFF 05126
Boots Systemic Weed and Grass Killer Ready To Use	The Boots Company Plc	MAFF 05028
Glypho	Zeneca Garden Care	MAFF 06441
Glypho Gun!	Zeneca Garden Care	MAFF 06443
Great Mills Systemic Action Weedkiller Ready to Use	Great Mills (Retail) Ltd	MAFF 07080
Greenscape Ready To Use Weed Killer	Monsanto Plc	MAFF 04676
Greenscape Weedkiller	Monsanto Plc	MAFF 04321
Homebase Systemic Action Weedkiller	Homebase Ltd	MAFF 06622
Homebase Systemic Action Weedkiller Ready to Use	Homebase Ltd	MAFF 06623
Murphy Tumbleweed	Fisons Plc	MAFF 04008
Murphy Tumbleweed Gel	Fisons Plc	MAFF 05186
Murphy Tumbleweed Gel	Murphy Home & Garden Products	MAFF 04009
Murphy Tumbleweed Spray	Fisons Plc	MAFF 04015
New Formula B & Q Complete Weedkiller	B & Q Plc	MAFF 06100
New Formula Greenscape	Monsanto Agricultural Co	MAFF 06051
New Improved Leaf Action Roundup Brushkiller Ready To Use	Monsanto Plc	MAFF 06760

* These products are "approved for agricultural use". For further details refer to page vii.
% These products are subject to the staged revocation procedure. For further details refer to page vii.
A These products are approved for use in or near water. For further details refer to page vii.

AMATEUR PRODUCTS

Product Name	Marketing Company	Reg. No.

1030 Glyphosate (Herbicide)—continued

New Improved Leaf Action Roundup Weedkiller Ready To Use	Monsanto Plc	MAFF 06272
Roundup Brushkiller	Monsanto Plc	MAFF 05755
Roundup Brushkiller Ready To Use	Monsanto Plc	MAFF 05832
Roundup GC	Monsanto Plc	MAFF 05538
Roundup Micro	Monsanto Plc	MAFF 05918
Roundup Ready To Use	Monsanto Plc	MAFF 06501
Roundup Tab	Monsanto Plc	MAFF 05917
Tough Weed Gun!	Zeneca Garden Care	MAFF 06442
Tough Weed Killer	Zeneca Garden Care	MAFF 06440
Woolworth's Weedkiller	Woolworths Plc	MAFF 07108

1031 Grease (Insecticide)

PBI Boltac Grease Bands	Pan Britannica Industries Ltd	MAFF 00288

1032 Heptenophos + *Permethrin* (Insecticide)

Murphy Systemic Action Insecticide	Fisons Plc	MAFF 05580
Murphy Tumblebug	Fisons Plc	MAFF 03637

1033 4-Indol-3-ylbutyric acid (Herbicide)

Fisons Clearcut II	Fisons Plc	MAFF 06316

1034 4-Indol-3-ylbutyric acid + *Thiram* + *1-Naphthylacetic acid* (Fungicide)(Herbicide)

Boots Hormone Rooting Powder	The Boots Company Plc	MAFF 01067

1035 Lindane (Insecticide)

Doff Ant Killer	Doff Portland Ltd	MAFF 00739
Doff Gamma BHC Dust	Doff Portland Ltd	MAFF 04868
Murphy Gamma BHC Dust	Fisons Plc	MAFF 04006

1036 Lindane + *Pyrethrins* (Insecticide)

% Secto Greenfly and Garden Insect Spray	Secto Co Ltd	MAFF 03841

1037 Lindane + *Rotenone* + *Thiram* (Fungicide)(Insecticide)

% PBI Hexyl	Pan Britannica Industries Ltd	MAFF 02650

1038 Malathion (Insecticide)

% Ban-Mite	Johnsons Veterinary Products Ltd	MAFF 02676
Duramitex	Harkers Ltd	MAFF 02512
Malathion Greenfly Killer	Pan Britannica Industries Ltd	MAFF 01247

---

* These products are "approved for agricultural use". For further details refer to page vii.
% These products are subject to the staged revocation procedure. For further details refer to page vii.
A These products are approved for use in or near water. For further details refer to page vii.

AMATEUR PRODUCTS

Product Name	Marketing Company	Reg. No.

1038 Malathion *(Insecticide)—continued*

Murphy Malathion Dust	Fisons Plc	MAFF 03972
Murphy Malathion Liquid	Fisons Plc	MAFF 03971

1039 Malathion + *Permethrin* *(Insecticide)*

Bio Crop Saver	Pan Britannica Industries Ltd	MAFF 03969

1040 Maleic hydrazide *(Herbicide)*

Stop Gro G8	Botanical Developments	MAFF 05923
Stop Gro G8	Synchemicals Ltd	MAFF 02029

1041 Mancozeb *(Fungicide)*

PBI Dithane 945	Pan Britannica Industries Ltd	MAFF 00718

1042 MCPA + *Amitrole* *(Herbicide)*

%	Murphy Problem Weeds Killer	Fisons Plc	MAFF 05295

1043 MCPA + *Amitrole* + *Simazine* *(Herbicide)*

%	Fisons Path Weeds Killer	Fisons Plc	MAFF 02700

1044 MCPA + *Dicamba* + *Dichlorprop* *(Herbicide)*

	B & Q Liquid Weed and Feed	B & Q Plc	MAFF 05293
	Boots Nettle and Bramble Weedkiller	The Boots Company Plc	MAFF 03455
	Groundclear	Pan Britannica Industries Ltd	MAFF 05953
	Groundclear Spot	Pan Britannica Industries Ltd	MAFF 05950
	Lawnsman Liquid Weed and Feed	Imperial Chemical Industries Plc	MAFF 03610
	Liquid Weed and Feed	ICI Garden and Professional Products	MAFF 05869
	Liquid Weed and Feed	Zeneca Garden Care	MAFF 06887
%	Ready To Use Groundclear	May & Baker Garden Care	MAFF 05704
	Woolworth Liquid Lawn Feed and Weed	Woolworths Plc	MAFF 07060

1045 MCPA + *Dicamba* + *Mecoprop* *(Herbicide)*

Bio Weed Pencil	Pan Britannica Industries Ltd	MAFF 04054
Biospot	Pan Britannica Industries Ltd	MAFF 05071
Fisons Evergreen Feed and Weed Liquid	Fisons Plc	MAFF 05664
Fisons Lawncare Liquid	Fisons Plc	MAFF 03625
Homebase Lawn Feed and Weed liquid	Homebase Ltd	MAFF 06086

---

* These products are "approved for agricultural use". For further details refer to page vii.
% These products are subject to the staged revocation procedure. For further details refer to page vii.
A These products are approved for use in or near water. For further details refer to page vii.

AMATEUR PRODUCTS

Product Name	Marketing Company	Reg. No.

1046 MCPA + *Dichlorprop* (Herbicide)

% Boots Lawn Weed and Feed Soluble Powder	The Boots Company Plc	MAFF 02403
Doff Lawn Weed and Feed Soluble Powder	Doff Portland Ltd	MAFF 05708
J Arthur Bower's Granular Feed and Weed	Sinclair Horticulture & Leisure Ltd	MAFF 06040
% J Arthur Bower's Granular Lawn Feed and Weed	Sinclair Horticulture & Leisure Ltd	MAFF 04480
% J Arthur Bowers Lawn Feed and Weed	Sinclair Horticulture & Leisure Ltd	MAFF 04186
Wilko Soluble Lawn Food and Weedkiller	Wilkinson Home & Garden Stores	MAFF 04391

1047 MCPA + *Dichlorprop* + *Ferrous sulphate* (Herbicide)

% J Arthur Bower's Feed and Weed Mosskiller	Sinclair Horticulture & Leisure Ltd	MAFF 04188
J Arthur Bower's Granular Feed, Weed and Mosskiller	Sinclair Horticulture & Leisure Ltd	MAFF 06041
% J Arthur Bower's Granular Feed, Weed and Mosskiller	Sinclair Horticulture & Leisure Ltd	MAFF 04473
J Arthur Bower's Granular Feed, Weed and Mosskiller	William Sinclair Horticulture Ltd	MAFF 07042
Wilko Lawn Feed, Weed and Mosskiller	Wilkinson Home & Garden Stores	MAFF 04602

1048 MCPA + *Ferrous sulphate* + *Mecoprop* (Herbicide)

Fisons Evergreen Extra	Fisons Plc	MAFF 03890
Great Mills Lawn Feed, Weed and Mosskiller	Great Mills (Retail) Ltd	MAFF 06109
Homebase Lawn Feed Weed and Mosskiller	Homebase Ltd	MAFF 06108
Murphy Lawn Feed, Weed and Mosskiller	Murphy Home & Garden Products	MAFF 07058
Premier Lawn Feed, Weed and Mosskiller	Premier Way Ltd	MAFF 07154
Woolworths Lawn Feed, Weed and Mosskiller	Woolworths Plc	MAFF 06724

1049 MCPA + *Mecoprop* (Herbicide)

Fisons Evergreen 90	Fisons Plc	MAFF 03131
Fisons Evergreen Feed and Weed	Fisons Plc	MAFF 05906
Gardenstore Lawn Feed and Weed	Texas Homecare Ltd	MAFF 07360
Homebase Lawn Feed and Weed	Homebase Ltd	MAFF 05614
Murphy Lawn Feed and Weed	Murphy Home & Garden Products	MAFF 07029

* These products are "approved for agricultural use". For further details refer to page vii.
% These products are subject to the staged revocation procedure. For further details refer to page vii.
A These products are approved for use in or near water. For further details refer to page vii.

AMATEUR PRODUCTS

Product Name	Marketing Company	Reg. No.

1050 Mecoprop + Benazolin + 2,4-D + Dicamba + Dichlorophen + Dichlorprop *(Herbicide)*

Boots Total Lawn Treatment	The Boots Company Plc	MAFF 00293
Vitax Green Up Lawn Feed 'N' Weed Plus Mosskiller	Vitax Ltd	MAFF 05570

1051 Mecoprop + 2,4-D *(Herbicide)*

	Asda Lawn Weed and Feed	Asda Stores Ltd	MAFF 04761
	Elliott's Turf Feed and Weed	Thomas Elliott Ltd	MAFF 05046
	Gem Lawn Weed and Feed	Gem Gardening	MAFF 04486
	Gem Lawn Weed and Feed 4	Gem Gardening	MAFF 04487
%	Green Up Lawn Spot Weedkiller	Vitax Ltd	MAFF 05086
	J. Arthur Bower's Lawn Food with Weedkiller	Sinclair Horticulture & Leisure Ltd	MAFF 04301
%	Lawn Feed and Weed Granules	May & Baker Garden Care	MAFF 05263
	Lawn Feed and Weed Granules	Pan Britannica Industries Ltd	MAFF 05902
%	Lawn Feed and Weed Granules	Rhone-Poulenc Agriculture Ltd	MAFF 05343
%	Lawn Spot Weed Granules	May & Baker Garden Care	MAFF 05291
	Lawn Spot Weed Granules	Pan Britannica Industries Ltd	MAFF 05903
%	Spot Weeder	Rhone-Poulenc Agriculture Ltd	MAFF 05347
	Spraydex Lawn Spot Weeder	Spraydex Ltd	MAFF 03141
	Supergreen Double (Feed and Weed)	Pan Britannica Industries Ltd	MAFF 05949
	Supergreen Double (Feed and Weed)	Rhone-Poulenc Agriculture Ltd	MAFF 05348
	Supertox	Rhone-Poulenc Agriculture Ltd	MAFF 05350
	Supertox Lawn Weedkiller	Pan Britannica Industries Ltd	MAFF 05948
	Supertox Spot	Pan Britannica Industries Ltd	MAFF 05951
%	Supertox Spot Weeder	May & Baker Garden Care	MAFF 05168
%	Verdant F6 Combined Fertilizer and Weedkiller	T Parker & Sons (Turf Management) Ltd	MAFF 05200
	Verdant No. 6 Feed and Weed	T Parker & Sons (Turf Management) Ltd	MAFF 05047
	Verdone 2	Imperial Chemical Industries Plc	MAFF 03271
	Verdone 2	Zeneca Garden Care	MAFF 06853
	Wilko Lawn Feed 'n' Weed	Wilkinson Home & Garden Stores	MAFF 04403
	Wilko Lawn Food and Weedkiller	Wilkinson Home & Garden Stores	MAFF 04302

1052 Mecoprop + 2,4-D + Dicamba *(Herbicide)*

New Formulation SBK Brushwood Killer	Vitax Ltd	MAFF 05043

1053 Mecoprop + 2,4-D + Dichlorprop *(Herbicide)*

Boots Kill-A-Weed	The Boots Company Plc	MAFF 03523
Boots Lawn Weedkiller	The Boots Company Plc	MAFF 02677

---

* These products are "approved for agricultural use". For further details refer to page vii.
% These products are subject to the staged revocation procedure. For further details refer to page vii.
A These products are approved for use in or near water. For further details refer to page vii.

AMATEUR PRODUCTS

Product Name	Marketing Company	Reg. No.

1054 Mecoprop + 2,4-D + Ferrous sulphate (Herbicide)

Asda Lawn Feed and Weed Moss Killer	Asda Stores Ltd	MAFF 03819
Gem Lawn Weed and Feed + Mosskiller	Gem Gardening	MAFF 04488
Green Up Feed and Weed Plus Mosskiller	Vitax Ltd	MAFF 05491
Green Up Feed and Weed Plus Mosskiller	Vitax Ltd	MAFF 06513
Notcutts Feed and Weed with Mosskiller	Notcutts Garden Centres Ltd	MAFF 05124
% Supergreen Feed, Weed and Mosskiller	May & Baker Garden Care	MAFF 05349
Supergreen Triple (Feed, Weed and Mosskiller)	Pan Britannica Industries Ltd	MAFF 05955
Vitagrow Feed and Weed with Mosskiller	Vitagrow (Fertilisers) Ltd	MAFF 05098
Weed 'N' Feed Extra	Steetley Chemicals Ltd	MAFF 03450
Weed 'N' Feed Extra	Vitax Ltd	MAFF 06506

1055 Mecoprop + *Dicamba* (Herbicide)

Elliott's Touchweeder	Thomas Elliott Ltd	MAFF 06252

1056 Mecoprop + *Dicamba* + *Dichlorprop* (Herbicide)

New Supertox	Pan Britannica Industries Ltd	MAFF 06128

1057 Mecoprop + *Dicamba* + *MCPA* (Herbicide)

Bio Weed Pencil	Pan Britannica Industries Ltd	MAFF 04054
Biospot	Pan Britannica Industries Ltd	MAFF 05071
Fisons Evergreen Feed and Weed Liquid	Fisons Plc	MAFF 05664
Fisons Lawncare Liquid	Fisons Plc	MAFF 03625
Homebase Lawn Feed and Weed liquid	Homebase Ltd	MAFF 06086

1058 Mecoprop + *Ferrous sulphate* + *MCPA* (Herbicide)

Fisons Evergreen Extra	Fisons Plc	MAFF 03890
Great Mills Lawn Feed, Weed and Mosskiller	Great Mills (Retail) Ltd	MAFF 06109
Homebase Lawn Feed Weed and Mosskiller	Homebase Ltd	MAFF 06108
Murphy Lawn Feed, Weed and Mosskiller	Murphy Home & Garden Products	MAFF 07058

---

* These products are "approved for agricultural use". For further details refer to page vii.
% These products are subject to the staged revocation procedure. For further details refer to page vii.
A These products are approved for use in or near water. For further details refer to page vii.

AMATEUR PRODUCTS

Product Name	Marketing Company	Reg. No.

1058 Mecoprop + *Ferrous sulphate* + *MCPA* *(Herbicide)—continued*

Premier Lawn Feed, Weed and Mosskiller	Premier Way Ltd	MAFF 07154
Woolworths Lawn Feed, Weed and Mosskiller	Woolworths Plc	MAFF 06724

1059 Mecoprop + *MCPA* *(Herbicide)*

Fisons Evergreen 90	Fisons Plc	MAFF 03131
Fisons Evergreen Feed and Weed	Fisons Plc	MAFF 05906
Gardenstore Lawn Feed and Weed	Texas Homecare Ltd	MAFF 07360
Homebase Lawn Feed and Weed	Homebase Ltd	MAFF 05614
Murphy Lawn Feed and Weed	Murphy Home & Garden Products	MAFF 07029

1060 Mecoprop-P + *2,4-D* *(Herbicide)*

Green Up Lawn Spot Weedkiller	Vitax Ltd	MAFF 06028
Proctors Lawn Weed and Feed	H & T Proctor	MAFF 04669
Vitax 'Green Up' Granular Lawn Feed and Weed	Vitax Ltd	MAFF 06158

1061 Metaldehyde *(Miscellaneous)*

	Aro Slug Killer Blue Mini Pellets	Makro Self Service Wholesalers Ltd	MAFF 06289
	B & Q Slug Killer Blue Mini Pellets	B & Q Plc	MAFF 05607
%	B & Q Slug Killer Blue Mini Pellets	B & Q Plc	MAFF 04387
	Co-op Garden Maker Slug Killer Pellets	Co-Operative Wholesale Society Ltd	MAFF 04689
	Devcol Morgan 3% Metaldehyde Slug Killer	Devcol Morgan Ltd	MAFF 07113
	Do It All Slug Killer Pellets	Do-It-All Ltd	MAFF 04895
	Doff Slugoids Slugkiller Blue Mini Pellets	Doff Portland Ltd	MAFF 00744
	Fellside Green Slug Killer Mini Pellets	Doff Portland Ltd	MAFF 05004
	Focus Slug Killer Pellets	Focus DIY Ltd	MAFF 06027
	Great Mills Slug Killer Blue Mini Pellets	Great Mills (Retail) Ltd	MAFF 06088
	Homebase Slug Killer Blue Mini Pellets	Sainsbury's Homebase House and Garden Centres	MAFF 06410
	ICI Slug Pellets	Garden and Professional Products	MAFF 06429
	Murphy Slug Tape	Fisons Plc	MAFF 05114
	Murphy Slugit Liquid	Fisons Plc	MAFF 03633
	Murphy Slugits	Fisons Plc	MAFF 05308
	Murphy Slugits	Murphy Home & Garden Products	MAFF 03634
	PBI Slug Mini Pellets	Pan Britannica Industries Ltd	MAFF 02611

* These products are "approved for agricultural use". For further details refer to page vii.
% These products are subject to the staged revocation procedure. For further details refer to page vii.
A These products are approved for use in or near water. For further details refer to page vii.

AMATEUR PRODUCTS

Product Name	Marketing Company	Reg. No.
1061 Metaldehyde *(Miscellaneous)—continued*		
Portland Brand Slug Killer Blue Mini Pellets	Doff Portland Ltd	MAFF 06411
Slug Pellets	Zeneca Garden Care	MAFF 06835
Slug Xtra	Imperial Chemical Industries Plc	MAFF 05870
Slug Xtra	Zeneca Garden Care	MAFF 06822
Thrifty Pack Blue Mini Slug Killer Pellets	David Morgan (Nottingham) Ltd	MAFF 06430
Vapona Slug and Snail Killer Pellets	Sara Lee Household & Personal Care	MAFF 06923
Wilko Slug Killer Blue Mini Pellets	Wilkinson Home & Garden Stores	MAFF 05032
Wilko Slug Killer Blue Mini Pellets	Wilkinson Home & Garden Stores	MAFF 05608
Woolworths Slug Killer Pellets	Woolworths Plc	MAFF 06924
Woolworths Slug Pellets	Woolworths Plc	MAFF 07038
1062 Methiocarb *(Miscellaneous)*		
Slug Gard	Pan Britannica Industries Ltd	MAFF 01963
1063 Methyl nonyl ketone *(Vertebrate control)*		
Get Off My Garden	Get-Off My Garden Ltd	MAFF 04573
Get Off My Garden	Pet and Garden Manufacturing Plc	MAFF 06614
1064 Methyl nonyl ketone + ***Citronella oil*** *(Vertebrate control)*		
Secto Keep Off	Secto Co Ltd	MAFF 04960
1065 Myclobutanil *(Fungicide)*		
Systhane	Pan Britannica Industries Ltd	MAFF 04522
Systhane	Pan Britannica Industries Ltd	MAFF 04523
1066 Naphthalene *(Vertebrate control)*		
Scent Off Buds	Synchemicals Ltd	MAFF 02907
Scent Off Pellets	Vitax Ltd	MAFF 01888
1067 1-Naphthylacetic acid *(Herbicide)*		
Strike 2	May & Baker Garden Care	MAFF 04849
Strike 2	Pan Britannica Industries Ltd	MAFF 05952
1068 1-Naphthylacetic acid + ***Captan*** *(Fungicide)(Herbicide)*		
Doff Hormone Rooting Powder	Doff Portland Ltd	MAFF 01065
Murphy Hormone Rooting Powder	Fisons Plc	MAFF 03618
% New Strike	May & Baker Garden Care	MAFF 04430
New Strike	Pan Britannica Industries Ltd	MAFF 05956
Rooting Powder	Vitax Ltd	MAFF 06334

* These products are "approved for agricultural use". For further details refer to page vii.
% These products are subject to the staged revocation procedure. For further details refer to page vii.
A These products are approved for use in or near water. For further details refer to page vii.

AMATEUR PRODUCTS

Product Name	Marketing Company	Reg. No.

1069 1-Naphthylacetic acid + *Dichlorophen* *(Herbicide)*
    Bio Roota   Pan Britannica Industries Ltd   MAFF 00271

1070 1-Naphthylacetic acid + *Thiram* *(Fungicide)(Herbicide)*
%   Green Fingers Hormone Rooting Powder   Secto Co Ltd   MAFF 01001

1071 1-Naphthylacetic acid +Thiram + *4-Indol-3-ylbutyric acid* *(Fungicide)(Herbicide)*
    Boots Hormone Rooting Powder   The Boots Company Plc   MAFF 01067

1072 2-Naphthyloxyacetic acid *(Herbicide)*
    Synchemicals Tomato Setting Spray   Synchemicals Ltd   MAFF 03735

1073 Orange peel oil + *Garlic oil* + *Orange pith oil* *(Vertebrate control)*
    Growing Success Cat Repellent   Growing Success Organics Ltd   MAFF 06262

1074 Orange pith oil + *Garlic oil* + *Orange peel oil* *(Vertebrate control)*
    Growing Success Cat Repellent   Growing Success Organics Ltd   MAFF 06262

1075 Paraquat + *Amitrole* + *Diquat* + *Simazine* *(Herbicide)*
    Pathclear   ICI Garden and Professional Products   MAFF 01546
    Pathclear   Zeneca Garden Care   MAFF 06845

1076 Paraquat + *Diquat* *(Herbicide)*
    Weedol   Imperial Chemical Industries Plc   MAFF 02357
    Weedol   Zeneca Garden Care   MAFF 06863

1077 Penconazole *(Fungicide)*
    Murphy Tumbleblite II   Fisons Plc   MAFF 06751
    Murphy Tumbleblite II Ready to Use   Fisons Plc   MAFF 06752

1078 Pepper *(Vertebrate control)*
    PBI Pepper Dust   Pan Britannica Industries Ltd   MAFF 01569
    Pepper Dust   Synchemicals Ltd   MAFF 01570
    Secto Pepper Dust   Secto Co Ltd   MAFF 05052

1079 Permethrin *(Herbicide)*
    Fisons Insect Spray For Houseplant   Fisons Plc   MAFF 02521

---

*   These products are "approved for agricultural use". For further details refer to page vii.
%   These products are subject to the staged revocation procedure. For further details refer to page vii.
A   These products are approved for use in or near water. For further details refer to page vii.

AMATEUR PRODUCTS

Product Name	Marketing Company	Reg. No.

1080 Permethrin (Insecticide)

Bio Flydown	Pan Britannica Industries Ltd	MAFF 00267
Bio Sprayday	Pan Britannica Industries Ltd	MAFF 00272
Boots Caterpillar and Whitefly Killer	The Boots Company Plc	MAFF 00290
Darmycel Agarifume Smoke Generator	Octavius Hunt Ltd	MAFF 02508
Fumite Whitefly Smoke Cone	Zeneca Garden Care	MAFF 06834
% Murphy Permethrin Whitefly Smoke	Fisons Plc	MAFF 05657
Murphy Zap Cap General Insecticide	Fisons Plc	MAFF 03923
Picket	ICI Garden and Professional Products	MAFF 01590
Picket	Zeneca Garden Care	MAFF 06846

1081 Permethrin + *Bioallethrin* (Insecticide)

Floracid	Perycut Insectengun Ltd	MAFF 04577
Floracid	Perycut Insectengun Ltd	MAFF 06798
Longer Lasting Bug Gun	ICI Garden and Professional Products	MAFF 06137
Longer Lasting Bug Gun	Zeneca Garden Care	MAFF 06916
New Spraydex Greenfly Killer	Spraydex Ltd	MAFF 04847
Spraydex Greenfly Killer	Spraydex Ltd	MAFF 06587

1082 Permethrin + *Carbendazim* + *Copper oxychloride* + *Sulphur* (Fungicide)(Insecticide)

% Bio Multiveg	Pan Britannica Industries Ltd	MAFF 03116

1083 Permethrin + *Dimethoate* (Insecticide)

Bio Long Last	Pan Britannica Industries Ltd	MAFF 00269

1084 Permethrin + *Fenarimol* (Fungicide)(Insecticide)

Murphy Zap Cap Combined Insecticide and Fungicide	Fisons Plc	MAFF 03924

1085 Permethrin + *Heptenophos* (Insecticide)

Murphy Systemic Action Insecticide	Fisons Plc	MAFF 05580
Murphy Tumblebug	Fisons Plc	MAFF 03637

1086 Permethrin + *Malathion* (Insecticide)

Bio Crop Saver	Pan Britannica Industries Ltd	MAFF 03969

1087 Permethrin + *Sulphur* + *Triforine* (Fungicide)(Insecticide)

Bio Multirose	Pan Britannica Industries Ltd	MAFF 05716
% Bio Multirose	Pan Britannica Industries Ltd	MAFF 02541

---

* These products are "approved for agricultural use". For further details refer to page vii.
% These products are subject to the staged revocation procedure. For further details refer to page vii.
A These products are approved for use in or near water. For further details refer to page vii.

AMATEUR PRODUCTS

Product Name	Marketing Company	Reg. No.

1088 Permethrin + Tetramethrin *(Insecticide)*
Nippon Ant and Crawling Insect Killer	Synchemicals Ltd	MAFF 03039

1089 Phenothrin + Tetramethrin *(Insecticide)*
Pesguard House and Plant Spray	Sumitomo Chemical (UK) Plc	MAFF 01580

1090 Pirimicarb *(Insecticide)*
Rapid Aerosol	Imperial Chemical Industries Plc	MAFF 01689
Rapid Aerosol	Zeneca Garden Care	MAFF 06847
Rapid Greenfly Killer	Imperial Chemical Industries Plc	MAFF 01690

1091 Pirimicarb + Bupirimate + Triforine *(Fungicide)(Insecticide)*
Roseclear	Imperial Chemical Industries Plc	MAFF 01826
Roseclear	Zeneca Garden Care	MAFF 06849

1092 Pirimiphos-methyl *(Insecticide)*
Ant Powder	Woolworths Plc	MAFF 06977
Antkiller Dust	Imperial Chemical Industries Plc	MAFF 00101
Antkiller Dust	Zeneca Garden Care	MAFF 06865
B & Q Antkiller Dust	B & Q Plc	MAFF 05830
Fumite General Purpose Insecticide Smoke Cone	ICI Garden Products	MAFF 00932
Fumite General Purpose Insecticide Smoke Cone	Zeneca Garden Care	MAFF 06833
Sybol	Imperial Chemical Industries Plc	MAFF 05448
Sybol	Zeneca Garden Care	MAFF 06888
Sybol Dust	Imperial Chemical Industries Plc	MAFF 05665
Sybol Dust	Zeneca Garden Care	MAFF 06851

1093 Pirimiphos-methyl + Pyrethrins *(Insecticide)*
Kerispray	Imperial Chemical Industries Plc	MAFF 02653
Kerispray	Zeneca Garden Care	MAFF 06840

1094 Pirimiphos-methyl + Resmethrin + Tetramethrin *(Insecticide)*
Sybol Aerosol	Imperial Chemical Industries Plc	MAFF 05449
Sybol Aerosol	Zeneca Garden Care	MAFF 06850

1095 Primicarb *(Insecticide)*
Rapid Greenfly Killer	Zeneca Garden Care	MAFF 06848

1096 Propiconazole *(Fungicide)*
Tumbleblite	Fisons Plc	MAFF 04691

---

* These products are "approved for agricultural use". For further details refer to page vii.
% These products are subject to the staged revocation procedure. For further details refer to page vii.
A These products are approved for use in or near water. For further details refer to page vii.

AMATEUR PRODUCTS

Product Name	Marketing Company	Reg. No.

1097 Pyrethrins *(Insecticide)*

	Product Name	Marketing Company	Reg. No.
	Aquablast Bug Spray	Agropharm Ltd	MAFF 03461
	B & Q Complete Insecticide Spray	B & Q Plc	MAFF 06964
	B & Q Fruit and Vegetable Insecticide Spray	B & Q Plc	MAFF 05766
%	B & Q Fruit and Vegetable Insecticide Spray	B & Q Plc	MAFF 04981
	B & Q House Plant Insecticide Spray	B & Q Plc	MAFF 05828
	B & Q Houseplant Insect Spray	B & Q Plc	MAFF 06103
	B & Q Houseplant Insecticide Spray	B & Q Plc	MAFF 07032
	B & Q Insecticide Spray For Roses and Flowers	Zeneca Garden Care	MAFF 05875
	B & Q Insecticide Spray for Fruit and Vegetables	B & Q Plc	MAFF 05829
	B & Q Rose and Flower Insecticide Spray	B & Q Plc	MAFF 05769
%	B & Q Rose and Flower Insecticide Spray	B & Q Plc	MAFF 04980
	Bio Friendly Anti-Ant Duster	Pan Britannica Industries Ltd	MAFF 05100
	Bug Gun!	Imperial Chemical Industries Plc	MAFF 05965
	Bug Gun!	Imperial Chemical Industries Plc	MAFF 05966
	Bug Gun!	Zeneca Garden Care	MAFF 06836
	Bug Gun!	Zeneca Garden Care	MAFF 06837
%	Bug-Gun	Imperial Chemical Industries Plc	MAFF 03076
	Co-op Garden Maker Rose and Flower Insecticide Spray	Co-Operative Wholesale Society Ltd	MAFF 04998
	Co-op Garden Maker Rose and Flower Insecticide Spray	Co-Operative Wholesale Society Ltd	MAFF 05609
	Devcol All Purpose Natural Insecticide Spray	Devcol Ltd	MAFF 05802
	Do-It-All Fruit and Vegetable Insecticide Spray	W H Smith Do-It-All Ltd	MAFF 04995
	Do-It-All Rose and Flower Insecticide Spray	Do-It-All Ltd	MAFF 05765
	Do-It-All Rose and Flower Insecticide Spray	W H Smith Do-It-All Ltd	MAFF 04984
	Do-it-All Fruit and Vegetable Insecticide Spray	Do-It-All Ltd	MAFF 05767
	Doff 'All in One' Insecticide Spray	Doff Portland Ltd	MAFF 06069
	Doff Fruit and Vegetable Insecticide Spray	Doff Portland Ltd	MAFF 04040
	Doff Houseplant Insecticide Spray	Doff Portland Ltd	MAFF 06066
	Doff Rose and Flower Insecticide Spray	Doff Portland Ltd	MAFF 04041
	Doff Greenfly Killer	Doff Portland Ltd	MAFF 07030

* These products are "approved for agricultural use". For further details refer to page vii.
% These products are subject to the staged revocation procedure. For further details refer to page vii.
A These products are approved for use in or near water. For further details refer to page vii.

AMATEUR PRODUCTS

Product Name	Marketing Company	Reg. No.
1097 Pyrethrins *(Insecticide)—continued*		
Homebase Pest Gun	Sainsbury's Homebase House and Garden Centres	MAFF 06962
Fellside Green Fruit and Vegetable Insect Spray	Doff Portland Ltd	MAFF 05008
Fellside Green Rose and Flower Insect Spray	Doff Portland Ltd	MAFF 05007
Fisons Natural House Plant Insect Spray	Fisons Plc	MAFF 06323
Fisons Nature's Answer to Insect Pests	Fisons Plc	MAFF 06936
Fisons Nature's Answer to Insect Pests on Flowers, Fruit and Vegetables	Fisons Plc	MAFF 05137
Focus Fruit and Vegetable Insecticide Spray	Focus DIY Ltd	MAFF 06067
Focus Rose and Flower Insecticide Spray	Focus DIY Ltd	MAFF 06068
Garden Centre Pest Spray for Fruit and Vegetables	Gateway Food Markets Ltd	MAFF 05006
Garden Centre Pest Spray for Roses and Flowers	Gateway Food Markets Ltd	MAFF 05005
Great Mills Complete Insecticide Spray	Great Mills (Retail) Ltd	MAFF 07031
Great Mills Fruit and Vegetable Insecticide Spray	Great Mills (Retail) Ltd	MAFF 04983
Great Mills Fruit and Vegetable Insecticide Spray	Great Mills (Retail) Ltd	MAFF 05772
Great Mills Rose and Flower Insecticide Spray	Great Mills (Retail) Ltd	MAFF 04982
Great Mills Rose and Flower Insecticide Spray	Great Mills (Retail) Ltd	MAFF 05771
Keri Insect Spray	ICI Garden and Professional Products	MAFF 06155
Keri Insect Spray	Zeneca Garden Care	MAFF 06885
Murphy Bugmaster	Fisons Plc	MAFF 06324
Natures 'Bug Gun!'	Zeneca Garden Care	MAFF 07077
PY Spray Insect Killer	Synchemicals Ltd	MAFF 05082
PBI Anti-Ant Duster	Pan Britannica Industries Ltd	MAFF 00098
Py Garden Insect Killer	Py Spray Co Ltd	MAFF 02810
% Py Garden Insecticide	Py Spray Co Ltd	MAFF 02811
Py Garden Insecticide	Vitax Ltd	MAFF 06395
Py Powder	Synchemicals Ltd	MAFF 05542
Py Powder	Zeneca Professional Products	MAFF 06886
Py Spray Garden Insect Killer	Vitax Ltd	MAFF 06085

* These products are "approved for agricultural use". For further details refer to page vii.
% These products are subject to the staged revocation procedure. For further details refer to page vii.
A These products are approved for use in or near water. For further details refer to page vii.

AMATEUR PRODUCTS

Product Name	Marketing Company	Reg. No.
1097 Pyrethrins *(Insecticide)—continued*		
Py Spray Insect Killer	Vitax Ltd	MAFF 05543
Secto Nature Care Garden Insect Powder	Secto Co Ltd	MAFF 06300
Secto Nature Care Garden Insect Spray	Secto Co Ltd	MAFF 06263
Secto Nature Care Houseplant Insect Spray	Secto Co Ltd	MAFF 06264
Vapona All-In-One	Sara Lee Household & Personal Care	MAFF 06965
% Vapona House and Garden Plant Insect Killer	Ashe Consumer Products Ltd	MAFF 05089
Vapona House and Garden Plant Insect Killer	Sara Lee Household & Personal Care	MAFF 05996
Wilko Fruit and Vegetable Insecticide Spray	Wilkinson Home & Garden Stores	MAFF 05770
Wilko Multi-Purpose Insecticide Spray	Wilkinson Home & Garden Stores	MAFF 06963
Wilko Rose and Flower Insecticide Spray	Wilkinson Home & Garden Stores	MAFF 04996
Wilko Rose and Flower Insecticide Spray	Wilkinson Home & Garden Stores	MAFF 05768
Woolworths Complete Insecticide Spray	Woolworths Plc	MAFF 06961
1098 Pyrethrins + *Lindane* *(Insecticide)*		
% Secto Greenfly and Garden Insect Spray	Secto Co Ltd	MAFF 03841
1099 Pyrethrins + *Pirimiphos-methyl* *(Insecticide)*		
Kerispray	Imperial Chemical Industries Plc	MAFF 02653
Kerispray	Zeneca Garden Care	MAFF 06840
1100 Pyrethrins + *Resmethrin* *(Insecticide)*		
House Plant Pest Killer	Vitax Ltd	MAFF 06432
% Rentokil Blackfly & Greenfly Killer	Rentokil Ltd	MAFF 02984
% Rentokil Greenhouse and Garden Insect Killer	Rentokil Ltd	MAFF 03689
1101 Quassia *(Vertebrate control)*		
Bird-Away	Fisons Plc	MAFF 05547
Cat Away	Fisons Plc	MAFF 05549
Cat Off	Fieldspray Division, Nilco Chemical Co Ltd	MAFF 05479
% Cat Off	Fieldspray Ltd	MAFF 04400

* These products are "approved for agricultural use". For further details refer to page vii.
% These products are subject to the staged revocation procedure. For further details refer to page vii.
A These products are approved for use in or near water. For further details refer to page vii.

AMATEUR PRODUCTS

Product Name	Marketing Company	Reg. No.

1101 Quassia *(Vertebrate control)—continued*

	Dog Away	Fisons Plc	MAFF 05548
	Dog Off	Fieldspray Division, Nilco Chemical Co Ltd	MAFF 05490
%	Dog Off	Fieldspray Ltd	MAFF 04401
	Garden Hoppit	Fieldspray Ltd	MAFF 04398
	Garden Hoppit Ready to Use	Fieldspray Division, Nilco Chemical Co Ltd	MAFF 05478
%	Garden Hoppit Ready-For-Use	Fieldspray Ltd	MAFF 04399

1102 Resmethrin + *Pirimiphos-methyl* + *Tetramethrin* *(Insecticide)*

Sybol Aerosol	Imperial Chemical Industries Plc	MAFF 05449
Sybol Aerosol	Zeneca Garden Care	MAFF 06850

1103 Resmethrin + *Pyrethrins* *(Insecticide)*

	House Plant Pest Killer	Vitax Ltd	MAFF 06432
%	Rentokil Blackfly & Greenfly Killer	Rentokil Ltd	MAFF 02984
%	Rentokil Greenhouse and Garden Insect Killer	Rentokil Ltd	MAFF 03689

1104 Rotenone *(Insecticide)*

%	Abol Derris Dust	Imperial Chemical Industries Plc	MAFF 02651
	Bio Friendly Insect Spray	Pan Britannica Industries Ltd	MAFF 05148
	Bio Friendly Pest Duster	Pan Britannica Industries Ltd	MAFF 06811
	Bio Liquid Derris Plus	Pan Britannica Industries Ltd	MAFF 07059
	Derris Dust	Vitax Ltd	MAFF 00676
	Derris Dust	Vitax Ltd	MAFF 05452
	Doff Derris Dust	Doff Portland Ltd	MAFF 00740
	ICI Derris Dust	Imperial Chemical Industries Plc	MAFF 05999
	Murphy Derris Dust	Fisons Plc	MAFF 04005
	PBI Liquid Derris	Pan Britannica Industries Ltd	MAFF 01214
	Wasp Exterminator	Battle Hayward & Bower Ltd	MAFF 06333
	Zeneca Derris Dust	Zeneca Garden Care	MAFF 06779

1105 Rotenone + *Carbaryl* *(Insecticide)*

Boots Garden Insect Powder with Derris	The Boots Company Plc	MAFF 02849

1106 Rotenone + *Lindane* + *Thiram* *(Fungicide)(Insecticide)*

%	PBI Hexyl	Pan Britannica Industries Ltd	MAFF 02650

---

* These products are "approved for agricultural use". For further details refer to page vii.
% These products are subject to the staged revocation procedure. For further details refer to page vii.
A These products are approved for use in or near water. For further details refer to page vii.

AMATEUR PRODUCTS

Product Name	Marketing Company	Reg. No.

1107 Rotenone + Sulphur *(Fungicide)(Insecticide)*

Bio Back To Nature Pest & Disease Duster	Pan Britannica Industries Ltd	MAFF 00265
% Bio Friendly Pest and Disease Duster	Pan Britannica Industries Ltd	MAFF 04887

1108 Simazine *(Herbicide)*

% Murphy Weedex	Murphy Chemicals Ltd	MAFF 02352

1109 Simazine + Amitrole *(Herbicide)*

% Path and Drive Weedkiller	May & Baker Garden Care	MAFF 04762
Path and Drive Weedkiller	Pan Britannica Industries Ltd	MAFF 05958
% Path and Drive Weedkiller	Rhone-Poulenc Agriculture Ltd	MAFF 05351
% Super Weedex	Fisons Plc	MAFF 02054

1110 Simazine + Amitrole + 2,4-D + Diuron *(Herbicide)*

Hytrol	Agrichem (International) Ltd	MAFF 04540

1111 Simazine + Amitrole + Diquat + Paraquat *(Herbicide)*

Pathclear	ICI Garden and Professional Products	MAFF 01546
Pathclear	Zeneca Garden Care	MAFF 06845

1112 Simazine + Amitrole + MCPA *(Herbicide)*

% Fisons Path Weeds Killer	Fisons Plc	MAFF 02700

1113 Simazine + Diuron *(Herbicide)*

Total Weedkiller Granules	Pan Britannica Industries Ltd	MAFF 05946
% Total Weedkiller Granules	Rhone-Poulenc Agriculture Ltd	MAFF 05352

1114 Sodium chlorate *(Herbicide)*

Barrettine Sodium Chlorate (Fire Suppressed) Weedkiller	Cooke's Chemicals (Sales) Ltd	MAFF 06617
Battle, Hayward and Bower Sodium Chlorate Weedkiller with Fire Depressant Battle	Hayward & Bower Ltd	MAFF 05876
Blanchard's Sodium Chlorate Weedkiller (Fire Suppressed)	Blanchard Martin and Simmonds Ltd	MAFF 05649
Cooke's Liquid Sodium Chlorate Weedkiller	Cooke's Chemicals (Sales) Ltd	MAFF 04280
Cookes Sodium Chlorate Weedkiller with Fire Depressant	Cooke's Chemicals (Sales) Ltd	MAFF 04281
Devcol – Sodium Chlorate Weedkiller	Devcol Ltd	MAFF 05656
Devcol Path Weedkiller	Devcol Ltd	MAFF 06580
Doff Path Weedkiller	Doff Portland Ltd	MAFF 07044

* These products are "approved for agricultural use". For further details refer to page vii.
% These products are subject to the staged revocation procedure. For further details refer to page vii.
A These products are approved for use in or near water. For further details refer to page vii.

AMATEUR PRODUCTS

Product Name	Marketing Company	Reg. No.

1114 Sodium chlorate *(Herbicide)—continued*

	Product Name	Marketing Company	Reg. No.
	Doff Sodium Chlorate Weedkiller	Doff Portland Ltd	MAFF 00500
	Focus Sodium Chlorate Weedkiller	Focus DIY Ltd	MAFF 06000
	Gem Sodium Chlorate Weedkiller	Joseph Metcalf Ltd	MAFF 04159
%	Gibbs-Palmer Sodium Chlorate	Gibbs-Palmer Ltd	MAFF 04378
	Great Mills Sodium Chlorate Weedkiller	Great Mills (Retail) Ltd	MAFF 07078
	Homebase Sodium Chlorate Weedkiller	Sainsbury's Homebase House and Garden Centres	MAFF 06620
	Murphy Sodium Chlorate	Fisons Plc	MAFF 03635
	Sodium Chlorate	Imperial Chemical Industries Plc	MAFF 01973
	Wilko Sodium Chlorate Weedkiller	Wilkinson Home & Garden Stores	MAFF 06281

1115 Sodium tetraborate *(Insecticide)*

	Product Name	Marketing Company	Reg. No.
%	Nippon Ant Destroyer Liquid	Synchemicals Ltd	MAFF 01502

1116 Sulphur *(Fungicide)*

	Product Name	Marketing Company	Reg. No.
%	Bio-Friendly Sulphur Dust	Pan Britannica Industries Ltd	MAFF 04974
	Fisons Nature's Answer to Powdery Mildew and Scab	Fisons Plc	MAFF 05143
	Green Sulphur	Synchemicals Ltd	MAFF 01007
	Green Sulphur	Vitax Ltd	MAFF 05782
	Phostrogen Safer's Ready-To-Use Garden Fungicide	Phostrogen Ltd	MAFF 04342
	Safer's Natural Garden Fungicide	Koppert (UK) Ltd	MAFF 06298
%	Safer's Natural Garden Fungicide	Koppert (UK) Ltd	MAFF 03569
	Sulphur Candles	Battle Hayward & Bower Ltd	MAFF 02039
	Yellow Sulphur	Synchemicals Ltd	MAFF 02372
	Yellow Sulphur	Vitax Ltd	MAFF 05783

1117 Sulphur *(Vertebrate control)*

	Product Name	Marketing Company	Reg. No.
	Murphy Mole Smoke	Fisons Plc	MAFF 03615

1118 Sulphur + *Fatty acids* *(Fungicide)(Insecticide)*

	Product Name	Marketing Company	Reg. No.
	Fisons Nature's Answer Fungicide and Insect Killer	Fisons Plc	MAFF 06767

1119 Sulphur + *Permethrin* + *Triforine* *(Fungicide)(Insecticide)*

	Product Name	Marketing Company	Reg. No.
	Bio Multirose	Pan Britannica Industries Ltd	MAFF 05716
%	Bio Multirose	Pan Britannica Industries Ltd	MAFF 02541

---

* These products are "approved for agricultural use". For further details refer to page vii.
% These products are subject to the staged revocation procedure. For further details refer to page vii.
A These products are approved for use in or near water. For further details refer to page vii.

AMATEUR PRODUCTS

Product Name	Marketing Company	Reg. No.

1120 Sulphur + *Rotenone* *(Fungicide)(Insecticide)*

	Bio Back To Nature Pest & Disease Duster	Pan Britannica Industries Ltd	MAFF 00265
%	Bio Friendly Pest and Disease Duster	Pan Britannica Industries Ltd	MAFF 04887

1121 Sulphur + *Carbendazim* + *Copper oxychloride* + *Permethrin* *(Fungicide)(Insecticide)*

%	Bio Multiveg	Pan Britannica Industries Ltd	MAFF 03116

1122 Tar oils *(Insecticide)*

	Jeyes Fluid	Jeyes Ltd	MAFF 04606
	Murphy Mortegg	Fisons Plc	MAFF 03616

1123 2-3-6 TBA + *2,4-D* *(Herbicide)*

	Touchweeder	Thomas Elliott Ltd	MAFF 02864

1124 Terbutryn *(Herbicide)*

	A Algae Kit	Ciba Agriculture	MAFF 04545
	A Blanc-kit	Intercel (UK)	MAFF 04546

1125 Tetramethrin + *Permethrin* *(Insecticide)*

	Nippon Ant and Crawling Insect Killer	Synchemicals Ltd	MAFF 03039

1126 Tetramethrin + *Phenothrin* *(Insecticide)*

	Pesguard House and Plant Spray	Sumitomo Chemical (UK) Plc	MAFF 01580

1127 Tetramethrin + *Pirimiphos-methyl* + *Resmethrin* *(Insecticide)*

	Sybol Aerosol	Imperial Chemical Industries Plc	MAFF 05449
	Sybol Aerosol	Zeneca Garden Care	MAFF 06850

1128 Thiophanate-methyl *(Fungicide)*

%	Fungus Fighter	May & Baker Agrochemicals	MAFF 00952
	Liquid Club Root Control	Pan Britannica Industries Ltd	MAFF 05957
%	Systemic Fungicide Liquid	May & Baker Garden Care	MAFF 04971

1129 Thiram + *4-Indol-3-ylbutyric acid* + *1-Naphthylacetic acid* *(Fungicide)(Herbicide)*

	Boots Hormone Rooting Powder	The Boots Company Plc	MAFF 01067

1130 Thiram + *Lindane* + *Rotenone* *(Fungicide)(Insecticide)*

%	PBI Hexyl	Pan Britannica Industries Ltd	MAFF 02650

---

* These products are "approved for agricultural use". For further details refer to page vii.
% These products are subject to the staged revocation procedure. For further details refer to page vii.
A These products are approved for use in or near water. For further details refer to page vii.

AMATEUR PRODUCTS

Product Name	Marketing Company	Reg. No.

1131 Thiram + *1-Naphthylacetic acid* *(Fungicide)(Herbicide)*

% Green Fingers Hormone Rooting Powder	Secto Co Ltd	MAFF 01001

1132 Triclopyr *(Herbicide)*

Murphy Nettlemaster	Fisons Plc	MAFF 06757

1133 Triforine + *Bupirimate* *(Fungicide)*

Nimrod T	Imperial Chemical Industries Plc	MAFF 03982
Nimrod T	Zeneca Garden Products	MAFF 06843

1134 Triforine + *Bupirimate* + *Pirimicarb* *(Fungicide)(Insecticide)*

Roseclear	Imperial Chemical Industries Plc	MAFF 01826
Roseclear	Zeneca Garden Care	MAFF 06849

1135 Triforine + *Permethrin* + *Sulphur* *(Fungicide)(Insecticide)*

Bio Multirose	Pan Britannica Industries Ltd	MAFF 05716
% Bio Multirose	Pan Britannica Industries Ltd	MAFF 02541

---

* These products are "approved for agricultural use". For further details refer to page vii.
% These products are subject to the staged revocation procedure. For further details refer to page vii.
A These products are approved for use in or near water. For further details refer to page vii.

3
PSD PRODUCT TRADE NAME INDEX

The number after the Trade Name gives the Active Ingredient Code Number under which the Active Ingredient in the product occurs in the Professional or Amateur Sections.

100 Plus *718*
2,4-D 50% *111*
200 Plus *500*
200-Plus *500*
3C Chlormequat 460 *62*
3C Chlormequat 600 *62*
5 Star Chlormequat *62*
50-50 Liquid Mosskiller *530*
5C Chlormequat Plus *62*
6 X Slug Killer *927*
ASCE 3748 *505*
AAprotect *889*
AAterra WP *547*
ABM Chlormequat 40 *62*
ABM Chlormequat 67.5 *62*
Abol Derris Dust *1104*
Accord *152*
Ace-Sodium Chlorate (Fire Suppressed) Weedkiller *419*
Acquit *751*
Acrobat *533*
Acrobat MZ *534*
Actellic Dust *837*
Actellic Smoke Generator 10 *837*
Actellic Smoke Generator No 20 *837*
Actellifog *837*
Actrilawn 10 *247*
Acumen *26*
Adition *173*
Adjust *62*
Advance *45*
Advizor *60*
Advizor S *60*
Advocate *662*
Afalon *290*
Afalon EC *290*
Afugan *671*
Agri Carbaryl 85 WP *761*
Agri HBN 400 *46*
Agrichem 2,4-D *111*
Agrichem DB Plus *129*
Agrichem Flowable Thiram *706*
Agrichem Mancozeb 80 *609*
Agrichem Maneb 80 *621*
Agrichem Maneb Dry Flowable *621*
Agrichem MCPA 25 *301*
Agrichem MCPA 50 *301*
Agricorn 500 *301*
Agricorn D *111*
Agritox 50 *301*
Agroxone 50 *301*
Aitken's Lawn Sand Plus *157*
Albrass *398*
Albrass SC *398*
Alfacron 10 WP *754*

Alfadex *839*
Alfracon 50 WP *754*
Algae Kit *1124*
Alicep *55*
Aliette *581*
Alistell *128*
Allicur *686*
Ally *360*
Ally WSB *360*
Alpha Atrazine 50 SC *12*
Alpha Atrazine 50 WP *12*
Alpha Biotril Plus *46*
Alpha Briotril *46*
Alpha Briotril Plus 19/19 *46*
Alpha Bromacil 80 WP *34*
Alpha Captan 49 Flowable *479*
Alpha Captan 50 WP *479*
Alpha Captan 83 WP *479*
Alpha Chloralose (Pure) *868*
Alpha Chloralose 4% Ready Mixed *868*
Alpha Chlormequat 460 *62*
Alpha Chlorpyrifos 48 EC *766*
Alpha Chlortoluron 500 *69*
Alpha Dicofol 18.5 EC *780*
Alpha Isoproturon 500 *267*
Alpha Isoproturon 650 *267*
Alpha Linuron 50 WP *290*
Alpha Linuron 50 SC *290*
Alpha Pentagan *62*
Alpha Pentagan Extra *62*
Alpha Prometryne 50 WP *395*
Alpha Prometryne 80 WP *395*
Alpha Propachlor 50 SC *398*
Alpha Propachlor 65 WP *398*
Alpha Simazine 50 SC *412*
Alpha Simazine 50 WP *412*
Alpha Simazine 80 WP *412*
Alpha Simazol *8*
Alpha Simazol T *8*
Alpha Terbalin 35 SC *438*
Alpha Terbalin 350 CS *438*
Alpha Terbutryne 50 SC *433*
Alpha Terbutryne 50 WP *433*
Alpha Trifluralin 48 EC *463*
Alto 100 SL *523*
Alto Combi *487*
Alto Eco *526*
Alto Elite *509*
Alto Major *528*
Alutal *870*
Ambush *829*
Ambush C *772*
Amcide *9*
Amcide *935*
Amizol-D *7*

Amos CMPP   *325*
Amos Cypermethrin 10   *772*
Amos Dimethoate   *784*
Amos Gro-Stop   *83*
Amos Maneb 80   *621*
Amos Maneb 800 WP   *621*
Amos Metaldehyde   *915*
Amos Micro-Sulphur   *677*
Amos Talunex   *870*
Angle 567 SC   *109*
Ant Powder   *1092*
Antec Durakil   *797*
Antergon MH 180   *298*
Antkiller Dust   *1092*
Antracol   *670*
Apache   *236*
Apex   *751*
Apex 5E   *823*
Aphox   *834*
Aplan   *523*
Apollo 50 SC   *770*
Applaud   *759*
Apron 350 FS   *631*
Apron 350 FS (Without Stickers)   *631*
Apron Combi 453 FS   *637*
Apron Combi FS   *637*
Apron T69 WS   *636*
Apron T69 WS (Without Stickers)   *636*
Aquablast Bug Spray   *1097*
Ardent   *174*
Arelon   *267*
Arelon WDG   *267*
Arena Plus   *498*
Argylene   *422*
Armillatox   *972*
Aro Slug Killer Blue Mini Pellets   *1061*
Arotex Extra   *62*
Arpal Non Selex Powder   *419*
Arresin   *363*
Arsenal   *242*
Arsenal 50   *242*
Arsenal 50 F   *242*
Ascot 480 FS   *703*
Asda Lawn Feed and Weed Moss Killer   *983*
Asda Lawn Sand   *1022*
Asda Lawn Weed and Feed   *984*
Ashlade 460 CCC   *62*
Ashlade 5C   *62*
Ashlade 700 5C   *62*
Ashlade 700 CCC   *62*
Ashlade CP   *61*
Ashlade Carbendazim Flowable   *484*
Ashlade Chlorotoluron 500   *69*
Ashlade Cyclops   *508*
Ashlade Cypermethrin 10 EC   *772*

Ashlade Dimethoate   *784*
Ashlade Flint   *297*
Ashlade Flotin   *564*
Ashlade Flotin 2   *564*
Ashlade Linuron FL   *290*
Ashlade Mancarb FL   *492*
Ashlade Mancarb Plus   *486*
Ashlade Mancozeb FL   *609*
Ashlade Mancozeb Flowable   *609*
Ashlade Maneb FL   *621*
Ashlade Maneb FL   *629*
Ashlade SMC Flowable   *517*
Ashlade Simazine 50 FL   *412*
Ashlade Solace   *521*
Ashlade Sulphur FL   *677*
Ashlade Summit   *438*
Ashlade Trimaran   *463*
Aspire   *378*
Asset   *17*
Astix   *341*
Astix K   *341*
Astrol   *47*
Asulox   *11*
Atlacide Extra Dusting Powder   *14*
Atlacide Soluble Powder Weedkiller   *419*
Atladox HI   *124*
Atlas 2,4-D   *111*
Atlas 3C:645 Chlormequat   *62*
Atlas 5C Chlormequat   *62*
Atlas Atrazine   *12*
Atlas Brown   *82*
Atlas CMPP   *325*
Atlas Chlormequat 46   *62*
Atlas Chlormequat 460:46   *62*
Atlas Chlormequat 640 5C   *62*
Atlas Chlormequat 670   *62*
Atlas Chlormequat 700   *62*
Atlas Chlormequat 730   *62*
Atlas CIPC 40   *75*
Atlas Dimethoate   *784*
Atlas Electrum   *58*
Atlas Gold   *80*
Atlas Herbon Pabrac   *75*
Atlas Indigo   *83*
Atlas Janus   *297*
Atlas Linuron   *290*
Atlas MCPA   *301*
Atlas Minerva   *41*
Atlas Orange   *398*
Atlas Pink C   *78*
Atlas Prometryne 50 WP   *395*
Atlas Propachlor   *398*
Atlas Protall   *267*
Atlas Protrum K   *386*
Atlas Quintacel   *62*

Atlas Red 79
Atlas Sheriff 767
Atlas Silver 56
Atlas Simazine 412
Atlas Solan 40 384
Atlas Somon 421
Atlas Steward 809
Atlas Sulphur 80 FL 677
Atlas Tecgran 100 426
Atlas Tecnazene 6% Dust 426
Atlas Terbine 62
Atlas Thor 193
Atlas Tricol 62
Atlas Trifluralin 463
Atraflow 12
Atrinal 175
Auger 267
Aura 750 EC 553
Autumn Kite 280
Autumn and Winter Toplawn 958
Avadex BW 446
Avadex BW 15G 446
Avadex BW 480 446
Avadex BW Granular 446
Avenge 2 169
Aventox SC 418
B & Q Ant Killer 940
B & Q Antkiller Dust 1092
B & Q Complete Insecticide Spray 1097
B & Q Complete Weedkiller 1030
B & Q Complete Weedkiller Ready To Use 1030
B & Q Fruit and Vegetable Insecticide Spray 1097
B & Q Garden Fungicide Spray 969
B & Q Granular Weed and Feed 978
B & Q House Plant Insecticide Spray 1097
B & Q Houseplant Insect Spray 1097
B & Q Houseplant Insecticide Ready-to-Use 1018
B & Q Houseplant Insecticide Spray 1097
B & Q Insecticide Spray For Roses and Flowers 1097
B & Q Insecticide Spray for Fruit and Vegetables 1097
B & Q Lawn Feed and Weed 981
B & Q Lawn Spot Weeder 981
B & Q Lawn Weedkiller 981
B & Q Lawnweed Spray 978
B & Q Liquid Weed and Feed 992
B & Q Mosskiller for Lawns 1022
B & Q Natural Pest Killer 1018
B & Q Rose and Flower Insecticide Spray 1097
B & Q Slug Killer Blue Mini Pellets 1061
B & Q Triple Action Lawn Care 979
B & Q Woodlice killer 940

B-Nine 131
BL500 75
Bactospeine WP 890
Bactospeine WP 938
Ban-Mite 817
Ban-Mite 1038
Banlene Plus 146
Banvel 4S 135
Banvel M 145
Banvel MP 146
Banvel P 148
Banvel T 152
Banweed 369
Barclay Bezant 472
Barclay Calypso 774
Barclay Carbosect 762
Barclay Champion 56
Barclay Chlormequat 72.5 62
Barclay Claddagh 56
Barclay Clinch 766
Barclay Corrib 505
Barclay Corrib 500 505
Barclay Cypersect XL 772
Barclay Dart 236
Barclay Desiquat 180
Barclay Dodex 539
Barclay Eyetak 658
Barclay Gallup 236
Barclay Gallup Amenity 236
Barclay Guideline 267
Barclay Holdup 62
Barclay Holdup 460 62
Barclay Hurler 221
Barclay Keeper 200 193
Barclay Manzeb 80 609
Barclay Manzeb 80W 609
Barclay Manzeb Flow 629
Barclay Metaza 354
Barclay Pirate 409
Barclay Pirimisect 834
Barclay Piza 500 407
Barclay Proton 267
Barclay Shandon 523
Barclay Total 373
Barclay Winner 216
Barleyquat B 62
Barrettine Sodium Chlorate (Fire Suppressed) Weedkiller 1114
BAS 03729H 341
BAS 421F 553
BAS 438 510
BAS 46402F 561
BAS 48500F 556
BAS 91580F 686
Basagran 24

Basamid *905*
BASF 2,4-D Ester 480 *111*
BASF 3C Chlormequat *62*
BASF 3C Chlormequat 600 *62*
BASF 3C Chlormequat 720 *62*
BASF 3C Chlormequat 750 *62*
BASF Bravo 500 *505*
BASF Dimethoate 40 *784*
BASF Maneb *621*
BASF MCPA Amine 50 *301*
BASF Phorate *832*
BASF Turf Protectant Fungicide *741*
BASF Turf Systemic Fungicide *484*
Bastion T *229*
Basudin 40WP *778*
Basudin 5FG *778*
Battal FL *484*
Battle, Hayward and Bower Sodium Chlorate Weedkiller with Fire Depressant *1114*
Bavistin *484*
Bavistin DF *484*
Bavistin FL *484*
Bay UK 292 *546*
Bayer UK 300 *686*
Bayer UK 407 WG *197*
Bayfidan *720*
Bayleton *718*
Bayleton 5 *718*
Bayleton BM *500*
Baytan *586*
Baytan Flowable *586*
Baytan IM *585*
Baythroid *771*
BB Chlorothalonil *505*
Beetomax *386*
Beetup *386*
Bellmac Plus *319*
Bellmac Straight *321*
Benazalox *18*
Benlate *941*
Benlate Fungicide *474*
Berelex *233*
Beret 050 FS *549*
Beret Extra 060 FS *550*
Besiege *521*
Betanal Compact *133*
Betanal Congress *132*
Betanal E *386*
Betanal Maestro *198*
Betanal Montage *198*
Betanal Progress *132*
Betanal Quadrant *198*
Betanal Quorum *133*
Betanal Rostrum *133*
Betanal Tandem *198*

Betanal Ultima *132*
Betapal Concentrate *368*
Betaren 120 *386*
Betosip *386*
Betosip 114 *386*
Bettaquat B *62*
Better DF *56*
Better Flowable *56*
BH 2,4-D Ester 50 *111*
BH CMPP/2,4-D *122*
BH CMPP Extra *325*
BH MCPA 75 *301*
BH Prefix D *154*
Binab T *891*
Bingo *557*
Bio 'BT' Caterpillar Killer *938*
Bio Back To Nature Pest & Disease Duster *1107*
Bio Catapult *926*
Bio Crop Saver *1039*
Bio Flydown *1080*
Bio Friendly Anti-Ant Duster *1097*
Bio Friendly Insect Spray *1104*
Bio Friendly Pest Duster *1104*
Bio Friendly Pest Pistol *1018*
Bio Friendly Pest and Disease Duster *1107*
Bio Lawn Weed Killer *978*
Bio Liquid Derris Plus *1104*
Bio Long Last *1012*
Bio Moss Killer *997*
Bio Multirose *1087*
Bio Multiveg *962*
Bio Pest Pistol *1018*
Bio Roota *999*
Bio Speedweed *1017*
Bio Sprayday *1080*
Bio Weed Pencil *994*
Bio-Friendly Sulphur Dust *1116*
Biobit Flowable Concentrate *890*
Biospot *994*
Bird-Away *1101*
Birlane 24 *765*
Birlane Granules *765*
Birlane Liquid Seed Treatment *765*
Blanc-kit *1124*
Blanchard's Sodium Chlorate Weedkiller (Fire Suppressed) *1114*
Blex *837*
Bofix *96*
Bolda FL *493*
Bombardier *505*
Bonzi *371*
Boots Ant Killer *948*
Boots Caterpillar and Whitefly Killer *1080*
Boots Garden Fungicide *961*
Boots Garden Insect Powder with Derris *960*

Boots Greenfly and Blackfly Killer *1011*
Boots Hormone Rooting Powder *1034*
Boots Kill-A-Weed *982*
Boots Lawn Moss Killer and Fertiliser *1022*
Boots Lawn Weed and Feed Soluble
 Powder *1006*
Boots Lawn Weedkiller *982*
Boots Nettle and Bramble Weedkiller *992*
Boots Systemic Weed and Grass Killer *1030*
Boots Systemic Weed and Grass Killer Ready To
 Use *1030*
Boots Total Lawn Treatment *939*
Bordeaux Mixture *965*
Bordeaux Mixture *966*
Borocil 1.5 *34*
Borocil K *36*
Bos MH 180 *298*
Botrilex *674*
BR Destral *5*
Brabant PCNB 20 % Dust *674*
Brasoran 50 WP *15*
Bravo 500 *505*
Bravo 720 *505*
Bravocarb *485*
Brays Emulsion *903*
Brestan 60 *563*
Brestan 60 SP *563*
Brimstone Plus *657*
Brittox *49*
Broadshot *118*
Brodifacoum *872*
Brodifacoum Bait *872*
Brodifacoum Bait Blocks *872*
Brodifacoum Sewer Bait *872*
Brom-O-Gas *901*
Bromadeth Rat and Mouse Killer *950*
Bronox *296*
Bug Gun! *1097*
Bug-Gun *1097*
BUK 92300F *686*
Burtolin *298*
Bushwacker Spray *425*
Butisan S *354*
Bygran F *426*
Bygran S *687*
CIA Dimethoate *784*
CIA Diquat-200 *180*
CIA Fentin-475-WP *564*
CIA MCPA 500 *301*
CIA Paraquat 200 *373*
CIA Propyzamide 500 *407*
CX 171 *197*
Calidan *490*
Calirus *473*
Calixin *729*

Camco Ant Powder *940*
Camco Insect Powder *940*
Campbell's Beetup *383*
Campbell's Bellmac Plus *319*
Campbell's Bellmac Straight *321*
Campbell's CMPP *325*
Campbell's Carbendazim 50% Flowable *484*
Campbell's CIPC 40% *75*
Campbell's DSM *777*
Campbell's DB Straight *125*
Campbell's Destox *111*
Campbell's Dimethoate 40 *784*
Campbell's Dioweed 50 *111*
Campbell's Disulfoton FE 10 *786*
Campbell's Field Marshall *146*
Campbell's Grassland Herbicide *146*
Campbell's Linuron 45% Flowable *290*
Campbell's MC Flowable *492*
Campbell's MCPA 25 *301*
Campbell's Metham Sodium *916*
Campbell's New Camppex *120*
Campbell's Nico Soap *825*
Campbell's Phorate *832*
Campbell's Rapier *407*
Campbell's Redipon *158*
Campbell's Redlegor *129*
Campbell's Trifluralin *463*
Campbell's Trifluron *297*
Campbell's X-Spor SC *621*
Campbells MCPA 50 *301*
Campbells Redipon Extra *166*
Cantor *280*
Captan 83 *479*
Capture *42*
Carbate Flowable *484*
Carbetamex *53*
Casoron G *154*
Casoron G-4 *996*
Casoron G4 *154*
Casoron G4 *996*
Casoron GSR *154*
Castaway Plus *815*
Cat Away *1101*
Cat Off *1101*
CCC 700 *62*
CDA CMPP/2,4-D *122*
CDA Castaway Plus *815*
CDA Dicotox Extra *111*
CDA Mildothane *704*
CDA Regulox K *298*
CDA Rovral *600*
CDA Simflow Plus *8*
CDA Spasor *236*
CDA Supertox 30 *122*
Celt *22*

Centex 419
Ceratotect 690
Cercobin Liquid 704
Cereline 476
Cereline Flowable 476
Cerevax 504
Cerevax Extra 502
Cerone 65
Certan 890
Certrol-PA 165
Challenge 234
Challenge 2 234
Chandor 297
Check Turf II 349
Checkmate 411
Cheetah R 200
Cheetah Super 201
Chempak Lawn Sand 1022
Chempak Slug Killer 927
Chemtech Cypermethrin 10 EC 772
Chemtech Nuarimol 642
Childion 781
Chiltern Chlorothalonil 500 505
Chiltern Hardy 915
Chiltern Isoproturon 267
Chipko Diuron 80 183
Chipman Diuron Flowable 183
Chipman Garlon 4 455
Chlorophos 963
Chloropicrin Fumigant 900
Chlortoluron 500 69
Chorus 216
Chrysotek Beige 245
Chryzotop Green 245
Cirrus 386
Citadel 216
Clarion 236
Clarosan 1FG 433
Claymore 242
Clayton Alpha-Cyper 751
Clayton Benzeb 472
Clayton Cyperten 772
Clayton Fencer 200
Clayton Fenican IPU 173
Clayton Fluroxypyr 221
Clayton Flutriafol 573
Clayton Glyphosate 236
Clayton Glyphosate 360 236
Clayton Krypton 521
Clayton Lenacil 80W 286
Clayton Meteor 772
Clayton Metsulfuron 360
Clayton Pirimicarb 50 SG 834
Clayton Propel 407
Clayton Propiconazole 664

Clayton Quatrow 180
Clayton Swath 236
Clayton Tebucon 683
Clean Up 972
Clean-Up-360 236
Cleanacres PDR 675 62
Cleanrun 122
Clear-Up 236
Clearway 8
Cleaval 108
Clenecorn 325
Clinafarm Spray 591
Clortosip 505
Clout 1
Clovacorn Extra 128
Clovotox 325
Club 820
CMPP 60 325
Co-op Garden Maker Ant Killer 940
Co-op Garden Maker Rose and Flower
  Insecticide Spray 1097
Co-op Garden Maker Slug Killer Pellets 1061
Colstar 556
Com-Trol 122
Combinex 656
Comet 354
Commando 215
Comodor 600 424
Compass 603
Compete 20 217
Compete 5 217
Compete Forte 218
Compete Mix 20 PVA 219
Compete Mix A 220
Competitor 218
Compitox Extra 325
Compo Benodanil 473
Condox 148
Conquest 235
Contact 75 505
Contest 751
Contrast 488
Cooke's Liquid Sodium Chlorate
  Weedkiller 1114
Cooke's Professional Sodium Chlorate Weedkiller
  with Fire Depressant 419
Cooke's Weedclear 419
Cookes Sodium Chlorate Weedkiller with Fire
  Depressant 1114
Corbel 553
Corbel CL 510
Corral 216
Corry's Fruit Tree Grease 944
Cosmic FL 494
Cossack 215

Cougar *173*
Country Fresh Disinfectant *191*
Coupler SC *92*
Crackdown *774*
Crackshot *440*
Crimson 514 FW *71*
Crimson 514 SC *71*
Croneton *792*
Crop Phenmedipham *386*
Cropsafe Dimethoate 40 *784*
Cropsafe MCPB *321*
Croptex Amber *398*
Croptex Bronze *384*
Croptex Chrome *79*
Croptex Fungex *514*
Croptex Pewter *75*
Croptex Steel *421*
Crossfire *766*
Crusader S *40*
CSC 2-Aminobutane *470*
Cubisan *911*
Cudgel *802*
Cultar *371*
Cuprokylt *516*
Cuprokylt L *516*
Cuprosana H *516*
Curb *869*
Curb (Garden Pack) *926*
Curzate M *521*
Cutlass *1010*
Cyanamid D50 *111*
Cyanamid DD Soil Fumigant *906*
Cyclone *579*
Cyclops *508*
Cymag *885*
Cymbush *772*
Cyperkill 10 *772*
Cyperkill 25 *772*
Cyperkill 5 *772*
Cypertox *772*
Cytro-Lane *818*
DAL-CT *499*
DAL CT *499*
DP 353 *230*
Dacamox 5 G *856*
Daconil Turf *505*
Dacthal W 75 *84*
Dacthal W-75 *84*
Dacthal W75 *84*
Dagger *240*
Danadim 400 *784*
Danadim Dimethoate 40 *784*
Darlingtons Diazinon Granules *778*
Darmycel Agarifume Smoke Generator *1080*
Darmycel Dichlorvos *779*

Dash *234*
Dazide *131*
Dazomet 98 *905*
Deadline Place Packs *873*
Debut *280*
Decade 500 EC *560*
Decimate *85*
Decis *774*
Decis Quick *775*
Decoy *820*
Deeweed *928*
Deftor *357*
Deloxil *46*
Deloxil 400 *46*
Delsanex Chlorate Weedkiller *419*
Delsene 50 DF *484*
Delta Insect Powder *940*
Deosan Chlorate Weedkiller *419*
Dequiman MZ *609*
Derosal Liquid *484*
Derosal WDG *484*
Derris Dust *1104*
Dethmor Plusbait *887*
Dethmor Warfarin 5 *887*
Detia Gas EX-B *895*
Detia Gas EX-T *895*
Deuce *382*
Devcol - Sodium Chlorate Weedkiller *1114*
Devcol 928 Insecticide Concentrate *1018*
Devcol All Purpose Natural Insecticide Spray *1097*
Devcol Ant Killer *940*
Devcol Green/Blackfly Killer *1018*
Devcol Houseplant Insecticide Ready For Use *1018*
Devcol Liquid Derris *848*
Devcol Morgan 3% Metaldehyde Slug Killer *1061*
Devcol Natural Pest Gun *1018*
Devcol Natural Pest Killer *1018*
Devcol Path Weedkiller *1114*
Devcol Pest Killer *1018*
Devcol Whitefly Insecticide *1018*
Devrinol *369*
Dextrone X *373*
Dexuron *190*
Di Farmon *148*
Dicofen *797*
Dicotox Extra *111*
Dicurane 500 FW *69*
Dicurane 500 SC *69*
Dicurane 700 SC *69*
Dicurane Combi 540 FW *73*
Dicurane Combi 540 SC *73*
Dicurane Duo 446 FW *29*

Dicurane Duo 446 SC *29*
Dicurane Duo 495 FW *29*
Dicurane Duo 495 SC *29*
Digermin *463*
Dimanquat *62*
Dimex TWK *419*
Dimilin ODC-45 *783*
Dimilin WP *783*
Dipel *890*
Dipel *938*
Dipterex 80 *866*
Diquat 200 *180*
Disyston FE-10 *786*
Disyston P-10 *786*
Dithane 945 *609*
Dithane Dry Flowable *609*
Dithane Superflo *609*
Dithianon Flowable *536*
Ditrac All-Weather Bait Bar *880*
Ditrac All-Weather Blox *880*
Ditrac Rat and Mouse Bait *880*
Ditrac Super Blox All-Weather *880*
Diuron 50 FL *183*
Diuron 80 FL *183*
Diuron 80% WP *183*
Divora *523*
Do It All Ant Killer *940*
Do It All Slug Killer Pellets *1061*
Do-It-All Fruit and Vegetable Insecticide Spray *1097*
Do-It-All Rose and Flower Insecticide Spray *1097*
Do-it-All Fruit and Vegetable Insecticide Spray *1097*
Docklene *146*
Dockmaster *143*
Doff 'All in One' Insecticide Spray *1097*
Doff Agricultural Slug Killer with Animal Repellent *915*
Doff Ant Control Powder *940*
Doff Ant Killer *1035*
Doff Derris Dust *1104*
Doff Fruit and Vegetable Insecticide Spray *1097*
Doff Gamma BHC Dust *1035*
Doff Garden Insect Powder *1021*
Doff Greenfly Killer *1097*
Doff Hormone Rooting Powder *956*
Doff Horticultural Slug Killer Blue Mini Pellets *915*
Doff Houseplant Insecticide Spray *1097*
Doff Lawn Feed and Weed *981*
Doff Lawn Mosskiller and Fertilizer *1022*
Doff Lawn Spot Weeder *981*
Doff Lawn Weed and Feed Soluble Powder *1006*
Doff Metaldehyde Slug Killer Mini Pellets *915*
Doff Nettle Gun *981*
Doff New Formula Lawn Weedkiller *981*
Doff Path Weedkiller *1114*
Doff Plant Disease Control *961*
Doff Rose and Flower Insecticide Spray *1097*
Doff Slugoids Slugkiller Blue Mini Pellets *1061*
Doff Sodium Chlorate Weedkiller *419*
Doff Sodium Chlorate Weedkiller *1114*
Doff Systemic Insecticide *1011*
Doff Total Path Weedkiller *928*
Doff Wasp Nest Killer *940*
Doff Woodlice Killer *940*
Dog Away *1101*
Dog Off *883*
Dog Off *1101*
Dorado *672*
Dorin *727*
Dormone *111*
Dosaflo *357*
Double H Lawn Feed and Weed *973*
Doubledown *788*
Doublet *47*
Dow Shield *87*
Doxstar *231*
Drat *876*
Drat Bait *876*
Drat Rat & Mouse Bait *876*
Draza *820*
Draza 2 *820*
Draza Plus *820*
Draza ST *917*
Du-Ter 50 *564*
Du-ter 50 *564*
Duet *347*
DUK 110 *444*
DUK 118 *440*
DUK 51 *557*
DUK 747 *568*
DUK-880 *289*
Duomo *505*
Duplosan *341*
Duplosan New System CMPP *341*
Duramitex *1038*
Dursban 4 *766*
Dursban 5G *766*
Dyfonate 10G *802*
Dynamec *748*
Dyrene *471*
Dyrene 720 *471*
Eagle *2*
Early Impact *489*
Echo *349*
Eclipse *541*
Effect *218*

Ekatin 857
Elliot's Easy-Weed 978
Elliot's Lawn Sand 210
Elliot's Mosskiller 210
Elliott's Touchweeder 995
Elliott's Turf Feed and Weed 984
Elocril 50 WP 807
Elvaron 529
Elvaron WG 529
Elvitox 820
Embark 349
Empal 301
Encore 278
Endorats 876
Endox 148
Endspray 205
Endspray II 357
Enforcer 156
Enide 50 W 179
Enide 50W 179
Ensign 542
Epic 540
Epox 820
Eptam 6E 192
Escar-go 3 915
Escar-go 6 915
Escort 97
Estrad 219
Estrad Duplo 219
Ethrel C 65
Everest 410
Evidence 776
EXP 4005 170
EXP 8506 33
Extratect Flowable 599
F 238 538
FCL Metsulfuron-Methyl-20 360
FD 4058 567
FP107 915
Fairy Ring Destroyer 739
Falcon 401
Fanfare 469 FW 276
Fanfare 469 SC 276
Fargro Chlormequat 62
Fargro Thuricide HP 890
Farm Fly Spray 10 WP 754
Farmatin 560 562
Farmatin 560 564
Farmon 2,4-D 111
Farmon Condox 148
Farmon PDQ 181
Fastac 751
Favour 600 FW 638
Favour 600 SC 638
Fazor 298

FCC Rapier 407
FCC Topcorn Extra 108
Fellside Green Fruit and Vegetable Insect Spray 1097
Fellside Green Rose and Flower Insect Spray 1097
Fellside Green Slug Killer Mini Pellets 1061
Fernex Granules 836
Ferrax 545
Ferrax IM 544
Fertosan Slug and Snail Powder 927
Fettel 149
Ficam Insect Powder 940
Ficam ULV 755
Field Marshal 147
Fielder 827
Finish 488
Fisons Autumn Extra 1022
Fisons Basilex 717
Fisons Clearcut II 1033
Fisons Clover-kil 981
Fisons Evergreen 90 1049
Fisons Evergreen Extra 1026
Fisons Evergreen Feed and Weed 1049
Fisons Evergreen Feed and Weed Liquid 994
Fisons Filex 662
Fisons Greenmaster Autumn 210
Fisons Greenmaster Extra 320
Fisons Greenmaster Mosskiller 210
Fisons Helarion 915
Fisons Insect Spray For Houseplant 1079
Fisons Lawn Sand 1022
Fisons Lawn Spot Weeder 978
Fisons Lawncare Liquid 994
Fisons Mosskil Extra 1022
Fisons Natural House Plant Insect Spray 1097
Fisons Nature's Answer Fungicide and Insect Killer 1019
Fisons Nature's Answer Organic Insecticide RTU 1018
Fisons Nature's Answer to Insect Pests 1097
Fisons Nature's Answer to Insect Pests on Flowers, Fruit and Vegetables 1097
Fisons Nature's Answer to Powdery Mildew and Scab 1116
Fisons Octave 658
Fisons Path Weeds Killer 933
Fisons Ready-to-Use Clover-kil 981
Fisons Ready-to-Use Lawn Weedkiller 981
Fisons Tritox 146
Fisons Turfclear 484
Fisons Water On Lawn Pest Killer 958
Fisons Water-on Lawn Weedkiller 981
Fix Up Slow Release Fly Killer 1007
Flarepath 642

Flexidor *281*
Flexidor 125 *281*
Floracid *943*
Focus Ant Killer *940*
Focus Fruit and Vegetable Insecticide Spray *1097*
Focus Lawn Weedkiller *981*
Focus Rose and Flower Insecticide Spray *1097*
Focus Slug Killer Pellets *1061*
Focus Sodium Chlorate Weedkiller *1114*
Folicur *683*
Folicur EC *683*
Folimat *826*
Folio 575 FW *512*
Folio 575 SC *512*
Fongarid 25 WP *587*
Fonofos Seed Treatment *802*
For-ester *111*
Force ST *852*
Format *87*
Fortrol *102*
FR 1001 *46*
FR 1433 *603*
FR 1442 *30*
Framolene *152*
Freeway *183*
Frowncide *567*
Fubol 58 WP *617*
Fubol 75 WP *617*
Fumispore *650*
Fumite Dicloran Smoke *531*
Fumite Dicofol Smoke *780*
Fumite General Purpose Insecticide Smoke Cone *1092*
Fumite Lindane 10 *809*
Fumite Lindane 40 *809*
Fumite Lindane Pellets *809*
Fumite Permethrin Smoke Generators *829*
Fumite Pirimiphos Methyl Smoke *837*
Fumite Propoxur Smoke *838*
Fumite Whitefly Smoke Cone *1080*
Fumitoxin *752*
Fumyl-O-Gas *918*
Fungaflor *591*
Fungaflor C *591*
Fungaflor Smoke *591*
Fungazil 100 SL *591*
Funginex *739*
Fungus Fighter *1128*
Fusarex 10% Granules *687*
Fusarex 6% Dust *687*
Fusilade 250 EW *216*
Fusilade 5 *216*
Fusion *572*
Fydulan *130*

Fytospore *521*
Galben M *472*
Gallery *281*
Gallery 125 *281*
Galtak 50 SC *16*
Gamma Col *809*
Gamma HCH Dust *809*
Gamma-Col Turf *809*
Gammalex *480*
Gammasan 30 *809*
Garden Centre Pest Spray for Fruit and Vegetables *1097*
Garden Centre Pest Spray for Roses and Flowers *1097*
Garden Hoppit *1101*
Garden Hoppit Ready to Use *1101*
Garden Hoppit Ready-For-Use *1101*
Gardenstore Lawn Feed and Weed *1049*
Garlon 2 *455*
Garlon 4 *455*
Garnet *684*
Garvox 3G *755*
Gastrotox 6G Pellets *915*
Gaucho *806*
Geest Fumite 300 MK2 *829*
Gem Lawn Sand *1022*
Gem Lawn Weed and Feed *984*
Gem Lawn Weed and Feed + Mosskiller *983*
Gem Lawn Weed and Feed 4 *984*
Gem Sodium Chlorate Weedkiller *419*
Gem Sodium Chlorate Weedkiller *1114*
Gemini *561*
Genesis *922*
Genie *568*
Gesagard 50 WP *395*
Gesaprim 500 FW *12*
Gesaprim 500 SC *12*
Gesatop 50 WP *412*
Gesatop 500 FW *412*
Gesatop 500 SC *412*
Get Off My Garden *1063*
Gibbs-Palmer Sodium Chlorate *1114*
Gladiator *56*
Gladiator DF *56*
Glint 500 EC *560*
Glyfonex *236*
Glyfosate - 360 *236*
Glypho *1030*
Glypho Gun! *1030*
Glyphogan *236*
Glytex *284*
Golden Malrin Fly Bait *822*
Goliath *386*
Goltix Triple *197*
Goltix WG *352*

Good Life Lawn Feed and Weed   *973*
Graminon 500 SC   *267*
Graminon 500SC   *267*
Gramonol 5   *365*
Gramonol Five   *365*
Gramoxone 100   *373*
Grapple   *216*
Grasp   *445*
Grazon 90   *100*
Great Mills Ant Killer   *940*
Great Mills Complete Insecticide Spray   *1097*
Great Mills Fast Action Weedkiller Ready to Use   *1029*
Great Mills Fruit and Vegetable Insecticide Spray   *1097*
Great Mills Lawn Feed, Weed and Mosskiller   *1026*
Great Mills Lawn Spot Weeder   *981*
Great Mills Rose and Flower Insecticide Spray   *1097*
Great Mills Slug Killer Blue Mini Pellets   *1061*
Great Mills Sodium Chlorate Weedkiller   *1114*
Great Mills Systemic Action Weedkiller Ready to Use   *1030*
Great Mills Woodlice Killer   *940*
Green Fingers Hormone Rooting Powder   *1070*
Green Sulphur   *1116*
Green Up Feed and Weed Plus Mosskiller   *983*
Green Up Lawn Feed and Weed   *978*
Green Up Lawn Spot Weedkiller   *984*
Green Up Lawn Spot Weedkiller   *985*
Green Up Mossfree   *1022*
Green Up Weedfree Lawn Weedkiller   *978*
Green Up Weedfree Spot Weedkiller for Lawns   *978*
Greenco GR1   *1018*
Greenco GR3   *1018*
Greenco GR3 Pest Jet   *1018*
Greenscape Ready To Use Weed Killer   *1030*
Greenscape Weedkiller   *1030*
Greenshield   *485*
Greensward   *979*
Grey Squirrel Liquid Conc   *887*
Grey Squirrel Liquid Concentrate   *887*
Gro-Tard II   *349*
Groundclear   *992*
Groundclear Spot   *992*
Grovex R15 Liquid Warfarin   *887*
Grovex R18 Sewer Bait   *887*
Grovex Warfarin 5   *887*
Grovex Warfarin 5 (Inert Base)   *887*
Grovex Zinc Phosphide   *888*
Growing Success Cat Repellent   *1028*
Growing Success Slug Killer   *927*
Guardian   *508*

Guardsman B   *869*
Guardsman L Crop Spray   *869*
Guardsman M   *869*
Guardsman STP   *869*
HOE 39866 SL06   *234*
HY-CARB   *484*
HY-D   *111*
HY-MCPA   *301*
HY-TL   *703*
HY-VIC   *703*
Hallmark   *808*
Halo   *511*
Halo 300   *511*
Halophen RE 49   *530*
Harlequin 500 FW   *279*
Harlequin 500 SC   *279*
Harmony M   *362*
Harvest   *234*
Hawk 440 SC   *174*
HCH 50% Tracking Dust   *882*
Headland Addstem   *484*
Headland Charge   *325*
Headland Dephend   *386*
Headland Dual   *492*
Headland Inorganic Liquid Copper   *516*
Headland Relay   *146*
Headland Spear   *301*
Headland Spirit   *621*
Headland Staff   *111*
Headland Sulphur   *677*
Headland Swift   *62*
Helmsman   *62*
Helosate   *236*
Hemoxone   *166*
Herbasan   *386*
Herrisol   *146*
Hickson Phenmedipham   *386*
Hickstor 10   *687*
Hickstor 3   *426*
Hickstor 5   *426*
Hickstor 6   *426*
Hickstor 6 Plus MBC   *52*
Hilite   *236*
Hispor 45WP   *497*
Hobby   *496*
Hoe 39866 SL06   *234*
Hoechst Diuron   *183*
Hoegrass   *167*
Hoegrass 280   *167*
Hoegrass EW   *167*
Holdfast D   *151*
Holdup   *236*
Homebase Ant Killer   *940*
Homebase Contact Action Weedkiller   *1029*
Homebase Contact Action Weedkiller Ready-to-Use Sprayer   *1029*

Homebase Lawn Feed Weed and Mosskiller *1026*
Homebase Lawn Feed and Mosskiller *1022*
Homebase Lawn Feed and Weed *1049*
Homebase Lawn Feed and Weed liquid *994*
Homebase Natural Pest Killer *1018*
Homebase Pest Gun *1097*
Homebase Slug Killer Blue Mini Pellets *1061*
Homebase Sodium Chlorate Weedkiller *1114*
Homebase Systemic Action Weedkiller *1030*
Homebase Systemic Action Weedkiller Ready to Use *1030*
Homebase Woodlice Killer *940*
Hoppit *883*
Hortag Carbotec *52*
Hortag Cypermethrin 10 EC *772*
Hortag Tecnacarb *52*
Hortag Tecnazene 10% Granules *687*
Hortag Tecnazene Double Dust *687*
Hortag Tecnazene Plus *689*
Hortag Tecnazene Potato Dust *687*
Hortag Tecnazene Potato Granules *687*
Hortag Zineb Wettable *743*
Hortichem 2-Aminobutane *470*
Hortichem Spraying Oil *824*
Hostaquick *804*
Hostathion *865*
Hotspur *95*
House Plant Pest Killer *1100*
Hy-Flo *706*
Hy-TL *703*
Hyban *148*
Hydon *37*
Hydraguard *608*
Hygrass *148*
Hykeep *690*
Hymec *325*
Hymush *690*
Hyperkil *874*
Hyprone *146*
Hyquat *62*
Hyquat 70 *62*
Hyquat 75 *62*
Hysede FL *607*
Hystore 10 *426*
Hysward *146*
Hytane 500 SC *267*
Hytane 500FW *267*
Hytec *426*
Hytec 6 *426*
Hytec Super *428*
Hytrol *930*
Hyvar X *34*
Hyzon *56*
I T Glyphosate *236*

IPSO *276*
Ibis *65*
ICI Derris Dust *1104*
ICI Insecticidal Soap *1018*
ICI Mosskiller For Lawns *1022*
ICI Slug Pellets *1061*
Impact *573*
Impact Excel *511*
Impact Excel 375 *511*
Improved Golden Malrin Fly Bait *822*
Interlates Chlormequat 46 *62*
Invader *534*
Invicta *32*
Ipicombi TL *297*
Ipifluor *463*
Ipithoate *784*
Ipitrax *12*
IPU 500 *267*
IPU 500 Herbicide *267*
Iso Cornox 57 *325*
Isoguard *267*
Isoproturon 500 *267*
Isoproturon 553 *267*
J Arthur Bower's Feed and Weed Mosskiller *1005*
J Arthur Bower's Granular Feed and Weed *1006*
J Arthur Bower's Granular Feed, Weed and Mosskiller *1005*
J Arthur Bower's Granular Lawn Feed and Weed *1006*
J Arthur Bower's Lawn Sand *1022*
J Arthur Bowers Lawn Feed and Weed *1006*
J Arthur Bowers Lawn Sand *1022*
J. Arthur Bower's Lawn Food with Weedkiller *984*
Janus *297*
Jasper *727*
Javelin *173*
Javelin Gold *173*
Jester *559*
Jeyes Fluid *1122*
Jolt *278*
Jubilee *360*
Jubilee 20DF *360*
Jupital *505*
Kapitol *481*
Kapitol WDG *481*
Karamate Dry Flo *609*
Karamate N *609*
Karate Mouse Bait *955*
Karate Ready To Use Rat and Mouse Bait *876*
Karate Ready To Use Rodenticide Sachets *876*
Karathane Liquid *535*
Karathane WP *535*
Karmex *183*

Keetak 553
Kelthane 780
Kemifam E 386
Kemiron 193
Kemiron Flow 193
Kemkill Sewer Bait 0.025% 887
Kemkill Sewer Bait 0.05% 887
Kerb 50 W 407
Kerb Flo 407
Kerb Granules 407
Keri Insect Spray 1097
Kerispray 1093
Ki-Hara IPU 500 267
Killgercide Ready To Use Cut Wheat Bait 878
Killgerm Rat Rod 878
Killgerm Rat Rods 878
Killgerm Ratak Cut Wheat Rat Bait 878
Killgerm Ratak Rat Pellets 878
Killgerm Ratak Whole Wheat Rat Bait 878
Killgerm Sakarat Special 876
Killgerm Sewercide Cut Wheat Rat Bait 887
Killgerm Sewercide Whole Wheat Rat Bait 887
Killgerm Wax Bait 878
Kilmol 887
Klearwell Alphachloralose Bait 868
Klearwell Blue Warfarin 5 887
Klearwell Drat Rodent Poison 876
Klearwell Oiled Warfarin Sewer Bait 887
Klearwell Ready Mixed Warfarin 887
Klerat 872
Klerat Mouse Tube 872
Klerat Wax Blocks 872
Knave 789
Knot Out 281
Kombat WDG 483
Kombat WDG 491
Koppert De Moss 199
Kotol FS 809
Krenite 232
Krovar 1 36
Kumulus DF 677
Kumulus S 677
KW Chlormequat 600 62
Landgold Aldicarb 10G 749
Landgold Benazolin Plus 18
Landgold Carbosulfan 10G 763
Landgold Chlorothalonil 50 505
Landgold CMPP F 325
Landgold Cycloxydim 110
Landgold DFF 625 173
Landgold Diquat 180
Landgold Ethofumesate 200 193
Landgold FF550 173
Landgold Fenoxaprop 200
Landgold Fenpropimorph 750 553

Landgold Fluazifop-P 216
Landgold Fluroxypyr 221
Landgold Flusilazole 400 568
Landgold Flutriafol 573
Landgold Glyphosate 360 236
Landgold Isoproturon 267
Landgold Isoproturon FC 267
Landgold Isoproturon FL 267
Landgold Isoproturon SC 267
Landgold Lambda-C 808
Landgold Mecoprop-P 341
Landgold Mecoprop-P 600 341
Landgold Metamitron 352
Landgold Metsulfuron 360
Landgold Paraquat 373
Landgold Pirimicarb 50 834
Landgold Propiconazole 664
Landgold Propyzamide 50 407
Landgold Triadimefon 25 718
Lannate 20L 821
Lanslide 288
Laser 110
Lawn Builder Plus Weed Control 978
Lawn Feed and Weed Granules 984
Lawn Spot Weed Granules 984
Lawn Weed Gun 978
Lawnsman Liquid Weed and Feed 992
Lawnsman Weed and Feed 978
Leap 648
Legacy 567
Legend 552
Legumex Extra 20
Legumex Extra 21
Lektan 352
Lentagran 45 WP 409
Lentagran 50% WP 409
Lentagran WP 409
Lentipur CL 500 69
Lever Industrial Sodium Chlorate Weedkiller with Fire Depressant 419
Lexone 70 DF 359
Leyclene 43
Lindane 20 809
Lindane Flowable 809
Lindex Plus FS Seed Treatment 558
Linnet 297
Linuron 50 290
Linuron Flowable 290
Liquid Club Root Control 1128
Liquid Curb Crop Spray 869
Liquid Derris 848
Liquid Linuron 290
Liquid Weed and Feed 992
Lo-Gran 20WG 447
Loft 3

Longer Lasting Bug Gun *943*
Longlife Cleanrun *122*
Longlife plus *114*
Lontrel 100 *87*
Lontrel Plus *93*
Lorate 20DF *360*
Lorsban T *766*
Ludorum *69*
Ludorum 700 *69*
Luxan CMPP *325*
Luxan Chloridazon *56*
Luxan CMPP 600 *325*
Luxan Cypermethrin *772*
Luxan Dimethoate *784*
Luxan Gro-Stop *83*
Luxan Mancozeb Flowable *609*
Luxan Maneb 80 *621*
Luxan Maneb 800 WP *621*
Luxan MCPA 50 *301*
Luxan Metaldehyde *915*
Luxan Micro-Sulphur *677*
Luxan Phenmedipham *386*
Luxan Talunex *870*
Lyric *568*
MON 35010 *236*
MON 44068 Pro *236*
MON 52276 *236*
MTM CIPC 40 *75*
MTM DB Straight *125*
MTM Dimethoate 40 *784*
MTM Disulfoton P10 *786*
MTM Diuron 80 *183*
MTM Eminent *108*
MTM Grassland Herbicide *147*
MTM Malathion 60 *817*
MTM Phorate *832*
MTM Sugar Beet Herbicide *80*
MTM Sulphur Flowable *677*
MTM Trifluralin *463*
Magnum *59*
Mainstay *505*
Malathion Greenfly Killer *1038*
Mallard 750 EC *551*
Mandops Barleyquat B *62*
Mandops Bettaquat B *62*
Mandops Chlormequat 460 *62*
Mandops Chlormequat 700 *62*
Mandops Helestone *62*
Maneb 80 *621*
Manex *629*
Manipulator *62*
Mantis 250 EC *664*
Mantle 425 EC *560*
Manzate *621*
Manzate 200 *609*

Manzate 200 DF *609*
Manzate 200 PI *609*
Marks 2,4-D-A *111*
Marks 2,4-DB *125*
Marks 2,4-DB Extra *129*
Marks MCPB *321*
Marks MCPA 50A *301*
Marks MCPA P30 *301*
Marks MCPA S.25 *301*
Marks MCPA SP *301*
Marks Mecoprop BAI *341*
Marks Mecoprop E *325*
Marks Mecoprop K *325*
Marks PCF Beet Herbicide *77*
Marks Polytox-K *158*
Marks Polytox-M *119*
Marshal 10 G *763*
Mascot Clearing *473*
Mascot Cloverkiller *325*
Mascot Cloverkiller-P *341*
Mascot Contact Turf Fungicide *741*
Mascot Highway WP *8*
Mascot Mosskiller *156*
Mascot Selective Weedkiller *122*
Mascot Selective-P *123*
Mascot Simazine 2 % Granular *412*
Mascot Super Selective *146*
Mascot Super Selective-P *147*
Mascot Systemic Turf Fungicide *484*
Masterspray *51*
Match *102*
Matrikerb *99*
Maxicrop Mosskiller and Conditioner *210*
Maxicrop Mosskiller and Lawn Tonic *1022*
Maxim *484*
Mazide Selective *144*
Mazin *621*
MC Flowable *492*
MCPA 25% *301*
Mebrom 100 *918*
Mebrom 98 *918*
Medo *972*
Mega D *111*
Mega M *301*
Meld *572*
Meothrin *799*
Merit *382*
Metarex *915*
Metarex RG *915*
Metasystox R *828*
Meteor *64*
Methoxone *325*
Methyl Bromide 100% *918*
Methyl Bromide 98% *901*
Microcarb Suspendable Powder *761*

PSD PRODUCT NAME INDEX

Microsul Flowable Sulphur   *677*
Microthiol Special   *677*
Midstream   *180*
Mifaslug   *915*
Mifatox   *777*
Mildothane Liquid   *704*
Mildothane Turf Liquid   *704*
Milgo E   *543*
Militan   *546*
Milstem   *543*
Miros DF   *505*
Mirvale 500 HN   *75*
Missile   *671*
Mission   *410*
Mistral   *553*
Mistral CT   *510*
Mitac 20   *753*
Mocap 10G   *793*
Mogul   *236*
Monarch   *381*
Monceren   *654*
Monceren DS   *654*
Monceren FS   *654*
Monceren IM   *598*
Monceren IM Flowable   *598*
Moot   *528*
Morestan   *673*
Morgan's Blue Mini Slug Killer Pellets   *915*
Mortegg Emulsion   *850*
Moss Control Plus Lawn Fertilizer   *1022*
Mossgun   *997*
Mosskiller For Lawns   *1022*
Mouser   *949*
Mowchem   *349*
MSS 2,4-D Amine   *111*
MSS 2,4-D Ester   *111*
MSS 2,4-DP   *158*
MSS 2,4-DP + MCPA   *166*
MSS 2,4-DB + MCPA   *129*
MSS Aminotriazole 80% WP   *3*
MSS Aminotriazole Technical   *3*
MSS Atrazine 50 FL   *12*
MSS Atrazine 80 WP   *12*
MSS CMPP   *325*
MSS Chlormequat 40   *62*
MSS Chlormequat 460   *62*
MSS Chlormequat 60   *62*
MSS Chlormequat 70   *62*
MSS CIPC 40 EC   *75*
MSS CIPC 5 G   *75*
MSS CIPC 50 LF   *75*
MSS CIPC 50 M   *75*
MSS Diuron 50 FL   *183*
MSS Diuron 50FL   *183*
MSS Iprofile   *267*

MSS MH 18   *298*
MSS MCPA 50   *301*
MSS MCPB + MCPA   *319*
MSS Mircam   *150*
MSS Mircam Plus   *147*
MSS Mircell   *62*
MSS Mirprop   *341*
MSS Optica   *341*
MSS Simazine 50 FL   *412*
MSS Simazine 80 WP   *412*
MSS Simazine/Aminotriazole 43% FL   *8*
MSS Sugar Beet Herbicide   *77*
MSS Sulphur 80   *677*
MSS Sulphur 80 WP   *677*
MSS Trifluralin 48 EC   *463*
Multi-W FL   *492*
Murbenine   *588*
Murbenine Plus   *589*
Murfume Grain Store Smoke   *809*
Murfume Lindane Smoke   *809*
Murfume Permethrin Smoke   *829*
Murfume Propoxur Smoke   *838*
Murphy Bugmaster   *1097*
Murphy Clover-Kil   *981*
Murphy Derris Dust   *1104*
Murphy Gamma BHC Dust   *1035*
Murphy Hormone Rooting Powder   *956*
Murphy Lawn Feed and Moss Killer   *1022*
Murphy Lawn Feed and Weed   *1049*
Murphy Lawn Feed, Weed and Mosskiller   *1026*
Murphy Lawn Pest Killer   *957*
Murphy Lawn Pest Killer   *958*
Murphy Lawn Weedkiller   *981*
Murphy Lawn Weedkiller and Lawn Tonic   *981*
Murphy Malathion Dust   *1038*
Murphy Malathion Liquid   *1038*
Murphy Mole Smoke   *1117*
Murphy Mortegg   *1122*
Murphy Nettlemaster   *1132*
Murphy Path Weed Killer   *928*
Murphy Permethrin Whitefly Smoke   *1080*
Murphy Problem Weeds Killer   *932*
Murphy Slug Tape   *1061*
Murphy Slugit Liquid   *1061*
Murphy Slugits   *1061*
Murphy Sodium Chlorate   *1114*
Murphy Super Mosskiller Ready To Use   *997*
Murphy Systemic Action Fungicide   *961*
Murphy Systemic Action Insecticide   *1032*
Murphy Traditional Copper Fungicide   *967*
Murphy Tumbleblite II   *1077*
Murphy Tumbleblite II Ready to Use   *1077*
Murphy Tumblebug   *1032*
Murphy Tumbleweed   *1030*
Murphy Tumbleweed Gel   *1030*

Murphy Tumbleweed Spray 1030
Murphy Ultra-Weed 1029
Murphy Weedex 1108
Murphy Weedmaster 1029
Murphy Weedmaster Ready-to-use Sprayer 1029
Murphy Zap Cap Combined Insecticide and Fungicide 1020
Murphy Zap Cap General Insecticide 1080
Murphy Zap Cap Systemic Fungicide 961
Murvin 85 761
Musketeer 265
Muster 236
Muster LA 236
Mycotal 892
NaTA 423
Narsty 869
NATA 423
Natural Bug Gun 1018
Nature's Answer 938
Natures 'Bug Gun!' 1097
Natures Friends Bactospeine 938
Nebulin 687
Neminfest 297
Nemispor 609
Nemolt 851
Neokil 878
Neosorexa Bait 878
Neosorexa Concentrate 878
Neosorexa Liquid Concentrate 878
Neosorexa Ratpacks 878
Neporex 2 SG 773
New 5C Cycocel 62
New Arena Plus 52
New Arena TBZ 6 428
New Atraflow Plus 4
New Bandock 146
New Estermone 114
New Formula B & Q Complete Weedkiller 1030
New Formula Greenscape 1030
New Formulation SBK Brushwood Killer 980
New Formulation Weed and Brush Killer 117
New Formulation Weed and Brushkiller 117
New Hickstor 6 426
New Hickstor 6 Plus MBC 52
New Hickstor TBZ 6 428
New Hystore 687
New Improved Leaf Action Roundup Brushkiller Ready To Use 1030
New Improved Leaf Action Roundup Weedkiller Ready To Use 1030
New Kotol 809
New Murbetex Fl 56
New Simflow 412
New Spraydex Greenfly Killer 943

New Squadron 492
New Strike 956
New Supertox 993
Nex 762
Nicotine 40% Shreds 825
Nimrod 477
Nimrod T 478
Nimrod T 953
Nippon Ant Destroyer Liquid 1115
Nippon Ant Killer Liquid 947
Nippon Ant and Crawling Insect Killer 1088
Nomix 2,4-D Herbicide 111
Nomix Turf Selective Herbicide 122
Nomix Turf Selective LC Herbicide 122
Nomix-Chipman Mosskiller 156
Nortron 193
Notcutts Feed and Weed with Mosskiller 983
Novosol Flowable Concentrate 890
Nuvan 500 EC 779
Octave 658
Octolan 509
Oncol 10 G 756
Oncol 10G 756
Opal 552
Opogard 500 SC 432
Opogard 500FW 432
Optica 341
Optimol 819
Optimol 915
Opus 540
Opus Plus 542
Opus Team 541
Oracle 277
Orchard Herbicide 7
Osprey 58 WP 617
Osprey 58WP 617
Outlaw 236
Oxytril CM 46
Oyster 173
P A Dimethoate 40 784
PA Chlormequat 400 62
PA Chlormequat 460 62
PA Propachlor Flowable 398
PA Weedmaster 56
PCQ Rat and Mouse Bait 880
PDQ 181
PDR 675 62
PP005 216
PP007 216
PP630 508
PY Spray Insect Killer 1097
Pacer 489
Palette 489
Pallitop 641
Palormone D 111

Panacide M   156
Panacide TS   156
Panacide Technical Solution   156
Pancil T   644
Panoctine   588
Panoctine Plus   589
Panther   173
Parable   181
Pasturol   146
Pasturol D   147
Patafol Plus   618
Path and Drive Weedkiller   934
Path and Shrub Guard   996
Pathclear   931
Patrol   551
PBI Anti-Ant Duster   1097
PBI Arbrex Pruning Compound   945
PBI Boltac Grease Bands   1031
PBI Bromophos   951
PBI Cheshunt Compound   969
PBI Dithane 945   1041
PBI Fenitrothion   1021
PBI Hexyl   1037
PBI Liquid Derris   1104
PBI Pepper Dust   1078
PBI Racumin Mouse Bait   971
PBI Racumin Rat Bait   971
PBI Slug Mini Pellets   1061
PBI Supercarb   961
PBI Toplawn   978
PBI Velvas   1022
Peaweed   397
Penncozeb   609
Penncozeb Flowable   609
Penncozeb WDG   609
Pentac Aquaflow   782
Pepper Dust   1078
Percept   46
Perflan 80W   425
Permasect 10 EC   829
Permasect 25 EC   829
Permit   829
Pesguard House and Plant Spray   1089
Pest-Pel   878
Phantom   834
Phenoxylene 50   301
Phosfleur 1.5   74
Phosfleur 10% Liquid   74
Phostek   752
Phostek   870
Phostoxin I   752
Phostrogen House Plant Insecticide   1018
Phostrogen Safer's Garden Insecticide Concentrate   1018
Phostrogen Safer's Ready-To-Use Fruit and Vegetable Insecticide   1018
Phostrogen Safer's Ready-To-Use Garden Fungicide   1116
Phostrogen Safer's Ready-To-Use House Plant Insecticide   1018
Phostrogen Safer's Ready-To-Use Rose and Flower Insecticide   1018
Phostrogen Soluble Lawn Moss Killer and Lawn Tonic   1022
Phyto IPU   267
Phyto-CCC   62
Phyto-Chloridazon   56
Phyto-Medipham   386
Phyto-Toluron   69
Picket   1080
Pigeon Loft Spray   838
Pilot   410
Pinnacle 400   241
Pirimicarb 50 WG   834
Pirimor   834
Pistol 25   386
Plant Pin   954
Plantvax 20   649
Plantvax 75   649
Plover 250EC   532
Podquat   62
Pointer   573
Polycote Select   636
Polycote Universal   631
Polymone X   119
Polyram   746
Polysect Insecticide   942
Polysect Insecticide Ready To Use   942
Pommetrol M   83
Portland Brand Ant Killer   940
Portland Brand Slug Killer Blue Mini Pellets   1061
Portland Glyphosate 480   236
Portman Betalion   386
Portman Chlormequat 700   62
Portman Chlortoluron   69
Portman Dimethoate EC   784
Portman Glider   236
Portman Glyphosate   236
Portman Glyphosate 360   236
Portman Isotop   267
Portman Mandate 80   609
Portman Pirimicarb   834
Portman Propachlor 50 FL   398
Portman Supaquat   62
Portman Trifluralin   463
Portman Trifluralin 480   463
Portman Weedmaster   56
Post-Kite   265
Power Blade   827
Power CMPP   325

Power Carbendazim 50   *484*
Power Chase   *763*
Power Chloridazon   *56*
Power Chlortoluron 500 FC   *69*
Power Countdown   *352*
Power Demo H   *834*
Power Glyphosate 360   *236*
Power Gro-Stop   *83*
Power Maneb 80   *621*
Power Metribuzin 70   *359*
Power Micro-Sulphur   *677*
Power Paraquat   *373*
Power Prime   *216*
Power Prochloraz 400   *658*
Power Swing   *267*
PP 321   *808*
PP 321 Oilseed Rape Summer Pests   *808*
PP 450   *573*
PP Captan 80WG   *479*
PP Captan 83   *479*
Pradone Plus   *54*
Prebane 500 FW   *433*
Prebane 500 SC   *433*
Precede   *181*
Prefix D   *154*
Prelude 20 LF   *658*
Prelude 20LF   *658*
Premier Autumn Lawn Feed with Mosskiller   *1022*
Premier Lawn Feed, Weed and Mosskiller   *1026*
Primatol SE 500 FW   *8*
Primatol SE 500 SC   *8*
Proctors Lawn Weed and Feed   *985*
Profalon   *81*
Promark Diuron   *183*
Prospect   *440*
Protector   *492*
Pulsar   *27*
Puma X   *203*
Punch   *568*
Punch C   *488*
Pursuit   *2*
Py Garden Insect Killer   *1097*
Py Garden Insecticide   *1097*
Py Powder   *1097*
Py Spray Garden Insect Killer   *1097*
Py Spray Insect Killer   *1097*
Pynosect 30 Fogging Solution   *840*
Pynosect 30 Water Miscible   *840*
Pyramin DF   *56*
Pyramin FL   *56*
Quad CMPP 600   *325*
Quadrangle Chlormequat 700   *62*
Quadrangle Cyper 10   *772*
Quadrangle Gunner   *215*

Quadrangle Hinge   *484*
Quadrangle Maneb Flowable   *621*
Quadrangle Manex   *629*
Quadrangle MCPA 50   *301*
Quadrangle Mini Slug Pellets   *915*
Quadrangle Onslaught   *297*
Quadrangle Pistol 400   *386*
Quadrangle Quadban   *147*
Quadrangle Super-Tin 4L   *564*
Quantum   *452*
Quartz   *352*
Quartz BL   *42*
Quickstep   *31*
Quintil 500   *267*
Quintozene Wettable Powder   *674*
R H Maneb 80   *621*
RH-3866 ST   *640*
RP 1001   *46*
RP 283   *31*
RP 4169   *32*
RP 83/8   *49*
RP Chlormequat 40   *62*
RP Chlormequat 67.5   *62*
Rabbit Smear Liquid   *871*
Racumin Master Mix   *877*
Racumin Rat Bait   *877*
Racumin Tracking Powder   *877*
Radar   *664*
Radspor FL   *539*
Rampart   *762*
Ramrod 20 Granular   *398*
Ramrod Flowable   *398*
Rapid Aerosol   *1090*
Rapid Greenfly Killer   *1090*
Rapid Greenfly Killer   *1095*
Rapier   *407*
Rappor Liquid Seed Treatment   *588*
Rappor Plus Liquid Seed Treatment   *589*
Rat-Eyre Rat and Mouse Bait   *876*
Ratak   *878*
Ratak   *1008*
Ratak Wax Blocks   *878*
Rataway   *878*
Rataway Bait Bags   *878*
Rataway W   *878*
Raven   *348*
Ravine   *588*
Ravine Plus   *589*
Raxil   *683*
Raxil S   *685*
RCR Zinc Phosphide   *888*
Ready To Use Groundclear   *992*
Rearguard   *517*
Recoil   *619*
Redipon   *158*

Reglone 180
Regulex 233
Regulox K 298
Reldan 50 769
Remtal SC 418
Renardine 946
Renardine 72/2 946
Renovator 115
Rentokil Alphachloralose 868
Rentokil Alphachloralose Concentrate 868
Rentokil Alphakil 925
Rentokil Alphalard 868
Rentokil Biotrol 887
Rentokil Biotrol Concentrate 887
Rentokil Biotrol Plus Outdoor Rat Killer 950
Rentokil Biotrol Plus Outdoor Rat Killer (for service use only) 873
Rentokil Blackfly & Greenfly Killer 1100
Rentokil Bromard 873
Rentokil Bromatrol 873
Rentokil Bromatrol Contact Dust 873
Rentokil Deadline 873
Rentokil Deadline Contact Dust 873
Rentokil Deadline Liquid Concentrate 873
Rentokil Deerat Concentrate 874
Rentokil Degesch Phostoxin Plates 914
Rentokil Fentrol 878
Rentokil Fentrol Contact Dust 878
Rentokil Fentrol Gel 878
Rentokil Greenhouse and Garden Insect Killer 1100
Rentokil Grey Squirrel Bait 887
Rentokil Lindane Contact Dust 882
Rentokil Liquid Bromatrol 873
Rentokil Liquid Warfarin Concentrate 887
Rentokil Methyl Bromide 918
Rentokil Mouse Dust 882
Rentokil Path & Patio Weedkiller 929
Rentokil Phostoxin 752
Rentokil Rodine C 873
Rentokil Rodine C (For Home Garden Use) 950
Rentokil Total Mouse Killer System 950
Reply 102
Repulse 505
Resistone PC 191
Rhizopon A Powder 244
Rhizopon A Tablets 244
Rhizopon AA Powder (1%) 245
Rhizopon AA Powder (0.5%) 245
Rhizopon AA Powder (2%) 245
Rhizopon AA Tablets 245
Rhizopon B Powder (0.1%) 366
Rhizopon B Powder (0.2%) 366
Rhizopon B Tablets 366
Ridento Ready to use Rat Bait 876

Ridomil MBC 60 WP 495
Ridomil Plus 50 WP 518
Rimidin 548
Ringer 729
Ringmaster 649
Ripcord 772
Ripost 522
Ripost Pepite 522
Ripost WDG 522
Rite Weed Carbendazim 484
Ritefeed 2,4-D Amine 111
Ritefeed Dichlorophen 156
Ritefeed Simazine/Aminotriazole 43% FL 8
Rival 238
Rizolex 717
Rizolex 50 WP 717
Rizolex Flowable 717
Robinson's Thiram 60 706
Rodent Cake 880
Rodentex Rat Killer 887
Rodentex Universal Rat Bait 887
Rodextra Mouse Bait 868
Ronilan 741
Ronilan DF 741
Ronilan FL 741
Ronstar 2G 370
Ronstar Liquid 370
Root-Out 935
Rooting Powder 956
Roseclear 952
Rotalin 290
Roundup 236
Roundup A 236
Roundup Biactive 236
Roundup Biactive Dry 236
Roundup Brushkiller 1030
Roundup Brushkiller Ready To Use 1030
Roundup Four 80 236
Roundup GC 1030
Roundup Micro 1030
Roundup Pro 236
Roundup Pro Biactive 236
Roundup Ready To Use 1030
Roundup Tab 1030
Roundup Two 40 236
Rout 876
Rout Rat Packs 876
Rover DF 505
Rovral Flo 600
Rovral Green 600
Rovral Liquid FS 600
Rovral WP 600
Roxion 784
Royal MH 180 298
Rubigan 548

Ruby 684
Ruby Rat 876
SAN 619 523
SAN 703 509
SHL Lawn Sand 210
SHL Lawn Sand Plus 157
SHL Turf Feed and Weed 166
SHL Turf and Weed and Mosskiller 164
SP 283 31
STR 301 497
Sabina 500 SC 267
Sabre 267
Sabre WDG 267
Safer's Insecticidal Soap Ready to Use 1018
Safer's Natural Garden Fungicide 1116
Safers Insecticidal Soap 795
Sakarat Concentrate 887
Sakarat Ready To Use (Cut Wheat Base) 887
Sakarat Ready To Use (Whole Wheat Base) 887
Sakarat X Ready To Use Warfarin Rat Bait 887
Sakarat X Warfarin Concentrate 887
Salvo 567
Salvo 905
Sambarin 312.5 FW 513
Sambarin 312.5 SC 513
Sambarin TP 513
San 735 528
Sanction 568
Sapecron 10 FG 765
Sapecron 240 EC 765
Sapir 834
Saprol 739
Satis 15 WP 220
Satisfar 794
Satisfar Dust 794
Savall 841
Savona 795
Savona Ready to Use 1018
Savona Ready-For Use 1018
Savona Rose Spray 1018
Scattercarb 761
Scent Off Buds 1066
Scent Off Pellets 1066
Scoot 926
Scorpio 19
Scorpio 400 23
Scuttle 886
Scythe 373
Scythe LC 373
Seconal 884
Secto Ant and Crawling Insect Powder 940
Secto Greenfly and Garden Insect Spray 1036
Secto Keep Off 964
Secto Nature Care Garden Insect Powder 1097
Secto Nature Care Garden Insect Spray 1097

Secto Nature Care Houseplant Insect Spray 1097
Secto Pepper Dust 1078
Secto Wasp Killer Powder 940
Sectovap Greenhouse Pest Killer 1007
Sedanox Granules 765
Seedox SC 755
Select-Trol 122
Selective Weedkiller 122
Seloxone 98
Semeron 25 WP 134
Senate 437
Sencorex WG 359
Sentinel 2 398
Sentry 750
Septal WDG 491
Septico Slug Killer 927
Seradix 1 245
Seradix 2 245
Seradix 3 245
Seradix No 2 245
Seradix No 3 245
Seritox 50 166
Seritox Turf 166
Setter 33 21
Sevin 85 Sprayable 761
Sewarin Extra 887
Sewarin Extra Ready For Use Bait 887
Sewarin P 887
Sheen 550 EC 552
Shell D50 111
Shell DD Soil Fumigant 906
Sheriff 767
Shirlan 567
Shogun 100 EC 401
Sibutol 475
Sibutol Flowable 475
Sickle 44
Silvacur 684
Silvapron D 111
Simazine SC 412
Simflow 412
Simflow Plus 8
Sinbar 429
Sipcam UK Rover 500 505
Sistan 916
Skirmish 495 SC 285
SL 471 200 EC 672
Slaymor 873
Slaymor Bait Bags 873
Slaymor Liquid Concentrate 873
Slug Destroyer 915
Slug Gard 1062
Slug Pellets 915
Slug Pellets 1061

Slug Xtra *1061*
Snailaway *970*
Sobrom BM 100 *918*
Sobrom BM 98 *918*
Sodium Chlorate *1114*
Sodium Chlorate (Fire suppressed) Weedkiller *419*
Sodium chlorate *419*
Solfa *677*
Soltair *182*
Solvigran 10 *786*
Sonar *802*
Sorex Golden Fly Bait *821*
Sorex Warfarin 250 ppm Rat Bait *887*
Sorex Warfarin 500 ppm Rat Bait *887*
Sorex Warfarin Rat Bait *887*
Sorex Warfarin Sewer Bait *887*
Sorexa CD *875*
Sorexa CD Concentrate *875*
Sorexa D *878*
Sorexa D Mouse Killer *1008*
Sorexa Plus *887*
Sovereign 330 EC *378*
Spannit *766*
Spannit Granules *766*
Sparkle 45 WP *497*
Spasor *236*
Spearhead *94*
Spectron *59*
Speedway *373*
Speedway Liquid *373*
Spinnaker *720*
Spitfire *106*
Splendor *445*
Sporgon 50 WP *658*
Sportak *658*
Sportak 45 *658*
Sportak 45 HF *658*
Sportak Alpha *496*
Sportak Delta 460 *527*
Sportak Focus HF *658*
Sportak HF *658*
Sportak Sierra HF *658*
Spot Weeder *984*
Spraydex General Purpose Fungicide *969*
Spraydex Greenfly Killer *943*
Spraydex Lawn Spot Weeder *984*
Sprayon Inorganic Liquid Copper *516*
Springcorn Extra *146*
Sprint *559*
Sprint HF *559*
Spudweed *397*
Squadron *492*
Stabilan 5C *62*
Stabilan 750 *62*

Stacato *236*
Stalker *173*
Stampede *236*
Standon Aldicarb 10 G *749*
Standon Carbosulfan 10G *763*
Standon Chorothalonil 50 *505*
Standon CMPP F *325*
Standon Cymoxanil Extra *521*
Standon Cypermethrin *772*
Standon Deltamethrin *774*
Standon Diquat *180*
Standon Ethofumesate 200 *193*
Standon Fenpropimorph 750 *553*
Standon Fluazifop-P *216*
Standon Fluroxypyr *221*
Standon Flusilazole Plus *488*
Standon Flutriafol *573*
Standon Glyphosate 360 *236*
Standon Lambda-C *808*
Standon Mecoprop-P *341*
Standon Metazachlor 50 *354*
Standon Metsulfuron *360*
Standon Paraquat *373*
Standon Pirimicarb 50 *834*
Standon Pirimicarb H *834*
Standon Propiconazole *664*
Standon Propyzamide 50 *407*
Standon Triadimefon 25 *718*
Standon Tridemorph 750 *729*
Standup 700 *62*
Stantion *65*
Starane 2 *221*
Starane Super *230*
Starter Flowable *56*
Status *46*
Stay-Kleen *107*
Stay-Off *926*
Stefes 7G *352*
Stefes Alphacypermethrin *751*
Stefes Biggles *410*
Stefes Blight Spray *521*
Stefes C-Flo *484*
Stefes CCC *62*
Stefes Carbendazim Flo *484*
Stefes Carbofuran *762*
Stefes CCC 640 *62*
Stefes CCC 700 *62*
Stefes CCC 720 *62*
Stefes Chloridazon *56*
Stefes Complete *236*
Stefes Cypermethrin *772*
Stefes Cypermethrin 2 *772*
Stefes Deltamethrin *774*
Stefes Diquat *180*
Stefes Flamprop *215*

Stefes Fluroxypyr *221*
Stefes Forte *386*
Stefes Fumat *193*
Stefes Glyphosate *236*
Stefes IPU *267*
Stefes K2 *62*
Stefes Kickdown *236*
Stefes Kickdown 2 *236*
Stefes Lenacil *286*
Stefes Maneb DF *621*
Stefes Mecoprop-P *341*
Stefes Mecoprop-P2 *341*
Stefes Medipham *386*
Stefes Mepiquat *68*
Stefes Metamitron *352*
Stefes Paraquat *373*
Stefes Pirimicarb *834*
Stefes Pride *407*
Stefes Restore *664*
Stefes Slayer *216*
Stefes Stance *65*
Stefes Toluron *69*
Stellox 380 EC *46*
Stellox 60 WG *46*
Stempor DG *484*
Stempor WG *484*
Sterilite Hop Defoliant *10*
Sterilite Tar Oil Winter Wash 60 % Stock Emulsion *682*
Sterilite Tar Oil Winter Wash 80 % Miscible Quality *682*
Stetson *236*
Stexal *228*
Sting *236*
Sting CT *236*
Stirrup *236*
Stoller Flowable Sulphur *677*
Stomp 400 SC *378*
Stop Gro G8 *1040*
Storite Clear Liquid *690*
Storite Flowable *690*
Storite SS *689*
Storm *881*
Storm Mouse Bait Blocks *881*
Strate *63*
Stratos *110*
Strike 2 *1067*
Suffix *28*
Sulfosate *236*
Sulphur Candles *1116*
Sumi-Alpha *791*
Sumicidin *801*
Super Mosstox *156*
Super Verdone *116*
Super Weedex *934*
Super-Tin 4L *564*
Super-flor 6 % Metaldehyde Slug Killer Mini Pellets *915*
Supercarb *484*
Supergreen Double (Feed and Weed) *984*
Supergreen Feed, Weed and Mosskiller *121*
Supergreen Feed, Weed and Mosskiller *983*
Supergreen Triple (Feed, Weed and Mosskiller) *983*
Supertox *984*
Supertox 30 *122*
Supertox Lawn Weedkiller *984*
Supertox Spot *984*
Supertox Spot Weeder *984*
Suscon Green Soil Insecticide *766*
Swipe 560 EC *49*
Sybol *1092*
Sybol Aerosol *1094*
Sybol Dust *1092*
Sydex *122*
Syford *111*
Synchemicals Couch and Grass Killer *986*
Synchemicals Mazide 25 *298*
Synchemicals Mazide Selective *144*
Synchemicals Tomato Setting Spray *1072*
Synergol *153*
Syngran *412*
Synox *266*
Sypex *63*
Sypex M *67*
Systemic Fungicide Liquid *1128*
Systhane *1065*
Systhane 40W *640*
Systhane 6 Flo *640*
Systhane 6W *640*
Systol M *521*
TWK Total Weedkiller *419*
Tachigaren 70 WP *590*
Takron *56*
Taktic *753*
Taktic Building Sprays *753*
Talgard *64*
Talisman *69*
Talon *766*
Talstar *757*
Talunex *870*
Tattoo *620*
Taylor's Lawn Sand *210*
Teal *50*
Teal G *50*
Tecto Dust *690*
Tecto Flowable Turf Fungicide *690*
Tecto Wettable Powder *690*
Tedion V-18 EC *853*
Telone 2000 *907*

Telone II 907
Temik 10G 749
Tempo 295
Tern 750 EC 551
Terpal 65
Terpal 67
Terpal C 63
Terraclor 20 D 674
Terrathion Granules 832
Terset 48
Tesco Lawn Feed 'n' Weed 978
Tetralex Plus 146
Texas Ant killer 940
Thinsec 761
Thiodan 20 EC 790
Thiodan 35 EC 790
Thiovit 677
Thrifty Pack Blue Mini Slug Killer Pellets 1061
Thripstick 774
Throttle 762
Thuricide HP 890
Tiara 501
Tigress 168
Tilt 250 EC 664
Tilt Turbo 475 EC 669
Timbrel 455
Tipoff 366
Tiptor 527
Titan 62
Titan 670 62
Tol 7 69
Tolkan Liquid 267
Tolkan Turbo 173
Tombel 843
Tomcat All-Weather Blox 880
Tomcat All-Weather Rodent Bar 880
Tomcat Rat and Mouse Bait 880
Top Cop 519
Top Farm CMPP 325
Top Farm Carbendazim - 435 484
Top Farm Chlormequat 640 62
Top Farm Chlorothalonil 500 505
Top Farm Dimethoate 784
Top Farm Diquat-200 180
Top Farm Ethofumesate 200 193
Top Farm Fentin-475-WP 564
Top Farm IPU 500 267
Top Farm MCPA 500 301
Top Farm Paraquat 373
Top Farm Paraquat 200 373
Top Farm Propiconazole 250 664
Top Farm Propyzamide 500 407
Top Farm Toluron 500 86
Topas 100EC 651
Topas C 50WP 482

Topas D275 FW 537
Topas D275 SC 537
Toppel 10 772
Topshot 25
Tordon 101 124
Tordon 22K 392
Toro 69
Torque 796
Total Weedkiller Granules 1016
Totem 72
Totril 247
Touchdown 236
Touchdown LA 236
Touche 189
Touchweeder 974
Tough Weed Gun! 1030
Tough Weed Killer 1030
Townex Mouse Poison 868
Townex Sachets 877
Tracker 135
Trapper 43
Treflan 463
Tri-farmon FL 297
Tribunil 355
Tribunil WG 355
Tribute 146
Trifolex-Tra 319
Triforine Liquid Seed Treatment 739
Trigard 463
Triherbicide CIPC 75
Trik 6
Trimangol 80 621
Trimangol WDG 621
Trimanzone PVA 566
Trimanzone WP 566
Triminol 642
Tripart 5C 62
Tripart Accendo 352
Tripart Arena 10 G 426
Tripart Arena 10G 687
Tripart Arena 3 426
Tripart Arena 3 687
Tripart Arena 5G 426
Tripart Arena 6% 426
Tripart Arena Plus 52
Tripart Audax 766
Tripart Beta 386
Tripart Beta 2 386
Tripart Brevis 62
Tripart Brevis 2 62
Tripart Chlormequat 460 62
Tripart Culmus 69
Tripart Defensor FL 484
Tripart Faber 505
Tripart Gladiator 2 56

PSD PRODUCT NAME INDEX

Tripart Imber *677*
Tripart Legion *492*
Tripart MCPA 50 *301*
Tripart Mini Slug Pellets *915*
Tripart New Arena 6 *426*
Tripart Obex *621*
Tripart Pugil *267*
Tripart Ratio *281*
Tripart Sentinel *398*
Tripart Trifluralin 48 EC *463*
Tripart Victor *486*
Triple Action Grass Hopper *979*
Triple Action Grasshopper *979*
Triple XXX *878*
Triplen *463*
Triplen Combi *297*
Tristar *463*
Trithac *630*
Tritoftorol *743*
Trojan SC *56*
Trooper *215*
Tropotox *321*
Tropotox Plus *319*
Trump *278*
Trustan *522*
Trustan WDG *522*
Tumbleblite *1096*
Turbair Acaricide *781*
Turbair Flydown *839*
Turbair Grain Store Insecticide *798*
Turbair Kilsect Short Life Grade *839*
Turbair Kilsect Super *758*
Turbair Permethrin *829*
Turbair Resmethrin Extra *844*
Turbair Rovral *600*
Turbair Super Flydown *839*
Turbair Systemic Insecticide *784*
Turbair Zineb *743*
Twin-Tak *47*
Twinspan *768*
Twister *896*
Twister Flow *896*
Ultra-sonic *238*
Ultrafaber *505*
Unicrop 6 % Mini Slug Pellets *915*
Unicrop Atrazine 50 *12*
Unicrop Dimethoate 40 *784*
Unicrop Fenitrothion 50 *797*
Unicrop Flowable Atrazine *12*
Unicrop Flowable Diuron *183*
Unicrop Flowable Mancozeb *609*
Unicrop Flowable Maneb *621*
Unicrop Flowable Simazine *412*
Unicrop Leatherjacket Pellets *809*
Unicrop Mancozeb *609*

Unicrop Maneb *621*
Unicrop Maneb 80 *621*
Unicrop Manguard DG *621*
Unicrop Simazine 50 *412*
Unicrop Thianosan DG *706*
Unicrop Zineb *743*
Unistar CMPP *325*
Unistar CMPP *341*
Unistar Ethephon 480 *65*
Unistar Glyfosate 360 *236*
Unistar Glyphosate 360 *236*
Unistar Pirimicarb 500 *834*
Upgrade *63*
Vangard *386*
Vantage *30*
Vapona All-In-One *1097*
Vapona House and Garden Plant Insect Killer *1097*
Vapona Slug and Snail Killer Pellets *1061*
Vapona Woodlice Killer *940*
Varmint *60*
Vassgro Cypermethrin Insecticide *772*
Vassgro Mini Slug Pellets *915*
Velpar Liquid *239*
Venzar Flowable *286*
Venzar Weedkiller *286*
Verdant F6 Combined Fertilizer and Weedkiller *984*
Verdant No 12 Mosskiller *210*
Verdant No. 6 Feed and Weed *984*
Verdant No.7 Lawn Sand *210*
Verdone 2 *984*
Verdone CDA *122*
Vertalec *892*
Vincit IM *578*
Vincit L *580*
Vindex *39*
Viper *447*
Vitaflo Extra *502*
Vitagrow Feed and Weed with Mosskiller *983*
Vitagrow Lawn Sand *210*
Vitavax *504*
Vitavax Extra *502*
Vitavax RS *503*
Vitax 'Green Up' Granular Lawn Feed and Weed *985*
Vitax Green Up Lawn Feed 'N' Weed Plus Mosskiller *939*
Vitax Lawn Sand *1022*
Vitax Micro Gran 2 *210*
Vitax Turf Tonic *210*
Vitesse *490*
Vitigran *516*
Vizor *286*
Vydate 10G *827*

199

Warefog 25 *75*
Warfarin 0.025% Ready For Use (Cereal Base) *887*
Warfarin 0.05% Ready For Use (Cereal Base) *887*
Warfarin 0.5% Master Mix (Cereal Base) *887*
Warfarin 0.5% Master Mix (Talc Base) *887*
Warfarin 1% Master Mix (Cereal Base) *887*
Warfarin 1% Master Mix (Talc Base) *887*
Warfarin 1% Tracking Dust (Talc Base) *887*
Warfarin Concentrate *887*
Warfarin Master Mix 0.5% (Cereal Base) *887*
Warfarin Master Mix 1.0% (Cereal Base) *887*
Warfarin RTU *887*
Warfarin Ready For Use 0.025% *887*
Warfarin Ready For Use 0.05% *887*
Warfarin Ready Mixed Bait *887*
Warfarin Sewer Bait *887*
Warfarin Sewer Treatment *887*
Warfarin Soluble 1% *887*
Wasp Exterminator *1104*
WBC Systemic Aphicide *857*
Weed 'N' Feed Extra *983*
Weed Out *924*
Weed-out Couchgrass Killer *924*
Weedazol - TL *3*
Weedazol Total *6*
Weedazol-TL *3*
Weedex S 2 FG *412*
Weedol *1014*
Weedone LV4 *111*
Wetcol 3 Copper Fungicide *515*
Whip X *203*
Widgeon 750 EC *553*
Wildcat *201*
Wilko Ant Destroyer *940*
Wilko Fruit and Vegetable Insecticide Spray *1097*
Wilko Lawn Feed 'n' Weed *984*
Wilko Lawn Feed, Weed and Mosskiller *1005*
Wilko Lawn Food and Weedkiller *984*
Wilko Lawn Sand *1022*
Wilko Lawn Spot Weeder *981*
Wilko Lawn Weedkiller *981*
Wilko Multi-Purpose Insecticide Spray *1097*
Wilko Path Weedkiller *928*
Wilko Rose and Flower Insecticide Spray *1097*
Wilko Slug Killer Blue Mini Pellets *1061*
Wilko Sodium Chlorate Weedkiller *1114*
Wilko Soluble Lawn Food and Weedkiller *1006*
Wilko Woodlice Killer *940*
Wireworm FS Seed Treatment *809*
Wireworm Liquid Seed Treatment *809*
Wizzard *216*
Woolworth Lawn Weedkiller *981*

Woolworth Liquid Lawn Feed and Weed *992*
Woolworth's Weedkiller *1030*
Woolworths Ant Killer *940*
Woolworths Complete Insecticide Spray *1097*
Woolworths Lawn Feed, Weed and Mosskiller *1026*
Woolworths Lawn Spot Weeder *981*
Woolworths Slug Killer Pellets *1061*
Woolworths Slug Pellets *1061*
X-Spor SC *621*
XL All Insecticide *825*
XL All Nicotine 95% *825*
Yaltox *762*
Yellow Sulphur *1116*
Zeneca Derris Dust *1104*
Zennapron *122*
Zinc Phosphide *888*
Zodiac TX *173*
Zolone Liquid *833*
ZP Rodent Bait *888*
Zulu 550 EC *552*

4

PSD ACTIVE INGREDIENT INDEX

The number after the Active Ingredient gives the Ingredient Code Number from the Professional or Amateur Sections.

Abamectin 748
Aldicarb 749, 893
Aldicarb
  + Lindane 750, 894
Alloxydim-sodium 1, 924
Alphachloralose 868, 925
Alphacypermethrin 751
Aluminium ammonium sulphate 869, 926
Aluminium phosphide 752, 870, 895
Aluminium sulphate 927
Amidosulfuron 2
2-Aminobutane 470
Amitraz 753
Amitrole 3
  + Atrazine 4, 928
  + Atrazine + 2,4-D 929
  + Bromacil + Diuron 5
  + 2,4-D + Diuron 6
  + 2,4-D + Diuron + Simazine 930
  + Diquat + Paraquat + Simazine 931
  + Diuron 7
  + MCPA 932
  + MCPA + Simazine 933
  + Simazine 8, 934
Ammonium sulphamate 9, 935
Anilazine 471
Anthracene oil 10
Asulam 11
Atrazine 12
  + Amitrole 13, 936
  + Amitrole + 2,4-D 937
  + Sodium chlorate 14
Azamethiphos 754
Aziprotryne 15
Bacillus thuringiensis 890, 938
Benalaxyl
  + Mancozeb 472
Benazolin 16
+ Bromoxynil + Ioxynil 17
  + Clopyralid 18
  + Clopyralid + Dimefuron 19
  + 2,4-D + Dicamba + Dichlorophen +
      Dichlorprop + Mecoprop 939
  + 2,4-D + MCPA 20
  + 2,4-DB + MCPA 21
  + Dicamba + Dichlorprop 22
  + Dimefuron 23
Bendiocarb 755, 940
Benfuracarb 756
Benodanil 473
Benomyl 474, 941
Bentazone 24
  + Cyanazine + 2,4-DB 25
  + MCPA + MCPB 26
  + MCPB 27

Benzoylprop-ethyl 28
Bifenox
  + Chlorotoluron 29
  + Cyanazine 30
  + Dicamba 31
  + Isoproturon 32
  + Mecoprop 33
Bifenthrin 757, 942
Bioallethrin
  + Permethrin 943
Bitertanol
  + Fuberidazole 475
  + Fuberidazole + Triadimenol 476
Bitumen 944, 945
Bone oil 871, 946
Borax 947
  + Carbaryl 948
Brodifacoum 872, 949
Bromacil 34
  + Amitrole + Diuron 35
  + Diuron 36
  + Picloram 37
Bromadiolone 873, 950
Bromophos 951
  + Resmethrin 758
Bromoxynil
  + Benazolin + Ioxynil 38
  + Clopyralid 39
  + Clopyralid + Fluroxypyr + Ioxynil 40
  + Dichlorprop + Ioxynil + MCPA 41
  + Diflufenican + Ioxynil 42
  + Ethofumesate + Ioxynil 43
  + Fluroxypyr 44
  + Fluroxypyr + Ioxynil 45
  + Ioxynil 46
  + Ioxynil + Isoproturon 47
  + Ioxynil + Isoproturon + Mecoprop 48
  + Ioxynil + Mecoprop 49
  + Ioxynil + Triasulfuron 50
  + Ioxynil + Trifluralin 51
Bupirimate 477
  + Pirimicarb + Triforine 952
  + Triforine 478, 953
Buprofezin 759
Butoxycarboxim 954
Calciferol 874
+ Difenacoum 875, 955
Captan 479
  + Lindane 480, 760
  + 1-Naphthylacetic acid 956
  + Nuarimol 481
  + Penconazole 482
Carbaryl 761, 896, 957, 958
  + Borax 959
  + Rotenone 960

Carbendazim 484, 897, 961
 + Mancozeb 483
 + Chlorothalonil 485
 + Chlorothalonil + Maneb 486
 + Copper oxychloride + Permethrin + Sulphur 962
 + Cyproconazole 487
 + Flusilazole 488
 + Flutriafol 489
 + Iprodione 490
 + Mancozeb 491
 + Maneb 492
 + Maneb + Sulphur 493
 + Maneb + Tridemorph 494
 + Metalaxyl 495
 + Prochloraz 496
 + Propiconazole 497
 + Tecnazene 498, 52
 + Thiram 499
 + Triadimefon 500
 + Triadimenol 501
Carbetamide 53
 + Dimefuron 54
Carbofuran 762, 898
Carbosulfan 763, 899
Carboxin
 + Imazalil + Thiabendazole 502
 + Lindane + Thiram 503, 764
 + Thiabendazole 504
Chlorbufam
 + Chloridazon 55
Chlorfenvinphos 765
Chloridazon 56
 + Chlorbufam 57
 + Chlorpropham + Fenuron + Propham 58
 + Ethofumesate 59
 + Lenacil 60
 + Propachlor 61
Chlormequat 62
 + 2-Chloroethylphosphonic acid 63
 + Imazaquin 64
2-Chloroethylphosphonic acid 65
 + Chlormequat 66
 + Mepiquat 67
 + Mepiquat chloride 68
Chlorophacinone 876
Chloropicrin 900
 + Methyl bromide 901
Chlorothalonil 505
 + Carbendazim 506
 + Carbendazim + Maneb 507
 + Cymoxanil 508
 + Cyproconazole 509
 + Fenpropimorph 510

Chlorothalonil–*continued*
 + Flutriafol 511
 + Metalaxyl 512
 + Propiconazole 513
Chlorotoluron 69
 + Bifenox 70
 + Isoxaben 71
 + Pendimethalin 72
 + Trifluralin 73
Chlorphonium 74
Chlorpropham 75
 + Chloridazon + Fenuron + Propham 76
 + Cresylic acid + Fenuron + Propham 77, 902
 + Diuron + Propham 78
 + Fenuron 79
 + Fenuron + Propham 80
 + Linuron 81
 + Pentanochlor 82
 + Propham 83
Chlorpyrifos 766
 + Diazinon 963
 + Dimethoate 767
 + Disulfoton 768
Chlorpyrifos-methyl 769
Chlorthal dimethyl 84
 + Propachlor 85
Chlortoluron 86
Citronella oil
 + Methyl nonyl ketone 964
Clofentezine 770
Clopyralid 87
 + Benazolin 88
 + Benazolin + Dimefuron 89
 + Bromoxynil 90
 + Bromoxynil + Fluroxypyr + Ioxynil 91
 + Cyanazine 92
 + Dichlorprop + MCPA 93
 + Diflufenican + MCPA 94
 + Fluroxypyr + Ioxynil 95
 + Fluroxypyr + MCPA 96
 + Ioxynil 97
 + Mecoprop 98
 + Propyzamide 99
 + Triclopyr 100
Copper 965
Copper ammonium carbonate 514
Copper complex - bordeaux 515, 966
Copper oxychloride 516, 967
 + Carbendazim + Permethrin + Sulphur 968
 + Maneb + Sulphur 517
 + Metalaxyl 518
Copper sulphate 969
 + Sulphur 519
 + Ferrous sulphate 970

Coumatetralyl 877, 971
Cresylic acid 903, 972
   + Chlorpropham + Fenuron + Propham 101, 904
Cyanazine 102
   + Bentazone + 2,4-DB 103
   + Bifenox 104
   + Clopyralid 105
   + Fluroxypyr 106
   + Linuron 107
   + Mecoprop 108
   + Terbuthylazine 109
Cycloxydim 110
Cyfluthrin 771
Cymoxanil
   + Chlorothalonil 520
   + Mancozeb 521
   + Mancozeb + Oxadixyl 522
Cypermethrin 772
Cyproconazole 523
   + Carbendazim 524
   + Chlorothalonil 525
   + Mancozeb 526
   + Prochloraz 527
Cyproconazole
   + Tridemorph 528
Cyromazine 773
2,4-D 111, 973
   + 2-3-6 TBA 974
   + Amitrole + Atrazine 975
   + Amitrole + Diuron 112
   + Amitrole + Diuron + Simazine 976
   + Benazolin + Dicamba + Dichlorophen + Dichlorprop + Mecoprop 977
   + Benazolin + MCPA 113
   + Dicamba 114, 978
   + Dicamba + Ferrous sulphate 115, 979
   + Dicamba + Ioxynil 116
   + Dicamba + Mecoprop 117, 980
   + Dicamba + Triclopyr 118
   + Dichlorprop 119, 981
   + Dichlorprop + MCPA + Mecoprop 120
   + Dichlorprop + Mecoprop 982
   + Ferrous sulphate + Mecoprop 121, 983
   + Mecoprop 122, 984
   + Mecoprop-P 123, 985
   + Picloram 124
2,4-DB 125
   + Benazolin + MCPA 126
   + Bentazone + Cyanazine 127
   + Linuron + MCPA 128
   + MCPA 129
Dalapon 986
Dalapon
   + Dichlobenil 130

Daminozide 131
Dazomet 905
Deltamethrin 774
   + Heptenophos 775
   + Pirimicarb 776
Demeton-S-methyl 777
Desmedipham
   + Ethofumesate + Phenmedipham 132
   + Phenmedipham 133
Desmetryne 134
Diazinon 778
   + Chlorpyrifos 987
Dicamba 135
   + Benazolin + 2,4-D + Dichlorophen + Dichlorprop + Mecoprop 988
   + Benazolin + Dichlorprop 136
   + Bifenox 137
   + 2,4-D 138, 989
   + 2,4-D + Ferrous sulphate 139, 990
   + 2,4-D + Ioxynil 140
   + 2,4-D + Mecoprop 141, 991
   + 2,4-D + Triclopyr 142
   + Dichlorprop + MCPA 992
   + Dichlorprop + Mecoprop 993
   + Maleic hydrazide 143
   + Maleic hydrazide + MCPA 144
   + MCPA 145
   + MCPA + Mecoprop 146, 994
   + MCPA + Mecoprop-P 147
   + Mecoprop 148, 995
   + Mecoprop + Triclopyr 149
   + Mecoprop-P 150
   + Paclobutrazol 151
   + Triasulfuron 152
Dichlobenil 154, 996
   + Dalapon 155
Dichlofluanid 529
Dichlorophen 156, 530, 997
   + Benazolin + 2,4-D + Dicamba + Dichlorprop + Mecoprop 998
   + Ferrous sulphate 157
   + 4-Indol-3-ylbutyric acid + 1-Naphthylacetic acid 153
   + 1-Naphthylacetic acid 999
1,2-Dichloropropane
   + 1,3-Dichloropropene 906
1,3-Dichloropropene 907
   + 1,2-Dichloropropane 908
Dichlorprop 158
   + Benazolin + 2,4-D + Dicamba + Dichlorophen + Mecoprop 1000
   + Benazolin + Dicamba 159
   + Bromoxynil + Ioxynil + MCPA 160
   + Clopyralid + MCPA 161
   + 2,4-D 162, 1001

Dichlorprop—*continued*
 + 2,4-D + MCPA + Mecoprop   *163*
 + 2,4-D + Mecoprop   *1002*
 + Dicamba + MCPA   *1003*
 + Dicamba + Mecoprop   *1004*
 + Ferrous sulphate + MCPA   *164, 1005*
 + Ioxynil + MCPA   *165*
 + MCPA   *166, 1006*
Dichlorvos   *779, 1007*
Diclofop-methyl   *167*
 + Fenoxaprop-P-ethyl   *168*
Dicloran   *531*
Dicofol   *780*
 + Tetradifon   *781*
Dienochlor   *782*
Difenacoum   *878, 1008*
 + Calciferol   *879, 1009*
Difenoconazole   *532*
Difenzoquat   *169*
Diflubenzuron   *783*
Diflufenican   *170*
 + Bromoxynil + Ioxynil   *171*
 + Clopyralid + MCPA   *172*
 + Isoproturon   *173*
 + Trifluralin   *174*
Dikegulac   *175, 1010*
Dimefuron
 + Benazolin   *176*
 + Benazolin + Clopyralid   *177*
 + Carbetamide   *178*
Dimethoate   *784, 1011*
 + Chlorpyrifos   *785*
 + Permethrin   *1012*
Dimethomorph   *533*
 + Mancozeb   *534*
Dinocap   *535*
Diphacinone   *880*
Diphenamid   *179*
Diquat   *180*
 + Amitrole + Paraquat + Simazine   *1013*
 + Paraquat   *181, 1014*
 + Paraquat + Simazine   *182*
Disulfoton   *786*
 + Chlorpyrifos   *787*
 + Fonofos   *788*
 + Quinalphos   *789*
Dithianon   *536*
 + Penconazole   *537*
Diuron   *183*
 + Amitrole   *184*
 + Amitrole + Bromacil   *185*
 + Amitrole + 2,4-D   *186*
 + Amitrole + 2,4-D + Simazine   *1015*
 + Bromacil   *187*

Diuron—*continued*
 + Chlorpropham + Propham   *188*
 + Glyphosate   *189*
 + Paraquat   *190*
 + Simazine   *1016*
Dodecylbenzyl trimethyl ammonium chloride   *191*
Dodemorph   *538*
Dodine   *539*
Endosulfan   *790*
Epoxiconazole   *540*
 + Fenpropimorph   *541*
 + Tridemorph   *542*
EPTC   *192*
Esfenvalerate   *791*
Ethiofencarb   *792*
Ethirimol   *543*
 + Flutriafol + Imazalil + Thiabendazole   *544*
 + Flutriafol + Thiabendazole   *545*
 + Fuberidazole + Triadimenol   *546*
Ethofumesate   *193*
 + Bromoxynil + Ioxynil   *194*
 + Chloridazon   *195*
 + Desmedipham + Phenmedipham   *196*
 + Metamitron + Phenmedipham   *197*
 + Phenmedipham   *198*
Ethoprophos   *793, 909*
Etridiazole   *547*
Etrimfos   *794*
Fatty acids   *199, 795, 1017, 1018*
 + Sulphur   *1019*
Fenarimol   *548*
 + Permethrin   *1020*
Fenbutatin oxide   *796*
Fenitrothion   *797, 1021*
 + Permethrin + Resmethrin   *798*
Fenoxaprop-ethyl   *200*
Fenoxaprop-P-ethyl   *201*
 + Diclofop-methyl   *202*
 + Isoproturon   *203*
Fenpiclonil   *549*
 + Imazalil   *550*
Fenpropathrin   *799*
Fenpropidin   *551*
 + Propiconazole   *552*
Fenpropimorph   *553*
 + Chlorothalonil   *554*
 + Epoxiconazole   *555*
 + Flusilazole   *556*
 + Flusilazole + Tridemorph   *557*
 + Lindane + Thiram   *558, 800*
 + Prochloraz   *559*
 + Propiconazole   *560*
 + Tridemorph   *561*

Fentin Hydroxide 562
Fentin acetate
  + Maneb   563
Fentin hydroxide   564
  + Metoxuron   205, 565
Fenuron
  + Chloridazon + Chlorpropham + Propham   206
  + Chlorpropham   207
  + Chlorpropham + Cresylic acid + Propham   208, 910
  + Chlorpropham + Propham   209
Fenvalerate   801
Ferbam
  + Maneb + Zineb   566
Ferrous sulphate   210, 1022
  + 2,4-D + Dicamba   211, 1023
  + 2,4-D + Mecoprop   212, 1024
  + Dichlorophen   213
  + Dichlorprop + MCPA   214, 1025
  + MCPA + Mecoprop   1026
  + Copper sulphate   1027
Flamprop-M-isopropyl   215
Flocoumafen   881
Fluazifop-P-butyl   216
Fluazinam   567
Fluoroglycofen-ethyl   217
  + Isoproturon   218
  + Mecoprop-P   219
  + Triasulfuron   220
Fluroxypyr   221
  + Bromoxynil   222
  + Bromoxynil + Clopyralid + Ioxynil   223
  + Bromoxynil + Ioxynil   224
  + Clopyralid + Ioxynil   225
  + Clopyralid + MCPA   226
  + Cyanazine   227
  + Ioxynil   228
  + Mecoprop-P   229
  + Thifensulfuron-methyl + Tribenuron-methyl   230
  + Triclopyr   231
Flusilazole   568
  + Carbendazim   569
  + Fenpropimorph   570
  + Fenpropimorph + Tridemorph   571
  + Tridemorph   572
Flutriafol   573
  + Carbendazim   574
  + Chlorothalonil   575
  + Ethirimol + Imazalil + Thiabendazole   576
  + Ethirimol + Thiabendazole   577
  + Imazalil + Thiabendazole   578
  + Iprodione   579
  + Thiabendazole   580

Fonofos   802
  + Disulfoton   803
Formaldehyde   911
Fosamine-ammonium   232
Fosetyl-aluminium   581
Fuberidazole
  + Bitertanol   582
  + Bitertanol + Triadimenol   583
  + Ethirimol + Triadimenol   584
  + Imazalil + Triadimenol   585
  + Triadimenol   586
Furalaxyl   587
Garlic oil
  + Orange peel oil + Orange pith oil   1028
Gibberellins   233
Glufosinate ammonium   234, 1029
  + Monolinuron   235
Glyphosate   236, 1030
  + Diuron   237
  + Simazine   238
Grease   1031
Guazatine   588
  + Imazalil   589
Heptenophos   804
  + Deltamethrin   805
  + Permethrin   1032
Hexazinone   239
Hymexazol   590
Imazalil   591
  + Carboxin + Thiabendazole   592
  + Ethirimol + Flutriafol + Thiabendazole   593
  + Fenpiclonil   594
  + Flutriafol + Thiabendazole   595
  + Fuberidazole + Triadimenol   596
  + Guazatine   597
  + Pencycuron   598
  + Thiabendazole   599
Imazamethabenz-methyl   240
  + Isoproturon   241
Imazapyr   242
Imazaquin
  + Chlormequat   243
Imidacloprid   806
Indol-3-ylacetic acid   244
4-Indol-3-ylbutyric acid   245, 1033
  + Dichlorophen + 1-Naphthylacetic acid   246
  + Thiram + 1-Naphthylacetic acid   1034
Iodofenphos   807
Ioxynil   247
  + Benazolin + Bromoxynil   248
  + Bromoxynil   249
  + Bromoxynil + Clopyralid + Fluroxypyr   250
  + Bromoxynil + Dichlorprop + MCPA   251
  + Bromoxynil + Diflufenican   252
  + Bromoxynil + Ethofumesate   253

PSD ACTIVE INGREDIENT INDEX

Ioxynil–*continued*
  + Bromoxynil + Fluroxypyr   254
  + Bromoxynil + Isoproturon   255
  + Bromoxynil + Isoproturon + Mecoprop   256
  + Bromoxynil + Mecoprop   257
  + Bromoxynil + Triasulfuron   258
  + Bromoxynil + Trifluralin   259
  + Clopyralid   260
  + Clopyralid + Fluroxypyr   261
  + 2,4-D + Dicamba   262
  + Dichlorprop + MCPA   263
  + Fluroxypyr   264
  + Isoproturon + Mecoprop   265
  + Mecoprop   266

Iprodione   600
  + Carbendazim   601
  + Flutriafol   602
  + Thiophanate-methyl   603

Isoproturon   267
  + Bifenox   268
  + Bromoxynil + Ioxynil   269
  + Bromoxynil + Ioxynil + Mecoprop   270
  + Diflufenican   271
  + Fenoxaprop-P-ethyl   272
  + Fluoroglycofen-ethyl   273
  + Imazamethabenz-methyl   274
  + Ioxynil + Mecoprop   275
  + Isoxaben   276
  + Metsulfuron-methyl   277
  + Pendimethalin   278
  + Simazine   279
  + Trifluralin   280

Isoxaben   281
  + Chlorotoluron   282
  + Isoproturon   283
  + Methabenzthiazuron   284
  + Terbuthylazine   285

Lambda cyhalothrin   808
Lenacil   286
  + Chloridazon   287
  + Linuron   288
  + Phenmedipham   289

Lindane   809, 882, 1035
  + Aldicarb   810, 912
  + Captan   604, 811
  + Carboxin + Thiram   605, 812
  + Fenpropimorph + Thiram   606, 813
  + Pyrethrins   1036
  + Rotenone + Thiram   1037
  + Thiabendazole + Thiram   607, 814
  + Thiophanate-methyl   815, 913
  + Thiram   608, 816

Linuron   290
  + Chlorpropham   291
  + Cyanazine   292

Linuron–*continued*
  + 2,4-DB + MCPA   293
  + Lenacil   294
  + Terbutryn   295
  + Trietazine   296
  + Trifluralin   297

Magnesium phosphide   914
Malathion   817, 1038
  + Permethrin   1039

Maleic hydrazide   298, 1040
  + Dicamba   299
  + Dicamba + MCPA   300

Mancozeb   609, 1041
  + Benalaxyl   610
  + Carbendazim   611
  + Carbendazim   612
  + Cymoxanil   613
  + Cymoxanil + Oxadixyl   614
  + Cyproconazole   615
  + Dimethomorph   616
  + Metalaxyl   617
  + Ofurace   618
  + Oxadixyl   619
  + Propamocarb hydrochloride   620

Maneb   621
  + Carbendazim   622
  + Carbendazim + Chlorothalonil   623
  + Carbendazim + Sulphur   624
  + Carbendazim + Tridemorph   625
  + Copper oxychloride + Sulphur   626
  + Fentin acetate   627
  + Ferbam + Zineb   628
  + Zinc   629
  + Zineb   630

MCPA   301
  + Amitrole   1042
  + Amitrole + Simazine   1043
  + Benazolin + 2,4-D   302
  + Benazolin + 2,4-DB   303
  + Bentazone + MCPB   304
  + Bromoxynil + Dichlorprop + Ioxynil   305
  + Clopyralid + Dichlorprop   306
  + Clopyralid + Diflufenican   307
  + Clopyralid + Fluroxypyr   308
  + 2,4-D + Dichlorprop + Mecoprop   309
  + 2,4-DB   310
  + 2,4-DB + Linuron   311
  + Dicamba   312
  + Dicamba + Dichlorprop   1044
  + Dicamba + Maleic hydrazide   313
  + Dicamba + Mecoprop   314, 1045
  + Dicamba + Mecoprop-P   315
  + Dichlorprop   316, 1046
  + Dichlorprop + Ferrous sulphate   317, 1047
  + Dichlorprop + Ioxynil   318

MCPA–continued
+ Ferrous sulphate + Mecoprop   1048
+ MCPB   319
+ Mecoprop   320, 1049
MCPB   321
+ Bentazone   322
+ Bentazone + MCPA   323
+ MCPA   324
Mecoprop   325
+ Benazolin + 2,4-D + Dicamba + Dichlorophen + Dichlorprop   1050
+ Bifenox   326
+ Bromoxynil + Ioxynil   327
+ Bromoxynil + Ioxynil + Isoproturon   328
+ Clopyralid   329
+ Cyanazine   330
+ 2,4-D   331, 1051
+ 2,4-D + Dicamba   332, 1052
+ 2,4-D + Dichlorprop   1053
+ 2,4-D + Dichlorprop + MCPA   333
+ 2,4-D + Ferrous sulphate   334, 1054
+ Dicamba   335, 1055
+ Dicamba + Dichlorprop   1056
+ Dicamba + MCPA   336, 1057
+ Dicamba + Triclopyr   337
+ Ferrous sulphate + MCPA   1058
+ Ioxynil   338
+ Ioxynil + Isoproturon   339
+ MCPA   340, 1059
Mecoprop-P   341
+ 2,4-D   342, 1060
+ Dicamba   343
+ Dicamba + MCPA   344
+ Fluoroglycofen-ethyl   345
+ Fluroxypyr   346
+ Thifensulfuron-methyl   347
+ Triasulfuron   348
Mefluidide   349
Mephosfolan   818
Mepiquat
+ 2-Chloroethylphosphonic acid   350
Mepiquat chloride
+ 2-Chloroethylphosphonic acid   351
Metalaxyl   631
+ Carbendazim   632
+ Chlorothalonil   633
+ Copper oxychloride   634
+ Mancozeb   635
+ Thiabendazole   636
+ Thiabendazole + Thiram   637
+ Thiram   638
Metaldehyde   819, 915, 1061
Metam-sodium   916
Metamitron   352
+ Ethofumesate + Phenmedipham   353

Metazachlor   354
Methabenzthiazuron   355
+ Isoxaben   356
Methiocarb   820, 917, 1062
Methomyl   821
+ (Z)-9-Tricosene   822
Methoprene   823
Methyl bromide   918
+ Chloropicrin   919
Methyl nonyl ketone   1063
+ Citronella oil   1064
Metoxuron   357
+ Fentin hydroxide   358, 639
Metribuzin   359
Metsulfuron-methyl   360
+ Isoproturon   361
+ Thifensulfuron-methyl   362
Mineral oil   824
Monolinuron   363
+ Glufosinate ammonium   364
+ Paraquat   365
Myclobutanil   640, 1065
Naphthalene   1066
1-Naphthylacetic acid   366, 1067
+ Captan   1068
+ Dichlorophen   1069
+ Dichlorophen + 4-Indol-3-ylbutyric acid   367
+ Thiram   1070
+ hiram + 4-Indol-3-ylbutyric acid   1071
2-Naphthyloxyacetic acid   368, 1072
Napropamide   369
Nicotine   825
Nitrothal-isopropyl
+ Zineb-ethylene thiuram disulphide adduct   641
Nuarimol   642
+ Captan   643
Octhilinone   644
Ofurace
+ Mancozeb   645
Ometohate   826
Orange peel oil
+ Garlic oil + Orange pith oil   1073
Orange pith oil
+ Garlic oil + Orange peel oil   1074
Oxadiazon   370
Oxadixyl
+ Cymoxanil + Mancozeb   646
+ Mancozeb   647
+ Thiabendazole + Thiram   648
Oxamyl   827, 920
Oxycarboxin   649
Oxydemeton-methyl   828
Paclobutrazol   371
+ Dicamba   372

Parahydroxyphenylsalicylamide 650
Paraquat 373
 + Amitrole + Diquat + Simazine 1075
 + Diquat 374, 1076
 + Diquat + Simazine 375
 + Diuron 376
 + Monolinuron 377
Penconazole 651, 1077
 + Captan 652
 + Dithianon 653
Pencycuron 654
 + Imazalil 655
Pendimethalin 378
 + Chlorotoluron 379
 + Isoproturon 380
 + Prometryn 381
 + Simazine 382
Penmedipham 383
Pentanochlor 384
 + Chlorpropham 385
Pepper 1078
Permethrin 829, 1079, 1080
 + Bioallethrin 1081
 + Carbendazim + Copper oxychloride + Sulphur 1082
 + Dimethoate 1083
 + Fenarimol 1084
 + Fenitrothion + Resmethrin 830
 + Heptenophos 1085
 + Malathion 1086
 + Sulphur + Triforine 1087
 + Tetramethrin 1088
 + Thiram 656, 831
Phenmedipham 386
 + Desmedipham 387
 + Desmedipham + Ethofumesate 388
 + Ethofumesate 389
 + Ethofumesate + Metamitron 390
 + Lenacil 391
Phenothrin + Tetramethrin 1089
Phorate 832
Phosalone 833
Picloram 392
 + Bromacil 393
 + 2,4-D 394
Pirimicarb 834, 1090
 + Bupirimate + Triforine 1091
 + Deltamethrin 835
Pirimiphos-ethyl 836
Pirimiphos-methyl 837, 1092
 + Pyrethrins 1093
 + Resmethrin + Tetramethrin 1094
Potassium sorbate
 + Sodium metabisulphite + Sodium propionate 657

Primicarb 1095
Prochloraz 658
 + Carbendazim 659
 + Cyproconazole 660
 + Fenpropimorph 661
Prometryn 395
 + Pendimethalin 396
 + Terbutryn 397
Propachlor 398
 + Chloridazon 399
 + Chlorthal dimethyl 400
Propamocarb hydrochloride 662
 + Mancozeb 663
Propaquizafop 401
Propham
 + Chloridazon + Chlorpropham + Fenuron 402
 + Chlorpropham 403
 + Chlorpropham + Cresylic acid + Fenuron 404, 921
 + Chlorpropham + Diuron 405
 + Chlorpropham + Fenuron 406
Propiconazole 664, 1096
 + Carbendazim 665
 + Chlorothalonil 666
 + Fenpropidin 667
 + Fenpropimorph 668
 + Tridemorph 669
Propineb 670
Propoxur 838
Propyzamide 407
 + Clopyralid 408
Pyrazophos 671
Pyrethrins 839, 1097
 + Lindane 1098
 + Pirimiphos-methyl 1099
 + Resmethrin 840, 1100
Pyridate 409
Pyrifenox 672
Quassia 883, 1101
Quinalphos 841
 + Disulfoton 842
 + Thiometon 843
Quinomethionate 673
Quintozene 674
Quizalofop-ethyl 410
Resmethrin 844
 + Bromophos 845
 + Fenitrothion + Permethrin 846
 + Pirimiphos-methyl + Tetramethrin 1102
 + Pyrethrins 847, 1103
Rotenone 848, 1104
 + Carbaryl 1105
 + Lindane + Thiram 1106
 + Sulphur 1107

Seconal *884*
Sethoxydim *411*
Simazine *412, 1108*
  + Amitrole *413, 1109*
  + Amitrole + 2,4-D + Diuron *1110*
  + Amitrole + Diquat + Paraquat *1111*
  + Amitrole + MCPA *1112*
  + Diquat + Paraquat *414*
  + Diuron *1113*
  + Glyphosate *415*
  + Isoproturon *416*
  + Pendimethalin *417*
  + Trietazine *418*
Sodium chlorate *419, 1114*
  + Atrazine *420*
Sodium cyanide *885*
Sodium metabisulphite
  + Potassium sorbate + Sodium propionate *675*
Sodium monochloroacetate *421*
Sodium propionate
  + Potassium sorbate + Sodium metabisulphite *676*
Sodium silver thiosulphate *422*
Sodium tetraborate *1115*
Sulphonated cod liver oil *886*
Sulphur *677, 849, 1116, 1117*
  + Carbendazim + Copper oxychloride + Permethrin *1121*
  + Carbendazim + Maneb *678*
  + Copper oxychloride + Maneb *679*
  + Copper sulphate *680*
  + Fatty acids *1118*
  + Permethrin + Triforine *1119*
  + Rotenone *1120*
2,3,6-TBA
  + 2,4-D *1123*
Tar oils *682, 850, 1122*
TCA *423*
Tebuconazole *683*
  + Triadimenol *684*
  + Triazoxide *685*
  + Tridemorph *686*
Tebutam *424*
Tebuthiuron *425*
Tecnazene *426, 687*
  + Carbendazim *427, 688*
  + Thiabendazole *428, 689*
Teflubenzuron *851*
Tefluthrin *852*
Terbacil *429*
Terbuthylazine
  + Cyanazine *430*
  + Isoxaben *431*
  + Terbutryn *432*

Terbutryn *433, 1124*
  + Linuron *434*
  + Prometryn *435*
  + Terbuthylazine *436*
  + Trietazine *437*
  + Trifluralin *438*
Tetradifon *853*
  + Dicofol *854*
Tetramethrin
  + Permethrin *1125*
  + Phenothrin *1126*
  + Pirimiphos-methyl + Resmethrin *1127*
Thiabendazole *690*
  + Carboxin *691*
  + Carboxin + Imazalil *692*
  + Ethirimol + Flutriafol *693*
  + Ethirimol + Flutriafol + Imazalil *694*
  + Flutriafol *695*
  + Flutriafol + Imazalil *696*
  + Imazalil *697*
  + Lindane + Thiram *698, 855*
  + Metalaxyl *699*
  + Metalaxyl + Thiram *700*
  + Oxadixyl + Thiram *701*
  + Tecnazene *439, 702*
  + Thiram *703*
Thifensulfuron-methyl *440*
  + Fluroxypyr + Tribenuron-methyl *441*
  + Mecoprop-P *442*
  + Metsulfuron-methyl *443*
  + Tribenuron-methyl *444*
Thiodicarb *922*
Thiofanox *856*
Thiometon *857*
  + Quinalphos *858*
Thiophanate-methyl *704, 1128*
  + Iprodione *705*
  + Lindane *859, 923*
Thiram *706*
  + Carbendazim *707*
  + Carboxin + Lindane *708, 860*
  + Fenpropimorph + Lindane *709, 861*
  + Lindane *710, 862*
  + Lindane + Rotenone *1129*
  + Lindane + Thiabendazole *711, 863*
  + Metalaxyl *712*
  + Metalaxyl + Thiabendazole *713*
  + 1-Naphthylacetic acid *1130*
  + Oxadixyl + Thiabendazole *714*
  + Permethrin *715, 864*
  + Thiabendazole *716*
Tolclofos-methyl *717*
Tralkoxydim *445*
Tri-allate *446*

Triadimefon 718
+ Carbendazim 719
Triadimenol 720
+ Bitertanol + Fuberidazole 721
+ Carbendazim 722
+ Ethirimol + Fuberidazole 723
+ Fuberidazole 724
+ Fuberidazole + Imazalil 725
+ Tebuconazole 726
+ Tridemorph 727
Triasulfuron 447
+ Bromoxynil + Ioxynil 448
+ Dicamba 449
+ Fluoroglycofen-ethyl 450
+ Mecoprop-P 451
Triazophos 865
Triazoxide
+ Tebuconazole 728
Tribenuron-methyl 452
+ Fluroxypyr + Thifensulfuron-methyl 453
+ Thifensulfuron-methyl 454
Trichlorfon 866
Trichoderma viride 891
Triclopyr 455, 1131
+ Clopyralid 456
+ 2,4-D + Dicamba 457
+ Dicamba + Mecoprop 458
+ Fluroxypyr 459
(Z)-9-Tricosene
+ Methomyl 867
Tridemorph 729
+ Carbendazim + Maneb 730
+ Cyproconazole 731
+ Epoxiconazole 732
+ Fenpropimorph 733

Tridemorph–*continued*
+ Fenpropimorph + Flusilazole 734
+ Flusilazole 735
+ Propiconazole 736
+ Tebuconazole 737
+ Triadimenol 738
Trietazine
+ Linuron 460
+ Simazine 461
+ Terbutryn 462
Trifluralin 463
+ Bromoxynil + Ioxynil 464
+ Chlorotoluron 465
+ Diflufenican 466
+ Isoproturon 467
+ Linuron 468
+ Terbutryn 469
Triforine 739
+ Bupirimate 740, 1132
+ Bupirimate + Pirimicarb 1133
+ Permethrin + Sulphur 1134
Verticillium lecanii 892
Vinclozolin 741
Warfarin 887
Zinc
+ Maneb 742
Zinc phosphide 888
Zineb 743
+ Ferbam + Maneb 744
+ Maneb 745
Zineb-ethylene thiuram disulphide adduct 746
+ Nitrothal-isopropyl 747
Ziram 889

PART B

HSE Registered Products

4

1
WOOD PRESERVATIVES

Wood Preservatives

1 2-(THIOCYANOMETHYLTHIO) BENZOTHIAZOLE

Product Name	Marketing Company	Use	Reg. No
Bayer WPC-3	Bayer Plc	WI	HSE 4969
Bayer WPC-4	Bayer Plc	WI	HSE 4970
BL 1111	Buckman Laboratories Ltd	WI	HSE 4764
Crysolite Glaramara Concentrated Timber Treatment	Crysolite Protective Coatings	WI	HSE 5443
Fentex NP-UF	Protim Solignum Ltd	WI	HSE 5069
Paneltone	Rentokil Ltd	WI	HSE 4253

2 2-(THIOCYANOMETHYLTHIO) BENZOTHIAZOLE and BORIC ACID

Product Name	Marketing Company	Use	Reg. No
Protim Fentex Europa 1 RFU	Protim Solignum Ltd	WI	HSE 5119
Protim Fentex Europa I	Protim Solignum Ltd	WI	HSE 4984

3 2-(THIOCYANOMETHYLTHIO) BENZOTHIAZOLE and BORIC ACID and METHYLENE BIS (THIOCYANATE)

Product Name	Marketing Company	Use	Reg. No
Fentex Elite	Protim Solignum Ltd	WI	HSE 5532

4 2-(THIOCYANOMETHYLTHIO) BENZOTHIAZOLE and METHYLENE BIS (THIOCYANATE)

Product Name	Marketing Company	Use	Reg. No
* Aquatect Wood Preservative Stain	Aqua Coatings	WI	HSE 3566
Busan 1009	Buckman Laboratories SA	WI	HSE 3887
Busan 1009	Buckman Laboratories SA	WI	HSE 5256
Celbrite TC	Rentokil Ltd	WI	HSE 3921
* Gallwey BMC	Fosroc Ltd	WI	HSE 3534
Hickson Antiblu 3739	Hickson Timber Products Ltd	WI	HSE 4841
Mect	Buckman Laboratories SA	WI	HSE 3888
Mect	Buckman Laboratories SA	WI	HSE 5255
* Protek Blue	Ashby Timber Treatments	WI	HSE 3017
* Protek Wood Protection (FT Grade 20:1)	Ashby Timber Treatments	WI	HSE 3016
* Protek Wood Protection (FT Grade 9:1)	Ashby Timber Treatments	WI	HSE 3015
* Protim Stainguard	Fosroc Ltd	WI	HSE 3533

Product Name	Marketing Company	Use	Reg. No

4 2-(THIOCYANOMETHYLTHIO) BENZOTHIAZOLE and METHYLENE BIS (THIOCYANATE)—continued

Product Name	Marketing Company	Use	Reg. No
Protim Stainguard	Protim Solignum Ltd	WI	HSE 5009
Timbercol Preservative Concentrate	Timbercol Industries Ltd	WI	HSE 3124

5 2-METHYL-4-ISOTHIAZOLIN-3-ONE and 5-CHLORO-2-METHYL-4-ISOTHIAZOLIN-3-ONE

Product Name	Marketing Company	Use	Reg. No
* Biomek	Mechema Chemicals Ltd	WI	HSE 3091
Celkil 90	Rentokil Ltd	WI	HSE 4275
Celkil 90	Rentokil Ltd	WI	HSE 5428
* Gallwey SA	Fosroc Ltd	WI	HSE 3702
Kathon 886F	Rohm and Haas (UK) Ltd	WI	HSE 4866
Kathon 886F	Rohm and Haas (UK) Ltd	WI	HSE 5431
Laporte Mould-Ex	Laporte Wood Preservation	WI	HSE 3953
Laporte Mould-Ex	Laporte Wood Preservation	WI	HSE 5430
Tanamix 3743	Hickson Timber Products Ltd	WI	HSE 4852
Tanamix 3743	Hickson Timber Products Ltd	WI	HSE 5429

6 2-PHENYLPHENOL

Product Name	Marketing Company	Use	Reg. No
Barrettine WSP Wood Treatment	Barrettine (Products) Ltd	WA	HSE 3092
Basiment 560	Venilia Ltd	WP	HSE 4411
Croda Wood Preserver	Croda Hydrocarbons Ltd	WA WP	HSE 5382
Preventol OF	Bayer (UK) Ltd	WP	HSE 4033
Preventol OF	Bayer Plc	WI WP	HSE 5200
Protim Joinery Lining 280	Protim Solignum Ltd	WP	HSE 5135

7 2-PHENYLPHENOL and BENZALKONIUM CHLORIDE

Product Name	Marketing Company	Use	Reg. No
Chel Dry Rot Killer for Masonry and Brickwork	Chelec Ltd	WP	HSE 3067

8 2-PHENYLPHENOL and CYPERMETHRIN

Product Name	Marketing Company	Use	Reg. No
Croda Timber Protector	Croda Hydrocarbons Ltd	WA WP	HSE 5369

WOOD PRESERVATIVES

Product Name	Marketing Company	Use	Reg. No

9 2-PHENYLPHENOL and PERMETHRIN

Product Name	Marketing Company	Use	Reg. No
Brunol OPA	Stanhope Chemical Products Ltd	WP	HSE 4473

10 3-IODO-2-PROPYNYL-N-BUTYL CARBAMATE

Product Name	Marketing Company	Use	Reg. No
Bio-Kil Board Preservative	Bio-Kil Chemicals Ltd	WI WP	HSE 4899
Conductive Pilt 80 RFU	PPG Industries (UK) Ltd	WI	HSE 4500
Deepkill F	Sovereign Chemical Industries Ltd	WI WP	HSE 5250
Pilt 80 RFU	PPG Industries (UK) Ltd	WI	HSE 4877
Pilt NF4 RFU	PPG Industries (UK) Ltd	WI	HSE 5541
Polyphase Emulsifiable Concentrate	Troy Chemical Company BV	WI	HSE 3938
Polyphase Solvent Based Concentrate	Troy Chemical Company BV	WP	HSE 3939
Polyphase Solvent Based Ready For Use	Troy Chemical Company BV	WA	HSE 3943
Polyphase Water Based Concentrate	Troy Chemical Company BV	WP	HSE 3944
Polyphase Water Based Ready For Use	Troy Chemical Company BV	WA	HSE 3945
Protim FDR 250	Protim Solignum Ltd	WI	HSE 5095
Protim JP 250	Protim Solignum Ltd	WI	HSE 5105
Remecology Spirit Based Fungicide R5	Remtox Chemicals Ltd	WP	HSE 4803
Remtox AQ Fungicide R7	Remtox (Chemicals) Ltd	WI WP	HSE 5410
Remtox Dry Rot Paint	Remtox Chemicals Ltd	WA SA	HSE 5187
Remtox Fungicide Microemulsion M7	Remtox (Chemicals) Ltd	WI WP	HSE 5419
Remtox Fungicide Paste K6	Remtox (Chemicals) Ltd	WI WP	HSE 5400
Remtox Microactive Fungicide W7	Remtox Chemicals Ltd	WP	HSE 5590
Safeguard Fungicidal Micro Emulsifiable Concentrate	Safeguard Chemicals Ltd	WP SP	HSE 5648
SOV AQ Micro F	Sovereign Chemical Industries Ltd	WI WP	HSE 5223
Sovac F	Sovereign Chemical Industries Ltd	WI WP	HSE 5233
Sovac FWR	Sovereign Chemical Industries Ltd	WI WP	HSE 5344

Product Name	Marketing Company	Use	Reg. No

10 3-IODO-2-PROPYNYL-N-BUTYL CARBAMATE—continued

Sovaq Micro F	Sovereign Chemical Industries Ltd	WI WP	HSE 5668
Sovereign AQF	Sovereign Chemical Industries Ltd	WI WP	HSE 5229

11 3-IODO-2-PROPYNYL-N-BUTYL CARBAMATE and CYPERMETHRIN

Microtech Dual Purpose AQ	Cementone Beaver Ltd	WP	HSE 4994
Microtech Dual Purpose AQ	Cementone Beaver Ltd	WP	HSE 5453
* Microtech Dual Purpose AQ	Cementone-Beaver Ltd	WP	HSE 4518
Water-Based Wood Preserver (Concentrate)	Cementone-Beaver Ltd	WA WP	HSE 5630
Water-Based Wood Preserver (Concentrate)	Cementone-Beaver Ltd	WA	HSE 5515
Water-Based Wood Preserver	Cementone-Beaver Ltd	WA WP	HSE 4902

12 3-IODO-2-PROPYNYL-N-BUTYL CARBAMATE and DIALKYLDIMETHYL AMMONIUM CHLORIDE

Hickson NP-1	Hickson Timber Products Ltd	WI	HSE 5401

13 3-IODO-2-PROPYNYL-N-BUTYL CARBAMATE and GAMMA-HCH

# Sadolin Sadovac 35	Sadolin (UK) Ltd	WI	HSE 4358

14 3-IODO-2-PROPYNYL-N-BUTYL CARBAMATE and PERMETHRIN

Brunol PP	Stanhope Chemical Products Ltd	WI WP	HSE 5414
Brunol SPI	Stanhope Chemical Products Ltd	WI WP	HSE 5415
Deepkill	Sovereign Chemical Industries Ltd	WA	HSE 4226
Deepkill	Sovereign Chemical Industries Ltd	WI WP	HSE 5238
Fungicide/Insecticide Microemulsion M9	Remtox (Chemicals) Ltd	WI WP	HSE 5711
Protim 250	Protim Solignum Ltd	WI	HSE 5070
Protim 250 WR	Protim Solignum Ltd	WI	HSE 5071

WOOD PRESERVATIVES

Product Name	Marketing Company	Use	Reg. No
14 3-IODO-2-PROPYNYL-N-BUTYL CARBAMATE and PERMETHRIN—continued			
Remecology Timber Preservative Paste	Remtox (Chemicals) Ltd	WA	HSE 4416
Remtox AQ Fungicide/Insecticide R9	Remtox (Chemicals) Ltd	WI WP	HSE 5370
Remtox Dual Purpose Paste K9	Remtox (Chemicals) Ltd	WI WP	HSE 5399
Remtox Fungicide/Insecticide Microemulsion M9	Remtox (Chemicals) Ltd	WI WP	HSE 5360
Remtox Spirit Based F/I K7	Remtox (Chemicals) Ltd	WI WP	HSE 5391
Sadolin Base No 561–2611	Sadolin (UK) Ltd	WA	HSE 4330
Sadolin Base No 561–2611	Sadolin Nobel UK Ltd	WA WP	HSE 5649
SOV AQ Micro F/I	Sovereign Chemical Industries Ltd	WI WP	HSE 5222
Sovac F/I	Sovereign Chemical Industries Ltd	WI WP	HSE 5232
Sovac F/I WR	Sovereign Chemical Industries Ltd	WI WP	HSE 5300
Sovaq Micro F/I	Sovereign Chemical Industries Ltd	WI WP	HSE 5669
Sovereign AQ F/I	Sovereign Chemical Industries Ltd	WI WP	HSE 5221
Tripaste PP	Triton Chemical Manufacturing Company Ltd	WA	HSE 4595
Woodtreat 25	Stanhope Chemical Products Ltd	WP	HSE 5413
15 5-CHLORO-2-METHYL-4-ISOTHIAZOLIN-3-ONE and 2-METHYL-4-ISOTHIAZOLIN-3-ONE			
* Biomek	Mechema Chemicals Ltd	WI	HSE 3091
Celkil 90	Rentokil Ltd	WI	HSE 4275
Celkil 90	Rentokil Ltd	WI	HSE 5428
* Gallwey SA	Fosroc Ltd	WI	HSE 3702
Kathon 886F	Rohm and Haas (UK) Ltd	WI	HSE 4866
Kathon 886F	Rohm and Haas (UK) Ltd	WI	HSE 5431
Laporte Mould-Ex	Laporte Wood Preservation	WI	HSE 3953
Laporte Mould-Ex	Laporte Wood Preservation	WI	HSE 5430
Tanamix 3743	Hickson Timber Products Ltd	WI	HSE 4852
Tanamix 3743	Hickson Timber Products Ltd	WI	HSE 5429

WOOD PRESERVATIVES

Product Name	Marketing Company	Use	Reg. No

16 ACYPETACS COPPER

Product Name	Marketing Company	Use	Reg. No
Cuprinol Wood Preserver Green S	Cuprinol Ltd	WA	HSE 4697

17 ACYPETACS COPPER and ACYPETACS ZINC and PERMETHRIN

Product Name	Marketing Company	Use	Reg. No
Cuprisol P	Cuprinol Ltd	WI	HSE 3634
Protim 800P	Protim Solignum Ltd	WI	HSE 5702

18 ACYPETACS ZINC

Product Name	Marketing Company	Use	Reg. No
Cuprinol Wet and Dry Rot Killer for Timber(S)	Cuprinol Ltd	WA	HSE 3589
Cuprinol Wood Preserver	Cuprinol Ltd	WA	HSE 4698
Cuprinol Wood Preserver Clear S	Cuprinol Ltd	WA	HSE 4708
# Cuprinol Wood Preserver Dark Oak S	Cuprinol Ltd	WA	HSE 4730
Cuprisol F	Cuprinol Ltd	WI	HSE 3734
Cuprisol XQD	Cuprinol Ltd	WI	HSE 3733
Exterior Wood Preserver S	Cuprinol Ltd	WA	HSE 4716
Fungicidal Preservative	Protim Solignum Ltd	WA WI WP	HSE 5489
* Protim FDR 800	Fosroc Ltd	WI	HSE 4492
Protim FDR 800	Protim Solignum Ltd	WI	HSE 4963
* Protim JP 800	Fosroc Ltd	WI	HSE 4491
Protim JP 800	Protim Solignum Ltd	WI	HSE 4987
* Protim Paste 800 F	Fosroc Ltd	WP	HSE 4497
Protim Paste 800 F	Protim Solignum Ltd	WI WP	HSE 4989
Wickes Exterior Wood Preserver	Wickes Building Supplies Ltd	WA WP	HSE 5697

19 ACYPETACS ZINC and DICHLOFLUANID

Product Name	Marketing Company	Use	Reg. No
Cuprinol Decorative Wood Preserver	Cuprinol Ltd	WA	HSE 3156
Cuprinol Decorative Wood Preserver Red Cedar	Cuprinol Ltd	WA	HSE 4696
Cuprinol Low Odour Decorative Wood Preserver	Cuprinol Ltd	WA WI WP	HSE 5343
Cuprinol Low Odour Decorative Wood Preserver Red Cedar	Cuprinol Ltd	WA WI WP	HSE 5342
Cuprinol Magnatreat F	Cuprinol Ltd	WI	HSE 3752

WOOD PRESERVATIVES

Product Name	Marketing Company	Use	Reg. No
19 **ACYPETACS ZINC and DICHLOFLUANID**—continued			
Cuprinol Magnatreat XQD	Cuprinol Ltd	WI	HSE 3751
Cuprinol Preservative Base	Cuprinol Ltd	WA WP	HSE 4929
20 **ACYPETACS ZINC and PERMETHRIN**			
# Cedasol Ready To Use (2310)	Hickson Timber Products Ltd	WI	HSE 3975
Cuprinol 5 Star Complete Wood Treatment S	Cuprinol Ltd	WA	HSE 4710
Cuprinol Combination Grade S	Cuprinol Ltd	WA	HSE 3588
Cuprinol Low Odour 5 Star Complete Wood Treatment	Cuprinol Ltd	WA WI WP	HSE 5445
Cuprisol FN	Cuprinol Ltd	WI	HSE 3632
Cuprisol WR	Cuprinol Ltd	WI	HSE 3633
Cuprisol XQD Special	Cuprinol Ltd	WI	HSE 3732
Cut End	Protim Solignum Ltd	WA WI WP	HSE 5490
* Protim 800	Fosroc Ltd	WI	HSE 4490
Protim 800	Protim Solignum Ltd	WI	HSE 4991
* Protim 800 C	Fosroc Ltd	WI	HSE 4494
Protim 800 C	Protim Solignum Ltd	WI	HSE 4990
* Protim 800 C Oil Brown	Fosroc Ltd	WI	HSE 4488
* Protim 800 CWR	Fosroc Ltd	WI	HSE 4493
Protim 800 CWR	Protim Solignum Ltd	WI	HSE 4993
* Protim 800 WR	Fosroc Ltd	WI	HSE 4495
Protim 800 WR	Protim Solignum Ltd	WI	HSE 4992
* Protim Brown 800	Fosroc Ltd	WI	HSE 4486
Protim Curative 800	Fosroc Ltd	WA	HSE 4489
* Protim Paste 800	Fosroc Ltd	WP	HSE 4485
Protim Paste 800	Protim Solignum Ltd	WI WP	HSE 4986
Protim WR 800	Fosroc Ltd	WA	HSE 4487
Universal	Protim Solignum Ltd	WA WI WP	HSE 5491
Wickes All Purpose Wood Treatment	Wickes Building Supplies Ltd	WA WP	HSE 5696

WOOD PRESERVATIVES

Product Name	Marketing Company	Use	Reg. No
21 ACYPETACS ZINC and PERMETHRIN and ACYPETACS COPPER			
Cuprisol P	Cuprinol Ltd	WI	HSE 3634
Protim 800P	Protim Solignum Ltd	WI	HSE 5702
22 ALKYLARYLTRIMETHYL AMMONIUM CHLORIDE			
Anti-Mould Solution	Macpherson Paints Ltd	WA SA	HSE 4873
Barrettine Timberguard	Barrettine (Products) Ltd	WA	HSE 3089
Langlow Timbershield	Langlow Products Ltd	WA	HSE 4394
* Resistone PC	ABM Chemicals Ltd	WP SP	HSE 3774
Timberdip	Langlow Products Ltd	WI WP	HSE 4393
* Timberglow	Cotech Services Ltd	WA	HSE 3079
Timberguard 1 Plus 7 Concentrate	Barrettine (Products) Ltd	WP	HSE 3090
Timberguard 1 Plus 7 Concentrate	Barrettine Products Ltd	WI WP	HSE 5211
23 ALKYLARYLTRIMETHYL AMMONIUM CHLORIDE and DISODIUM OCTABORATE			
Timberglow Concentrate	Cotech Services Ltd	WP	HSE 3078
Timberglow Extra	Cotech Services Ltd	WA	HSE 3077
24 ALKYLARYLTRIMETHYL AMMONIUM CHLORIDE and TRIBUTYLTIN OXIDE			
# Killgerm Masonry Sterilant	Killgerm Chemicals Ltd	WP SP	HSE 3125
25 ALKYLTRIMETHYL AMMONIUM CHLORIDE and SODIUM TETRABORATE			
Sinesto B	Finnmex Co	WI	HSE 5136
* Sinesto B	KYMI Chemicals	WI	HSE 4878
26 AMMONIUM BIFLUORIDE and SODIUM DICHROMATE and SODIUM FLUORIDE			
Rentex	Rentokil Ltd	WI	HSE 4756
27 ARSENIC PENTOXIDE and CHROMIUM TRIOXIDE and COPPER OXIDE			
Celcure AO	Rentokil Ltd	WI	HSE 4220
* Celcure CCA Type C	Rentokil Ltd	WI	HSE 4673
Celcure CCA Type C	Rentokil Ltd	WI	HSE 5104

WOOD PRESERVATIVES

Product Name	Marketing Company	Use	Reg. No

27 ARSENIC PENTOXIDE and CHROMIUM TRIOXIDE and COPPER OXIDE— continued

Product Name	Marketing Company	Use	Reg. No
Laporte CCA AWPA Type C	Laporte Wood Preservation	WI	HSE 3179
Laporte CCA Oxide Type 1	Laporte Wood Preservation	WI	HSE 3177
Laporte CCA Oxide Type 2	Laporte Wood Preservation	WI	HSE 3178
* Mekure T1 Oxide	Mechema Chemicals Ltd	WI	HSE 4079
* Mekure T2 Oxide	Mechema Chemicals Ltd	WI	HSE 4080
Protim CCA Oxide – Type II	Fosroc Ltd	WI	HSE 4463
Protim CCA Oxide 50	Protim Solignum Ltd	WI	HSE 5537
Protim CCA Oxide 58	Protim Solignum Ltd	WI	HSE 5686
Protim CCA Oxide 72	Protim Solignum Ltd	WI	HSE 5594
Tanalith 3302	Hickson Timber Products Ltd	WI	HSE 5098
Tanalith 3313	Hickson Timber Products Ltd	WI	HSE 4422
Tanalith C3310	Hickson Timber Products Ltd	WI	HSE 4669
Tanalith Oxide C3309	Hickson Timber Products Ltd	WI	HSE 4668
Tanalith Oxide C3314	Hickson Timber Products Ltd	WI	HSE 4774

28 ARSENIC PENTOXIDE and COPPER SULPHATE and SODIUM DICHROMATE

Product Name	Marketing Company	Use	Reg. No
Celcure A Concentrate	Rentokil Ltd	WI	HSE 5458
Celcure A Fluid 10	Rentokil Ltd	WI	HSE 3764
Celcure A Fluid 6	Rentokil Ltd	WI	HSE 3155
Celcure A Paste	Rentokil Ltd	WI	HSE 4523
Injecta CCA-C	Injecta APS	WI	HSE 4447
Kemira CCA Type BS	Kemira Kemwood AB	WI	HSE 3879
Kemwood CCA Type BS	Laporte Kemwood AB	WI	HSE 5534
Laporte CCA Type 1	Laporte Wood Preservation	WI	HSE 3175
Laporte CCA Type 2	Laporte Wood Preservation	WI	HSE 3176
Laporte Permawood CCA	Laporte Wood Preservation	WI	HSE 3529
* Meksol	Mechema Chemicals Ltd	WI	HSE 3994
* Mekure	Mechema Chemicals Ltd	WI	HSE 4352
* Mekure T2	Mechema Chemicals Ltd	WI	HSE 4350
* Protim CCA Salts Type 2	Fosroc Ltd	WI	HSE 4462
Protim CCA Salts Type 2	Protim Solignum Ltd	WI	HSE 4972
Tanalith 3357	Hickson Timber Products Ltd	WI	HSE 4431
Tanalith CL (3354)	Hickson Timber Products Ltd	WI	HSE 4196

WOOD PRESERVATIVES

Product Name	Marketing Company	Use	Reg. No
28 ARSENIC PENTOXIDE and COPPER SULPHATE and SODIUM DICHROMATE— continued			
Tanalith CP 3353	Hickson Timber Products Ltd	WI	HSE 4667
Tecca P2	Tecca Ltd	WI	HSE 3958
29 AZACONAZOLE			
# Rodewod 10 OL	Janssen Pharmaceutica NV	WA	HSE 3840
Rodewod 50 SL	Janssen Pharmaceutica NV	WA	HSE 3839
Safetray SL	Progress Products	WP	HSE 5464
30 AZACONAZOLE and DICHLOFLUANID			
Xyladecor Matt U 404	Venilia Ltd	WA	HSE 4445
Xylamon Primer Dipping Stain U 415	Venilia Ltd	WP	HSE 4444
31 AZACONAZOLE and PERMETHRIN			
Fongix SE Total Treatment for Wood	Liberon Waxes Ltd	WA SA	HSE 5610
Fongix SE Total Treatment for Wood	V33 UK	WA SA	HSE 4288
Woodworm Killer and Rot Treatment	Liberon Waxes Ltd	WA SA	HSE 4826
Xylamon Brown U 101 C	Venilia Ltd	WA	HSE 4443
Xylamon Curative U 152 G/H	Venilia Ltd	WP	HSE 4446
32 BENZALKONIUM CHLORIDE			
Algae Remover	Sadolin (UK) Ltd	WA SA	HSE 4781
Basilit Bauholz-KD	Venilia Ltd	WI	HSE 4356
Basiment NT	Kay Metzeler Ltd	WI	HSE 3614
Glen Wood Care Wood Preservative	Glen Wood Care	WA	HSE 4096
Gloquat RP	Rhone-Poulenc Chemicals	WP SP	HSE 4815
# Timbertreat Wood Preservative	Esmi Chemicals Ltd	WA	HSE 3188
# Vulcanite Timbertreat Wood Preservative	Vulcanite Ltd	WA	HSE 3187
33 BENZALKONIUM CHLORIDE and 2-PHENYLPHENOL			
Chel Dry Rot Killer for Masonry and Brickwork	Chelec Ltd	WP	HSE 3067

WOOD PRESERVATIVES

Product Name	Marketing Company	Use	Reg. No

34 BENZALKONIUM CHLORIDE and BORIC ACID

Product Name	Marketing Company	Use	Reg. No
Microguard Mouldicidal Wood Preserver	Permagard Products Ltd	WA SA	HSE 5100

35 BENZALKONIUM CHLORIDE and BORIC ACID and DIALKYLDIMETHYL AMMONIUM CHLORIDE and METHYLENE BIS (THIOCYANATE)

Product Name	Marketing Company	Use	Reg. No
Celbrite MT	Rentokil Ltd	WI	HSE 4550

36 BENZALKONIUM CHLORIDE and CYPERMETHRIN

Product Name	Marketing Company	Use	Reg. No
# Timbertreat Multi Purpose Preservative	Esmi Chemicals	WA	HSE 3518
# Timbertreat Wood Preservative – Clear	Vulcanite Ltd	WA	HSE 3186
# Timbertreat Woodworm and Dry Rot Killer	Esmi Chemicals Ltd	WA	HSE 3183
# Timbertreat Woodworm and Dry Rot Killer	Vulcanite Ltd	WA	HSE 3184
# Timbertreat Woodworm and Dry Rot Killer – Water Based	Esmi Chemicals Ltd	WA	HSE 3182
# Vulcanite Timbertreat Multi Purpose Preservative	Vulcanite Ltd	WA	HSE 3519
# Vulcanite Timbertreat Wood Preservative Clear	Vulcanite Limited	WA	HSE 3185
# Vulcanite Timbertreat Woodworm and Dry Rot Killer Water Based	Vulcanite Ltd	WA	HSE 3181

37 BENZALKONIUM CHLORIDE and CYPERMETHRIN and DIALKYLDIMETHYL AMMONIUM CHLORIDE

Product Name	Marketing Company	Use	Reg. No
Gainserv 684	Industrial Chemical Company (Preston) Ltd	WA	HSE 3996

38 BENZALKONIUM CHLORIDE and DIALKYLDIMETHYL AMMONIUM CHLORIDE

Product Name	Marketing Company	Use	Reg. No
ABL Aqueous Wood Preserver Concentrate 1:9	Advanced Bitumens Ltd	WP	HSE 4199
Aqueous Wood Preserver	Advanced Bitumens Ltd	WA	HSE 4727
Celbrite M	Rentokil Ltd	WI	HSE 4535
Celbronze B	Rentokil Ltd	WP	HSE 4549

38 BENZALKONIUM CHLORIDE and DIALKYLDIMETHYL AMMONIUM CHLORIDE—continued

Product Name	Marketing Company	Use	Reg. No
Colourfast Protector	Rentokil Ltd	WA	HSE 4609
Sterilising Fluid	Rentokil Ltd	WP SP	HSE 4605

39 BENZALKONIUM CHLORIDE and DISODIUM OCTABORATE

Product Name	Marketing Company	Use	Reg. No
Boracol 10 Rh	Channelwood Preservations Ltd	WA SA	HSE 4104
Boracol 20 Rh	Channelwood Preservations Ltd	WP	HSE 4019
Boracol B8.5 Rh Mouldicide/Wood Preservative	Channelwood Preservations Ltd	WA SA	HSE 4809
Deepwood 20 Inorganic Boron Wood Preservative	Safeguard Chemicals Ltd	WP	HSE 5514
Deepwood 50 Inorganic Boron Wood Preservative Paste	Safeguard Chemicals Ltd	WP SP	HSE 5596

40 BENZALKONIUM CHLORIDE and GAMMA-HCH

Product Name	Marketing Company	Use	Reg. No
Biokil	Jaymar Chemicals	WP	HSE 4247

41 BENZALKONIUM CHLORIDE and PERMETHRIN

Product Name	Marketing Company	Use	Reg. No
Biokil Emulsion	Jaymar Chemicals	WP	HSE 4562

42 BENZALKONIUM CHLORIDE and TRIBUTYLTIN OXIDE

Product Name	Marketing Company	Use	Reg. No
# Nubex Fungicide QT Concentrate	Albright and Wilson Ltd	WP SP	HSE 3166

43 BORIC ACID

Product Name	Marketing Company	Use	Reg. No
Basilit B 85	Kay-Metzeler Ltd	WI	HSE 3813
Celbor M	Rentokil Ltd	WI	HSE 4912
Control Fluid FB	Rentokil Ltd	WP	HSE 5323
Dricon	Hickson Timber Products Ltd	WI	HSE 5688
Dricon	Lambson Ltd	WI	HSE 3069
Dricon	Lambson Speciality Chemicals Ltd	WI	HSE 5315
New Cut 'N' Spray	Rentokil Ltd	WA	HSE 4187
Rentokil Dry Rot Fluid (E)	Rentokil Ltd	WP SP	HSE 3997

WOOD PRESERVATIVES

Product Name	Marketing Company	Use	Reg. No
43 BORIC ACID—continued			
Rentokil Dry Rot Fluid (E) for Bonded Warehouses	Rentokil Ltd	WP SP	HSE 3986
Woodworm Fluid FB	Rentokil Ltd	WP	HSE 4251
44 BORIC ACID and 2-(THIOCYANOMETHYLTHIO) BENZOTHIAZOLE			
Protim Fentex Europa 1 RFU	Protim Solignum Ltd	WI	HSE 5119
Protim Fentex Europa I	Protim Solignum Ltd	WI	HSE 4984
45 BORIC ACID and BENZALKONIUM CHLORIDE			
Microguard Mouldicidal Wood Preserver	Permagard Products Ltd	WA SA	HSE 5100
46 BORIC ACID and CHROMIUM ACETATE and COPPER SULPHATE and SODIUM DICHROMATE			
* Celcure CB Salts	Rentokil Ltd	WI	HSE 4536
Celgard CF	Rentokil Ltd	WI	HSE 4608
47 BORIC ACID and CHROMIUM TRIOXIDE and COPPER OXIDE			
Tanalith (3419) CBC	Hickson Timber Products Ltd	WI	HSE 4022
48 BORIC ACID and COPPER CARBONATE HYDROXIDE			
Tanalith 3487	Hickson Timber Products Ltd	WI	HSE 5305
49 BORIC ACID and COPPER CARBONATE HYDROXIDE and TEBUCONAZOLE			
Tanalith 3485	Hickson Timber Products Ltd	WI	HSE 5562
50 BORIC ACID and COPPER OXIDE			
Tanalith 3422	Hickson Timber Products Ltd	WI	HSE 4403
51 BORIC ACID and COPPER SULPHATE			
Ensele 3426	Hickson Timber Products Ltd	WP	HSE 4364
Ensele 3427	Hickson Timber Products Ltd	WA	HSE 5053

WOOD PRESERVATIVES

Product Name	Marketing Company	Use	Reg. No
52 BORIC ACID and COPPER SULPHATE and POTASSIUM DICHROMATE			
Tecca CCB 1	Tecca Ltd	WI	HSE 5584
* Wolmanit CB	BASF (United Kingdom) Ltd	WI	HSE 3108
* Wolmanit CB-P	BASF (United Kingdom) Ltd	WI	HSE 3110
53 BORIC ACID and COPPER SULPHATE and SODIUM DICHROMATE			
Celcure CB90	Rentokil Ltd	WI	HSE 5117
Tanalith CBC Paste 3402	Hickson Timber Products Ltd	WI	HSE 3833
* Wolmanit CB-A	BASF (United Kingdom) Ltd	WI	HSE 3109
54 BORIC ACID and DIALKYLDIMETHYL AMMONIUM CHLORIDE and METHYLENE BIS (THIOCYANATE) and BENZALKONIUM CHLORIDE			
Celbrite MT	Rentokil Ltd	WI	HSE 4550
55 BORIC ACID and METHYLENE BIS (THIOCYANATE)			
Hickson Antiblu 3737	Hickson Timber Products Ltd	WI	HSE 4745
Hickson Antiblu 3738	Hickson Timber Products Ltd	WI	HSE 4746
56 BORIC ACID and METHYLENE BIS (THIOCYANATE) and 2-(THIOCYANOMETHYLTHIO) BENZOTHIAZOLE			
Fentex Elite	Protim Solignum Ltd	WI	HSE 5532
57 BORIC ACID and PERMETHRIN			
Microguard Fl Concentrate	Permagard Products Ltd	WI WP	HSE 5101
58 BORIC ACID and PERMETHRIN and ZINC OCTOATE			
Celpruf BZP	Rentokil Ltd	WI	HSE 4190
Celpruf BZP WR	Rentokil Ltd	WI	HSE 4191
Premium Grade Wood Treatment	Rentokil Ltd	WA	HSE 4193
59 BORIC ACID and SODIUM TETRABORATE			
Celbor	Rentokil Ltd	WI	HSE 4614
PC-K	Peter Cox Preservation	WP	HSE 4409
Pyrolith 3505 Ready to Use	Hickson Timber Products Ltd	WI	HSE 3636

WOOD PRESERVATIVES

Product Name	Marketing Company	Use	Reg. No
60 BORIC ACID and ZINC OCTOATE			
Celpruf BZ	Rentokil Ltd	WI	HSE 4188
Celpruf BZ WR	Rentokil Ltd	WI	HSE 4189
Dry Rot Fluid (D) EC	Rentokil Ltd	WP	HSE 4194
Rentokil Dry Rot Paste (D)	Rentokil Ltd	WP	HSE 3987
61 CARBENDAZIM			
* Ronseal Fencelife	Sterling-Winthrop Group Ltd	WA	HSE 3130
62 CARBENDAZIM and FURMECYCLOX			
# Bio Woody	Pan Britannica Industries Ltd	WA	HSE 4638
63 CARBENDAZIM and TRIBUTYLTIN OXIDE and ZINC NAPHTHENATE			
# Dulux Preservative Wood Primer	ICI Paints Division	WA	HSE 3826
# Dulux Weathershield Exterior Preservative Primer	ICI Paints Division	WA	HSE 3819
# Dulux Weathershield Preservative Basecoat	ICI Paints Division	WA	HSE 4026
64 CHROMIUM ACETATE and COPPER SULPHATE and SODIUM DICHROMATE			
Celcure B	Rentokil Ltd	WI	HSE 4541
Celcure O	Rentokil Ltd	WI	HSE 4539
* Mekseal	Mechema Chemicals Ltd	WI	HSE 4351
65 CHROMIUM ACETATE and COPPER SULPHATE and SODIUM DICHROMATE and BORIC ACID			
* Celcure CB Salts	Rentokil Ltd	WI	HSE 4536
Celgard CF	Rentokil Ltd	WI	HSE 4608
66 CHROMIUM TRIOXIDE and COPPER OXIDE and ARSENIC PENTOXIDE			
Celcure AO	Rentokil Ltd	WI	HSE 4220
* Celcure CCA Type C	Rentokil Ltd	WI	HSE 4673
Celcure CCA Type C	Rentokil Ltd	WI	HSE 5104
Laporte CCA AWPA Type C	Laporte Wood Preservation	WI	HSE 3179

WOOD PRESERVATIVES

Product Name	Marketing Company	Use	Reg. No
66 CHROMIUM TRIOXIDE and COPPER OXIDE and ARSENIC PENTOXIDE— continued			
Laporte CCA Oxide Type 1	Laporte Wood Preservation	WI	HSE 3177
Laporte CCA Oxide Type 2	Laporte Wood Preservation	WI	HSE 3178
* Mekure T1 Oxide	Mechema Chemicals Ltd	WI	HSE 4079
* Mekure T2 Oxide	Mechema Chemicals Ltd	WI	HSE 4080
Protim CCA Oxide – Type II	Fosroc Ltd	WI	HSE 4463
Protim CCA Oxide 50	Protim Solignum Ltd	WI	HSE 5537
Protim CCA Oxide 58	Protim Solignum Ltd	WI	HSE 5686
Protim CCA Oxide 72	Protim Solignum Ltd	WI	HSE 5594
Tanalith 3302	Hickson Timber Products Ltd	WI	HSE 5098
Tanalith 3313	Hickson Timber Products Ltd	WI	HSE 4422
Tanalith C3310	Hickson Timber Products Ltd	WI	HSE 4669
Tanalith Oxide C3309	Hickson Timber Products Ltd	WI	HSE 4668
Tanalith Oxide C3314	Hickson Timber Products Ltd	WI	HSE 4774
67 CHROMIUM TRIOXIDE and COPPER OXIDE and BORIC ACID			
Tanalith (3419) CBC	Hickson Timber Products Ltd	WI	HSE 4022
68 CHROMIUM TRIOXIDE and COPPER OXIDE and DISODIUM OCTABORATE			
Celcure CB Paste	Rentokil Ltd	WI	HSE 4537
69 COAL TAR CREOSOTE			
ABL Brown Creosote	Advanced Bitumens Ltd	WA	HSE 3859
B & Q Creosote	B & Q Plc	WA	HSE 4101
Bartoline Dark and Light Creosote	Bartoline Ltd	WA	HSE 4459
* Blended Coal Tar Creosote	Croda Hydrocarbons Ltd	WA	HSE 4293
Blended Coal Tar Creosote	John Astley and Sons Ltd	WA	HSE 4789
BS 144 Creosote	Lancashire Tar Distillers Ltd	WI	HSE 3144
BS 144 Creosote	Lanstar Ltd	WI	HSE 5189
BS 144 Creosote	Oakmere Technical Services Ltd	WI	HSE 3712
Builders Mate Creosote	Builders Mate Ltd	WA WP	HSE 5110
Carbo Creosote	Talke Chemical Co Ltd	WA	HSE 4362

WOOD PRESERVATIVES

Product Name	Marketing Company	Use	Reg. No
69 **COAL TAR CREOSOTE**—continued			
Co-op Creosote	Co-operative Wholesale Society Ltd	WA	HSE 3639
Coal Tar Creosote	Creohaul and Company	WA	HSE 3898
Coal Tar Creosote	Croda Hydrocarbons Ltd	WA	HSE 4294
Coal Tar Creosote	Hardmans of Hull Ltd	WA	HSE 4917
Coal Tar Creosote	South Western Tar Distilleries Ltd	WP	HSE 4574
Coal Tar Creosote	William Mathwin and Son (Newcastle) Ltd	WA	HSE 4602
Coal Tar Creosote Blend	Laycocks Agricultural Chemists	WA	HSE 4450
* Creosote	Ashby Timber Treatments	WA	HSE 3021
Creosote	Blanchard Martin and Simmonds Ltd	WA	HSE 3884
Creosote	C W Wastnage Ltd	WA WP	HSE 5267
Creosote	Dee Oil (Midlands) Ltd	WA	HSE 3602
Creosote	G and B Fuels Ltd	WA	HSE 4469
Creosote	Great Mills (Retail) Ltd	WA WP	HSE 5666
Creosote	Liver Grease Oil and Chemical Company Ltd	WA	HSE 4467
Creosote	RK and J Jones	WA WI WP	HSE 4384
* Creosote	Robert McBride Homecare Ltd	WA	HSE 3142
Creosote	Strathclyde Chemical Company Ltd	WA	HSE 4525
Creosote	T K Bird Ltd	WA WI WP	HSE 4903
Creosote Blend	Croda Hydrocarbons Ltd	WA WI WP	HSE 4948
Creosote Blend	Davan Industrial Ltd	WA WI WP	HSE 5302
* Creosote Blend	James D Johnson and Company Ltd	WA	HSE 3530
Creosote Blend MK1 (Medium Dark)	South Western Tar Distilleries Ltd	WA	HSE 4588
Creosote Blend MK3 (Light Golden)	South Western Tar Distilleries Ltd	WA	HSE 4590

WOOD PRESERVATIVES

Product Name	Marketing Company	Use	Reg. No
69 COAL TAR CREOSOTE—continued			
Creosote Blended Wood Preservative	Rye Oil Ltd	WA	HSE 3730
Creosote BS 144	Coalite Chemicals Division	WI	HSE 3811
* Creosote BS 144 (2)	Printar Industries Ltd	WI	HSE 4533
Creosote BS 144 (3)	Printar Industries Ltd	WP	HSE 4534
Creosote BS 144 Type III	Coalite Chemicals Division	WA	HSE 3812
Dark Brown Creosote	Coal Products Ltd	WA	HSE 3741
Dark Creosote Emulsion	E W Middleton and Sons	WA	HSE 4289
* Four Seasons Creosote	William Pinchin and Company	WA	HSE 4715
Greenhills Creosote	Greenhills (Wessex) Ltd	WA	HSE 4205
Hickson Timbercare 2511 Brown	Hickson Timber Products Ltd	WI	HSE 4666
Homebase Creosote	Homebase Ltd	WA	HSE 3672
Jewson Creosote	Jewson Ltd	WA	HSE 3761
Langlow Creosote	Langlow Products Ltd	WA	HSE 4639
Leyland Creosote	Leyland Paint Co	WA	HSE 4296
Medium Brown Creosote	C W Wastnage Ltd	WA	HSE 4691
Middletons Creosote	Biotech Environmental Services Ltd	WA	HSE 4441
# Morrisons Creosote	Morrisons Supermarkets Plc	WA	HSE 3086
Nitromors Creosote	Kalon Group Plc	WA WP	HSE 4879
Ovoline 275 Golden Creosote	Bretts Oils Ltd	WA	HSE 3651
Solignum Dark Brown	Protim Solignum Ltd	WA WI WP	HSE 5023
* Solignum Dark Brown	Solignum Ltd	WA	HSE 4316
Solignum Medium Brown	Protim Solignum Ltd	WA WI WP	HSE 5022
* Solignum Medium Brown	Solignum Ltd	WA	HSE 4317
Super Andyman Creosote	AMB Products Ltd	WA	HSE 4211
* Texas Creosote	Texas Homecare Ltd	WA	HSE 3074
Texas Creosote	Texas Homecare Ltd	WA	HSE 4297
# Timbrol Creosote Blend	Technical and Manufacturing Services Ltd	WA WP	HSE 3084
Wickes Creosote	Wickes Building Supplies Ltd	WA	HSE 4565
Wilko Creosote	T K Bird Ltd	WA	HSE 5043

WOOD PRESERVATIVES

Product Name	Marketing Company	Use	Reg. No
69 COAL TAR CREOSOTE—continued			
Wood Preservative Type 3	Printar Industries Ltd	WP	HSE 4532
Wood Preservative Type 3	SWTD Hertford	WI WP	HSE 5252
70 COAL TAR CREOSOTE and CREOSOTE			
* Arborsan 4	Lancashire Tar Distillers Ltd	WA	HSE 4453
* Arborsan 6	Lancashire Tar Distillers Ltd	WA	HSE 4452
71 COPPER CARBONATE HYDROXIDE and BORIC ACID			
Tanalith 3487	Hickson Timber Products Ltd	WI	HSE 5305
72 COPPER CARBONATE HYDROXIDE and TEBUCONAZOLE and BORIC ACID			
Tanalith 3485	Hickson Timber Products Ltd	WI	HSE 5562
73 COPPER NAPHTHENATE			
Barkep Wood Preserver Green	Bartoline Ltd	WA	HSE 4458
Barrettine Green Wood Preserver	Barrettine Products Ltd	WA	HSE 4159
Bio-Kil Cunap Pole Wrap	Bio-Kil Chemicals Ltd	WP	HSE 5341
Blackfriars Green Wood Preserver	E Parsons and Sons Ltd	WA	HSE 4209
Carbo Wood Preservative	Talke Chemical Company Ltd	WA	HSE 3820
Copper Naphthenate Solution to BS 5056	Cuprinol Ltd	WA	HSE 4711
Croda Green Wood Preserver	Croda Hydrocarbons Ltd	WA WP	HSE 5372
Flag Brand Wood Preservative Green	C W Wastnage Ltd	WA	HSE 3073
* Gainserv 681	Industrial Chemical Company (Preston) Ltd	WA	HSE 3955
Glen Wood Care Green	Glen Wood Care	WA	HSE 4095
# Granyte Farmcare Green Wood Preservative	Granyte Surface Coatings Plc	WA WP	HSE 3524
Green Plus Wood Preserver	Ultrabond Ltd	WA WP	HSE 5040
Green Wood Preservative	Advanced Bitumens Ltd	WA	HSE 4728
Green Wood Preservative	Strathbond Ltd	WP	HSE 4660
Langlow Wood Preservative Green	Langlow Products Ltd	WA	HSE 4395
Larsen Green Wood Preservative	Larsen Manufacturing Ltd	WA	HSE 3708

WOOD PRESERVATIVES

Product Name	Marketing Company	Use	Reg. No
73 COPPER NAPHTHENATE—continued			
Leyland Timbrene Green Environmental Formula	Leyland Paint Company	WA	HSE 4799
Preservative for Wood Green	Rentokil Ltd	WA	HSE 4613
* Protim Green WR	Fosroc Ltd	WA	HSE 4435
Protim Green WR	Protim Solignum Ltd	WA WI WP	HSE 4976
Supergrade Wood Preserver Green	Rentokil Ltd	WP	HSE 3717
Teamac Woodtec Green	Teal and Mackrill Ltd	WA	HSE 4429
Timber Preservative Green	Antel-Tridex	WP	HSE 4847
Timbermate Green	Lanstar Coatings Ltd	WA	HSE 4401
# Timbertreat Green Preservative	Esmi Chemicals Ltd	WA WP	HSE 3147
Timbrene Green Wood Preserver	Kalon Group Plc	WA WP	HSE 4967
# Vulcanite Timbertreat Green Preservative	Vulcanite Ltd	WA WP	HSE 3148
Wood Preserver Green	Cementone-Beaver Ltd	WA	HSE 4927

74 COPPER NAPHTHENATE and DISODIUM OCTABORATE

Product Name	Marketing Company	Use	Reg. No
Bio-Kil SR Pole Wrap	Bio-Kil Chemicals Ltd	WP	HSE 5675

75 COPPER NAPHTHENATE and GAMMA-HCH and TRIBUTYLTIN OXIDE

Product Name	Marketing Company	Use	Reg. No
# Decor-8 Wood Preservative Green	Decor-8 Ltd	WA WP	HSE 3065
# Fads Wood Preservative Green	Fads AG Stanley Ltd	WA WP	HSE 3037
# Great Mills Wood Preservative Green	Great Mills	WA WP	HSE 3041
# Homecharm Wood Preservative Green	Homecharm Ltd	WA WP	HSE 3061
# Payless Wood Preservative Green	Payless DIY Ltd	WA WP	HSE 3049
# POB Wood Preservative Green	POB Savident Ltd	WA WP	HSE 3025
# SPL Wood Preservative Green	Silver Paint and Lacquer Ltd	WA WP	HSE 3057
# Texas Wood Preservative Green	Texas Homecare Ltd	WA WP	HSE 3033
# Trend Wood Preservative Green	Trend Voluntary Group Services Ltd	WA WP	HSE 3045
# Wickes Wood Preservative Green	Wickes Building Supplies Ltd	WA WP	HSE 3029
# Wilco Wood Preservative Green	Wilkinson Group of Companies	WA WP	HSE 3053

WOOD PRESERVATIVES

Product Name	Marketing Company	Use	Reg. No

76 COPPER NAPHTHENATE and PERMETHRIN

Product Name	Marketing Company	Use	Reg. No
Green	Protim Solignum Ltd	WA WI WP	HSE 5483
* Protim Green E	Fosroc Ltd	WA	HSE 3799
Protim Green E	Protim Solignum Ltd	WA WI WP	HSE 4975
Solignum Green	Protim Solignum Ltd	WA	HSE 5020
* Solignum Green	Solignum Ltd	WA	HSE 3900

77 COPPER NAPHTHENATE and TRI (HEXYLENE GLYCOL) BIBORATE

Product Name	Marketing Company	Use	Reg. No
Dark Green Timber Preservative 2	Palace Chemicals Ltd	WA	HSE 4321
Dark Green Wood Preservative	Palace Chemicals Ltd	WA WI WP	HSE 5243
Kingfisher Wood Preservative	Kingfisher Chemicals Ltd	WP	HSE 4011
Pale Green Timber Preservative 2	Palace Chemicals Ltd	WA	HSE 4320
Pale Green Wood Preservative	Palace Chemicals Ltd	WA WI WP	HSE 5241
Sovereign Timber Preservative Dark Green	Sovereign Chemical Industries Ltd	WP	HSE 3808
Sovereign Timber Preservative Pale Green	Sovereign Chemical Industries Ltd	WP	HSE 3806

78 COPPER NAPHTHENATE and ZINC OCTOATE

Product Name	Marketing Company	Use	Reg. No
Dulux Weathershield Timber Preservative	ICI Plc	WA	HSE 3805
Weathershield Exterior Timber Preservative	ICI Paints Division	WA	HSE 4915

79 COPPER OXIDE and ARSENIC PENTOXIDE and CHROMIUM TRIOXIDE

Product Name	Marketing Company	Use	Reg. No
Celcure AO	Rentokil Ltd	WI	HSE 4220
* Celcure CCA Type C	Rentokil Ltd	WI	HSE 4673
Celcure CCA Type C	Rentokil Ltd	WI	HSE 5104
Laporte CCA AWPA Type C	Laporte Wood Preservation	WI	HSE 3179
Laporte CCA Oxide Type 1	Laporte Wood Preservation	WI	HSE 3177
Laporte CCA Oxide Type 2	Laporte Wood Preservation	WI	HSE 3178
* Mekure T1 Oxide	Mechema Chemicals Ltd	WI	HSE 4079

Product Name	Marketing Company	Use	Reg. No

79 COPPER OXIDE and ARSENIC PENTOXIDE and CHROMIUM TRIOXIDE—continued

Product Name	Marketing Company	Use	Reg. No
* Mekure T2 Oxide	Mechema Chemicals Ltd	WI	HSE 4080
Protim CCA Oxide – Type II	Fosroc Ltd	WI	HSE 4463
Protim CCA Oxide 50	Protim Solignum Ltd	WI	HSE 5537
Protim CCA Oxide 58	Protim Solignum Ltd	WI	HSE 5686
Protim CCA Oxide 72	Protim Solignum Ltd	WI	HSE 5594
Tanalith 3302	Hickson Timber Products Ltd	WI	HSE 5098
Tanalith 3313	Hickson Timber Products Ltd	WI	HSE 4422
Tanalith C3310	Hickson Timber Products Ltd	WI	HSE 4669
Tanalith Oxide C3309	Hickson Timber Products Ltd	WI	HSE 4668
Tanalith Oxide C3314	Hickson Timber Products Ltd	WI	HSE 4774

80 COPPER OXIDE and BORIC ACID

Product Name	Marketing Company	Use	Reg. No
Tanalith 3422	Hickson Timber Products Ltd	WI	HSE 4403

81 COPPER OXIDE and BORIC ACID and CHROMIUM TRIOXIDE

Product Name	Marketing Company	Use	Reg. No
Tanalith (3419) CBC	Hickson Timber Products Ltd	WI	HSE 4022

82 COPPER OXIDE and DISODIUM OCTABORATE and CHROMIUM TRIOXIDE

Product Name	Marketing Company	Use	Reg. No
Celcure CB Paste	Rentokil Ltd	WI	HSE 4537

83 COPPER SALT OF SYNTHETIC CARBOXYLIC ACIDS (C8-C12)

Product Name	Marketing Company	Use	Reg. No
* Cuprinol WP CPTF 3	Cuprinol Ltd	WI	HSE 4694

84 COPPER SULPHATE and BORIC ACID

Product Name	Marketing Company	Use	Reg. No
Ensele 3426	Hickson Timber Products Ltd	WP	HSE 4364
Ensele 3427	Hickson Timber Products Ltd	WA	HSE 5053

85 COPPER SULPHATE and POTASSIUM DICHROMATE and BORIC ACID

Product Name	Marketing Company	Use	Reg. No
Tecca CCB 1	Tecca Ltd	WI	HSE 5584
* Wolmanit CB	BASF (United Kingdom) Ltd	WI	HSE 3108
* Womanit CB-P	BASF (United Kingdom) Ltd	WI	HSE 3110

WOOD PRESERVATIVES

Product Name	Marketing Company	Use	Reg. No

86 COPPER SULPHATE and SODIUM DICHROMATE

* Ensele 3424	Hickson Timber Products Ltd	WI	HSE 4670
Laporte Cut End Preservative	Laporte Wood Preservation Ltd	WI	HSE 3528

87 COPPER SULPHATE and SODIUM DICHROMATE and ARSENIC PENTOXIDE

Celcure A Concentrate	Rentokil Ltd	WI	HSE 5458
Celcure A Fluid 10	Rentokil Ltd	WI	HSE 3764
Celcure A Fluid 6	Rentokil Ltd	WI	HSE 3155
Celcure A Paste	Rentokil Ltd	WI	HSE 4523
Injecta CCA-C	Injecta APS	WI	HSE 4447
Kemira CCA Type BS	Kemira Kemwood AB	WI	HSE 3879
Kemwood CCA Type BS	Laporte Kemwood AB	WI	HSE 5534
Laporte CCA Type 1	Laporte Wood Preservation	WI	HSE 3175
Laporte CCA Type 2	Laporte Wood Preservation	WI	HSE 3176
Laporte Permawood CCA	Laporte Wood Preservation	WI	HSE 3529
* Meksol	Mechema Chemicals Ltd	WI	HSE 3994
* Mekure	Mechema Chemicals Ltd	WI	HSE 4352
* Mekure T2	Mechema Chemicals Ltd	WI	HSE 4350
* Protim CCA Salts Type 2	Fosroc Ltd	WI	HSE 4462
Protim CCA Salts Type 2	Protim Solignum Ltd	WI	HSE 4972
Tanalith 3357	Hickson Timber Products Ltd	WI	HSE 4431
Tanalith CL (3354)	Hickson Timber Products Ltd	WI	HSE 4196
Tanalith CP 3353	Hickson Timber Products Ltd	WI	HSE 4667
Tecca P2	Tecca Ltd	WI	HSE 3958

88 COPPER SULPHATE and SODIUM DICHROMATE and BORIC ACID

Celcure CB90	Rentokil Ltd	WI	HSE 5117
Tanalith CBC Paste 3402	Hickson Timber Products Ltd	WI	HSE 3833
* Wolmanit CB-A	BASF (United Kingdom) Ltd	WI	HSE 3109

89 COPPER SULPHATE and SODIUM DICHROMATE and BORIC ACID and CHROMIUM ACETATE

* Celcure CB Salts	Rentokil Ltd	WI	HSE 4536
Celgard CF	Rentokil Ltd	WI	HSE 4608

WOOD PRESERVATIVES

Product Name	Marketing Company	Use	Reg. No
90 COPPER SULPHATE and SODIUM DICHROMATE and CHROMIUM ACETATE			
Celcure B	Rentokil Ltd	WI	HSE 4541
Celcure O	Rentokil Ltd	WI	HSE 4539
* Mekseal	Mechema Chemicals Ltd	WI	HSE 4351
91 COPPER VERSATATE			
* Cuprinol WP CPTF 2	Cuprinol Ltd	WI	HSE 4695
92 CREOSOTE			
Arborsan 3 Creosote	Lanstar Ltd	WI	HSE 5159
Arborsan 4 Creosote	Lanstar Ltd	WA WI WP	HSE 5160
Arborsan 6 Creosote	Lanstar Ltd	WA WI WP	HSE 5158
B and Q Creosote	B and Q Ltd	WA WP	HSE 5093
* B and Q Creosote	B and Q Ltd	WA	HSE 3526
Barrettine Creosote	Barrettine (Products) Ltd	WA	HSE 3146
Chel Creosote	Chelec Ltd	WA	HSE 4200
Chelec Creosote	Chelec Ltd	WA WP	HSE 5412
Co-op Creosote	Co-operative Wholesale Society Ltd	WA WP	HSE 5073
Creosote	Clarkes Products (Weyhill) Ltd	WA	HSE 3111
Creosote	Fosroc Ltd	WA	HSE 3729
# Creosote	Great Mills (Retail) Ltd	WA	HSE 4071
Creosote	Hilbre Building Chemicals Ltd	WA WP	HSE 5301
Creosote	Langlow Products Ltd	WA	HSE 4034
Creosote	Palace Chemicals Ltd	WA	HSE 3841
* Creosote	Readymix Drypack Ltd	WA	HSE 3608
Creosote	South Western Tar Distilleries Ltd	WP	HSE 4575
Creosote	Spectar Ltd	WA WI WP	HSE 5495
Creosote	Suffolk & Essex Supplies Ltd	WA WP	HSE 3141
Creosote (Light Brown and Dark Brown)	PLA Products Ltd	WA WP	HSE 5121

4

239

WOOD PRESERVATIVES

Product Name	Marketing Company	Use	Reg. No
92 **CREOSOTE**—continued			
Creosote 131C	South Western Tar Distilleries Ltd	WP	HSE 4576
Creosote Blend	Langlow Products Ltd	WA WP	HSE 4392
Creosote Coke Oven Oil	South Western Tar Distilleries Ltd	WP	HSE 4586
Creosote Emulsion	Oakmere Technical Services Ltd	WI	HSE 5345
Creosote MK2	South Western Tar Distilleries Ltd	WP	HSE 4589
DIY Time Creosote	Nurdin and Peacock	WA WP	HSE 5314
Golden Creosote	A-Chem Ltd	WA	HSE 4334
Golden Creosote	C W Wastnage Ltd	WA	HSE 4690
Homebase Creosote	Homebase Ltd	WA WP	HSE 5054
Jewson Creosote	Jewson Ltd	WA WP	HSE 5072
Kalon Creosote	Kalon Group Plc	WA WP	HSE 5496
Lanstar Creosote	Lanstar Coatings Ltd	WA	HSE 4374
Larsen Creosote	Larsen Manufacturing Ltd	WA WI WP	HSE 4930
Leyland Creosote	Leyland Paint Company	WA WP	HSE 5074
McDougall Rose Hi-Life Creosote	McDougall Rose Ltd	WA	HSE 4308
Nitromors Creosote	Kalon Group Plc	WA WP	HSE 5129
Nut Brown Creosote	A-Chem Ltd	WA	HSE 4335
Oakmere Creosote Type 2	Oakmere Technical Services Ltd	WA	HSE 4390
* Payless Creosote	Payless DIY Ltd	WA	HSE 4309
Signpost Creosote	MacPherson Paints Ltd	WA	HSE 4507
Solignum Fencing Fluid	Protim Solignum Ltd	WI	HSE 5016
* Solignum Fencing Fluid	Solignum Ltd	WI	HSE 4318
Solignum Gold	Protim Solignum Ltd	WA	HSE 5024
* Solignum Gold	Solignum Ltd	WA	HSE 4305
Southdown Creosote	C Brewer and Sons Ltd	WA	HSE 4310
# Tesco Dark Brown Creosote	Tesco Ltd	WA WP	HSE 3538
# Tesco Light Brown Creosote	Tesco Ltd	WA WP	HSE 3539
Texas Creosote	Texas Homecare Ltd	WA WP	HSE 5092

WOOD PRESERVATIVES

Product Name	Marketing Company	Use	Reg. No
92 CREOSOTE—continued			
Travis Perkins Creosote	Travis Perkins Trading Company Ltd	WA WP	HSE 4941
Valspar Creosote	Akzo Coatings Plc	WA WP	HSE 5469
Woodman Creosote and Light Brown Creosote	J H Woodman	WA WI WP	HSE 5120
93 CREOSOTE and COAL TAR CREOSOTE			
* Arborsan 4	Lancashire Tar Distillers Ltd	WA	HSE 4453
* Arborsan 6	Lancashire Tar Distillers Ltd	WA	HSE 4452
94 CREOSOTE and TC OIL			
* Arborsan 3	Lancashire Tar Distillers Ltd	WI	HSE 3143
95 CYPERMETHRIN			
Cementone Woodworm Killer	Cementone-Beaver Ltd	WA	HSE 3850
Crown Woodworm Concentrate	Crown Chemicals Ltd	WP	HSE 5470
Cyperguard Woodworm Killer	Safeguard Chemicals Ltd	WP	HSE 3890
Devatern 0.5L	Sorex Ltd	WA	HSE 3874
Devatern 1.0 L	Sorex Ltd	WP	HSE 3876
Devatern EC	Sorex Ltd	WP	HSE 3875
Gainserv 686	Industrial Chemical Company (Preston) Ltd	WA	HSE 3971
Green Range Woodworm Killer	Cementone-Beaver Ltd	WP	HSE 3711
Green Range Woodworm Killer AQ	Cementone-Beaver Ltd	WP	HSE 3688
Hickson Antiborer 3767	Hickson Timber Products Ltd	WI	HSE 3583
Killgerm Woodworm Killer	Killgerm Chemicals Ltd	WP	HSE 3126
Microtech Woodworm Killer AQ	Cementone Beaver Ltd	WP	HSE 5452
* Microtech Woodworm Killer AQ	Cementone-Beaver Ltd	WP	HSE 4481
Microtech Woodworm Killer AQ	Cementone-Beaver Ltd	WP	HSE 5062
# Ness Woodworm Killer	Cementone-Beaver Ltd	WP	HSE 3905
# Ness Woodworm Killer AQ	Cementone-Beaver Ltd	WP	HSE 3906
# Nubex Emulsion Concentrate (Low Odour)	Albright and Wilson Ltd	WP	HSE 3616

WOOD PRESERVATIVES

Product Name	Marketing Company	Use	Reg. No
95 CYPERMETHRIN—continued			
Nubex Emulsion Concentrate C (Low Odour)	Nubex Ltd	WP	HSE 3880
Nubex Emulsion Concentrate C (Low Odour)	Nubex Ltd	WP	HSE 5268
Palace Microfine Insecticide Concentrate	Palace Chemicals Ltd	WA WI WP	HSE 5550
PC-H/3	Peter Cox Preservation	WP	HSE 4408
PCX 12/P Concentrate	Peter Cox Preservation	WP	HSE 4406
PCX-12/P	Peter Cox Preservation	WP	HSE 4407
Protim Insecticidal Emulsion C	Fosroc Ltd	WI WP	HSE 3777
Protim Woodworm Killer C	Fosroc Ltd	WA WI WP	HSE 3778
Water Based Woodworm Killer	Cementone-Beaver Ltd	WA	HSE 4874
Woodworm Killer	Oakmere Technical Services Ltd	WA	HSE 4371
Woodworm Killer	Palace Chemicals Ltd	WA	HSE 3769
96 CYPERMETHRIN and 2-PHENYLPHENOL			
Croda Timber Protector	Croda Hydrocarbons Ltd	WA WP	HSE 5369
97 CYPERMETHRIN and 3-IODO-2-PROPYNYL-N-BUTYL CARBAMATE			
Microtech Dual Purpose AQ	Cementone Beaver Ltd	WP	HSE 4994
Microtech Dual Purpose AQ	Cementone Beaver Ltd	WP	HSE 5453
* Microtech Dual Purpose AQ	Cementone-Beaver Ltd	WP	HSE 4518
Water-Based Wood Preserver (Concentrate)	Cementone-Beaver Ltd	WA WP	HSE 5630
Water-Based Wood Preserver (Concentrate)	Cementone-Beaver Ltd	WA	HSE 5515
Water-Based Wood Preserver	Cementone-Beaver Ltd	WA WP	HSE 4902
98 CYPERMETHRIN and BENZALKONIUM CHLORIDE			
# Timbertreat Multi Purpose Preservative	ESMI Chemicals	WA	HSE 3518
# Timbertreat Wood Preservative – Clear	Vulcanite Ltd	WA	HSE 3186

WOOD PRESERVATIVES

Product Name	Marketing Company	Use	Reg. No

98 CYPERMETHRIN and BENZALKONIUM CHLORIDE—continued

# Timbertreat Woodworm and Dry Rot Killer	ESMI Chemicals Ltd	WA	HSE 3183
# Timbertreat Woodworm and Dry Rot Killer	Vulcanite Ltd	WA	HSE 3184
# Timbertreat Woodworm and Dry Rot Killer – Water Based	ESMI Chemicals Ltd	WA	HSE 3182
# Vulcanite Timbertreat Multi Purpose Preservative	Vulcanite Ltd	WA	HSE 3519
# Vulcanite Timbertreat Wood Preservative Clear	Vulcanite Limited	WA	HSE 3185
# Vulcanite Timbertreat Woodworm and Dry Rot Killer Water Based	Vulcanite Ltd	WA	HSE 3181

99 CYPERMETHRIN and DIALKYLDIMETHYL AMMONIUM CHLORIDE and BENZALKONIUM CHLORIDE

Gainserv 684	Industrial Chemical Company (Preston) Ltd	WA	HSE 3996

100 CYPERMETHRIN and DICHLOFLUANID

Universal Wood Preservative	Oakmere Technical Services Ltd	WA	HSE 4170

101 CYPERMETHRIN and PENTACHLOROPHENOL

# Devatern 1.0 LP	Sorex Ltd	WP	HSE 3877

102 CYPERMETHRIN and TEBUCONAZOLE

Celpruf TZC	Rentokil Ltd	WI	HSE 5722

103 CYPERMETHRIN and TRI(HEXYLENE GLYCOL) BIBORATE

Cementone Multiplus	Cementone-Beaver Ltd	WA WP	HSE 3849
Crown Fungicide Insecticide Concentrate	Crown Chemicals Ltd	WP	HSE 5463
# Cyperguard BC Wood Treatment	Safeguard Chemicals Ltd	WP	HSE 3889
Devatern Wood Preserver	Sorex Ltd	WP	HSE 4073
Ecology Fungicide Insecticide Concentrate	Palace Chemicals Ltd	WP	HSE 3792

WOOD PRESERVATIVES

Product Name	Marketing Company	Use	Reg. No

103 CYPERMETHRIN and TRI(HEXYLENE GLYCOL) BIBORATE—continued

Product Name	Marketing Company	Use	Reg. No
Fungicide Insecticide	Palace Chemicals Ltd	WA	HSE 3795
Gainpaste	Industrial Chemical Company (Preston) Ltd	WA	HSE 4547
Green Range Dual Purpose AQ	Cementone Beaver Ltd	WP	HSE 3725
Green Range Wykamol Plus	Cementone Beaver Ltd	WP	HSE 3731
Killgerm Wood Protector	Killgerm Chemicals Ltd	WP	HSE 3127
Langlow Clear Wood Preserver	Langlow Products Ltd	WA	HSE 3802
# Ness Dual Purpose	Cementone Beaver Ltd	WP	HSE 3768
# Ness Dual Purpose AQ	Cementone Beaver Ltd	WP	HSE 3767
Nubex Emulsion Concentrate CB (Low Odour)	Nubex Ltd	WP	HSE 3637
Nubex Emulsion Concentrate CB (Low Odour)	Nubex Ltd	WP	HSE 5328
Nubex Woodworm All Purpose CB	Nubex Ltd	WP	HSE 3189
Palace Microfine Fungicide Insecticide (Concentrate)	Palace Chemicals Ltd	WI WP	HSE 5421
* Payless Wood Preserver – Clear	Payless DIY Ltd	WA	HSE 3831
PC-H/4	Peter Cox Preservation	WA	HSE 3740
PCX-122	Peter Cox Preservation	WP	HSE 3999
PCX-122 Concentrate	Peter Cox Preservation	WP	HSE 4000
Protim CBC	Fosroc Ltd	WA	HSE 3798
Timbermate Clear Wood Preservative	Lanstar Coatings Ltd	WA	HSE 3787

104 CYPERMETHRIN and ZINC OCTOATE

Product Name	Marketing Company	Use	Reg. No
Palavac Industrial Timber Preservative	Palace Chemicalts Ltd	WI WP	HSE 5494

105 CYPERMETHRIN and ZINC VERSATATE

Product Name	Marketing Company	Use	Reg. No
Dipsar G R	Cementone-Beaver Ltd	WI	HSE 4449
* Green Range Dipsar	Cementone Beaver Ltd	WI	HSE 4186

106 DIALKYLDIMETHYL AMMONIUM CHLORIDE

Product Name	Marketing Company	Use	Reg. No
Mould Inhibitor	Castle Products	WA SA	HSE 3582

Product Name	Marketing Company	Use	Reg. No

107 DIALKYLDIMETHYL AMMONIUM CHLORIDE and 3-IODO-2-PROPYNYL-N-BUTYL CARBAMATE

Hickson NP-1	Hickson Timber Products Ltd	WI	HSE 5401

108 DIAKYLDIMETHYL AMMONIUM CHLORIDE and BENZALKONIUM CHLORIDE

ABL Aqueous Wood Preserver Concentrate 1:9	Advanced Bitumens Ltd	WP	HSE 4199
Aqueous Wood Preserver	Advanced Bitumens Ltd	WA	HSE 4727
Celbrite M	Rentokil Ltd	WI	HSE 4535
Celbronze B	Rentokil Ltd	WP	HSE 4549
Colourfast Protector	Rentokil Ltd	WA	HSE 4609
Sterilising Fluid	Rentokil Ltd	WP SP	HSE 4605

109 DIALKYLDIMETHYL AMMONIUM CHLORIDE and BENZALKONIUM CHLORIDE and CYPERMETHRIN

Gainserv 684	Industrial Chemical Company (Preston) Ltd	WA	HSE 3996

110 DIALKYLDIMETHYL AMMONIUM CHLORIDE and METHYLENE BIS (THIOCYANATE) and BENZALKONIUM CHLORIDE and BORIC ACID

Celbrite MT	Rentokil Ltd	WI	HSE 4550

111 DICHLOFLUANID

ABL Wood Preservative (D)	Advanced Bitumens Ltd	WA	HSE 3744
Cedarwood Protector	Rentokil Ltd	WA	HSE 4587
Cedarwood Special	Cuprinol Ltd	WA WI WP	HSE 4729
Cuprinol Hardwood Basecoat	Cuprinol Ltd	WA WI WP	HSE 5015
* Cuprinol Hardwood Basecoat	Cuprinol Ltd	WI	HSE 4721
Curpinol Hardwood Basecoat Meranti	Cuprinol Ltd	WI	HSE 4707
Cuprinol Preservative-Wood Hardener	Cuprinol Ltd	WA	HSE 3590
Dualprime F	Cuprinol Ltd	WI	HSE 4706
* Exterior Ronseal Satin Wood Finish	Sterling Winthrop Group Ltd	WA	HSE 3081

WOOD PRESERVATIVES

Product Name	Marketing Company	Use	Reg. No
111 DICHLOFLUANID—continued			
Flexarb Timber Coating	MacPherson Paints Ltd	WA	HSE 4861
Hardwood Protector	Rentokil Ltd	WA	HSE 4580
Impra-Color	Glen Products Ltd	WA WP	HSE 5403
International Intertox Blue Peter	International Paint Ltd	WA	HSE 5217
Langlow Wood Preservative	Langlow Products Ltd	WA WI WP	HSE 5548
Langlow Wood Preserver Formulation A	Langlow Products Ltd	WA	HSE 3753
Lister Teak Dressing	Green Brothers Ltd	WA	HSE 4628
New Formula Cedarwood	Cuprinol Ltd	WA	HSE 4699
* Payless Wood Preserver	Payless DIY Ltd	WA	HSE 3830
Permalene Satin Wood Stain	C Stabler Ltd	WA	HSE 3644
* Ronseal Trade High Build Preservative Woodstain	Sterling-Winthrop Group Ltd	WA	HSE 3922
Teak Oil	Rentokil Ltd	WA	HSE 4612
Timberlife	Palace Chemicals Ltd	WA	HSE 4501
Timberlife Extra	Palace Chemicals Ltd	WA WP	HSE 5352
Ultrabond Wood Preservative	Ultrabond Ltd	WA	HSE 4343
Wood Preservative	Langlow Products Ltd	WA	HSE 4600
Wood Preservative	Oakmere Technical Services Ltd	WA	HSE 4370
112 DICHLOFLUANID and ACYPETACS ZINC			
Cuprinol Decorative Wood Preserver	Cuprinol Ltd	WA	HSE 3156
Cuprinol Decorative Wood Preserver Red Cedar	Cuprinol Ltd	WA	HSE 4696
Cuprinol Low Odour Decorative Wood Preserver	Cuprinol Ltd	WA WI WP	HSE 5343
Cuprinol Low Odour Decorative Wood Preserver Red Cedar	Cuprinol Ltd	WA WI WP	HSE 5342
Cuprinol Magnatreat F	Cuprinol Ltd	WI	HSE 3752
Cuprinol Magnatreat XQD	Cuprinol Ltd	WI	HSE 3751
Cuprinol Preservative Base	Cuprinol Ltd	WA WP	HSE 4929

WOOD PRESERVATIVES

Product Name	Marketing Company	Use	Reg. No
113 DICHLOFLUANID and AZACONAZOLE			
Xyladecor Matt U 404	Venilia Ltd	WA	HSE 4445
Xylamon Primer Dipping Stain U 415	Venilia Ltd	WP	HSE 4444
114 DICHLOFLUANID and CYPERMETHRIN			
Universal Wood Preservative	Oakmere Technical Services Ltd	WA	HSE 4170
115 DICHLOFLUANID and FURMECYCLOX			
# Nitromors Timbrene Clear	Nitromors Ltd	WA	HSE 3647
# Xylamon Primer Dipping Stain U411	Kay Metzeler Ltd	WP	HSE 3747
116 DICHLOFLUANID and FURMECYCLOX and GAMMA-HCH			
# Fungol Primer 55	BASF (UK) Ltd	WA WP	HSE 3198
# Wolvac 55	BASF (UK) Ltd	WI	HSE 3199
117 DICHLOFLUANID and FURMECYCLOX and PERMETHRIN			
# Nitromors Timbrene Supreme	Nitromors Ltd	WA	HSE 3646
# Xyladecor Matt U-4010	Kay Metzeler Ltd	WA WI WP	HSE 3140
118 DICHLOFLUANID and FURMECYCLOX and PERMETHRIN and TRIBUTYLTIN OXIDE			
# OS Color WR	Ostermann and Scheiwe GMBH	WA	HSE 3160
119 DICHLOFLUANID and PERMETHRIN			
Impra-Color	Glen Products	WA WP	HSE 3870
120 DICHLOFLUANID and PERMETHRIN and TEBUCONAZOLE			
Impra-Holzschutzgrund (Primer)	Glen Products Ltd	WI WP	HSE 5735
121 DICHLOFLUANID and PERMETHRIN and TRI(HEXYLENE GLYCOL) BIBORATE			
Barrettine New Universal Fluid D	Barrettine Products Ltd	WA	HSE 4004

WOOD PRESERVATIVES

Product Name	Marketing Company	Use	Reg. No
121 DICHLOFLUANID and PERMETHRIN and TRI(HEXYLENE GLYCOL) BIBORATE—continued			
Blackfriars Gold Star Clear	E Parsons and Sons Ltd	WA	HSE 4149
Leyland Timbrene Supreme Environmental Formula	Leyland Paint Company	WA	HSE 4800
Nitromors Timbrene Supreme Environmental Formula	Henkel Home Improvements	WA	HSE 4107
Timbrene Supreme	Kalon Group Plc	WA WP	HSE 4995
122 DICHLOFLUANID and PERMETHRIN and ZINC OCTOATE			
OS Color Wood Stain and Preservative	Ostermann and Scheiwe GmbH	WA	HSE 5531
OS Color WR	Ostermann and Scheiwe GmbH	WA	HSE 5364
123 DICHLOFLUANID and TRI(HEXYLENE GLYCOL) BIBORATE			
ABL Universal Woodworm Killer DB	Advanced Bitumens Ltd	WA	HSE 3858
Barrettine New Wood Preserver	Barrettine Products Ltd	WA	HSE 4038
Blackfriars New Wood Preserver	E Parsons and Sons Ltd	WA	HSE 4150
Double Action Timber Preservative for Doors	International Paint Ltd	WA	HSE 3707
Double Action Wood Preservative	International Paint Ltd	WA	HSE 3706
Nitromors Timbrene Clear Environmental Formula	Henkel Home Improvements	WA	HSE 4106
Ranch Preservative	International Paint Ltd	WA	HSE 4085
* Ronseal Trade Low Build Preservative Woodstain	Sterling-Winthrop Group Ltd	WA	HSE 3959
* Ronseal Trade Preservative Woodstain Basecoat and Colour Harmoniser	Sterling-Winthrop Group Ltd	WA	HSE 3923
Sadolin New Base	Sadolin UK Ltd	WA	HSE 4341
Ultrabond Woodworm Killer	Ultrabond Ltd	WA	HSE 4344
124 DICHLOFLUANID and TRI(HEXYLENE GLYCOL) BIBORATE and ZINC NAPHTHENATE			
# Dulux Exterior Preservative Basecoat	ICI Paints Division	WA	HSE 4035

WOOD PRESERVATIVES

Product Name	Marketing Company	Use	Reg. No

124 DICHLOFLUANID and TRI(HEXYLENE GLYCOL) BIBORATE and ZINC NAPHTHENATE—continued

Product Name	Marketing Company	Use	Reg. No
Weathershield Exterior Preservative Basecoat	ICI Paints Division	WA WP	HSE 5245
Weathershield Exterior Preservative Basecoat	ICI Paints Division	WA	HSE 4036
Weathershield Preservative Primer	ICI Paints Division	WA WP	HSE 5244
Weathershield Preservative Primer	ICI Paints Division	WA	HSE 4013

125 DICHLOFLUANID and TRI(HEXYLENE GLYCOL) BIBORATE and ZINC OCTOATE

Product Name	Marketing Company	Use	Reg. No
Timber Preservative Clear TFP7	Johnstone's Paints Plc	WA	HSE 4016

126 DICHLOFLUANID and TRIBUTYLTIN OXIDE

Product Name	Marketing Company	Use	Reg. No
AA155/00 Industrial Wood Preservative	Becker Paint Ltd	WI	HSE 3099
AA155/03 Industrial Wood Preservative	Becker Paint Ltd	WI	HSE 3100
Celpruf CP Special	Rentokil Ltd	WI	HSE 4582
Hickson Woodex	Hickson Timber Products Ltd	WI	HSE 5405
Industrial Wood Preservative AA 155	Becker-Acroma Ltd	WI	HSE 4629
* Protim Cedar	Fosroc Ltd	WI	HSE 4544
Protim Cedar	Protim Solignum Ltd	WI	HSE 5042
* Protim R Coloured Plus	Fosroc Ltd	WI	HSE 4391
Protim Solignum Softwood Basestain CS	Protim Solignum Ltd	WI	HSE 5647
# Wilko Red Cedar Wood Preserver	Wilkinson Home and Garden Stores	WA WP	HSE 3598
Wood Preservative AA 155/00	Becker-Acroma Ltd	WI	HSE 5308
Wood Preservative AA 155/03	Becker-Acroma Ltd	WI	HSE 5310
Wood Preservative AA155	Becker-Acroma Ltd	WI	HSE 5311
# Woodex Intra	Hygaea Colours and Varnishes Ltd	WA	HSE 3159

127 DICHLOFLUANID and ZINC NAPHTHENATE

Product Name	Marketing Company	Use	Reg. No
Masterstroke Wood Preserver	Akzo-Nobel Ltd	WA	HSE 5689

WOOD PRESERVATIVES

Product Name	Marketing Company	Use	Reg. No
128 DICHLOFLUANID and ZINC OCTOATE			
Aquaseal Wood Preserver	Aquaseal Ltd	WA	HSE 4252
Exterior Preservative Primer	Windeck Paints Ltd	WA	HSE 3857
Febwood Wood Preserver	FEB Ltd	WA	HSE 4255
Great Mills Exterior Preservative Primer	Great Mills (Retail) Ltd	WA	HSE 3704
Texas Exterior Wood Preserver	Texas Homecare	WA	HSE 4128
Timbercare Microporous Exterior Preservative	Manders Paints Ltd	WA WP	HSE 5157
Wilko Exterior System Preservative Primer	Wilkinson Home and Garden Stores	WA	HSE 3686
129 DISODIUM OCTABORATE			
BioKil Boron Paste	Bio-Kil Chemicals Ltd	WP	HSE 3866
Biokil B40 Past Wood Preservative	Biokil Chemicals Ltd	WP	HSE 4161
Biokil Timbor Rods	Biokil Chemicals Ltd	WP	HSE 4502
Boracol 20	Channelwood Preservations Ltd	WP	HSE 4018
Boracol B40	Channelwood Preservations Ltd	WP	HSE 4788
Celgard FP	Rentokil Ltd	WI	HSE 4757
Dry Pin	Window Care Systems Ltd	WI WP	HSE 5144
Pandrol Timbershield Rods	Pandrol International Ltd	WP	HSE 3901
Permadip 9 Concentrate	Permagard Products Ltd	WP	HSE 4160
Protek 9 Star Wood Protection	Proteck Products (Sun Europa) Ltd	WA	HSE 3104
Protek Double 9 Star Wood Protection	Protek Products (Sun Europa) Ltd	WA	HSE 3101
Protek Shedstar	Protek Products (Sun Europa) Ltd	WA	HSE 3103
Protek Wood Protection Fencegrade	Ashby Timber Treatments	WA	HSE 3018
Protek Wood Protection Fencegrade 9:1	Ashby Timber Treatments	WA	HSE 3019
Protek Wood Protection Shed Grade	Ashby Timber Treatments	WA	HSE 3020
Protek Woodstar	Protek Products (Sun Europa) Ltd	WA	HSE 3102
* Protim B10	Fosroc Ltd	WA	HSE 4531

WOOD PRESERVATIVES

Product Name	Marketing Company	Use	Reg. No

129 DISODIUM OCTABORATE—continued

Product Name	Marketing Company	Use	Reg. No
Protim B10	Protim Solignum Ltd	WA WP	HSE 4971
Remtox Borocol 20 Wood Preservative	Remtox Chemicals Ltd	WP	HSE 4094
Remtox Boron Rods	Remtox (Chemicals) Ltd	WP	HSE 5560
Ronseal Wood Preservative Tablets	Roncraft	WA WP	HSE 5271
Ronseal Wood Preservative Tablets	Sterling Roncraft	WA	HSE 4324
Safeguard Antiflame 4050 WD Wood Preservative	Safeguard Chemicals Ltd	WP	HSE 5656
Sovereign Timbor Rod	Sovereign Chemical Industries Ltd	WA	HSE 4627
Super Andy Man Wood Protection	AMB Products Ltd	WA	HSE 3535
Timbertex PI	Marcher Chemicals Ltd	WP	HSE 3611
Timbor	Borax Consolidated Ltd	WI WP	HSE 5687
Timbor	Borax Consolidated Ltd	WI	HSE 3621
Timbor Paste	Rentokil Ltd	WI	HSE 5402

130 DISODIUM OCTABORATE and ALKYLARYLTRIMETHYL AMMONIUM CHLORIDE

Product Name	Marketing Company	Use	Reg. No
Timberglow Concentrate	Cotech Services Ltd	WP	HSE 3078
Timberglow Extra	Cotech Services Ltd	WA	HSE 3077

131 DISODIUM OCTABORATE and BENZALKONIUM CHLORIDE

Product Name	Marketing Company	Use	Reg. No
Boracol 10 Rh	Channelwood Preservations Ltd	WA SA	HSE 4104
Boracol 20 Rh	Channelwood Preservations Ltd	WP	HSE 4019
Boracol B8.5 Rh Mouldicide/Wood Preservative	Channelwood Preservations Ltd	WA SA	HSE 4809
Deepwood 20 Inorganic Boron Wood Preservative	Safeguard Chemicals Ltd	WP	HSE 5514
Deepwood 50 Inorganic Boron Wood Preservative Paste	Safeguard Chemicals Ltd	WP SP	HSE 5596

132 DISODIUM OCTABORATE and CHROMIUM TRIOXIDE and COPPER OXIDE

Product Name	Marketing Company	Use	Reg. No
Celcure CB Paste	Rentokil Ltd	WI	HSE 4537

Product Name	Marketing Company	Use	Reg. No

133 DISODIUM OCTABORATE and COPPER NAPHTHENATE

Product Name	Marketing Company	Use	Reg. No
Bio-Kil SR Pole Wrap	Bio-Kil Chemicals Ltd	WP	HSE 5675

134 DISODIUM OCTABORATE and SODIUM PENTACHLOROPHENOXIDE

Product Name	Marketing Company	Use	Reg. No
Gainserv 140	Industrial Chemical Company (Preston) Ltd	WP	HSE 3963
Gainserv Concentrate	Industrial Chemical Company (Preston) Ltd	WP	HSE 4601
Gainserv Polymeric	Industrial Chemical Company (Preston) Ltd	WP	HSE 3964
* Gallwey ABS	Fosroc Ltd	WI	HSE 4529
Gallwey ABS	Protim Solignum Ltd	WI	HSE 5014
Protim Fentex Green Concentrate	Protim Solignum Ltd	WI	HSE 5122
Protim Fentex Green RFU	Protim Solignum Ltd	WI	HSE 5123
* Protim Fentex M	Fosroc Ltd	WI WP	HSE 3515
Protim Fentex M	Protim Solignum Ltd	WI	HSE 4973
Protim Kleen II	Fosroc Ltd	WI	HSE 3973
Protim Kleen II	Protim Solignum Ltd	WI	HSE 5203

135 DODECYLAMINE LACTATE and DODECYLAMINE SALICYLATE

Product Name	Marketing Company	Use	Reg. No
Environmental Woodrot Treatment	SCI (Building Products)	WI WP SP	HSE 5530
Fungicidal Wash	Polybond Ltd	WA SA	HSE 3601
Fungicidal Wash	Smyth-Morris	WP SP	HSE 3004
MacPherson Anti-Mould Solution	MacPherson Paints Ltd	WP SP	HSE 4791
Metalife Fungicidal Wash	Metalife International Ltd	WA SA	HSE 5406
Mouldrid	S and K Maintenance Products	WA SA	HSE 3671
Nubex WDR	Nubex Ltd	WP SP	HSE 3810
Nuodex 87	Durham Chemicals	WP SP	HSE 4744
Permoglaze Micatex Fungicidal Treatment	AKZO Coatings Plc	WP SP	HSE 5249
* Weathershield Fungicidal Solution	ICI Paints	WA SA	HSE 4918
Weathershield Fungicidal Wash	ICI Paints Division	WA SA	HSE 5134

WOOD PRESERVATIVES

Product Name	Marketing Company	Use	Reg. No

136 DODECYLAMINE LACTATE and DODECYLAMINE SALICYLATE and PERMETHRIN

Kingfisher Timber Paste	Kingfisher Chemicals Ltd	WP	HSE 4327

137 DODECYLAMINE SALICYLATE and DODECYLAMINE LACTATE

Environmental Woodrot Treatment	SCI (Building Products)	WI WP SP	HSE 5530
Fungicidal Wash	Polybond Ltd	WA SA	HSE 3601
Fungicidal Wash	Smyth-Morris	WP SP	HSE 3004
MacPherson Anti-Mould Solution	MacPherson Paints Ltd	WP SP	HSE 4791
Metalife Fungicidal Wash	Metalife International Ltd	WA SA	HSE 5406
Mouldrid	S and K Maintenance Products	WA SA	HSE 3671
Nubex WDR	Nubex Ltd	WP SP	HSE 3810
Nuodex 87	Durham Chemicals	WP SP	HSE 4744
Permoglaze Micatex Fungicidal Treatment	Akzo Coatings Plc	WP SP	HSE 5249
* Weathershield Fungicidal Solution	ICI Paints	WA SA	HSE 4918
Weathershield Fungicidal Wash	ICI Paints Division	WA SA	HSE 5134

138 DODECYLAMINE SALICYLATE and PERMETHRIN and DODECYLAMINE LACTATE

Kingfisher Timber Paste	Kingfisher Chemicals Ltd	WP	HSE 4327

139 FURMECYCLOX and CARBENDAZIM

# Bio Woody	Pan Britannica Industries Ltd	WA	HSE 4638

140 FURMECYCLOX and DICHLOFLUANID

# Nitromors Timbrene Clear	Nitromors Ltd	WA	HSE 3647
# Xylamon Primer Dipping Stain U411	Kay Metzeler Ltd	WP	HSE 3747

141 FURMECYCLOX and GAMMA-HCH and DICHLOFLUANID

# Fungol Primer 55	BASF (UK) Ltd	WA WP	HSE 3198
# Wolvac 55	BASF (UK) Ltd	WI	HSE 3199

WOOD PRESERVATIVES

Product Name	Marketing Company	Use	Reg. No

142 FURMECYCLOX and PERMETHRIN

* Xylamon Brown U1011	Kay Metzeler Ltd	WA	HSE 3746

143 FURMECYCLOX and PERMETHRIN and DICHLOFLUANID

# Nitromors Timbrene Supreme	Nitromors Ltd	WA	HSE 3646
# Xyladecor Matt U-4010	Kay Metzeler Ltd	WA WI WP	HSE 3140

144 FURMECYCLOX and PERMETHRIN and TRIBUTYLTIN OXIDE and DICHLOFLUANID

# OS Color WR	Ostermann and Scheiwe GmbH	WA	HSE 3160

145 GAMMA-HCH

Fumite Lindane Generator Size 10	Octavius Hunt Ltd	WA IA	HSE 4713
Fumite Lindane Generator Size 40	Octavius Hunt Ltd	WA IA	HSE 4714
* Gammexane Smoke Generator No 22	ICI Garden and Professional Products	WP IP	HSE 4822
# Nubex Emulsion Grade L	Nubex Ltd	WP	HSE 3656
# Palace Woodworm Killer	Palace Chemicals Ltd	WA WP	HSE 3586
# PC-I	Peter Cox Preservation Ltd	WA	HSE 3736
Protim Woodworm Killer	Fosroc Ltd	WA	HSE 4545
* Roofguard Smoke Generator	Octavius Hunt Ltd	WA IA	HSE 4712
Wax Polish	Rentokil Ltd	WA	HSE 4465
Woodworm Fluid	Rentokil Ltd	WA	HSE 4603
Woodworm Fluid for Bonded Warehouses and Distilleries	Rentokil Ltd	WP	HSE 4520
Woodworm Fluid Minimum Odour	Rentokil Ltd	WP	HSE 4521
Woodworm Paste	Rentokil Ltd	WP	HSE 4522

146 GAMMA-HCH and 3-IODO-2-PROPYNYL-N-BUTYL CARBAMATE

# Sadolin Sadovac 35	Sadolin (UK) Ltd	WI	HSE 4358

147 GAMMA-HCH and BENZALKONIUM CHLORIDE

Biokil	Jaymar Chemicals	WP	HSE 4247

WOOD PRESERVATIVES

Product Name	Marketing Company	Use	Reg. No

148 GAMMA-HCH and DICHLOFLUANID and FURMECYCLOX

# Fungol Primer 55	BASF (UK) Ltd	WA WP	HSE 3198
# Wolvac 55	BASF (UK) Ltd	WI	HSE 3199

149 GAMMA-HCH and PENTACHLOROPHENOL

# Ness Mayonnaise Wood Treatment	Cementone-Beaver Ltd	WP	HSE 3907
Protim 200C	Fosroc Ltd	WI	HSE 3670
Supergrade Wood Preserver	Rentokil Ltd	WP	HSE 4641
Supergrade Wood Preserver Black	Rentokil Ltd	WP	HSE 3713
Supergrade Wood Preserver Brown	Rentokil Ltd	WP	HSE 3715
Supergrade Wood Preserver Clear	Rentokil Ltd	WP	HSE 3716
Water Repellent Pink Primer	Rentokil Ltd	WP	HSE 4551
Woodtreat	Stanhope Chemical Products Ltd	WP	HSE 4475

150 GAMMA-HCH and PENTACHLOROPHENOL and TRIBUTYLTIN OXIDE

Celpruf PK	Rentokil Ltd	WI	HSE 4540
Celpruf PK WR	Rentokil Ltd	WI	HSE 4538
Lar-Vac 100	Larsen Manufacturing Ltd	WI	HSE 3723
# Preservative For Wood Black	Rentokil Ltd	WA	HSE 3714
* Protim 80	Fosroc Ltd	WI	HSE 3550
Protim 80	Protim Solignum Ltd	WI	HSE 5029
* Protim 80 C	Fosroc Ltd	WI	HSE 3548
Protim 80 C	Protim Solignum Ltd	WI	HSE 5036
* Protim 80 CWR	Fosroc Ltd	WI	HSE 3547
Protim 80 CWR	Protim Solignum Ltd	WI	HSE 5031
Protim 80 Oil Brown	Fosroc Ltd	WI	HSE 3516
* Protim 80 WR	Fosroc Ltd	WI	HSE 3549
Protim 80 WR	Protim Solignum Ltd	WI	HSE 5030
Protim 80C Oil Brown	Fosroc Ltd	WI	HSE 3517
* Protim Brown	Fosroc Ltd	WI	HSE 3551
Protim Brown	Protim Solignum Ltd	WI	HSE 5000
Protim Grade Basic	Fosroc Ltd	WI	HSE 3544
Protim Grade Basic C	Fosroc Ltd	WI	HSE 3543

WOOD PRESERVATIVES

Product Name	Marketing Company	Use	Reg. No
150 GAMMA-HCH and PENTACHLOROPHENOL and TRIBUTYLTIN OXIDE—continued			
Protim Grade Basic CWR	Fosroc Ltd	WI	HSE 3545
Protim Grade Basic WR	Fosroc Ltd	WI	HSE 3546
# Wilko Clear Wood Preserver	Wilkinson Home and Garden Stores	WA WP	HSE 3599
151 GAMMA-HCH and PENTACHLOROPHENOL and ZINC NAPHTHENATE			
Protim 200	Fosroc Ltd	WI	HSE 3669
152 GAMMA-HCH and PENTACHLOROPHENYL LAURATE			
Brunol ATP	Stanhope Chemical Products Ltd	WP	HSE 4498
Mystox BTL	Stanhope Chemical Products Ltd	WP	HSE 4425
Mystox BTV	Stanhope Chemical Products Ltd	WP	HSE 4424
* PLA Products Woodworm Killer	PLA Products (Chemicals) Ltd	WP	HSE 4634
Protim WB12	Fosroc Ltd	WI WP	HSE 3138
153 GAMMA-HCH and PENTACHLOROPHENYL LAURATE and TRIBUTYLTIN OXIDE			
# Larsen Clear Wood Preservative	Larsen Manufacturing Ltd	WA	HSE 3771
154 GAMMA-HCH and TRI(HEXYLENE GLYCOL) BIBORATE			
Cube F and I Concentrate/Grade BL	Kleeneze Ltd	WP	HSE 3095
Cube Solvent F and I Fluid/Grade BL	Kleeneze Ltd	WA	HSE 3094
# Nubex Emulsion LB (Low Odour)	Nubex Ltd	WP	HSE 3664
155 GAMMA-HCH and TRI(HEXYLENE GLYCOL) BIBORATE and TRIBUTYLTIN OXIDE			
# Timber Preservative Solvent Based Fungicidal Insecticide 30	Ambersil Ltd	WP	HSE 3006

WOOD PRESERVATIVES

Product Name	Marketing Company	Use	Reg. No

156 GAMMA-HCH and TRIBUTYLTIN OXIDE

Product Name	Marketing Company	Use	Reg. No
# Aqueous Fungicide—Insecticide Concentrate	Palace Chemicals Ltd	WP	HSE 3129
# Aqueous Fungicide-Insecticide	Palace Chemicals Ltd	WA WP	HSE 3128
Cedasol 2304 RTU	Hickson Timber Products Ltd	WI	HSE 3151
Cedasol 2320	Hickson Timber Products Ltd	WI	HSE 4365
# Cleartreat	South Bucks Estates Ltd	WP	HSE 3131
# Decor-8 Wood Preservative Brown	Decor-8 Ltd	WA WP	HSE 3063
# Decor-8 Wood Preservative Clear	Decor-8 Ltd	WA WP	HSE 3066
# Decor-8 Wood Preservative Red Cedar	Decor-8 Ltd	WA WP	HSE 3064
Dipsar	Cementone Beaver Ltd	WI	HSE 3604
# Fads Wood Preservative Brown	Fads AG Stanley Ltd	WA WP	HSE 3035
# Fads Wood Preservative Clear	Fads AG Stanley Ltd	WA WP	HSE 3038
# Fads Wood Preservative Red Cedar	Fads AG Stanley Ltd	WA WP	HSE 3036
# Fungicide Insecticide Wood Preservative	Palace Chemicals Ltd	WA	HSE 3587
# Great Mills Wood Preservative Brown	Great Mills	WA WP	HSE 3039
# Great Mills Wood Preservative Clear	Great Mills	WA WP	HSE 3042
# Great Mills Wood Preservative Red Cedar	Great Mills	WA WP	HSE 3040
Hickson Timbercare WRQD	Hickson Timber Products Ltd	WI	HSE 3594
# Homecharm Wood Preservative Brown	Homecharm Ltd	WA WP	HSE 3059
# Homecharm Wood Preservative Clear	Homecharm Ltd	WA WP	HSE 3062
# Homecharm Wood Preservative Red Cedar	Homecharm Ltd	WA WP	HSE 3060
# Kingston Dual Purpose Fluid	Kingston Chemicals	WP	HSE 3641
# Laporte Permatreat	Laporte Wood Preservation	WI	HSE 3699
# Laporte Permatreat Cut End Preservative	Laporte Wood Preservation	WI	HSE 3700
# Laporte Permatreat With Water Repellant	Laporte Wood Preservation	WI	HSE 3701
# Larsen Brown Wood Preservative	Larsen Manufacturing Ltd	WA	HSE 3772
# Nubex Standard Emulsion Concentrate LT	Nubex Ltd	WP	HSE 3653

156 GAMMA-HCH and TRIBUTYLTIN OXIDE—continued

Product Name	Marketing Company	Use	Reg. No
# Nubex Woodworm All Purpose LT	Nubex Ltd	WA	HSE 3655
# Payless Wood Preservative Brown	Payless DIY Ltd	WA WP	HSE 3047
# Payless Wood Preservative Clear	Payless DIY Ltd	WA WP	HSE 3050
# Payless Wood Preservative Red Cedar	Payless DIY Ltd	WA WP	HSE 3048
# Pob Wood Preservative Brown	Pob Savident Ltd	WA WP	HSE 3023
# Pob Wood Preservative Clear	Pob Savident Ltd	WA WP	HSE 3026
# Pob Wood Preservative Red Cedar	Pob Savident Ltd	WA WP	HSE 3024
* Protim 210	Fosroc Ltd	WI	HSE 3755
Protim 210	Protim Solignum Ltd	WI	HSE 5033
* Protim 210 C	Fosroc Ltd	WI	HSE 3757
Protim 210 C	Protim Solignum Ltd	WI	HSE 5038
* Protim 210 CWR	Fosroc Ltd	WI	HSE 3758
Protim 210 CWR	Protim Solginum Ltd	WI	HSE 5035
* Protim 210 WR	Fosroc Ltd	WI	HSE 3756
Protim 210 WR	Protim Solignum Ltd	WI	HSE 5039
* Protim 23 WR	Fosroc Ltd	WI	HSE 3869
Protim 23 WR	Protim Solignum Ltd	WI	HSE 5037
Protim 90	Fosroc Ltd	WI	HSE 4564
Protim 90	Protim Solignum Ltd	WI	HSE 5276
* Protim Paste	Fosroc Ltd	WP	HSE 4748
Protim Paste	Protim Solignum Ltd	WP	HSE 4979
Protim TWR	Fosroc Ltd	WI	HSE 3542
# SPL Wood Preservative Brown	Silver Paint and Lacquer Ltd	WA WP	HSE 3055
# SPL Wood Preservative Clear	Silver Paint and Lacquer Ltd	WA WP	HSE 3058
# SPL Wood Preservative Red Cedar	Silver Paint and Lacquer Ltd	WA WP	HSE 3056
# Texas Wood Preservative Brown	Texas Homecare Ltd	WA WP	HSE 3031
# Texas Wood Preservative Red Cedar	Texas Homecare Ltd	WA WP	HSE 3032
# Timber Preservative Solvent Based Fungicidal Insecticide 10	Ambersil Ltd	WP	HSE 3005
# Timber Treat	Palace Chemicals Ltd	WP	HSE 3570
# Trend Wood Preservative Brown	Trend Voluntary Group Services Ltd	WA WP	HSE 3043
# Trend Wood Preservative Clear	Trend Voluntary Group Services Ltd	WA WP	HSE 3046

WOOD PRESERVATIVES

Product Name	Marketing Company	Use	Reg. No

156 GAMMA-HCH and TRIBUTYLTIN OXIDE—continued

Product Name	Marketing Company	Use	Reg. No
# Trend Wood Preservative Red Cedar	Trend Voluntary Group Services Ltd	WA WP	HSE 3044
Vacsol MWR Concentrate 2203	Hickson Timber Products Ltd	WI	HSE 3170
Vacsol MWR Ready To Use 2204	Hickson Timber Products Ltd	WI	HSE 3167
Vacsol P 2304 RTU	Hickson Timber Products Ltd	WI	HSE 4597
Vacsol WR Concentrate 2115	Hickson Timber Products Ltd	WI	HSE 3169
Vacsol WR Ready To Use 2116	Hickson Timber Products Ltd	WI	HSE 3168
# Wickes Wood Preservative Brown	Wickes Building Supplies Ltd	WA WP	HSE 3027
# Wickes Wood Preservative Clear	Wickes Building Supplies Ltd	WA	HSE 3030
# Wickes Wood Preservative Red Cedar	Wickes Building Supplies Ltd	WA WP	HSE 3028
# Wilco Wood Preservative Brown	Wilkinson Group of Companies	WA WP	HSE 3051
# Wilco Wood Preservative Clear	Wilkinson Group of Companies	WA WP	HSE 3054
# Wilco Wood Preservative Red Cedar	Wilkinson Group of Companies	WA WP	HSE 3052

157 GAMMA-HCH and TRIBUTYLTIN OXIDE and COPPER NAPHTHENATE

Product Name	Marketing Company	Use	Reg. No
# Decor-8 Wood Preservative Green	Decor-8 Ltd	WA WP	HSE 3065
# Fads Wood Preservative Green	Fads AG Stanley Ltd	WA WP	HSE 3037
# Great Mills Wood Preservative Green	Great Mills	WA WP	HSE 3041
# Homecharm Wood Preservative Green	Homecharm Ltd	WA WP	HSE 3061
# Payless Wood Preservative Green	Payless DIY Ltd	WA WP	HSE 3049
# Pob Wood Preservative Green	Pob Savident Ltd	WA WP	HSE 3025
# SPL Wood Preservative Green	Silver Paint and Lacquer Ltd	WA WP	HSE 3057
# Texas Wood Preservative Green	Texas Homecare Ltd	WA WP	HSE 3033
# Trend Wood Preservative Green	Trend Voluntary Group Services Ltd	WA WP	HSE 3045
# Wickes Wood Preservative Green	Wickes Building Supplies Ltd	WA WP	HSE 3029
# Wilco Wood Preservative Green	Wilkinson Group of Companies	WA WP	HSE 3053

WOOD PRESERVATIVES

Product Name	Marketing Company	Use	Reg. No

158 METHYLENE BIS (THIOCYANATE)

# Basiment 540	Kay Metzeler Ltd	WI	HSE 3615
Timbercol Preservative (MBT) Concentrate	Timbercol Industries Ltd	WI	HSE 3892
Timbertone S7 Preservative	Timbertone	WI	HSE 4118

159 METHYLENE BIS (THIOCYANATE) and 2-(THIOCYANOMETHYLTHIO) BENZOTHIAZOLE

* Aquatect Wood Preservative Stain	Aqua Coatings	WI	HSE 3566
Busan 1009	Buckman Laboratories SA	WI	HSE 3887
Busan 1009	Buckman Laboratories SA	WI	HSE 5256
Celbrite TC	Rentokil Ltd	WI	HSE 3921
* Gallwey BMC	Fosroc Ltd	WI	HSE 3534
Hickson Antiblu 3739	Hickson Timber Products Ltd	WI	HSE 4841
Mect	Buckman Laboratories SA	WI	HSE 3888
Mect	Buckman Laboratories SA	WI	HSE 5255
* Protek Blue	Ashby Timber Treatments	WI	HSE 3017
* Protek Wood Protection (FT Grade 20:1)	Ashby Timber Treatments	WI	HSE 3016
* Protek Wood Protection (FT Grade 9:1)	Ashby Timber Treatments	WI	HSE 3015
* Protim Stainguard	Fosroc Ltd	WI	HSE 3533
Protim Stainguard	Protim Solignum Ltd	WI	HSE 5009
Timbercol Preservative Concentrate	Timbercol Industries Ltd	WI	HSE 3124

160 METHYLENE BIS (THIOCYANATE) and 2-(THIOCYANOMETHYLTHIO) BENZOTHIAZOLE and BORIC ACID

Fentex Elite	Protim Solignum Ltd	WI	HSE 5532

161 METHYLENE BIS (THIOCYANATE) and BENZALKONIUM CHLORIDE and BORIC ACID and DIALKYLDIMETHYL AMMONIUM CHLORIDE

Celbrite MT	Rentokil Ltd	WI	HSE 4550

162 METHYLENE BIS (THIOCYANATE) and BORIC ACID

Hickson Antiblu 3737	Hickson Timber Products Ltd	WI	HSE 4745
Hickson Antiblu 3738	Hickson Timber Products Ltd	WI	HSE 4746

Product Name	Marketing Company	Use	Reg. No

163 OXINE-COPPER

Mitrol PQ 8	Kemira Kemwood AB	WI	HSE 4023
Mitrol PQ 8	Laporte Kemwood AB	WI	HSE 5533

164 PENTACHLOROPHENOL

# Dry Rot and Wet Rot Fluid	Rentokil Ltd	WA SA	HSE 4604
Gainserv 680	Industrial Chemical Company (Preston) Ltd	WP	HSE 3993
Gainserve 682	Industrial Chemical Company (Preston) Ltd	WP	HSE 3956
Gainserv 683	Industrial Chemical Company (Preston) Ltd	WP	HSE 3957
# Pro AM Timbertreat Brown	Fosroc Ltd	WA	HSE 3580
Protim Exterior Brown	Fosroc Ltd	WP	HSE 4530
Protim GC	Fosroc Ltd	WI	HSE 4552
Protim GC	Protim Solignum Ltd	WI	HSE 5202

165 PENTACHLOROPHENOL and CYPERMETHRIN

# Devatern 1.0 LP	Sorex Ltd	WP	HSE 3877

166 PENTACHLOROPHENOL and GAMMA-HCH

# Ness Mayonnaise Wood Treatment	Cementone-Beaver Ltd	WP	HSE 3907
Protim 200C	Fosroc Ltd	WI	HSE 3670
Supergrade Wood Preserver	Rentokil Ltd	WP	HSE 4641
Supergrade Wood Preserver Black	Rentokil Ltd	WP	HSE 3713
Supergrade Wood Preserver Brown	Rentokil Ltd	WP	HSE 3715
Supergrade Wood Preserver Clear	Rentokil Ltd	WP	HSE 3716
Water Repellent Pink Primer	Rentokil Ltd	WP	HSE 4551
Woodtreat	Stanhope Chemical Products Ltd	WP	HSE 4475

167 PENTACHLOROPHENOL and TRIBUTYLTIN OXIDE

Celpruf JP	Rentokil Ltd	WI	HSE 4548
Celpruf JP WR	Rentokil Ltd	WI	HSE 4543
* Protim FDR-H	Fosroc Ltd	WI	HSE 4438

WOOD PRESERVATIVES

Product Name	Marketing Company	Use	Reg. No
167 **PENTACHLOROPHENOL and TRIBUTYLTIN OXIDE**—continued			
Protim FDR-H	Protim Solignum Ltd	WI	HSE 4985
* Protim Joinery Lining	Fosroc Ltd	WP	HSE 4426
Protim Joinery Lining	Protim Solignum Ltd	WP	HSE 4977
* Protim JP	Fosroc Ltd	WI	HSE 4326
Protim JP	Protim Solignum Ltd	WI	HSE 4978
* Protim R Clear	Fosroc Ltd	WI	HSE 3541
Protim R Clear	Protim Solignum Ltd	WI	HSE 5013
168 **PENTACHLOROPHENOL and TRIBUTYLTIN OXIDE and GAMMA-HCH**			
Celpruf PK	Rentokil Ltd	WI	HSE 4540
Celpruf PK WR	Rentokil Ltd	WI	HSE 4538
Lar-Vac 100	Larsen Manufacturing Ltd	WI	HSE 3723
# Preservative for Wood Black	Rentokil Ltd	WA	HSE 3714
* Protim 80	Fosroc Ltd	WI	HSE 3550
Protim 80	Protim Solignum Ltd	WI	HSE 5029
* Protim 80 C	Fosroc Ltd	WI	HSE 3548
Protim 80 C	Protim Solignum Ltd	WI	HSE 5036
* Protim 80 CWR	Fosroc Ltd	WI	HSE 3547
Protim 80 CWR	Protim Solignum Ltd	WI	HSE 5031
Protim 80 Oil Brown	Fosroc Ltd	WI	HSE 3516
* Protim 80 WR	Fosroc Ltd	WI	HSE 3549
Protim 80 WR	Protim Solignum Ltd	WI	HSE 5030
Protim 80 C Oil Brown	Fosroc Ltd	WI	HSE 3517
* Protim Brown	Fosroc Ltd	WI	HSE 3551
Protim Brown	Protim Solignum Ltd	WI	HSE 5000
Protim Grade Basic	Fosroc Ltd	WI	HSE 3544
Protim Grade Basic C	Fosroc Ltd	WI	HSE 3543
Protim Grade Basic CWR	Fosroc Ltd	WI	HSE 3545
Protim Grade Basic WR	Fosroc Ltd	WI	HSE 3546
# Wilko Clear Wood Preserver	Wilkinson Home and Garden Stores	WA WP	HSE 3599
169 **PENTACHLOROPHENOL and ZINC NAPHTHENATE and GAMMA-HCH**			
Protim 200	Fosroc Ltd	WI	HSE 3669

WOOD PRESERVATIVES

Product Name	Marketing Company	Use	Reg. No

170 PENTACHLOROPHENYL LAURATE

# Blackfriar Wood Preserver	E Parsons and Sons Ltd	WA	HSE 3998
# Chel Wood Preserver	Chelec Ltd	WA	HSE 3745
# Chel Wood Preserver/Woodworm Dry Rot Killer	Chelec Ltd	WA	HSE 3068
# Timberplus	Nitromors Ltd	WA	HSE 3605

171 PENTACHLOROPHENYL LAURATE and GAMMA-HCH

Brunol ATP	Stanhope Chemical Products Ltd	WP	HSE 4498
Mystox BTL	Stanhope Chemical Products Ltd	WP	HSE 4425
Mystox BTV	Stanhope Chemical Products Ltd	WP	HSE 4424
* PLA Products Woodworm Killer	PLA Products (Chemicals) Ltd	WP	HSE 4634
Protim WB12	Fosroc Ltd	WI WP	HSE 3138

172 PENTACHLOROPHENYL LAURATE and TRIBUTYLTIN OXIDE

* Larsen Joinery Grade	Larsen Manufacturing Ltd	WI	HSE 3709

173 PENTACHLOROPHENYL LAURATE and TRIBUTYLTIN OXIDE and GAMMA-HCH

# Larsen Clear Wood Preservative	Larsen Manufacturing Ltd	WA	HSE 3771

174 PERMETHRIN

Antel Woodworm Killer (P) Concentrate (Water Dilutable)	Antel-Tridex	WP	HSE 4044
# Aquakill Ecology	Rentacure	WP	HSE 3881
Barrettine New Woodworm Fluid	Barrettine Products Ltd	WA WI WP	HSE 3622
Barrettine New Woodworm Killer	Barrettine Products Ltd	WA	HSE 4005
Brunol PC	Stanhope Chemical Products Ltd	WP	HSE 3863
Brunol PY	Stanhope Chemical Products Ltd	WP	HSE 4476
Cuprinol Insecticidal Emulsion Concentrate	Cuprinol Ltd	WP	HSE 4709

WOOD PRESERVATIVES

Product Name	Marketing Company	Use	Reg. No
174 **PERMETHRIN**—continued			
Cuprinol Low Odour Woodworm Killer	Cuprinol Ltd	WA WI WP	HSE 5449
Cuprinol Woodworm Killer S	Cuprinol Ltd	WA	HSE 4693
Cuprinol Woodworm Killer S (Aerosol)	Cuprinol Ltd	WA	HSE 4700
Deepkill I	Sovereign Chemical Industries Ltd	WI WP	HSE 5239
Deepwood "Clear" Insecticide Emulsion Concentrate	Safeguard Chemicals Ltd	WI WP	HSE 4950
Deepwood 8 Micro Emulsifiable Insecticide Concentrate	Safeguard Chemicals Ltd	WI WP	HSE 5664
Deepwood Standard Emulsion Concentrate	Safeguard Chemicals Ltd	WP	HSE 5665
Environmental Timber Treatment	SCI (Building Products)	WP	HSE 5451
Environmental Timber Treatment Paste	SCI (Building Products)	WP	HSE 5529
* Heritage Woodworm Killer	Heritage Preservation	WP	HSE 4171
Hickson Antiborer 3768	Hickson Timber Products Ltd	WI	HSE 3149
Impra-Sanol	Glen Products Ltd	WA	HSE 4325
Leyland Timbrene Woodworm Killer	Leyland Paint Company	WA WP	HSE 4825
Low Odour Woodworm Killer	Cuprinol Ltd	WA WP	HSE 5435
Microguard Permethrin Concentrate	Permagard Products Ltd	WI WP	HSE 5102
Microguard Woodworm Fluid	Permagard Products Ltd	WA WI WP	HSE 5103
* Nitromors Timbrene Woodworm Killer	Nitromors Ltd	WA	HSE 3645
Permethrin WW Conc	Permagard Products Ltd	WP	HSE 4020
# Permolite Low Odour Concentrate P	Nubex Ltd	WP	HSE 3667
Protim AQ	Fosroc Ltd	WI	HSE 3650
Protim AQ	Protim Solignum Ltd	WI	HSE 5201
Protim Aquachem-Insecticidal Emulsion P	Protim Solignum Ltd	WA WP	HSE 5161
Protim Insecticidal Emulsion 8	Protim Solignum Ltd	WI WP	HSE 5701
* Protim Insecticidal Emulsion P	Fosroc Ltd	WP	HSE 3775
Protim Insecticidal Emulsion P	Protim Solignum Ltd	WI WP	HSE 5008
* Protim Paste P	Fosroc Ltd	WA WP	HSE 3838
Protim Paste P	Protim Solignum Ltd	WA WP	HSE 4980

WOOD PRESERVATIVES

Product Name	Marketing Company	Use	Reg. No
174 **PERMETHRIN**—continued			
* Protim Woodworm Killer P	Fosroc Ltd	WA	HSE 4594
Protim Woodworm Killer P	Protim Solignum Ltd	WA WI WP	HSE 5011
Remecology Insecticide R8	Remtox (Chemicals) Ltd	WP	HSE 4233
Remecology Spirit Based Insecticide R6	Remtox Chemicals Ltd	WP	HSE 4836
Remtox Insecticide Microemulsion M8	Remtox (Chemicals) Ltd	WI WP	HSE 5325
Remtox Insecticide Microemulsion M8	Remtox (Chemicals) Ltd	WI WP	HSE 5710
Remtox Insecticide Paste R4	Remtox (Chemicals) Ltd	WI WP	HSE 5389
Remox Microactive Insecticide W8	Remtox (Chemicals) Ltd	WP	HSE 5555
Rentokil Woodworm Killer	Rentokil Ltd	WA	HSE 4208
Rentokil Woodworm Treatment	Rentokil Ltd	WA	HSE 3911
Safeguard Deepwood 1 Insecticide Emulsion Concentrate	Safeguard Chemicals Ltd	WP	HSE 4132
Safeguard Deepwood I Insecticide Emulsion Concentrate	Safeguard Chemicals Ltd	WP	HSE 5304
Safeguard Deepwood IV Woodworm Killer	Safeguard Chemicals Ltd	WP	HSE 3946
Safeguard Woodworm Killer	Safeguard Chemicals Ltd	WP	HSE 4243
Solignum Woodworm Killer	Protim Solignum Ltd	WA WI WP	HSE 5488
Solignum Woodworm Killer	Protim Solignum Ltd	WA	HSE 5002
* Solignum Woodworm Killer	Solignum Ltd	WA	HSE 3822
Solignum Woodworm Killer Concentrate	Protim Solignum Ltd	WI WP	HSE 5466
Solignum Woodworm Killer Concentrate P	Protim Solignum Ltd	WP	HSE 5004
* Solignum Woodworm Killer Concentrate P	Solignum Ltd	WP	HSE 4307
Solignum Woodworm Killer Trade	Protim Solignum Ltd	WI WP	HSE 5003
* Solignum Woodworm Killer Trade	Solignum Ltd	WP	HSE 3823
Sov AQ Micro I	Sovereign Chemical Industries Ltd	WI WP	HSE 5220
Sovac I	Sovereign Chemical Industries Ltd	WI WP	HSE 5230
Sovaq Micro I	Sovereign Chemical Industries Ltd	WI WP	HSE 5663

WOOD PRESERVATIVES

Product Name	Marketing Company	Use	Reg. No
174 **PERMETHRIN**—continued			
Sovereign Aqueous Insecticide	Sovereign Chemical Industries Ltd	WP	HSE 4620
Timbrene Woodworm Killer	Kalon Group Plc	WA WP	HSE 4996
Trimethrin 20S	Triton Chemical Manufacturing Company Ltd	WA	HSE 4621
Trimethrin 2AQ	Triton Chemical Manufacturing Company Ltd	WP	HSE 4625
Trimethrin 3AQ	Triton Chemical Manufacturing Company Ltd	WP	HSE 4626
Trimethrin 3OS	Triton Chemical Manufacturing Company Ltd	WA	HSE 4622
Tritec	Triton Chemical Manufacturing Company Ltd	WP	HSE 5424
Wickes Woodworm Killer	Wickes Building Supplies Ltd	WA WP	HSE 5692
Woodworm Fluid (B) EC	Rentokil Ltd	WP	HSE 4213
Woodworm Fluid (B) Emulsion Concentrate	Rentokil Ltd	WP	HSE 4726
Woodworm Fluid (B) for Bonded Warehouses	Rentokil Ltd	WP	HSE 4725
Woodworm Fluid B	Rentokil Ltd	WP	HSE 4743
# Woodworm Fluid FP	Rentokil Ltd	WP	HSE 4250
Woodworm Fluid FP	Rentokil Ltd	WP	HSE 4577
Woodworm Fluid Z Emulsion Concentrate	Rentokil Ltd	WP	HSE 4264
Woodworm Furniture Polish	Rentokil Ltd	WA	HSE 3954
Woodworm Killer	Langlow Products Ltd	WA	HSE 3691
Woodworm Killer	Protim Solignum Ltd	WA WI WP	HSE 5487
Woodworm Killer (Grade P)	Barrettine Products Ltd	WA WI WP	HSE 5213
Woodworm Paste B	Rentokil Ltd	WP	HSE 4578
Woodworm Roof Void Paste	Rentokil Ltd	WA	HSE 3765
Woodworm Treatment Spray	Rentokil Ltd	WA	HSE 5619
175 **PERMETHRIN and 2-PHENYLPHENOL**			
Brunol OPA	Stanhope Chemical Products Ltd	WP	HSE 4473

176 PERMETHRIN and 3-IODO-2-PROPYNYL-N-BUTYL CARBAMATE

Product Name	Marketing Company	Use	Reg. No
Brunol PP	Stanhope Chemical Products Ltd	WI WP	HSE 5414
Brunol SPI	Stanhope Chemical Products Ltd	WI WP	HSE 5415
Deepkill	Sovereign Chemical Industries Ltd	WA	HSE 4226
Deepkill	Sovereign Chemical Industries Ltd	WI WP	HSE 5238
Fungicide/Insecticide Microemulsion M9	Remtox (Chemicals) Ltd	WI WP	HSE 5711
Protim 250	Protim Solignum Ltd	WI	HSE 5070
Protim 250 WR	Protim Solignum Ltd	WI	HSE 5071
Remecology Timber Preservative Paste	Remtox (Chemicals) Ltd	WA	HSE 4416
Remtox AQ Fungicide/Insecticide R9	Remtox (Chemicals) Ltd	WI WP	HSE 5370
Remtox Dual Purpose Paste K9	Remtox (Chemicals) Ltd	WI WP	HSE 5399
Remtox Fungicide/Insecticide Microemulsion M9	Remtox (Chemicals) Ltd	WI WP	HSE 5360
Remtox Spirit Based F/I K7	Remtox (Chemicals) Ltd	WI WP	HSE 5391
Sadolin Base No 561-2611	Sadolin (UK) Ltd	WA	HSE 4330
Sadolin Base No 561-2611	Sadolin Nobel UK Ltd	WA WP	HSE 5649
Sov AQ Micro F/I	Sovereign Chemical Industries Ltd	WI WP	HSE 5222
Sovac F/I	Sovereign Chemical Industries Ltd	WI WP	HSE 5232
Sovac F/I WR	Sovereign Chemical Industries Ltd	WI WP	HSE 5300
Sovaq Micro F/I	Sovereign Chemical Industries Ltd	WI WP	HSE 5669
Sovereign AQ F/I	Sovereign Chemical Industries Ltd	WI WP	HSE 5221
Tripaste PP	Triton Chemical Manufacturing Company Ltd	WA	HSE 4595
Woodtreat 25	Stanhope Chemical Products Ltd	WP	HSE 5413

WOOD PRESERVATIVES

Product Name	Marketing Company	Use	Reg. No

177 PERMETHRIN and ACYPETACS COPPER and ACYPETACS ZINC

Product Name	Marketing Company	Use	Reg. No
Cuprisol P	Cuprinol Ltd	WI	HSE 3634
Protim 800P	Protim Solignum Ltd	WI	HSE 5702

178 PERMETHRIN and ACYPETACS ZINC

Product Name	Marketing Company	Use	Reg. No
# Cedasol Ready To Use (2310)	Hickson Timber Products Ltd	WI	HSE 3975
Cuprinol 5 Star Complete Wood Treatment S	Cuprinol Ltd	WA	HSE 4710
Cuprinol Combination Grade S	Cuprinol Ltd	WA	HSE 3588
Cuprinol Low Odour 5 Star Complete Wood Treatment	Cuprinol Ltd	WA WI WP	HSE 5445
Cuprisol FN	Cuprinol Ltd	WI	HSE 3632
Cuprisol WR	Cuprinol Ltd	WI	HSE 3633
Cuprisol XQD Special	Cuprinol Ltd	WI	HSE 3732
Cut End	Protim Solignum Ltd	WA WI WP	HSE 5490
* Protim 800	Fosroc Ltd	WI	HSE 4490
Protim 800	Protim Solignum Ltd	WI	HSE 4991
* Protim 800 C	Fosroc Ltd	WI	HSE 4494
Protim 800 C	Protim Solignum Ltd	WI	HSE 4990
* Protim 800 C Oil Brown	Fosroc Ltd	WI	HSE 4488
* Protim 800 CWR	Fosroc Ltd	WI	HSE 4493
Protim 800 CWR	Protim Solignum Ltd	WI	HSE 4993
* Protim 800 WR	Fosroc Ltd	WI	HSE 4495
Protim 800 WR	Protim Solignum Ltd	WI	HSE 4992
* Protim Brown 800	Fosroc Ltd	WI	HSE 4486
Protim Curative 800	Fosroc Ltd	WA	HSE 4489
* Protim Paste 800	Fosroc Ltd	WP	HSE 4485
Protim Paste 800	Protim Solignum Ltd	WI WP	HSE 4986
Protim WR 800	Fosroc Ltd	WA	HSE 4487
Universal	Protim Solignum Ltd	WA WI WP	HSE 5491
Wickes All Purpose Wood Treatment	Wickes Building Supplies Ltd	WA WP	HSE 5696

179 PERMETHRIN and AZACONAZOLE

Product Name	Marketing Company	Use	Reg. No
Fongix SE Total Treatment for Wood	Liberon Waxes Ltd	WA SA	HSE 5610
Fongix SE Total Treatment for Wood	V33 UK	WA SA	HSE 4288

WOOD PRESERVATIVES

Product Name	Marketing Company	Use	Reg. No
179 PERMETHRIN and AZACONAZOLE—continued			
Woodworm Killer and Rot Treatment	Liberon Waxes Ltd	WA SA	HSE 4826
Xylamon Brown U 101 C	Venilia Ltd	WA	HSE 4443
Xylamon Curative U 152 G/H	Venilia Ltd	WP	HSE 4446
180 PERMETHRIN and BENZALKONIUM CHLORIDE			
Biokil Emulsion	Jaymar Chemicals	WP	HSE 4562
181 PERMETHRIN and BORIC ACID			
Microguard Fl Concentrate	Permagard Products Ltd	WI WP	HSE 5101
182 PERMETHRIN and COPPER NAPHTHENATE			
Green	Protim Solignum Ltd	WA WI WP	HSE 5483
* Protim Green E	Fosroc Ltd	WA	HSE 3799
Protim Green E	Protim Solignum Ltd	WA WI WP	HSE 4975
Soligum Green	Protim Solignum Ltd	WA	HSE 5020
* Solignum Green	Solignum Ltd	WA	HSE 3900
183 PERMETHRIN and DICHLOFLUANID			
Impra-Color	Glen Products	WA WP	HSE 3870
184 PERMETHRIN and DICHLOFLUANID and FURMECYCLOX			
# Nitromors Timbrene Supreme	Nitromors Ltd	WA	HSE 3646
# Xyladecor Matt U-4010	Kay Metzeler Ltd	WA WI WP	HSE 3140
185 PERMETHRIN and DODECYLAMINE LACTATE and DODECYLAMINE SALICYLATE			
Kingfisher Timber Paste	Kingfisher Chemicals Ltd	WP	HSE 4327
186 PERMETHRIN and FURMECYCLOX			
# Xylamon Brown U1011	Kay Metzeler Ltd	WA	HSE 3746

Product Name	Marketing Company	Use	Reg. No

187 PERMETHRIN and PROPICONAZOLE

Protim 340	Protim Solignum Ltd	WI	HSE 5510
Protim 340 WR	Protim Solignum Ltd	WI	HSE 5509
Wocosen 12 OL	Janssen Pharmaceutica NV	WA WI WP	HSE 5404

188 PERMETHRIN and TEBUCONAZOLE

Brunol ATP New	Stanhope Chemical Products Ltd	WI WP	HSE 5620
Brunol STP	Stanhope Chemical Products Ltd	WI WP	HSE 5621

189 PERMETHRIN and TEBUCONAZOLE and DICHLOFLUANID

Impra-Holzschutzgrund (Primer)	Glen Products Ltd	WI WP	HSE 5735

190 PERMETHRIN and TRI(HEXYLENE GLYCOL) BIBORATE

Antel Dual Purpose (BP) Concentrate (Water Dilutable)	Antel Tridex	WP	HSE 4046
Barrettine New Universal Fluid	Barrettine Products Ltd	WA WI WP	HSE 3623
Brunol PBO	Stanhope Chemical Products Ltd	WP	HSE 4474
Brunol Special P	Stanhope Chemical Products Ltd	WP	HSE 3864
Chel-Wood Preserver/Woodworm Dry Rot Killer BP	Chelec Ltd	WA	HSE 4897
Deepwood Fl Dual Purpose Emulsion Concentrate	Safeguard Chemicals Ltd	WI WP	HSE 5087
* Domexyl Paste Wood Preservative	Domeel Ltd	WP	HSE 4311
Ecology Fungicide Insecticide Aqueous (Concentrate) Wood Preservative Dual Purpose	Kingfisher Chemicals Ltd	WP	HSE 4039
Insecticide Fungicide Wood Preservative	Kingfisher Chemicals Ltd	WP	HSE 4048
Larsen Wood Preservative Clear 2 and Brown 2	Larsen Manufacturing Ltd	WA	HSE 4287
Larsen Woodworm Killer 2	Larsen Manufacturing Ltd	WA	HSE 4285
Mar-Kil S	Marcher Chemicals Ltd	WI WP	HSE 5234
Mar-Kil W	Marcher Chemicals Ltd	WI WP	HSE 5219

WOOD PRESERVATIVES

Product Name	Marketing Company	Use	Reg. No
190 **PERMETHRIN and TRI (HEXYLENE GLYCOL) BIBORATE**—continued			
* Payless Wood Preserver—Universal	Payless DIY Ltd	WA	HSE 3828
Perma AQ Dual Purpose	Triton Perma Industries Ltd	WP	HSE 5617
Permapaste PE	Triton Perma Industries Ltd	WP	HSE 5616
Permatreat Paste	Permagard Products Ltd	WP	HSE 4075
Permethrin F and I Concentrate	Permagard Products Ltd	WP	HSE 4008
* Protim CDB	Fosroc Ltd	WA	HSE 4197
# Protim CDB	Fosroc Ltd	WP	HSE 3638
Protim CDB	Protim Solignum Ltd	WA WI WP	HSE 4999
Protim Curative GP	Fosroc Ltd	WA	HSE 4279
* Protim WR 260	Fosroc Ltd	WA	HSE 4130
Protim WR 260	Protim Solignum Ltd	WA WI WP	HSE 4997
Remecology Fungicide Insecticide R9	Remtox (Chemicals) Ltd	WP	HSE 4249
Remecology Spirit Based K7	Remtox (Chemicals) Ltd	WP	HSE 4248
Safeguard BP Dual Purpose Wood Preservative	Safeguard Chemicals Ltd	WP	HSE 4230
Safeguard BP O/S Wood Preservative	Safeguard Chemicals Ltd	WP	HSE 3107
Safeguard Deepwood II Dual Purpose Emulsion Concentrate	Safeguard Chemicals Ltd	WP	HSE 4133
Safeguard Deepwood II Dual Purpose Emulsion Concentrate	Safeguard Chemicals Ltd	WP	HSE 5303
Safeguard Deepwood III Timber Treatment	Safeguard Chemicals Ltd	WA	HSE 3937
Safeguard Deepwood Paste	Safeguard Chemicals Ltd	WP	HSE 4110
Saturin E10	British Building and Engineering Appliances Plc	WP	HSE 3896
Saturin E5	British Building and Engineering Appliances Plc	WA	HSE 3897
Solignum Remedial Concentrate	Protim Solignum Ltd	WP	HSE 5021
* Solignum Remedial Concentrate	Solignum Ltd	WP	HSE 4169
Solignum Remedial Fluid PB	Protim Solignum Ltd	WP	HSE 5017
* Solignum Remedial Fluid PB	Solignum Ltd	WP	HSE 4306
Solignum Remedial PB	Protim Solignum Ltd	WA WI WP	HSE 5607
Sovereign Aqueous Fungicide/Insecticide 2	Sovereign Chemical Industries Ltd	WP	HSE 4619

WOOD PRESERVATIVES

Product Name	Marketing Company	Use	Reg. No
190 PERMETHRIN and TRI (HEXYLENE GLYCOL) BIBORATE—continued			
Sovereign Insecticide/Fungicide 2	Sovereign Chemical Industries Ltd	WP	HSE 4649
Trimethrin AQ Plus	Triton Chemical Manufacturing Company Ltd	WP	HSE 3522
Trimethrin OS Plus	Triton Chemical Manufacturing Company Ltd	WA WP	HSE 3521
Tritec Plus	Triton Chemical Manufacturing Company Ltd	WP	HSE 5425
Universal Fluid (Grade PB)	Barrettine Products Ltd	WA WI WP	HSE 5212
Universal Wood Preserver	Langlow Products Ltd	WA	HSE 3692
Woodworm Killer	PLA Products Ltd	WA WP	HSE 5028
191 PERMETHRIN and TRI(HEXYLENE GLYCOL) BIBORATE and DICHLOFLUANID			
Barrettine New Universal Fluid D	Barrettine Products Ltd	WA	HSE 4004
Blackfriars Gold Star Clear	E Parsons and Sons Ltd	WA	HSE 4149
Leyland Timbrene Supreme Environmental Formula	Leyland Paint Company	WA	HSE 4800
Nitromors Timbrene Supreme Environmental Formula	Henkel Home Improvements	WA	HSE 4107
Timbrene Supreme	Kalon Group PLC	WA WF	HSE 4995
192 PERMETHRIN and TRIBUTYLTIN NAPHTHENATE			
Celpruf TNM	Rentokil Ltd	WI	HSE 4704
Celpruf TNM WR	Rentokil Ltd	WI	HSE 4703
* Protim 230	Fosroc Ltd	WI	HSE 4559
Protim 230	Protim Solignum Ltd	WI	HSE 5044
* Protim 230 WR	Fosroc Ltd	WI	HSE 4560
Protim 230 WR	Protim Solignum Ltd	WI	HSE 5045
Vacsol 2622/2623 WR	Hickson Timber Products Ltd	WI	HSE 4645
Vacsol 2625/2625 WR 2:1 Concentrate	Hickson Timber Products Ltd	WI	HSE 5329
Vacsol 2713/2714	Hickson Timber Products Ltd	WI	HSE 4808
Vacsol 2716/2717 2:1 Concentrate	Hickson Timber Products Ltd	WI	HSE 5441

193 PERMETHRIN and TRIBUTYLTIN OXIDE and DICHLOFLUANID and FURMECYCLOX

Product Name	Marketing Company	Use	Reg. No
# OS Color WR	Ostermann and Scheiwe GmbH	WA	HSE 3160

194 PERMETHRIN and TRIBUTYLTIN PHOSPHATE

Product Name	Marketing Company	Use	Reg. No
# Celpruf TPM	Rentokil Ltd	WI	HSE 4028
# Celpruf TPM WR	Rentokil Ltd	WI	HSE 4029
# Protim 240	Fosroc Ltd	WI	HSE 3860
# Protim 240 C	Fosroc Ltd	WI	HSE 3988
# Protim 240 CWR	Fosroc Ltd	WI	HSE 3972
# Protim 240 WR	Fosroc Ltd	WI	HSE 3861
# Timbercare WRQD Ready to Use	Hickson Timber Products Ltd	WI	HSE 3679
# Vacsele	Hickson Timber Products Ltd	WI	HSE 3678
# Vacsol MWR Ready for Use	Hickson Timber Products Ltd	WI	HSE 3677
# Vacsol WR Ready to Use	Hickson Timber Products Ltd	WI	HSE 3676

195 PERMETHRIN and ZINC OCTOATE

Product Name	Marketing Company	Use	Reg. No
Chel-Wood Preserver/Woodworm Dry Rot Killer	Chelec Ltd	WA	HSE 4172
Cuprinol Difusol S	Cuprinol Ltd	WP	HSE 3750
# Green Range Mayonnaise	Cementone-Beaver Ltd	WP	HSE 3801
Green Range Timber Treatment Paste	Cementone-Beaver Ltd	WP	HSE 4651
Larvac 300	Larsen Manufacturing Ltd	WI	HSE 4461
Palace Timbertreat Ecology	Palace Chemicals Ltd	WP	HSE 4931
Premium Wood Treatment	Rentokil Ltd	WA SA	HSE 5350
Rentokil Dual Purpose Fluid	Rentokil Ltd	WP	HSE 4074
Rentokil Wood Preservative	Rentokil Ltd	WA	HSE 4195
Solignum Colourless	Protim Solignum Ltd	WA WI WP	HSE 5549
# Solignum Remedial Mayonnaise	Solignum Ltd	WP	HSE 3899
Solignum Wood Preservative Paste	Protim Solignum Ltd	WP	HSE 5001
* Solignum Wood Preservative Paste	Solignum Ltd	WP	HSE 4082
* Trimethrin 6	Triton Chemical Manufacturing Company Ltd	WA	HSE 3918

Product Name	Marketing Company	Use	Reg. No

195 PERMETHRIN and ZINC OCTOATE—continued

Product Name	Marketing Company	Use	Reg. No
* Triton Tripaste	Triton Chemical Manufacturing Company Ltd	WA	HSE 3927
Wood Preservative Clear	Rentokil Ltd	WA	HSE 4077
Woodtreat BP	Stanhope Chemical Products Ltd	WP	HSE 3085

196 PERMETHRIN and ZINC OCTOATE and BORIC ACID

Product Name	Marketing Company	Use	Reg. No
Celpruf BZP	Rentokil Ltd	WI	HSE 4190
Celpruf BZP WR	Rentokil Ltd	WI	HSE 4191
Premium Grade Wood Treatment	Rentokil Ltd	WA	HSE 4193

197 PERMETHRIN and ZINC OCTOATE and DICHLOFLUANID

Product Name	Marketing Company	Use	Reg. No
OS Color Wood Stain and Preservative	Ostermann and Scheiwe GmbH	WA	HSE 5531
OS Color WR	Ostermann and Scheiwe GmbH	WA	HSE 5364

198 PERMETHRIN and ZINC VERSATATE

Product Name	Marketing Company	Use	Reg. No
Cedasol Ready To Use (2306)	Hickson Timber Products Ltd	WI	HSE 4117
Celpruf ZOP	Rentokil Ltd	WI	HSE 4681
Celpruf ZOP WR	Rentokil Ltd	WI	HSE 4682
Imersol 2410	Hickson Timber Products Ltd	WI	HSE 4674
* Imersol WRQD Ready to Use (2523)	Hickson Timber Products Ltd	WI	HSE 3969
Protim 220	Fosroc Ltd	WI	HSE 3912
Protim 220	Protim Solignum Ltd	WI	HSE 5198
Protim 220 CWR	Fosroc Ltd	WA	HSE 3995
Protim 220 CWR	Protim Solignum Ltd	WA WI	HSE 5196
Protim 220 WR	Fosroc Ltd	WI	HSE 3913
Protim 220 WR	Protim Solignum Ltd	WI	HSE 5195
Protim Curative Z	Fosroc Ltd	WA	HSE 4268
* Protim Paste 220	Fosroc Ltd	WP	HSE 4058
Protim Paste 220	Protim Solignum Ltd	WI WP	HSE 4983
Protim WR 220	Fosroc Ltd	WA	HSE 4113
Solignum Universal	Protim Solignum Ltd	WA WI WP	HSE 5019

WOOD PRESERVATIVES

Product Name	Marketing Company	Use	Reg. No
198 PERMETHRIN and ZINC VERSATATE—continued			
* Solignum Universal	Solignum Ltd	WA	HSE 4147
Vacsele 2611	Hickson Timber Products Ltd	WP	HSE 4675
Vacsele P2312	Hickson Timber Products Ltd	WP	HSE 4363
Vacsele Ready to Use (2605)	Hickson Timber Products Ltd	WP	HSE 3976
Vacsol (2:1 Conc) 2711/2712	Hickson Timber Products Ltd	WI	HSE 4685
Vacsol 2709/2710	Hickson Timber Products Ltd	WI	HSE 4678
Vacsol 2:1 Concentrate	Hickson Timber Products Ltd	WI	HSE 3977
Vacsol P (2310)	Hickson Timber Products Ltd	WI	HSE 4377
Vacsol Ready to Use	Hickson Timber Products Ltd	WI	HSE 3978
Vacsol WR (2:1 Conc) 2614/2615	Hickson Timber Products Ltd	WI	HSE 4679
Vacsol WR 2612/2613	Hickson Timber Products Ltd	WI	HSE 4676
Vacsol WR 2:1 Concentrate	Hickson Timber Products Ltd	WI	HSE 3824
Vacsol WR Ready to Use	Hickson Timber Products Ltd	WI	HSE 3825
199 PIRIMIPHOS-METHYL			
Actellic 25 EC	Zeneca Public Health	WP IAH IFSP IP	HSE 4880
200 POTASSIUM 2-PHENYLPHENOXIDE			
Timbercol Preservative Concentrate	Timbercol Industries Ltd	WI	HSE 3868
201 POTASSIUM DICHROMATE and BORIC ACID and COPPER SULPHATE			
Tecca CCB 1	Tecca Ltd	WI	HSE 5584
* Wolmanit CB	BASF (United Kingdom) Ltd	WI	HSE 3108
* Wolmanit CB-P	BASF (United Kingdom) Ltd	WI	HSE 3110
202 PROPICONAZOLE			
Wocosen S	Janssen Pharmaceutica NV	WA WI WP	HSE 5394
203 PROPICONAZOLE and PERMETHRIN			
Protim 340	Protim Solignum Ltd	WI	HSE 5510
Protim 340 WR	Protim Solignum Ltd	WI	HSE 5509
Wocosen 12 OL	Janssen Pharmaceutica NV	WA WI WP	HSE 5404

WOOD PRESERVATIVES

Product Name	Marketing Company	Use	Reg. No
204 SODIUM 2-PHENYLPHENOXIDE			
Brunosol Concentrate	Stanhope Chemical Products Ltd	WP SP	HSE 4472
Coaltec 50	Coalite Chemicals Division	WP SP	HSE 4761
Coaltec 50TD	Coalite Chemicals Division	WA SA	HSE 4762
Cube Concentrate Fluid for Dry Rot	Kleeneze Ltd	WP SP	HSE 3083
Cube Fluid for Dry Rot	Kleeneze Ltd	WA SA	HSE 3082
Fence Protector	Barrettine Products Ltd	WA WI WP	HSE 5208
Fence 'n' Shed Concentrate	Lanstar Coatings Ltd	WI	HSE 3106
Mar-Cide	Marcher Chemicals Ltd	WI WP SP	HSE 5631
Plantsafe Autumn Gold	RK and J Jones	WA WP	HSE 4383
Timbertex Pl/2	Marcher Chemcials Ltd	WI WP	HSE 5628
Valspar Shed and Fence Preservative	Macpherson Paints Ltd	WA	HSE 4865
Valspar Timberguard Shed and Fence Preservative	Macpherson Paints Ltd	WA	HSE 4867
W.S.P. Wood Treatment	Barrettine Products Ltd	WA WI WP	HSE 5209
Wickes Shed and Fence Treatment	Macpherson Paints Ltd	WA	HSE 4849
Wickes Wood Preserver	Macpherson Paints Ltd	WA	HSE 4850
205 SODIUM DICHROMATE and ARSENIC PENTOXIDE and COPPER SULPHATE			
Celcure A Concentrate	Rentokil Ltd	WI	HSE 5458
Celcure A Fluid 10	Rentokil Ltd	WI	HSE 3764
Celcure A Fluid 6	Rentokil Ltd	WI	HSE 3155
Celcure A Paste	Rentokil Ltd	WI	HSE 4523
Injecta CCA-C	Injecta APS	WI	HSE 4447
Kemira CCA Type BS	Kemira Kemwood AB	WI	HSE 3879
Kemwood CCA Type BS	Laporte Kemwood AB	WI	HSE 5534
Laporte CCA Type 1	Laporte Wood Preservation	WI	HSE 3175
Laporte CCA Type 2	Laporte Wood Preservation	WI	HSE 3176
Laporte Permawood CCA	Laporte Wood Preservation	WI	HSE 3529
* Meksol	Mechema Chemicals Ltd	WI	HSE 3994
* Mekure	Mechema Chemicals Ltd	WI	HSE 4352
* Mekure T2	Mechema Chemicals Ltd	WI	HSE 4350
* Protim CCA Salts Type 2	Fosroc Ltd	WI	HSE 4462

WOOD PRESERVATIVES

Product Name	Marketing Company	Use	Reg. No

205 SODIUM DICHROMATE and ARSENIC PENTOXIDE and COPPER SULPHATE—continued

Protim CCA Salts Type 2	Protim Solignum Ltd	WI	HSE 4972
Tanalith 3357	Hickson Timber Products Ltd	WI	HSE 4431
Tanalith CL (3354)	Hickson Timber Products Ltd	WI	HSE 4196
Tanalith CP 3353	Hickson Timber Products Ltd	WI	HSE 4667
Tecca P2	Tecca Ltd	WI	HSE 3958

206 SODIUM DICHROMATE and BORIC ACID and CHROMIUM ACETATE and COPPER SULPHATE

* Celcure CB Salts	Rentokil Ltd	WI	HSE 4536
Celgard CF	Rentokil Ltd	WI	HSE 4608

207 SODIUM DICHROMATE and BORIC ACID and COPPER SULPHATE

Celcure CB90	Rentokil Ltd	WI	HSE 5117
Tanalith CBC Paste 3402	Hickson Timber Products Ltd	WI	HSE 3833
* Wolmanit CB-A	BASF (United Kingdom) Ltd	WI	HSE 3109

208 SODIUM DICHROMATE and CHROMIUM ACETATE and COPPER SULPHATE

Celcure B	Rentokil Ltd	WI	HSE 4541
Celcure O	Rentokil Ltd	WI	HSE 4539
* Mekseal	Mechema Chemicals Ltd	WI	HSE 4351

209 SODIUM DICHROMATE and COPPER SULPHATE

* Ensele 3424	Hickson Timber Products Ltd	WI	HSE 4670
Laporte Cut End Preservative	Laporte Wood Preservation Ltd	WI	HSE 3528

210 SODIUM DICHROMATE and SODIUM FLUORIDE and AMMONIUM BIFLUORIDE

Rentex	Rentokil Ltd	WI	HSE 4756

211 SODIUM FLUORIDE and AMMONIUM BIFLUORIDE and SODIUM DICHROMATE

Rentex	Rentokil Ltd	WI	HSE 4756

WOOD PRESERVATIVES

Product Name	Marketing Company	Use	Reg. No
212 SODIUM PENTACHLOROPHENOXIDE			
Celbrite Nap 100	Rentokil Ltd	WP	HSE 4542
* Clarkes Woodcare	Clarkes Products (Weyhill) Ltd	WP	HSE 3531
# Fen-Tan	Forest Fencing Ltd	WA	HSE 3773
Gainserv 685	Industrial Chemical Company (Preston) Ltd	WP SP	HSE 3960
# Mosgo	Agrichem Ltd	WA SA	HSE 4336
Mosgo P	Agrichem Ltd	WP SP	HSE 4337
# Nubex Fungicide FS15	Nubex Ltd	WP SP	HSE 3666
* Protim Fentex WR	Fosroc Ltd	WI	HSE 4672
Protim Fentex WR	Protim Solignum Ltd	WI	HSE 4974
* Protim Panelguard	Fosroc Ltd	WI	HSE 4553
Protim Panelguard	Protim Solignum Ltd	WI	HSE 5010
* Protim Plug Compound	Fosroc Ltd	WP SP	HSE 4555
Protim Plug Compound	Protim Solignum Ltd	WP SP	HSE 5012
213 SODIUM PENTACHLOROPHENOXIDE and DISODIUM OCTABORATE			
Gainserv 140	Industrial Chemical Company (Preston) Ltd	WP	HSE 3963
Gainserv Concentrate	Industrial Chemical Company (Preston) Ltd	WP	HSE 4601
Gainserv Polymeric	Industrial Chemical Company (Preston) Ltd	WP	HSE 3964
* Gallwey ABS	Fosroc Ltd	WI	HSE 4529
Gallwey ABS	Protim Solignum Ltd	WI	HSE 5014
Protim Fentex Green Concentrate	Protim Solignum Ltd	WI	HSE 5122
Protim Fentex Green RFU	Protim Solignum Ltd	WI	HSE 5123
* Protim Fentex M	Fosroc Ltd	WI WP	HSE 3515
Protim Fentex M	Protim Solignum Ltd	WI	HSE 4973
Protim Kleen II	Fosroc Ltd	WI	HSE 3973
Protim Kleen II	Protim Solignum Ltd	WI	HSE 5203
214 SODIUM TETRABORATE			
Ensele 3430	Hickson Timber Products Ltd	WP	HSE 4671

Product Name	Marketing Company	Use	Reg. No

214 SODIUM TETRABORATE—continued

# Protect A Fence	Solihull Fencing Company Ltd	WA	HSE 3786

215 SODIUM TETRABORATE and ALKYLTRIMETHYL AMMONIUM CHLORIDE

Sinesto B	Finnmex Co	WI	HSE 5136
* Sinesto B	KYMI Chemicals	WI	HSE 4878

216 SODIUM TETRABORATE and BORIC ACID

Celbor	Rentokil Ltd	WI	HSE 4614
PC-K	Peter Cox Preservation	WP	HSE 4409
Pyrolith 3505 Ready To Use	Hickson Timber Products Ltd	WI	HSE 3636

217 TC OIL and CREOSOTE

* Arborsan 3	Lancashire Tar Distillers Ltd	WI	HSE 3143

218 TEBUCONAZOLE

Bayer WPC 2–1.5	Bayer Plc	WI WP	HSE 5501
Bayer WPC 2–25	Bayer Plc	WI WP	HSE 5500
Bayer WPC 2–25–SB	Bayer Plc	WI WP	HSE 5502

219 TEBUCONAZOLE and BORIC ACID and COPPER CARBONATE HYDROXIDE

Tanalith 3485	Hickson Timber Products Ltd	WI	HSE 5562

220 TEBUCONAZOLE and CYPERMETHRIN

Celpruf TZC	Rentokil Ltd	WI	HSE 5722

221 TEBUCONAZOLE and DICHLOFLUANID and PERMETHRIN

Impra-Holzschutzgrund (Primer)	Glen Products Ltd	WI WP	HSE 5735

222 TEBUCONAZOLE and PERMETHRIN

Brunol ATP New	Stanhope Chemical Products Ltd	WI WP	HSE 5620

WOOD PRESERVATIVES

Product Name	Marketing Company	Use	Reg. No
222 TEBUCONAZOLE and PERMETHRIN—continued			
Brunol STP	Stanhope Chemical Products Ltd	WI WP	HSE 5621
223 TRI (HEXYLENE GLYCOL) BIBORATE			
ABL Wood Preservative (B)	Advanced Bitumens Ltd	WA	HSE 3743
Barrettine New Preserver	Barrettine Products Ltd	WA WI WP	HSE 3624
Brown Timber Preservative 2	Palace Chemicals Ltd	WA	HSE 4323
Brown Wood Preservative	Palace Chemicals Ltd	WA WI WP	HSE 5242
Clear Timber Preservative 2	Palace Chemicals Ltd	WA	HSE 4322
Clear Wood Preservative	Palace Chemicals Ltd	WA WI WP	HSE 5240
* Cube Preserver—Grade B—Light Brown	Kleeneze Ltd	WA WI WP	HSE 3610
Green Range Fungicidal Solvent	Cementone-Beaver Ltd	WP	HSE 3827
Kingfisher Wood Preservative	Kingfisher Chemicals Ltd	WP	HSE 4012
Langlow Wood Preserver Formulation B	Langlow Products Ltd	WA	HSE 3754
Larsen Joinery Grade 2	Larsen Manufacturing Ltd	WI WP	HSE 4286
# Ness Fungicidal Solvent	Cemetone-Beaver Ltd	WP	HSE 3904
Palace Base Coat Wood Preservative	Palace Chemicals Ltd	WA WI WP	HSE 5351
Palace Microfine Fungicide Concentrate	Palace Chemicals Ltd	WI WP	HSE 5551
* Payless Wood Preserver—Red Cedar	Payless DIY Ltd	WA	HSE 3829
PC-F/B	Peter Cox Preservation	WA	HSE 3739
PC-XJ/2	Peter Cox Preservation	WA	HSE 3737
* Protim 260 F	Fosroc Ltd	WA	HSE 4105
Protim 260 F	Protim Solignum Ltd	WA WI WP	HSE 4998
Protim Curative F	Fosroc Ltd	WA	HSE 4267
Protim Injection Fluid	Fosroc Ltd	WA	HSE 4204
Safeguard Deepwood Fungicide	Safeguard Chemicals Ltd	WA	HSE 4183
Sovereign AQ/FT	Sovereign Chemical Industries Ltd	WI WP	HSE 5433
Sovereign Timber Preservative	Sovereign Chemical Industries Ltd	WP	HSE 3807
Wood Preserver	Langlow Products Ltd	WA	HSE 3690

WOOD PRESERVATIVES

Product Name	Marketing Company	Use	Reg. No
224 TRI (HEXYLENE GLYCOL) BIBORATE and COPPER NAPHTHENATE			
Dark Green Timber Preservative 2	Palace Chemicals Ltd	WA	HSE 4321
Dark Green Wood Preservative	Palace Chemicals Ltd	WA WI WP	HSE 5243
Kingfisher Wood Preservative	Kingfisher Chemicals Ltd	WP	HSE 4011
Pale Green Timber Preservative 2	Palace Chemicals Ltd	WA	HSE 4320
Pale Green Wood Preservative	Palace Chemicals Ltd	WA WI WP	HSE 5241
Sovereign Timber Preservative	Sovereign Chemical Industries Ltd	WP	HSE 3808
Sovereign Timber Preservative	Sovereign Chemical Industries Ltd	WP	HSE 3806
225 TRI (HEXYLENE GLYCOL) BIBORATE and CYPERMETHRIN			
Cementone Multiplus	Cementone-Beaver Ltd	WA WP	HSE 3849
Crown Fungicide Insecticide Concentrate	Crown Chemicals Ltd	WP	HSE 5463
# Cyperguard BC Wood Treatment	Safeguard Chemicals Ltd	WP	HSE 3889
Devatern Wood Preserver	Sorex Ltd	WP	HSE 4073
Ecology Fungicide Insecticide Concentrate	Palace Chemicals Ltd	WP	HSE 3792
Fungicide Insecticide	Palace Chemicals Ltd	WA	HSE 3795
Gainpaste	Industrial Chemical Company (Preston) Ltd	WA	HSE 4547
Green Range Dual Purpose AQ	Cementone Beaver Ltd	WP	HSE 3725
Green Range Wykamol Plus	Cementone Beaver Ltd	WP	HSE 3731
Killgerm Wood Protector	Killgerm Chemicals Ltd	WP	HSE 3127
Langlow Clear Wood Preserver	Langlow Products Ltd	WA	HSE 3802
# Ness Dual Purpose	Cementone Beaver Ltd	WP	HSE 3768
# Ness Dual Purpose AQ	Cementone Beaver Ltd	WP	HSE 3767
Nubex Emulsion Concentrate CB (Low Odour)	Nubex Ltd	WP	HSE 3637
Nubex Emulsion Concentrate CB (Low Odour)	Nubex Ltd	WP	HSE 5328
Nubex Woodworm All Purpose CB	Nubex Ltd	WP	HSE 3189
Palace Microfine Fungicide Insecticide (Concentrate)	Palace Chemicals Ltd	WI WP	HSE 5421
* Payless Wood Preserver—Clear	Payless DIY Ltd	WA	HSE 3831
PC-H/4	Peter Cox Preservation	WA	HSE 3740

WOOD PRESERVATIVES

Product Name	Marketing Company	Use	Reg. No
225 TRI (HEXYLENE GLYCOL) BIBORATE and CYPERMETHRIN—continued			
PCX-122	Peter Cox Preservation	WP	HSE 3999
PCX-122 Concentrate	Peter Cox Preservation	WP	HSE 4000
Protim CBC	Fosroc Ltd	WA	HSE 3798
Timbermate Clear Wood Preservative	Lanstar Coatings Ltd	WA	HSE 3787
226 TRI (HEXYLENE GLYCOL) BIBORATE and DICHLOFLUANID			
ABL Universal Woodworm Killer DB	Advanced Bitumens Ltd	WA	HSE 3858
Barrettine New Wood Preserver	Barrettine Products Ltd	WA	HSE 4038
Blackfriars New Wood Preserver	E Parsons and Sons Ltd	WA	HSE 4150
Double Action Timber Preservative For Doors	International Paint Ltd	WA	HSE 3707
Double Action Wood Preservative	International Paint Ltd	WA	HSE 3706
Nitromors Timbrene Clear Environmental Formula	Henkel Home Improvements	WA	HSE 4106
Ranch Preservative	International Paint Ltd	WA	HSE 4085
* Ronseal Trade Low Build Preservative Woodstain	Sterling-Winthrop Group Ltd	WA	HSE 3959
* Ronseal Trade Preservative Woodstain Basecoat and Colour Harmoniser	Sterling-Winthrop Group Ltd	WA	HSE 3923
Sadolin New Base	Sadolin UK Ltd	WA	HSE 4341
Ultrabond Woodworm Killer	Ultrabond Ltd	WA	HSE 4344
227 TRI (HEXYLENE GLYCOL) BIBORATE and DICHLOFLUANID and PERMETHRIN			
Barrettine New Universal Fluid D	Barrettine Products Ltd	WA	HSE 4004
Blackfriars Gold Star Clear	E Parsons and Sons Ltd	WA	HSE 4149
Leyland Timbrene Supreme Environmental Formula	Leyland Paint Company	WA	HSE 4800
Nitromors Timbrene Supreme Environmental Formula	Henkel Home Improvements	WA	HSE 4107
Timbrene Supreme	Kalon Group Plc	WA WP	HSE 4995
228 TRI (HEXYLENE GLYCOL) BIBORATE and GAMMA-HCH			
Cube F and I Concentrate/Grade BL	Kleeneze Ltd	WP	HSE 3095

WOOD PRESERVATIVES

Product Name	Marketing Company	Use	Reg. No

228 TRI (HEXYLENE GLYCOL) BIBORATE and GAMMA-HCH—continued

Product Name	Marketing Company	Use	Reg. No
Cube Solvent F and I Fluid/Grade BL	Kleeneze Ltd	WA	HSE 3094
# Nubex Emulsion LB (Low Odour)	Nubex Ltd	WP	HSE 3664

229 TRI (HEXYLENE GLYCOL) BIBORATE and PERMETHRIN

Product Name	Marketing Company	Use	Reg. No
Antel Dual Purpose (BP) Concentrate (Water Dilutable)	Antel-Tridex	WP	HSE 4046
Barrettine New Universal Fluid	Barrettine Products Ltd	WA WI WP	HSE 3623
Brunol PBO	Stanhope Chemical Products Ltd	WP	HSE 4474
Brunol Special P	Stanhope Chemical Products Ltd	WP	HSE 3864
Chel-Wood Preserver/Woodworm Dry Rot Killer BP	Chelec Ltd	WA	HSE 4897
Deepwood FI Dual Purpose Emulsion Concentrate	Safeguard Chemicals Ltd	WI WP	HSE 5087
* Domexyl Paste Wood Preservative	Domeel Ltd	WP	HSE 4311
Ecology Fungicide Insecticide Aqueous (Concentrate) Wood Preservative Dual Purpose	Kingfisher Chemicals Ltd	WP	HSE 4039
Insecticide Fungicide Wood Preservative	Kingfisher Chemicals Ltd	WP	HSE 4048
Larsen Wood Preservative Clear 2 and Brown 2	Larsen Manufacturing Ltd	WA	HSE 4287
Larsen Woodworm Killer 2	Larsen Manufacturing Ltd	WA	HSE 4285
Mar-Kil S	Marcher Chemicals Ltd	WI WP	HSE 5234
Mar-Kil W	Marcher Chemicals Ltd	WI WP	HSE 5219
* Payless Wood Preserver—Universal	Payless DIY Ltd	WA	HSE 3828
Perma AQ Dual Purpose	Triton Perma Industries Ltd	WP	HSE 5617
Permapaste PB	Triton Perma Industries Ltd	WP	HSE 5616
Permatreat Paste	Permagard Products Ltd	WP	HSE 4075
Permethrin P and I Concentrate	Permagard Products Ltd	WP	HSE 4008
* Protim CDB	Fosroc Ltd	WA	HSE 4197
# Protim CDB	Fosroc Ltd	WP	HSE 3638
Protim CDB	Protim Solignum Ltd	WA WI WP	HSE 4999
Protim Curative GP	Fosroc Ltd	WA	HSE 4279
* Protim WR 260	Fosroc Ltd	WA	HSE 4130

229 TRI (HEXYLENE GLYCOL) BIBORATE and PERMETHRIN—continued

Product Name	Marketing Company	Use	Reg. No
Protim WR 260	Protim Solignum Ltd	WA WI WP	HSE 4997
Remecology Fungicide Insecticide R9	Remtox (Chemicals) Ltd	WP	HSE 4249
Remecology Spirit Based K7	Remtox (Chemicals) Ltd	WP	HSE 4248
Safeguard BP Dual Purpose Wood Preservative	Safeguard Chemicals Ltd	WP	HSE 4230
Safeguard BP O/S Wood Preservative	Safeguard Chemicals Ltd	WP	HSE 3107
Safeguard Deepwood II Dual Purpose Emulsion Concentrate	Safeguard Chemicals Ltd	WP	HSE 4133
Safeguard Deepwood II Dual Purpose Emulsion Concentrate	Safeguard Chemicals Ltd	WP	HSE 5303
Safeguard Deepwood III Timber Treatment	Safeguard Chemicals Ltd	WA	HSE 3937
Safeguard Deepwood Paste	Safeguard Chemicals Ltd	WP	HSE 4110
Saturin E10	British Building and Engineering Appliances Plc	WP	HSE 3896
Saturin E5	British Building and Engineering Appliances Plc	WA	HSE 3897
Solignum Remedial Concentrate	Protim Solignum Ltd	WP	HSE 5021
* Solignum Remedial Concentrate	Solignum Ltd	WP	HSE 4169
Solignum Remedial Fluid PB	Protim Solignum Ltd	WP	HSE 5017
* Solignum Remedial Fluid PB	Solignum Ltd	WP	HSE 4306
Solignum Remedial PB	Protim Solignum Ltd	WA WI WP	HSE 5607
Sovereign Aqueous Fungicide/ Insecticide 2	Sovereign Chemical Industries Ltd	WP	HSE 4619
Sovereign Insecticide/Fungicide 2	Sovereign Chemical Industries Ltd	WP	HSE 4649
Trimethrin AQ Plus	Triton Chemical Manufacturing Company Ltd	WP	HSE 3522
Trimethrin OS Plus	Triton Chemical Manufacturing Company Ltd	WA WP	HSE 3521
Tritec Plus	Triton Chemical Manufacturing Company Ltd	WP	HSE 5425
Universal Fluid (Grade PB)	Barrettine Products Ltd	WA WI WP	HSE 5212
Universal Wood Preserver	Langlow Products Ltd	WA	HSE 3692
Woodworm Killer	PLA Products Ltd	WA WP	HSE 5028

WOOD PRESERVATIVES

Product Name	Marketing Company	Use	Reg. No

230 TRI (HEXYLENE GLYCOL) BIBORATE and TRIBUTYLTIN OXIDE and GAMMA-HCH

Product Name	Marketing Company	Use	Reg. No
# Timber Preservative Solvent Based Fungicidal Insecticide 30	Ambersil Ltd	WP	HSE 3006

231 TRI (HEXYLENE GLYCOL) BIBORATE and ZINC NAPHTHENATE and DICHLOFLUANID

Product Name	Marketing Company	Use	Reg. No
# Dulux Exterior Preservative Basecoat	ICI Paints Division	WA	HSE 4035
Weathershield Exterior Preservative Basecoat	ICI Paints Division	WA WP	HSE 5245
Weathershield Exterior Preservative Basecoat	ICI Paints Division	WA	HSE 4036
Weathershield Preservative Primer	ICI Paints Division	WA WP	HSE 5244
Weathershield Preservative Primer	ICI Paints Division	WA	HSE 4013

232 TRI (HEXYLENE GLYCOL) BIBORATE and ZINC OCTOATE and DICHLOFLUANID

Product Name	Marketing Company	Use	Reg. No
Timber Preservative Clear TFP7	Johnstone's Paints Plc	WA	HSE 4016

233 TRIBUTYLTIN NAPHTHENATE

Product Name	Marketing Company	Use	Reg. No
Celpruf Primer TN	Rentokil Ltd	WI	HSE 5378
Celpruf TN	Rentokil Ltd	WI	HSE 4701
Celpruf TN WR	Rentokil Ltd	WI	HSE 4702
* Protim FDR 230	Fosroc Ltd	WI	HSE 4563
Protim FDR 230	Protim Solignum Ltd	WI	HSE 5046
Vacsol 2652/2653 JWR	Hickson Timber Products Ltd	WI	HSE 4646
Vacsol 2746/2747 J	Hickson Timber Products Ltd	WI	HSE 4807

234 TRIBUTYLTIN NAPHTHENATE and PERMETHRIN

Product Name	Marketing Company	Use	Reg. No
Celpruf TNM	Rentokil Ltd	WI	HSE 4704
Celpruf TNM WR	Rentokil Ltd	WI	HSE 4703
* Protim 230	Fosroc Ltd	WI	HSE 4559
Protim 230	Protim Solignum Ltd	WI	HSE 5044
* Protim 230 WR	Fosroc Ltd	WI	HSE 4560
Protim 230 WR	Protim Solignum Ltd	WI	HSE 5045

WOOD PRESERVATIVES

Product Name	Marketing Company	Use	Reg. No
234 TRIBUTYLTIN NAPHTHENATE and PERMETHRIN—continued			
Vacsol 2622/2623 WR	Hickson Timber Products Ltd	WI	HSE 4645
Vacsol 2625/2626 WR 2:1 Concentrate	Hickson Timber Products Ltd	WI	HSE 5329
Vacsol 2713/2714	Hickson Timber Products Ltd	WI	HSE 4808
Vacsol 2716/2717 2:1 Concentrate	Hickson Timber Products Ltd	WI	HSE 5441
235 TRIBUTYLTIN OXIDE			
Celpruf Primer	Rentokil Ltd	WI	HSE 3796
# Conducive Pilt RFU	PPG Industries (UK) Ltd	WI	HSE 4402
# Febwood Brown	FEB Ltd	WA WP	HSE 3011
# Febwood Exterior Wood Preserver	FEB Ltd	WI	HSE 3627
# Febwood Green	FEB Ltd	WA	HSE 3012
# Febwood Preservative WP3	FEB Ltd	WA WP	HSE 3010
# Flag Brand Wood Preservative	C W Wastnage Ltd	WA	HSE 3071
# PC-F/A	Peter Cox Preservation Ltd	WA	HSE 3738
* Protim 215 PP	Fosroc Ltd	WI	HSE 4769
Protim 215 PP	Protim Solignum Ltd	WI	HSE 5034
Protim FDR 210	Protim Solignum Ltd	WI	HSE 5278
* Protim JP 210	Fosroc Ltd	WI	HSE 3934
Protim JP 210	Protim Solignum Ltd	WI	HSE 5032
* Protim R Coloured	Fosroc Ltd	WI	HSE 4257
* Protim R Coloured	Protim Solignum Ltd	WI	HSE 5005
# Protim R Coloured 333	Fosroc Ltd	WI	HSE 4427
# Sadolin Sadovac No 561–2392	Sadolin (UK) Ltd	WI	HSE 3643
# Scanvac Cedar, No 561–2508	Sadolin Paints Ltd	WI	HSE 3105
# Solvent Black Preserver	Cotech Services Ltd	WA WI WP	HSE 3080
# Texas Exterior Wood Preserver	Texas Homecare Ltd	WI	HSE 3628
Vacsol 2234 J Conc	Hickson Timber Products Ltd	WI	HSE 4665
Vacsol J RTU	Hickson Timber Products Ltd	WI	HSE 3152
Vacsol JWR Concentrate	Hickson Timber Products Ltd	WI	HSE 3153
Vacsol JWR RTU	Hickson Timber Products Ltd	WI	HSE 3154
Vacsol P Ready To Use (2334)	Hickson Timber Products Ltd	WI	HSE 4203
# Wilko Exterior Wood Preserver	Wilkinson Home and Garden Stores	WA WP	HSE 3600

WOOD PRESERVATIVES

Product Name	Marketing Company	Use	Reg. No

235 TRIBUTYLTIN OXIDE—continued

Wood Preservative—AA155/01	Becker-Acroma Ltd	WI	HSE 5090
Wood Preservative AA 155/01	Becker-Acroma Ltd	WI	HSE 5309
# Wudfil Wet Rot Treatment	Wudcare Products	WA	HSE 3917

236 TRIBUTYLTIN OXIDE and ALKYLARYLTRIMETHYL AMMONIUM CHLORIDE

# Killgerm Masonry Sterilant	Killgerm Chemicals Ltd	WP SP	HSE 3125

237 TRIBUTYLTIN OXIDE and BENZALKONIUM CHLORIDE

# Nubex Fungicide QT Concentrate	Albright and Wilson Ltd	WP SP	HSE 3166

238 TRIBUTYLTIN OXIDE and COPPER NAPHTHENATE and GAMMA-HCH

# Decor-8 Wood Preservative Green	Decor-8 Ltd	WA WP	HSE 3065
# Fads Wood Preservative Green	Fads AG Stanley Ltd	WA WP	HSE 3037
# Great Mills Wood Preservative Green	Great Mills	WA WP	HSE 3041
# Homecharm Wood Preservative Green	Homecharm Ltd	WA WP	HSE 3061
# Payless Wood Preservative Green	Payless DIY Ltd	WA WP	HSE 3049
# POB Wood Preservative Green	POB Savident Ltd	WA WP	HSE 3025
# SPL Wood Preservative Green	Silver Paint and Lacquer Ltd	WA WP	HSE 3057
# Texas Wood Preservative Green	Texas Homecare Ltd	WA WP	HSE 3033
# Trend Wood Preservative Green	Trend Voluntary Group Services Ltd	WA WP	HSE 3045
# Wickes Wood Preservative Green	Wickes Building Supplies Ltd	WA WP	HSE 3029
# Wilco Wood Preservative Green	Wilkinson Group of Companies	WA WP	HSE 3053

239 TRIBUTYLTIN OXIDE and DICHLOFLUANID

AA155/00 Industrial Wood Preservative	Becker Paint Ltd	WI	HSE 3099
AA155/03 Industrial Wood Preservative	Becker Paint Ltd	WI	HSE 3100

WOOD PRESERVATIVES

Product Name	Marketing Company	Use	Reg. No

239 TRIBUTYLTIN OXIDE and DICHLOFLUANID—continued

Product Name	Marketing Company	Use	Reg. No
Celpruf CP Special	Rentokil Ltd	WI	HSE 4582
Hickson Woodex	Hickson Timber Products Ltd	WI	HSE 5405
Industrial Wood Preservative AA 155	Becker-Acroma Ltd	WI	HSE 4629
* Protim Cedar	Fosroc Ltd	WI	HSE 4544
Protim Cedar	Protim Solignum Ltd	WI	HSE 5042
* Protim R Coloured Plus	Fosroc Ltd	WI	HSE 4391
Protim Solignum Softwood Basestain CS	Protim Solignum Ltd	WI	HSE 5647
# Wilko Red Cedar Wood Preserver	Wilkinson Home and Garden Stores	WA WP	HSE 3598
Wood Preservative AA 155/00	Becker-Acroma Ltd	WI	HSE 5308
Wood Preservative AA 155/03	Becker-Acroma Ltd	WI	HSE 5310
Wood Preservative AA155	Becker-Acroma Ltd	WI	HSE 5311
# Woodex Intra	Hygaea Colours and Varnishes Ltd	WA	HSE 3159

240 TRIBUTYLTIN OXIDE and DICHLOFLUANID and FURMECYCLOX and PERMETHRIN

Product Name	Marketing Company	Use	Reg. No
# OS Color WR	Ostermann and Scheiwe GmbH	WA	HSE 3160

241 TRIBUTYLTIN OXIDE and GAMMA-HCH

Product Name	Marketing Company	Use	Reg. No
# Aqueous Fungicide-Insecticide Concentrate	Palace Chemicals Ltd	WP	HSE 3129
# Aqueous Fungicide-Insecticide	Palace Chemicals Ltd	WA WP	HSE 3128
Cedasol 2304 RTU	Hickson Timber Products Ltd	WI	HSE 3151
Cedasol 2320	Hickson Timber Products Ltd	WI	HSE 4365
# Cleartreat	South Bucks Estates Ltd	WP	HSE 3131
# Decor-8 Wood Preservative Brown	Decor-8 Ltd	WA WP	HSE 3063
# Decor-8 Wood Preservative Clear	Decor-8 Ltd	WA WP	HSE 3066
# Decor-8 Wood Preservative Red Cedar	Decor-8 Ltd	WA WP	HSE 3064
Dipsar	Cementone Beaver Ltd	WI	HSE 3604
# Fads Wood Preservative Brown	Fads AG Stanley Ltd	WA WP	HSE 3035
# Fads Wood Preservative Clear	Fads AG Stanley Ltd	WA WP	HSE 3038

WOOD PRESERVATIVES

Product Name	Marketing Company	Use	Reg. No
241 **TRIBUTYLTIN OXIDE and GAMMA-HCH**—continued			
# Fads Wood Preservative Red Cedar	Fads AG Stanley Ltd	WA WP	HSE 3036
# Fungicide Insecticide Wood Preservative	Palace Chemicals Ltd	WA	HSE 3587
# Great Mills Wood Preservative Brown	Great Mills	WA WP	HSE 3039
# Great Mills Wood Preservative Clear	Great Mills	WA WP	HSE 3042
# Great Mills Wood Preservative Red Cedar	Great Mills	WA WP	HSE 3040
Hickson Timbercare WRQD	Hickson Timber Products Ltd	WI	HSE 3594
# Homecharm Wood Preservative Brown	Homecharm Ltd	WA WP	HSE 3059
# Homecharm Wood Preservative Clear	Homecharm Ltd	WA WP	HSE 3062
# Homecharm Wood Preservative Red Cedar	Homecharm Ltd	WA WP	HSE 3060
# Kingston Dual Purpose Fluid	Kingston Chemicals	WP	HSE 3641
# Laporte Permatreat	Laporte Wood Preservation	WI	HSE 3699
# Laporte Permatreat Cut End Preservative	Laporte Wood Preservation	WI	HSE 3700
# Laporte Permatreat With Water Repellant	Laporte Wood Preservation	WI	HSE 3701
# Larsen Brown Wood Preservative	Larsen Manufacturing Ltd	WA	HSE 3772
# Nubex Standard Emulsion Concentrate LT	Nubex Ltd	WP	HSE 3653
# Nubex Woodworm All Purpose LT	Nubex Ltd	WA	HSE 3655
# Payless Wood Preservative Brown	Payless DIY Ltd	WA WP	HSE 3047
# Payless Wood Preservative Clear	Payless DIY Ltd	WA WP	HSE 3050
# Payless Wood Preservative Red Cedar	Payless DIY Ltd	WA WP	HSE 3048
# POB Wood Preservative Brown	POB Savident Ltd	WA WP	HSE 3023
# POB Wood Preservative Clear	POB Savident Ltd	WA WP	HSE 3026
# POB Wood Preservative Red Cedar	POB Savident Ltd	WA WP	HSE 3024
* Protim 210	Fosroc Ltd	WI	HSE 3755
Protim 210	Protim Solignum Ltd	WI	HSE 5033
* Protim 210 C	Fosroc Ltd	WI	HSE 3757
Protim 210 C	Protim Solignum Ltd	WI	HSE 5038

WOOD PRESERVATIVES

Product Name	Marketing Company	Use	Reg. No
241 TRIBUTYLTIN OXIDE and GAMMA-HCH—continued			
* Protim 210 CWR	Fosroc Ltd	WI	HSE 3758
Protim 210 CWR	Protim Solignum Ltd	WI	HSE 5035
* Protim 210 WR	Fosroc Ltd	WI	HSE 3756
Protim 210 WR	Protim Solignum Ltd	WI	HSE 5039
* Protim 23 WR	Fosroc Ltd	WI	HSE 3869
Protim 23 WR	Protim Solignum Ltd	WI	HSE 5037
Protim 90	Fosroc Ltd	WI	HSE 4564
Protim 90	Protim Solignum Ltd	WI	HSE 5276
* Protim Paste	Fosroc Ltd	WP	HSE 4748
Protim Paste	Protim Solignum Ltd	WP	HSE 4979
Protim TWR	Fosroc Ltd	WI	HSE 3542
# SPL Wood Preservative Brown	Silver Paint and Lacquer Ltd	WA WP	HSE 3055
# SPL Wood Preservative Clear	Silver Paint and Lacquer Ltd	WA WP	HSE 3058
# SPL Wood Preservative Red Cedar	Silver Paint and Lacquer Ltd	WA WP	HSE 3056
# Texas Wood Preservative Brown	Texas Homecare Ltd	WA WP	HSE 3031
# Texas Wood Preservative Red Cedar	Texas Homecare Ltd	WA WP	HSE 3032
# Timber Preservative Solvent Based Fungicidal Insecticide 10	Ambersil Ltd	WP	HSE 3005
# Timber Treat	Palace Chemicals Ltd	WP	HSE 3570
# Trend Wood Preservative Brown	Trend Voluntary Group Services Ltd	WA WP	HSE 3043
# Trend Wood Preservative Clear	Trend Voluntary Group Services Ltd	WA WP	HSE 3046
# Trend Wood Preservative Red Cedar	Trend Voluntary Group Services Ltd	WA WP	HSE 3044
Vacsol MWR Concentrate 2203	Hickson Timber Products Ltd	WI	HSE 3170
Vacsol MWR Ready To Use 2204	Hickson Timber Products Ltd	WI	HSE 3167
Vacsol P 2304 RTU	Hickson Timber Products Ltd	WI	HSE 4597
Vacsol WR Concentrate 2115	Hickson Timber Products Ltd	WI	HSE 3169
Vacsol WR Ready To Use 2116	Hickson Timber Products Ltd	WI	HSE 3168
# Wickes Wood Preservative Brown	Wickes Building Supplies Ltd	WA WP	HSE 3027
# Wickes Wood Preservative Clear	Wickes Building Supplies Ltd	WA	HSE 3030
# Wickes Wood Preservative Red Cedar	Wickes Building Supplies Ltd	WA WP	HSE 3028

WOOD PRESERVATIVES

Product Name	Marketing Company	Use	Reg. No

241 TRIBUTYLTIN OXIDE and GAMMA-HCH—continued

# Wilco Wood Preservative Brown	Wilkinson Group Of Companies	WA WP	HSE 3051
# Wilco Wood Preservative Clear	Wilkinson Group Of Companies	WA WP	HSE 3054
# Wilco Wood Preservative Red Cedar	Wilkinson Group Of Companies	WA WP	HSE 3052

242 TRIBUTYLTIN OXIDE and GAMMA-HCH and PENTACHLOROPHENOL

Celpruf PK	Rentokil Ltd	WI	HSE 4540
Celpruf PK WR	Rentokil Ltd	WI	HSE 4538
Lar-Vac 100	Larsen Manufacturing Ltd	WI	HSE 3723
# Preservative For Wood Black	Rentokil Ltd	WA	HSE 3714
* Protim 80	Fosroc Ltd	WI	HSE 3550
Protim 80	Protim Solignum Ltd	WI	HSE 5029
* Protim 80 C	Fosroc Ltd	WI	HSE 3548
Protim 80 C	Protim Solignum Ltd	WI	HSE 5036
* Protim 80 CWR	Fosroc Ltd	WI	HSE 3547
Protim 80 CWR	Protim Solignum Ltd	WI	HSE 5031
Protim 80 Oil Brown	Fosroc Ltd	WI	HSE 3516
* Protim 80 WR	Fosroc Ltd	WI	HSE 3549
Protim 80 WR	Protim Solignum Ltd	WI	HSE 5030
Protim 80C Oil Brown	Fosroc Ltd	WI	HSE 3517
* Protim Brown	Fosroc Ltd	WI	HSE 3551
Protim Brown	Protim Solignum Ltd	WI	HSE 5000
Protim Grade Basic	Fosroc Ltd	WI	HSE 3544
Protim Grade Basic C	Fosroc Ltd	WI	HSE 3543
Protim Grade Basic CWR	Fosroc Ltd	WI	HSE 3545
Protim Grade Basic WR	Fosroc Ltd	WI	HSE 3546
# Wilko Clear Wood Preserver	Wilkinson Home and Garden Stores	WA WP	HSE 3599

243 TRIBUTYLTIN OXIDE and GAMMA-HCH and PENTACHLOROPHENYL LAURATE

# Larsen Clear Wood Preservative	Larsen Manufacturing Ltd	WA	HSE 3771

WOOD PRESERVATIVES

Product Name	Marketing Company	Use	Reg. No

244 TRIBUTYLTIN OXIDE and GAMMA-HCH and TRI (HEXYLENE GLYCOL) BIBORATE

# Timber Preservative Solvent Based Fungicidal Insecticide 30	Ambersil Ltd	WP	HSE 3006

245 TRIBUTYLTIN OXIDE and PENTACHLOROPHENOL

Celpruf JP	Rentokil Ltd	WI	HSE 4548
Celpruf JP WR	Rentokil Ltd	WI	HSE 4543
* Protim FDR-H	Fosroc Ltd	WI	HSE 4438
Protim FDR-H	Protim Solignum Ltd	WI	HSE 4985
* Protim Joinery Lining	Fosroc Ltd	WP	HSE 4426
Protim Joinery Lining	Protim Solignum Ltd	WP	HSE 4977
* Protim JP	Fosroc Ltd	WI	HSE 4326
Protim JP	Protim Solignum Ltd	WI	HSE 4978
* Protim R Clear	Fosroc Ltd	WI	HSE 3541
Protim R Clear	Protim Solignum Ltd	WI	HSE 5013

246 TRIBUTYLTIN OXIDE and PENTACHLOROPHENYL LAURATE

* Larsen Joinery Grade	Larsen Manufacturing Ltd	WI	HSE 3709

247 TRIBUTYLTIN OXIDE and ZINC NAPHTHENATE and CARBENDAZIM

# Dulux Preservative Wood Primer	ICI Paints Division	WA	HSE 3826
# Dulux Weathershield Exterior Preservative Primer	ICI Paints Divsion	WA	HSE 3819
# Dulux Weathershield Preservative Basecoat	ICI Paints Division	WA	HSE 4026

248 TRIBUTYLTIN PHOSPHATE

# Celpruf TP	Rentokil Ltd	WI	HSE 3984
# Celpruf TP WR	Rentokil Ltd	WI	HSE 3985
# Protim FDR 240	Fosroc Ltd	WI	HSE 3941
# Protim JP 240	Fosroc Ltd	WI	HSE 3940
# Vacsol J Ready To Use	Hickson Timber Products Ltd	WI	HSE 3680
# Vacsol JWR Ready To Use	Hickson Timber Products Ltd	WI	HSE 3681

WOOD PRESERVATIVES

Product Name	Marketing Company	Use	Reg. No
249 TRIBUTYLTIN PHOSPHATE and PERMETHRIN			
# Celpruf TPM	Rentokil Ltd	WI	HSE 4028
# Celpruf TPM WR	Rentokil Ltd	WI	HSE 4029
# Protim 240	Fosroc Ltd	WI	HSE 3860
# Protim 240 C	Fosroc Ltd	WI	HSE 3988
# Protim 240 CWR	Fosroc Ltd	WI	HSE 3972
# Protim 240 WR	Fosroc Ltd	WI	HSE 3861
# Timbercare WRQD Ready To Use	Hickson Timber Products Ltd	WI	HSE 3679
# Vacsele	Hickson Timber Products Ltd	WI	HSE 3678
# Vacsol MWR Ready For Use	Hickson Timber Products Ltd	WI	HSE 3677
# Vacsol WR Ready To Use	Hickson Timber Products Ltd	WI	HSE 3676
250 ZINC NAPHTHENATE			
Barkep Wood Preserver	Bartoline Ltd	WA	HSE 4457
Carbo Wood Preservative	Talke Chemical Company Ltd	WA	HSE 3821
Cover Plus Clear Wood Preservative	Macpherson Paints Ltd	WA	HSE 4517
Flexarb Clear Preservative	Macpherson Paints Ltd	WA WP	HSE 3770
Spencer Wood Preservative	Spencer Coatings Ltd	WA	HSE 3947
Teamac Wood Preservative	Teal and Mackrill Ltd	WA	HSE 4428
# TRC Water Repellent Wood Preserver	Texas Refinery Corporation of Canada Ltd	WP	HSE 3687
Wood Preservative	C W Wastnage Ltd	WA WI WP	HSE 5284
251 ZINC NAPHTHENATE and CARBENDAZIM and TRIBUTYLTIN OXIDE			
# Dulux Preservative Wood Primer	ICI Paints Division	WA	HSE 3826
# Dulux Weathershield Exterior Preservative Primer	ICI Paints Division	WA	HSE 3819
# Dulux Weathershield Preservative Basecoat	ICI Paints Division	WA	HSE 4026
252 ZINC NAPHTHENATE and DICHLOFLUANID			
Masterstroke Wood Preserver	Akzo-Nobel Ltd	WA	HSE 5689

WOOD PRESERVATIVES

Product Name	Marketing Company	Use	Reg. No

253 ZINC NAPHTHENATE and DICHLOFLUANID and TRI(HEXYLENE GLYCOL) BIBORATE

# Dulux Exterior Preservative Basecoat	ICI Paints Division	WA	HSE 4035
Weathershield Exterior Preservative Basecoat	ICI Paints Division	WA WP	HSE 5245
Weathershield Exterior Preservative Basecoat	ICI Paints Division	WA	HSE 4036
Weathershield Preservative Primer	ICI Paints Division	WA WP	HSE 5244
Weathershield Preservative Primer	ICI Paints Division	WA	HSE 4013

254 ZINC NAPHTHENATE and GAMMA-HCH and PENTACHLOROPHENOL

Protim 200	Fosroc Ltd	WI	HSE 3669

255 ZINC OCTOATE

B and Q Clear Wood Preserver	B and Q Ltd	WA	HSE 3013
B and Q Exterior Wood Preservative	B and Q Ltd	WA	HSE 4705
B and Q Formula Wood Preserver	B and Q Ltd	WA	HSE 3014
Co-op Exterior Wood Preserver	Co-Operative Wholesale Society Ltd	WA	HSE 3640
Crown Timber Preservative	Crown Chemicals Ltd	WP	HSE 5442
# Homework All Purpose Wood Preserver	Kalon Group Plc	WA	HSE 3087
# Homework Exterior Wood Preservative	Kalon Group Plc	WA	HSE 3088
Leyland Timbrene Environmental Formula	Leyland Paint Company	WA	HSE 4801
Leyland Universal Preservative Base	Kalon Group Plc	WA WP	HSE 3760
Nitromors Timbrene Environmental Formula Wood Preservative	Henkel Home Improvements	WA	HSE 4108
PLA Wood Preserver	PLA Products (Chemicals) Ltd	WA	HSE 4021
PLA Wood Preserver	PLA Products Ltd	WA WI WP	HSE 5340
Premier Pro-Tec(s) Environmental Formula Wood Preservative	Premier Decorative Products	WA	HSE 4689
Rentokil Dry Rot and Wet Rot Treatment	Rentokil Ltd	WA SA	HSE 4378
Ronseal Low Odour Wood Preserver	Roncraft	WA	HSE 4919

WOOD PRESERVATIVES

Product Name	Marketing Company	Use	Reg. No

255 ZINC OCTOATE—continued

Product Name	Marketing Company	Use	Reg. No
Square Deal Deep Protection Wood Preserver	Texas Homecare Ltd	WA	HSE 4640
* Square Deal Wood Preserver	Texas Homecare Ltd	WA	HSE 3693
Timber Preservative	Antel-Tridex	WP	HSE 4846
Timbrene Wood Preserver	Kalon Group Plc	WA WP	HSE 4966
Trade Ronseal Low Odour Wood Preserver	Roncraft	WI WP	HSE 4932

256 ZINC OCTOATE and BORIC ACID

Product Name	Marketing Company	Use	Reg. No
Celpruf BZ	Rentokil Ltd	WI	HSE 4188
Celpruf BZ WR	Rentokil Ltd	WI	HSE 4189
Dry Rot Fluid (D) EC	Rentokil Ltd	WP	HSE 4194
Rentokil Dry Rot Paste (D)	Rentokil Ltd	WP	HSE 3987

257 ZINC OCTOATE and BORIC ACID and PERMETHRIN

Product Name	Marketing Company	Use	Reg. No
Celpruf BZP	Rentokil Ltd	WI	HSE 4190
Celpruf BZP WR	Rentokil Ltd	WI	HSE 4191
Premium Grade Wood Treatment	Rentokil Ltd	WA	HSE 4193

258 ZINC OCTOATE and COPPER NAPHTHENATE

Product Name	Marketing Company	Use	Reg. No
Dulux Weathershield Timber Preservative	ICI Plc	WA	HSE 3805
Weathershield Exterior Timber Preservative	ICI Paints Division	WA	HSE 4915

259 ZINC OCTOATE and CYPERMETHRIN

Product Name	Marketing Company	Use	Reg. No
Palavac Industrial Timber Preservative	Palace Chemicals Ltd	WI WP	HSE 5494

260 ZINC OCTOATE and DICHLOFLUANID

Product Name	Marketing Company	Use	Reg. No
Aquaseal Wood Preserver	Aquaseal Ltd	WA	HSE 4252
Exterior Preservative Primer	Windeck Paints Ltd	WA	HSE 3857
Febwood Wood Preserver	FEB Ltd	WA	HSE 4255

Product Name	Marketing Company	Use	Reg. No

260 ZINC OCTOATE and DICHLOFLUANID—continued

Great Mills Exterior Preservative Primer	Great Mills (Retail) Ltd	WA	HSE 3704
Texas Exterior Wood Preserver	Texas Homecare	WA	HSE 4128
Timbercare Microporous Exterior Preservative	Manders Paints Ltd	WA WP	HSE 5157
Wilko Exterior System Preservative Primer	Wilkinson Home and Garden Stores	WA	HSE 3686

261 ZINC OCTOATE and DICHLOFLUANID and PERMETHRIN

OS Color Wood Stain and Preservative	Ostermann and Scheiwe GmbH	WA	HSE 5531
OS Color WR	Ostermann and Scheiwe GmbH	WA	HSE 5364

262 ZINC OCTOATE and DICHLOFLUANID and TRI (HEXYLENE GLYCOL) BIBORATE

Timber Preservative Clear TFP7	Johnstone's Paints Plc	WA	HSE 4016

263 ZINC OCTOATE and PERMETHRIN

Chel-Wood Preserver/Woodworm Dry Rot Killer	Chelec Ltd	WA	HSE 4172
Cuprinol Difusol S	Cuprinol Ltd	WP	HSE 3750
# Green Range Mayonnaise	Cementone-Beaver Ltd	WP	HSE 3801
Green Range Timber Treatment Paste	Cementone-Beaver Ltd	WP	HSE 4651
Larvac 300	Larsen Manufacturing Ltd	WI	HSE 4461
Palace Timbertreat Ecology	Palace Chemicals Ltd	WP	HSE 4931
Premium Wood Treatment	Rentokil Ltd	WA SA	HSE 5350
Rentokil Dual Purpose Fluid	Rentokil Ltd	WP	HSE 4074
Rentokil Wood Preservative	Rentokil Ltd	WA	HSE 4195
Solignum Colourless	Protim Solignum Ltd	WA WI WP	HSE 5549
# Solignum Remedial Mayonnaise	Solignum Ltd	WP	HSE 3899
Solignum Wood Preservative Paste	Protim Solignum Ltd	WP	HSE 5001
* Solignum Wood Preservative Paste	Solignum Ltd	WP	HSE 4082
* Trimethrin 6	Triton Chemical Manufacturing Company Ltd	WA	HSE 3918

Product Name	Marketing Company	Use	Reg. No

263 ZINC OCTOATE and PERMETHRIN—continued

* Triton Tripaste	Triton Chemical Manufacturing Company Ltd	WA	HSE 3927
Wood Preservative Clear	Rentokil Ltd	WA	HSE 4077
Woodtreat BP	Stanhope Chemical Products Ltd	WP	HSE 3085

264 ZINC SALT OF SYNTHETIC (C8-C12) CARBOXYLIC ACIDS

* Cuprinol WP CPTF 5	Cuprinol Ltd	WI	HSE 4718

265 ZINC VERSATATE

Celpruf Primer ZV	Rentokil Ltd	WI WP	HSE 5379
Celpruf ZO	Rentokil Ltd	WI	HSE 4683
Celpruf ZO WR	Rentokil Ltd	WI	HSE 4684
* Cuprinol WP CPTF 4	Cuprinol Ltd	WI	HSE 4719
Protim Curative ZF	Fosroc Ltd	WA	HSE 4207
Protim JP 220	Fosroc Ltd	WI	HSE 3992
Protim JP 220	Protim Solignum Ltd	WI	HSE 5197
* Protim Paste 220 F	Fosroc Ltd	WP	HSE 4102
Protim Paste 220 F	Protim Solignum Ltd	WI WP	HSE 4988
Vacsol J (2:1 Conc) 2744/2745	Hickson Timber Products Ltd	WI	HSE 4687
Vacsol J 2742/2743	Hickson Timber Products Ltd	WI	HSE 4686
Vacsol J 2:1 Concentrate	Hickson Timber Products Ltd	WI	HSE 3965
Vacsol J Ready To Use	Hickson Timber Products Ltd	WI	HSE 3967
Vacsol JWR (2:1 Conc) 2642/2643	Hickson Timber Products Ltd	WI	HSE 4677
Vasol JWR 2640/2641	Hickson Timber Products Ltd	WI	HSE 4680
Vacsol JWR 2:1 Concentrate	Hickson Timber Products Ltd	WI	HSE 3966
Vacsol JWR Ready To Use	Hickson Timber Products Ltd	WI	HSE 3968
Vacsol P Ready To Use (2335)	Hickson Timber Products Ltd	WI	HSE 4116
Wood Preserver	Cementone Beaver Ltd	WA	HSE 4376

266 ZINC VERSATATE and CYPERMETHRIN

Dipsar G R	Cementone-Beaver Ltd	WI	HSE 4449
* Green Range Dipsar	Cementone Beaver Ltd	WI	HSE 4186

WOOD PRESERVATIVES

Product Name	Marketing Company	Use	Reg. No
267 ZINC VERSATATE and PERMETHRIN			
Cedasol Ready To Use (2306)	Hickson Timber Products Ltd	WI	HSE 4117
Celpruf Zop	Rentokil Ltd	WI	HSE 4681
Celpruf Zop WR	Rentokil Ltd	WI	HSE 4682
Imersol 2410	Hickson Timber Products Ltd	WI	HSE 4674
* Imersol WRQD Ready To Use (2523)	Hickson Timber Products Ltd	WI	HSE 3969
Protim 220	Fosroc Ltd	WI	HSE 3912
Protim 220	Protim Solignum Ltd	WI	HSE 5198
Protim 220 CWR	Fosroc Ltd	WA	HSE 3995
Protim 220 CWR	Protim Solignum Ltd	WA WI WP	HSE 5196
Protim 220 WR	Fosroc Ltd	WI	HSE 3913
Protim 220 WR	Protim Solignum Ltd	WI	HSE 5195
Protim Curative Z	Fosroc Ltd	WA	HSE 4268
* Protim Paste 220	Fosroc Ltd	WP	HSE 4058
Protim Paste 220	Protim Solignum Ltd	WI WP	HSE 4983
Protim WR 220	Fosroc Ltd	WA	HSE 4113
Solignum Universal	Protim Solignum Ltd	WA WI WP	HSE 5019
* Solignum Universal	Solignum Ltd	WA	HSE 4147
Vacsele 2611	Hickson Timber Products Ltd	WP	HSE 4675
Vacsele P2312	Hickson Timber Products Ltd	WP	HSE 4363
Vacsele Ready To Use (2605)	Hickson Timber Products Ltd	WP	HSE 3976
Vacsol (2:1 Conc) 2711/2712	Hickson Timber Products Ltd	WI	HSE 4685
Vacsol 2709/2710	Hickson Timber Products Ltd	WI	HSE 4678
Vacsol 2:1 Concentrate	Hickson Timber Products Ltd	WI	HSE 3977
Vacsol P (2310)	Hickson Timber Products Ltd	WI	HSE 4377
Vacsol Ready To use	Hickson Timber Products Ltd	WI	HSE 3978
Vacsol WR (2:1 Conc) 2614/2615	Hickson Timber Products Ltd	WI	HSE 4679
Vacsol WR 2612/2613	Hickson Timber Products Ltd	WI	HSE 4676
Vacsol WR 2:1 Concentrate	Hickson Timber Products Ltd	WI	HSE 3824
Vacsol WR Ready To Use	Hickson Timber Products Ltd	WI	HSE 3825

2
SURFACE BIOCIDES

Product Name	Marketing Company	Use	Reg. No

Surface Biocides

268 2-PHENYLPHENOL

# Biochem Masonry Fungicide	Biochem Ltd	SP	HSE 4114

269 2-PHENYLPHENOL and BENZALKONIUM CHLORIDE

Chelwash	Chelec Ltd	SA	HSE 3096
Deadly Nightshade	Andura Textured Masonry Coatings Ltd	SP	HSE 4895
Earnshaws Fungal Wash	E Earnshaw and Company (1965) Ltd	SA	HSE 4256
Fungicidal Algaecidal Bacteriacidal Wash	Ultrabond Ltd	SA	HSE 4216
Fungicidal Solution FL.2	Manders Paints Ltd	SA	HSE 4245
Fungicidal Wash	C W Wastnage Ltd	SA	HSE 3072
Fungicidal Wash Solution	Johnstones Paints Plc	SA SP	HSE 3776
Fungiguard	Spaldings Ltd	SP	HSE 5582
# Iscosan Fungicidal Solution	Spencer (Aberdeen) Plc	SA SP	HSE 3705
Johnstone's Fungicidal Wash	Johnstone's Paints Plc	SA SP	HSE 5678
# Marley Patio Cleaner	Marley Adhesives	SA	HSE 3832
Microstel	Microbicide Ltd	SP	HSE 4198
Nitromors Mould Remover	Henkel Home Improvements and Adhesive Products	SA	HSE 5507
Permacide Masonry Fungicide	Triton Perma Industries Ltd	SP	HSE 5570
Permagard FWS	Permagard Products Ltd	SP	HSE 5485
Permarock Fungicidal Wash	Permarock Products Ltd	SP	HSE 5193
POB Mould Treatment	The Robert McBride Group Ltd	SA	HSE 4076
Spencer Fungicidal Treatment	Spencer Coatings Ltd	SP	HSE 4155
Tetra Concentrated Mould Cleaner	Tetrosyl (Building Products) Ltd	SA	HSE 4206
Z. 144 Fungicidal Wash	Crosbie Coatings Ltd	SA SP	HSE 3003

SURFACE BIOCIDES

Product Name	Marketing Company	Use	Reg. No
270 3-IODO-2-PROPYNYL-N-BUTYL CARBAMATE			
Fungicidal Wall Solution	Sovereign Chemical Industries Ltd	SP	HSE 5387
Remtox Dry Rot FWS.	Remtox Chemicals Ltd	SP	HSE 5188
Remtox Dry Rot Paint	Remtox Chemicals Ltd	SA WA	HSE 5187
Remtox Fungicidal Wall Solution RS	Remtox (Chemicals) Ltd	SP	HSE 5484
Remtox Microactive FWS W6	Remtox Chemicals Ltd	SP	HSE 5589
Safeguard Fungicidal Micro Emulsifiable Concentrate	Safeguard Chemicals Ltd	SP WF	HSE 5648
271 ALKYLARYLTRIMETHYL AMMONIUM CHLORIDE			
Abicide 82	Langlow Products Ltd	SA	HSE 4470
Albany Fungicidal Wash	C Brewer and Sons Ltd	SA	HSE 4471
Anti-Mould Solution	Macpherson Paints Ltd	SA WA	HSE 4873
Beeline Fungicidal Wash	Ward Bekker Ltd	SA	HSE 4224
Betterware Anti-Fungus and Mildew Wipe	Betterware Sales Ltd	SA	HSE 4043
Betterware Fungus and Mildew Killer	Betterware Sales Ltd	SA	HSE 4049
Bio-Kil Scrub Out Black Mould/Refill	Bio-Kil Chemicals Ltd	SA	HSE 4810
# Biocidal Wash	Liquid Plastics Ltd	SP	HSE 4001
Fungishield Sterilising Solution Concentrate GS36	Glixtone Ltd	SA	HSE 4283
Fungishield Sterilising Solution GS37	Glixtone Ltd	SA	HSE 4284
Kibes Sterilising Solution Concentrate GS 36	Kibes (UK) Insulation Ltd	SA SP	HSE 5493
Mould Cleaner	J H Woodman Ltd	SA	HSE 4291
* Resistone PC	ABM Chemicals Ltd	SP WP	HSE 3774
* Wickes Fungicidal Wash	Wickes Building Supplies Ltd	SA SP	HSE 3122
272 ALKYLARYLTRIMETHYL AMMONIUM CHLORIDE and BORIC ACID			
Kleeneze Anti-Mould Spray	Kleeneze Ltd	SA	HSE 4599
273 ALKYLARYLTRIMETHYL AMMONIUM CHLORIDE and DISODIUM OCTABORATE			
Boracol 10RH Surface Biocide	Channelwood Preservations Ltd	SP	HSE 4911
Remtox Borocol 10 RH Masonry Biocide	Remtox Chemicals Ltd	SP	HSE 4100

SURFACE BIOCIDES

Product Name	Marketing Company	Use	Reg. No

274 ALKYLARYLTRIMETHYL AMMONIUM CHLORIDE and PERMETHRIN and PYRETHRINS

* Roxem C	Horton Hygiene Co Ltd	SP IP	HSE 3123

275 ALKYLARYLTRIMETHYL AMMONIUM CHLORIDE and TRIBUTYLTIN OXIDE

# Killgerm Masonry Sterilant	Killgerm Chemicals Ltd	SP WP	HSE 3125

276 AZACONAZOLE and PERMETHRIN

Fongix SE Total Treatment for Wood	Liberon Waxes Ltd	SA WA	HSE 5610
Fongix SE Total Treatment for Wood	V33 UK	SA WA	HSE 4288
Woodworm Killer and Rot Treatment	Liberon Waxes Ltd	SA WA	HSE 4826

277 BENZALKONIUM CHLORIDE

A-Zygo 3 Sterilising Solution	Premier Condensation and Mould Control Ltd	SP	HSE 5368
Algae Remover	Sadolin (UK) Ltd	SA WA	HSE 4781
Bedclear	Fargro Ltd	SP	HSE 5179
Bio-Kil Dentolite Solution	Bio-Kil Chemicals Ltd	SA	HSE 4593
BN Algae Remover	Morgan Nehra Holdings Ltd	SA SP	HSE 5381
BN Mosskiller	Morgan Nehra Holdings Ltd	SA SP	HSE 5080
Conc Quat	Semitec Ltd	SP	HSE 4904
Cuprinol Cuprotect Exterior Fungicidal Wash	Cuprinol Ltd	SA	HSE 4610
Cuprinol Cuprotect Exterior Fungicide	Cuprinol Ltd	SA SP	HSE 5526
Cuprinol Cuprotect Fungicidal Patio Cleaner	Cuprinol Ltd	SA	HSE 4606
Cuprinol Cuprotect Patio Cleaner	Cuprinol Ltd	SA SP	HSE 5527
Cuprotect Exterior Fungicide	Cuprinol Ltd	SA	HSE 4165
Cuprotect Patio Cleaner	Cuprinol Ltd	SA	HSE 4164
# ESMI CA: 20 Concentrated Algicide	ESMI Chemicals Ltd	SP	HSE 3113
# ESMI Mouldicide for Walls, Ceilings and Cupboards	ESMI Chemicals Ltd	SA	HSE 3114
Fungicidal Solution	Cementone-Beaver Ltd	SA SP	HSE 4928
Gloquat RP	Rhone-Poulenc Chemicals	SP WP	HSE 4815
Homebase Fungicidal Wash	Homebase Ltd	SA SP	HSE 4860

SURFACE BIOCIDES

Product Name	Marketing Company	Use	Reg. No
277 **BENZALKONIUM CHLORIDE**—continued			
Homebase Mould Cleaner	Homebase Ltd	SA SP	HSE 5707
Howes Olympic Algaecide	Killgerm Chemicals Ltd	SP	HSE 4845
Interior Mould Remover Wipes	Cadismark Ltd	SA	HSE 5695
Leyland Sterilisation Wash	Leyland Paint Company	SP	HSE 4751
LPL Biocidal Wash	Liquid Plastics Ltd	SP	HSE 4479
Moss-Cure	Cementone-Beaver Ltd	SP	HSE 4478
Mould Killer	Cadismark Ltd	SA	HSE 5047
* Mycofen Concentrate	Mycofen Systems Ltd	SA	HSE 3915
* Mycospray	Mycofen Systems Ltd	SA	HSE 4010
Paramos	Chemsearch	SP	HSE 4524
* Payless Fungicidal Wash	Payless DIY Ltd	SA	HSE 3909
Remtox Remwash Extra	Remtox Chemicals Ltd	SP	HSE 5178
Snowcem Algicide	Snowcem PMC Ltd	SA SP	HSE 4806
Square Deal Fungicidal Solution	Texas Homecare Ltd	SA	HSE 4644
* Square Deal Fungicidal Wash	Texas Homecare Ltd	SA	HSE 3759
# Vulcanite CA: 20 Concentrated Algicide	Vulcanite Ltd	SP	HSE 3112
# Vulcanite Mouldicide for Walls, Ceilings and Cupboards	Vulcanite Ltd	SA	HSE 3115
278 **BENZALKONIUM CHLORIDE and 2-PHENYLPHENOL**			
Chelwash	Chelec Ltd	SA	HSE 3096
Deadly Nightshade	Andura Textured Masonry Coatings Ltd	SP	HSE 4895
Earnshaws Fungal Wash	E Earnshaw and Company (1965) Ltd	SA	HSE 4256
Fungicidal Algaecidal Bacteriacidal Wash	Ultrabond Ltd	SA	HSE 4216
Fungicidal Solution FL.2	Manders Paints Ltd	SA	HSE 4245
Fungicidal Wash	C W Wastnage Ltd	SA	HSE 3072
Fungicidal Wash Solution	Johnstone's Paints Plc	SA SP	HSE 3776
Fungiguard	Spaldings Ltd	SP	HSE 5582
# Iscosan Fungicidal Solution	Spencer (Aberdeen) Plc	SA SP	HSE 3705
Johnstone's Fungicidal Wash	Johnstone's Paints Plc	SA SP	HSE 5678
# Marley Patio Cleaner	Marley Adhesives	SA	HSE 3832

SURFACE BIOCIDES

Product Name	Marketing Company	Use	Reg. No
278 **BENZALKONIUM CHLORIDE and 2-PHENYLPHENOL**—continued			
Microstel	Microbicide Ltd	SP	HSE 4198
Nitromors Mould Remover	Henkel Home Improvements and Adhesive Products	SA	HSE 5507
Permacide Masonry Fungicide	Triton Perma Industries Ltd	SP	HSE 5570
Permagard FWS	Permagard Products Ltd	SP	HSE 5485
Permarock Fungicidal Wash	Permarock Products Ltd	SP	HSE 5193
POB Mould Treatment	The Robert McBride Group Ltd	SA	HSE 4076
Spencer Fungicidal Treatment	Spencer Coatings Ltd	SP	HSE 4155
Tetra Concentrated Mould Cleaner	Tetrosyl (Building Products) Ltd	SA	HSE 4206
Z.144 Fungicidal Wash	Crosbie Coatings Ltd	SA SP	HSE 3003
279 **BENZALKONIUM CHLORIDE and BORIC ACID**			
Microguard Mouldicidal Wood Preserver	Permagard Products Ltd	SA WA	HSE 5100
280 **BENZALKONIUM CHLORIDE and CARBENDAZIM and DIALKYLDIMETHYL AMMONIUM CHLORIDE**			
Algotox	May and Baker Garden Care	SA	HSE 4901
Exterior Mouldicide	Rentokil Ltd	SA	HSE 4913
Mould Cure	Rentokil Ltd	SA	HSE 4914
Rentokil Mould Cure Spray	Rentokil Ltd	SA	HSE 4227
281 **BENZALKONIUM CHLORIDE and DIALKYLDIMETHYL AMMONIUM CHLORIDE**			
Sterilising Fluid	Rentokil Ltd	SP WP	HSE 4605
282 **BENZALKONIUM CHLORIDE and DISODIUM OCTABORATE**			
Boracol 10 Rh	Channelwood Preservations Ltd	SA WA	HSE 4104
Boracol B8.5 RH Mouldicide/Wood Preservative	Channelwood Preservations Ltd	SA WA	HSE 4809
Cuprinol No More Mould Fungicidal Spray	Cuprinol Ltd	SA	HSE 4162
Cuprotect Fungicidal Spray	Cuprinol Ltd	SA	HSE 4221

SURFACE BIOCIDES

Product Name	Marketing Company	Use	Reg. No
282 BENZALKONIUM CHLORIDE and DISODIUM OCTABORATE—continued			
Cuprotect Interior Mould Killer	Cuprinol Ltd	SA	HSE 4163
Deepflow 11 Inorganic Boron Masonary Biocide	Safeguard Chemicals Ltd	SP	HSE 5528
Deepwood 50 Inorganic Boron Wood Preservative Paste	Safeguard Chemicals Ltd	SP WF	HSE 5596
283 BENZALKONIUM CHLORIDE and NAPHTHALENE and PERMETHRIN and PYRETHRINS			
Roxem D	Horton Hygiene Company	SP IP	HSE 5107
284 BENZALKONIUM CHLORIDE and TRIBUTYLTIN OXIDE			
# Fungaside	Anti Pollution Chemicals Ltd	SP	HSE 3584
# Nubex Fungicide QT Concentrate	Albright and Wilson Ltd	SP WP	HSE 3166
285 BORIC ACID			
Rentokil Dry Rot Fluid (E)	Rentokil Ltd	SP WP	HSE 3997
Rentokil Dry Rot Fluid (E) for Bonded Warehouses	Rentokil Ltd	SP WP	HSE 3986
286 BORIC ACID and ALKYLARYLTRIMETHYL AMMONIUM CHLORIDE			
Kleeneze Anti-Mould Spray	Kleeneze Ltd	SA	HSE 4599
287 BORIC ACID and BENZALKONIUM CHLORIDE			
Microguard Mouldicidal Wood Preserver	Permagard Products Ltd	SA WA	HSE 5100
288 CARBENDAZIM and DIALKYLDIMETHYL AMMONIUM CHLORIDE and BENZALKONIUM CHLORIDE			
Algotox	May and Baker Garden Care	SA	HSE 4901
Exterior Mouldicide	Rentokil Ltd	SA	HSE 4913
Mould Cure	Rentokil Ltd	SA	HSE 4914
Rentokil Mould Cure Spray	Rentokil Ltd	SA	HSE 4227

4

SURFACE BIOCIDES

Product Name	Marketing Company	Use	Reg. No
289 DIALKYLDIMETHYL AMMONIUM CHLORIDE			
Bioclean 911	Biotech Environmental Services Ltd	SA SP	HSE 5192
Clearmold D	Mould Growth Consultants Ltd	SA	HSE 3931
Halophane Bonding Solution	Mould Growth Consultants Ltd	SP	HSE 3929
Halophen BM1165L	Mould Growth Consultants Ltd	SP	HSE 3930
Lichenite	Mould Growth Consultants Ltd	SP	HSE 3936
Mosscheck	Biotech Environmental Services Ltd	SA	HSE 4870
Mould Inhibitor	Castle Products	SA WA	HSE 3582
Mouldcheck Barrier	Biotech Environmental Services Ltd	SA	HSE 4506
Mouldcheck Barrier 2	Biotech Environmental Services Ltd	SA SP	HSE 4856
# RLT Clearmold Spray	Mould Growth Consultants Ltd	SA	HSE 3932
RLT Clearmould Spray	Mould Growth Consultants Ltd	SA SP	HSE 4824
RLT Halophen	Mould Growth Consultants Ltd	SA SP	HSE 4081
# RLT Halophen	Mould Growth Consultants Ltd	SP	HSE 3814
RLT Halophen DS	Mould Growth Consultants Ltd	SA SP	HSE 4483
290 DIALKYLDIMETHYL AMMONIUM CHLORIDE and BENZALKONIUM CHLORIDE			
Sterilising Fluid	Rentokil Ltd	SP WP	HSE 4605
291 DIALKYLDIMETHYL AMMONIUM CHLORIDE and BENZALKONIUM CHLORIDE and CARBENDAZIM			
Algotox	May and Baker Garden Care	SA	HSE 4901
Exterior Mouldicide	Rentokil Ltd	SA	HSE 4913
Mould Cure	Rentokil Ltd	SA	HSE 4914
Rentokil Mould Cure Spray	Rentokil Ltd	SA	HSE 4227

SURFACE BIOCIDES

Product Name	Marketing Company	Use	Reg. No
292 DICHLOROPHEN			
Bactdet D	Mould Growth Consultants Ltd	SP	HSE 3809
Denso Mouldshield Biocidal Cleanser	Winn and Coales (Denso) Ltd	SA	HSE 4731
Denso Mouldshield Surface Biocide	Winn and Coales (Denso) Ltd	SP	HSE 4732
Fungo	Dax Products Ltd	SA	HSE 4768
Halodec	Mould Growth Consultants Ltd	SP	HSE 4454
Halophane 105	Mould Growth Consultants Ltd	SP	HSE 3817
Halophane 106M	Mould Growth Consultants Ltd	SP	HSE 3818
Halophane No.1 Aerosol	Mould Growth Consultants Ltd	SP	HSE 3815
Halophane No.3 Aerosol	Mould Growth Consultants Ltd	SP	HSE 3816
Mouldcheck Spray	Biotech Environmental Services Ltd	SA	HSE 4093
Mouldcheck Sterilizer	Biotech Environmental Services Ltd	SA	HSE 4505
# Mycodet	Mycofen Systems Ltd	SA SP	HSE 3585
* Mycodet	Mycofen Systems Ltd	SA	HSE 4276
* Mycofen Barrier	Mycofen Systems Ltd	SP	HSE 3916
Nipacide DP30	Nipa Laboratories Ltd	SP	HSE 4925
Panacide M	Coalite Chemicals	SP	HSE 3853
Panacide M21	Coalite Chemicals	SP	HSE 4923
Panaclean 736	Coalite Chemicals	SP	HSE 5075
RLT Bactdet	Mould Growth Consultants Ltd	SA	HSE 3928
SP.153 Sterilising Detergent Wash	Joseph Mason Plc	SP	HSE 3920
SP.154 Fungicidal and Bactericidal Treatment	Joseph Mason Plc	SP	HSE 3919

SURFACE BIOCIDES

Product Name	Marketing Company	Use	Reg. No
293 DISODIUM OCTABORATE and ALKYLARYLTRIMETHYL AMMONIUM CHLORIDE			
Boracol 10 RH Surface Biocide	Channelwood Preservations Ltd	SP	HSE 4911
Remtox Borocol 10 RH Masonry Biocide	Remtox Chemicals Ltd	SP	HSE 4100
294 DISODIUM OCTABORATE and BENZALKONIUM CHLORIDE			
Boracol 10 Rh	Channelwood Preservations Ltd	SA WA	HSE 4104
Boracol B8.5 RH Mouldicide/Wood Preservative	Channelwood Preservations Ltd	SA WA	HSE 4809
Cuprinol No More Mould Fungicidal Spray	Cuprinol Ltd	SA	HSE 4162
Cuprotect Fungicidal Spray	Cuprinol Ltd	SA	HSE 4221
Cuprotect Interior Mould Killer	Cuprinol Ltd	SA	HSE 4163
Deepflow 11 Inorganic Boron Masonry Biocide	Safeguard Chemicals Ltd	SP	HSE 5528
Deepwood 50 Inorganic Boron Wood Preservative Paste	Safeguard Chemicals Ltd	SP WP	HSE 5596
295 DODECYLAMINE LACTATE and DODECYLAMINE SALICYLATE			
Anti Fungus Wash	Craig and Rose Plc	SP	HSE 4375
Biocide G	Nutec Chemicals Ltd	SA	HSE 3561
Blackfriar Anti Mould Solution 195/9	E Parsons and Sons Ltd	SA	HSE 4231
# BMS Biowash	Suber Industries Ltd	SP	HSE 3520
Crown Fungicidal Wash	Crown Berger Limited	SP	HSE 3569
Dacrylate Fungicidal Wash Solution	Dacrylate Paints Ltd	SA SP	HSE 3675
* Dulux Weathershield Fungicidal Solution	ICI Plc	SA SP	HSE 3742
Environmental Woodrot Treatment	SCI (Building Products)	SP WI WP	HSE 5530
Febflex Fungicide	FEB Ltd	SA SP	HSE 3009
Fungicidal Wash	Polybond Ltd	SA WA	HSE 3601
Fungicidal Wash	Smyth-Morris	SP WP	HSE 3004
Green Range Murosol 20	Cementone-Beaver Ltd	SP	HSE 4456
Hyperion Mould Inhibiting Solution	Hird Hastie Paints Ltd	SP	HSE 4412

SURFACE BIOCIDES

Product Name	Marketing Company	Use	Reg. No
295 **DODECYLAMINE LACTATE and DODECYLAMINE SALICYLATE**—continued			
Kingfisher Fungicidal Wall Solution	Kingfisher Chemicals Ltd	SP	HSE 4014
Larsen Concentrated Algicide	Larsen Manufacturing Ltd	SP	HSE 4968
* Linotol Fungicidal Wash	Linotol Products Ltd	SA	HSE 4223
M-Tec Biocide	M-Protective Coatings Ltd	SP	HSE 4381
Macpherson Anti-Mould Solution	Macpherson Paints Ltd	SP WP	HSE 4791
Macpherson Antimould Solution (RFU)	Akzo Coatings Plc	SA SP	HSE 5190
Metalife Fungicidal Wash	Metalife International Ltd	SA WA	HSE 5406
Mouldrid	S and K Maintenance Products	SA WA	HSE 3671
Mrs Clear Mould Fungicidal Wash	Mr (Polymer Cement Products) Ltd	SP	HSE 3620
Nubex WDR	Nubex Ltd	SP WP	HSE 3810
Nuodex 87	Durham Chemicals	SP WP	HSE 4744
Palace Mould Remover	Palace Chemicals Ltd	SA SP	HSE 5204
Permadex Masonry Fungicide Concentrate	Permagard Products Ltd	SP	HSE 4007
Permoglaze Micatex Fungicidal Treatment	Akzo Coatings Plc	SP WP	HSE 5249
Polycell Mould Cleaner	Polycell Products Ltd	SA SP	HSE 3612
* Protim Wall Solution II	Fosroc Ltd	SA	HSE 4158
# Protim Wall Solution II	Fosroc Ltd	SP	HSE 3837
Protim Wall Solution II	Protim Solignum Ltd	SA SP	HSE 5006
* Protim Wall Solution II Concentrate	Fosroc Ltd	SP	HSE 4060
Protim Wall Solution II Concentrate	Protim Solignum Ltd	SP	HSE 5007
Safeguard Deepwood Surface Biocide Concentrate	Safeguard Chemicals Ltd	SP	HSE 4616
Safeguard Mould and Moss Killer	Safeguard Chemicals Ltd	SA	HSE 4607
Solignum Dry Rot Killer	Protim Solignum Ltd	SA SP	HSE 5025
* Solignum Dry Rot Killer	Solignum Ltd	SA	HSE 3924
Solignum Dry Rot Killer Concentrate	Solignum Ltd	SP	HSE 3925
Solignum Fungicide	Protim Solignum Ltd	SA SP	HSE 5026
* Solignum Fungicide	Solignum Ltd	SA	HSE 3926
Thompson's Interior and Exterior Fungicidal Spray	Roncraft	SA	HSE 5626
Thompson's Interior Mould Killer	Roncraft	SA	HSE 5624

SURFACE BIOCIDES

Product Name	Marketing Company	Use	Reg. No

295 **DODECYLAMINE LACTATE and DODECYLAMINE SALICYLATE**—continued

Torkill Fungicidal Solution 'W'	Tor Coatings Ltd	SP	HSE 4246
Unisil S Silicone Waterproofing Solution	United Paints Ltd	SP	HSE 4758
Unitas Fungicidal Wash-Exterior (Solvent Based)	United Paints Ltd	SP	HSE 4759
Unitas Fungicidal Wash-Interior (Water Based)	United Paints Ltd	SP	HSE 4760
Vallance Fungicidal Wash	Siroflex Ltd	SA SP	HSE 5375
* Weathershield Fungicidal Solution	ICI Paints	SA WA	HSE 4918
Weathershield Fungicidal Wash	ICI Paints Division	SA WA	HSE 5134

296 **DODECYLAMINE LACTATE and DODECYLAMINE SALICYLATE and TRIBUTYLTIN OXIDE**

# Algicide E	Nutec Chemicals Ltd	SA	HSE 3562

297 **DODECYLAMINE LACTATE and DODECYLAMINE SALICYLATE and ZINC OCTOATE**

Algicide E	Nutec Chemicals Ltd	SA	HSE 4214

298 **DODECYLAMINE SALICYLATE and DODECYLAMINE LACTATE**

Anti Fungus Wash	Craig and Rose Plc	SP	HSE 4375
Biocide G	Nutec Chemicals Ltd	SA	HSE 3561
Blackfriar Anti Mould Solution 195/9	E Parsons and Sons Ltd	SA	HSE 4231
# BMS Biowash	Suber Industries Ltd	SP	HSE 3520
Crown Fungicidal Wash	Crown Berger Limited	SP	HSE 3569
Dacrylate Fungicidal Wash Solution	Dacrylate Paints Ltd	SA SP	HSE 3675
* Dulux Weathershield Fungicidal Solution	ICI Plc	SA SP	HSE 3742
Environmental Woodrot Treatment	SCI (Building Products)	SP WI WP	HSE 5530
Febflex Fungicide	FEB Ltd	SA SP	HSE 3009
Fungicidal Wash	Polybond Ltd	SA WA	HSE 3601
Fungicidal Wash	Smyth-Morris	SP WP	HSE 3004
Green Range Murosol 20	Cementone-Beaver Ltd	SP	HSE 4456
Hyperion Mould Inhibiting Solution	Hird Hastie Paints Ltd	SP	HSE 4412

SURFACE BIOCIDES

Product Name	Marketing Company	Use	Reg. No
298 **DODECYLAMINE SALICYLATE and DODECYLAMINE LACTATE**—continued			
Kingfisher Fungicidal Wall Solution	Kingfisher Chemicals Ltd	SP	HSE 4014
Larsen Concentrated Algicide	Larsen Manufacturing Ltd	SP	HSE 4968
* Linotol Fungicidal Wash	Linotol Products Ltd	SA	HSE 4223
M-Tec Biocide	M-Protective Coatings Ltd	SP	HSE 4381
Macpherson Anti-Mould Solution	Macpherson Paints Ltd	SP WP	HSE 4791
Macpherson Antimould Solution (RFU)	Akzo Coatings Plc	SA SP	HSE 5190
Metalife Fungicidal Wash	Metalife International Ltd	SA WA	HSE 5406
Mouldrid	S and K Maintenance Products	SA WA	HSE 3671
Mrs Clear Mould Fungicidal Wash	Mr (Polymer Cement Products) Ltd	SP	HSE 3620
Nubex WDR	Nubex Ltd	SP WP	HSE 3810
Nuodex 87	Durham Chemicals	SP WP	HSE 4744
Palace Mould Remover	Palace Chemicals Ltd	SA SP	HSE 5204
Permadex Masonry Fungicide Concentrate	Permagard Products Ltd	SP	HSE 4007
Permoglaze Micatex Fungicidal Treatment	Akzo Coatings Plc	SP WP	HSE 5249
Polycell Mould Cleaner	Polycell Products Ltd	SA SP	HSE 3612
* Protim Wall Solution II	Fosroc Ltd	SA	HSE 4158
# Protim Wall Solution II	Fosroc Ltd	SP	HSE 3837
Protim Wall Solution II	Protim Solignum Ltd	SA SP	HSE 5006
* Protim Wall Solution II Concentrate	Fosroc Ltd	SP	HSE 4060
Protim Wall Solution II Concentrate	Protim Solignum Ltd	SP	HSE 5007
Safeguard Deepwood Surface Biocide Concentrate	Safeguard Chemicals Ltd	SP	HSE 4616
Safeguard Mould and Moss Killer	Safeguard Chemicals Ltd	SA	HSE 4607
Solignum Dry Rot Killer	Protim Solignum Ltd	SA SP	HSE 5025
* Solignum Dry Rot Killer	Solignum Ltd	SA	HSE 3924
Solignum Dry Rot Killer Concentrate	Solignum Ltd	SP	HSE 3925
Solignum Fungicide	Protim Solignum Ltd	SA SP	HSE 5026
* Solignum Fungicide	Solignum Ltd	SA	HSE 3926
Thompson's Interior and Exterior Fungicidal Spray	Roncraft	SA	HSE 5626
Thompson's Interior Mould Killer	Roncraft	SA	HSE 5624

SURFACE BIOCIDES

Product Name	Marketing Company	Use	Reg. No
298 **DODECYLAMINE SALICYLATE and DODECYLAMINE LACTATE**—continued			
Torkill Fungicidal Solution 'W'	Tor Coatings Ltd	SP	HSE 4246
Unisil S Silicone Waterproofing Solution	United Paints Ltd	SP	HSE 4758
Unitas Fungicidal Wash-Exterior (Solvent Based)	United Paints Ltd	SP	HSE 4759
Unitas Fungicidal Wash-Interior (Water Based)	United Paints Ltd	SP	HSE 4760
Vallance Fungicidal Wash	Siroflex Ltd	SA SP	HSE 5375
* Weathershield Fungicidal Solution	ICI Paints	SA WA	HSE 4918
Weathershield Fungicidal Wash	ICI Paints Division	SA WA	HSE 5134
299 **DODECYLAMINE SALICYLATE and TRIBUTYLTIN OXIDE and DODECYLAMINE LACTATE**			
# Algicide E	Nutec Chemicals Ltd	SA	HSE 3562
300 **DODECYLAMINE SALICYLATE and ZINC OCTOATE and DODECYLAMINE LACTATE**			
Algicide E	Nutec Chemicals Ltd	SA	HSE 4214
301 **NAPHTHALENE and PERMETHRIN and PYRETHRINS and BENZALKONIUM CHLORIDE**			
Roxem D	Horton Hygiene Company	SP IP	HSE 5107
302 **PENTACHLOROPHENOL**			
# Dry Rot and Wet Rot Fluid	Rentokil Ltd	SA WA	HSE 4604
303 **PERMETHRIN and AZACONAZOLE**			
Fongix SE Total Treatment for Wood	Liberon Waxes Ltd	SA WA	HSE 5610
Fongix SE Total Treatment for Wood	V33 UK	SA WA	HSE 4288
Woodworm Killer and Rot Treatment	Liberon Waxes Ltd	SA WA	HSE 4826
304 **PERMETHRIN and PYRETHRINS and ALKYLARYLTRIMETHYL AMMONIUM CHLORIDE**			
* Roxem C	Horton Hygiene Co Ltd	SP IP	HSE 3123

SURFACE BIOCIDES

Product Name	Marketing Company	Use	Reg. No

305 **PERMETHRIN and PYRETHRINS and BENZALKONIUM CHLORIDE and NAPHTHALENE**

Roxem D	Horton Hygiene Company	SP IP	HSE 5107

306 **PERMETHRIN and ZINC OCTOATE**

Premium Wood Treatment	Rentokil Ltd	SA WA	HSE 5350

307 **PYRETHRINS and ALKYLARYLTRIMETHYL AMMONIUM CHLORIDE and PERMETHRIN**

* Roxem C	Horton Hygiene Co Ltd	SP IP	HSE 3123

308 **PYRETHRINS and PERMETHRIN and NAPHTHALENE and BENZALKONIUM CHLORIDE**

Roxem D	Horton Hygiene Company	SP IP	HSE 5107

309 **SODIUM 2-PHENYLPHENOXIDE**

Aqueous Fungicidal Irrigation Fluid	Palace Chemicals Ltd	SA SP	HSE 3794
Aqueous Fungicidal Irrigation Fluid Concentrate	Palace Chemicals Ltd	SP	HSE 3793
Brunosol Concentrate	Stanhope Chemical Products Ltd	SP WP	HSE 4472
Coaltec 50	Coalite Chemicals Division	SP WP	HSE 4761
Coaltec 50TD	Coalite Chemicals Division	SA WA	HSE 4762
Cube Concentrate Fluid for Dry Rot	Kleeneze Ltd	SP WP	HSE 3083
Cube Fluid for Dry Rot	Kleeneze Ltd	SA WA	HSE 3082
Cuprinol Dry Rot Killer S for Brickwork and Masonry	Cuprinol Ltd	SA	HSE 4720
Dry Rot Treatment for Masonry	Cementone-Beaver Ltd	SA	HSE 3983
Fungicidal Wash	Croda Hydrocarbons Ltd	SA SP	HSE 5089
Green Range Fungicidal Concentrate	Cementone Beaver Ltd	SP	HSE 3689
Mar-Cide	Marcher Chemicals Ltd	SP WI WP	HSE 5631
* Ness Fungicidal Concentrate	Cementone Beaver Ltd	SP	HSE 3766
PC-D	Peter Cox Preservation	SP	HSE 4405
PC-D	Peter Cox Preservation	SP	HSE 5719

SURFACE BIOCIDES

Product Name	Marketing Company	Use	Reg. No
309 SODIUM 2-PHENYLPHENOXIDE—continued			
PC-D Concentrate	Peter Cox Preservation	SP	HSE 4404
PC-D Concentrate	Peter Cox Preservation	SP	HSE 5720
PLA Products Dry Rot Killer	PLA Products (Chemicals) Ltd	SA	HSE 4635
PLA Products Dry Rot Killer	PLA Products Ltd	SA SP	HSE 5339
PLA Products Fungicidal Wash	PLA Products Ltd	SA	HSE 5373
Remtox FWS (Low Odour)	Remtox Chemicals Ltd	SP	HSE 3882
Safeguard Fungicidal Wall Solution	Safeguard Chemicals Ltd	SP	HSE 3903
Saturin E30	British Building and Engineering Appliances Plc	SP	HSE 3902
Solignum Anti Fungi Concentrate	Protim Solignum Ltd	SP	HSE 5018
* Solignum Anti Fungi Concentrate	Solignum Ltd	SP	HSE 4232
# Sovereign Fungicidal Wall Solution	Sovereign Chemical Industries Ltd	SP	HSE 3132
Sovereign Fungicidal Wall Solution	Sovereign Chemical Industries Ltd	SP	HSE 3933
Trisol 21	Triton Chemical Manufacturing Company Ltd	SP	HSE 3789
310 SODIUM HYPOCHLORITE			
Brolac Fungicidal Solution	Crown Berger Ltd	SA	HSE 4413
Crown Trade Fungicidal Solution	Crown Berger Ltd	SA	HSE 5076
Crown Trade Stronghold Fungicidal Solution	Crown Berger Ltd	SP	HSE 5383
Do It All Fungicidal Solution	W H Smith Do It All	SA	HSE 3075
Fads Homestyle Fungicidal Solution	Fads AG Stanley Ltd	SA	HSE 4152
Fungicidal Wash	Rentokil Ltd	SA	HSE 4615
Laura Ashley Home Fungicidal Wash	Laura Ashley Ltd	SA	HSE 4192
# Magicote Fungicidal Concentrate	Crown Berger Europe Ltd	SA	HSE 3949
Magicote Masonry Fungicidal Solution	Crown Berger Ltd	SA	HSE 3948
Palace Fungicidal Wash	Palace Chemicals Ltd	SA	HSE 4091
Rainstopper Deadly Nightshade Biowash	Andura Textured Masonry Coatings Ltd	SP	HSE 4894
Sadolin Bioclean 979-9020	Sadolin (UK) Ltd	SA	HSE 4637
Sandtex Fungicide	Akzo Coatings Plc	SA SP	HSE 5258

SURFACE BIOCIDES

Product Name	Marketing Company	Use	Reg. No
311 SODIUM PENTACHLOROPHENOXIDE			
# Biochem Masonry Gel	Biochem Ltd	SP	HSE 4103
Gainserv 685	Industrial Chemical Company (Preston) Ltd	SP WP	HSE 3960
# Mosgo	Agrichem Ltd	SA WA	HSE 4336
Mosgo P	Agrichem Ltd	SP WP	HSE 4337
# Nubex Fungicide FS15	Nubex Ltd	SP WP	HSE 3666
* Protim Plug Compound	Fosroc Ltd	SP WP	HSE 4555
Protim Plug Compound	Protim Solignum Ltd	SP WP	HSE 5012
312 TRIBUTYLTIN OXIDE and ALKYLARYLTRIMETHYL AMMONIUM CHLORIDE			
# Killgerm Masonry Sterilant	Killgerm Chemicals Ltd	SP WP	HSE 3125
313 TRIBUTYLTIN OXIDE and BENZALKONIUM CHLORIDE			
# Fungaside	Anti Pollution Chemicals Ltd	SP	HSE 3584
# Nubex Fungicide QT Concentrate	Albright and Wilson Ltd	SP WP	HSE 3166
314 TRIBUTYLTIN OXIDE and DODECYLAMINE LACTATE and DODECYLAMINE SALICYLATE			
# Algicide E	Nutec Chemicals Ltd	SA	HSE 3562
315 ZINC OCTOATE			
Rentokil Dry Rot and Wet Rot Treatment	Rentokil Ltd	SA WA	HSE 4378
316 ZINC OCTOATE and DODECYLAMINE LACTATE and DODECYLAMINE SALICYLATE			
Algicide E	Nutec Chemicals Ltd	SA	HSE 4214
317 ZINC OCTOATE and PERMETHRIN			
Premium Wood Treatment	Rentokil Ltd	SA WA	HSE 5350

3
INSECTICIDES

Insecticides

Product Name	Marketing Company	Use	Reg. No

318 1, 4-DICHLOROBENZENE

Jertox Moth Crystals	Thornton and Ross Ltd	IA	HSE 5056
Moth Repellent	Boots Company Plc	IA	HSE 5079

319 1, 4-DICHLOROBENZENE and DICHLORVOS

Secto Moth Killer	Secto Company Ltd	IA	HSE 4430

320 1, 4-DICHLOROBENZENE and NAPHTHALENE

Mothaks	Sara Lee Household and Personal Care (UK) Ltd	IA	HSE 5124
Vapona Mothaks	Ashe Ltd	IA	HSE 5682

321 ALKYLARYLTRIMETHYL AMMONIUM CHLORIDE and PERMETHRIN and PYRETHRINS

* Roxem C	Horton Hygiene Co Ltd	IP SP	HSE 3123

322 ALLETHRIN

Spira No Bite Outdoor Mosquito Coils	Travel Accessories (UK) Ltd	IA	HSE 4421

323 ALPHACYPERMETHRIN

Fendona 1.5 SC	Sorex Ltd	IFSP IP	HSE 4092
# Fendona 25 SC	Sorex Ltd	IFSP IP	HSE 3161
Fendona 6SC	Sorex Ltd	IFSP IP	HSE 4455
Fendona 6SC	Sorex Ltd	IP	HSE 3133
Fendona ASC	Rentokil Ltd	IFSP IP	HSE 4946
Fendona Lacquer	Sorex Ltd	IA IP	HSE 4278
Fendona WP	Sorex Ltd	IP	HSE 4299
Littac	Sorex Ltd	IAH IFSP IP	HSE 5176

INSECTICIDES

Product Name	Marketing Company	Use	Reg. No
324 AZAMETHIPHOS			
AL63 Crawling Insect Killer	Roussel Uclaf Environmental Health Ltd	IFSP IP	HSE 5450
Azamethiphos Fly Spray	Rentokil Ltd	IP	HSE 4466
325 BACILLUS THURINGIENSIS VAR ISRAELENSIS			
Bactimos Flowable Concentrate	Koppert (UK) Ltd	IP	HSE 4792
Bactimos Wettable Powder	Koppert (UK) Ltd	IP	HSE 4793
Skeetal Flowable Concentrate	Novo Enzyme Products Ltd	IP	HSE 3603
Skeetal Flowable Concentrate (Aerial)	Novo Enzyme Products Ltd	IP	HSE 3878
Teknar HP-D	Killgerm Chemicals Ltd	IP	HSE 5355
326 BACILLUS THURINGIENSIS VAR ISRAELENSIS (SEROTYPE H14) FERMENTER PRODUCT 600			
Skeetal Flowable Concentrate	Novo Nordisk Bioindustries UK Ltd	IP	HSE 5706
327 BENDIOCARB			
Bendiocarb Dusting Powder	Rentokil Ltd	IP	HSE 5125
Bendiocarb Wettable Powder	Rentokil Ltd	IP	HSE 5416
Camco Ant Powder	Cambridge Animal and Public Health Ltd	IA	HSE 5218
FICAM 20W	Camco	IFSP IP	HSE 3682
Ficam D	Cambridge Animal and Public Health Ltd	IP	HSE 4829
Ficam W	Cambridge Animal and Public Health Ltd	IP	HSE 4831
Ficam W	Camco	IP	HSE 5390
# Ficam Wasp and Hornet Killer	Cambridge Animal and Public Health Ltd	IP	HSE 3886
Ficam Wasp and Hornet Spray	Cambridge Animal and Public Health Ltd	IP	HSE 5253
Ficam Wasp and Hornet Spray	Camco	IP	HSE 4229
Murphy Kil-Ant Powder	Fisons Plc Horticulture Division	IA	HSE 5237
* Raid Ant Bait C	Johnson Wax Ltd	IA	HSE 3935

INSECTICIDES

Product Name	Marketing Company	Use	Reg. No
327 BENDIOCARB—continued			
Rentokil Ant and Insect Powder Professional	Rentokil Ltd	IP	HSE 5386
Secto Ant Bait	Secto Company Ltd	IA	HSE 5088
Wasp Next Killer Professional	Rentokil Ltd	IP	HSE 5651
328 BENDIOCARB and PYRETHRINS			
Ficam Plus	Cambridge Animal and Public Health Ltd	IP	HSE 4830
329 BENDIOCARB and TETRAMETHRIN			
Camco Insect Spray	Camco	IA	HSE 4222
Devcol Houshold Insect Spray	Devcol Ltd	IA	HSE 5512
330 BENZALKONIUM CHLORIDE and NAPHTHALENE and PERMETHRIN and PYRETHRINS			
Roxem D	Horton Hygiene Company	IP SP	HSE 5107
331 BIOALLETHRIN			
Boots Electric Mosquito Killer	Boots Plc	IA	HSE 3871
Boots UK Flying Insect Killer	Boots Company Plc	IA	HSE 4145
Buzz-Off	Ross Consumer Electronics	IA	HSE 3854
Buzz-Off 2	Ross Consumer Electronics	IA	HSE 3855
First Class Moquito Killer Travel Pack	Superdrug Plc	IA	HSE 4078
Globol Pyrethrum Electrical Evaporator	Global Chemicals (UK) Ltd	IA	HSE 4439
Haden Mosquito and Flying Insect Killer	D H Haden Ltd	IA	HSE 5716
Lyvia Mosquito Killer	Lyvia Ltd	IA	HSE 4212
Mosqui – Go Electric	Jack Rogers and Co Ltd	IA	HSE 4083
Mosquito Killer Travel Pack	Culmstock Ltd	IA	HSE 4926
Mosquito Repellent	Shopfield Ltd	IA	HSE 4653
Nippon Flying Insect Killer Tablets	Vitax Ltd	IA	HSE 5199
Pif Paf Mosquito Mats	Roussel Uclaf Environmental Health Ltd	IA	HSE 5156

INSECTICIDES

Product Name	Marketing Company	Use	Reg. No
331 BIOALLETHRIN—continued			
Pif Paf Mosquito Mats	Wellcome Foundation Ltd	IA	HSE 4261
Shelltox Mat 1	Temana International Ltd	IA	HSE 4740
Spira "No Bite" Mosquito Killer	Travel Accessories (UK) Ltd	IA	HSE 3727
Stradz Mosquito Killer	Stradz Ltd	IA	HSE 4144
Wahl Envoyage Mosquito Killer	Wahl Europe Ltd	IA	HSE 5108
Woolworth Mosquito Killer	F W Woolworth Ltd	IA	HSE 4210
332 BIOALLETHRIN and BIORESMETHRIN			
* Coopers Fly Spray BB (Ready To Use)	Coopers Animal Health Ltd	IAH IP	HSE 4041
Pybuthrin 33 BB	Roussel Uclaf Environmental Health Ltd	IFSP IP	HSE 5162
Pybuthrin 33 BB	Wellcome Foundation Ltd	IFSP IP	HSE 4886
333 BIOALLETHRIN and PERMETHRIN			
Ant Gun! 2	Zeneca Garden Care	IA IP	HSE 5184
Ant Gun! 2	Zeneca Garden Centre	IA IP	HSE 5629
Antkiller Spray 2	Zeneca Agrochemicals Limited	IA IP	HSE 5185
Crawling Insect And Ant Killer	Rentokil Ltd	IA	HSE 4650
Creepy Crawly Gun!	Zeneca Garden Care	IA IP	HSE 5183
Fleegard	Bayer Plc	IA	HSE 5564
# Forest Friends Crawling Pest Control	Pharmaceuticals International (UK) Ltd	IA	HSE 4389
# Forest Friends Flying Pest Control	Pharmaceuticals International (UK) Ltd	IA IP	HSE 4379
# Greenerway Crawling Pest Control	Pharmaceuticals International (UK) Ltd	IA	HSE 4380
# Greenerway Flying Pest Control	Pharmaceuticals International (UK) Ltd	IA	HSE 4388
Insectrol	Rentokil Ltd	IA	HSE 4568
Insectrol Professional	Rentokil Ltd	IP IFSP	HSE 4624
Insektigun	Spraydex Ltd	IA IAH IFSP	HSE 3002
Nippon Fly Killer Pads	Vitax Ltd	IA	HSE 5426
# Odex Fly Spray	Odex Ltd	IA IP	HSE 4146

INSECTICIDES

Product Name	Marketing Company	Use	Reg. No
333 BIOALLETHRIN and PERMETHRIN—continued			
PC Insect Killer	Perycut Insectengun Ltd	IA IP	HSE 4415
# PC Pest Control Spray	Pharmaceuticals International (UK) Ltd	IA	HSE 4244
Perycut Cockroach Mat	Perycut Chemie AG	IA IP	HSE 5467
Pif Paf Crawling Insect Killer	Roussel Uclaf Environmental Health Ltd	IFSP IP	HSE 5153
Pif Paf Crawling Insect Killer Aerosol	Wellcome Foundation Ltd	IA IFSP IP	HSE 3070
Pif Paf Flying Insect Killer	Roussel Uclaf Environmental Health Ltd	IA IFSP IP	HSE 5151
Pif Paf Flying Insect Killer Aerosol	Roussel Uclaf Environmental Health Ltd	IA IFSP IP	HSE 5152
Pif Paf Flying Insect Killer Aerosol	Wellcome Foundation Ltd	IA IFSP IP	HSE 4567
Pybuthrin Fly Killer	Roussel Uclaf Environmental Health Ltd	IA IFSP IP	HSE 5141
Pybuthrin Fly Spray	Roussel Uclaf Environmental Health Ltd	IA	HSE 5116
Sainsbury's Crawling Insect and Ant Killer	J Sainsbury Plc	IA	HSE 5231
Spraydex Ant & Insect Killer	Pan Britannica Industries Ltd	IA	HSE 5645
Spraydex Houseplant Spray	Spraydex Ltd	IA	HSE 3540
Spraydex Insect Killer	Spraydex Ltd	IA	HSE 3834
Vapona Ant And Crawling Insect Killer	Ashe Consumer Products Ltd	IA IP	HSE 4167
Vapona Ant and Crawling Insect Killer	Ashe Consumer Products Ltd	IA IP	HSE 5206
Vapona Ant and Crawling Insect Killer	Ashe Consumer Products Ltd	IA	HSE 3596
Vapona Fly and Wasp Killer	Ashe Consumer Products Ltd	IA	HSE 4843
Vapona Fly and Wasp Killer	Ashe Consumer Products Ltd	IA	HSE 5205
Wellcome Environmental Health Fly Spray	Wellcome Foundation Ltd	IA	HSE 4835
Wellcome Fly Killer	Wellcome Foundation Ltd	IA IFSP IP	HSE 4834
Wellcome Pif Paf Crawling Insect Killer	Wellcome Foundation Ltd	IFSP IP	HSE 4168
Wellcome Pif Paf Flying Insect Killer	Wellcome Foundation Ltd	IA IFSP IP	HSE 4510

INSECTICIDES

Product Name	Marketing Company	Use	Reg. No

333 BIOALLETHRIN and PERMETHRIN—continued

Zap Pest Control	Delsol Ltd	IA IP	HSE 5068

334 BIOALLETHRIN and PERMETHRIN and PYRETHRINS

Supabug Crawling Insect Killer	Sharpstow International Homecare Products Ltd	IA	HSE 4794

335 BIOALLETHRIN and S-BIOALLETHRIN

Actomite	G D Searle and Co Ltd	IA	HSE 4182

336 BIORESMETHRIN

Biosol RTU	Microsol	IFSP IP	HSE 5306
Blade	Chemsearch	IP	HSE 4839
Safe Kill RTU	Friendly Systems	IFSP IP	HSE 5563

337 BIORESMETHRIN and BIOALLETHRIN

* Coopers Fly Spray BB (Ready To Use)	Coopers Animal Health Ltd	IAH IP	HSE 4041
Pybuthrin 33 BB	Roussel Uclaf Environmental Health Ltd	IFSP IP	HSE 5162
Pybuthrin 33 BB	Wellcome Foundation Ltd	IFSP IP	HSE 4886

338 BIORESMETHRIN and S-BIOALLETHRIN

* Coopers Fly Spray EB (Ready To Use)	Coopers Animal Health Ltd	IAH IP	HSE 4042

339 BORIC ACID

Baracaf Cockroach Control Sticker	Baracaf	IFSP IP	HSE 3180
Baracaf Cockroach Control Sticker (Domestic)	European Marketing Systems (GB) Ltd	IA IP	HSE 4964
Boric Acid Concentrate	Rentokil Ltd	IP	HSE 4420
Boric Acid Powder	Rentokil Ltd	IP	HSE 4373
Instasective	Instafoam & Fibre Ltd	IP	HSE 5109
Killgerm Boric Acid Powder	Killgerm Chemicals Ltd	IP	HSE 4527
# Raid Ant Bait (NZ)	Johnson Wax Ltd	IA	HSE 4002
Roachbuster	Roachbuster (UK) Ltd	IA IP	HSE 5215

INSECTICIDES

Product Name	Marketing Company	Use	Reg. No
340 CARBARYL			
Ant and Crawling Insect Powder	May and Baker Garden Care	IA	HSE 4804
Ant and Insect Powder	Rentokil Ltd	IA	HSE 4556
Killgerm Carbaryl 5% Dust	Killgerm Chemicals Ltd	IFSP IP	HSE 3157
Rentokil Wasp Nest Killer	Rentokil Ltd	IA	HSE 4557
Sevin D	Rhone-Poulenc Environmental Products	IP	HSE 4598
Wasp Nest Destroyer	May and Baker Garden Care	IA	HSE 4784
Wasp Nest Destroyer	Pan Britannica Industries Ltd	IA	HSE 5456
341 CETO-STEARYL DIETHOXYLATE and OLEYL MONOETHOXYLATE			
Larvex-100	Accotec Ltd	IP	HSE 5393
Larvex-15	Accotec Ltd	IP	HSE 5091
342 CHLORPYRIFOS			
Dursban 4TC	Dowelanco Ltd	IFSP IP	HSE 3097
Dursban LO	Dowelanco Ltd	IP	HSE 4771
Empire 20	Dowelanco Ltd	IFSP IP	HSE 4844
Killgerm Terminate	Killgerm Chemicals Ltd	IFSP IP	HSE 3553
Raid Ant Bait	Johnson Wax Ltd	IA	HSE 5585
343 CHLORPYRIFOS and CYPERMETHRIN and PYRETHRINS			
New Tetracide	Killgerm Chemicals Ltd	IP	HSE 3724
344 CHLORPYRIFOS and TETRAMETHRIN			
Raid Wasp Nest Destroyer	Johnson Wax Ltd	IA	HSE 5597
345 CHLORPYRIFOS-METHYL			
Reldan 50 EC	Dowelanco Ltd	IP	HSE 4875
Smite	Roussel Uclaf Environmental Health Ltd	IP	HSE 5142
Smite	Wellcome Foundation Ltd	IP	HSE 4891

INSECTICIDES

Product Name	Marketing Company	Use	Reg. No
346 CHLORPYRIFOS METHYL and PERMETHRIN and PYRETHRINS			
Multispray	Roussel Uclaf Environmental Health Ltd	IFSP IP	HSE 5165
Multispray	Wellcome Foundation Ltd	IFSP IP	HSE 4893
347 CITRONELLA OIL			
Aztec BBQ Patio Candles	Chartan Aldred Ltd	IA	HSE 5613
BBQ Fly Repellent Terracotta Pot Candle	Eclipse Candles	IA	HSE 5627
BBQ Fly Repellent Candle	Eclipse Candles	IA	HSE 5625
BBQ Patio Candles	Sherwood Promark Ltd	IA	HSE 5556
Citromax Citronella Insect Repellent Liquid Candle Oil	Lamplight Farms Inc	IA	HSE 5423
348 CYPERMETHRIN			
B and Q New Formula Ant Killer Spray	B and Q Plc	IA	HSE 5606
Cymperator	Killgerm Chemicals Ltd	IFSP IP	HSE 3970
Cyperkill 10	Mitchell Cotts Chemicals Ltd	IP	HSE 4025
Cyperkill 10 WP	Mitchell Cotts Chemicals Ltd	IFSP IP	HSE 3649
Cypermethrin 10% EC	Killgerm Chemicals Ltd	IFSP IP	HSE 3136
Cypermethrin 10% WP	Killgerm Chemicals Ltd	IFSP IP	HSE 3137
Cypermethrin Lacquer	Killgerm Chemicals Ltd	IFSP IP	HSE 3164
Cypermethrin PH-10EC	Zeneca Public Health	IP	HSE 4961
Demon 40 WP	ICI Plc	IP	HSE 3174
Demon 40 WP	Zeneca Public Health	IFSP IP	HSE 5457
Doff Ant Killer Spray	Doff Portland Ltd	IA	HSE 5588
Great Mills Ant Killer Spray	Great Mills (Retail) Ltd	IA	HSE 5572
Homebase Antkiller Spray	Sainsbury's Homebase House and Garden Centres	IA	HSE 5573
Murphy Kil-Ant Ready To Use	Fisons Plc Horticulture Division	IA	HSE 5168
New Siege II	Chemsearch	IA IFSP IP	HSE 5638
Pyrasol C RTU	Microsol	IA IFSP IP	HSE 5336
Pyrasol CP	Microsol	IFSP IP	HSE 5478

INSECTICIDES

Product Name	Marketing Company	Use	Reg. No
348 CYPERMETHRIN—continued			
Ready Kill	Friendly Systems	IA IFSP IP	HSE 5566
Siege	Chemsearch	IP	HSE 4840
Siege II	Chemsearch	IA IFSP IP	HSE 5251
Texas Ant Gun	Texas Homecare	IA	HSE 5679
Tinocide Insecticidal Lacquer	Phenoglaze Coatings Ltd	IP	HSE 4484
Vapona Antpen	Ashe Consumer Products Ltd	IA IP	HSE 3592
Vapona Flypen	Ashe Consumer Products Ltd	IA	HSE 3606
Wilko Ant Killer Spray	Wilkinson Home and Garden Stores	IA	HSE 5583
349 CYPERMETHRIN and METHOPRENE			
Killgerm Precor ULV III	Killgerm Chemicals Ltd	IFSP IP	HSE 3523
350 CYPERMETHRIN and PYRETHRINS			
Bolt Crawling Insect Killer	Johnson Wax Ltd	IA	HSE 3609
Raid Cockroach Killer Formula 2	Johnson Wax Ltd	IA	HSE 4239
351 CYPERMETHRIN and PYRETHRINS and CHLORPYRIFOS			
New Tetracide	Killgerm Chemicals Ltd	IP	HSE 3724
352 CYPERMETHRIN and TETRAMETHRIN			
Cyperkill Plus WP	Mitchell Cotts Chemicals Ltd	IFSP IP	HSE 4855
New Vapona Ant Killer	Ashe Consumer Products Ltd	IA	HSE 4298
Raid Ant and Crawling Insect Killer	Johnson Wax Ltd	IA	HSE 3001
Raid Residual Crawling Insect Killer	Johnson Wax Ltd	IA IP	HSE 4584
S C Johnson Raid Ant and Cockroach Killer	Johnson Wax Ltd	IA	HSE 5216
# Shelltox Cockroach and Crawling Insect Killer 2	Temana International Ltd	IA IP	HSE 4030
Shelltox Cockroach and Crawling Insect Killer 3	Temana International Ltd	IA IP	HSE 4031
Shelltox Cockroach and Crawling Insect Killer 4	Temana International Ltd	IA	HSE 4739

INSECTICIDES

Product Name	Marketing Company	Use	Reg. No
352 CYPERMETHRIN and TETRAMETHRIN—continued			
Super Shelltox Crawling Insect Killer	Temana International Ltd	IA	HSE 5454
Vapona Ant and Crawling Insect Killer Aerosol	Ashe Ltd	IA	HSE 5282
353 CYPERMETHRIN and TETRAMETHRIN and d-ALLETHRIN			
"Bop" Flying and Crawling Insect Killer	Robert McBride Ltd – Exports	IA IP	HSE 4136
"Bop" Flying and Crawling Insect Killer (Water Based)	Robert McBride Ltd – Exports	IA IP	HSE 4137
"Bop" Flying Insect Killer	Robert McBride Ltd – Exports	IA IP	HSE 4141
"Bop" Flying Insect Killer (MC)	Robert McBride Ltd – Exports	IA IP	HSE 4140
Bop Flying and Crawling Insect Killer (MC)	Robert McBride Ltd – Exports	IA IP	HSE 4135
354 CYPERMETHRIN and d-ALLETHRIN			
Shelltox Crawling Insect Killer Liquid Spray	Temana International Ltd	IA IP	HSE 3565
Vapona Ant and Crawling Insect Spray	Ashe Consumer Products Ltd	IA	HSE 3593
355 DELTAMETHRIN			
Crackdown	Roussel Uclaf Environmental Health Ltd	IFSP IP	HSE 5097
Crackdown	Wellcome Foundation Ltd	IFSP IP	HSE 4889
356 DIAZINON			
B and Q Ant Killer Lacquer	B and Q Ltd	IA	HSE 4750
Dethlac Insecticidal Lacquer	Gerhardt Pharmaceuticals Ltd	IA	HSE 4423
Doff Antlak	Doff Portland Ltd	IA	HSE 3591
Knox Out 2 FM	Rentokil Ltd	IAH IFSP IP	HSE 4464
Secto Ant and Crawling Insect Lacquer	Secto Company Ltd	IA	HSE 4400
Secto Kil-A-Line	Secto Company Ltd	IA	HSE 3631

INSECTICIDES

Product Name	Marketing Company	Use	Reg. No
356 DIAZINON—continued			
Wilko Ant Killer Lacquer	Wilkinson Home and Garden Stores	IA	HSE 4965
357 DIAZINON and DICHLORVOS			
Vijurrax Spray	Vijusa UK Ltd	IA	HSE 5717
358 DIAZINON and PYRETHRINS			
Ant Gun!	ICI Garden and Professional Products	IA	HSE 4981
B and Q Antkiller Spray	B and Q Plc	IA	HSE 5114
359 DICHLOROPHEN and DICHLORVOS and TETRAMETHRIN			
Wintox	Mould Growth Consultants Ltd	IP	HSE 4236
365 DICHLORVOS			
Boots Moth Killer	Boots Company Plc	IA	HSE 5543
Boots Slow Release Fly and Wasp Killer	Boots Company Plc	IA	HSE 5545
Boots Slow Release Fly Killer	Boots Company Plc	IA	HSE 4922
Boots Small Space Fly and Moth Killer	Boots Company Plc	IA	HSE 5272
Boots Small Space Moth and Fly Killer	Boots Company Plc	IA	HSE 4573
Cromessol Fly-Away	Cromessol Ltd	IA	HSE 4338
Culmstock Mothproofer	Culmstock Ltd	IA	HSE 4027
Culmstock Slow Release Fly Killer	Culmstock Ltd	IA	HSE 3961
DDVP (Toxicant) Strip	International Pheromone Systems Ltd	IFSP IP	HSE 5517
Fly Killer	Rentokil Ltd	IA	HSE 5676
Flying Insect Killer	Culmstock Ltd	IA	HSE 4442
Flykil	Calmic Service UK Ltd	IA	HSE 4129
Freshways Slow Release Insect Killer	Freshways of York	IA	HSE 4618
Funnel Trap Insecticidal Strip	Agrisense BCS Ltd	IP	HSE 4292

INSECTICIDES

Product Name	Marketing Company	Use	Reg. No
365 **DICHLORVOS**—continued			
Globol Small Space Fly and Moth Strip	Globol Chemicals (UK) Ltd	IA	HSE 4692
Industrial Hygiene Controllable Cassette Insect killer	Industrial Hygiene Company	IA	HSE 4617
Jeyes Expel Moth Killer	Jeyes Group Plc	IA	HSE 5504
Jeyes Expel Slow Release Fly and Wasp Killer	Jeyes Group Plc	IA	HSE 5505
Jeyes Kontrol Moth Killer	J Sainsbury Plc	IA	HSE 5546
Jeyes Kontrol Slow Release Fly and Wasp Killer	J Sainsbury Plc	IA	HSE 5544
Jeyes Slow Release Fly Killer Controllable Cassette	Jeyes Ltd	IA	HSE 5259
Jeyes Small Space Fly and Moth Strip	Jeyes Ltd	IA	HSE 5289
Kleenoff Slow Release Fly Killer Controllable Cassette	Jeyes Ltd	IA	HSE 5261
Kleenoff Small Space Fly and Moth Strip	Jeyes Ltd	IA	HSE 5288
Kontrol Kitchen Size Fly Killer	Secto Company Ltd	IA	HSE 5598
Lloyds Supersave Slow Release Fly and Wasp Killer	Lloyds Chemist	IA	HSE 4920
Lloyds Supersave Small Space Fly and Moth Strip	Lloyds Chemist	IA	HSE 4921
Moth and Fly Killer	Rentokil Ltd	IA	HSE 5677
Russco Strips	Russell Fine Chemicals Ltd	IFSP IP	HSE 5654
Secto Fly Killer Living Room Size	Secto Company Ltd	IA	HSE 4397
Secto Mini-Space Insect Killers	Secto Company Ltd	IA	HSE 4398
Secto Slow Release Fly Killer Kitchen Size	Secto Company Ltd	IA	HSE 4399
Slow Release Fly Killer Controllable Cassette	Globol Chemicals (UK) Ltd	IA	HSE 4339
Superdrug Slow Release Fly Killer	Superdrug Plc	IA	HSE 3962
Superdrug Small Space Fly/Moth Strip	Superdrug Plc	IA	HSE 4068
Teepol Products Vapona Fly Killer	Teepol Products Ltd	IA IP	HSE 3673
Vapona Fly Killer	Ashe Consumer Products Ltd	IA	HSE 4717
Vapona Moth Killer	Ashe Consumer Products Ltd	IA	HSE 4724

INSECTICIDES

Product Name	Marketing Company	Use	Reg. No

365 DICHLORVOS—continued

Vapona Professional Cockroach Killer	Ashe Ltd	IA	HSE 5685
Vapona Small Space Fly Killer	Ashe Consumer Products Ltd	IA	HSE 4723

361 DICHLORVOS and 1, 4-DICHLOROBENZENE

Secto Moth Killer	Secto Company Ltd	IA	HSE 4430

362 DICHLORVOS and DIAZINON

Vijurrax Spray	Vijusa UK Ltd	IA	HSE 5717

363 DICHLORVOS and IODOFENPHOS

Defest Flea Free	CIBA Animal Health	IA	HSE 4331
Defest Flea Free	Sherleys Ltd	IA	HSE 5715
Nuvan Staykil	CIBA Animal Health	IA IP	HSE 4017

364 DICHLORVOS and PERMETHRIN and S-BIOALLETHRIN

Pif Paf Insecticide	Roussel Uclaf Environmental Health Ltd	IA IFSP IP	HSE 5150
Pif Paf Insecticide	Wellcome Foundation Ltd	IA IFSP IP	HSE 3748

365 DICHLORVOS and PERMETHRIN and TETRAMETHRIN

Laser Insect Killer	Sharpstow International Homecare Products Ltd	IA	HSE 4910
Supaswat Insect Killer	Sharpstow International Homecare Products Ltd	IA	HSE 4630

366 DICHLORVOS and PERMETHRIN and TETRAMETHRIN and d-ALLETHRIN

Bop	Robert McBride Ltd—Exports	IA	HSE 4773

367 DICHLORVOS and TETRAMETHRIN

Shelltox Extra Flykiller	Temana International Ltd	IA IP	HSE 3696
Shelltox Flykiller	Temana International Ltd	IA	HSE 3162
Vapona Fly and Wasp Killer Spray	Ashe Consumer Products Ltd	IA	HSE 4907
Vapona Wasp and Fly Killer	Ashe Ltd	IA	HSE 5281

INSECTICIDES

Product Name	Marketing Company	Use	Reg. No
368 DICHLORVOS and TETRAMETHRIN and DICHLOROPHEN			
Wintox	Mould Growth Consultants Ltd	IP	HSE 4236
369 DIMETHOATE and TETRAMETHRIN			
Killgerm Dimethoate Extra	Killgerm Chemicals Ltd	IFSP IP	HSE 4633
370 ENCAPSULATED PERMETHRIN and PERMETHRIN			
Kudos	Roussel Uclaf Environmental Health Ltd	IP	HSE 5263
Kudos	Wellcome Foundation Ltd	IP	HSE 3910
376 FENITROTHION			
Antec E-C Kill	Antec International Ltd	IAH IFSP IP	HSE 5468
Demise	Sorex Ltd	IFSP IP	HSE 5084
Fenitrothion Dusting Powder	Rentokil Ltd	IP	HSE 4372
Fenitrothion Emulsion Concentrate	Rentokil Ltd	IFSP IP	HSE 4783
Fenitrothion Wettable Powder	Rentokil Ltd	IFSP IP	HSE 4827
Killgerm Fenitrothion 40 WP	Killgerm Chemicals Ltd	IAH IFSP IP	HSE 4858
Killgerm Fenitrothion 50 EC	Killgerm Chemicals Ltd	IAH IFSP IP	HSE 4722
Micromite	Micro-Biologicals Ltd	IFSP IP	HSE 4480
Sectacide 50 EC	Anglian Farm Supplies Ltd	IAH IFSP IP	HSE 4939
* Sectacide 50 EC	Anglian Farm Supplies Ltd	IFSP IP	HSE 3139
Sumithion 20% MC	Sumitomo Chemical (UK) Plc	IP	HSE 4905
372 FENITROTHION and PERMETHRIN and RESMETHRIN			
Turbair Beetle Killer	Pan Britannica Industries Ltd	IFSP IP	HSE 4648
373 FENITROTHION and PERMETHRIN and TETRAMETHRIN			
Motox	HVM International Ltd	IA	HSE 5622

INSECTICIDES

Product Name	Marketing Company	Use	Reg. No
374 FENITROTHION and TETRAMETHRIN			
Ant and Crawling Insect Killer	Rentokil Ltd	IA	HSE 4642
Big D Cockroach and Crawling Insect Killer	Domestic Fillers Ltd	IA	HSE 4098
Big D Rocket Fly Killer	Domestic Fillers Ltd	IA	HSE 3552
Doom Ant and Crawling Insect Killer Aerosol	NAPA Products Ltd	IA	HSE 4753
Flying Insect Killer	Keen (World Marketing) Ltd	IA	HSE 4313
Flying Insect Killer Faster Knockdown	Keen (World Marketing) Ltd	IA	HSE 4312
Killgerm Fenitrothion-Pyrethroid Concentrate	Killgerm Chemicals Ltd	IFSP IP	HSE 4623
Tox Exterminating Fly and Wasp Killer	Keen (World Marketing) Ltd	IA	HSE 4314
375 GAMMA-HCH			
Doom Ant and Insect Powder	NAPA Products Ltd	IA	HSE 4570
Fumite Lindane Generator Size 10	Octavius Hunt Ltd	IA WA	HSE 4713
Fumite Lindane Generator Size 40	Octavius Hunt Ltd	IA WA	HSE 4714
Fumite Lindane Pellet No 3	Octavius Hunt Ltd	IAH IFSP IP	HSE 4733
Fumite Lindane Pellet No 4	Octavius Hunt Ltd	IAH IFSP IP	HSE 4734
* Gammexane Smoke Generator No 22	ICI Garden and Professional Products	IP WP	HSE 4822
* Roofguard Smoke Generator	Octavius Hunt Ltd	IA WA	HSE 4712
376 GAMMA-HCH and TETRAMETHRIN			
Doom Flea Killer	NAPA Products Ltd	IA	HSE 4572
Doom Moth Proofer Aerosol	NAPA Products Ltd	IA	HSE 4571
377 HYDRAMETHYLNON			
Maxforce Bait Station	Roussel Uclaf Environmental Health Ltd	IFSP IP	HSE 5371
Maxforce Bait Stations	Wellcome Foundation Ltd	IFSP IP	HSE 4154
Maxforce Gel	Roussel Uclaf Environmental Health Ltd	IFSP IP IAH	HSE 5365

INSECTICIDES

Product Name	Marketing Company	Use	Reg. No
377 HYDRAMETHYLNON—continued			
Maxforce Gel	Wellcome Foundation Ltd	IFSP IP	HSE 4153
Maxforce Pharaoh's Ant Killer	Roussel Uclaf Environmental Health Ltd	IFSP IP	HSE 5082
378 HYDROPRENE			
Protrol	Sandoz Speciality Pest Control Ltd	IFSP IP	HSE 5333
Protrol	Zoecon Corporation	IFSP IP	HSE 4908
379 HYDROPRENE and PYRETHRINS			
Protrol Plus ULV	Zoecon Corporation	IFSP IP	HSE 4909
380 IODOFENPHOS			
Cockroach Bait	Rentokil Ltd	IP	HSE 4385
Iodofenphos Granular Bait	Rentokil Ltd	IP	HSE 4387
* Nuvanol N 500 FW	Ciba-Geigy Agrochemicals	IAH IFSP IP	HSE 4357
Nuvanol N 500 SC	Ciba Animal Health	IAH IFSP IP	HSE 4951
Rentokil Iodofenphos Gel	Rentokil Ltd	IFSP IP	HSE 4359
Waspex	Rentokil Ltd	IP	HSE 4386
381 IODOFENPHOS and DICHLORVOS			
Defest Flea Free	Ciba Animal Health	IA	HSE 4331
Defest Flea Free	Sherleys Ltd	IA	HSE 5715
Nuvan Staykil	Ciba Animal Health	IA IP	HSE 4017
382 LAMBDA-CYHALOTHRIN			
Icon 2.5 EC	Zeneca Public Health	IP	HSE 5614
383 METHOPRENE			
Pharorid	Sandoz Speciality Pest Control Ltd	IP	HSE 5332
Pharorid	Zoecon Corporation	IP	HSE 4823

INSECTICIDES

Product Name	Marketing Company	Use	Reg. No

384 METHOPRENE and CYPERMETHRIN

Killgerm Precor ULV III	Killgerm Chemicals Ltd	IFSP IP	HSE 3523

385 METHOPRENE and PERMETHRIN

Acclaim Plus	Sanofi Animal Health Ltd	IA	HSE 3883
Precor Plus Premise Spray	Zoecon Corporation	IA	HSE 3885
Precor ULV II	Killgerm Chemicals Ltd	IFSP IP	HSE 3191
# Siphotrol Premise Spray	Zoecon Corporation	IA	HSE 3567
Siphotrol Premise Spray	Zoecon Corporation	IA	HSE 3568
Zodiac Household Spray	Labtec Animal Health	IA	HSE 4937

386 METHOPRENE and PYRETHRINS

Canovel Pet Bedding and Household Spray	Smithkline Beecham Animal Health	IA	HSE 5330
Killgerm Precor RTU	Killgerm Chemicals Ltd	IFSP IP	HSE 3165
Killgerm Precor ULV I	Killgerm Chemicals Ltd	IFSP IP	HSE 3098

387 NAPHTHALENE

Dragon Brand Moth Balls	R A Davies and Partners	IA	HSE 5385
Jertox Moth Balls	Thornton and Ross Ltd	IA	HSE 5057
Phernal Brand Moth Balls	Harrow Drug Company	IA	HSE 5740

388 NAPHTHALENE and 1, 4-DICHLOROBENZENE

Mothaks	Sara Lee Household and Personal Care (UK) Ltd	IA	HSE 5124
Vapona Mothaks	Ashe Ltd	IA	HSE 5682

389 NAPHTHALENE and PERMETHRIN and PYRETHRINS and BENZALKONIUM CHLORIDE

Roxem D	Horton Hygiene Company	IP SP	HSE 5107

390 NATAMYCIN

Tymasil	Brocades (Great Britain) Ltd	IA	HSE 3625

INSECTICIDES

Product Name	Marketing Company	Use	Reg. No
391 **OLEYL MONOETHOXYLATE and CETO-STEARYL DIETHOXYLATE**			
Larvex-100	Accotec Ltd	IP	HSE 5393
Larvex-15	Accotec Ltd	IP	HSE 5091
392 **PERMETHRIN**			
Amogas Ant Killer	Thornton and Ross Ltd	IA	HSE 4265
Amogas Ant Killer	Thornton and Ross Ltd	IA	HSE 5126
* Arrow Residual Powerkill	Arrow Chemicals Ltd	IP	HSE 4003
Boots Ant Killer Powder	Boots Company Plc	IA	HSE 5525
Chirton Ant and Insect Killer	Ronson Plc	IA	HSE 3835
* Coopers Stomoxin P	Coopers Animal Health Ltd	IAH IP	HSE 3007
Coopex 25% EC	Roussel Uclaf Environmental Health Ltd	IFSP IP	HSE 5147
Coopex 25% EC	Wellcome Foundation Ltd	IFSP IP	HSE 3194
Coopex Insect Powder	Roussel Uclaf Environmental Health Ltd	IAH IFSP IP	HSE 5052
Coopex Maxi Smoke Generator	Roussel Uclaf Environmental Health Ltd	IAH IFSP IP	HSE 5131
Coopex Maxi Smoke Generator	Wellcome Foundation Ltd	IAH IFSP IP	HSE 4561
Coopex Mini Smoke Generator	Roussel Uclaf Environmental Health Ltd	IAH IFSP IP	HSE 5130
Coopex Mini Smoke Generator	Wellcome Foundation Ltd	IAH IFSP IP	HSE 4554
Coopex Smoke Generator	Roussel Uclaf Environmental Health Ltd	IAH IFSP IP	HSE 5132
Coopex Smoke Generator	Wellcome Foundation Ltd	IAH IFSP IP	HSE 4888
Coopex WP	Roussel Uclaf Environmental Health Ltd	IFSP IP	HSE 5096
Coopex WP	Wellcome Foundation Ltd	IFSP IP	HSE 4890
Delta Insect Powder	Middle East Aerosol and Detergent Company Ltd	IA	HSE 4664
Flea Kill from Bob Martin. Insecticidal Carpet and Funiture Deodoriser	Bob Martin Company	IA	HSE 5569
Fresh-A-Pet Insecticidal Rug and Carpet Freshener	Loxley Medical	IA	HSE 4109

INSECTICIDES

Product Name	Marketing Company	Use	Reg. No
392 **PERMETHRIN**—continued			
Gold Label Kennel and Stable Powder	Stockcare Ltd	IA	HSE 5699
Hartz Rid Flea Carpet Freshener	Thomas Cork SML	IA	HSE 5392
Household Flea Spray from Bob Martin	Bob Martin Company	IA	HSE 3914
Jeyes Expel Ant Killer Powder	Jeyes Group Plc	IA	HSE 5524
Jeyes Kontrol Ant Killer Powder	Jeyes Group Plc	IA	HSE 5520
Johnson's Carpet Flea Guard	Johnson's Veterinary Products Ltd	IA	HSE 4943
* Johnson's Flea Guard	Johnson's Veterinary Products Ltd	IA	HSE 3865
# Killoff Ant Killer Powder	The Kleenoff Company	IA	HSE 4062
Lanosol P	Microsol	IP	HSE 5477
Lanosol RTU	Microsol	IFSP IP	HSE 5307
# MAFU Ant and Insect Powder	Bayer (UK) Ltd	IA	HSE 3173
Micrapor Environmental Flea Spray	Petlife	IA	HSE 5444
Mosiguard Shield	Medical Advisory Services for Travellers Abroad	IA	HSE 4515
Muscatrol	Rentokil Ltd	IP	HSE 4579
Nippon Ant Killer Powder	Vitax Ltd	IA	HSE 4333
Nippon Ant Killer Powder	Vitax Ltd	IA	HSE 5499
Nippon Ready For Use Ant and Crawling Insect Killer	Vitax Ltd	IA	HSE 4328
Nomad Residex P	Nomad Travel Pharmacy	IA	HSE 5177
One Shot Space Spray	Roussel Uclaf Environmental Health Ltd	IP	HSE 5086
Outright Household Flea Spray	Outright Pet Care Products	IA	HSE 5623
Peripel	Roussel Uclaf Environmental Health Ltd	IP	HSE 5265
Peripel	Wellcome Foundation Ltd	IP	HSE 3800
Peripel 55	Roussel Uclaf Environmental Health Ltd	IP	HSE 5658
Permasect 0.5 Dust	Mitchell Cotts Chemicals Ltd	IFSP IP	HSE 3134
Permasect 10 WP	Mitchell Cotts Chemicals Ltd	IFSP IP	HSE 4854
Permasect Powder	Mitchell Cotts Chemicals Ltd	IA	HSE 5498
Permethrin	Lifesystems Ltd	IA	HSE 5612

INSECTICIDES

Product Name	Marketing Company	Use	Reg. No
392 PERMETHRIN—continued			
Permethrin Dusting Powder	Rentokil Ltd	IP	HSE 4688
Permethrin PH – 25WP	Zeneca Public Health	IP	HSE 4956
Permethrin Wettable Powder	Rentokil Ltd	IP	HSE 4581
# Pestrin	Hughes and Hughes Ltd	IP	HSE 3684
Pestrin	Hughes and Hughes Ltd	IP	HSE 4499
Pif Paf Crawling Insect Powder	Roussel Uclaf Environmental Health Ltd	IA IFSP IP	HSE 5155
Pif Paf Crawling Insect Powder	Wellcome Foundation Ltd	IA IFSP IP	HSE 4770
Pynosect 6	Mitchell Cotts Chemicals Ltd	IP	HSE 4528
Pynosect PCO	Mitchell Cotts Chemicals Ltd	IFSP IP	HSE 4566
Raid Mothproofer	Johnson Wax Ltd	IA	HSE 5646
Residex P	Agropharm Ltd	IP	HSE 5170
Residual Powerkill	Yule Catto Consumer Chemicals Ltd	IP	HSE 5180
Secto Extra Strength Insect Killer	Secto Company Ltd	IA	HSE 4329
Secto Flea Free Insecticidal Rug and Carpet Freshener	Secto Co Ltd	IA	HSE 4151
Sherley's Rug-De-Bug	Sherley's Pet Care CIBA-Geigy Animal Health	IA	HSE 5128
Stapro Insecticide	Stapro Ltd	IP	HSE 4526
Vapona Ant and Crawling Insect Powder	Ashe Consumer Products Ltd	IA	HSE 3595
Vapona Ant and Crawling Insect Powder	Ashe Consumer Products Ltd	IA	HSE 5207
Vapona Carpet and Household Flea Powder	Ashe Ltd	IA	HSE 5650
* Vapona Moth Proofer	Ashe Consumer Products Ltd	IA	HSE 3597
Vitax Kontrol Ant Killer Powder	Vitax Ltd	IA	HSE 5593
Z-Stop Anti-Wasp Strip	Thames Laboratories Ltd	IA	HSE 3532
393 PERMETHRIN and BIOALLETHRIN			
Ant Gun! 2	Zeneca Garden Care	IA IP	HSE 5184
Ant Gun! 2	Zeneca Garden Care	IA IP	HSE 5629
Antkiller Spray 2	Zeneca Agrochemicals Limited	IA IP	HSE 5185

INSECTICIDES

Product Name	Marketing Company	Use	Reg. No
393 **PERMETHRIN and BIOALLETHRIN**—continued			
Crawling Insect and Ant Killer	Rentokil Ltd	IA	HSE 4650
Creepy Crawly Gun!	Zeneca Garden Care	IA IP	HSE 5183
Fleegard	Bayer Plc	IA	HSE 5564
# Forest Friends Crawling Pest Control	Pharmaceuticals International (UK) Ltd	IA	HSE 4389
# Forest Friends Flying Pest Control	Pharmaceuticals International (UK) Ltd	IA IP	HSE 4379
# Greenerway Crawling Pest Control	Pharmaceuticals International (UK) Ltd	IA	HSE 4380
# Greenerway Flying Pest Control	Pharmaceuticals International (UK) Ltd	IA	HSE 4388
Insectrol	Rentokil Ltd	IA	HSE 4568
Insectrol Professional	Rentokil Ltd	IFSP IP	HSE 4624
Insektigun	Spraydex Ltd	IA IAH IFSP	HSE 3002
Nippon Fly Killer Pads	Vitax Ltd	IA	HSE 5426
# Odex Fly Spray	Odex Ltd	IA IP	HSE 4146
PC Insect Killer	Perycut Insectengun Ltd	IA IP	HSE 4415
# PC Pest Control Spray	Pharmaceuticals International (UK) Ltd	IA	HSE 4244
Perycut Cockroach Mat	Perycut Chemie AG	IA IP	HSE 5467
Pif Paf Crawling Insect Killer	Roussel Uclaf Environmental Health Ltd	IFSP IP	HSE 5153
Pif Paf Crawling Insect Killer Aerosol	Wellcome Foundation Ltd	IA IFSP IP	HSE 3070
Pif Paf Flying Insect Killer	Roussel Uclaf Environmental Health Ltd	IA IFSP IP	HSE 5151
Pif Paf Flying Insect Killer Aerosol	Roussel Uclaf Environmental Health Ltd	IA IFSP IP	HSE 5152
Pif Paf Flying Insect Killer Aerosol	Wellcome Foundation Ltd	IA IFSP IP	HSE 4567
Pybuthrin Fly Killer	Roussel Uclaf Environmental Health Ltd	IA IFSP IP	HSE 5141
Pybuthrin Fly Spray	Roussel Uclaf Environmental Health Ltd	IA	HSE 5116
Sainsbury's Crawling Insect and Ant Killer	J Sainsbury Plc	IA	HSE 5231

INSECTICIDES

Product Name	Marketing Company	Use	Reg. No
393 PERMETHRIN and BIOALLETHRIN—continued			
Spraydex Ant & Insect Killer	Pan Britannica Industries Ltd	IA	HSE 5645
Spraydex Houseplant Spray	Spraydex Ltd	IA	HSE 3540
Spraydex Insect Killer	Spraydex Ltd	IA	HSE 3834
Vapona Ant and Crawling Insect Killer	Ashe Consumer Products Ltd	IA IP	HSE 4167
Vapona Ant and Crawling Insect Killer	Ashe Consumer Products Ltd	IA IP	HSE 5206
Vapona Ant and Crawling Insect Killer	Ashe Consumer Products Ltd	IA	HSE 3596
Vapona Fly and Wasp Killer	Ashe Consumer Products Ltd	IA	HSE 4843
Vapona Fly and Wasp Killer	Ashe Consumer Products Ltd	IA	HSE 5205
Wellcome Environmental Health Fly Spray	Wellcome Foundation Ltd	IA	HSE 4835
Wellcome Fly Killer	Wellcome Foundation Ltd	IA IFSP IP	HSE 4834
Wellcome Pif Paf Crawling Insect Killer	Wellcome Foundation Ltd	IFSP IP	HSE 4168
Wellcome Pif Paf Flying Insect Killer	Wellcome Foundation Ltd	IA IFSP IP	HSE 4510
Zap Pest Control	Delsol Ltd	IA IP	HSE 5068
394 PERMETHRIN and ENCAPSULATED PERMETHRIN			
Kudos	Roussel Uclaf Environmental Health Ltd	IP	HSE 5263
Kudos	Wellcome Foundation Ltd	IP	HSE 3910
395 PERMETHRIN and METHOPRENE			
Acclaim Plus	Sanofi Animal Health Ltd	IA	HSE 3883
Precor Plus Premise Spray	Zoecon Corporation	IA	HSE 3885
Precor ULV II	Killgerm Chemicals Ltd	IFSP IP	HSE 3191
# Siphotrol Premise Spray	Zoecon Corporation	IA	HSE 3567
Siphotrol Premise Spray	Zoecon Corporation	IA	HSE 3568
Zodiac Household Spray	Labtec Animal Health	IA	HSE 4937
396 PERMETHRIN and PYRETHRINS			
* Coopers Head To Tail Animal Bedding Spray	Coopers Animal Health Ltd	IA	HSE 3008

INSECTICIDES

Product Name	Marketing Company	Use	Reg. No
396 **PERMETHRIN and PYRETHRINS**—continued			
Detia Crawling Insect Spray	K-D Pest Control Products	IA	HSE 3710
Flamil Finale	Flore-Chemie GmbH	IP	HSE 5083
Household Flea Spray	Johnson's Veterinary Products Ltd	IA	HSE 3908
Insecticidal Aerosol	Aerosols International Ltd	IA	HSE 5518
Johnson's Household Flea Spray	Johnson's Veterinary Products Ltd	IA	HSE 5703
Kybosh	Pan Britannica Industries Ltd	IA	HSE 3022
Secto Household Flea Spray	Secto Co Ltd	IA	HSE 4131
397 **PERMETHRIN and PYRETHRINS and ALKYLARYLTRIMETHYL AMMONIUM CHLORIDE**			
* Roxem C	Horton Hygiene Co Ltd	IP SP	HSE 3123
398 **PERMETHRIN and PYRETHRINS and BENZALKONIUM CHLORIDE and NAPHTHALENE**			
Roxem D	Horton Hygiene Company	IP SP	HSE 5107
399 **PERMETHRIN and PYRETHRINS and BIOALLETHRIN**			
Supabug Crawling Insect Killer	Sharpstow International Homecare Products Ltd	IA	HSE 4794
400 **PERMETHRIN and PYRETHRINS and CHLORPYRIFOS-METHYL**			
Multispray	Roussel Uclaf Environmental Health Ltd	IFSP IP	HSE 5165
Multispray	Wellcome Foundation Ltd	IFSP IP	HSE 4893
401 **PERMETHRIN and RESMETHRIN and FENITROTHION**			
Turbair Beetle Killer	Pan Britannica Industries Ltd	IFSP IP	HSE 4648
402 **PERMETHRIN and S-BIOALLETHRIN**			
* Aqua Reslin	Wellcome Foundation Ltd	IFSP IP	HSE 3195
Aqua Reslin Premium	Roussel Uclaf Environmental Health Ltd	IFSP IP	HSE 5172
Aqua Reslin Premium	Wellcome Foundation Ltd	IFSP IP	HSE 4087

INSECTICIDES

Product Name	Marketing Company	Use	Reg. No
402 **PERMETHRIN and S-BIOALLETHRIN**—continued			
Aqua Reslin Super	Roussel Uclaf Environmental Health Ltd	IFSP IP	HSE 5175
Aqua Reslin Super	Wellcome Foundation Ltd	IFSP IP	HSE 4088
Coopex S25% EC	Roussel Uclaf Environmental Health Ltd	IFSP IP	HSE 5146
Coopex S25% EC	Wellcome Foundation Ltd	IFSP IP	HSE 3525
Imperator Fog S 18/6	Zeneca Public Health	IP	HSE 4960
Imperator Fog Super S 30/15	Zeneca Public Health	IP	HSE 4955
Imperator ULV S 6/3	Zeneca Public Health	IP	HSE 4954
Imperator ULV Super S 10/4	Zeneca Public Health	IP	HSE 4953
Neopybuthrin Premium	Roussel Uclaf Environmental Health Ltd	IFSP IP	HSE 5149
Neopybuthrin S200Q	Wellcome Foundation Ltd	IFSP IP	HSE 3192
* Neopybuthrin S9E	Wellcome Foundation Ltd	IFSP IP	HSE 3193
Neopybuthrin Super	Roussel Uclaf Environmental Health Ltd	IFSP IP	HSE 5148
Neopybuthrin Super	Wellcome Foundation Ltd	IFSP IP	HSE 4086
Resigen	Roussel Uclaf Environmental Health Ltd	IFSP IP	HSE 5143
Resigen	Wellcome Foundation Ltd	IFSP IP	HSE 4887
* Reslin 25P	Wellcome Foundation Ltd	IFSP IP	HSE 3190
# Reslin 25S	Wellcome Foundation Ltd	IFSP IP	HSE 3197
* Reslin 25S	Wellcome Foundation Ltd	IFSP IP	HSE 4258
# Reslin 25SE	Wellcome Foundation Ltd	IFSP IP	HSE 3196
Reslin 25SE	Wellcome Foundation Ltd	IFSP IP	HSE 4259
Reslin Premium	Roussel Uclaf Environmental Health Ltd	IFSP IP	HSE 5174
Reslin Super	Roussel Uclaf Environmental Health Ltd	IFSP IP	HSE 5173
Reslin Super	Wellcome Foundation Ltd	IFSP IP	HSE 4089
* Wellcome Crawling Insect Killer	Wellcome Foundation Ltd	IA IFSP IP	HSE 4166
403 **PERMETHRIN and S-BIOALLETHRIN and DICHLORVOS**			
Pif Paf Insecticide	Roussel Uclaf Environmental Health Ltd	IA IFSP IP	HSE 5150

INSECTICIDES

Product Name	Marketing Company	Use	Reg. No
403 PERMETHRIN and S-BIOALLETHRIN and DICHLORVOS—continued			
Pif Paf Insecticide	Wellcome Foundation Ltd	IA IFSP IP	HSE 3748
404 PERMETHRIN and S-BIOALLETHRIN and TETRAMETHRIN			
Pif Paf Crawling Insect Killer Aerosol	Roussel Uclaf Environmental Health Ltd	IA IFSP IP	HSE 5595
405 PERMETHRIN and S-HYDROPRENE			
Protrol Plus Crack and Crevice Aerosol	Sandoz Speciality Pest Control Ltd	IA IFSP IP	HSE 5384
Protrol Plus Crack and Crevice Aerosol	Zoecon Corporation	IA IFSP IP	HSE 4906
406 PERMETHRIN and S-METHOPRENE			
Vet-Kem Pump Spray	Sanofi Animal Health Ltd	IA	HSE 5475
Zodiac and Pump Spray	Grosvenor Pet Health Ltd	IA	HSE 5472
407 PERMETHRIN and TETRAMETHRIN			
"Bop" Crawling Insect Killer	Robert McBride Ltd –Exports	IA IP	HSE 4138
"Bop" Crawling Insect Killer (MC)	Robert McBride Ltd –Exports	IA IP	HSE 4139
Bolt Ant and Crawling Insect Killer	Johnson Wax Ltd	IA	HSE 3619
Ciba-Geigy Ko Fly and Wasp Killer	Ciba-Geigy Plc	IA	HSE 3726
Corry's Fly and Wasp Killer	Vitax Ltd	IA	HSE 5322
D-Stroy	Future Developments (Manufacturing) Ltd	IP	HSE 5411
Delta Cockroach and Ant Spray	Middle East Aerosol and Detergent Company Ltd	IA	HSE 3950
Doom Tropical Strength Fly and Wasp Killer Aerosol	NAPA Products Ltd	IA	HSE 4752
Dragon Flykiller	Zeneca Public Health	IA	HSE 4952
Fly and Wasp Killer	Dimex Ltd	IP	HSE 5050
Fly and Wasp Killer	Rentokil Ltd	IA IP	HSE 4643
Fly Spray	Merton Cleaning Supplies	IP	HSE 5112
Hail Flying Insect Killer	International Consumer Products Ltd	IA	HSE 5568
* Household Flea Spray	Boots Company Plc	IA	HSE 4661

INSECTICIDES

Product Name	Marketing Company	Use	Reg. No
407 PERMETHRIN and TETRAMETHRIN—continued			
KO2 Selkil	Selden Research Ltd	IP	HSE 4828
Nippon Ant and Crawling Insect Killer	Vitax Ltd	IA	HSE 4414
Nippon Fly Killer Spray	Vitax Ltd	IA	HSE 4254
Nippon Killaquer Crawling Insect Killer	Arrow Chemicals	IFSP IP	HSE 3694
Nisa Fly and Wasp Killer	Nisa	IA	HSE 4962
* Pestkill Insect Spray	Phillips Yeast Products Ltd	IA	HSE 4353
Purge Insect Destroyer	Aztec Chemicals Ltd	IP	HSE 5048
Pynosect 10	Mitchell Cotts Chemicals Ltd	IP	HSE 3171
Pynosect Extra Fog	Mitchell Cotts Chemicals Ltd	IAH IFSP IP	HSE 3172
Raid Ant and Roach Killer	Johnson Wax Ltd	IA	HSE 5182
Residroid	Sharpstow International Homecare Products Ltd	IAH IP	HSE 4736
Sactif Flying Insect Killer	Lever Industrial Ltd	IP	HSE 4924
Sainsbury's Fly and Wasp Killer	J Sainsbury Plc	IA	HSE 5335
Sanmex Fly and Wasp Killer	The British Products Sanmex Company Ltd	IA	HSE 4006
Sanmex Supakil Insecticide	The British Product Sanmex Company Ltd	IA	HSE 4009
Scorpio Fly Spray for Flying Insects	Jangro Ltd	IP	HSE 5049
Secto Fly Killer	Secto Company Ltd	IA	HSE 3862
Shelltox Insect Killer	Temana International Ltd	IA IP	HSE 3697
Shelltox Insect Killer 2	Temana International Ltd	IA IP	HSE 4061
Stop Insect Killer	Temana International Ltd	IA IP	HSE 3163
Superdrug Fly Killer	Superdrug Stores Plc	IA	HSE 4185
408 PERMETHRIN and TETRAMETHRIN and DICHLORVOS			
Laser Insect Killer	Sharpstow International Homecare Products Ltd	IA	HSE 4910
Supaswat Insect Killer	Sharpstow International Homecare Products Ltd	IA	HSE 4630
409 PERMETHRIN and TETRAMETHRIN and FENITROTHION			
Motox	HVM International Ltd	IA	HSE 5622

INSECTICIDES

Product Name	Marketing Company	Use	Reg. No
410 **PERMETHRIN and TETRAMETHRIN and d-ALLETHRIN**			
Hail Plus Crawling Insect Killer	International Consumer Products Ltd	IA	HSE 5567
New Super Raid Insecticide	Johnson Wax Ltd	IA IP	HSE 4583
411 **PERMETHRIN and TETRAMETHRIN and d-ALLETHRIN and DICHLORVOS**			
Bop	Robert McBride Ltd –Exports	IA	HSE 4773
412 **PERMETHRIN and d-ALLETHRIN**			
New Vapona Fly and Wasp Killer	Ashe Consumer Products Ltd	IA	HSE 4201
Raid Yardguard	Johnson Wax Ltd	IA	HSE 5459
413 **PHOXIM**			
* Mafu Ant Killer	Bayer (UK) Ltd	IA	HSE 3145
414 **PIRIMIPHOS-METHYL**			
Actellic 25 EC	Zeneca Public Health	IAH IFSP IP WP	HSE 4880
Actellic Dust	Zeneca Public Health	IFSP IP	HSE 4881
# Actellic M20	Fosroc Ltd	IP	HSE 3093
# Actellic M20	ICI Plc	IP	HSE 3527
415 **PIRIMIPHOS-METHYL and RESMETHRIN and TETRAMETHRIN**			
Waspend	Zeneca Garden Care	IA IP	HSE 4857
416 **POTASSIUM SALTS OF FATTY ACIDS**			
Devcol Liquid Ant Killer	Devcol Morgan Ltd	IA	HSE 5518
Devcol Wasp Killer	Devcol Morgan Ltd	IA	HSE 5519
Greenco Ant Killer	Septico Organic Products	IA	HSE 4260
Natural Ant Gun	Zeneca Garden Care	IA	HSE 4916
Natural Wasp Gun	ICI Garden and Professional Products	IA	HSE 5327
Natural Wasp Killer	Septico Organic Products	IA	HSE 5326

INSECTICIDES

Product Name	Marketing Company	Use	Reg. No
417 PRALLETHRIN			
ETOC Liquid Vaporiser	Sumitomo Chemical (UK) Plc	IA	HSE 5618
418 PROPOXUR			
* Blattenex 20	Bayer UK Ltd	IP	HSE 4434
Killgerm Propoxur 20 EC	Killgerm Chemicals Ltd	IFSP IP	HSE 4546
419 PYRETHRINS			
Bolt Dry Flying Insect Killer	Johnson Wax Ltd	IA	HSE 3581
* Bolt Natural Flying Insect Killer	Johnson Wax Ltd	IA	HSE 3607
Boots Pump Action Fly and Wasp Killer	Boots Company Plc	IA	HSE 5292
Boots Pump Action Fly and Wasp Killer	Boots Company Plc	IA	HSE 5547
# Cooper Fly Spray	Wellcome Foundation Ltd	IP	HSE 3867
Coopers Fly Spray N	Pitman-Moore Ltd	IAH IP	HSE 5377
Coopers Fly Spray N (Ready to Use)	Coopers Animal Health Ltd	IAH IP	HSE 4040
Dairy Fly Spray	Battle, Hayward and Bower Ltd	IAH IP	HSE 5579
Days Farm Fly Spray	Day and Sons (Crewe) Ltd	IP	HSE 4749
Detia Pyrethrum Spray	K-D Pest Control Products	IA	HSE 3695
Devcol Universal Pest Powder	Devcol Ltd	IA	HSE 4864
Drione	Roussel Uclaf Environmental Health Ltd	IFSP IP	HSE 5254
Drione	Wellcome Foundation Ltd	IFSP IP	HSE 4656
Flamil Finale Super	Flore-Chemie GmbH	IA IP	HSE 5615
* Flit Flying and Crawling Insect Killer	Agropharm Ltd	IA IP	HSE 4821
Insect Killer Fortefog	Agropharm Ltd	IA IP	HSE 4820
Globol Shake and Spray Insect Killer	Globol Chemicals (UK) Ltd	IA	HSE 4883
Jeyes Expel Pump Action Fly and Wasp Killer	Jeyes Group Plc	IA	HSE 5506
Jeyes Flying Insect Spray	Jeyes Ltd	IA	HSE 5248
Karate Dairy Fly Spray	Lever Industrial Ltd	IAH IP	HSE 5611
Killgerm Pyrethrum Spray	Killgerm Chemicals Ltd	IAH IFSP IP	HSE 4636
Killgerm ULV 400	Killgerm Chemicals Ltd	IAH IFSP IP	HSE 4838
Kleenoff Flying Insect Spray	Jeyes Ltd	IA	HSE 5247

INSECTICIDES

Product Name	Marketing Company	Use	Reg. No
419 **PYRETHRINS**—continued			
Konk 1(B) Flying Insect Killer	Nolan Chemicals Ltd	IAH IFSP IP	HSE 5486
Konk I	Airguard Control Inc	IAH IFSP IP	HSE 5085
Patriot Flying and Crawling Insect Killer	Agropharm Ltd	IA IP	HSE 4959
Pif Paf Fly Spray	Roussel Uclaf Environmental Health Ltd	IFSP IP	HSE 5154
Pif Paf Fly Spray	Wellcome Foundation Ltd	IFSP IP	HSE 4514
Prevent	Agropharm Ltd	IA	HSE 4851
Pybuthrin 2/16	Roussel Uclaf Environmental Health Ltd	IAH IFSP IP	HSE 5115
Pybuthrin 2/16	Wellcome Foundation Ltd	IAH IFSP IP	HSE 4885
Pybuthrin 33	Roussel Uclaf Environmental Health Ltd	IFSP IP	HSE 5106
Pybuthrin 33	Wellcome Foundation Ltd	IFSP IP	HSE 4892
* Pybuthrin ULV	Wellcome Foundation Ltd	IAH IFSP IP	HSE 4234
Pyrematic Flying Insect Killer	Colmart Marketing Ltd	IAH IFSP IP	HSE 4735
Raid Dry Natural Flying Insect Killer	Johnson Wax Ltd	IA	HSE 4235
Residex	Agropharm Ltd	IA	HSE 4811
SC Johnson Flying Insect Killer	Johnson Wax Ltd	IA	HSE 5094
Superdrug Fly and Wasp Killer Pump Spray	Superdrug Stores Plc	IA	HSE 5171
Swak Natural	Technical Concepts International Ltd	IAH IFSP IP	HSE 4982
Trilanco Fly Spray	Trilanco	IP	HSE 4558
Vapona Green Arrow Fly and Wasp Killer	Ashe Consumer Products Ltd	IA	HSE 4460
Vapona House and Plant Fly Aerosol	Ashe Ltd	IA	HSE 5235

420 **PYRETHRINS and ALKYLARYLTRIMETHYL AMMONIUM CHLORIDE and PERMETHRIN**

* Roxem C	Horton Hygiene Co Ltd	IP SP	HSE 3123

421 **PYRETHRINS and BENDIOCARB**

Ficam Plus	Cambridge Animal and Public Health Ltd	IP	HSE 4830

INSECTICIDES

Product Name	Marketing Company	Use	Reg. No
422 PYRETHRINS and BIOALLETHRIN and PERMETHRIN			
Supabug Crawling Insect Killer	Sharpstow International Homecare Products Ltd	IA	HSE 4794
423 PYRETHRINS and CHLORPYRIFOS and CYPERMETHRIN			
New Tetracide	Killgerm Chemicals Ltd	IP	HSE 3724
424 PYRETHRINS and CHLORPYRIFOS-METHYL and PERMETHRIN			
Multispray	Roussel Uclaf Environmental Health Ltd	IFSP IP	HSE 5165
Multispray	Wellcome Foundation Ltd	IFSP IP	HSE 4893
425 PYRETHRINS and CYPERMETHRIN			
Bolt Crawling Insect Killer	Johnson Wax Ltd	IA	HSE 3609
Raid Cockroach Killer Formula 2	Johnson Wax Ltd	IA	HSE 4239
426 PYRETHRINS and DIAZINON			
Ant Gun!	ICI Garden and Professional Products	IA	HSE 4981
B and Q Antkiller Spray	B and Q Plc	IA	HSE 5114
427 PYRETHRINS and HYDROPRENE			
Protrol Plus ULV	Zoecon Corporation	IFSP IP	HSE 4909
428 PYRETHRINS and METHOPRENE			
Canovel Pet Bedding and Household Spray	Smithkline Beecham Animal Health	IA	HSE 5330
Killgerm Precor RTU	Killgerm Chemicals Ltd	IFSP IP	HSE 3165
Killgerm Precor ULV I	Killgerm Chemicals Ltd	IFSP IP	HSE 3098
429 PYRETHRINS and PERMETHRIN			
* Coopers Head to Tail Animal Bedding Spray	Coopers Animal Health Ltd	IA	HSE 3008
Detia Crawling Insect Spray	K-D Pest Control Products	IA	HSE 3710
Flamil Finale	Flore-Chemie GmbH	IP	HSE 5083

INSECTICIDES

Product Name	Marketing Company	Use	Reg. No

429 PYRETHRINS and PERMETHRIN—continued

Product Name	Marketing Company	Use	Reg. No
Household Flea Spray	Johnson's Veterinary Products Ltd	IA	HSE 3908
Insecticidal Aerosol	Aerosols International Ltd	IA	HSE 5118
Johnson's Household Flea Spray	Johnson's Veterinary Products Ltd	IA	HSE 5703
Kybosh	Pan Britannica Industries Ltd	IA	HSE 3022
Secto Household Flea Spray	Secto Co Ltd	IA	HSE 4131

430 PYRETHRINS and PERMETHRIN and NAPHTHALENE and BENZALKONIUM CHLORIDE

Product Name	Marketing Company	Use	Reg. No
Roxem D	Horton Hygiene Company	IP SP	HSE 5107

431 PYRETHRINS and RESMETHRIN

Product Name	Marketing Company	Use	Reg. No
Rentokil Houseplant Insect Killer	Rentokil Ltd	IA	HSE 4433

432 PYRETHRINS and S-METHOPRENE

Product Name	Marketing Company	Use	Reg. No
Acclaim	Sanofi Animal Health Ltd	IA	HSE 5473
Vet-Kem Acclaim	Sanofi Animal Health Ltd	IA	HSE 5474
Zodiac and Household Flea Spray	Grosvenor Pet Health Ltd	IA	HSE 5471

433 RESMETHRIN and FENITROTHION and PERMETHRIN

Product Name	Marketing Company	Use	Reg. No
Turbair Beetle Killer	Pan Britannica Industries Ltd	IFSP IP	HSE 4648

434 RESMETHRIN and PYRETHRINS

Product Name	Marketing Company	Use	Reg. No
Rentokil Houseplant Insect Killer	Rentokil Ltd	IA	HSE 4433

435 RESMETHRIN and TETRAMETHRIN

Product Name	Marketing Company	Use	Reg. No
151 Dry Fly and Wasp Killer	SB Ltd	IA	HSE 5169
Apex Fly-Killer	Apex Industrial Chemicals Ltd	IA IP	HSE 4300
Chem-Kil	Inter-Chem	IA IP	HSE 4266
Chirton Dry Fly and Wasp Killer	Ronson Plc	IA	HSE 5287
Chirton Fly and Wasp Killer	Ronson Plc	IA	HSE 3836
Chromessol Insect Killer	Cromessol Company Ltd	IA	HSE 5067

INSECTICIDES

Product Name	Marketing Company	Use	Reg. No
435 **RESMETHRIN and TETRAMETHRIN**—continued			
Fly Killer	Kalon Agricultural Division	IA	HSE 4817
Hospital Flying Insect Spray	Kalon Chemicals	IA	HSE 4652
Janisolve Fly Killer	Johmar Enterprises Company	IA	HSE 4228
Penetone Fly Killer	Penetone Ltd	IA	HSE 4072
Pynosect 2	Mitchell Cotts Chemicals Ltd	IP	HSE 4654
Pynosect 4	Mitchell Cotts Chemicals Ltd	IP	HSE 4655
Quantum Hygiene Fly Killer	Quantum Hygiene	IA IP	HSE 5066
Swat	Solvitol Ltd	IA IP	HSE 3564
Swat A	Solvitol Ltd	IP	HSE 5051
Trilanco Fly Killer	Trilanco	IA	HSE 4816
Wasp Nest Destroyer	Sorex Ltd	IP	HSE 4410
436 **RESMETHRIN and TETRAMETHRIN and PIRIMIPHOS-METHYL**			
Waspend	Zeneca Garden Care	IA IP	HSE 4857
437 **S-BIOALLETHRIN and BIOALLETHRIN**			
Actomite	G D Searle and Co Ltd	IA	HSE 4182
438 **S-BIOALLETHRIN and BIORESMETHRIN**			
* Coopers Fly Spray EB (Ready To Use)	Coopers Animal Health Ltd	IAH IP	HSE 4042
439 **S-BIOALLETHRIN and DICHLORVOS and PERMETHRIN**			
Pif Paf Insecticide	Roussel Uclaf Environmental Health Ltd	IA IFSP IP	HSE 5150
Pif Paf Insecticide	Wellcome Foundation Ltd	IA IFSP IP	HSE 3748
440 **S-BIOALLETHRIN and PERMETHRIN**			
* Aqua Reslin	Wellcome Foundation Ltd	IFSP IP	HSE 3195
Aqua Reslin Premium	Roussel Uclaf Environmental Health Ltd	IFSP IP	HSE 5172
Aqua Reslin Premium	Wellcome Foundation Ltd	IFSP IP	HSE 4087

INSECTICIDES

Product Name	Marketing Company	Use	Reg. No
440 S-BIOALLETHRIN and PERMETHRIN—continued			
Aqua Reslin Super	Roussel Uclaf Environmental Health Ltd	IFSP IP	HSE 5175
Aqua Reslin Super	Wellcome Foundation Ltd	IFSP IP	HSE 4088
Coopex S25% EC	Roussel Uclaf Environmental Health Ltd	IFSP IP	HSE 5146
Coopex S25% EC	Wellcome Foundation Ltd	IFSP IP	HSE 3525
Imperator Fog S 18/6	Zeneca Public Health	IP	HSE 4960
Imperator Fog Super S 30/15	Zeneca Public Health	IP	HSE 4955
Imperator ULV S 6/3	Zeneca Public Health	IP	HSE 4954
Imperator ULV Super S 10/4	Zeneca Public Health	IP	HSE 4953
Neopybuthrin Premium	Roussel Uclaf Environmental Health Ltd	IFSP IP	HSE 5149
Neopybuthrin S200Q	Wellcome Foundation Ltd	IFSP IP	HSE 3192
* Neopybuthrin S9E	Wellcome Foundation Ltd	IFSP IP	HSE 3193
Neopybuthrin Super	Roussel Uclaf Environmental Health Ltd	IFSP IP	HSE 5148
Neopybuthrin Super	Wellcome Foundation Ltd	IFSP IP	HSE 4086
Resigen	Roussel Uclaf Environmental Health Ltd	IFSP IP	HSE 5143
Resigen	Wellcome Foundation Ltd	IFSP IP	HSE 4887
* Reslin 25P	Wellcome Foundation Ltd	IFSP IP	HSE 3190
# Reslin 25S	Wellcome Foundation Ltd	IFSP IP	HSE 3197
* Reslin 25S	Wellcome Foundation Ltd	IFSP IP	HSE 4258
# Reslin 25SE	Wellcome Foundation Ltd	IFSP IP	HSE 3196
Reslin 25SE	Wellcome Foundation Ltd	IFSP IP	HSE 4259
Reslin Premium	Roussel Uclaf Environmental Health Ltd	IFSP IP	HSE 5174
Reslin Super	Roussel Uclaf Environmental Health Ltd	IFSP IP	HSE 5173
Reslin Super	Wellcome Foundation Ltd	IFSP IP	HSE 4089
* Wellcome Crawling Insect Killer	Wellcome Foundation Ltd	IA IFSP IP	HSE 4166
441 S-BIOALLETHRIN and TETRAMETHRIN and PERMETHRIN			
Pif Paf Crawling Insect Killer Aerosol	Roussel Uclaf Environmental Health Ltd	IA IFSP IP	HSE 5595

INSECTICIDES

Product Name	Marketing Company	Use	Reg. No
442 S-HYDROPRENE and PERMETHRIN			
Protrol Plus Crack and Crevice Aerosol	Sandoz Speciality Pest Control Ltd	IA IFSP IP	HSE 5384
Protrol Plus Crack and Crevice Aerosol	Zoecon Corporation	IA IFSP IP	HSE 4906
443 S-METHOPRENE			
Dianex	Sandoz Speciality Pest Control Ltd	IP	HSE 5709
Pharorid S	Sandoz Speciality Pest Control Ltd	IP	HSE 5479
444 S-METHOPRENE and PERMETHRIN			
Vet-Kem Pump Spray	Sanofi Animal Health Ltd	IA	HSE 5475
Zodiac and Pump Spray	Grosvenor Pet Health Ltd	IA	HSE 5472
445 S-METHOPRENE and PYRETHRINS			
Acclaim	Sanofi Animal Health Ltd	IA	HSE 5473
Vet-Kem Acclaim	Sanofi Animal Health Ltd	IA	HSE 5474
Zodiac and Household Flea Spray	Grosvenor Pet Health Ltd	IA	HSE 5471
446 TETRAMETHRIN			
Big D Fly and Wasp Killer	D F Marketing Ltd	IA	HSE 5422
Bolt Flying Insect Killer	Johnson Wax Ltd	IA	HSE 3791
Boots Fly and Wasp Killer	Boots Company Plc	IA	HSE 5632
Di-Fly	Lever Industrial Ltd	IA	HSE 4509
Dry Fly Killer	Domestic Fillers Ltd	IA	HSE 4237
Fly and Wasp Killer	Lloyds Chemists Plc	IA	HSE 5639
Jeyes Expel Fly and Wasp Killer	Jeyes Group Plc	IA	HSE 5516
Jeyes Kontrol Fly and Wasp Killer	Jeyes Group Plc	IA	HSE 5359
* Johnson Wax Raid Fly and Wasp Killer	Johnson Wax Ltd	IA	HSE 4238
Killgerm Py-Kill W	Killgerm Chemicals Ltd	IAH IFSP IP	HSE 4632
Kleeneze Fly Killer	Kleeneze Ltd	IA	HSE 4225
Odex Fly Spray	Odex Ltd	IA	HSE 4814

INSECTICIDES

Product Name	Marketing Company	Use	Reg. No
446 **TETRAMETHRIN**—continued			
Premiere Fly and Wasp Killer	Premiere Products	IA IP	HSE 4516
Provence Fly and Wasp Killer	Haventrail Ltd	IA	HSE 4382
Pyrakill Flying Insect Spray	Forward Chemicals Ltd	IA	HSE 5357
Raid Flying Insect Killer	Johnson Wax Ltd	IA IP	HSE 4585
Red Can Fly Killer	Yellow Can Company Ltd	IA	HSE 4508
SC Johnson Raid Fly and Wasp Killer	Johnson Wax Ltd	IA	HSE 5113
St Michael Flyspray	Marks and Spencer Plc	IA	HSE 4177
Super Raid Insect Killer Formula 2	Johnson Wax Ltd	IA	HSE 4240
447 **TETRAMETHRIN and BENDIOCARB**			
Camco Insect Spray	Camco	IA	HSE 4222
Devcol Household Insect Spray	Devcol Ltd	IA	HSE 5512
448 **TETRAMETHRIN and CHLORPYRIFOS**			
Raid Wasp Nest Destroyer	Johnson Wax Ltd	IA	HSE 5597
449 **TETRAMETHRIN and CYPERMETHRIN**			
Cyperkill Plus WP	Mitchell Cotts Chemicals Ltd	IFSP IP	HSE 4855
New Vapona Ant Killer	Ashe Consumer Products Ltd	IA	HSE 4298
Raid Ant and Crawling Insect Killer	Johnson Wax Ltd	IA	HSE 3001
Raid Residual Crawling Insect Killer	Johnson Wax Ltd	IA IP	HSE 4584
SC Johnson Raid Ant and Cockroach Killer	Johnson Wax Ltd	IA	HSE 5216
# Shelltox Cockroach and Crawling Insect Killer 2	Temana International Ltd	IA IP	HSE 4030
Shelltox Cockroach and Crawling Insect Killer 3	Temana International Ltd	IA IP	HSE 4031
Shelltox Cockroach and Crawling Insect Killer 4	Temana International Ltd	IA	HSE 4739
Super Shelltox Crawling Insect Killer	Temana International Ltd	IA	HSE 5454
Vapona Ant and Crawling Insect Killer Aerosol	Ashe Ltd	IA	HSE 5282

INSECTICIDES

Product Name	Marketing Company	Use	Reg. No

450 **TETRAMETHRIN and DICHLOROPHEN and DICHLORVOS**

Wintox	Mould Growth Consultants Ltd	IP	HSE 4236

451 **TETRAMETHRIN and DICHLORVOS**

Shelltox Extra Flykiller	Temana International Ltd	IA IP	HSE 3696
Shelltox Flykiller	Temana International Ltd	IA	HSE 3162
Vapona Fly and Wasp Killer Spray	Ashe Consumer Products Ltd	IA	HSE 4907
Vapona Wasp and Fly Killer	Ashe Ltd	IA	HSE 5281

452 **TETRAMETHRIN and DICHLORVOS and PERMETHRIN**

Laser Insect Killer	Sharpstow International Homecare Products Ltd	IA	HSE 4910
Supaswat Insect Killer	Sharpstow International Homecare Products Ltd	IA	HSE 4630

453 **TETRAMETHRIN and DIMETHOATE**

Killgerm Dimethoate Extra	Killgerm Chemicals Ltd	IFSP IP	HSE 4633

454 **TETRAMETHRIN and FENITROTHION**

Ant and Crawling Insect Killer	Rentokil Ltd	IA	HSE 4642
Big D Cockroach and Crawling Insect Killer	Domestic Fillers Ltd	IA	HSE 4098
Big D Rocket Fly Killer	Domestic Fillers Ltd	IA	HSE 3552
Doom Ant and Crawling Insect Killer Aerosol	Napa Products Ltd	IA	HSE 4753
Flying Insect Killer	Keen (World Marketing) Ltd	IA	HSE 4313
Flying Insect Killer Faster Knockdown	Keen (World Marketing) Ltd	IA	HSE 4312
Killgerm Fenitrothion-Pyrethroid Concentrate	Killgerm Chemicals Ltd	IFSP IP	HSE 4623
Tox Exterminating Fly and Wasp Killer	Keen (World Marketing) Ltd	IA	HSE 4314

455 **TETRAMETHRIN and FENITROTHION and PERMETHRIN**

Motox	HVM International Ltd	IA	HSE 5622

INSECTICIDES

Product Name	Marketing Company	Use	Reg. No
456 TETRAMETHRIN and GAMMA-HCH			
Doom Flea Killer	Napa Products Ltd	IA	HSE 4572
Doom Moth Proofer Aerosol	Napa Products Ltd	IA	HSE 4571
457 TETRAMETHRIN and PERMETHRIN			
"Bop" Crawling Insect Killer	Robert McBride Ltd –Exports	IA IP	HSE 4138
"Bop" Crawling Insect Killer (MC)	Robert McBride Ltd –Exports	IA IP	HSE 4139
Bolt Ant and Crawling Insect Killer	Johnson Wax Ltd	IA	HSE 3619
Ciba-Geigy Ko Fly and Wasp Killer	Ciba-Geigy Plc	IA	HSE 3726
Corry's Fly and Wasp Killer	Vitax Ltd	IA	HSE 5322
D-Stroy	Future Developments (Manufacturing) Ltd	IP	HSE 5411
Delta Cockroach and Ant Spray	Middle East Aerosol and Detergent Company Ltd	IA	HSE 3950
Doom Tropical Strength Fly and Wasp Killer Aerosol	Napa Products Ltd	IA	HSE 4752
Dragon Flykiller	Zeneca Public Health	IA	HSE 4952
Fly and Wasp Killer	Dimex Ltd	IP	HSE 5050
Fly and Wasp Killer	Rentokil Ltd	IA IP	HSE 4643
Fly Spray	Merton Cleaning Supplies	IP	HSE 5112
Hail Flying Insect Killer	International Consumer Products Ltd	IA	HSE 5568
* Household Flea Spray	Boots Company Plc	IA	HSE 4661
K02 Selkil	Selden Research Ltd	IP	HSE 4828
Nippon Ant and Crawling Insect Killer	Vitax Ltd	IA	HSE 4414
Nippon Fly Killer Spray	Vitax Ltd	IA	HSE 4254
Nippon Killaquer Crawling Insect Killer	Arrow Chemicals	IFSP IP	HSE 3694
Nisa Fly and Wasp Killer	Nisa	IA	HSE 4962
* Pestkill Insect Spray	Phillips Yeast Products Ltd	IA	HSE 4353
Purge Insect Destroyer	Aztec Chemicals Ltd	IP	HSE 5048
Pynosect 10	Mitchell Cotts Chemicals Ltd	IP	HSE 3171
Pynosect Extra Fog	Mitchell Cotts Chemicals Ltd	IAH IFSP IP	HSE 3172
Raid Ant and Roach Killer	Johnson Wax Ltd	IA	HSE 5182

INSECTICIDES

Product Name	Marketing Company	Use	Reg. No

457 TETRAMETHRIN and PERMETHRIN—continued

Product Name	Marketing Company	Use	Reg. No
Residroid	Sharpstow International Homecare Products Ltd	IAH IP	HSE 4736
Sactif Flying Insect Killer	Lever Industrial Ltd	IP	HSE 4924
Sainsbury's Fly and Wasp Killer	J Sainsbury Plc	IA	HSE 5335
Sanmex Fly and Wasp Killer	The British Products Sanmex Company Ltd	IA	HSE 4006
Sanmex Supakil Insecticide	The British Product Sanmex Company Ltd	IA	HSE 4009
Scorpio Fly Spray for Flying Insects	Jangro Ltd	IP	HSE 5049
Secto Fly Killer	Secto Company Ltd	IA	HSE 3862
Shelltox Insect Killer	Temana International Ltd	IA IP	HSE 3697
Shelltox Insect Killer 2	Temana International Ltd	IA IP	HSE 4061
Stop Insect Killer	Temana International Ltd	IA IP	HSE 3163
Superdrug Fly Killer	Superdrug Stores Plc	IA	HSE 4185

458 TETRAMETHRIN and PERMETHRIN and S-BIOALLETHRIN

Product Name	Marketing Company	Use	Reg. No
Pif Paf Crawling Insect Killer Aerosol	Roussel Uclaf Environmental Health Ltd	IA IFSP IP	HSE 5595

459 TETRAMETHRIN and PIRIMIPHOS-METHYL and RESMETHRIN

Product Name	Marketing Company	Use	Reg. No
Waspend	Zeneca Garden Care	IA IP	HSE 4857

460 TETRAMETHRIN and RESMETHRIN

Product Name	Marketing Company	Use	Reg. No
151 Dry Fly and Wasp Killer	SB Ltd	IA	HSE 5169
Apex Fly-Killer	Apex Industrial Chemicals Ltd	IA IP	HSE 4300
Chem-Kil	Inter-Chem	IA IP	HSE 4266
Chirton Dry Fly and Wasp Killer	Ronson Plc	IA	HSE 5287
Chirton Fly and Wasp Killer	Ronson Plc	IA	HSE 3836
Cromessol Insect Killer	Cromessol Company Ltd	IA	HSE 5067
Fly Killer	Kalon Agricultural Division	IA	HSE 4817
Hospital Flying Insect Spray	Kalon Chemicals	IA	HSE 4652
Janisolve Fly Killer	Johmar Enterprises Company	IA	HSE 4228
Penetone Fly Killer	Penetone Ltd	IA	HSE 4072

INSECTICIDES

Product Name	Marketing Company	Use	Reg. No

460 TETRAMETHRIN and RESMETHRIN—continued

Pynosect 2	Mitchell Cotts Chemicals Ltd	IP	HSE 4654
Pynosect 4	Mitchell Cotts Chemicals Ltd	IP	HSE 4655
Quantum Hygiene Fly Killer	Quantum Hygiene	IA IP	HSE 5066
Swat	Solvitol Ltd	IA IP	HSE 3564
Swat A	Solvitol Ltd	IP	HSE 5051
Trilanco Fly Killer	Trilanco	IA	HSE 4816
Wasp Nest Destroyer	Sorex Ltd	IP	HSE 4410

461 TETRAMETHRIN and d-ALLETHRIN

Boots Dry Fly and Wasp Killer	Boots Company Plc	IA	HSE 4037
Boots Dry Fly and Wasp Killer 3	Boots Company Plc	IA	HSE 4241
Fly, Wasp and Mosquito Killer	Rentokil Ltd	IA	HSE 4882
New Vapona Fly Killer Dry Formulation	Ashe Consumer Products Ltd	IA	HSE 4202
Raid Fly and Wasp Killer	Johnson Wax Ltd	IA	HSE 3000
Sainsbury's Fly, Wasp and Mosquito Killer	J Sainsbury Plc	IA	HSE 5164

462 TETRAMETHRIN and d-ALLETHRIN and CYPERMETHRIN

"Bop" Flying and Crawling Insect Killer	Robert McBride Ltd –Exports	IA IP	HSE 4136
"Bop" Flying and Crawling Insect Killer (Water Based)	Robert McBride Ltd –Exports	IA IP	HSE 4137
"Bop" Flying Insect Killer	Robert McBride Ltd –Exports	IA IP	HSE 4141
"Bop" Flying Insect Killer (MC)	Robert McBride Ltd –Exports	IA IP	HSE 4140
Bop Flying and Crawling Insect Killer (MC)	Robert McBride Ltd –Exports	IA IP	HSE 4135

463 TETRAMETHRIN and d-ALLETHRIN and DICHLORVOS and PERMETHRIN

Bop	Robert McBride Ltd –Exports	IA	HSE 4773

464 TETRAMETHRIN and d-ALLETHRIN and PERMETHRIN

Hail Plus Crawling Insect Killer	International Consumer Products Ltd	IA	HSE 5567
New Super Raid Insecticide	Johnson Wax Ltd	IA IP	HSE 4583

INSECTICIDES

Product Name	Marketing Company	Use	Reg. No
465 TETRAMETHRIN and d-PHENOTHRIN			
"Flak"	Robert McBride Ltd –Exports	IA IP	HSE 4143
Big D Ant and Crawling Insect Killer	Domestic Fillers Ltd	IA	HSE 4097
Boots Ant and Crawling Insect Killer	Boots Company Plc	IA	HSE 5554
Boots Flea Spray	Boots Company Plc	IA	HSE 5522
Bop	Robert McBride Ltd –Exports	IA IP	HSE 4142
Brimpex ULV 1500	Brimpex Metal Treatments	IAH IFSP IP	HSE 4942
Delta Fly Spray	Middle East Aerosol and Detergent Company Ltd	IA	HSE 4663
Deosan Fly Spray	Deosan Ltd	IP	HSE 5246
Envar 6/14 EC	Agrimar (UK) Plc	IP	HSE 4354
Envar 6/14 Oil	Agrimar (UK) Plc	IP	HSE 4355
Goodboy Household Flea Spray	Armitage Brothers Ltd	IA	HSE 4090
Jeyes Expel Ant and Crawling Insect Killer	Jeyes Group Plc	IA	HSE 5552
Jeyes Expel Flea Killer	Jeyes Group Plc	IA	HSE 5523
Jeyes Kontrol Ant and Crawling Insect Killer	Jeyes Group Plc	IA	HSE 5553
Jeyes Kontrol Flea Killer	Jeyes Group Plc	IA	HSE 5521
Killgerm ULV 500	Killgerm Chemicals Ltd	IAH IFSP IP	HSE 4647
Killoff Aerosol Insect Killer	Ferosan Products	IA	HSE 3974
Pesguard NS 6/14 EC	Sumitomo Chemical (UK) Plc	IFSP IP	HSE 4360
Pesguard NS 6/14 Oil	Sumitomo Chemical (UK) Plc	IFSP IP	HSE 4361
Pesguard OBA F 7305 C	Sumitomo Chemical (UK) Plc	IA	HSE 4064
Pesguard OBA F 7305 D	Sumitomo Chemical (UK) Plc	IA	HSE 4065
Pesguard OBA F 7305 E	Sumitomo Chemical (UK) Plc	IA	HSE 4066
Pesguard OBA F 7305 F	Sumitomo Chemical (UK) Plc	IA	HSE 4067
Pesguard OBA F 7305 G	Sumitomo Chemical (UK) Plc	IA	HSE 4063
Pesguard OBA F 7305 B	Sumitomo Chemical (UK) Plc	IA	HSE 4419
Pesguard WBA F-2656	Sumitomo Chemical (UK) Plc	IA	HSE 4263
Pesguard WBA F-2692	Sumitomo Chemical (UK) Plc	IA	HSE 4262
Pestkill Household Spray	Phillips Yeast Products Ltd	IA	HSE 4938
Py Kill 25 Plus	Killgerm Chemicals Ltd	IAH IFSP IP	HSE 4900

INSECTICIDES

Product Name	Marketing Company	Use	Reg. No
465 **TETRAMETHRIN and d-PHENOTHRIN**—continued			
Raid House and Plant	Johnson Wax Ltd	IA	HSE 5432
S C Johnson Raid	Johnson Wax Ltd	IA	HSE 5181
Sergeant's Dust Mite Patrol Injector and Spray	Conagra Pet Products Company	IA IP	HSE 5513
Sergeant's Dust Mite Patrol Spray	Conagra Pet Products Company	IA	HSE 5041
Sergeant's Insect Patrol	Conagra Pet Products Company	IA	HSE 5397
Shelltox Flying Insect Killer 2	Temana International Ltd	IA IP	HSE 4032
Shelltox Flying Insect Killer 3	Temana International Ltd	IA	HSE 4741
Shelltox Super Flying Insect Killer	Temana International Ltd	IA	HSE 4790
Sorex Fly Spray RTU	Sorex Ltd	IP	HSE 5718
Supaswat Insect Killer	Sharpstow International Homecare Products Ltd	IA	HSE 5691
Super Fly Spray	Sorex Ltd	IA	HSE 4468
Super Raid II	Johnson Wax Ltd	IA	HSE 4015
Super Shelltox Flying Insect Killer	Temana International Ltd	IA	HSE 5440
Vapona House and Plant Fly Spray	Ashe Ltd	IA	HSE 5236
Vapona Wasp and Fly Killer Spray	Ashe Ltd	IA	HSE 5280
Vitapet Flearid Household Spray	Seven Seas Ltd	IA	HSE 5434
466 **TRICHLORPHON**			
Boots Ant Trap	Boots Company Plc	IA	HSE 5542
Detia Ant Bait	K-D Pest Control Products	IA	HSE 3803
Jeyes Expel Ant Trap	Jeyes Group Plc	IA	HSE 5503
Vapona Ant Trap	Ashe Consumer Products Ltd	IA	HSE 3804
467 **d-ALLETHRIN**			
Floret Fast Knock Down	Reckitt and Colman Products Ltd	IA	HSE 4396
Gelert Mosquito Repellent	Bryncir Products Ltd	IA	HSE 5465
Jeyes Expel Plug-In Flying Insect Killer	Jeyes Ltd	IA	HSE 5653
Jeyes Kontrol Plug-In Flying Insect Killer	Jeyes Ltd	IA	HSE 5652

INSECTICIDES

Product Name	Marketing Company	Use	Reg. No
467 d-ALLETHRIN—continued			
Moskil Mosquito Repellent	I and M Steiner Ltd	IA	HSE 4342
Pynamin Forte Mat 120	Sumitomo Chemical (UK) Plc	IA	HSE 4417
Shelltox Mat 2	Temana International Ltd	IA	HSE 4742
Vapona Plug-In Flying Insect Killer	Ashe Ltd	IA	HSE 5708
468 d-ALLETHRIN and CYPERMETHRIN			
Shelltox Crawling Insect Killer Liquid Spray	Temana International Ltd	IA IP	HSE 3565
Vapona Ant and Crawling Insect Spray	Ashe Consumer Products Ltd	IA	HSE 3593
469 d-ALLETHRIN and CYPERMETHRIN and TETRAMETHRIN			
"Bop" Flying and Crawling Insect Killer	Robert McBride Ltd –Exports	IA IP	HSE 4136
"Bop" Flying and Crawling Insect Killer (Water Based)	Robert McBride Ltd –Exports	IA IP	HSE 4137
"Bop" Flying Insect Killer	Robert McBride Ltd –Exports	IA IP	HSE 4141
"Bop" Flying Insect Killer (MC)	Robert McBride Ltd –Exports	IA IP	HSE 4140
Bop Flying and Crawling Insect Killer (MC)	Robert McBride Ltd –Exports	IA IP	HSE 4135
470 d-ALLETHRIN and DICHLORVOS and PERMETHRIN and TETRAMETHRIN			
Bop	Robert McBride Ltd –Exports	IA	HSE 4773
471 d-ALLETHRIN and PERMETHRIN			
New Vapona and Wasp Killer	Ashe Consumer Products Ltd	IA	HSE 4201
Raid Yardguard	Johnson Wax Ltd	IA	HSE 5459
472 d-ALLETHRIN and PERMETHRIN and TETRAMETHRIN			
Hail Plus Crawling Insect Killer	International Consumer Products Ltd	IA	HSE 5567
New Super Raid Insecticide	Johnson Wax Ltd	IA IP	HSE 4583

INSECTICIDES

Product Name	Marketing Company	Use	Reg. No
473 d-ALLETHRIN and TETRAMETHRIN			
Boots Dry Fly and Wasp Killer	Boots Company Plc	IA	HSE 4037
Boots Dry Fly and Wasp Killer 3	Boots Company Plc	IA	HSE 4241
Fly, Wasp and Mosquito Killer	Rentokil Ltd	IA	HSE 4882
New Vapona Fly Killer Dry Formulation	Ashe Consumer Products Ltd	IA	HSE 4202
Raid Fly and Wasp Killer	Johnson Wax Ltd	IA	HSE 3000
Sainsbury's Fly, Wasp and Mosquito Killer	J Sainsbury Plc	IA	HSE 5164
474 d-ALLETHRIN and d-PHENOTHRIN			
Fly and Ant Killer	Rentokil Ltd	IA	HSE 5561
Jeyes Crawling Insect Spray	Jeyes Ltd	IA	HSE 5270
Kleenoff Crawling Insect Spray	Jeyes Ltd	IA	HSE 5269
Kwik Insecticide	Hand Associates Ltd	IA	HSE 4290
Nippon Ready For Use Fly Killer Spray	Vitax Ltd	IA	HSE 4631
Pedigree Exelpet Bedding Spray	Thomas's	IA	HSE 5427
Pesguard PS 102	Sumitomo Chemical (UK) Plc	IA	HSE 5312
Pesguard PS 102A	Sumitomo Chemical (UK) Plc	IA	HSE 4173
Pesguard PS 102B	Sumitomo Chemical (UK) Plc	IA	HSE 4174
Pesguard PS 102C	Sumitomo Chemical (UK) Plc	IA	HSE 4175
Pesguard PS 102D	Sumitomo Chemical (UK) Plc	IA	HSE 4176
Secto Ant and Crawling Insect Spray	Secto Company Ltd	IA	HSE 5714
Target Flying Insect Killer	Reckitt and Colman Products Ltd	IA	HSE 4178
Whiskas Bedding Pest Control Spray	Thomas's	IA	HSE 4596
Whiskas Exelpet Bedding Spray	Thomas's	IA	HSE 4837
475 d-PHENOTHRIN			
Aircraft Aerosol Insect Control	Sumitomo Chemical (UK) Plc	IP	HSE 5408
# Canovel – Vac Attack Fragrant Insecticidal Carpet Freshener	Beecham Animal Health Ltd	IA	HSE 4069
Insecticide Aerosol Single Dose for Aircraft Disinsection	Aerosols International Ltd	IP	HSE 3797

INSECTICIDES

Product Name	Marketing Company	Use	Reg. No
475 d-PHENOTHRIN—continued			
# Oko Pet Care Insecticidal Carpet Freshener	Oko (Pet Products) Ltd	IA	HSE 4340
One-Shot Aircraft Aerosol Insect Control	Sumitomo Chemical (UK) Plc	IP	HSE 5407
# Pet Fresh Insecticidal Carpet Freshener	The Kleenoff Company	IA	HSE 4070
Phillips Pestkill Karpet Killer	Phillips Yeast Products Ltd	IA	HSE 5604
Sergeants Car Patrol	Conagra Pet Products Company	IA	HSE 4503
Sergeants Carpet Patrol	Conagra Pet Products Company	IA	HSE 4763
Sergeants Dust Mite Patrol	Conagra Pet Products Company	IA	HSE 4862
# Sergeants Mite Patrol	Conagra Pet Products Company	IA	HSE 4772
Sergeants Rug Patrol	Conagra Pet Products Company	IA	HSE 4504
Sumithrin 10 Sec	Sumitomo Chemical Company Ltd	IP	HSE 3762
Sumithrin 10 Sec Carpet Treatment	Sumitomo Chemical (UK) Plc	IA	HSE 5398
Wellcome Multishot Aircraft Aerosol	Wellcome Foundation Ltd	IP	HSE 3990
Wellcome One-Shot Aircraft Aerosol	Wellcome Foundation Ltd	IP	HSE 3989
Willo-Zawb	Willows Francis Veterinary	IA	HSE 3076
Willodorm	Willows Francis Veterinary	IA	HSE 4782
476 d-PHENOTHRIN and TETRAMETHRIN			
"Flak"	Robert McBride Ltd –Exports	IA IP	HSE 4143
Big D Ant and Crawling Insect Killer	Domestic Fillers Ltd	IA	HSE 4097
Boots Ant and Crawling Insect Killer	Boots Company Plc	IA	HSE 5554
Boots Flea Spray	Boots Company Plc	IA	HSE 5522
Bop	Robert McBride Ltd –Exports	IA IP	HSE 4142
Brimpex ULV 1500	Brimpex Metal Treatments	IAH IFSP IP	HSE 4942
Delta Fly Spray	Middle East Aerosol and Detergent Company Ltd	IA	HSE 4663
Deosan Fly Spray	Deosan Ltd	IP	HSE 5246
Envar 6/14 EC	Agrimar (UK) Plc	IP	HSE 4354

INSECTICIDES

Product Name	Marketing Company	Use	Reg. No
476 d-PHENOTHRIN and TETRAMETHRIN—continued			
Envar 6/14 Oil	Agrimar (UK) Plc	IP	HSE 4355
Goodboy Household Flea Spray	Armitage Brothers Ltd	IA	HSE 4090
Jeyes Expel Ant and Crawling Insect Killer	Jeyes Group Plc	IA	HSE 5552
Jeyes Expel Flea Killer	Jeyes Group Plc	IA	HSE 5523
Jeyes Kontrol Ant and Crawling Insect Killer	Jeyes Group Plc	IA	HSE 5553
Jeyes Kontrol Flea Killer	Jeyes Group Plc	IA	HSE 5521
Killgerm ULV 500	Killgerm Chemicals Ltd	IAH IFSP IP	HSE 4647
Killoff Aerosol Insect Killer	Ferosan Products	IA	HSE 3974
Pesguard NS 6/14 EC	Sumitomo Chemical (UK) Plc	IFSP IP	HSE 4360
Pesguard NS 6/14 Oil	Sumitomo Chemical (UK) Plc	IFSP IP	HSE 4361
Pesguard OBA F 7305 C	Sumitomo Chemical (UK) Plc	IA	HSE 4064
Pesguard OBA F 7305 D	Sumitomo Chemical (UK) Plc	IA	HSE 4065
Pesguard OBA F 7305 E	Sumitomo Chemical (UK) Plc	IA	HSE 4066
Pesguard OBA F 7305 F	Sumitomo Chemical (UK) Plc	IA	HSE 4067
Pesguard OBA F 7305 G	Sumitomo Chemical (UK) Plc	IA	HSE 4063
Pesguard OBA F 7305 B	Sumitomo Chemical (UK) Plc	IA	HSE 4419
Pesguard WBA F-2656	Sumitomo Chemical (UK) Plc	IA	HSE 4263
Pesguard WBA F-2692	Sumitomo Chemical (UK) Plc	IA	HSE 4262
Pestkill Household Spray	Phillips Yeast Products Ltd	IA	HSE 4938
PY Kill 25 Plus	Killgerm Chemicals Ltd	IAH IFSP IP	HSE 4900
Raid House and Plant	Johnson Wax Ltd	IA	HSE 5432
S C Johnson Raid	Johnson Wax Ltd	IA	HSE 5181
Sergeant's Dust Mite Patrol Injector and Spray	Conagra Pet Products Company	IA IP	HSE 5513
Sergeant's Dust Mite Patrol Spray	Conagra Pet Products Company	IA	HSE 5041
Sergeant's Insect Patrol	Conagra Pet Products Company	IA	HSE 5397
Shelltox Flying Insect Killer 2	Temana International Ltd	IA IP	HSE 4032
Shelltox Flying Insect Killer 3	Temana International Ltd	IA	HSE 4741
Shelltox Super Flying Insect Killer	Temana International Ltd	IA	HSE 4790
Sorex Fly Spray RTU	Sorex Ltd	IP	HSE 5718

INSECTICIDES

Product Name	Marketing Company	Use	Reg. No
476 d-PHENOTHRIN and TETRAMETHRIN—continued			
Supaswat Insect Killer	Sharpstow International Homecare Products Ltd	IA	HSE 5691
Super Fly Spray	Sorex Ltd	IA	HSE 4468
Super Raid II	Johnson Wax Ltd	IA	HSE 4015
Super Shelltox Flying Insect Killer	Temana International Ltd	IA	HSE 5440
Vapona House and Plant Fly Spray	Ashe Ltd	IA	HSE 5236
Vapona Wasp and Fly Killer Spray	Ashe Ltd	IA	HSE 5280
Vitapet Flearid Household Spray	Seven Seas Ltd	IA	HSE 5434
477 d-PHENOTHRIN and d-ALLETHRIN			
Fly and Ant Killer	Rentokil Ltd	IA	HSE 5561
Jeyes Crawling Insect Spray	Jeyes Ltd	IA	HSE 5270
Kleenoff Crawling Insect Spray	Jeyes Ltd	IA	HSE 5269
Kwik Insecticide	Hand Associates Ltd	IA	HSE 4290
Nippon Ready For Use Fly Killer Spray	Vitax Ltd	IA	HSE 4631
Pedigree Exelpet Bedding Spray	Thomas's	IA	HSE 5427
Pesguard PS 102	Sumitomo Chemical (UK) Plc	IA	HSE 5312
Pesguard PS 102A	Sumitomo Chemical (UK) Plc	IA	HSE 4173
Pesguard PS 102B	Sumitomo Chemical (UK) Plc	IA	HSE 4174
Pesguard PS 102C	Sumitomo Chemical (UK) Plc	IA	HSE 4175
Pesguard PS 102D	Sumitomo Chemical (UK) Plc	IA	HSE 4176
Secto Ant and Crawling Insect Spray	Secto Company Ltd	IA	HSE 5714
Target Flying Insect Killer	Reckitt and Colman Products Ltd	IA	HSE 4178
Whiskas Bedding Pest Control Spray	Thomas's	IA	HSE 4596
Whiskas Exelpet Bedding Spray	Thomas's	IA	HSE 4837
478 d-PHENOTHRIN and d-TETRAMETHRIN			
Pesguard WBA F2714	Sumitomo Chemical (UK) Plc	IA	HSE 4418
479 d-TETRAMETHRIN and d-PHENOTHRIN			
Pesguard WBA F2714	Sumitomo Chemical (UK) Plc	IA	HSE 4418

4
ANTIFOULING PRODUCTS

ANTIFOULING PRODUCTS

Product Name	Marketing Company	Use	Reg. No

Antifouling Products

480 2,3,5,6-TETRACHLORO-4-(METHYL SULPHONYL) PYRIDINE

# Chemcrest	Chemflake International Ltd	AQA AVA AVP	HSE 3245

481 2,3,5,6-TETRACHLORO-4-(METHYL SULPHONYL) PYRIDINE and 2-METHYLTHIO-4-TERTIARY-BUTYLAMINO-6-CYCLOPROPYLAMINO-S-TRIAZINE and CUPROUS OXIDE and CUPROUS THIOCYANATE and ZINC OXIDE

Envoy TF 400	W and J Leigh and Company	AVA AVP	HSE 4432
Envoy TF 500	W and J Leigh and Company	AVA AVP	HSE 5599
Meridian MP40 Antifouling	Deangate Marine Products	AVA AVP	HSE 5027

482 2,3,5,6-TETRACHLORO-4-(METHYL SULPHONYL) PYRIDINE and CUPROUS OXIDE

Envoy TF 200	W and J Leigh and Company	AVA AVP	HSE 4111
Grassline ABL Anti-Fouling Type M397	W and J Leigh and Company	AVA AVP	HSE 3470
Grassline TF Anti-Fouling Type M396	W and J Leigh and Company	AVA AVP	HSE 3462

483 2,4,5,6-TETRACHLORO ISOPHTHALONITRILE

Flexgard IV Waterbase Preservative	Flexabar Corporation	AQA	HSE 3321
Flexgard V Waterbase Preservative	Flexabar Corporation	AQA	HSE 3320

484 2,4,5,6-TETRACHLORO ISOPHTHALONITRILE and 4,5-DICHLORO-2-N-OCTYL-4-ISOTHIAZOLIN-3-ONE and CUPROUS OXIDE and ZINC OXIDE

Seatender 15	Camrex Chugoku Ltd	AVA AVP	HSE 5348
TFA 10 HG	Camrex Chugoku Ltd	AVA AVP	HSE 5366
TFA 10G	Camrex Chugoku Ltd	AVA AVP	HSE 5347

485 2,4,5,6-TETRACHLORO ISOPHTHALONITRILE and COPPER SULPHATE

Flexgard VI Waterbase Preservative	Flexabar Corporation	AQA	HSE 3319

ANTIFOULING PRODUCTS

Product Name	Marketing Company	Use	Reg. No
486 2,4,5,6-TETRACHLORO ISOPHTHALONITRILE and CUPROUS OXIDE			
# Hempel's Antifouling Tin Free 741GB	Hempel Paints Ltd	AVA AVP	HSE 3327
487 2,4,5,6-TETRACHLORO ISOPHTHALONITRILE and CUPROUS OXIDE and DICHLOROPHENYL DIMETHYLUREA and ZINC OXIDE			
Seatender 12	Camrex Chugoku Ltd	AVA AVP	HSE 5324
TFA 10	Camrex Chugoku Ltd	AVA AVP	HSE 5346
TFA 10 H	Camrex Chugoku Ltd	AVA AVP	HSE 5358
488 2,4,5,6-TETRACHLORO ISOPHTHALONITRILE and CUPROUS OXIDE and ZINC NAPHTHENATE			
Teamac Killa Copper Plus	Teal and Mackrill Ltd	AVA AVP	HSE 3495
489 2,4,5,6-TETRACHLORO ISOPHTHALONITRILE and CUPROUS OXIDE and ZINC OXIDE			
Hempel's Antifouling Classic 7654C	Hempel Paints Ltd	AVA AVP	HSE 5298
Hempel's Antifouling Combic 7199C	Hempel Paints Ltd	AVA AVP	HSE 5296
Hempel's Antifouling Nautic 7190C	Hempel Paints Ltd	AVA AVP	HSE 5294
TFA 10 LA	Camrex Chugoku Ltd	AVA AVP	HSE 5361
490 2,4,5,6-TETRACHLORO ISOPHTHALONITRILE and CUPROUS THIOCYANATE and ZINC OXIDE			
Gummipaint A/F	Veneziani Spa	AVA AVP	HSE 4934
491 2-(THIOCYANOMETHYLTHIO) BENZOTHIAZOLE			
Lynx Antifouling	Blakes Marine Paints Ltd	AVA AVP	HSE 3618
# Net Clean Beta	Gramos Chemicals International Ltd	AQA	HSE 3322
492 2-(THIOCYANOMETHYLTHIO) BENZOTHIAZOLE and 2-METHYLTHIO-4-TERTIARY-BUTYLAMINO-6-CYCLOPROPYLAMINO-S-TRIAZINE and DICHLOROPHENYL DIMETHYLUREA and ZINC OXIDE and ZINC PYRITHIONE			
A2 Antifouling	Nautix SA	AVA AVP	HSE 4832
A2 Teflon Antifouling	Nautix SA	AVA AVP	HSE 4833

ANTIFOULING PRODUCTS

Product Name	Marketing Company	Use	Reg. No

493 2-(THIOCYANOMETHYLTHIO) BENZOTHIAZOLE and CUPROUS OXIDE

Product Name	Marketing Company	Use	Reg. No
A3 Antifouling	Nautix SA	AVA AVP	HSE 4367
A3 Teflon Antifouling	Nautix SA	AVA AVP	HSE 4368
* Hempel's Antifouling Tin Free 740GB	Hempel Paints Ltd	AVA AVP	HSE 3326
* Hempel's Antifouling Tin Free 752GB	Hempel Paints Ltd	AVA AVP	HSE 3334
Titan Tin Free	Blakes Marine Paints Ltd	AVA AVP	HSE 3556

494 2-METHYLTHIO-4 TERTIARY-BUTYLAMINO-6-CYCLOPROPYLAMINO-S-TRIAZINE and CUPROUS OXIDE and ZINC OXIDE

Product Name	Marketing Company	Use	Reg. No
Algicide Red Antifouling	Blakes Marine Paints	AVA AVP	HSE 5738

495 2-METHYLTHIO-4-TERTIARY and ZINC OXIDE

Product Name	Marketing Company	Use	Reg. No
Broads Red Antifouling	Blakes Marine Paints	AVA AVP	HSE 5736

496 2-METHYLTHIO-4-TERTIARY-BUTYLAMINO-6-CYCLOPROPYLAMINO-S-TRIAZINE and COPPER METAL

Product Name	Marketing Company	Use	Reg. No
VC 17M Tropicana	Extensor AB	AVA AVP	HSE 4218

497 2-METHYLTHIO-4-TERTIARY-BUTYLAMINO-6-CYCLOPROPYLAMINO-S-TRIAZINE and COPPER RESINATE and CUPROUS OXIDE

Product Name	Marketing Company	Use	Reg. No
# Sigmaplane Tin Free AF SO 0672	Sigma Coatings BV	AVA AVP	HSE 3873

498 2-METHYLTHIO-4-TERTIARY-BUTYLAMINO-6-CYCLOPROPYLAMINO-S-TRIAZINE and COPPER RESINATE and CUPROUS OXIDE and ZINC OXIDE

Product Name	Marketing Company	Use	Reg. No
Sigmaplane Ecol Antifouling	Sigma Coatings BV	AVA AVP	HSE 4348

499 2-METHYLTHIO-4-TERTIARY-BUTYLAMINO-6-CYCLOPROPYLAMINO-S-TRIAZINE and CUPROUS OXIDE

Product Name	Marketing Company	Use	Reg. No
Antifouling Seamate HB 33 Dark Red Non-Tin	Jotun-Henry Clark Ltd	AVA AVP	HSE 3422
Antifouling Seamate HB 33 Light Red Non-Tin	Jotun-Henry Clark Ltd	AVA AVP	HSE 3423
Antifouling Seamate HB 66 Dark Red Non-Tin	Jotun-Henry Clark Ltd	AVA AVP	HSE 3420

ANTIFOULING PRODUCTS

Product Name	Marketing Company	Use	Reg. No

499 2-METHYLTHIO-4-TERTIARY-BUTYLAMINO-6-CYCLOPROPYLAMINO-S-TRIAZINE and CUPROUS OXIDE—continued

Product Name	Marketing Company	Use	Reg. No
Antifouling Seamate HB 66 Light Red Non-Tin	Jotun-Henry Clark Ltd	AVA AVP	HSE 3421
Aquaspeed	Blakes Marine Paints Ltd	AVA AVP	HSE 4511
Challenger Antifouling	Blakes Marine Paints Ltd	AVA AVP	HSE 4099
Hempel's Antifouling Classic Tin Free 7654	Hempel Paints Ltd	AVA AVP	HSE 3720
Hempel's Antifouling Forte Tin Free 7625	Hempel Paints Ltd	AVA AVP	HSE 3351
Hempel's Antifouling Olympic Tin Free 7154	Hempel Paints Ltd	AVA AVP	HSE 3718
Hempel's Antifouling Tin Free 745GB	Hempel Paints Ltd	AVA AVP	HSE 3331
Hempel's Antifouling Tin Free 750GB	Hempel Paints Ltd	AVA AVP	HSE 3332
* Hempel's Antifouling Tin Free 7661	Hempel Paints Ltd	AVA AVP	HSE 3335
Interspeed Antifouling BW0900 Series	International Paint Ltd	AVA AVP	HSE 5636
* Interspeed Extra BWA750 Red	International Paint Ltd	AVA AVP	HSE 4282
# Interspeed Super BWA900 Red	International Paint Ltd	AVA AVP	HSE 4304
Interspeed System 2 BRA143 Brown	International Paint Ltd	AVA AVP	HSE 4301
Interviron BQA450 Series	International Paint Ltd	AVA AVP	HSE 4657
* Interviron BQA750 Series (Base)	International Paint Ltd	AVA AVP	HSE 4477
Interviron Super Tin-Free Polishing Antifouling BQ0400 Series	International Paint Ltd	AVA AVP	HSE 5637
# Micron (Gull White)	International Paint Plc	AVA AVP	HSE 3394
Micron CSC Dover White	International Paint Ltd	AVA AVP	HSE 3557
Patente Laxe	Teais SA	AVA AVP	HSE 5497
Raffaello Plus Tin Free	Veneziani Spa	AVA AVP	HSE 3642
VC Offshore 602	Extensor AB	AVA AVP	HSE 3508
VC Offshore Extra Strength	Extensor AB	AVA AVP	HSE 4777

ANTIFOULING PRODUCTS

Product Name	Marketing Company	Use	Reg. No

500 2-METHYLTHIO-4-TERTIARY-BUTYLAMINO-6-CYCLOPROPYLAMINO-S-TRIAZINE and CUPROUS OXIDE and CUPROUS THIOCYANATE and ZINC OXIDE and 2,3,5,6-TETRACHLORO-4-(METHYL SULPHONYL) PYRIDINE

Product Name	Marketing Company	Use	Reg. No
Envoy TF 400	W and J Leigh and Company	AVA AVP	HSE 4432
Envoy TF 500	W and J Leigh and Company	AVA AVP	HSE 5599
Meridian MP40 Antifouling	Deangate Marine Products	AVA AVP	HSE 5027

501 2-METHYLTHIO-4-TERTIARY-BUTYLAMINO-6-CYCLOPROPYLAMINO-S-TRIAZINE and CUPROUS OXIDE and DICHLOROPHENYL DIMETHYLUREA and ZINC OXIDE

Product Name	Marketing Company	Use	Reg. No
A3 Antifouling 072015	Nautix SA	AVA AVP	HSE 5224
A4 Antifouling 072017	Nautix SA	AVA AVP	HSE 5226
Le Marin	Nautix SA	AVA AVP	HSE 5480

502 2-METHYLTHIO-4-TERTIARY-BUTYLAMINO-6-CYCLOPROPYLAMINO-S-TRIAZINE and CUPROUS OXIDE and ZINC OXIDE

Product Name	Marketing Company	Use	Reg. No
Even TF	Veneziani Spa	AVA AVP	HSE 5166
Hempel's Antifouling Classic 7611 Red (Tin Free) 5000	Hempel Paints Ltd	AVA AVP	HSE 5064
Hempel's Antifouling Classic 76540	Hempel Paints Ltd	AVA AVP	HSE 5291
Hempel's Antifouling Combic 71990	Hempel Paints Ltd	AVA AVP	HSE 5600
Hempel's Antifouling Combic 71992	Hempel Paints Ltd	AVA AVP	HSE 5601
Hempel's Antifouling Combic Tin Free 71990	Hempel Paints Ltd	AVA AVP	HSE 5266
Hempel's Antifouling Mille Dynamic	Hempel Paints Ltd	AVA	HSE 4440
Hempel's Antifouling Nautic 71900	Hempel Paints Ltd	AVA AVP	HSE 5290
Hempel's Antifouling Nautic 71902	Hempel Paints Ltd	AVA AVP	HSE 5605
Hempel's Antifouling Nautic Tin Free 7190	Hempel Paints Ltd	AVA AVP	HSE 4869
Hempel's Antifouling Olympic HI-7661	Hempel Paints Ltd	AVA AVP	HSE 4898
Hempel's Hard Racing 76480	Hempel Paints Ltd	AVA AVP	HSE 5538
Hempel's Mille Dynamic 71700	Hempel Paints Ltd	AVA AVP	HSE 5574
Hempels Antifouling Bravo Tin Free 7610	Hempel Paints Ltd	AVA AVP	HSE 4482
Long Life T.F.	Veneziani Spa	AVA AVP	HSE 4936

Product Name	Marketing Company	Use	Reg. No

503 2-METHYLTHIO-4-TERTIARY-BUTYLAMINO-6-CYCLOPROPYLAMINO-S-TRIAZINE and CUPROUS THIOCYANATE

Product Name	Marketing Company	Use	Reg. No
AL–27	International Paint Ltd	AVA AVP	HSE 3613
Aquaspeed White	Blakes Marine Paints Ltd	AVA AVP	HSE 4513
Boot Top Plus (Gull White)	International Paint Ltd	AVA AVP	HSE 3389
Broads Sweetwater Antifouling	Blakes Marine Paints Ltd	AVA AVP	HSE 3222
* Cruiser Superior (White)	International Paint Ltd	AVA AVP	HSE 3784
Hempel's Tin Free Hard Racing 7648	Hempel Paints Ltd	AVA AVP	HSE 3367
Interspeed 2000	International Paint Ltd	AVA AVP	HSE 4148
* Interspeed 2000 (White)	International Paint Plc	AVA AVP	HSE 3780
MPX	International Paint Ltd	AVA AVP	HSE 4818
* MPX (White)	International Paint Ltd	AVA AVP	HSE 3782
Prop-N-Drive	Extensor AB	AVA AVP	HSE 3509
Propeller T.F.	Veneziani Spa	AVA AVP	HSE 5660
Tiger White	Blakes Marine Paints Ltd	AVA AVP	HSE 4512
Tigerline Antifouling	Blakes Marine Paints Ltd	AVA AVP	HSE 3842
VC 17-Victory Antifouling	Extensor AB	AVA AVP	HSE 3506
* VC Offshore	Extensor AB	AVA AVP	HSE 3843
VC Offshore	Extensor AB	AVA AVP	HSE 4779
VC Prop-O-Drev	Extensor AB	AVA AVP	HSE 4217

504 2-METHYLTHIO-4-TERTIARY-BUTYLAMINO-6-CYCLOPROPYLAMINO-S-TRIAZINE and CUPROUS THIOCYANATE and DICHLOFLUANID and ZINC OXIDE

Product Name	Marketing Company	Use	Reg. No
Hempel's Antifouling Mille Dynamic 717 GB	Hempel Paints Ltd	AVA AVP	HSE 4437

505 2-METHYLTHIO-4-TERTIARY-BUTYLAMINO-6-CYCLOPROPYLAMINO-S-TRIAZINE and CUPROUS THIOCYANATE and DICHLOROPHENYL DIMETHYLUREA and ZINC OXIDE

Product Name	Marketing Company	Use	Reg. No
A3 Antifouling 072016	Nautix SA	AVA AVP	HSE 5225
A3 Teflon Antifouling 062015	Nautix SA	AVA AVP	HSE 5481
A4 Antifouling 072018	Nautix SA	AVA AVP	HSE 5227
A4 Teflon Antifouling 062017	Nautix SA	AVA AVP	HSE 5482

ANTIFOULING PRODUCTS

Product Name	Marketing Company	Use	Reg. No

506 2-METHYLTHIO-4-TERTIARY-BUTYLAMINO-6-CYCLOPROPYLAMINO-S-TRIAZINE and CUPROUS THIOCYANATE and ZINC OXIDE

Even TF Light Grey	Veneziani Spa	AVA AVP	HSE 5167
Hempel's Hard Racing 76380	Hempel Paints Ltd	AVA AVP	HSE 5540
Hempel's Mille Alu 71601	Hempel Paints Ltd	AVA AVP	HSE 5577
Hempel's Mille Dynamic 71600	Hempel Paints Ltd	AVA AVP	HSE 5576
Raffaello Alloy	Veneziani Spa	AVA AVP	HSE 5492

507 2-METHYLTHIO-4-TERTIARY-BUTYLAMINO-6-CYCLOPROPYLAMINO-S-TRIAZINE and CUPROUS THIOCYANATE and ZINEB

# Interspeed 2000	International Paint Plc	AVA AVP	HSE 3781
* MPX	International Paint Ltd	AVA	HSE 3783

508 2-METHYLTHIO-4-TERTIARY-BUTYLAMINO-6-CYCLOPROPYLAMINO-S-TRIAZINE and DICHLOROPHENYL DIMETHYLUREA and ZINC OXIDE and ZINC PYRITHIONE

A7 Teflon Antifouling	Nautix SA	AVA AVP	HSE 4945
A7 Teflon Antifouling 072019	Nautix SA	AVA AVP	HSE 5228

509 2-METHYLTHIO-4-TERTIARY-BUTYLAMINO-6-CYCLOPROPYLAMINO-S-TRIAZINE and DICHLOROPHENYL DIMETHYLUREA and ZINC OXIDE and ZINC PYRITHIONE and 2-(THIOCYANOMETHYLTHIO) BENZOTHIAZOLE

A2 Antifouling	Nautix SA	AVA AVP	HSE 4832
A2 Teflon Antifouling	Nautix SA	AVA AVP	HSE 4833

510 2-METHYLTHIO-4-TERTIARY-BUTYLAMINO-6-CYCLOPROPYLAMINO-S-TRIAZINE and DICHLOROPHENYL DIMETHYLUREA and ZINC PYRITHIONE

A6 Antifouling	Nautix SA	AVA AVP	HSE 4944

511 2-METHYLTHIO-4-TERTIARY-BUTYLAMINO-6-CYCLOPROPYLAMINO-S-TRIAZINE and THIRAM

Broads Freshwater Antifouling	Blakes Marine Paints Ltd	AVA AVP	HSE 3221
Lynx Metal Free Antifouling	Blakes Marine Paints Ltd	AVA AVP	HSE 3555
Trawler Tin-Free	Blakes Marine Paints Ltd	AVA AVP	HSE 3235

ANTIFOULING PRODUCTS

Product Name	Marketing Company	Use	Reg. No

512 2-METHYLTHIO-4-TERTIARY-BUTYLAMINO-6-CYCLOPROPYLAMINO-S-TRIAZINE and THIRAM and TRIBUTYLTIN METHACRYLATE and TRIBUTYLTIN OXIDE

| # Commercial Copolymer Extra Antifouling | Blakes Marine Paints Ltd | AVP | HSE 3225 |

513 2-METHYLTHIO-4-TERTIARY-BUTYLAMINO-6-CYCLOPROPYLAMINO-S-TRIAZINE and TRIBUTYLTIN METHACRYLATE and ZINC OXIDE

| Antifouling Alusea | Jotun-Henry Clark Ltd | AVP | HSE 4569 |

514 2-METHYLTHIO-4-TERTIARY-BUTYLAMINO-6-CYCLOPROPYLAMINO-S-TRIZINE and CUPROUS OXIDE

| Micron CSC 200 Series | International Paint Ltd | AVA AVP | HSE 5732 |

515 2-METHYLTHIO-4-TERTIARY-BUTYLAMINO-6-CYCLOPROPYLAMINO-S-TRIAZINE and CUPROUS THIOCYANATE

| Cruiser Superior 100 Series | International Paint Ltd | AVA AVP | HSE 5723 |

516 2-METHYLTHIO-4-TERTIARY-BUTYLAMINO-6-CYCLOPROPYLAMINO-S-TRIAZINE and CUPROUS OXIDE and ZINC OXIDE

| Broads Black Antifouling | Blakes Marine Paints | AVA AVP | HSE 5739 |
| Hard Racing Antifouling | Blakes Marine Paints | AVA AVP | HSE 5704 |

517 2-METHYLTHIO-4-TERTIARY-BUTYLAMINO-6-CYCLOPROPYLAMINO-S-TRIAZINE and CUPROUS THIOCYANATE and ZINC OXIDE

| Titan FGA Antifouling White | Blakes Marine Paints | AVA AVP | HSE 5680 |

518 2-METHYLTHIO-4-TERTIARY-BUTYLAMINO-6-CYCLOPROPYLAMINO-S-TRIAZINE and CUPROUS OXIDE and ZINC OXIDE

| Titan FGA Antifouling | Blakes Marine Paints | AVA AVP | HSE 5681 |

519 2-METHYLTHIO-4-TERTIARY-BUTYLAMINO-6-CYCLOPROPYLAMINO-S-TRIAZINE and CUPROUS THIOCYANATE and ZINC OXIDE

| Hard Racing Antifouling White | Blakes Marine Paints | AVA AVP | HSE 5705 |

ANTIFOULING PRODUCTS

Product Name	Marketing Company	Use	Reg. No

520 4,5-DICHLORO-2-N-OCTYL -4-ISOTHIAZOLIN-3-ONE and CUPROUS OXIDE

Product Name	Marketing Company	Use	Reg. No
Hempel's Antifouling Tin Free 743GB	Hempel Paints Ltd	AVA AVP	HSE 3329
Hempel's Antifouling Tin Free 751GB	Hempel Paints Ltd	AVA AVP	HSE 3333
Hempel's Antifouling Tin Free 7662	Hempel Paints Ltd	AVA AVP	HSE 3336
Hempel's Tin Free Antifouling 7626	Hempel Paints Ltd	AVA AVP	HSE 3337
Interviron Super Tin-Free Polishing Antifouling BQ0420 Series	International Paint Ltd	AVA AVP	HSE 5642

521 4,5-DICHLORO-2-N-OCTYL-4-ISOTHIAZOLIN-3-ONE and CUPROUS OXIDE and TRIBUTYLTIN METHACRYLATE and TRIBUTYLTIN OXIDE

Product Name	Marketing Company	Use	Reg. No
Intersmooth Hisol BF0270/950/970 Series	International Paint Ltd	AVP	HSE 5641

522 4,5-DICHLORO-2-N-OCTYL -4-ISOTHIAZOLIN-3-ONE and CUPROUS OXIDE and ZINC OXIDE

Product Name	Marketing Company	Use	Reg. No
Antifouling Seavictor 50	Jotun-Henry Clark Ltd	AVA AVP	HSE 4958
Hempel's Antifouling Classic 7654E	Hempel Paints Ltd	AVA AVP	HSE 5286
Hempel's Antifouling Combic 7199E	Hempel Paints Ltd	AVA AVP	HSE 5277
Hempel's Antifouling Nautic 7190E	Hempel Paints Ltd	AVA AVP	HSE 5283
Hempel's Antifouling Nautic Tin Free 7190E	Hempel Paints Ltd	AVA AVP	HSE 5145

523 4,5-DICHLORO-2-N-OCTYL -4-ISOTHIAZOLIN-3-ONE and CUPROUS OXIDE and ZINC OXIDE and 2,4,5,6-TETRACHLORO ISOPHTHALONITRILE

Product Name	Marketing Company	Use	Reg. No
Seatender 15	Camrex Chugoku Ltd	AVA AVP	HSE 5348
TFA 10 HG	Camrex Chugoku Ltd	AVA AVP	HSE 5366
TFA 10G	Camrex Chugoku Ltd	AVA AVP	HSE 5347

524 4,5-DICHLORO-2-N-OCTYL-4-ISOTHIAZOLIN-3-ONE and CUPROUS OXIDE

Product Name	Marketing Company	Use	Reg. No
Micron CSC 300 Series	International Paint Ltd	AVA AVP	HSE 5724

525 4-CHLORO-META-CRESOL and CUPROUS OXIDE

Product Name	Marketing Company	Use	Reg. No
# Mebon Tropical Anti-Fouling Paint 6-2-01	Mebon Paints Ltd	AVA AVP	HSE 3475

ANTIFOULING PRODUCTS

Product Name	Marketing Company	Use	Reg. No

526 ARSENIC TRIOXIDE and COPPER NAPHTHENATE and CUPROUS OXIDE

# Crodec Antifouling Super	Croda Paints Ltd	AVA AVP	HSE 3294

527 ARSENIC TRIOXIDE and CUPROUS OXIDE

# Craig and Rose Green Antifouling	Craig and Rose Plc	AVA AVP	HSE 3292
# Craig and Rose Red Oxide Antifouing	Craig and Rose Plc	AVA AVP	HSE 3293

528 CIS 1-(3-CHLOROALLYL)-3,5,7-TRIAZA-1-AZONIA ADAMANTANE CHLORIDE

# Chemlife	Chemflake International Ltd	AQA AVA AVP	HSE 3250

529 COPPER BRONZE POWDER

# Copper/Bronze Antifouling Powder AF26	Desoto Titanine Plc	AVA AVP	HSE 3298

530 COPPER METAL

# Bronze	International Paint Ltd	AVA AVP	HSE 3391
Copperbot	Corrosion Defence Ltd	AVA AVP	HSE 5262
Crystic Copperclad 70PA	Scott Bader Co Ltd	AVA AVP	HSE 3479
Miricoat A.F. Coating	Miricoat Ltd	AVP	HSE 5587
* Ruwa Bronze Bottom Paint	Sikkens BV	AVA AVP	HSE 3512
* Scomet	Metallisation Service Ltd	AVA AVP	HSE 3477
VC 17M EP-Antifouling	Extensor AB	AVA AVP	HSE 3318
VC17M	Extensor AB	AVA AQA AVP	HSE 4780

531 COPPER METAL and 2-METHYLTHIO-4-TERTIARY-BUTYLAMINO-6-CYCLOPROPYLAMINO-S-TRIAZINE

VC 17M Tropicana	Extensor AB	AVA AVP	HSE 4218

532 COPPER METAL and TRIBUTYLTIN METHACRYLATE and TRIBUTYLTIN OXIDE

# Hempel's Antifouling Mille Copper 7671	Hempel Paints Ltd	AVP	HSE 3354

ANTIFOULING PRODUCTS

Product Name	Marketing Company	Use	Reg. No
533 COPPER METAL and ZINC OXIDE			
# AF29 Desoto Antifouling	Desoto Titanine Plc	AVA AVP	HSE 3893
Amercoat 67E	Ameron BV.	AVA AVP	HSE 3201
Amercoat 70E	Ameron BV.	AQA AVA AVP	HSE 3202
Amercoat 70ESP	Ameron BV.	AQA AVA AVP	HSE 3203
534 COPPER NAPHTHENATE			
# Net Clean Gamma	Gramos Chemicals International Ltd	AQA	HSE 3323
535 COPPER NAPHTHENATE and CUPROUS OXIDE and ARSENIC TRIOXIDE			
# Crodec Antifouling Super	Croda Paints Ltd	AVA AVP	HSE 3294
536 COPPER NAPHTHENATE and CUPROUS OXIDE and DICHLOFLUANID and ZINC NAPHTHENATE and ZINC OXIDE			
Teamac Killa Copper Plus	Teal and Mackrill Ltd	AVA AVP	HSE 4659
537 COPPER RESINATE and CUPROUS OXIDE			
Double Shield Antifouling	Llewellyn Ryland Ltd	AVA AVP	HSE 3471
538 COPPER RESINATE and CUPROUS OXIDE and 2-METHYLTHIO-4-TERTIARY-BUTYLAMINO-6-CYCLOPROPYLAMINO-S-TRIAZINE			
# Sigmaplane Tin Free AF SO 0672	Sigma Coatings BV	AVA AVP	HSE 3873
539 COPPER RESINATE and CUPROUS OXIDE and DICHLOROPHENYL DIMETHYLUREA			
# Sigmaplane Tin Free AF SO 0671	Sigma Coatings BV	AVA AVP	HSE 3872
540 COPPER RESINATE and CUPROUS OXIDE and DICHLOROPHENYL DIMETHYLUREA and ZINC OXIDE			
* Sigmaplane Ecol AF First Coat	Sigma Coatings BV	AVA AVP	HSE 4349

ANTIFOULING PRODUCTS

Product Name	Marketing Company	Use	Reg. No

541 COPPER RESINATE and CUPROUS OXIDE and TRIBUTYLTIN FLUORIDE

* Lily Antifouling Super Tropical	Sigma Coatings BV	AVP	HSE 3485

542 COPPER RESINATE and CUPROUS OXIDE and TRIBUTYLTIN FLUORIDE and ZINC OXIDE

* Lily Antifouling IV	Sigma Coatings BV	AVP	HSE 3484
* Sigma Antifouling CR	Sigma Coatings BV	AVP	HSE 3483
* Sigma Antifouling IV	Sigma Coatings BV	AVP	HSE 4347

543 COPPER RESINATE and CUPROUS OXIDE and TRIBUTYLTIN FLUORIDE and ZINC OXIDE and ZINEB

* Sigma Pilot Antifouling LL	Sigma Coatings BV	AVA	HSE 3481

544 COPPER RESINATE and CUPROUS OXIDE and ZINC OXIDE and 2-METHYLTHIO-4-TERTIARY-BUTYLAMINO-6-CYCLOPROPYLAMINO-S-TRIAZINE

Sigmaplane Ecol Antifouling	Sigma Coatings BV	AVA AVP	HSE 4348

545 COPPER RESINATE and CUPROUS OXIDE and ZINC OXIDE and ZINEB

* Sigma Pilot Antifouling LL/TF	Sigma Coatings BV	AVA AVP	HSE 3480
Sigma Pilot Ecol Antifouling	Sigma Coatings Ltd	AVA AVP	HSE 4933

546 COPPER SULPHATE and 2,4,5,6-TETRACHLORO ISOPHTHALONITRILE

Flexgard VI Waterbase Preservative	Flexabar Corporation	AQA	HSE 3319

547 CUPROUS OXIDE

Admiralty Antifouling (TO TS10240)	Sigma Coatings BV	AVA AVP	HSE 3490
Algicide Antifouling	Blakes Marine Paints Ltd	AVA AVP	HSE 3219
Amercoat 275	Ameron BV	AVA AVP	HSE 3204
Amercoat 277	Ameron BV	AVA AVP	HSE 3205
Anti-Fouling Paint 161P (Red and Chocolate TO TS10240)	United Paints Ltd	AVA AVP	HSE 3503
Antifouling Paint 161P (Red and Chocolate TO TS10240)	International Paint Ltd	AVP	HSE 3401
# Antifouling Paint 161P (TO TS10240)	Desoto Titanine Plc	AVA AVP	HSE 3308

547 CUPROUS OXIDE—continued

Product Name	Marketing Company	Use	Reg. No
Antifouling Sargasso Non-Tin	Jotun-Henry Clark Ltd	AVA AVP	HSE 3418
Antifouling Seaguardian	Jotun-Henry Clark Ltd	AVA AVP	HSE 3856
Antifouling Seamaster Non-Tin	Jotun-Henry Clark Ltd	AVA AVP	HSE 3416
* Antifouling Searose	Jotun-Henry Clark Ltd	AVA AVP	HSE 3685
Antifouling Seasafe	Jotun-Henry Clark Ltd	AVA AVP	HSE 3417
Antifouling Super Tropic	Jotun-Henry Clark Ltd	AVA AVP	HSE 3413
Antifouling Tropic Black	Jotun-Henry Clark Ltd	AVA AVP	HSE 3419
* Aquarius	International Paint Ltd	AVA AVP	HSE 3387
Blueline Copper SBA100	International Paint Ltd	AVA AVP	HSE 5140
# Britannic K504 Long Life Anti-Fouling	W and J Leigh and Company	AVA	HSE 3452
# Britannic M351 Anti-Fouling Extra	W and J Leigh and Company	AVA	HSE 3451
* Britannic M451 Antifouling Super	W and J Leigh and Company	AVA AVP	HSE 3763
Broads Antifouling	Blakes Marine Paints Ltd	AVA AVP	HSE 3220
# Camrex A/F 1	Camrex Ltd	AVA AVP	HSE 3236
# Camrex A/F 2	Camrex Ltd	AVA AVP	HSE 3237
# Camrex A/F 4	Camrex Ltd	AVA AVP	HSE 3244
# Chocolate Antifouling AF23 (TO TS10240)	Desoto Titanine Plc	AVA AVP	HSE 3297
Classica 3786/093 Red	Stoppani (UK) Ltd	AVA AVP	HSE 4797
* Coppercoat	International Paint Ltd	AVA AVP	HSE 3392
Copperpaint	International Paint Ltd	AVA AVP	HSE 4119
* Cruiser Premium (Light Grey)	International Paint Ltd	AVA AVP	HSE 4332
Devoe ABC 3 Black Antifouling	Devoe Coatings BV	AVA AVP	HSE 3315
Devoe ABC 3 Red Antifouling	Devoe Coatings BV	AVA AVP	HSE 3317
Envoy TF 100	W and J Leigh and Company	AVA AVP	HSE 3951
General Purpose Antifouling Paint	Spencer Coatings Ltd	AVA AVP	HSE 4156
Hard Racing	International Paint Ltd	AVA AVP	HSE 3393
Hempel's Antifouling 761GB	Hempel Paints Ltd	AVA AVP	HSE 3339
Hempel's Antifouling Nordic 7133	Hempel Paints Ltd	AVA AVP	HSE 3325
Hempel's Antifouling Paint 161P (Red and Chocolate TO TS10240)	Hempel Paints Ltd	AVA AVP	HSE 3355
Hempel's Tin Free Antifouling 744GB	Hempel Paints Ltd	AVA AVP	HSE 3330
Hempel's Tin Free Antifouling 7660	Hempel Paints Ltd	AVA AVP	HSE 3338

ANTIFOULING PRODUCTS

Product Name	Marketing Company	Use	Reg. No
547 **CUPROUS OXIDE**—continued			
Interclene Extra BAA100 Series	International Paint Ltd	AVA AVP	HSE 3371
Interclene Premium BCA300 Series	International Paint Ltd	AVA AVP	HSE 3372
Interclene Super BCA400 Series (BCA400 Red)	International Paint Ltd	AVA AVP	HSE 4084
Interclene Underwater Premium BCA468 Red	International Paint Ltd	AVA AVP	HSE 5059
International TBT Free Copolymer Antifouling BQA100 Series	International Paint Ltd	AVA AVP	HSE 3375
# Interspeed Super BJA600 Series	International Paint Ltd	AVA AVP	HSE 3380
# Interspeed System 2 BRA240 Series	International Paint Plc	AVA AVP	HSE 3383
Interspeed System 2 BR0142/240 Series	International Paint Ltd	AVA AVP	HSE 5634
# Kartini Antifouling AF24	Desoto Titanine Plc	AVA AVP	HSE 3300
Micron 400 Series	International Paint Ltd	AVA AVP	HSE 5728
Micron CSC 100 Series	International Paint Ltd	AVA AVP	HSE 5731
Norden Kobberstoff Red	Jotun-Henry Clark Ltd	AVA AVP	HSE 3415
Norimp 2000 Black	Jotun-Henry Clark Ltd	AQA	HSE 3404
Pilot Antifouling	Blakes Marine Paints Ltd	AVA AVP	HSE 3226
Puma Antifouling	Blakes Marine Paints Ltd	AVA AVP	HSE 3227
# Red Antifouling AF22 (TO TS10240)	Desoto Titanine Plc	AVA AVP	HSE 3303
Ruwa Vinyl Anti-Fouling TL	Sikkens BV	AVA AVP	HSE 3511
Shearwater Racing Antifouling	Blakes Marine Paints Ltd	AVA AVP	HSE 3228
Super Tropical Antifouling	Blakes Marine Paints Ltd	AVA AVP	HSE 3229
Super Tropical Extra Antifouling	Blakes Marine Paints Ltd	AVA AVP	HSE 3230
Teamac Standard Copper	Teal and Mackrill Ltd	AVA AVP	HSE 3496
Titan Tin Free Antifouling	Blakes Marine Paints Ltd	AVA AVP	HSE 3232
Trawler	International Paint Ltd	AVA AVP	HSE 3398
Tropical Super Service Antifouling Paint	Devoe Coatings BV	AVA AVP	HSE 3316
TS 10240 Antifouling ADA160 Series	International Paint Ltd	AVA AVP	HSE 3386
Unitas Antifouling Paint Chocolate	United Paints Ltd	AVA AVP	HSE 3499
Unitas Antifouling Paint Red	United Paints Ltd	AVA AVP	HSE 3498
Vinilstop 9926 Red	Stoppani (UK) Ltd	AVA AVP	HSE 4798
Vinyl AF	Chugoku Marine Paints (UK) Ltd	AVA AVP	HSE 3275

ANTIFOULING PRODUCTS

Product Name	Marketing Company	Use	Reg. No

547 CUPROUS OXIDE—continued

Waterways	International Paint Ltd	AVA AQA AVP	HSE 3399

548 CUPROUS OXIDE and 2,3,5,6-TETRACHLORO-4-(METHYL SULPHONYL) PYRIDINE

Envoy TF 200	W and J Leigh and Company	AVA AVP	HSE 4111
Grassline Abl Anti-Fouling Type M397	W and J Leigh and Company	AVA AVP	HSE 3470
Grassline TF Anti-Fouling Type M396	W and J Leigh and Company	AVA AVP	HSE 3462

549 CUPROUS OXIDE and 2,4,5,6-TETRACHLORO ISOPHTHALONITRILE

# Hempel's Antifouling Tin Free 741GB	Hempel Paints Ltd	AVA AVP	HSE 3327

550 CUPROUS OXIDE and 2-(THIOCYANOMETHYLTHIO) BENZOTHIAZOLE

A3 Antifouling	Nautix SA	AVA AVP	HSE 4367
A3 Teflon Antifouling	Nautix SA	AVA AVP	HSE 4368
* Hempel's Antifouling Tin Free 740GB	Hempel Paints Ltd	AVA AVP	HSE 3326
* Hempel's Antifouling Tin Free 752GB	Hempel Paints Ltd	AVA AVP	HSE 3334
Titan Tin Free	Blakes Marine Paints Ltd	AVA AVP	HSE 3556

551 CUPROUS OXIDE and 2-METHYLTHIO-4-TERTIARY-BUTYLAMINO-6-CYCLOPROPYLAMINO-S-TRIAZINE

Antifouling Seamate HB 33 Dark Red Non-Tin	Jotun-Henry Clark Ltd	AVA AVP	HSE 3422
Antifouling Seamate HB 33 Light Red Non-Tin	Jotun-Henry Clark Ltd	AVA AVP	HSE 3423
Antifouling Seamate HB 66 Dark Red Non-Tin	Jotun-Henry Clark Ltd	AVA AVP	HSE 3420
Antifouling Seamate HB 66 Light Red Non-Tin	Jotun-Henry Clark Ltd	AVA AVP	HSE 3421
Aquaspeed	Blakes Marine Paints Ltd	AVA AVP	HSE 4511
Challenger Antifouling	Blakes Marine Paints Ltd	AVA AVP	HSE 4099

Product Name	Marketing Company	Use	Reg. No

551 CUPROUS OXIDE and 2-METHLYTHIO-4-TERTIARY-BUTYLAMINO-6-CYCLOPROPYLAMINO-S-TRIAZINE—continued

Product Name	Marketing Company	Use	Reg. No
Hempel's Antifouling Classic Tin Free 7654	Hempel Paints Ltd	AVA AVP	HSE 3720
Hempel's Antifouling Forte Tin Free 7625	Hempel Paints Ltd	AVA AVP	HSE 3351
Hempel's Antifouling Olympic Tin Free 7154	Hempel Paints Ltd	AVA AVP	HSE 3718
Hempel's Antifouling Tin Free 745GB	Hempel Paints Ltd	AVA AVP	HSE 3331
Hempel's Antifouling Tin Free 750GB	Hempel Paints Ltd	AVA AVP	HSE 3332
* Hempel's Antifouling Tin Free 7661	Hempel Paints Ltd	AVA AVP	HSE 3335
Interspeed Antifouling BW0900 Series	International Paint Ltd	AVA AVP	HSE 5636
* Interspeed Extra BWA750 Red	International Paint Ltd	AVA AVP	HSE 4282
# Interspeed Super BWA900 Red	International Paint Ltd	AVA AVP	HSE 4304
Interspeed System 2 BRA143 Brown	International Paint Ltd	AVA AVP	HSE 4301
Interviron BQA450 Series	International Paint Ltd	AVA AVP	HSE 4657
* Interviron BQA750 Series (Base)	International Paint Ltd	AVA AVP	HSE 4477
Interviron Super Tin-Free Polishing Antifouling BQ0400	International Paint Ltd	AVA AVP	HSE 5637
# Micron (Gull White)	International Paint Plc	AVA AVP	HSE 3394
Micron CSC Dover White	International Paint Ltd	AVA AVP	HSE 3557
Patente Laxe	Teais SA	AVA AVP	HSE 5497
Raffaello Plus Tin Free	Veneziani Spa	AVA AVP	HSE 3642
VC Offshore 602	Extensor AB	AVA AVP	HSE 3508
VC Offshore Extra Strength	Extensor AB	AVA AVP	HSE 4777

552 CUPROUS OXIDE and 2-METHYLTHIO-4-TERTIARY-BUTYLAMINO-6-CYCLOPROPYLAMINO-S-TRIAZINE and COPPER RESINATE

Product Name	Marketing Company	Use	Reg. No
# Sigmaplane Tin Free AF SO 0672	Sigma Coatings BV	AVA AVP	HSE 3873

553 CUPROUS OXIDE and 2-METHYLTHIO-4-TERTIARY-BUTYLAMINO-6-CYCLOPROPYLAMINO-S-TRIAZINE

Product Name	Marketing Company	Use	Reg. No
Micron CSC 200 Series	International Paint Ltd	AVA AVP	HSE 5732

ANTIFOULING PRODUCTS

Product Name	Marketing Company	Use	Reg. No

554 CUPROUS OXIDE and 4,5-DICHLORO-2-N-OCTYL-4-ISOTHIAZOLIN-3-ONE

Hempel's Antifouling Tin Free 743GB	Hempel Paints Ltd	AVA AVP	HSE 3329
Hempel's Antifouling Tin Free 751GB	Hempel Paints Ltd	AVA AVP	HSE 3333
Hempel's Antifouling Tin Free 7662	Hempel Paints Ltd	AVA AVP	HSE 3336
Hempel's Tin Free Antifouling 7626	Hempel Paints Ltd	AVA AVP	HSE 3337
Interviron Super Tin-free Polishing Antifouling BQ0420 Series	International Paint Ltd	AVA AVP	HSE 5642

555 CUPROUS OXIDE and 4,5-DICHLORO-2-N-OCTYL-4-ISOTHIAZOLIN-3-ONE

Micron CSC 300 Series	International Paint Ltd	AVA AVP	HSE 5724

556 CUPROUS OXIDE and 4-CHLORO-META-CRESOL

# Mebon Tropical Anti-Fouling Paint 6-2-01	Mebon Paints Ltd	AVA AVP	HSE 3475

557 CUPROUS OXIDE and ARSENIC TRIOXIDE

# Craig and Rose Green Antifouling	Craig and Rose Plc	AVA AVP	HSE 3292
# Craig and Rose Red Oxide Antifouling	Craig and Rose Plc	AVA AVP	HSE 3293

558 CUPROUS OXIDE and ARSENIC TRIOXIDE and COPPER NAPHTHENATE

# Crodec Antifouling Super	Croda Paints Ltd	AVA AVP	HSE 3294

559 CUPROUS OXIDE and COPPER RESINATE

Double Shield Antifouling	Llewellyn Ryland Ltd	AVA AVP	HSE 3471

560 CUPROUS OXIDE and CUPROUS SULPHIDE

Admiralty Antifouling Black (TO TS10239)	Sigma Coatings BV	AVA AVP	HSE 3489
Black Anti-Fouling Paint 317 (TO TS10239)	United Paints Ltd	AVA AVP	HSE 3502
# Black Antifouling AF21 (TO TS10239)	Desoto Titanine Plc	AVA AVP	HSE 3295

ANTIFOULING PRODUCTS

Product Name	Marketing Company	Use	Reg. No

560 CUPROUS OXIDE and CUPROUS SULPHIDE—continued

# Black Antifouling Paint 317 (TO TS10239)	Desoto Titanine Plc	AVA AVP	HSE 3307
Black Antifouling Paint 317 (TO TS10239)	Hempel Paints Ltd	AVA AVP	HSE 3370
Hempel's Antifouling 762GB	Hempel Paints Ltd	AVA AVP	HSE 3340
# TS 10239 Antifouling	International Paint Ltd	AVA AVP	HSE 3385
Unitas Antifouling Paint Black	United Paints Ltd	AVA AVP	HSE 3501

561 CUPROUS OXIDE and CUPROUS THIOCYANATE and TRIBUTYLTIN METHACRYLATE and TRIBUTYLTIN OXIDE

# Hempel's Antifouling Mille 7678	Hempel Paints Ltd	AVP	HSE 3353

562 CUPROUS OXIDE and CUPROUS THIOCYANATE and ZINC OXIDE and 2,3,5,6-TETRACHLORO-4- (METHYL SULPHONYL) PYRIDINE and 2-METHYLTHIO-4-TERTIARY-BUTYLAMINO-6-CYCLOPROPYLAMINO-S-TRIAZINE

Envoy TF 400	W and J Leigh and Company	AVA AVP	HSE 4432
Envoy TF 500	W and J Leigh and Company	AVA AVP	HSE 5599
Meridian MP40 Antifouling	Deangate Marine Products	AVA AVP	HSE 5027

563 CUPROUS OXIDE and DICHLOFLUANID

Forcecontact AFUR514 (Base)	Forcecontact Ltd	AVA AVP	HSE 4876
Halcyon 5000 (Base)	Waterline	AVA AVP	HSE 5396
Halcyon 5000 (Base)	Waterline	AVA	HSE 5334
Hempel's Antifouling Rennot 7150	Hempel Paints Ltd	AQA	HSE 3364
Hempel's Antifouling Rennot 7177	Hempel Paints Ltd	AQA	HSE 3365
Hempel's Antifouling Tin Free 742GB	Hempel Paints Ltd	AVA AVP	HSE 3328
Seashield (Base)	Waterline	AVA AVP	HSE 5438
Slipstream Antifouling	United Paints Ltd	AVA AVP	HSE 3721
Tiger Tin Free Antifouling	Blakes Marine Paints Ltd	AVA AVP	HSE 3231

564 CUPROUS OXIDE and DICHLOFLUANID and ZINC NAPHTHENATE and ZINC OXIDE and COPPER NAPHTHENATE

Teamac Killa Copper Plus	Teal and Mackrill Ltd	AVA AVP	HSE 4659

ANTIFOULING PRODUCTS

Product Name	Marketing Company	Use	Reg. No

565 CUPROUS OXIDE and DICHLOFLUANID and ZINC OXIDE

Product Name	Marketing Company	Use	Reg. No
Dapaflow Antifoul	Dapa Services Ltd	AVA AVP	HSE 4754
Penguin Non-Stop	Marine and Industrial Sealants	AVA AVP	HSE 5671

566 CUPROUS OXIDE and DICHLOROPHENYL DIMETHYLUREA

Product Name	Marketing Company	Use	Reg. No
A4 Antifouling	Nautix SA	AVA AVP	HSE 4369
Aquarius Extra Strong	International Paint Ltd	AVA AVP	HSE 4280
Blueline Tropical SBA300	International Paint Ltd	AVA AVP	HSE 5139
Cruiser Premium	International Paint Ltd	AVA AVP	HSE 5127
Grafo Anti-Foul SW	Grafo Coatings Ltd	AVA AVP	HSE 5380
International Tin Free SPC BNA100 Series	International Paint Ltd	AVA AVP	HSE 5186
Intersmooth Hisol Tin Free BGA620 Series	International Paint Ltd	AVA AVP	HSE 4787
Intersmooth Tin Free BGA530 Series	International Paint Ltd	AVA AVP	HSE 4611
Interspeed Extra BWA500 Red	International Paint Ltd	AVA AVP	HSE 4303
* Interspeed Extra Strong	International Paint Ltd	AVA AVP	HSE 4215
Interspeed Extra Strong	International Paint Ltd	AVA AVP	HSE 4819
* Interspeed Premium BWA900 Red	International Paint Ltd	AVA AVP	HSE 4738
Interspeed Super BWA900 Red	International Paint Ltd	AVA AVP	HSE 4884
Interspeed Super BWA909 Black	International Paint Ltd	AVA AVP	HSE 5058
Interspeed System 2 BRA142 Brown	International Paint Ltd	AVA AVP	HSE 4302
Interswift Tin Free BQA400 Series	International Paint Ltd	AVA AVP	HSE 4842
Interswift Tin-Free Spc BTA540 Series	International Paint Ltd	AVA AVP	HSE 5078
* Interviron BQA400 Series	International Paint Ltd	AVA AVP	HSE 4658
Interviron Super BNA400 Series	International Paint Ltd	AVA AVP	HSE 5690
Interviron Super BQA400 Series	International Paint Ltd	AVA AVP	HSE 5409
Micron 500 Series	International Paint Ltd	AVA AVP	HSE 5729
Micron CSC	International Paint Ltd	AVA AVP	HSE 4775
Micron Plus Antifouling	International Paint Ltd	AVA AVP	HSE 5133
TFA–30	Chugoku Marine Paints (UK) Ltd	AVA AVP	HSE 3283
Waterways Antifouling	International Paint Plc	AVA AVP	HSE 5565

ANTIFOULING PRODUCTS

Product Name	Marketing Company	Use	Reg. No

567 CUPROUS OXIDE and DICHLOROPHENYL DIMETHYLUREA and COPPER RESINATE

# Sigmaplane Tin Free AF SO 0671	Sigma Coatings BV	AVA AVP	HSE 3872

568 CUPROUS OXIDE and DICHLOROPHENYL DIMETHYLUREA and TRIBUTYLTIN FLUORIDE and TRIBUTYLTIN OXIDE

# Hempel's Antifouling Hi Build 7600	Hempel Paints Ltd	AVP	HSE 3350

569 CUPROUS OXIDE and DICHLOROPHENYL DIMETHYLUREA and TRIBUTYLTIN METHACRYLATE and TRIBUTYLTIN OXIDE

Interswift BK0000/700 Series	International Paint Ltd	AVP	HSE 5640

570 CUPROUS OXIDE and DICHLOROPHENYL DIMETHYLUREA and TRIBUTYLTIN METHACRYLATE and TRIBUTYLTIN OXIDE and ZINC OXIDE

A11 Antifouling 072014	Nautix SA	AVP	HSE 5376

571 CUPROUS OXIDE and DICHLOROPHENYL DIMETHYLUREA and ZINC OXIDE

Hempel's Antifouling Classic 7654B	Hempel Paints Ltd	AVA AVP	HSE 5285
Hempel's Antifouling Combic 7199B	Hempel Paints Ltd	AVA AVP	HSE 5274
Hempel's Antifouling Nautic 7190B	Hempel Paints Ltd	AVA AVP	HSE 5273
Hempel's Bravo 7610A	Hempel Paints Ltd	AVA AVP	HSE 5603
Hempel's Hard Racing 7648A	Hempel Paints Ltd	AVA AVP	HSE 5535
Hempel's Mille Dynamic 7170A	Hempel Paints Ltd	AVA AVP	HSE 5536
Noa-Noa Rame	Stoppani (UK) Ltd	AVA AVP	HSE 4796
Seajet 033	Camrex Chugoku Ltd	AVA AVP	HSE 5331
Sigmaplane Ecol Antifouling	Sigma Coatings Ltd	AVA AVP	HSE 5670
TFA–10 LA Light/Dark	Chugoku Marine Paints (UK) Ltd	AVA AVP	HSE 3285
TFA–10 Light/Dark	Chugoku Marine Paints (UK) Ltd	AVA AVP	HSE 3286
TFA–20	Chugoku Marine Paints (UK) Ltd	AVA AVP	HSE 3284

ANTIFOULING PRODUCTS

Product Name	Marketing Company	Use	Reg. No

572 CUPROUS OXIDE and DICHLOROPHENYL DIMETHYLUREA and ZINC OXIDE and 2,4,5,6-TETRACHLORO ISOPHTHALONITRILE

Seatender 12	Camrex Chugoku Ltd	AVA AVP	HSE 5324
TFA 10	Camrex Chugoku Ltd	AVA AVP	HSE 5346
TFA 10 H	Camrex Chugoku Ltd	AVA AVP	HSE 5358

573 CUPROUS OXIDE and DICHLOROPHENYL DIMETHYLUREA and ZINC OXIDE and 2-METHYLTHIO-4-TERTIARY-BUTYLAMINO-6-CYCLOPROPYLAMINO-S-TRIAZINE

A3 Antifouling 072015	Nautix SA	AVA AVP	HSE 5224
A4 Antifouling 072017	Nautix SA	AVA AVP	HSE 5226
Le Marin	Nautix SA	AVA AVP	HSE 5480

574 CUPROUS OXIDE and DICHLOROPHENYL DIMETHYLUREA and ZINC OXIDE and COPPER RESINATE

* Sigmaplane Ecol AF First Coat	Sigma Coatings BV	AVA AVP	HSE 4349

575 CUPROUS OXIDE and FOLPET

# Leader	International Paint Plc	AVA AVP	HSE 3396

576 CUPROUS OXIDE and MANEB and TRIBUTYLTIN METHACRYLATE and ZINC OXIDE

Nu Wave A/F Flat Bottom	Kansai Paint Co Ltd	AVP	HSE 3447
Nu Wave A/F Vertical Bottom	Kansai Paint Co Ltd	AVP	HSE 3446
Rabamarine A/F No 2500 HS	Kansai Paint Co Ltd	AVP	HSE 3444
Rabamarine A/F No 2500M HS	Kansai Paint Co Ltd	AVP	HSE 3445

577 CUPROUS OXIDE and OXYTETRACYCLINE HYDROCHLORIDE

# Cobra	Adamarine Coatings	AVA AVP	HSE 3200

578 CUPROUS OXIDE and THIRAM

Envoy TF 300	W and J Leigh and Company	AVA AVP	HSE 3952

ANTIFOULING PRODUCTS

Product Name	Marketing Company	Use	Reg. No
579 CUPROUS OXIDE and TRIBUTYLTIN ACRYLATE and TRIBUTYLTIN OXIDE			
* Grassline ABL Anti-Fouling Type M336	W and J Leigh and Company	AVP	HSE 3463
* Grassline ABL Anti-Fouling Type M349	W and J Leigh and Company	AVP	HSE 3464
Grassline ABL Anti-Fouling Type M349 HS	W and J Leigh and Company	AVP	HSE 3465
* Grassline ABL Anti-Fouling Type M398	W and J Leigh and Company	AVP	HSE 3469
580 CUPROUS OXIDE and TRIBUTYLTIN ACRYLATE and ZINC OXIDE			
Amercoat 697	Ameron BV	AVP	HSE 3513
Amercoat 698 HS	Ameron BV	AVP	HSE 3214
581 CUPROUS OXIDE and TRIBUTYLTIN FLUORIDE			
* Chlorinated Rubber Antifouling AF19	Desoto Titanine Plc	AVP	HSE 3296
* Devchlor Chlorinated Rubber Antifouling	Devoe Coatings BV	AVP	HSE 3314
* Florida Antifouling AF13	Desoto Titanine Plc	AVP	HSE 3299
* Grassline XL Anti-Fouling Type K968B	W and J Leigh and Company	AVP	HSE 3454
* Grassline XLS Anti-Fouling Type M299	W and J Leigh and Company	AVP	HSE 3456
* Grassline XLS Antifouling Type M299	W and J Leigh and Company	AVP	HSE 3457
* Supertropical Antifouling AF5	Desoto Titanine Plc	AVP	HSE 3304
# Tropical Antifouling AF4	Desoto Titanine Plc	AVP	HSE 3305
* Vinyl Antifouling AF12	Desoto Titanine Plc	AVP	HSE 3306
582 CUPROUS OXIDE and TRIBUTYLTIN FLUORIDE and COPPER RESINATE			
* Lily Antifouling Super Tropical	Sigma Coatings BV	AVP	HSE 3485
583 CUPROUS OXIDE and TRIBUTYLTIN FLUORIDE and TRIBUTYLTIN OXIDE			
* Hempel's Antifouling Classic 7655	Hempel Paints Ltd	AVP	HSE 3346
# Hempel's Antifouling Forte 7620	Hempel Paints Ltd	AVP	HSE 3349
584 CUPROUS OXIDE and TRIBUTYLTIN FLUORIDE and TRIBUTYLTIN OXIDE and ZINC OXIDE			
* Hempel's Antifouling Olympic 7155	Hempel Paints Ltd	AVP	HSE 3719

ANTIFOULING PRODUCTS

Product Name	Marketing Company	Use	Reg. No

585 CUPROUS OXIDE and TRIBUTYLTIN FLUORIDE and ZINC NAPHTHENATE

* Teamac Killa	Teal and Mackrill Ltd	AVP	HSE 3492

586 CUPROUS OXIDE and TRIBUTYLTIN FLUORIDE and ZINC OXIDE

(ATMC 123) Amercoat 557	Ameron (UK) Ltd	AVP	HSE 3207
* Antifouling Seven Seas	Jotun-Henry Clark Ltd	AVP	HSE 3405

587 CUPROUS OXIDE and TRIBUTYLTIN FLUORIDE and ZINC OXIDE and COPPER RESINATE

* Lily Antifouling IV	Sigma Coatings BV	AVP	HSE 3484
* Sigma Antifouling CR	Sigma Coatings BV	AVP	HSE 3483
* Sigma Antifouling IV	Sigma Coatings BV	AVP	HSE 4347

588 CUPROUS OXIDE and TRIBUTYLTIN FLUORIDE and ZINC OXIDE and ZINEB

* Colturiet TCN Antifouling (Set and N Set)	Sigma Coatings BV	AVP	HSE 3488

589 CUPROUS OXIDE and TRIBUTYLTIN FLUORIDE and ZINC OXIDE AND ZINEB and COPPER RESINATE

* Sigma Pilot Antifouling LL	Sigma Coatings BV	AVA	HSE 3481

590 CUPROUS OXIDE and TRIBUTYLTIN METHACRYLATE

* AF Seaflo Z–100 LE–2 Light/Dark	Chugoku Marine Paints (UK) Ltd	AVP	HSE 3258
* AF Seaflo Z–100–2 Light/Dark	Chugoku Marine Paint (UK) Ltd	AVP	HSE 3252
AF Seaflo Z–200 C	Chugoku Marine Paints (UK) Ltd	AVP	HSE 3256
Devran MCP Antifouling Red	Devoe Coatings BV	AVP	HSE 3311
Devran MCP Antifouling Red/Brown	Devoe Coatings BV	AVP	HSE 3310
Takata LLL Antifouling	NOF Corporation	AVP	HSE 4050
Takata LLL Antifouling Hi-Solid	NOF Corporation	AVP	HSE 4052
Takata LLL Antifouling LS	NOF Corporation	AVP	HSE 4051
Takata LLL Antifouling LS Hi-Solid	NOF Corporation	AVP	HSE 4053
Takata LLL Antifouling NO 2001	NOF Corporation	AVP	HSE 4054

ANTIFOULING PRODUCTS

Product Name	Marketing Company	Use	Reg. No
591 CUPROUS OXIDE and TRIBUTYLTIN METHACRYLATE and TRIBUTYLTIN OXIDE			
AF Seaflo Z–100 HS–1	Camrex Chugoku Ltd	AVP	HSE 5318
AF Seaflo Z–100 HS–1 Light/Dark	Chugoku Marine Paints (UK) Ltd	AVP	HSE 3253
AF Seaflo Z–100 LE-HS–1	Camrex Chugoku Ltd	AVP	HSE 5313
AF Seaflo Z–100 LE–HS–1 Light/Dark	Chugoku Marine Paints (UK) Ltd	AVP	HSE 3259
# AF Seaflo Z–100–1 Light/Dark	Chugoku Marine Paints (UK) Ltd	AVP	HSE 3251
# Antifouling Seaflex	Jotun-Henry Clark Ltd	AVP	HSE 3428
Antifouling Seamate HB 66 Black	Jotun-Henry Clark Ltd	AVP	HSE 3409
Antifouling Seamate HB 99 Black	Jotun-Henry Clark Ltd	AVP	HSE 3410
Antifouling Seamate HB 99 Dark Red	Jotun-Henry Clark Ltd	AVP	HSE 3412
Antifouling Seamate HB 99 Light Red	Jotun-Henry Clark Ltd	AVP	HSE 3411
AF Seaflo Z–100 LE–1 Light/Dark	Chugoku Marine Paints (UK) Ltd	AVP	HSE 3257
Hempel's Antifouling Combic 7699	Hempel Paints Ltd	AVP	HSE 3348
Hempel's Antifouling Nautic HI 7691	Hempel Paints Ltd	AVP	HSE 3361
Hempel's Antifouling Nautic HI 7690	Hempel Paints Ltd	AVP	HSE 3360
Intersmooth SPC Antifouling BF0250 Series	International Paint Ltd	AVP	HSE 5635
Intersmooth SPC BFA090/BFA190 Series	International Paint Ltd	AVP	HSE 3377
592 CUPROUS OXIDE and TRIBUTYLTIN METHACRYLATE and TRIBUTYLTIN OXIDE and 4,5-DICHLORO-2-N-OCTYL-4-ISOTHIAZOLIN-3-ONE			
Intersmooth Hisol BF0270/950/970 Series	International Paint Ltd	AVP	HSE 5641
593 CUPROUS OXIDE and TRIBUTYLTIN METHACRYLATE and TRIBUTYLTIN OXIDE and TRIPHENYLTIN FLUORIDE			
* A1 Antifouling	Nautix SA	AVP	HSE 4366
# Hai-Hong's Antifouling Nautic HI–7695	Hai-Hong Marine Paint	AVP	HSE 3554

ANTIFOULING PRODUCTS

Product Name	Marketing Company	Use	Reg. No

593 CUPROUS OXIDE and TRIBUTYLTIN METHACRYLATE and TRIBUTYLTIN OXIDE and TRIPHENYLTIN FLUORIDE—continued

Product Name	Marketing Company	Use	Reg. No
# Hempel's Antifouling Mille MS7672	Hempel Paints Ltd	AVP	HSE 3356
# Hempel's Antifouling Nautic 7673	Hempel Paints Ltd	AVP	HSE 3357
# Hempel's Antifouling Nautic 7674	Hempel Paints Ltd	AVP	HSE 3358
* Hempel's Antifouling Nautic HI 7695	Hempel Paints Ltd	AVP	HSE 3362

594 CUPROUS OXIDE and TRIBUTYLTIN METHACRYLATE and TRIBUTYLTIN OXIDE and ZINC OXIDE

Product Name	Marketing Company	Use	Reg. No
# Ablative Antifouling AF28	Desoto Titanine Plc	AVP	HSE 3505
AF Seaflo Mark 2–1	Camrex Chugoku Ltd	AVP	HSE 5316
AF Seaflo Mark 2–1 Light/Dark	Chugoku Marine Paints (UK) Ltd	AVP	HSE 3261
# Antifouling Seaconomy	Jotun-Henry Clark Ltd	AVP	HSE 3438
Antifouling Seaconomy 200	Jotun-Henry Clark Ltd	AVP	HSE 4436
Antifouling Seaconomy 300	Jotun-Henry Clark Ltd	AVP	HSE 4272
# Antifouling Seaflex	Jotun-Henry Clark Ltd	AVP	HSE 3439
Antifouling Seamate HB 22	Jotun-Henry Clark Ltd	AVP	HSE 4271
Antifouling Seamate HB 33	Jotun-Henry Clark Ltd	AVP	HSE 4270
# Antifouling Seamate HB 33 DR	Jotun-Henry Clark Ltd	AVP	HSE 3440
# Antifouling Seamate HB 33 LR	Jotun-Henry Clark Ltd	AVP	HSE 3441
* Antifouling Seamate HB 66 Dark Red	Jotun-Henry Clark Ltd	AVP	HSE 3442
* Antifouling Seamate HB 66 Light Red	Jotun-Henry Clark Ltd	AVP	HSE 3443
Antifouling Seamate HB66	Jotun-Henry Clark Ltd	AVP	HSE 5698
# Camrex C–Clean EXA	Camrex Ltd	AVP	HSE 3240
# Camrex C–Clean HB EXA	Camrex Ltd	AVP	HSE 3241
# Camrex C–Clean HB LA	Camrex Ltd	AVP	HSE 3243
# Camrex C–Clean LA	Camrex Ltd	AVP	HSE 3242
Grassline ABL Antifouling Type M349	W and J Leigh and Company	AVP	HSE 5374

Product Name	Marketing Company	Use	Reg. No

595 CUPROUS OXIDE and TRIBUTYLTIN METHACRYLATE and TRIBUTYLTIN OXIDE and ZINC OXIDE and ZINEB

Product Name	Marketing Company	Use	Reg. No
Hempel's Antifouling Combic 7699B	Hempel Paints Ltd	AVP	HSE 5581
Hempel's Antifouling Nautic 7690B	Hempel Paints Ltd	AVP	HSE 5580
Hempel's Antifouling Nautic 7691B	Hempel Paints Ltd	AVP	HSE 5586

596 CUPROUS OXIDE and TRIBUTYLTIN METHACRYLATE and TRIBUTYLTIN OXIDE and ZINEB

Product Name	Marketing Company	Use	Reg. No
Blueline SBA900 Series	International Paint Ltd	AVP	HSE 5138
Intersmooth Hisol 2000 BFA270 Series	International Paint Ltd	AVP	HSE 3844
Intersmooth Hisol 9000 BFA970 Series	International Paint Plc	AVP	HSE 5461
Intersmooth Hisol BFA250/BFA900 Series	International Paint Ltd	AVP	HSE 3848
# Intersmooth Hisol BFA250/BFA900 Series	International Paint Plc	AVP	HSE 3378
Interswift BKA000/700 Series	International Paint Ltd	AVP	HSE 3384

597 CUPROUS OXIDE and TRIBUTYLTIN METHACRYLATE and ZINC METAL

Product Name	Marketing Company	Use	Reg. No
* AF Seaflow Z–100 HS–2 Light/Dark	Chugoku Marine Paints (UK) Ltd	AVP	HSE 3254
* AF Seaflo Z–100 LE–HS–2 Light/Dark	Chugoku Marine Paints (UK) Ltd	AVP	HSE 3260

598 CUPROUS OXIDE and TRIBUTYLTIN METHACRYLATE and ZINC METAL and ZINC OXIDE

Product Name	Marketing Company	Use	Reg. No
* AF Seaflo Mark 2–2 Light/Dark	Chugoku Marine Paints (UK) Ltd	AVP	HSE 3262
* AF Seaflo Mark 3 Light/Dark	Chugoku Marine Paints (UK) Ltd	AVP	HSE 3263

599 CUPROUS OXIDE and TRIBUTYLTIN METHACRYLATE and ZINC OXIDE and ZINEB

Product Name	Marketing Company	Use	Reg. No
Antifouling Seamate HB22	Jotun-Henry Clark Ltd	AVP	HSE 5363
Antifouling Seamate HB33	Jotun-Henry Clark Ltd	AVP	HSE 5362
Sigmaplane HA Antifouling	Sigma Coatings BV	AVP	HSE 4345
Sigmaplane HB	Sigma Coatings BV	AVP	HSE 3487

ANTIFOULING PRODUCTS

Product Name	Marketing Company	Use	Reg. No

600 CUPROUS OXIDE and TRIBUTYLTIN METHACRYLATE and ZINEB

# Sigma Pilot Antifouling TA	Sigma Coatings BV	AVP	HSE 3482
Sigmaplane TA Antifouling	Sigma Coatings BV	AVP	HSE 4346
# Sigmathrift Antifouling	Sigma Coatings BV	AVP	HSE 3486

601 CUPROUS OXIDE and TRIBUTYLTIN OXIDE

# Camrex A/F 3	Camrex Ltd	AVP	HSE 3238
# Camrex A/F 5	Camrex Ltd	AVP	HSE 3239
* Devchlor Antifouling Paint Red	Devoe Coatings BV	AVP	HSE 3309
* Devran 222 All Seasons Permanent	Devoe Coatings BV	AVP	HSE 3312
* Hempel's Antifouling Bravo 7610	Hempel Paints Ltd	AVP	HSE 3343
* Hempel's Antifouling Classic 7611	Hempel Paints Ltd	AVP	HSE 3344
* Hempel's Antifouling Classic 7633	Hempel Paints Ltd	AVP	HSE 3345
# Hempel's Antifouling Tropic 7644	Hempel Paints Ltd	AVP	HSE 3366
# High Performance Antifouling Paint	Spencer (Aberdeen) Plc	AVP	HSE 4157
* Interclene Extra BCA500 Series	International Paint Ltd	AVP	HSE 3374
* Interclene Super BCA400 Series	International Paint Ltd	AVP	HSE 3373
* Interspeed Extra BLA200 Series	International Paint Ltd	AVP	HSE 3381
* Interspeed Extra BJA450 Series	International Paint Ltd	AVP	HSE 3379
* Interspeed System 2 BRA140 Series	International Paint Ltd	AVP	HSE 3382
# Mebon Longlife Anti-Fouling (T) 6–2–09	Mebon Paints Ltd	AVP	HSE 3472
# Mebon Longlife Anti-Fouling 6–2–16	Mebon Paints Ltd	AVP	HSE 3473
# Mebon Longlife Anti-Fouling 6–2–16	Mebon Paints Ltd	AVP	HSE 3474
# Norden Kobberstoff Green	Jotun-Henry Clark Ltd	AVP	HSE 3432

602 CUPROUS OXIDE and TRIBUTYLTIN OXIDE and TRIPHENYLTIN FLUORIDE

# Hempel's Antifouling Classic 7677	Hempel Paints Ltd	AVP	HSE 3347

603 CUPROUS OXIDE and TRIBUTYLTIN OXIDE and ZINC OXIDE

# (ATMC 101) Amercoat 610	Ameron (UK) Ltd	AVP	HSE 3208
# (ATMC 102) Amercoat 611	Ameron (UK) Ltd	AVP	HSE 3209
# (ATMC 120) Amercoat 612	Ameron (UK) Ltd	AVP	HSE 3210
# (ATMC 121) Amercoat 613	Ameron (UK) Ltd	AVP	HSE 3211

ANTIFOULING PRODUCTS

Product Name	Marketing Company	Use	Reg. No

603 CUPROUS OXIDE and TRIBUTYLIN OXIDE and ZINC OXIDE—continued

# (ATMC 122) Amercoat 615	Ameron (UK) Ltd	AVP	HSE 3213
# Antifouling Equator KK	Jotun-Henry Clark Ltd	AVP	HSE 3431
# Antifouling Sargasso	Jotun-Henry Clark Ltd	AVP	HSE 3437
# Super Quality Antifouling	Spencer (Aberdeen) Plc	AVP	HSE 3491

604 CUPROUS OXIDE and TRIBUTYLTIN OXIDE and ZIRAM

* Devoe ABC 2 Antifouling	Devoe Coatings BV	AVP	HSE 3313

605 CUPROUS OXIDE and TRIBUTYLTIN POLYSILOXANE and ZINC OXIDE

* (ATMC 170) Amercoat 699	Ameron (UK) Ltd	AVP	HSE 3215

606 CUPROUS OXIDE and TRIPHENYLTIN FLUORIDE

* Envoy K926 High Performance Anti-Fouling	W and J Leigh and Company	AVP	HSE 3453
EXL-AF	Chugoku Marine Paints (UK) Ltd	AVP	HSE 3266
EXL-AF E	Chugoku Marine Paints (UK) Ltd	AVP	HSE 3267
* Grassline El Anti-Fouling Type L975	W and J Leigh and Company	AVP	HSE 3458
* Grassline XL3 Anti-Fouling Type L806 B	W and J Leigh and Company	AVP	HSE 3455

607 CUPROUS OXIDE and TRIPHENYLTIN FLUORIDE and ZINC OXIDE

* Rabamarine A/F No 100S	Kansai Paint Co Ltd	AVP	HSE 3448

608 CUPROUS OXIDE and TRIPHENYLTIN HYDROXIDE

(ATMC 135) Amercoat 614	Ameron (UK) Ltd	AVP	HSE 3212

609 CUPROUS OXIDE and TRIPHENYLTIN HYDROXIDE and ZINC OXIDE

AF Seaflo SP-1 Light/Dark	Chugoku Marine Paints (UK) Ltd	AVP	HSE 3269
AF Seaflo SP-2 Light/Dark	Chugoku Marine Paints (UK) Ltd	AVP	HSE 3268
* Rabamarine A/F No 1000 (A-Sol/B-Sol)	Kansai Paint Co Ltd	AVP	HSE 3449

ANTIFOULING PRODUCTS

Product Name	Marketing Company	Use	Reg. No

609 CUPROUS OXIDE and TRIPHENYLTIN OXIDE and ZINC OXIDE—continued

* Rabamarine A/F No 1000SP (A–Sol/B–Sol)	Kansai Paint Co Ltd	AVP	HSE 3450

610 CUPROUS OXIDE and ZINC NAPHTHENATE

Teamac Killa Copper	Teal and Mackrill Ltd	AVA AVP	HSE 3493
Teamac Killa Copper	Teal and Mackrill Ltd	AVA AVP	HSE 3494

611 CUPROUS OXIDE and ZINC NAPHTHENATE and 2,4,5,6-TETRACHLORO ISOPHTHALONITRILE

Teamac Killa Copper Plus	Teal and Mackrill Ltd	AVA AVP	HSE 3495

612 CUPROUS OXIDE and ZINC NAPHTHENATE and ZINC OXIDE

Teamac Super Tropical	Teal and Mackrill Ltd	AVA AVP	HSE 3497

613 CUPROUS OXIDE and ZINC OXIDE

(ATMC 129) Amercoat 279	Ameron (UK) Ltd	AVA AVP	HSE 3942
Amercoat 279	Ameron BV	AVA AVP	HSE 3206
Antifouling Seaguardian (Black and Blue)	Jotun-Henry Clark Ltd	AVP	HSE 4273
Antifouling Seavictor 40	Jotun-Henry Clark Ltd	AVA AVP	HSE 4957
Antifouling Tropic	Jotun-Henry Clark Ltd	AVA AVP	HSE 3414
Aquacleen	Mariner	AVA AVP	HSE 5667
Awlgrip Awlstar Gold Label Antifouling	Awlgrip NV	AVA AVP	HSE 5065
Biscon AF	Chugoku Marine Paints (UK) Ltd	AVA AVP	HSE 3272
C–Worthy	Benfleet Marine Wholesale	AVA AVP	HSE 5476
* Chugoku AF ST	Chugoku Marine Paints (UK) Ltd	AVA AVP	HSE 3270
* Chugoku AF T	Chugoku Marine Paints (UK) Ltd	AVA AVP	HSE 3271
Cobra V	Valiant Marine	AVA AVP	HSE 5194
Cooper's Copolymer Antifouling	Cooper's Marine Paints Ltd	AVA AVP	HSE 5609
# Copper Bottom Paint	Spencer (Aberdeen) Plc	AVA AVP	HSE 3504

ANTIFOULING PRODUCTS

Product Name	Marketing Company	Use	Reg. No
613 **CUPROUS OXIDE and ZINC OXIDE**—continued			
Cupron Plus T.F.	Veneziani Spa	AVA AVP	HSE 5661
Cupron T.F.	Veneziani Spa	AVA AVP	HSE 4935
Hempel's Copper Bottom Paint 7116	Hempel Paints Ltd	AVA AVP	HSE 4274
Marclear Antifouling	Marclear Marine Products Ltd	AVA AVP	HSE 5264
Marclear Antifouling	Marine Clearance	AVA AVP	HSE 4853
Noa-Noa Rame	Stoppani (UK) Ltd	AVA AVP	HSE 4795
* North Atlantic/Launching Antifouling AF2 Red	Desoto Titanine Plc	AVA AVP	HSE 3301
# North Atlantic/Launching Antifouling F2/801 Black	Desoto Titanine Plc	AVA AVP	HSE 3302
Penguin Racing	Marine and Industrial Sealants	AVA AVP	HSE 5673
Ravax AF	Camrex Chugoku Ltd	AVA AVP	HSE 5319
Ravax Anti-Fouling	Chugoku Marine Paints (UK) Ltd	AVA AVP	HSE 3291
Ravax Anti-Fouling HB	Chugoku Marine Paints (UK) Ltd	AVA AVP	HSE 3264
Ravax Anti-Fouling ND	Chugoku Marine Paints (UK) Ltd	AVA AVP	HSE 3265
Seatender 10	Camrex Chugoku Ltd	AVA AVP	HSE 5321
Seatender 7	Camrex Chugoku Ltd	AVA AVP	HSE 5320
Speedclean Antifouling	Mariner Paints	AVA AVP	HSE 5077
Superspeed	D R Margetson	AVA AVP	HSE 5191
Tiger Cruising	Blakes Marine Paints Ltd	AVA AVP	HSE 5099
Vinyl Antifouling 2000	Akzo Coatings BV	AVA AVP	HSE 5633
614 **CUPROUS OXIDE and ZINC OXIDE and 2,4,5,6-TETRACHLORO ISOPHTHALONITRILE**			
Hempel's Antifouling Classic 7654C	Hempel Paints Ltd	AVA AVP	HSE 5298
Hempel's Antifouling Combic 7199C	Hempel Paints Ltd	AVA AVP	HSE 5296
Hempel's Antifouling Nautic 7190C	Hempel Paints Ltd	AVA AVP	HSE 5294
TFA 10 LA	Camrex Chugoku Ltd	AVA AVP	HSE 5361

ANTIFOULING PRODUCTS

Product Name	Marketing Company	Use	Reg. No

615 CUPROUS OXIDE and ZINC OXIDE and 2,4,5,6-TETRACHLORO ISOPHTHALONITRILE and 4,5-DICHLORO-2-N-OCTYL -4-ISOTHIAZOLIN-3-ONE

Product Name	Marketing Company	Use	Reg. No
Seatender 15	Camrex Chugoku Ltd	AVA AVP	HSE 5348
TFA 10 HG	Camrex Chugoku Ltd	AVA AVP	HSE 5366
TFA 10G	Camrex Chugoku Ltd	AVA AVP	HSE 5347

616 CUPROUS OXIDE and ZINC OXIDE and 2-METHYLTHIO-4-TERTIARY-BUTYLAMINO-6-CYCLOPROPYLAMINO-S-TRIAZINE

Product Name	Marketing Company	Use	Reg. No
Algicide Red Antifouling	Blakes Marine Paints	AVA AVP	HSE 5738

617 CUPROUS OXIDE and ZINC OXIDE and 2-METHYLTHIO-4-TERTIARY-BUTYLAMINO-6-CYCLOPROPYLAMINO-S-TRIAZINE

Product Name	Marketing Company	Use	Reg. No
Even TF	Veneziani Spa	AVA AVP	HSE 5166
Hempel's Antifouling Classic 7611 Red (Tin Free) 5000	Hempel Paints Ltd	AVA AVP	HSE 5064
Hempel's Antifouling Classic 76540	Hempel Paints Ltd	AVA AVP	HSE 5291
Hempel's Antifouling Combic 71990	Hempel Paints Ltd	AVA AVP	HSE 5600
Hempel's Antifouling Combic 71992	Hempel Paints Ltd	AVA AVP	HSE 5601
Hempel's Antifouling Combic Tin Free 71990	Hempel Paints Ltd	AVA AVP	HSE 5266
Hempel's Antifouling Mille Dynamic	Hempel Paints Ltd	AVA	HSE 4440
Hempel's Antifouling Nautic 71900	Hempel Paints Ltd	AVA AVP	HSE 5290
Hempel's Antifouling Nautic 71902	Hempel Paints Ltd	AVA AVP	HSE 5605
Hempel's Antifouling Nautic Tin Free 7190	Hempel Paints Ltd	AVA AVP	HSE 4869
Hempel's Antifouling Olympic HI-7661	Hempel Paints Ltd	AVA AVP	HSE 4898
Hempel's Hard Racing 76480	Hempel Paints Ltd	AVA AVP	HSE 5538
Hempel's Mille Dynamic 71700	Hempel Paints Ltd	AVA AVP	HSE 5574
Hempels Antifouling Bravo Tin Free 7610	Hempel Paints Ltd	AVA AVP	HSE 4482
Long Life T.F.	Veneziani Spa	AVA AVP	HSE 4936

ANTIFOULING PRODUCTS

Product Name	Marketing Company	Use	Reg. No
618 CUPROUS OXIDE and ZINC OXIDE and 2-METHYLTHIO-4-TERTIARY-BUTYLAMINO-6-CYCLOPROPYLAMINO-S-TRIAZINE and COPPER RESINATE			
Sigmaplane Ecol Antifouling	Sigma Coatings BV	AVA AVP	HSE 4348
619 CUPROUS OXIDE and ZINC OXIDE and 2-METHYLTHIO-4-TERTIARY-BUTYLAMINO-6-CYCLOPROPYLAMINO-S-TRIAZINE			
Broads Black Antifouling	Blakes Marine Paints	AVA AVP	HSE 5739
Hard Racing Antifouling	Blakes Marine Paints	AVA AVP	HSE 5704
620 CUPROUS OXIDE and ZINC OXIDE and 2-METHYLTHIO-4-TERTIARY-BUTYLAMINO-6-CYCLOPROPYLAMINO- S-TRIAZINE			
Titan FGA Antifouling	Blakes Marine Paints	AVA AVP	HSE 5681
621 CUPROUS OXIDE and ZINC OXIDE and 4,5-DICHLORO-2-N-OCTYL-4-ISOTHIAZOLIN-3-ONE			
Antifouling Seavictor 50	Jotun-Henry Clark Ltd	AVA AVP	HSE 4958
Hempel's Antifouling Classic 7654E	Hempel Paints Ltd	AVA AVP	HSE 5286
Hempel's Antifouling Combic 7199E	Hempel Paints Ltd	AVA AVP	HSE 5277
Hempel's Antifouling Nautic 7190E	Hempel Paints Ltd	AVA AVP	HSE 5283
Hempel's Antifouling Nautic Tin Free 7190E	Hempel Paints Ltd	AVA AVP	HSE 5145
622 CUPROUS OXIDE and ZINC OXIDE and ZINEB and COPPER RESINATE			
* Sigma Pilot Antifouling LL/TF	Sigma Coatings BV	AVA AVP	HSE 3480
Sigma Pilot Ecol Antifouling	Sigma Coatings Ltd	AVA AVP	HSE 4933
623 CUPROUS OXIDE and ZINC OXIDE and ZIRAM			
Hempel's Antifouling Classic 7654A	Hempel Paints Ltd	AVA AVP	HSE 5297
Hempel's Antifouling Combic 7199A	Hempel Paints Ltd	AVA AVP	HSE 5295
Hempel's Antifouling Nautic 7190A	Hempel Paints Ltd	AVA AVP	HSE 5293
624 CUPROUS OXIDE and ZINC PYRITHIONE			
Micron 600 Series	International Paint Ltd	AVA AVP	HSE 5733
VC Offshore Extra 100 Series	International Paint Ltd	AVA AVP	HSE 5730
VC Offshore SP Antifouling	Extensor AB	AVA AVP	HSE 3507

4

ANTIFOULING PRODUCTS

Product Name	Marketing Company	Use	Reg. No
625 CUPROUS OXIDE and ZINEB			
Blueline SPC Tin Free SBA700 Series	International Paint Ltd	AVA AVP	HSE 5214
Equatorial	International Paint Ltd	AVA AVP	HSE 4121
Inter 100	International Paint Ltd	AVA AVP	HSE 4120
# International TBT Free Copolymer Antifouling BQA200 Series	International Paint Ltd	AVA AVP	HSE 3722
# Interspeed Extra (BWA 500 Red)	International Paint Ltd	AVA AVP	HSE 3845
Interspeed System 2 BRA140/ BRA240 Series	International Paint Ltd	AVA AVP	HSE 3847
Interviron BQA200 Series	International Paint Ltd	AVA AVP	HSE 3846
626 CUPROUS OXIDE and ZIRAM			
Awlstar Gold Label Anti-Fouling	Grow Group Incorporated	AVA AVP	HSE 3674
Cheetah Antifouling	Blakes Marine Paints Ltd	AVA AVP	HSE 3223
* Coppercoat	International Paint Ltd	AVA AVP	HSE 3560
* Cruiser Premium	International Paint Ltd	AVA AVP	HSE 4277
* Grassline TF Anti-Fouling Type M394	W and J Leigh and Company	AVA AVP	HSE 3460
# Micron	International Paint Plc	AVA AVP	HSE 3395
Micron CSC	International Paint Ltd	AVA AVP	HSE 3558
627 CUPROUS SULPHIDE and CUPROUS OXIDE			
Admiralty Antifouling Black (TO TS10239)	Sigma Coatings BV	AVA AVP	HSE 3489
Black Anti-Fouling Paint 317 (TO TS10239)	United Paints Ltd	AVA AVP	HSE 3502
# Black Antifouling AF21 (TO TS10239)	Desoto Titanine Plc	AVA AVP	HSE 3295
# Black Antifouling Paint 317 (TO TS10239)	Desoto Titanine Plc	AVA AVP	HSE 3307
Black Antifouling Paint 317 (TO TS10239)	Hempel Paints Ltd	AVA AVP	HSE 3370
Hempel's Antifouling 762GB	Hempel Paints Ltd	AVA AVP	HSE 3340
# TS 10239 Antifouling	International Paint Ltd	AVA AVP	HSE 3385
Unitas Antifouling Paint Black	United Paints Ltd	AVA AVP	HSE 3501

ANTIFOULING PRODUCTS

Product Name	Marketing Company	Use	Reg. No
628 CUPROUS THIOCYANATE			
Antifouling Broken White DL–2253	International Paint Ltd	AVA AVP	HSE 3400
Boot Top	International Paint Ltd	AVA AVP	HSE 3388
Hempel's Antifouling 763GB	Hempel Paints Ltd	AVA AVP	HSE 3537
# Sigma Anti-Fouling Broken White DL–2253	Sigma Coatings BV	AVA AVP	HSE 3536
Unitas Antifouling Paint White	United Paints Ltd	AVA AVP	HSE 3500
629 CUPROUS THIOCYANATE and 2-METHYLTHIO-4-TERTIARY-BUTYLAMINO-6-CYCLOPROPYLAMINO-S-TRIAZINE			
AL–27	International Paint Ltd	AVA AVP	HSE 3613
Aquaspeed White	Blakes Marine Paints Ltd	AVA AVP	HSE 4513
Boot Top Plus (Gull White)	International Paint Ltd	AVA AVP	HSE 3389
Broads Sweetwater Antifouling	Blakes Marine Paints Ltd	AVA AVP	HSE 3222
* Cruiser Superior (White)	International Paint Ltd	AVA AVP	HSE 3784
Hempel's Tin Free Hard Racing 7648	Hempel Paints Ltd	AVA AVP	HSE 3367
Interspeed 2000	International Paint Ltd	AVA AVP	HSE 4148
* Interspeed 2000 (White)	International Paint Plc	AVA AVP	HSE 3780
MPX	International Paint Ltd	AVA AVP	HSE 4818
* MPX (White)	International Paint Ltd	AVA AVP	HSE 3782
Prop-N-Drive	Extensor AB	AVA AVP	HSE 3509
Propeller T.F.	Veneziani Spa	AVA AVP	HSE 5660
Tiger White	Blakes Marine Paints Ltd	AVA AVP	HSE 4512
Tigerline Antifouling	Blakes Marine Paints Ltd	AVA AVP	HSE 3842
VC 17–Victory Antifouling	Extensor AB	AVA AVP	HSE 3506
* VC Offshore	Extensor AB	AVA AVP	HSE 3843
VC Offshore	Extensor AB	AVA AVP	HSE 4779
VC Prop-O-Drev	Extensor AB	AVA AVP	HSE 4217
630 CUPROUS THIOCYANATE and 2-METHYLTHIO-4-TERTIARY-BUTYLAMINO-6- CYCLOPROPYLAMINO-S-TRIAZINE			
Cruiser Superior 100 Series	International Paint Ltd	AVA AVP	HSE 5723

ANTIFOULING PRODUCTS

Product Name	Marketing Company	Use	Reg. No

631 CUPROUS THIOCYANATE and DICHLOFLUANID and ZINC OXIDE

Penguin Non-Stop White	Marine and Industrial Sealants	AVA AVP	HSE 5672

632 CUPROUS THIOCYANATE and DICHLOFLUANID and ZINC OXIDE and 2-METHYLTHIO-4-TERTIARY -BUTYLAMINO-6-CYCLOPROPYLAMINO -S-TRIAZINE

Hempel's Antifouling Mille Dynamic 717 GB	Hempel Paints Ltd	AVA AVP	HSE 4437

633 CUPROUS THIOCYANATE and DICHLOROPHENYL DIMETHYLUREA

Aquarius AL	International Paint Ltd	AVA AVP	HSE 4295
Cruiser Superior	International Paint Ltd	AVA AVP	HSE 4776
TFA–30 White	Chugoku Marine Paints (UK) Ltd	AVA AVP	HSE 3282
VC Aqua 12	Extensor AB	AVA AVP	HSE 4802

634 CUPROUS THIOCYANATE and DICHLOROPHENYL DIMETHYLUREA and TRIPHENYLTIN CHLORIDE

* Kamome FRP–DC 2	Chugoku Marine Paints (UK) Ltd	AVP	HSE 3290

635 CUPROUS THIOCYANATE and DICHLOROPHENYL DIMETHYLUREA and TRIPHENYLTIN CHLORIDE and ZINC OXIDE

* Kamome Colour 60–2	Chugoku Marine Paints (UK) Ltd	AVP	HSE 3276

636 CUPROUS THIOCYANATE and DICHLOROPHENYL DIMETHYLUREA and ZINC OXIDE

Hempel's Hard Racing 7638A	Hempel Paints Ltd	AVA AVP	HSE 5539
Hempel's Mille Alu 71602	Hempel Paints Ltd	AVA AVP	HSE 5578
Hempel's Mille Dynamic 7160A	Hempel Paints Ltd	AVA AVP	HSE 5575

ANTIFOULING PRODUCTS

Product Name	Marketing Company	Use	Reg. No

637 CUPROUS THIOCYANATE and DICHLOROPHENYL DIMETHYLUREA and ZINC OXIDE and 2-METHYLTHIO-4-TERTIARY-BUTYLAMINO-6-CYCLOPROPYLAMINO-S-TRIAZINE

Product Name	Marketing Company	Use	Reg. No
A3 Antifouling 072016	Nautix SA	AVA AVP	HSE 5225
A3 Teflon Antifouling 062015	Nautix SA	AVA AVP	HSE 5481
A4 Antifouling 072018	Nautix SA	AVA AVP	HSE 5227
A4 Teflon Antifouling 062017	Nautix SA	AVA AVP	HSE 5482

638 CUPROUS THIOCYANATE and MANEB and TRIBUTYLTIN MESO-DIBROMOSUCCINATE and TRIBUTYLTIN METHACRYLATE and TRIBUTYLTIN OXIDE and ZINC OXIDE

Product Name	Marketing Company	Use	Reg. No
# Marine Gold DX–1	Chugoku Marine Paints (UK) Ltd	AVP	HSE 3281

639 CUPROUS THIOCYANATE and TRIBUTYLTIN FLUORIDE

Product Name	Marketing Company	Use	Reg. No
* Kamome FRP–DC 1	Chugoku Marine Paints (UK) Ltd	AVP	HSE 3287

640 CUPROUS THIOCYANATE and TRIBUTYLTIN FLUORIDE and TRIBUTYLTIN OXIDE

Product Name	Marketing Company	Use	Reg. No
* Hempel's Antifouling 7650	Hempel Paints Ltd	AVP	HSE 3341
* Hempel's Antifouling 7659	Hempel Paints Ltd	AVP	HSE 3342

641 CUPROUS THIOCYANATE and TRIBUTYLTIN MESO-DIBROMOSUCCINATE and TRIBUTYLTIN OXIDE and ZINC OXIDE

Product Name	Marketing Company	Use	Reg. No
# Kamome Colour 60–1	Chugoku Marine Paints (UK) Ltd	AVP	HSE 3277

642 CUPROUS THIOCYANATE and TRIBUTYLTIN METHACRYLATE

Product Name	Marketing Company	Use	Reg. No
* AF Seaflo Z–200	Chugoku Marine Paints (UK) Ltd	AVP	HSE 3255
Norden Non-Stop Black	Jotun-Henry Clark Ltd	AVP	HSE 3433
Norden Non-Stop Blue	Jotun-Henry Clark Ltd	AVP	HSE 3434
Norden Non-Stop Red	Jotun-Henry Clark Ltd	AVP	HSE 3435
# Norden Non-Stop White	Jotun-Henry Clark Ltd	AVP	HSE 3436

ANTIFOULING PRODUCTS

Product Name	Marketing Company	Use	Reg. No
643 CUPROUS THIOCYANATE and TRIBUTYLTIN METHACRYLATE and TRIBUTYLTIN OXIDE			
Antifouling HB 66 Ocean Green	Jotun-Henry Clark Ltd	AVP	HSE 3408
* Antifouling Seaflex Sealinc Blue 110	Jotun-Henry Clark Ltd	AVP	HSE 3406
Antifouling Seamate HB Blue	Jotun-Henry Clark Ltd	AVP	HSE 3429
Antifouling Seamate HB Green	Jotun-Henry Clark Ltd	AVP	HSE 3430
# Antifouling Seamate HB Special Black	Jotun-Henry Clark Ltd	AVP	HSE 3407
Intersmooth Hisol SPC Antifouling BFA949 Red	International Paint Ltd	AVP	HSE 4949
Intersmooth SPC BFA040/BFA050 Series	International Paint Ltd	AVP	HSE 3376
644 CUPROUS THIOCYANATE and TRIBUTYLTIN METHACRYLATE and TRIBUTYLTIN OXIDE and CUPROUS OXIDE			
# Hempel's Antifouling Mille 7678	Hempel Paints Ltd	AVP	HSE 3353
645 CUPROUS THIOCYANATE and TRIBUTYLTIN METHACRYLATE and TRIBUTYLTIN OXIDE and TRIPHENYLTIN FLUORIDE			
# Hempel's Antifouling Mille 7670	Hempel Paints Ltd	AVP	HSE 3352
646 CUPROUS THIOCYANATE and TRIBUTYLTIN METHACRYLATE and TRIBUTYLTIN OXIDE and ZINC OXIDE			
Antifouling Seamate HB 33 BSL Blue	Jotun-Henry Clark Ltd	AVP	HSE 4269
Antifouling Seamate HB22 Roundel Blue	Jotun-Henry Clark Ltd	AVP	HSE 4872
Antifouling Seamate HB66 Black	Jotun-Henry Clark Ltd	AVP	HSE 4871
647 CUPROUS THIOCYANATE and TRIBUTYLTIN METHACRYLATE and TRIBUTYLTIN OXIDE and ZINEB			
Intersmooth Hisol BFA948 Orange	International Paint Ltd	AVP	HSE 4281
Micron 25 Plus	International Paint Ltd	AVP	HSE 3402
Superyacht Antifouling	International Paint Plc	AVP	HSE 5462

ANTIFOULING PRODUCTS

Product Name	Marketing Company	Use	Reg. No

648 CUPROUS THIOCYANATE and TRIBUTYLTIN METHACRYLATE and ZINC METAL and ZINC OXIDE

* Marine Gold DX White–2	Chugoku Marine Paints (UK) Ltd	AVP	HSE 3279

649 CUPROUS THIOCYANATE and TRIBUTYLTIN METHACRYLATE and ZINC OXIDE

* Marine Gold DX–2	Chugoku Marine Paints (UK) Ltd	AVP	HSE 3280

650 CUPROUS THIOCYANATE and TRIBUTYLTIN METHACRYLATE and ZINEB

Superyacht 800 Antifouling	International Paint Ltd	AVP	HSE 4778

651 CUPROUS THIOCYANATE and TRIBUTYLTIN METHACRYLATE and ZIRAM

Takata Seaqueen	Nippon Oil and Fats Co Ltd	AVP	HSE 4055

652 CUPROUS THIOCYANATE and TRIBUTYLTIN OXIDE

# Norden Durahart	Jotun-Henry Clark Ltd	AVP	HSE 3510

653 CUPROUS THIOCYANATE and TRIBUTYLTIN OXIDE and ZINEB

Cruiser Copolymer	International Paint Ltd	AVP	HSE 3514

654 CUPROUS THIOCYANATE and ZINC OXIDE and 2,3,5,6-TETRACHLORO-4-(METHYL SULPHONYL) PYRIDINE and 2-METHYLTHIO-4-TERTIARY-BUTYLAMINO-6-CYCLOPROPYLAMINO-S-TRIAZINE and CUPROUS OXIDE

Envoy TF 400	W and J Leigh and Company	AVA AVP	HSE 4432
Envoy TF 500	W and J Leigh and Company	AVA AVP	HSE 5599
Meridian MP40 Antifouling	Deangate Marine Products	AVA AVP	HSE 5027

655 CUPROUS THIOCYANATE and ZINC OXIDE and 2,4,5,6-TETRACHLORO ISOPHTHALONITRILE

Gummipaint A/F	Veneziani Spa	AVA AVP	HSE 4934

ANTIFOULING PRODUCTS

Product Name	Marketing Company	Use	Reg. No

656 CUPROUS THIOCYANATE and ZINC OXIDE and 2-METHYLTHIO-4-TERTIARY-BUTYLAMINO-6-CYCLOPROPYLAMINO-S-TRIAZINE

Even TF Light Grey	Veneziani Spa	AVA AVP	HSE 5167
Hempel's Hard Racing 76380	Hempel Paints Ltd	AVA AVP	HSE 5540
Hempel's Mille ALU 71601	Hempel Paints Ltd	AVA AVP	HSE 5577
Hempel's Mille Dynamic 71600	Hempel Paints Ltd	AVA AVP	HSE 5576
Raffaello Alloy	Veneziani Spa	AVA AVP	HSE 5492

657 CUPROUS THIOCYANATE and ZINC OXIDE and 2-METHYLTHIO-4-TERTIARY- BUTYLAMINO-6-CYCLOPROPYLAMINO -S-TRIAZINE

| Titan FGA Antifouling White | Blakes Marine Paints | AVA AVP | HSE 5680 |

658 CUPROUS THIOCYANATE and ZINC OXIDE and 2-METHYLTHIO-4-TERTIARY-BUTYLAMINO-6-CYCLOPROPYLAMINO-S-TRIAZINE

| Hard Racing Antifouling White | Blakes Marine Paints | AVA AVP | HSE 5705 |

659 CUPROUS THIOCYANATE and ZINEB

| * Cruiser Superior | International Paint Ltd | AVA AVP | HSE 3785 |

660 CUPROUS THIOCYANATE and ZINEB and 2-METHYLTHIO-4-TERTIARY-BUTYLAMINO-6-CYCLOPROPYLAMINO-S-TRIAZINE

| # Interspeed 2000 | International Paint Plc | AVA AVP | HSE 3781 |
| * MPX | International Paint Ltd | AVA | HSE 3783 |

661 CUPROUS THIOCYANATE and ZIRAM

| * Boot Top Plus | International Paint Ltd | AVA AVP | HSE 3390 |

662 DICHLOFLUANID

| Bayer AFC | Bayer Plc | AVA AVP | HSE 5163 |
| Bayer AFC | Bayer UK Ltd | AVA | HSE 3218 |

663 DICHLOFLUANID and CUPROUS OXIDE

Forcecontact AFUR514 (Base)	Forcecontact Ltd	AVA AVP	HSE 4876
Halcyon 5000 (Base)	Waterline	AVA AVP	HSE 5396
Halcyon 5000 (Base)	Waterline	AVA	HSE 5334

ANTIFOULING PRODUCTS

Product Name	Marketing Company	Use	Reg. No
663 DICHLOFLUANID and CUPROUS OXIDE—continued			
Hempel's Antifouling Rennot 7150	Hempel Paints Ltd	AQA	HSE 3364
Hempel's Antifouling Rennot 7177	Hempel Paints Ltd	AQA	HSE 3365
Hempel's Antifouling Tin Free 742GB	Hempel Paints Ltd	AVA AVP	HSE 3328
Seashield (Base)	Waterline	AVA AVP	HSE 5438
Slipstream Antifouling	United Paints Ltd	AVA AVP	HSE 3721
Tiger Tin Free Antifouling	Blakes Marine Paints Ltd	AVA AVP	HSE 3231
664 DICHLOFLUANID and ZINC NAPHTHENATE and ZINC OXIDE and COPPER NAPHTHENATE and CUPROUS OXIDE			
Teamac Killa Copper Plus	Teal and Mackrill Ltd	AVA AVP	HSE 4659
665 DICHLOFLUANID and ZINC OXIDE and 2-METHYLTHIO-4-TERTIARY-BUTYLAMINO-6-CYCLOPROPYLAMINO-S-TRIAZINE and CUPROUS THIOCYANATE			
Hempel's Antifouling Mille Dynamic 717 GB	Hempel Paints Ltd	AVA AVP	HSE 4437
666 DICHLOFLUANID and ZINC OXIDE and CUPROUS OXIDE			
Dapaflow Antifoul	Dapa Services Ltd	AVA AVP	HSE 4754
Penguin Non-Stop	Marine and Industrial Sealants	AVA AVP	HSE 5671
667 DICHLOFLUANID and ZINC OXIDE and CUPROUS THIOCYANATE			
Penguin Non-Stop White	Marine and Industrial Sealants	AVA AVP	HSE 5672
668 DICHLOROPHENYL DIMETHYLUREA and COPPER RESINATE and CUPROUS OXIDE			
# Sigmaplane Tin Free AF SO 0671	Sigma Coatings BV	AVA AVP	HSE 3872
669 DICHLOROPHENYL DIMETHYLUREA and CUPROUS OXIDE			
A4 Antifouling	Nautix SA	AVA AVP	HSE 4369
Aquarius Extra Strong	International Paint Ltd	AVA AVP	HSE 4280
Blueline Tropical SBA300	International Paint Ltd	AVA AVP	HSE 5139

ANTIFOULING PRODUCTS

Product Name	Marketing Company	Use	Reg. No
669 DICHLOROPHENYL DIMETHYLUREA and CUPROUS OXIDE—continued			
Cruiser Premium	International Paint Ltd	AVA AVP	HSE 5127
Grafo Anti-Foul SW	Grafo Coatings Ltd	AVA AVP	HSE 5380
International Tin Free SPC BNA100 Series	International Paint Ltd	AVA AVP	HSE 5186
Intersmooth Hisol Tin Free BGA620 Series	International Paint Ltd	AVA AVP	HSE 4787
Intersmooth Tin Free BGA530 Series	International Paint Ltd	AVA AVP	HSE 4611
Interspeed Extra BWA500 Red	International Paint Ltd	AVA AVP	HSE 4303
* Interspeed Extra Strong	International Paint Ltd	AVA AVP	HSE 4215
Interspeed Extra Strong	International Paint Ltd	AVA AVP	HSE 4819
* Interspeed Premium BWA900 Red	International Paint Ltd	AVA AVP	HSE 4738
Interspeed Super BWA900 Red	International Paint Ltd	AVA AVP	HSE 4884
Interspeed Super BWA909 Black	International Paint Ltd	AVA AVP	HSE 5058
Interspeed System 2 BRA142 Brown	International Paint Ltd	AVA AVP	HSE 4302
Interswift Tin Free BQA400 Series	International Paint Ltd	AVA AVP	HSE 4842
Interswift Tin-Free SPC BTA540 Series	International Paint Ltd	AVA AVP	HSE 5078
# Interviron BQA400 Series	International Paint Ltd	AVA AVP	HSE 4658
Interviron Super BQA400 Series	International Paint Ltd	AVA AVP	HSE 5690
Interviron Super BQA400 Series	International Paint Ltd	AVA AVP	HSE 5409
Micron 500 Series	International Paint Ltd	AVA AVP	HSE 5729
Micron CSC	International Paint Ltd	AVA AVP	HSE 4775
Micron Plus Antifouling	International Paint Ltd	AVA AVP	HSE 5133
TFA-30	Chugoku Marine Paints (UK) Ltd	AVA AVP	HSE 3283
Waterways Antifouling	International Paint Plc	AVA AVP	HSE 5565
670 DICHLOROPHENYL DIMETHYLUREA and CUPROUS THIOCYANATE			
Aquarius AL	International Paint Ltd	AVA AVP	HSE 4295
Cruiser Superior	International Paint Ltd	AVA AVP	HSE 4776
TFA-30 White	Chugoku Marine Paints (UK) Ltd	AVA AVP	HSE 3282
VC Aqua 12	Extensor AB	AVA AVP	HSE 4802

ANTIFOULING PRODUCTS

Product Name	Marketing Company	Use	Reg. No

671 DICHLOROPHENYL DIMETHYLUREA and TRIBUTYLTIN FLUORIDE and TRIBUTYLTIN METHACRYLATE and ZINC OXIDE

* Marine Gold DX White-1	Chugoku Marine Paints (UK) Ltd	AVP	HSE 3278

672 DICHLOROPHENYL DIMETHYLUREA and TRIBUTYLTIN FLUORIDE and TRIBUTYLTIN OXIDE and CUPROUS OXIDE

# Hempel's Antifouling Hi Build 7600	Hempel Paints Ltd	AVP	HSE 3350

673 DICHLOROPHENYL DIMETHYLUREA and TRIBUTYLTIN METHACRYLATE and TRIBUTYLTIN OXIDE and CURPROUS OXIDE

Interswift BK0000/700 Series	International Paint Ltd	AVP	HSE 5640

674 DICHLOROPHENYL DIMETHYLUREA and TRIBUTYLTIN METHACRYLATE and TRIBUTYLTIN OXIDE and ZINC OXIDE and CUPROUS OXIDE

A11 Antifouling 072014	Nautix SA	AVP	HSE 5376

675 DICHLOROPHENYL DIMETHYLUREA and TRIPHENYLTIN CHLORIDE

* Kamome FRP-60 2	Chugoku Marine Paints (UK) Ltd	AVP	HSE 3289

676 DICHLOROPHENYL DIMETHYLUREA and TRIPHENYLTIN CHLORIDE and CUPROUS THIOCYANATE

* Kamome FRP-DC 2	Chugoku Marine Paints (UK) Ltd	AVP	HSE 3290

677 DICHLOROPHENYL DIMETHYLUREA and TRIPHENYLTIN CHLORIDE and ZINC OXIDE and CUPROUS THIOCYANATE

Kamome Colour 60-2	Chugoku Marine Paints (UK) Ltd	AVP	HSE 3276

678 DICHLOROPHENYL DIMETHYLUREA and ZINC OXIDE

Aquashield	Silverblue Services Ltd	AVA AVP	HSE 3779

679 DICHLOROPHENYL DIMETHYLUREA and ZINC OXIDE and 2,4,5,6-TETRACHLORO ISOPHTHALONITRILE and CUPROUS OXIDE

Seatender 12	Camrex Chugoku Ltd	AVA AVP	HSE 5324

ANTIFOULING PRODUCTS

Product Name	Marketing Company	Use	Reg. No

679 **DICHLOROPHENYL DIMETHYLUREA and ZINC OXIDE AND 2,4,5,6-TETRACHLORO ISOPHTHALONITRILE and CUPROUS OXIDE**—continued

TFA 10	Camrex Chugoku Ltd	AVA AVP	HSE 5346
TFA 10 H	Camrex Chugoku Ltd	AVA AVP	HSE 5358

680 **DICHLOROPHENYL DIMETHYLUREA and ZINC OXIDE and 2-METHYLTHIO-4-TERTIARY-BUTYLAMINO-6-CYCLOPROPYLAMINO-S-TRIAZINE and CUPROUS OXIDE**

A3 Antifouling 072015	Nautix SA	AVA AVP	HSE 5224
A4 Antifouling 072017	Nautix SA	AVA AVP	HSE 5226
Le Marin	Nautix SA	AVA AVP	HSE 5480

681 **DICHLOROPHENYL DIMETHYLUREA and ZINC OXIDE and 2-METHYLTHIO-4-TERTIARY-BUTYLAMINO-6-CYCLOPROPYLAMINO-S-TRIAZINE and CUPROUS THIOCYANATE**

A3 Antifouling 072016	Nautix SA	AVA AVP	HSE 5225
A3 Teflon Antifouling 062015	Nautix SA	AVA AVP	HSE 5481
A4 Antifouling 072018	Nautix SA	AVA AVP	HSE 5227
A4 Teflon Antifouling 062017	Nautix SA	AVA AVP	HSE 5482

682 **DICHLOROPHENYL DIMETHYLUREA and ZINC OXIDE and COPPER RESINATE and CUPROUS OXIDE**

* Sigmaplane Ecol AF First Coat	Sigma Coatings BV	AVA AVP	HSE 4349

683 **DICHLOROPHENYL DIMETHYLUREA and ZINC OXIDE and CUPROUS OXIDE**

Hempel's Antifouling Classic 7654B	Hempel Paints Ltd	AVA AVP	HSE 5285
Hempel's Antifouling Combic 7199B	Hempel Paints Ltd	AVA AVP	HSE 5274
Hempel's Antifouling Nautic 7190B	Hempel Paints Ltd	AVA AVP	HSE 5273
Hempel's Bravo 7610A	Hempel Paints Ltd	AVA AVP	HSE 5603
Hempel's Hard Racing 7648A	Hempel Paints Ltd	AVA AVP	HSE 5535
Hempel's Mille Dynamic 7170A	Hempel Paints Ltd	AVA AVP	HSE 5536
Noa-Noa Rame	Stoppani (UK) Ltd	AVA AVP	HSE 4796
Seajet 033	Camrex Chugoku Ltd	AVA AVP	HSE 5331
Sigmaplane Ecol Antifouling	Sigma Coatings Ltd	AVA AVP	HSE 5670

ANTIFOULING PRODUCTS

Product Name	Marketing Company	Use	Reg. No

683 DICHLOROPHENYL DIMETHYLUREA and ZINC OXIDE and CUPROUS OXIDE—continued

TFA-10 LA Light/Dark	Chugoku Marine Paints (UK) Ltd	AVA AVP	HSE 3285
TFA-10 Light/Dark	Chugoku Marine Paints (UK) Ltd	AVA AVP	HSE 3286
TFA-20	Chugoku Marine Paints (UK) Ltd	AVA AVP	HSE 3284

684 DICHLOROPHENYL DIMETHYLUREA and ZINC OXIDE and CUPROUS THIOCYANATE

Hempel's Hard Racing 7638A	Hempel Paints Ltd	AVA AVP	HSE 5539
Hempel's Mille Alu 71602	Hempel Paints Ltd	AVA AVP	HSE 5578
Hempel's Mille Dynamic 7160A	Hempel Paints Ltd	AVA AVP	HSE 5575

685 DICHLOROPHENYL DIMETHYLUREA and ZINC OXIDE and ZINC PYRITHIONE and 2-(THIOCYANOMETHYLTHIO) BENZOTHIAZOLE and 2-METHYLTHIO-4-TERTIARY-BUTYLAMINO-6-CYCLOPROPYLAMINO-S-TRIAZINE

A2 Antifouling	Nautix SA	AVA AVP	HSE 4832
A2 Teflon Antifouling	Nautix SA	AVA AVP	HSE 4833

686 DICHLOROPHENYL DIMETHYLUREA and ZINC OXIDE and ZINC PYRITHIONE and 2-METHYLTHIO-4-TERTIARY-BUTYLAMINO-6-CYCLOPROPYLAMINO-S-TRIAZINE

A7 Teflon Antifouling	Nautix SA	AVA AVP	HSE 4945
A7 Teflon Antifouling 072019	Nautix SA	AVA AVP	HSE 5228

687 DICHLOROPHENYL DIMETHYLUREA and ZINC PYRITHIONE and 2-METHYLTHIO-4-TERTIARY-BUTYLAMINO-6-CYCLOPROPYLAMINO-S-TRIAZINE

A6 Antifouling	Nautix SA	AVA AVP	HSE 4944

688 FOLPET and CUPROUS OXIDE

# Leader	International Paint Plc	AVA AVP	HSE 3396

ANTIFOULING PRODUCTS

Product Name	Marketing Company	Use	Reg. No

689 MANEB and TRIBUTYLTIN MESO-DIBROMOSUCCINATE and TRIBUTYLTIN METHACRYLATE and TRIBUTYLTIN OXIDE and ZINC OXIDE and CUPROUS THIOCYANATE

# Marine Gold DX-1	Chugoku Marine Paints (UK) Ltd	AVP	HSE 3281

690 MANEB and TRIBUTYLTIN METHACRYLATE and ZINC OXIDE and CUPROUS OXIDE

Nu Wave A/F Flat Bottom	Kansai Paint Co Ltd	AVP	HSE 3447
Nu Wave A/F Vertical Bottom	Kansai Paint Co Ltd	AVP	HSE 3446
Rabamarine A/F No 2500 HS	Kansai Paint Co Ltd	AVP	HSE 3444
Rabamarine A/F No 2500M HS	Kansai Paint Co Ltd	AVP	HSE 3445

691 METHYLENE BIS (THIOCYANATE)

* Light Alloy	International Paint Ltd	AVA AVP	HSE 3397

692 NO RECOGNISED ACTIVE INGREDIENT

Bioclean	Chugoku Marine Paints (UK) Ltd	AVA AVP	HSE 3274
Bioclean DX	Camrex Chugoku Ltd	AQA AVA AVP	HSE 5317
Bioclean DX	Chugoku Marine Paints (UK) Ltd	AVA AVP	HSE 3273
# Chemflake	Chemflake International Ltd	AQA AVA	HSE 3246
* Chemglide	Chemflake International Ltd	AQA AVA AVP	HSE 3248
# Chemsilk	Chemflake International Ltd	AQA AVA AVP	HSE 3247
# Flex-O-Flake	Chemflake International Ltd	AQA AVA	HSE 3249
Interclene AQ HZA700 Series (Base)	International Paint Ltd	AQA AVA AVP	HSE 4765
Intersleek BXA 810/820	International Paint Ltd	AVA AVP	HSE 3403
Intersleek BXA560 Series (Base)	International Paint Ltd	AVA AVP	HSE 4785
Intersleek BXA580 Series (Base)	International Paint Ltd	AVA AVP	HSE 4786
Intersleek FCS HKA560 Series (Base)	International Paint Ltd	AQA AVA AVP	HSE 4767
Intersleek FCS HKA580 Series (Base)	International Paint Ltd	AQA AVA AVP	HSE 4766

ANTIFOULING PRODUCTS

Product Name	Marketing Company	Use	Reg. No

693 OXYTETRACYCLINE HYDROCHLORIDE and CUPROUS OXIDE

# Cobra	Adamarine Coatings	AVA AVP	HSE 3200

694 SODIUM HYPOCHLORITE

* Aqua-Tech Antifouling Berth	Aqua-Tech Marine Ltd	AVA AVP	HSE 3991

695 THIRAM and 2-METHYLTHIO-4-TERTIARY-BUTYLAMINO-6-CYCLOPROPYLAMINO-S-TRIAZINE

Broads Freshwater Antifouling	Blakes Marine Paints Ltd	AVA AVP	HSE 3221
Lynx Metal Free Antifouling	Blakes Marine Paints Ltd	AVA AVP	HSE 3555
Trawler Tin-free	Blakes Marine Paints Ltd	AVA AVP	HSE 3235

696 THIRAM and CUPROUS OXIDE

Envoy TF 300	W and J Leigh and Company	AVA AVP	HSE 3952

697 THIRAM and TRIBUTYLTIN METHACRYLATE and TRIBUTYLTIN OXIDE

# Commercial Copolymer Antifouling	Blakes Marine Paints Ltd	AVP	HSE 3224

698 THIRAM and TRIBUTYLTIN METHACRYLATE and TRIBUTYLTIN OXIDE and 2-METHYLTHIO-4-TERTIARY-BUTYLAMINO-6-CYCLOPROPYLAMINO-S-TRIAZINE

# Commercial Copolymer Extra Antifouling	Blakes Marine Paints Ltd	AVP	HSE 3225

699 THIRAM and TRIBUTYLTIN OLEATE and TRIBUTYLTIN TETRACHLOROPHTHALATE and TRIPHENYLTIN FLUORIDE

* Tropical Antifouling (Old Version)	Blakes Marine Paints Ltd	AVP	HSE 3233

700 THIRAM and TRIBUTYLTIN OXIDE and TRIPHENYLTIN FLUORIDE

# Tropical Antifouling (New Version)	Blakes Marine Paints Ltd	AVP	HSE 3234

701 TRIBUTYLTIN ACRYLATE

* Anti-Fouling Coating ARE3689	Ministry of Defence	AVP	HSE 3478

Product Name	Marketing Company	Use	Reg. No

702 **TRIBUTYLTIN ACRYLATE and TRIBUTYLTIN FLUORIDE**

* Grassline ABL Anti-Fouling Type M127	W and J Leigh and Company	AVP	HSE 3467
* Grassline ABL Anti-Fouling Type M190	W and J Leigh and Company	AVP	HSE 3466
* Grassline ABL Antifouling Type M129	W and J Leigh and Company	AVP	HSE 3468

703 **TRIBUTYLTIN ACRYLATE and TRIBUTYLTIN OXIDE and CUPROUS OXIDE**

* Grassline ABL Anti-Fouling Type M336	W and J Leigh and Company	AVP	HSE 3463
* Grassline ABL Anti-Fouling Type M349	W and J Leigh and Company	AVP	HSE 3464
Grassline ABL Anti-Fouling Type M349 HS	W and J Leigh and Company	AVP	HSE 3465
* Grassline ABL Anti-fouling Type M398	W and J Leigh and Company	AVP	HSE 3469

704 **TRIBUTYLTIN ACRYLATE and ZINC OXIDE and CUPROUS OXIDE**

Amercoat 697	Ameron BV	AVP	HSE 3513
Amercoat 698 HS	Ameron BV	AVP	HSE 3214

705 **TRIBUTYLTIN FLUORIDE**

* Envoy L840 Anti-Fouling (BSC 356)	W and J Leigh and Company	AVP	HSE 3459
* Hempel's Excelsior CR Antifouling 7643	Hempel Paints Ltd	AVP	HSE 3369

706 **TRIBUTYLTIN FLUORIDE and COPPER RESINATE and CUPROUS OXIDE**

* Lily Antifouling Super Tropical	Sigma Coatings BV	AVP	HSE 3485

707 **TRIBUTYLTIN FLUORIDE and CUPROUS OXIDE**

* Chlorinated Rubber Antifouling AF19	Desoto Titanine Plc	AVP	HSE 3296
* Devchlor Chlorinated Rubber Antifouling	Devoe Coatings BV	AVP	HSE 3314
* Florida Antifouling AF13	Desoto Titanine Plc	AVP	HSE 3299
* Grassline XL Anti-Fouling Type K968B	W and J Leigh and Company	AVP	HSE 3454

ANTIFOULING PRODUCTS

Product Name	Marketing Company	Use	Reg. No

707 TRIBUTYLTIN FLUORIDE and CUPROUS OXIDE—continued

* Grassline XLS Anti-Fouling Type M299	W and J Leigh and Company	AVP	HSE 3456
* Grassline XLS Antifouling Type M299	W and J Leigh and Company	AVP	HSE 3457
* Supertropical Antifouling AF5	Desoto Titanine Plc	AVP	HSE 3304
# Tropical Antifouling AF4	Desoto Titanine Plc	AVP	HSE 3305
* Vinyl Antifouling AF12	Desoto Titanine Plc	AVP	HSE 3306

708 TRIBUTYLTIN FLUORIDE and CUPROUS THIOCYANATE

* Kamome FRP-DC 1	Chugoku Marine Paints (UK) Ltd	AVP	HSE 3287

709 TRIBUTYLTIN FLUORIDE and TRIBUTYLTIN ACRYLATE

* Grassline ABL Anti-Fouling Type M127	W and J Leigh and Company	AVP	HSE 3467
* Grassline ABL Anti-Fouling Type M190	W and J Leigh and Company	AVP	HSE 3466
* Grassline ABL Antifouling Type M129	W and J Leigh and Company	AVP	HSE 3468

710 TRIBUTYLTIN FLUORIDE and TRIBUTYLTIN METHACRYLATE and ZINC OXIDE and DICHLOROPHENYL DIMETHYLUREA

* Marine Gold DX White-1	Chugoku Marine Paints (UK) Ltd	AVP	HSE 3278

711 TRIBUTYLTIN FLUORIDE and TRIBUTYLTIN OXIDE and CUPROUS OXIDE

* Hempel's Antifouling Classic 7655	Hempel Paints Ltd	AVP	HSE 3346
# Hempel's Antifouling Forte 7620	Hempel Paints Ltd	AVP	HSE 3349

712 TRIBUTYLTIN FLUORIDE and TRIBUTYLTIN OXIDE and CUPROUS OXIDE and DICHLOROPHENYL DIMETHYLUREA

# Hempel's Antifouling Hi Build 7600	Hempel Paints Ltd	AVP	HSE 3350

ANTIFOULING PRODUCTS

Product Name	Marketing Company	Use	Reg. No

713 TRIBUTYLTIN FLUORIDE and TRIBUTYLTIN OXIDE and CUPROUS THIOCYANATE

* Hempel's Antifouling 7650	Hempel Paints Ltd	AVP	HSE 3341
* Hempel's Antifouling 7659	Hempel Paints Ltd	AVP	HSE 3342

714 TRIBUTYLTIN FLUORIDE and TRIBUTYLTIN OXIDE and ZINC OXIDE and CUPROUS OXIDE

* Hempel's Antifouling Olympic 7155	Hempel Paints Ltd	AVP	HSE 3719

715 TRIBUTYLTIN FLUORIDE and ZINC NAPHTHENATE and CUPROUS OXIDE

* Teamac Killa	Teal and Mackrill Ltd	AVP	HSE 3492

716 TRIBUTYLTIN FLUORIDE and ZINC OXIDE and COPPER RESINATE and CUPROUS OXIDE

* Lily Antifouling IV	Sigma Coatings BV	AVP	HSE 3484
* Sigma Antifouling CR	Sigma Coatings BV	AVP	HSE 3483
* Sigma Antifouling IV	Sigma Coatings BV	AVP	HSE 4347

717 TRIBUTYLTIN FLUORIDE and ZINC OXIDE and CUPROUS OXIDE

(ATMC 123) Amercoat 557	Ameron (UK) Ltd	AVP	HSE 3207
* Antifouling Seven Seas	Jotun-Henry Clark Ltd	AVP	HSE 3405

718 TRIBUTYLTIN FLUORIDE and ZINC OXIDE and ZINEB and COPPER RESINATE and CUPROUS OXIDE

* Sigma Pilot Antifouling LL	Sigma Coatings BV	AVA	HSE 3481

719 TRIBUTYLTIN FLUORIDE and ZINC OXIDE and ZINEB and CUPROUS OXIDE

* Colturiet TCN Antifouling (Set and N Set)	Sigma Coatings BV	AVP	HSE 3488

720 TRIBUTYLTIN MESO-DIBROMOSUCCINATE

* Kamome FRP-60 1	Chugoku Marine Paints (UK) Ltd	AVP	HSE 3288

ANTIFOULING PRODUCTS

Product Name	Marketing Company	Use	Reg. No

721 **TRIBUTYLTIN MESO-DIBROMOSUCCINATE and TRIBUTYLTIN METHACRYLATE and TRIBUTYLTIN OXIDE and ZINC OXIDE and CUPROUS THIOCYANATE and MANEB**

# Marine Gold DX-1	Chugoku Marine Paints (UK) Ltd	AVP	HSE 3281

722 **TRIBUTYLTIN MESO-DIBROMOSUCCINATE and TRIBUTYLTIN OXIDE and ZINC OXIDE and CUPROUS THIOCYANATE**

# Kamome Colour 60-1	Chugoku Marine Paints (UK) Ltd	AVP	HSE 3277

723 **TRIBUTYLTIN METHACRYLATE and CUPROUS OXIDE**

* AF Seaflo Z-100 LE-2 Light/Dark	Chugoku Marine Paints (UK) Ltd	AVP	HSE 3258
* AF Seaflo Z-100-2 Light/Dark	Chugoku Marine Paints (UK) Ltd	AVP	HSE 3252
AF Seaflo Z-200 C	Chugoku Marine Paints (UK) Ltd	AVP	HSE 3256
Devran MCP Antifouling Red	Devoe Coatings BV	AVP	HSE 3311
Devran MCP Antifouling Red/Brown	Devoe Coatings BV	AVP	HSE 3310
Takata LLL Antifouling	NOF Corporation	AVP	HSE 4050
Takata LLL Antifouling Hi-Solid	NOF Corporation	AVP	HSE 4052
Takata LLL Antifouling LS	NOF Corporation	AVP	HSE 4051
Takata LLL Antifouling LS Hi-Solid	NOF Corporation	AVP	HSE 4053
Takata LLL Antifouling No 2001	NOF Corporation	AVP	HSE 4054

724 **TRIBUTYLTIN METHACRYLATE and CUPROUS THIOCYANATE**

* AF Seaflo Z-200	Chugoku Marine Paints (UK) Ltd	AVP	HSE 3255
Norden Non-Stop Black	Jotun-Henry Clark Ltd	AVP	HSE 3433
Norden Non-Stop Blue	Jotun-Henry Clark Ltd	AVP	HSE 3434
Norden Non-Stop Red	Jotun-Henry Clark Ltd	AVP	HSE 3435
# Norden Non-Stop White	Jotun-Henry Clark Ltd	AVP	HSE 3436

725 **TRIBUTYLTIN METHACRYLATE and TRIBUTYLTIN OXIDE**

* Hempel's Antifouling Clear 0777	Hempel Paints Ltd	AVP	HSE 3324

ANTIFOULING PRODUCTS

Product Name	Marketing Company	Use	Reg. No

726 TRIBUTYLTIN METHACRYLATE and TRIBUTYLTIN OXIDE and 2-METHYLTHIO-4-TERTIARY-BUTYLAMINO-6-CYCLOPROPYLAMINO-S-TRIAZINE and THIRAM

Product Name	Marketing Company	Use	Reg. No
# Commercial Copolymer Extra Antifouling	Blakes Marine Paints Ltd	AVP	HSE 3225

727 TRIBUTYLTIN METHACRYLATE and TRIBUTYLTIN OXIDE and 4,5-DICHLORO-2-N-OCTYL-4-ISOTHIAZOLIN-3-ONE and CUPROUS OXIDE

Product Name	Marketing Company	Use	Reg. No
Intersmooth Hisol BF0270/950/970 Series	International Paint Ltd	AVP	HSE 5641

728 TRIBUTYLTIN METHACRYLATE and TRIBUTYLTIN OXIDE and COPPER METAL

Product Name	Marketing Company	Use	Reg. No
# Hempel's Antifouling Mille Copper 7671	Hempel Paints Ltd	AVP	HSE 3354

729 TRIBUTYLTIN METHACRYLATE and TRIBUTYLTIN OXIDE and CUPROUS OXIDE

Product Name	Marketing Company	Use	Reg. No
AF Seaflo Z-100 HS-1	Camrex Chugoku Ltd	AVP	HSE 5318
AF Seaflo Z-100 HS-1 Light/Dark	Chugoku Marine Paints (UK) Ltd	AVP	HSE 3253
AF Seaflo Z-100 LE-HS-1	Camrex Chugoku Ltd	AVP	HSE 5313
AF Seaflo Z-100 LE-HS-1 Light/Dark	Chugoku Marine Paints (UK) Ltd	AVP	HSE 3259
# AF Seaflo Z-100-1 Light/Dark	Chugoku Marine Paints (UK) Ltd	AVP	HSE 3251
# Antifouling Seaflex	Jotun-Henry Clark Ltd	AVP	HSE 3428
Antifouling Seamate HB 66 Black	Jotun-Henry Clark Ltd	AVP	HSE 3409
Antifouling Seamate HB 99 Black	Jotun-Henry Clark Ltd	AVP	HSE 3410
Antifouling Seamate HB 99 Dark Red	Jotun-Henry Clark Ltd	AVP	HSE 3412
Antifouling Seamate HB 99 Light Red	Jotun-Henry Clark Ltd	AVP	HSE 3411
AF Seaflo Z-100 LE-1 Light/Dark	Chugoku Marine Paints (UK) Ltd	AVP	HSE 3257
Hempel's Antifouling Combic 7699	Hempel Paints Ltd	AVP	HSE 3348
Hempel's Antifouling Nautic HI7691	Hempel Paints Ltd	AVP	HSE 3361
Hempel's Antifouling Nautic HI7690	Hempel Paints Ltd	AVP	HSE 3360

ANTIFOULING PRODUCTS

Product Name	Marketing Company	Use	Reg. No

729 TRIBUTYLTIN METHACRYLATE and TRIBUTYLTIN OXIDE and CUPROUS OXIDE—continued

Intersmooth SPC Antifouling BFO250 Series	International Paint Ltd	AVP	HSE 5635
Intersmooth SPC BFA090/BFA190 Series	International Paint Ltd	AVP	HSE 3377

730 TRIBUTYLTIN METHACRYLATE and TRIBUTYLTIN OXIDE and CUPROUS OXIDE and CUPROUS THIOCYANATE

# Hempel's Antifouling Mille 7678	Hempel Paints Ltd	AVP	HSE 3353

731 TRIBUTYLTIN METHACRYLATE and TRIBUTYLTIN OXIDE and CUPROUS OXIDE and DICHLOROPHENYL DIMETHYLUREA

Interswift BK0000/700 Series	International Paint Ltd	AVP	HSE 5640

732 TRIBUTYLTIN METHACRYLATE and TRIBUTYLTIN OXIDE and CUPROUS THIOCYANATE

Antifouling HB 66 Ocean Green	Jotun-Henry Clark Ltd	AVP	HSE 3408
* Antifouling Seaflex Sealinc Blue 110	Jotun-Henry Clark Ltd	AVP	HSE 3406
Antifouling Seamate HB Blue	Jotun-Henry Clark Ltd	AVP	HSE 3429
Antifouling Seamate HB Green	Jotun-Henry Clark Ltd	AVP	HSE 3430
# Antifouling Seamate HB Special Black	Jotun-Henry Clark Ltd	AVP	HSE 3407
Intersmooth Hisol SPC Antifouling BFA949 Red	International Paint Ltd	AVP	HSE 4949
Intersmooth SPC BFA040/BFA050 Series	International Paint Ltd	AVP	HSE 3376

733 TRIBUTYLTIN METHACRYLATE and TRIBUTYLTIN OXIDE and THIRAM

# Commercial Copolymer Antifouling	Blakes Marine Paints Ltd	AVP	HSE 3224

734 TRIBUTYLTIN METHACRYLATE and TRIBUTYLTIN OXIDE and TRIPHENYLTIN FLUORIDE

# Hempel's Antifouling Nautic 7680	Hempel Paints Ltd	AVP	HSE 3359

ANTIFOULING PRODUCTS

Product Name	Marketing Company	Use	Reg. No
735 **TRIBUTYLTIN METHACRYLATE and TRIBUTYLTIN OXIDE and TRIPHENYLTIN FLUORIDE and CUPROUS OXIDE**			
* A1 Antifouling	Nautix SA	AVP	HSE 4366
# Hai-Hong's Antifouling Nautic HI-7695	Hai-Hong Marine Paint	AVP	HSE 3554
# Hempel's Antifouling Mille MS7672	Hempel Paints Ltd	AVP	HSE 3356
# Hempel's Antifouling Nautic 7673	Hempel Paints Ltd	AVP	HSE 3357
# Hempel's Antifouling Nautic 7674	Hempel Paints Ltd	AVP	HSE 3358
* Hempel's Antifouling Nautic HI7695	Hempel Paints Ltd	AVP	HSE 3362
736 **TRIBUTYLTIN METHACRYLATE and TRIBUTYLTIN OXIDE and TRIPHENYLTIN FLUORIDE and CUPROUS THIOCYANATE**			
# Hempel's Antifouling Mille 7670	Hempel Paints Ltd	AVP	HSE 3352
737 **TRIBUTYLTIN METHACRYLATE and TRIBUTYLTIN OXIDE and ZINC OXIDE and CUPROUS OXIDE**			
# Ablative Antifouling AF28	Desoto Titanine Plc	AVP	HSE 3505
AF Seaflo Mark 2-1	Camrex Chugoku Ltd	AVP	HSE 5316
AF Seaflo Mark 2-1 Light/Dark	Chugoku Marine Paints (UK) Ltd	AVP	HSE 3261
# Antifouling Seaconomy	Jotun-Henry Clark Ltd	AVP	HSE 3438
Antifouling Seaconomy 200	Jotun-Henry Clark Ltd	AVP	HSE 4436
Antifouling Seaconomy 300	Jotun-Henry Clark Ltd	AVP	HSE 4272
# Antifouling Seaflex	Jotun-Henry Clark Ltd	AVP	HSE 3439
Antifouling Seamate HB 22	Jotun-Henry Clark Ltd	AVP	HSE 4271
Antifouling Seamate HB 33	Jotun-Henry Clark Ltd	AVP	HSE 4270
# Antifouling Seamate HB 33 DR	Jotun-Henry Clark Ltd	AVP	HSE 3440
# Antifouling Seamate HB 33 LR	Jotun-Henry Clark Ltd	AVP	HSE 3441
* Antifouling Seamate HB 66 Dark Red	Jotun-Henry Clark Ltd	AVP	HSE 3442
* Antifouling Seamate HB 66 Light Red	Jotun-Henry Clark Ltd	AVP	HSE 3443
Antifouling Seamate HB66	Jotun-Henry Clark Ltd	AVP	HSE 5698
# Camrex C-Clean EXA	Camrex Ltd	AVP	HSE 3240
# Camrex C-Clean HB EXA	Camrex Ltd	AVP	HSE 3241
# Camrex C-Clean HB LA	Camrex Ltd	AVP	HSE 3243

Product Name	Marketing Company	Use	Reg. No

737 TRIBUTYLTIN METHACRYLATE and TRIBUTYLTIN OXIDE and ZINC OXIDE and CUPROUS OXIDE—continued

# Camrex C-Clean LA	Camrex Ltd	AVP	HSE 3242
Grassline ABL Antifouling Type M349	W and J Leigh and Company	AVP	HSE 5374

738 TRIBUTYLTIN METHACRYLATE and TRIBUTYLTIN OXIDE and ZINC OXIDE and CUPROUS OXIDE and DICHLOROPHENYL DIMETHYLUREA

A11 Antifouling 072014	Nautix SA	AVP	HSE 5376

739 TRIBUTYLTIN METHACRYLATE and TRIBUTYLTIN OXIDE and ZINC OXIDE and CUPROUS THIOCYANATE

Antifouling Seamate HB 33 BSL Blue	Jotun-Henry Clark Ltd	AVP	HSE 4269
Antifouling Seamate HB22 Roundel Blue	Jotun-Henry Clark Ltd	AVP	HSE 4872
Antifouling Seamate HB66 Black	Jotun-Henry Clark Ltd	AVP	HSE 4871

740 TRIBUTYLTIN METHACRYLATE and TRIBUTYLTIN OXIDE and ZINC OXIDE and CUPROUS THIOCYANATE and MANEB and TRIBUTYLTIN MESO-DIBROMOSUCCINATE

# Marine Gold DX-1	Chugoku Marine Paints (UK) Ltd	AVP	HSE 3281

741 TRIBUTYLIN METHACRYLATE and TRIBUTYLTIN OXIDE and ZINC OXIDE and ZINEB and CUPROUS OXIDE

Hempel's Antifouling Combic 7699B	Hempel Paints Ltd	AVP	HSE 5581
Hempel's Antifouling Nautic 7690B	Hempel Paints Ltd	AVP	HSE 5580
Hempel's Antifouling Nautic 7691B	Hempel Paints Ltd	AVP	HSE 5586

742 TRIBUTYLTIN METHACRYLATE and TRIBUTYLTIN OXIDE and ZINEB and CUPROUS OXIDE

Blueline SBA900 Series	International Paint Ltd	AVP	HSE 5138
Intersmooth Hisol 2000 BFA270 Series	International Paint Ltd	AVP	HSE 3844
Intersmooth Hisol 9000 BFA970 Series	International Paint Plc	AVP	HSE 5461

Product Name	Marketing Company	Use	Reg. No

742 TRIBUTYLTIN METHACRYLATE and TRIBUTYLTIN OXIDE and ZINEB and CUPROUS OXIDE—continued

Product Name	Marketing Company	Use	Reg. No
Intersmooth Hisol BFA250/BFA900 Series	International Paint Ltd	AVP	HSE 3848
# Intersmooth Hisol BFA250/BFA900 Series	International Paint Ltd	AVP	HSE 3378
Interswift BKA000/700 Series	International Paint Ltd	AVP	HSE 3384

743 TRIBUTYLTIN METHACRYLATE and TRIBUTYLTIN OXIDE and ZINEB and CUPROUS THIOCYANATE

Product Name	Marketing Company	Use	Reg. No
Intersmooth Hisol BFA948 Orange	International Paint Ltd	AVP	HSE 4281
Micron 25 Plus	International Paint Ltd	AVP	HSE 3402
Superyacht Antifouling	International Paint Plc	AVP	HSE 5462

744 TRIBUTYLTIN METHACRYLATE and TRIPHENYLTIN FLUORIDE

Product Name	Marketing Company	Use	Reg. No
* Antifouling Alusea Black	Jotun-Henry Clark Ltd	AVP	HSE 3424
* Antifouling Alusea Blue	Jotun-Henry Clark Ltd	AVP	HSE 3425
* Antifouling Alusea Red	Jotun-Henry Clark Ltd	AVP	HSE 3426
* Antifouling Alusea White	Jotun-Henry Clark Ltd	AVP	HSE 3427

745 TRIBUTYLTIN METHACRYLATE and ZINC METAL and CUPROUS OXIDE

Product Name	Marketing Company	Use	Reg. No
* AF Seaflo Z-100 HS-2 Light/Dark	Chugoku Marine Paints (UK) Ltd	AVP	HSE 3254
* AF Seaflo Z-100 LE-HS-2 Light/Dark	Chugoku Marine Paints (UK) Ltd	AVP	HSE 3260

746 TRIBUTYLTIN METHACRYLATE and ZINC METAL and ZINC OXIDE and CUPROUS OXIDE

Product Name	Marketing Company	Use	Reg. No
* AF Seaflo Mark 2-2 Light/Dark	Chugoku Marine Paints (UK) Ltd	AVP	HSE 3262
* AF Seaflo Mark 3 Light/Dark	Chugoku Marine Paints (UK) Ltd	AVP	HSE 3263

747 TRIBUTYLTIN METHACRYLATE and ZINC METAL and ZINC OXIDE and CUPROUS THIOCYANATE

Product Name	Marketing Company	Use	Reg. No
* Marine Gold DX White-2	Chugoku Marine Paints (UK) Ltd	AVP	HSE 3279

ANTIFOULING PRODUCTS

Product Name	Marketing Company	Use	Reg. No
748 TRIBUTYLTIN METHACRYLATE and ZINC OXIDE and 2-METHYLTHIO-4-TERTIARY-BUTYLAMINO-6-CYCLOPROPYLAMINO-S-TRIAZINE			
Antifouling Alusea	Jotun-Henry Clark Ltd	AVP	HSE 4569
749 TRIBUTYLTIN METHACRYLATE and ZINC OXIDE and CUPROUS OXIDE and MANEB			
Nu Wave A/F Flat Bottom	Kansai Paint Co Ltd	AVP	HSE 3447
Nu Wave A/F Vertical Bottom	Kansai Paint Co Ltd	AVP	HSE 3446
Rabamarine A/F No 2500 HS	Kansai Paint Co Ltd	AVP	HSE 3444
Rabamarine A/F No 2500M HS	Kansai Paint Co Ltd	AVP	HSE 3445
750 TRIBUTYLTIN METHACRYLATE and ZINC OXIDE and CUPROUS THIOCYANATE			
* Marine Gold DX-2	Chugoku Marine Paints (UK) Ltd	AVP	HSE 3280
751 TRIBUTYLTIN METHACRYLATE and ZINC OXIDE and DICHLOROPHENYL DIMETHYLUREA and TRIBUTYLTIN FLUORIDE			
* Marine Gold DX White-1	Chugoku Marine Paints (UK) Ltd	AVP	HSE 3278
752 TRIBUTYLTIN METHACRYLATE and ZINC OXIDE and ZINEB and CUPROUS OXIDE			
Antifouling Seamate HB22	Jotun-Henry Clark Ltd	AVP	HSE 5363
Antifouling Seamate HB33	Jotun-Henry Clark Ltd	AVP	HSE 5362
Sigmaplane HA Antifouling	Sigma Coatings BV	AVP	HSE 4345
Sigmaplane HB	Sigma Coatings BV	AVP	HSE 3487
753 TRIBUTYLTIN METHACRYLATE and ZINEB and CUPROUS OXIDE			
# Sigma Pilot Antifouling TA	Sigma Coatings BV	AVP	HSE 3482
Sigmaplane TA Antifouling	Sigma Coatings BV	AVP	HSE 4346
# Sigmathrift Antifouling	Sigma Coatings BV	AVP	HSE 3486
754 TRIBUTYLTIN METHACRYLATE and ZINEB and CUPROUS THIOCYANATE			
Superyacht 800 Antifouling	International Paint Ltd	AVP	HSE 4778

ANTIFOULING PRODUCTS

Product Name	Marketing Company	Use	Reg. No

755 TRIBUTYLTIN METHACRYLATE and ZIRAM and CUPROUS THIOCYANATE

Takata Seaqueen	Nippon Oil and Fats Co Ltd	AVP	HSE 4055

756 TRIBUTYLTIN OLEATE and TRIBUTYLTIN TETRACHLOROPHTHALATE and TRIPHENYLTIN FLUORIDE and THIRAM

* Tropical Antifouling (Old Version)	Blakes Marine Paints Ltd	AVP	HSE 3233

757 TRIBUTYLTIN OXIDE and 2-METHYLTHIO-4-TERTIARY-BUTYLAMINO-6-CYCLOPROPYLAMINO-S-TRIAZINE and THIRAM and TRIBUTYLTIN METHACRYLATE

# Commercial Copolymer Extra Antifouling	Blakes Marine Paints Ltd	AVP	HSE 3225

758 TRIBUTYLTIN OXIDE and 4,5,-DICHLORO-2-N-OCTYL-4-ISOTHIAZOLIN-3-ONE and CUPROUS OXIDE and TRIBUTYLTIN METHACRYLATE

Intersmooth Hisol BF0270/950/970 Series	International Paint Ltd	AVP	HSE 5641

759 TRIBUTYLTIN OXIDE and COPPER METAL and TRIBUTYLTIN METHACRYLATE

# Hempel's Antifouling Mille Copper 7671	Hempel Paints Ltd	AVP	HSE 3354

760 TRIBUTYLTIN OXIDE and CUPROUS OXIDE

# Camrex A/F 3	Camrex Ltd	AVP	HSE 3238
# Camrex A/F 5	Camrex Ltd	AVP	HSE 3239
* Devchlor Antifouling Paint Red	Devoe Coatings BV	AVP	HSE 3309
* Devran 222 All Seasons Permanent	Devoe Coatings BV	AVP	HSE 3312
* Hempel's Antifouling Bravo 7610	Hempel Paints Ltd	AVP	HSE 3343
* Hempel's Antifouling Classic 7611	Hempel Paints Ltd	AVP	HSE 3344
* Hempel's Antifouling Classic 7633	Hempel Paints Ltd	AVP	HSE 3345
# Hempel's Antifouling Tropic 7644	Hempel Paints Ltd	AVP	HSE 3366
# High Performance Antifouling Paint	Spencer (Aberdeen) Plc	AVP	HSE 4157
* Interclene Extra BCA500 Series	International Paint Ltd	AVP	HSE 3374
* Interclene Super BCA400 Series	International Paint Ltd	AVP	HSE 3373
* Interspeed Extra BLA200 Series	International Paint Ltd	AVP	HSE 3381

ANTIFOULING PRODUCTS

Product Name	Marketing Company	Use	Reg. No
760 TRIBUTYLTIN OXIDE and CUPROUS OXIDE—continued			
* Interspeed Extra Red BJA450 Series	International Paint Ltd	AVP	HSE 3379
* Interspeed System 2 BRA140 Series	International Paint Ltd	AVP	HSE 3382
# Mebon Longlife Anti-Fouling (T) 6-2-09	Mebon Paints Ltd	AVP	HSE 3472
# Mebon Longlife Anti-Fouling 6-2-16	Mebon Paints Ltd	AVP	HSE 3473
# Mebon Longlife Anti-Fouling 6-2-16	Mebon Paints Ltd	AVP	HSE 3474
# Norden Kobberstoff Green	Jotun-Henry Clark Ltd	AVP	HSE 3432
761 TRIBUTYLTIN OXIDE and CUPROUS OXIDE and CUPROUS THIOCYANATE and TRIBUTYLTIN METHACRYLATE			
# Hempel's Antifouling Mille 7678	Hempel Paints Ltd	AVP	HSE 3353
762 TRIBUTYLTIN OXIDE and CUPROUS OXIDE and DICHLOROPHENYL DIMETHYLUREA and TRIBUTYLTIN FLUORIDE			
# Hempel's Antifouling Hi Build 7600	Hempel Paints Ltd	AVP	HSE 3350
763 TRIBUTYLTIN OXIDE and CUPROUS OXIDE and DICHLOROPHENYL DIMETHYLUREA and TRIBUTYLTIN METHACRYLATE			
Interswift BK0000/700 Series	International Paint Ltd	AVP	HSE 5640
764 TRIBUTYLTIN OXIDE and CUPROUS OXIDE and TRIBUTYLTIN ACRYLATE			
* Grassline ABL Anti-Fouling Type M336	W and J Leigh and Company	AVP	HSE 3463
* Grassline ABL Anti-Fouling Type M349	W and J Leigh and Company	AVP	HSE 3464
Grassline ABL Anti-Fouling Type M349 HS	W and J Leigh and Company	AVP	HSE 3465
* Grassline ABL Anti-Fouling Type M398	W and J Leigh and Company	AVP	HSE 3469
765 TRIBUTYLTIN OXIDE and CUPROUS OXIDE and TRIBUTYLTIN FLUORIDE			
* Hempel's Antifouling Classic 7655	Hempel Paints Ltd	AVP	HSE 3346
# Hempel's Antifouling Forte 7620	Hempel Paints Ltd	AVP	HSE 3349

Product Name	Marketing Company	Use	Reg. No

766 TRIBUTYLTIN OXIDE and CUPROUS OXIDE and TRIBUTYLTIN METHACRYLATE

Product Name	Marketing Company	Use	Reg. No
AF Seaflo Z-100 HS-1	Camrex Chugoku Ltd	AVP	HSE 5318
AF Seaflo Z-100 HS-1 Light/Dark	Chugoku Marine Paints (UK) Ltd	AVP	HSE 3253
AF Seaflo Z-100 LE-HS-1	Camrex Chugoku Ltd	AVP	HSE 5313
AF Seaflo Z-100 LE-HS-1 Light/Dark	Chugoku Marine Paints (UK) Ltd	AVP	HSE 3259
# AF Seaflo Z-100-1 Light/Dark	Chugoku Marine Paints (UK) Ltd	AVP	HSE 3251
# Antifouling Seaflex	Jotun-Henry Clark Ltd	AVP	HSE 3428
Antifouling Seamate HB 66 Black	Jotun-Henry Clark Ltd	AVP	HSE 3409
Antifouling Seamate HB 99 Black	Jotun-Henry Clark Ltd	AVP	HSE 3410
Antifouling Seamate HB 99 Dark Red	Jotun-Henry Clark Ltd	AVP	HSE 3412
Antifouling Seamate HB 99 Light Red	Jotun-Henry Clark Ltd	AVP	HSE 3411
AF Seaflo Z-100 LE-1 Light/Dark	Chugoku Marine Paints (UK) Ltd	AVP	HSE 3257
Hempel's Antifouling Combic 7699	Hempel Paints Ltd	AVP	HSE 3348
Hempel's Antifouling Nautic HI 7691	Hempel Paints Ltd	AVP	HSE 3361
Hempel's Antifouling Nautic HI7690	Hempel Paints Ltd	AVP	HSE 3360
Intesmooth SPC Antifouling BF0250 Series	International Paint Ltd	AVP	HSE 5635
Intersmooth SPC BFA090/BFA190 Series	International Paint Ltd	AVP	HSE 3377

767 TRIBUTYLTIN OXIDE and CUPROUS THIOCYANATE

Product Name	Marketing Company	Use	Reg. No
# Norden Durahart	Jotun-Henry Clark Ltd	AVP	HSE 3510

768 TRIBUTYLTIN OXIDE and CUPROUS THIOCYANATE and TRIBUTYLTIN FLORIDE

Product Name	Marketing Company	Use	Reg. No
* Hempel's Antifouling 7650	Hempel Paints Ltd	AVP	HSE 3341
* Hempel's Antifouling 7659	Hempel Paints Ltd	AVP	HSE 3342

ANTIFOULING PRODUCTS

Product Name	Marketing Company	Use	Reg. No
769 TRIBUTYLTIN OXIDE and CUPROUS THIOCYANATE and TRIBUTYLTIN METHACRYLATE			
Antifouling HB 66 Ocean Green	Jotun-Henry Clark Ltd	AVP	HSE 3408
* Antifouling Seaflex Sealinc Blue 110	Jotun-Henry Clark Ltd	AVP	HSE 3406
Antifouling Seamate HB Blue	Jotun-Henry Clark Ltd	AVP	HSE 3429
Antifouling Seamate HB Green	Jotun-Henry Clark Ltd	AVP	HSE 3430
# Antifouling Seamate HB Special Black	Jotun-Henry Clark Ltd	AVP	HSE 3407
Intersmooth Hisol SPC Antifouling BFA949 Red	International Paint Ltd	AVP	HSE 4949
Intersmooth SPC BFA040/BFA050 Series	International Paint Ltd	AVP	HSE 3376
770 TRIBUTYLTIN OXIDE and THIRAM and TRIBUTYLTIN METHACRYLATE			
# Commercial Copolymer Antifouling	Blakes Marine Paints Ltd	AVP	HSE 3224
771 TRIBUTYLTIN OXIDE and TRIBUTYLTIN METHACRYLATE			
* Hempel's Antifouling Clear 0777	Hempel Paints Ltd	AVP	HSE 3324
772 TRIBUTYLTIN OXIDE and TRIPHENYLTIN FLUORIDE			
# Hempel's Antifouling Oceanic 7640	Hempel Paints Ltd	AVP	HSE 3363
* Hempel's Hard Racing Copper Free 7649	Hempel Paints Ltd	AVP	HSE 3368
773 TRIBUTYLTIN OXIDE and TRIPHENYLTIN FLUORIDE and CUPROUS OXIDE			
# Hempel's Antifouling Classic 7677	Hempel Paints Ltd	AVP	HSE 3347
774 TRIBUTYLTIN OXIDE and TRIPHENYLTIN FLUORIDE and CUPROUS OXIDE and TRIBUTYLTIN METHACRYLATE			
* A1 Antifouling	Nautix SA	AVP	HSE 4366
# Hai-Hong's Antifouling Nautic HI-7695	Hai-Hong Marine Paint	AVP	HSE 3554
# Hempel's Antifouling Mille MS7672	Hempel Paints Ltd	AVP	HSE 3356
# Hempel's Antifouling Nautic 7673	Hempel Paints Ltd	AVP	HSE 3357
# Hempel's Antifouling Nautic 7674	Hempel Paints Ltd	AVP	HSE 3358
* Hempel's Antifouling Nautic HI7695	Hempel Paints Ltd	AVP	HSE 3362

ANTIFOULING PRODUCTS

Product Name	Marketing Company	Use	Reg. No

775 TRIBUTYLTIN OXIDE and TRIPHENYLTIN FLUORIDE and CUPROUS THIOCYANATE and TRIBUTYLTIN METHACRYLATE

# Hempel's Antifouling Mille 7670	Hempel Paints Ltd	AVP	HSE 3352

776 TRIBUTYLTIN OXIDE and TRIPHENYLTIN FLUORIDE and THIRAM

# Tropical Antifouling (New Version)	Blakes Marine Paints Ltd	AVP	HSE 3234

777 TRIBUTYLTIN OXIDE and TRIPHENYLTIN FLUORIDE and TRIBUTYLTIN METHACRYLATE

* Hempel's Antifouling Nautic 7680	Hempel Paints Ltd	AVP	HSE 3359

778 TRIBUTYLTIN OXIDE and ZINC OXIDE and CUPROUS OXIDE

# (ATMC 101) Amercoat 610	Ameron (UK) Ltd	AVP	HSE 3208
# (ATMC 102) Amercoat 611	Ameron (UK) Ltd	AVP	HSE 3209
# (ATMC 120) Amercoat 612	Ameron (UK) Ltd	AVP	HSE 3210
# (ATMC 121) Amercoat 613	Ameron (UK) Ltd	AVP	HSE 3211
# (ATMC 122) Amercoat 615	Ameron (UK) Ltd	AVP	HSE 3213
# Antifouling Equator KK	Jotun-Henry Clark Ltd	AVP	HSE 3431
# Antifouling Sargasso	Jotun-Henry Clark Ltd	AVP	HSE 3437
# Super Quality Antifouling	Spencer (Aberdeen) Plc	AVP	HSE 3491

779 TRIBUTYLTIN OXIDE and ZINC OXIDE and CUPROUS OXIDE and DICHLOROPHENYL DIMETHYLUREA and TRIBUTYLTIN METHACRYLATE

A11 Antifouling 072014	Nautix SA	AVP	HSE 5376

780 TRIBUTYLTIN OXIDE and ZINC OXIDE and CUPROUS OXIDE and TRIBUTYLTIN FLUORIDE

* Hempel's Antifouling Olympic 7155	Hempel Paints Ltd	AVP	HSE 3719

781 TRIBUTYLTIN OXIDE and ZINC OXIDE and CUPROUS OXIDE and TRIBUTYLTIN METHACRYLATE

# Ablative Antifouling AF28	Desoto Titanine Plc	AVP	HSE 3505
* AF Seaflo Mark 2-1	Camrex Chugoku Ltd	AVP	HSE 5316
AF Seaflo Mark 2-1 Light/Dark	Chugoku Marine Paints (UK) Ltd	AVP	HSE 3261
# Antifouling Seaconomy	Jotun-Henry Clark Ltd	AVP	HSE 3438
Antifouling Seaconomy 200	Jotun-Henry Clark Ltd	AVP	HSE 4436

ANTIFOULING PRODUCTS

Product Name	Marketing Company	Use	Reg. No

781 **TRIBUTYLTIN OXIDE and ZINC OXIDE and CUPROUS OXIDE and TRIBUTYLTIN METHACRYLATE**—continued

Product Name	Marketing Company	Use	Reg. No
Antifouling Seaconomy 300	Jotun-Henry Clark Ltd	AVP	HSE 4272
# Antifouling Seaflex	Jotun-Henry Clark Ltd	AVP	HSE 3439
Antifouling Seamate HB 22	Jotun-Henry Clark Ltd	AVP	HSE 4271
Antifouling Seamate HB 33	Jotun-Henry Clark Ltd	AVP	HSE 4270
# Antifouling Seamate HB 33 DR	Jotun-Henry Clark Ltd	AVP	HSE 3440
# Antifouling Seamate HB 33 LR	Jotun-Henry Clark Ltd	AVP	HSE 3441
* Antifouling Seamate HB 66 Dark Red	Jotun-Henry Clark Ltd	AVP	HSE 3442
* Antifouling Seamate HB 66 Light Red	Jotun-Henry Clark Ltd	AVP	HSE 3443
Antifouling Seamate HB66	Jotun-Henry Clark Ltd	AVP	HSE 5698
# Camrex C-Clean EXA	Camrex Ltd	AVP	HSE 3240
# Camrex C-Clean HB EXA	Camrex Ltd	AVP	HSE 3241
# Camrex C-Clean HB LA	Camrex Ltd	AVP	HSE 3243
# Camrex C-Clean LA	Camrex Ltd	AVP	HSE 3242
Grassline ABL Antifouling Type M349	W and J Leigh and Company	AVP	HSE 5374

782 **TRIBUTYLTIN OXIDE and ZINC OXIDE and CUPROUS THIOCYANATE and MANEB and TRIBUTYLTIN MESO-DIBROMOSUCCINATE and TRIBUTYLTIN METHACRYLATE**

Product Name	Marketing Company	Use	Reg. No
# Marine Gold DX-1	Chugoku Marine Paints (UK) Ltd	AVP	HSE 3281

783 **TRIBUTYLTIN OXIDE and ZINC OXIDE and CUPROUS THIOCYANATE and TRIBUTYLTIN MESO-DIBROMOSUCCINATE**

Product Name	Marketing Company	Use	Reg. No
# Kamome Colour 60-1	Chugoku Marine Paints (UK) Ltd	AVP	HSE 3277

784 **TRIBUTYLTIN OXIDE and ZINC OXIDE and CUPROUS THIOCYANATE and TRIBUTYLTIN METHACRYLATE**

Product Name	Marketing Company	Use	Reg. No
Antifouling Seamate HB 33 BSL Blue	Jotun-Henry Clark Ltd	AVP	HSE 4269
Antifouling Seamate HB22 Roundel Blue	Jotun-Henry Clark Ltd	AVP	HSE 4872
Antifouling Seamate HB66 Black	Jotun-Henry Clark Ltd	AVP	HSE 4871

ANTIFOULING PRODUCTS

Product Name	Marketing Company	Use	Reg. No

785 TRIBUTYLTIN OXIDE and ZINC OXIDE and ZINEB and CUPROUS OXIDE and TRIBUTYLTIN METHACRYLATE

Hempel's Antifouling Combic 7699B	Hempel Paints Ltd	AVP	HSE 5581
Hempel's Antifouling Nautic 7690B	Hempel Paints Ltd	AVP	HSE 5580
Hempel's Antifouling Nautic 7691B	Hempel Paints Ltd	AVP	HSE 5586

786 TRIBUTYLTIN OXIDE and ZINEB and CUPROUS OXIDE and TRIBUTYLTIN METHACRYLATE

Blueline SBA900 Series	International Paint Ltd	AVP	HSE 5138
Intersmooth Hisol 2000 BFA270 Series	International Paint Ltd	AVP	HSE 3844
Intersmooth Hisol 9000 BFA970 Series	International Paint Plc	AVP	HSE 5461
Intersmooth Hisol BFA 250/BFA900 Series	International Paint Ltd	AVP	HSE 3848
# Intersmooth Hisol BFA250/BFA900 Series	International Paint Plc	AVP	HSE 3378
Interswift BKA000/700 Series	International Paint Ltd	AVP	HSE 3384

787 TRIBUTYLTIN OXIDE and ZINEB and CUPROUS THIOCYANATE

Cruiser Copolymer	International Paint Ltd	AVP	HSE 3514

788 TRIBUTYLTIN OXIDE and ZINEB and CUPROUS THIOCYANATE and TRIBUTYLTIN METHACRYLATE

Intersmooth Hisol BFA948 Orange	International Paint Ltd	AVP	HSE 4281
Micron 25 Plus	International Paint Ltd	AVP	HSE 3402
Superyacht Antifouling	International Paint Plc	AVP	HSE 5462

789 TRIBUTYLTIN OXIDE and ZIRAM and CUPROUS OXIDE

* Devoe ABC 2 Antifouling	Devoe Coatings BV	AVP	HSE 3313

790 TRIBUTYLTIN POLYSILOXANE and ZINC METAL

* (ATMC 172) Dimetcote 602	Ameron (UK) Ltd	AVP	HSE 3217

791 TRIBUTYLTIN POLYSILOXANE and ZINC OXIDE

* (ATMC 173) Dimetcote 601	Ameron (UK) Ltd	AVP	HSE 3216

ANTIFOULING PRODUCTS

Product Name	Marketing Company	Use	Reg. No
792 TRIBUTYLTIN POLYSILOXANE and ZINC OXIDE and CUPROUS OXIDE			
* (ATMC 170) Amercoat 699	Ameron (UK) Ltd	AVP	HSE 3215
793 TRIBUTYLTIN TETRACHLOROPHTHALATE and TRIPHENYLTIN FLUORIDE and THIRAM and TRIBUTYLTIN OLEATE			
* Tropical Antifouling (Old Version)	Blakes Marine Paints Ltd	AVP	HSE 3233
794 TRIPHENYLTIN CHLORIDE and CUPROUS THIOCYANATE and DICHLOROPHENYL DIMETHYLUREA			
* Kamome FRP-DC 2	Chugoku Marine Paints (UK) Ltd	AVP	HSE 3290
795 TRIPHENYLTIN CHLORIDE and DICHLOROPHENYL DIMETHYLUREA			
* Kamome FRP-60 2	Chugoku Marine Paints (UK) Ltd	AVP	HSE 3289
796 TRIPHENYLTIN CHLORIDE and ZINC OXIDE and CUPROUS THIOCYANATE and DICHLOROPHENYL DIMETHYLUREA			
* Kamome Colour 60-2	Chugoku Marine Paints (UK) Ltd	AVP	HSE 3276
797 TRIPHENYLTIN FLUORIDE and CUPROUS OXIDE			
* Envoy K926 High Performance Anti-Fouling	W and J Leigh and Company	AVP	HSE 3453
EXL-AF	Chugoku Marine Paints (UK) Ltd	AVP	HSE 3266
EXL-AF E	Chugoku Marine Paints (UK) Ltd	AVP	HSE 3267
* Grassline EL Anti-Fouling Type L975	W and J Leigh and Company	AVP	HSE 3458
* Grassline XL3 Anti-Fouling Type L806 B	W and J Leigh and Company	AVP	HSE 3455
798 TRIPHENYLTIN FLUORIDE and CUPROUS OXIDE and TRIBUTYLTIN METHACRYLATE and TRIBUTYLTIN OXIDE			
* A1 Antifouling	Nautix SA	AVP	HSE 4366
# Hai-Hong's Antifouling Nautic HI-7695	Hai-Hong Marine Paint	AVP	HSE 3554

ANTIFOULING PRODUCTS

Product Name	Marketing Company	Use	Reg. No

798 TRIPHENYLTIN FLUORIDE and CUPROUS OXIDE and TRIBUTYLTIN METHACRYLATE and TRIBUTYLTIN OXIDE—continued

# Hempel's Antifouling Mille MS7672	Hempel Paints Ltd	AVP	HSE 3356
# Hempel's Antifouling Nautic 7673	Hempel Paints Ltd	AVP	HSE 3357
# Hempel's Antifouling Nautic 7674	Hempel Paints Ltd	AVP	HSE 3358
* Hempel's Antifouling Nautic HI7695	Hempel Paints Ltd	AVP	HSE 3362

799 TRIPHENYLTIN FLUORIDE and CUPROUS OXIDE and TRIBUTYLTIN OXIDE

# Hempel's Antifouling Classic 7677	Hempel Paints Ltd	AVP	HSE 3347

800 TRIPHENYLTIN FLUORIDE and CUPROUS THIOCYANATE and TRIBUTYLTIN METHACRYLATE and TRIBUTYLTIN OXIDE

# Hempel's Antifouling Mille 7670	Hempel Paints Ltd	AVP	HSE 3352

801 TRIPHENYLTIN FLUORIDE and THIRAM and TRIBUTYLTIN OLEATE and TRIBUTYLTIN TETRACHLOROPHTHALATE

* Tropical Antifouling (Old Version)	Blakes Marine Paints Ltd	AVP	HSE 3233

802 TRIPHENYLTIN FLUORIDE and THIRAM and TRIBUTYLTIN OXIDE

# Tropical Antifouling (New Version)	Blakes Marine Paints Ltd	AVP	HSE 3234

803 TRIPHENYLTIN FLUORIDE and TRIBUTYLTIN METHACRYLATE

* Antifouling Alusea Black	Jotun-Henry Clark Ltd	AVP	HSE 3424
* Antifouling Alusea Blue	Jotun-Henry Clark Ltd	AVP	HSE 3425
* Antifouling Alusea Red	Jotun-Henry Clark Ltd	AVP	HSE 3426
* Antifouling Alusea White	Jotun-Henry Clark Ltd	AVP	HSE 3427

804 TRIPHENYLTIN FLUORIDE and TRIBUTYLTIN METHACRYLATE and TRIBUTYLTIN OXIDE

* Hempel's Antifouling Nautic 7680	Hempel Paints Ltd	AVP	HSE 3359

805 TRIPHENYLTIN FLUORIDE and TRIBUTYLTIN OXIDE

# Hempel's Antifouling Oceanic 7640	Hempel Paints Ltd	AVP	HSE 3363
* Hempel's Hard Racing Copper Free 7649	Hempel Paints Ltd	AVP	HSE 3368

ANTIFOULING PRODUCTS

Product Name	Marketing Company	Use	Reg. No
806 TRIPHENYLTIN FLUORIDE and ZINC OXIDE and CUPROUS OXIDE			
* Rabamarine A/F No 100s	Kansai Paint Co Ltd	AVP	HSE 3448
807 TRIPHENYLTIN HYDROXIDE and CUPROUS OXIDE			
(ATMC 135) Amercoat 614	Ameron (UK) Ltd	AVP	HSE 3212
808 TRIPHENYLTIN HYDROXIDE and ZINC OXIDE and CUPROUS OXIDE			
AF Seaflo SP-1 Light/Dark	Chugoku Marine Paints (UK) Ltd	AVP	HSE 3269
AF Seaflo SP-2 Light/Dark	Chugoku Marine Paints (UK) Ltd	AVP	HSE 3268
* Rabamarine A/F No 1000 (A-SOL/B-SOL)	Kansai Paint Co Ltd	AVP	HSE 3449
* Rabamarine A/F No 1000SP (A-SOL/B-SOL)	Kansai Paint Co Ltd	AVP	HSE 3450
809 ZINC METAL and CUPROUS OXIDE and TRIBUTYLTIN METHACRYLATE			
* AF Seaflo Z-100 HS-2 Light/Dark	Chugoku Marine Paints (UK) Ltd	AVP	HSE 3254
* AF Seaflo Z-100 LE-HS-2 Light/Dark	Chugoku Marine Paints (UK) Ltd	AVP	HSE 3260
810 ZINC METAL and TRIBUTYLTIN POLYSILOXANE			
* (ATMC 172) Dimetcote 602	Ameron (UK) Ltd	AVP	HSE 3217
811 ZINC METAL and ZINC OXIDE and CUPROUS OXIDE and TRIBUTYLTIN METHACRYLATE			
* AF Seaflo Mark 2-2 Light/Dark	Chugoku Marine Paints (UK) Ltd	AVP	HSE 3262
* AF Seaflo Mark 3 Light/Dark	Chugoku Marine Paints (UK) Ltd	AVP	HSE 3263
812 ZINC METAL and ZINC OXIDE and CUPROUS THIOCYANATE and TRIBUTYLTIN METHACRYLATE			
* Marine Gold DX White-2	Chugoku Marine Paints (UK) Ltd	AVP	HSE 3279

ANTIFOULING PRODUCTS

Product Name	Marketing Company	Use	Reg. No
813 ZINC NAPHTHENATE and 2,4,5,6-TETRACHLORO ISOPHTHALONITRILE and CUPROUS OXIDE			
Teamac Killa Copper Plus	Teal and Mackrill Ltd	AVA AVP	HSE 3495
814 ZINC NAPHTHENATE and CUPROUS OXIDE			
Teamac Killa Copper	Teal and Mackrill Ltd	AVA AVP	HSE 3493
Teamac Killa Copper	Teal and Mackrill Ltd	AVA AVP	HSE 3494
815 ZINC NAPHTHENATE and CUPROUS OXIDE and TRIBUTYLTIN FLUORIDE			
* Teamac Killa	Teal and Mackrill Ltd	AVP	HSE 3492
816 ZINC NAPHTHENATE and ZINC OXIDE and COPPER NAPHTHENATE and CUPROUS OXIDE and DICHLOFLUANID			
Teamac Killa Copper Plus	Teal and Mackrill Ltd	AVA AVP	HSE 4659
817 ZINC NAPHTHENATE and ZINC OXIDE and CUPROUS OXIDE			
Teamac Super Tropical	Teal and Mackrill Ltd	AVA AVP	HSE 3497
818 ZINC OXIDE and 2,3,5,6-TETRACHLORO-4-(METHYL SULPHONYL) PYRIDINE and 2-METHYLTHIO-4-TERTIARY-BUTYLAMINO-6-CYCLOPROPYLAMINO-S-TRIAZINE and CUPROUS OXIDE and CUPROUS THIOCYANATE			
Envoy TF 400	W and J Leigh and Company	AVA AVP	HSE 4432
Envoy TF 500	W and J Leigh and Company	AVA AVP	HSE 5599
Meridian MP40 Antifouling	Deangate Marine Products	AVA AVP	HSE 5027
819 ZINC OXIDE and 2,4,5,6-TETRACHLORO ISOPHTHALONITRILE and CUPROUS OXIDE			
Hempel's Antifouling Classic 7654C	Hempel Paints Ltd	AVA AVP	HSE 5298
Hempel's Antifouling Combic 7199C	Hempel Paints Ltd	AVA AVP	HSE 5296
Hempel's Antifouling Nautic 7190C	Hempel Paints Ltd	AVA AVP	HSE 5294
TFA 10 LA	Camrex Chugoku Ltd	AVA AVP	HSE 5361
820 ZINC OXIDE and 2,4,5,6-TETRACHLORO ISOPHTHALONITRILE and CUPROUS THIOCYANATE			
Gummipaint A/F	Veneziani Spa	AVA AVP	HSE 4934

ANTIFOULING PRODUCTS

Product Name	Marketing Company	Use	Reg. No

821 ZINC OXIDE and 2-METHYLTHIO-4-TERTIARY-BUTYLAMINO-6-CYCLOPROPYLAMINO-S-TRIAZINE and CUPROUS OXIDE

Product Name	Marketing Company	Use	Reg. No
Algicide Red Antifouling	Blakes Marine Paints	AVA AVP	HSE 5738

822 ZINC OXIDE and 2-METHYLTHIO-4-TERTIARY

Product Name	Marketing Company	Use	Reg. No
Broads Red Antifouling	Blakes Marine Paints	AVA AVP	HSE 5736

823 ZINC OXIDE and 2-METHYLTHIO-4-TERTIARY-BUTYLAMINO-6-CYCLOPROPYLAMINO-S-TRIAZINE and CUPROUS OXIDE

Product Name	Marketing Company	Use	Reg. No
Even TF	Veneziani Spa	AVA AVP	HSE 5166
Hempel's Antifouling Classic 7611 Red (Tin Free) 5000	Hempel Paints Ltd	AVA AVP	HSE 5064
Hempel's Antifouling Classic 76540	Hempel Paints Ltd	AVA AVP	HSE 5291
Hempel's Antifouling Combic 71990	Hempel Paints Ltd	AVA AVP	HSE 5600
Hempel's Antifouling Combic 71992	Hempel Paints Ltd	AVA AVP	HSE 5601
Hempel's Antifouling Combic Tin Free 71990	Hempel Paints Ltd	AVA AVP	HSE 5266
Hempel's Antifouling Mille Dynamic	Hempel Paints Ltd	AVA	HSE 4440
Hempel's Antifouling Nautic 71900	Hempel Paints Ltd	AVA AVP	HSE 5290
Hempel's Antifouling Nautic 71902	Hempel Paints Ltd	AVA AVP	HSE 5605
Hempel's Antifouling Nautic Tin Free 7190	Hempel Paints Ltd	AVA AVP	HSE 4869
Hempel's Antifouling Olympic HI-7661	Hempel Paints Ltd	AVA AVP	HSE 4898
Hempel's Hard Racing 76480	Hempel Paints Ltd	AVA AVP	HSE 5538
Hempel's Mille Dynamic 71700	Hempel Paints Ltd	AVA AVP	HSE 5574
Hempel's Antifouling Bravo Tin Free 7610	Hempel Paints Ltd	AVA AVP	HSE 4482
Long Life T.F.	Veneziani Spa	AVA AVP	HSE 4936

824 ZINC OXIDE and 2-METHYLTHIO-4-TERTIARY-BUTYLAMINO-6-CYCLOPROPYLAMINO-S-TRIAZINE and CUPROUS THIOCYANATE

Product Name	Marketing Company	Use	Reg. No
Even TF Light Grey	Veneziani Spa	AVA AVP	HSE 5167
Hempel's Hard Racing 76380	Hempel Paints Ltd	AVA AVP	HSE 5540
Hempel's Mille Alu 71601	Hempel Paints Ltd	AVA AVP	HSE 5577
Hempel's Mille Dynamic 71600	Hempel Paints Ltd	AVA AVP	HSE 5576
Raffaello Alloy	Veneziani Spa	AVA AVP	HSE 5492

ANTIFOULING PRODUCTS

Product Name	Marketing Company	Use	Reg. No

825 ZINC OXIDE and 2-METHYLTHIO-4-TERTIARY-BUTYLAMINO-6-CYCLOPROPYLAMINO-S-TRIAZINE and TRIBUTYLTIN METHACRYLATE

Antifouling Alusea	Jotun-Henry Clark Ltd	AVP	HSE 4569

826 ZINC OXIDE and 2-METHYLTHIO-4-TERTIARY-BUTYLAMINO-6-CYCLOPROPYLAMINO-S-TRIAZINE and CUPROUS OXIDE

Broads Black Antifouling	Blakes Marine Paints	AVA AVP	HSE 5739
Hard Racing Antifouling	Blakes Marine Paints	AVA AVP	HSE 5704

827 ZINC OXIDE and 2-METHYLTHIO-4-TERTIARY- BUTYLAMINO-6-CYCLOPROPYLAMINO-S-TRIAZINE and CUPROUS THIOCYANATE

Titan FGA Antifouling White	Blakes Marine Paints	AVA AVP	HSE 5680

828 ZINC OXIDE and 2-METHYLTHIO-4-TERTIARY- BUTYLAMINO-6-CYCLOPROPYLAMINO-S-TRIAZINE and CUPROUS OXIDE

Titan FGA Antifouling	Blakes Marine Paints	AVA AVP	HSE 5681

829 ZINC OXIDE and 2-METHYLTHIO-4-TERTIARY-BUTYLAMINO-6-CYCLOPROPYLAMINO-S-TRIAZINE and CUPROUS THIOCYANATE

Hard Racing Antifouling White	Blakes Marine Paints	AVA AVP	HSE 5705

830 ZINC OXIDE and 4,5-DICHLORO-2-N-OCTYL-4-ISOTHIAZOLIN-3-ONE and CUPROUS OXIDE

Antifouling Seavictor 50	Jotun-Henry Clark Ltd	AVA AVP	HSE 4958
Hempel's Antifouling Classic 7654E	Hempel Paints Ltd	AVA AVP	HSE 5286
Hempel's Antifouling Combic 7199E	Hempel Paints Ltd	AVA AVP	HSE 5277
Hempel's Antifouling Nautic 7190E	Hempel Paints Ltd	AVA AVP	HSE 5283
Hempel's Antifouling Nautic Tin Free 7190E	Hempel Paints Ltd	AVA AVP	HSE 5145

831 ZINC OXIDE and COPPER METAL

# AF29 Desoto Antifouling	Desoto Titanine Plc	AVA AVP	HSE 3893
Amercoat 67E	Ameron BV	AVA AVP	HSE 3201
Amercoat 70E	Ameron BV	AQA AVP AVA	HSE 3202
Amercoat 70ESP	Ameron BV	AQA AVA AVP	HSE 3203

ANTIFOULING PRODUCTS

Product Name	Marketing Company	Use	Reg. No

832 ZINC OXIDE and COPPER NAPHTHENATE and CUPROUS OXIDE and DICHLOFLUANID and ZINC NAPHTHENATE

Teamac Killa Copper Plus	Teal and Mackrill Ltd	AVA AVP	HSE 4659

833 ZINC OXIDE and COPPER RESINATE and CUPROUS OXIDE and TRIBUTYLTIN FLUORIDE

* Lily Antifouling IV	Sigma Coatings BV	AVP	HSE 3484
* Sigma Antifouling CR	Sigma Coatings BV	AVP	HSE 3483
* Sigma Antifouling IV	Sigma Coatings BV	AVP	HSE 4347

834 ZINC OXIDE and CUPROUS OXIDE

(ATMC 129) Amercoat 279	Ameron (UK) Ltd	AVA AVP	HSE 3942
Amercoat 279	Ameron BV	AVA AVP	HSE 3206
Antifouling Seaguardian (Black and Blue)	Jotun-Henry Clark Ltd	AVP	HSE 4273
Antifouling Seavictor 40	Jotun-Henry Clark Ltd	AVA AVP	HSE 4957
Antifouling Tropic	Jotun-Henry Clark Ltd	AVA AVP	HSE 3414
Aquacleen	Mariner	AVA AVP	HSE 5667
Awlgrip Awlstar Gold Label Antifouling	Awlgrip NV	AVA AVP	HSE 5065
Biscon AF	Chugoku Marine Paints (UK) Ltd	AVA AVP	HSE 3272
C-Worthy	Benfleet Marine Wholesale	AVA AVP	HSE 5476
* Chugoku AF ST	Chugoku Marine Paints (UK) Ltd	AVA AVP	HSE 3270
* Chugoku AF T	Chugoku Marine Paints (UK) Ltd	AVA AVP	HSE 3271
Cobra V	Valiant Marine	AVA AVP	HSE 5194
Cooper's Copolymer Antifouling	Cooper's Marine Paints Ltd	AVA AVP	HSE 5609
# Copper Bottom Paint	Spencer (Aberdeen) Plc	AVA AVP	HSE 3504
Cupron Plus T.F.	Veneziani Spa	AVA AVP	HSE 5661
Cupron T.F.	Veneziani Spa	AVA AVP	HSE 4935
Hempel's Copper Bottom Paint 7116	Hempel Paints Ltd	AVA AVP	HSE 4274
Marclear Antifouling	Marclear Marine Products Ltd	AVA AVP	HSE 5264
Marclear Antifouling	Marine Clearance	AVA AVP	HSE 4853
Noa-Noa Rame	Stoppani (UK) Ltd	AVA AVP	HSE 4795

435

ANTIFOULING PRODUCTS

Product Name	Marketing Company	Use	Reg. No
834 ZINC OXIDE and CUPROUS OXIDE—continued			
# North Atlantic/Launching Antifouling AF2 Red	Desoto Titanine Plc	AVA AVP	HSE 3301
# North Atlantic/Launching Antifouling F2/801 Black	Desoto Titanine Plc	AVA AVP	HSE 3302
Penguin Racing	Marine and Industrial Sealants	AVA AVP	HSE 5673
Ravax AF	Camrex Chugoku Ltd	AVA AVP	HSE 5319
Ravax Anti-Fouling	Chugoku Marine Paints (UK) Ltd	AVA AVP	HSE 3291
Ravax Anti-Fouling HB	Chugoku Marine Paints (UK) Ltd	AVA AVP	HSE 3264
Ravax Anti-Fouling ND	Chugoku Marine Paints (UK) Ltd	AVA AVP	HSE 3265
Seatender 10	Camrex Chugoku Ltd	AVA AVP	HSE 5321
Seatender 7	Camrex Chugoku Ltd	AVA AVP	HSE 5320
Speedclean Antifouling	Mariner Paints	AVA AVP	HSE 5077
Superspeed	D R Margetson	AVA AVP	HSE 5191
Tiger Cruising	Blakes Marine Paints Ltd	AVA AVP	HSE 5099
Vinyl Antifouling 2000	Akzo Coatings BV	AVA AVP	HSE 5633
835 ZINC OXIDE and CUPROUS OXIDE and 4,5-DICHLORO-2-N-OCTYL-4-ISOTHIAZOLIN-3-ONE and 2,4,5,6-TETRACHLORO ISOPHTHALONITRILE			
Seatender 15	Camrex Chugoku Ltd	AVA AVP	HSE 5348
TFA 10 HG	Camrex Chugoku Ltd	AVA AVP	HSE 5366
TFA 10G	Camrex Chugoku Ltd	AVA AVP	HSE 5347
836 ZINC OXIDE and CUPROUS OXIDE and COPPER RESINATE and 2-METHYLTHIO-4-TERTIARY-BUTYLAMINO-6-CYCLOPROPYLAMINO-S-TRIAZINE			
Sigmaplane Ecol Antifouling	Sigma Coatings BV	AVA AVP	HSE 4348
837 ZINC OXIDE and CUPROUS OXIDE and DICHLOFLUANID			
Dapaflow Antifoul	Dapa Services Ltd	AVA AVP	HSE 4754
Penguin Non-Stop	Marine and Industrial Sealants	AVA AVP	HSE 5671

ANTIFOULING PRODUCTS

Product Name	Marketing Company	Use	Reg. No
838 ZINC OXIDE and CUPROUS OXIDE and DICHLOROPHENYL DIMETHYLUREA			
Hempel's Antifouling Classic 7654B	Hempel Paints Ltd	AVA AVP	HSE 5285
Hempel's Antifouling Combic 7199B	Hempel Paints Ltd	AVA AVP	HSE 5274
Hempel's Antifouling Nautic 7190B	Hempel Paints Ltd	AVA AVP	HSE 5273
Hempel's Bravo 7610A	Hempel Paints Ltd	AVA AVP	HSE 5603
Hempel's Hard Racing 7648A	Hempel Paints Ltd	AVA AVP	HSE 5535
Hempel's Mille Dynamic 7170A	Hempel Paints Ltd	AVA AVP	HSE 5536
Noa-Noa Rame	Stoppani (UK) Ltd	AVA AVP	HSE 4796
Seajet 033	Camrex Chugoku Ltd	AVA AVP	HSE 5331
Sigmaplane Ecol Antifouling	Sigma Coatings Ltd	AVA AVP	HSE 5670
TFA-10 LA Light/Dark	Chugoku Marine Paints (UK) Ltd	AVA AVP	HSE 3285
TFA-10 Light/Dark	Chugoku Marine Paints (UK) Ltd	AVA AVP	HSE 3286
TFA-20	Chugoku Marine Paints (UK) Ltd	AVA AVP	HSE 3284
839 ZINC OXIDE and CUPROUS OXIDE and DICHLOROPHENYL DIMETHYLUREA and TRIBUTYLTIN METHACRYLATE and TRIBUTYLTIN OXIDE			
A11 Antifouling 072014	Nautix SA	AVP	HSE 5376
840 ZINC OXIDE and CUPROUS OXIDE and MANEB and TRIBUTYLTIN METHACRYLATE			
Nu Wave A/F Flat Bottom	Kansai Paint Co Ltd	AVP	HSE 3447
Nu Wave A/F Vertical Bottom	Kansai Paint Co Ltd	AVP	HSE 3446
Rabamarine A/F No 2500 HS	Kansai Paint Co Ltd	AVP	HSE 3444
Rabamarine A/F No 2500M HS	Kansai Paint Co Ltd	AVP	HSE 3445
841 ZINC OXIDE and CUPROUS OXIDE and TRIBUTYLTIN ACRYLATE			
Amercoat 697	Ameron BV	AVP	HSE 3513
Amercoat 698 HS	Ameron BV	AVP	HSE 3214

ANTIFOULING PRODUCTS

Product Name	Marketing Company	Use	Reg. No

842 ZINC OXIDE and CUPROUS OXIDE and TRIBUTYLTIN FLUORIDE

(ATMC 123) Amercoat 557	Ameron (UK) Ltd	AVP	HSE 3207
* Antifouling Seven Seas	Jotun-Henry Clark Ltd	AVP	HSE 3405

843 ZINC OXIDE and CUPROUS OXIDE and TRIBUTYLTIN FLUORIDE and TRIBUTYLTIN OXIDE

* Hempel's Antifouling Olympic 7155	Hempel Paints Ltd	AVP	HSE 3719

844 ZINC OXIDE and CUPROUS OXIDE and TRIBUTYLTIN METHACRYLATE and TRIBUTYLTIN OXIDE

# Ablative Antifouling AF28	Desoto Titanine Plc	AVP	HSE 3505
AF Seaflo Mark 2-1	Camrex Chugoku Ltd	AVP	HSE 5316
AF Seaflo Mark 2-1 Light/Dark	Chugoku Marine Paints (UK) Ltd	AVP	HSE 3261
# Antifouling Seaconomy	Jotun-Henry Clark Ltd	AVP	HSE 3438
Antifouling Seaconomy 200	Jotun-Henry Clark Ltd	AVP	HSE 4436
Antifouling Seaconomy 300	Jotun-Henry Clark Ltd	AVP	HSE 4272
# Antifouling Seaflex	Jotun-Henry Clark Ltd	AVP	HSE 3439
Antifouling Seamate HB 22	Jotun-Henry Clark Ltd	AVP	HSE 4271
Antifouling Seamate HB 33	Jotun-Henry Clark Ltd	AVP	HSE 4270
# Antifouling Seamate HB 33 DR	Jotun-Henry Clark Ltd	AVP	HSE 3440
# Antifouling Seamate HB 33 LR	Jotun-Henry Clark Ltd	AVP	HSE 3441
* Antifouling Seamate HB 66 Dark Red	Jotun-Henry Clark Ltd	AVP	HSE 3442
* Antifouling Seamate HB 66 Light Red	Jotun-Henry Clark Ltd	AVP	HSE 3443
Antifouling Seamate HB66	Jotun-Henry Clark Ltd	AVP	HSE 5698
# Camrex C-Clean EXA	Camrex Ltd	AVP	HSE 3240
# Camrex C-Clean HB EXA	Camrex Ltd	AVP	HSE 3241
# Camrex C-Clean HB LA	Camrex Ltd	AVP	HSE 3243
# Camrex C-Clean LA	Camrex Ltd	AVP	HSE 3242
Grassline ABL Antifouling Type M349	W and J Leigh and Company	AVP	HSE 5374

ANTIFOULING PRODUCTS

Product Name	Marketing Company	Use	Reg. No
845 ZINC OXIDE and CUPROUS OXIDE and TRIBUTYLTIN METHACRYLATE and ZINC METAL			
* AF Seaflo Mark 2-2 Light/Dark	Chugoku Marine Paints (UK) Ltd	AVP	HSE 3262
* AF Seaflo Mark 3 Light/Dark	Chugoku Marine Paints (UK) Ltd	AVP	HSE 3263
846 ZINC OXIDE and CUPROUS OXIDE and TRIBUTYLTIN OXIDE			
# (ATMC 101) Amercoat 610	Ameron (UK) Ltd	AVP	HSE 3208
# (ATMC 102) Amercoat 611	Ameron (UK) Ltd	AVP	HSE 3209
# (ATMC 120) Amercoat 612	Ameron (UK) Ltd	AVP	HSE 3210
# (ATMC 121) Amercoat 613	Ameron (UK) Ltd	AVP	HSE 3211
# (ATMC 122) Amercoat 615	Ameron (UK) Ltd	AVP	HSE 3213
# Antifouling Equator KK	Jotun-Henry Clark Ltd	AVP	HSE 3431
# Antifouling Sargasso	Jotun-Henry Clark Ltd	AVP	HSE 3437
# Super Quality Antifouling	Spencer (Aberdeen) Plc	AVP	HSE 3491
847 ZINC OXIDE and CUPROUS OXIDE and TRIBUTYLTIN POLYSILOXANE			
* (ATMC 170) Amercoat 699	Ameron (UK) Ltd	AVP	HSE 3215
848 ZINC OXIDE and CUPROUS OXIDE and TRIPHENYLTIN FLUORIDE			
* Rabamarine A/F No 100S	Kansai Paint Co Ltd	AVP	HSE 3448
849 ZINC OXIDE and CUPROUS OXIDE and TRIPHENYLTIN HYDROXIDE			
AF Seaflo SP-1 Light/Dark	Chugoku Marine Paints (UK) Ltd	AVP	HSE 3269
AF Seaflo SP-2 Light/Dark	Chugoku Marine Paints (UK) Ltd	AVP	HSE 3268
* Rabamarine A/F No 1000 (A-Sol/B-Sol)	Kansai Paint Co Ltd	AVP	HSE 3449
* Rabamarine A/F No 1000SP (A-Sol/B-Sol)	Kansai Paint Co Ltd	AVP	HSE 3450
850 ZINC OXIDE and CUPROUS OXIDE and ZINC NAPHTHENATE			
Teamac Super Tropical	Teal and Mackrill Ltd	AVA AVP	HSE 3497

ANTIFOULING PRODUCTS

Product Name	Marketing Company	Use	Reg. No
851 ZINC OXIDE and CUPROUS THIOCYANATE and DICHLOFLUANID			
Penguin Non-Stop White	Marine and Industrial Sealants	AVA AVP	HSE 5672
852 ZINC OXIDE and CUPROUS THIOCYANATE and DICHLOROPHENYL DIMETHYLUREA			
Hempel's Hard Racing 7638A	Hempel Paints Ltd	AVA AVP	HSE 5539
Hempel's Mille Alu 71602	Hempel Paints Ltd	AVA AVP	HSE 5578
Hempel's Mille Dynamic 7160A	Hempel Paints Ltd	AVA AVP	HSE 5575
853 ZINC OXIDE and CUPROUS THIOCYANATE and DICHLOROPHENYL DIMETHYLUREA and TRIPHENYLTIN CHLORIDE			
* Kamome Colour 60-2	Chugoku Marine Paints (UK) Ltd	AVP	HSE 3276
854 ZINC OXIDE and CUPROUS THIOCYANATE and MANEB and TRIBUTYLTIN MESO-DIBROMOSUCCINATE and TRIBUTYLTIN METHACRYLATE and TRIBUTYLTIN OXIDE			
# Marine Gold DX-1	Chugoku Marine Paints (UK) Ltd	AVP	HSE 3281
855 ZINC OXIDE and CUPROUS THIOCYANATE and TRIBUTYLTIN MESO-DIBROMOSUCCINATE and TRIBUTYLTIN OXIDE			
# Kamome Colour 60-1	Chugoku Marine Paints (UK) Ltd	AVP	HSE 3277
856 ZINC OXIDE and CUPROUS THIOCYANATE and TRIBUTYLTIN METHACRYLATE			
* Marine Gold DX-2	Chugoku Marine Paints (UK) Ltd	AVP	HSE 3280
857 ZINC OXIDE and CUPROUS THIOCYANATE and TRIBUTYLTIN METHACRYLATE and TRIBUTYLTIN OXIDE			
Antifouling Seamate HB 33 BSL Blue	Jotun-Henry Clark Ltd	AVP	HSE 4269
Antifouling Seamate HB22 Roundel Blue	Jotun-Henry Clark Ltd	AVP	HSE 4872
Antifouling Seamate HB66 Black	Jotun-Henry Clark Ltd	AVP	HSE 4871

ANTIFOULING PRODUCTS

Product Name	Marketing Company	Use	Reg. No

858 ZINC OXIDE and CUPROUS THIOCYANATE and TRIBUTYLTIN METHACRYLATE and ZINC METAL

* Marine Gold DX White-2	Chugoku Marine Paints (UK) Ltd	AVP	HSE 3279

859 ZINC OXIDE and DICHLOFLUANID and CUPROUS THIOCYANATE and 2-METHYLTHIO-4-TERTIARY-BUTYLAMINO-6-CYCLOPROPYLAMINO-S-TRIAZINE

Hempel's Antifouling Mille Dynamic 717 GB	Hempel Paints Ltd	AVA AVP	HSE 4437

860 ZINC OXIDE and DICHLOROPHENYL DIMETHYLUREA

Aquashield	Silverblue Services Ltd	AVA AVP	HSE 3779

861 ZINC OXIDE and DICHLOROPHENYL DIMETHYLUREA and CUPROUS OXIDE and 2,4,5,6-TETRACHLORO ISOPHTHALONITRILE

Seatender 12	Camrex Chugoku Ltd	AVA AVP	HSE 5324
TFA 10	Camrex Chugoku Ltd	AVA AVP	HSE 5346
TFA 10 H	Camrex Chugoku Ltd	AVA AVP	HSE 5358

862 ZINC OXIDE and DICHLOROPHENYL DIMETHYLUREA and CUPROUS OXIDE and 2-METHYLTHIO-4-TERTIARY-BUTYLAMINO-6-CYCLOPROPYLAMINO-S-TRIAZINE

A3 Antifouling 072015	Nautix SA	AVA AVP	HSE 5224
A4 Antifouling 072017	Nautix SA	AVA AVP	HSE 5226
Le Marin	Nautix SA	AVA AVP	HSE 5480

863 ZINC OXIDE and DICHLOROPHENYL DIMETHYLUREA and CUPROUS OXIDE and COPPER RESINATE

* Sigmaplane Ecol AF First Coat	Sigma Coatings BV	AVA AVP	HSE 4349

864 ZINC OXIDE and DICHLOROPHENYL DIMETHYLUREA and CUPROUS THIOCYANATE and 2-METHYLTHIO-4-TERTIARY-BUTYLAMINO-6-CYCLOPROPYLAMINO-S-TRIAZINE

A3 Antifouling 072016	Nautix SA	AVA AVP	HSE 5225
A3 Teflon Antifouling 062015	Nautix SA	AVA AVP	HSE 5481

ANTIFOULING PRODUCTS

Product Name	Marketing Company	Use	Reg. No

864 ZINC OXIDE and DICHLOROPHENYL DIMETHYLUREA and CUPROUS THIOCYANATE and 2-METHYLTHIO-4-TERTIARY-BUTYLAMINO-6-CYCLOPROPYLAMINO-S-TRIAZINE—continued

A4 Antifouling 072018	Nautix SA	AVA AVP	HSE 5227
A4 Teflon Antifouling 062017	Nautix SA	AVA AVP	HSE 5482

865 ZINC OXIDE and DICHLOROPHENYL DIMETHYLUREA and TRIBUTYLTIN FLUORIDE and TRIBUTYLTIN METHACRYLATE

* Marine Gold DX White-1	Chugoku Marine Paints (UK) Ltd	AVP	HSE 3278

866 ZINC OXIDE and TRIBUTYLTIN POLYSILOXANE

* (ATMC 173) Dimetcote 601	Ameron (UK) Ltd	AVP	HSE 3216

867 ZINC OXIDE and ZINC PYRITHIONE and 2-(THIOCYANOMETHYLTHIO) BENZOTHIAZOLE and 2-METHYLTHIO-4-TERTIARY-BUTYLAMINO-6-CYCLOPROPYLAMINO-S-TRIAZINE and DICHLOROPHENYL DIMETHYLUREA

A2 Antifouling	Nautix SA	AVA AVP	HSE 4832
A2 Teflon Antifouling	Nautix SA	AVA AVP	HSE 4833

868 ZINC OXIDE and ZINC PYRITHIONE and 2-METHYLTHIO-4-TERTIARY-BUTYLAMINO-6-CYCLOPROPYLAMINO-S-TRIAZINE and DICHLOROPHENYL DIMETHYLUREA

A7 Teflon Antifouling	Nautix SA	AVA AVP	HSE 4945
A7 Teflon Antifouling 072019	Nautix SA	AVA AVP	HSE 5228

869 ZINC OXIDE and ZINEB and COPPER RESINATE and CUPROUS OXIDE

* Sigma Pilot Antifouling LL/TF	Sigma Coatings BV	AVA AVP	HSE 3480
Sigma Pilot Ecol Antifouling	Sigma Coatings Ltd	AVA AVP	HSE 4933

870 ZINC OXIDE and ZINEB and COPPER RESINATE and CUPROUS OXIDE and TRIBUTYLTIN FLUORIDE

* Sigma Pilot Antifouling LL	Sigma Coatings BV	AVA	HSE 3481

871 ZINC OXIDE and ZINEB and CUPROUS OXIDE and TRIBUTYLTIN FLUORIDE

* Colturiet TCN Antifouling (Set and N Set)	Sigma Coatings BV	AVP	HSE 3488

ANTIFOULING PRODUCTS

Product Name	Marketing Company	Use	Reg. No
872 ZINC OXIDE and ZINEB and CUPROUS OXIDE and TRIBUTYLTIN METHACRYLATE			
Antifouling Seamate HB22	Jotun-Henry Clark Ltd	AVP	HSE 5363
Antifouling Seamate HB33	Jotun-Henry Clark Ltd	AVP	HSE 5362
Sigmaplane HA Antifouling	Sigma Coatings BV	AVP	HSE 4345
Sigmaplane HB	Sigma Coatings BV	AVP	HSE 3487
873 ZINC OXIDE and ZINEB and CUPROUS OXIDE and TRIBUTYLTIN METHACRYLATE and TRIBUTYLTIN OXIDE			
Hempel's Antifouling Combic 7699B	Hempel Paints Ltd	AVP	HSE 5581
Hempel's Antifouling Nautic 7690B	Hempel Paints Ltd	AVP	HSE 5580
Hempel's Antifouling Nautic 7691B	Hempel Paints Ltd	AVP	HSE 5586
874 ZINC OXIDE and ZIRAM and CUPROUS OXIDE			
Hempel's Antifouling Classic 7654A	Hempel Paints Ltd	AVA AVP	HSE 5297
Hempel's Antifouling Combic 7199A	Hempel Paints Ltd	AVA AVP	HSE 5295
Hempel's Antifouling Nautic 7190A	Hempel Paints Ltd	AVA AVP	HSE 5293
875 ZINC PYRITHIONE and 2-(THIOCYANOMETHYLTHIO) BENZOTHIAZOLE and 2-METHYLTHIO-4-TERTIARY-BUTYLAMINO-6-CYCLOPROPYLAMINO-S-TRIAZINE and DICHLOROPHENYL DIMETHYLUREA and ZINC OXIDE			
A2 Antifouling	Nautix SA	AVA AVP	HSE 4832
A2 Teflon Antifouling	Nautix SA	AVA AVP	HSE 4833
876 ZINC PYRITHIONE and 2-METHYLTHIO-4-TERTIARY-BUTYLAMINO-6-CYCLOPROPYLAMINO-S-TRIAZINE and DICHLOROPHENYL DIMETHYLUREA			
A6 Antifouling	Nautix SA	AVA AVP	HSE 4944
877 ZINC PYRITHIONE and CUPROUS OXIDE			
Micron 600 Series	International Paint Ltd	AVA AVP	HSE 5733
VC Offshore Extra 100 Series	International Paint Ltd	AVA AVP	HSE 5730
VC Offshore SP Antifouling	Extensor AB	AVA AVP	HSE 3507

ANTIFOULING PRODUCTS

Product Name	Marketing Company	Use	Reg. No

878 ZINC PYRITHIONE and ZINC OXIDE and DICHLOROPHENYL DIMETHYLUREA and 2-METHYLTHIO-4-TERTIARY-BUTYLAMINO-6-CYCLOPROPYLAMINO-S-TRIAZINE

Product Name	Marketing Company	Use	Reg. No
A7 Teflon Antifouling	Nautix SA	AVA AVP	HSE 4945
A7 Teflon Antifouling 072019	Nautix SA	AVA AVP	HSE 5228

879 ZINEB and 2-METHYLTHIO-4-TERTIARY-BUTYLAMINO-6-CYCLOPROPYLAMINO-S-TRIAZINE and CUPROUS THIOCYANATE

Product Name	Marketing Company	Use	Reg. No
# Interspeed 2000	International Paint Plc	AVA AVP	HSE 3781
* MPX	International Paint Ltd	AVA	HSE 3783

880 ZINEB and COPPER RESINATE and CUPROUS OXIDE and TRIBUTYLTIN FLUORIDE and ZINC OXIDE

Product Name	Marketing Company	Use	Reg. No
* Sigma Pilot Antifouling LL	Sigma Coatings BV	AVA	HSE 3481

881 ZINEB and CUPROUS OXIDE

Product Name	Marketing Company	Use	Reg. No
Blueline SPC Tin Free SBA700 Series	International Paint Ltd	AVA AVP	HSE 5214
Equatorial	International Paint Ltd	AVA AVP	HSE 4121
Inter 100	International Paint Ltd	AVA AVP	HSE 4120
# International TBT Free Copolymer Antifouling BQA200 Series	International Paint Ltd	AVA AVP	HSE 3722
# Interspeed Extra (BWA 500 Red)	International Paint Ltd	AVA AVP	HSE 3845
Interspeed System 2 BRA140/ BRA240 Series	International Paint Ltd	AVA AVP	HSE 3847
Interviron BQA200 Series	International Paint Ltd	AVA AVP	HSE 3846

882 ZINEB and CUPROUS OXIDE and TRIBUTYLTIN FLUORIDE and ZINC OXIDE

Product Name	Marketing Company	Use	Reg. No
* Colturiet TCN Antifouling (Set and N Set)	Sigma Coatings BV	AVP	HSE 3488

883 ZINEB and CUPROUS OXIDE and TRIBUTYLTIN METHACRYLATE

Product Name	Marketing Company	Use	Reg. No
# Sigma Pilot Antifouling TA	Sigma Coatings BV	AVP	HSE 3482
Sigmaplane TA Antifouling	Sigma Coatings BV	AVP	HSE 4346
* Sigmathrift Antifouling	Sigma Coatings BV	AVP	HSE 3486

ANTIFOULING PRODUCTS

Product Name	Marketing Company	Use	Reg. No

884 ZINEB and CUPROUS OXIDE and TRIBUTYLTIN METHACRYLATE and TRIBUTYLTIN OXIDE

Blueline SBA900 Series	International Paint Ltd	AVP	HSE 5138
Intersmooth Hisol 2000 BFA270 Series	International Paint Ltd	AVP	HSE 3844
Intersmooth Hisol 9000 BFA970 Series	International Paint Plc	AVP	HSE 5461
Intersmooth Hisol BFA250/BFA900 Series	International Paint Ltd	AVP	HSE 3848
# Intersmooth Hisol BFA250/BFA900 Series	International Paint Plc	AVP	HSE 3378
Interswift BKA000/700 Series	International Paint Ltd	AVP	HSE 3384

885 ZINEB and CUPROUS OXIDE and TRIBUTYLTIN METHACRYLATE and TRIBUTYLTIN OXIDE and ZINC OXIDE

Hempel's Antifouling Combic 7699B	Hempel Paints Ltd	AVP	HSE 5581
Hempel's Antifouling Nautic 7690B	Hempel Paints Ltd	AVP	HSE 5580
Hempel's Antifouling Nautic 7691B	Hempel Paints Ltd	AVP	HSE 5586

886 ZINEB and CUPROUS OXIDE and TRIBUTYLTIN METHACRYLATE and ZINC OXIDE

Antifouling Seamate HB22	Jotun-Henry Clark Ltd	AVP	HSE 5363
Antifouling Seamate HB33	Jotun-Henry Clark Ltd	AVP	HSE 5362
Sigmaplane HA Antifouling	Sigma Coatings BV	AVP	HSE 4345
Sigmaplane HB	Sigma Coatings BV	AVP	HSE 3487

887 ZINEB and CUPROUS THIOCYANATE

* Cruiser Superior	International Paint Ltd	AVA AVP	HSE 3785

888 ZINEB and CUPROUS THIOCYANATE and TRIBUTYLTIN METHACRYLATE

Superyacht 800 Antifouling	International Paint Ltd	AVP	HSE 4778

889 ZINEB and CUPROUS THIOCYANATE and TRIBUTYLTIN METHACRYLATE and TRIBUTYLTIN OXIDE

Intersmooth Hisol BFA948 Orange	International Paint Ltd	AVP	HSE 4281
Micron 25 Plus	International Paint Ltd	AVP	HSE 3402
Superyacht Antifouling	International Paint Plc	AVP	HSE 5462

ANTIFOULING PRODUCTS

Product Name	Marketing Company	Use	Reg. No

890 ZINEB and CUPROUS THIOCYANATE and TRIBUTYLTIN OXIDE

Cruiser Copolymer	International Paint Ltd	AVP	HSE 3514

891 ZINEB and ZINC OXIDE and CUPROUS OXIDE and COPPER RESINATE

* Sigma Pilot Antifouling LL/TF	Sigma Coatings BV	AVA AVP	HSE 3480
Sigma Pilot Ecol Antifouling	Sigma Coatings Ltd	AVA AVP	HSE 4933

892 ZIRAM and CUPROUS OXIDE

Awlstar Gold Label Anti-Fouling	Grow Group Incorporated	AVA AVP	HSE 3674
Cheetah Antifouling	Blakes Marine Paints Ltd	AVA AVP	HSE 3223
* Coppercoat	International Paint Ltd	AVA AVP	HSE 3560
* Cruiser Premium	International Paint Ltd	AVA AVP	HSE 4277
* Grassline TF Anti-Fouling Type M394	W and J Leigh and Company	AVA AVP	HSE 3460
# Micron	International Paint Plc	AVA AVP	HSE 3395
Micron CSC	International Paint Ltd	AVA AVP	HSE 3558

893 ZIRAM and CUPROUS OXIDE and TRIBUTYLTIN OXIDE

* Devoe ABC 2 Antifouling	Devoe Coatings BV	AVP	HSE 3313

894 ZIRAM and CUPROUS OXIDE and ZINC OXIDE

Hempel's Antifouling Classic 7654A	Hempel Paints Ltd	AVA AVP	HSE 5297
Hempel's Antifouling Combic 7199A	Hempel Paints Ltd	AVA AVP	HSE 5295
Hempel's Antifouling Nautic 7190A	Hempel Paints Ltd	AVA AVP	HSE 5293

895 ZIRAM and CUPROUS THIOCYANATE

* Boot Top Plus	International Paint Ltd	AVA AVP	HSE 3390

896 ZIRAM and CUPROUS THIOCYANATE and TRIBUTYLTIN METHACRYLATE

Takata Seaqueen	Nippon Oil and Fats Co Ltd	AVP	HSE 4055

5
HSE PRODUCT NAME INDEX

"Bop" Crawling Insect Killer   *407 457*
"Bop" Crawling Insect Killer (MC)   *407 457*
"Bop" Flying and Crawling Insect Killer   *353 462 469*
"Bop" Flying and Crawling Insect Killer (Water Based)   *353 462 469*
"Bop" Flying Insect Killer   *353 462 469*
"Bop" Flying Insect Killer (MC)   *353 462 469*
"Flak"   *465 476*
(ATMC 101) Amercoat 610   *603 778 846*
(ATMC 102) Amercoat 611   *603 778 846*
(ATMC 120) Amercoat 612   *603 778 846*
(ATMC 121) Amercoat 613   *603 778 846*
(ATMC 122) Amercoat 615   *603 778 846*
(ATMC 123) Amercoat 557   *586 717 842*
(ATMC 129) Amercoat 279   *613 834*
(ATMC 135) Amercoat 614   *608 807*
(ATMC 170) Amercoat 699   *605 792 847*
(ATMC 172) Dimetcote 602   *790 810*
(ATMC 173) Dimetcote 601   *791 866*
151 Dry Fly and Wasp Killer   *435 460*
A-Zygo 3 Sterilising Solution   *277*
A1 Antifouling   *593 735 774 798*
A11 Antifouling 072014   *570 674 738 779 839*
A2 Antifouling   *492 509 685 867 875*
A2 Teflon Antifouling   *492 509 685 867 875*
A3 Antifouling   *493 550*
A3 Antifouling 072015   *501 573 680 862*
A3 Antifouling 072016   *505 637 681 864*
A3 Teflon Antifouling   *493 550*
A3 Teflon Antifouling 062015   *505 637 681 864*
A4 Antifouling   *566 669*
A4 Antifouling 072017   *501 573 680 862*
A4 Antifouling 072018   *505 637 681 864*
A4 Teflon Antifouling 062017   *505 637 681 864*
A6 Antifouling   *510 687 876*
A7 Teflon Antifouling   *508 686 868 878*
A7 Teflon Antifouling 072019   *508 686 868 878*
AA155/00 Industrial Wood Preservative   *126 239*
AA155/03 Industrial Wood Preservative   *126 239*
Abicide 82   *271*
ABL Aqueous Wood Preserver Concentrate 1:9   *108 38*
ABL Brown Creosote   *69*
ABL Universal Woodworm Killer DB   *123 226*
ABL Wood Preservative (B)   *223*
ABL Wood Preservative (D)   *111*
Ablative Antifouling AF28   *594 737 781 844*
Acclaim   *432 445*
Acclaim Plus   *385 395*
Actellic 25 EC   *199 414*
Actellic Dust   *414*
Actellic M20   *414*
Actomite   *335 437*
Admiralty Antifouling (to TS10240)   *547*

Admiralty Antifouling Black (to TS10239)   *560 627*
AF Seaflo Mark 2-1   *594 737 781 844*
AF Seaflo Mark 2-1 Light/Dark   *594 737 781 844*
AF Seaflo Mark 2-2 Light/Dark   *598 746 811 845*
AF Seaflo Mark 3 Light/Dark   *598 746 811 845*
AF Seaflo SP-1 Light/Dark   *609 808 849*
AF Seaflo SP-2 Light/Dark   *609 808 849*
AF Seaflo Z-100 HS-1   *591 729 766*
AF Seaflo Z-100 HS-1 Light/Dark   *591 729 766*
AF Seaflo Z-100 HS-2 Light/Dark   *597 745 809*
AF Sealo Z-100 LE-1 Light/Dark   *591 729 766*
AF Seaflo Z-100 LE-2 Light/Dark   *590 723*
AF Seaflo Z-100 LE-HS-1   *591 729 766*
AF Seaflo Z-100 LE-HS-1 Light/Dark   *591 729 766*
AF Seaflo Z-100 LE-HS-2 Light/Dark   *597 745 809*
AF Seaflo Z-100-1 Light/Dark   *591 729 766*
AF Seaflo Z-100-2 Light/Dark   *590 723*
AF Seaflo Z-200   *642 724*
AF Seaflo Z-200 C   *590 723*
AF29 Desoto Antifouling   *533 831*
Aircraft Aerosol Insect Control   *475*
AL-27   *503 629*
AL63 Crawling Insect Killer   *324*
Albany Fungicidal Wash   *271*
Algae Remover   *277 32*
Algicide Antifouling   *547*
Algicide E   *296 297 299 300 314 316*
Algicide Red Antifouling   *494 616 821*
Algotox   *280 288 291*
Amercoat 275   *547*
Amercoat 277   *547*
Amercoat 279   *613 834*
Amercoat 67E   *533 831*
Amercoat 697   *580 704 841*
Amercoat 698 HS   *580 704 841*
Amercoat 70E   *533 831*
Amercoat 70ESP   *533 831*
Amogas Ant Killer   *392*
Ant and Crawling Insect Killer   *374 454*
Ant and Crawling Insect Powder   *340*
Ant and Insect Powder   *340*
Ant Gun!   *358 426*
Ant Gun! 2   *333 393*
Antec E-C Kill   *371*
Antel Dual Purpose (BP) Concentrate (Water Dilutable)   *190 229*
Antel Woodworm Killer (P) Concentrate (Water Dilutable)   *174*
Anti Fungus Wash   *295 298*
Anti-Fouling Coating ARE3689   *701*
Anti-Fouling Paint 161P (Red and Chocolate to TS10240)   *547*

HSE PRODUCT NAME INDEX

Anti-Mould Solution   *22 271*
Antifouling Alusea   *513 748 825*
Antifouling Alusea Black   *744 803*
Antifouling Alusea Blue   *744 803*
Antifouling Alusea Red   *744 803*
Antifouling Alusea White   *744 803*
Antifouling Broken White DL-2253   *628*
Antifouling Equator KK   *603 778 846*
Antifouling HB 66 Ocean Green   *643 732 769*
Antifouling Paint 161P (Red and Chocolate to TS10240)   *547*
Antifouling Paint 161P (to TS10240)   *547*
Antifouling Sargasso   *603 778 846*
Antifouling Sargasso Non-Tin   *547*
Antifouling Seaconomy   *594 737 781 844*
Antifouling Seaconomy 200   *594 737 781 844*
Antifouling Seaconomy 300   *594 737 781 844*
Antifouling Seaflex   *591 594 729 737 766 781 844*
Antifouling Seaflex Sealinc Blue 110   *643 732 769*
Antifouling Seaguardian   *547*
Antifouling Seaguardian (Black and Blue)   *613 834*
Antifouling Seamaster Non-Tin   *547*
Antifouling Seamate HB 22   *594 737 781 844*
Antifouling Seamate HB 33   *594 737 781 844*
Antifouling Seamate HB 33 BSL Blue   *646 739 784 857*
Antifouling Seamate HB 33 Dark Red Non-Tin   *499 551*
Antifouling Seamate HB 33 DR   *594 737 781 844*
Antifouling Seamate HB 33 Light Red Non-Tin   *499 551*
Antifouling Seamate HB 33 LR   *594 737 781 844*
Antifouling Seamate HB 66 Black   *591 729 766*
Antifouling Seamate HB 66 Dark Red   *594 737 781 844*
Antifouling Seamate HB 66 Dark Red Non-Tin   *499 551*
Antifouling Seamate HB 66 Light Red   *594 737 781 844*
Antifouling Seamate HB 66 Light Red Non-Tin   *499 551*
Antifouling Seamate HB 99 Black   *591 729 766*
Antifouling Seamate HB 99 Dark Red   *591 729 766*
Antifouling Seamate HB 99 Light Red   *591 729 766*
Antifouling Seamate HB Blue   *643 732 769*
Antifouling Seamate HB Green   *643 732 769*
Antifouling Seamate HB Special Black   *643 732 769*
Antifouling Seamate HB22   *599 752 872 886*
Antifouling Seamate HB22 Roundel Blue   *646 739 784 857*
Antifouling Seamate HB33   *599 752 872 886*
Antifouling Seamate HB66   *594 737 781 844*
Antifouling Seamate HB66 Black   *646 739 784 857*
Antifouling Searose   *547*
Antifouling Seasafe   *547*
Antifouling Seavictor 40   *613 834*
Antifouling Seavictor 50   *522 621 830*
Antifouling Seven Seas   *586 717 842*
Antifouling Super Tropic   *547*
Antifouling Tropic   *613 834*
Antifouling Tropic Black   *547*
Antkiller Spray 2   *333 393*
Apex Fly-Killer   *435 460*
Aqua Reslin   *402 440*
Aqua Reslin Premium   *402 440*
Aqua Reslin Super   *402 440*
Aqua-Tech Antifouling Berth   *694*
Aquacleen   *613 834*
Aquakill Ecology   *174*
Aquarius   *547*
Aquarius AL   *633 670*
Aquarius Extra Strong   *566 669*
Aquaseal Wood Preserver   *128 260*
Aquashield   *678 860*
Aquaspeed   *499 551*
Aquaspeed White   *503 629*
Aquatect Wood Preservative Stain   *159 4*
Aqueous Fungicidal Irrigation Fluid   *309*
Aqueous Fungicidal Irrigation Fluid Concentrate   *309*
Aqueous Fungicide-Insecticide Concentrate   *156 241*
Aqueous Fungicide-Insecticide   *156 241*
Aqueous Wood Preserver   *108 38*
Arborsan 3   *217 94*
Arborsan 3 Creosote   *92*
Arborsan 4   *70 93*
Arborsan 4 Creosote   *92*
Arborsan 6   *70 93*
Arborsan 6 Creosote   *92*
Arrow Residual Powerkill   *392*
Awlgrip Awlstar Gold Label Antifouling   *613 834*
Awlstar Gold Label Anti-Fouling   *626 892*
Azamethiphos Fly Spray   *324*
Aztec BBQ Patio Candles   *347*
B & Q Creosote   *69*
B and Q Ant Killer Lacquer   *356*
B and Q Antkiller Spray   *358 426*
B and Q Clear Wood Preserver   *255*
B and Q Creosote   *92*
B and Q Exterior Wood Preservative   *255*
B and Q Formula Wood Preserver   *255*

449

B and Q New Formula Ant Killer Spray 348
BBQ Fly Repellent Terracotta Pot Candle 347
BBQ Fly Repellent Candle 347
BBQ Patio Candles 347
Bactdet D 292
Bactimos Flowable Concentrate 325
Bactimos Wettable Powder 325
Baracaf Cockroach Control Sticker 339
Baracaf Cockroach Control Sticker (Domestic) 339
Barkep Wood Preserver 250
Barkep Wood Preserver Green 73
Barrettine Creosote 92
Barrettine Green Wood Preserver 73
Barrettine New Preserver 223
Barrettine New Universal Fluid 190 229
Barrettine New Universal Fluid D 121 191 227
Barrettine New Wood Preserver 123 226
Barrettine New Woodworm Fluid 174
Barrettine New Woodworm Killer 174
Barrettine Timberguard 22
Barrettine WSP Wood Treatment 6
Bartoline Dark and Light Creosote 69
Basilit B 85 43
Basilit Bauholz-KD 32
Basiment 540 158
Basiment 560 6
Basiment NT 32
Bayer AFC 662
Bayer WPC 2-1.5 218
Bayer WPC 2-25 218
Bayer WPC 2-25-SB 218
Bayer WPC-3 1
Bayer WPC-4 1
Bedclear 277
Beeline Fungicidal Wash 271
Bendiocarb Dusting Powder 327
Bendiocarb Wettable Powder 327
Betterware Anti-Fungus and Mildew Wipe 271
Betterware Fungus and Mildew Killer 271
Big D Ant and Crawling Insect Killer 465 476
Big D Cockroach and Crawling Insect Killer 374 454
Big D Fly and Wasp Killer 446
Big D Rocket Fly Killer 374 454
Bio Woody 139 62
Bio-Kil Board Preservative 10
Bio-Kil Boron Paste 129
Bio-Kil Cunap Pole Wrap 73
Bio-Kil Dentolite Solution 277
Bio-Kil Scrub Out Black Mould/Refill 271
Bio-Kil SR Pole Wrap 133 74
Biochem Masonry Fungicide 268
Biochem Masonry Gel 311
Biocidal Wash 271

Biocide G 295 298
Bioclean 692
Bioclean 911 289
Bioclean DX 692
Biokil 147 40
Biokil B40 Paste Wood Preservative 129
Biokil Emulsion 180 41
Biokil Timbor Rods 129
Biomek 15 5
Biosol RTU 336
Biscon AF 613 834
BL 1111 1
Black Anti-Fouling Paint 317 (to TS10239) 560 627
Black Antifouling AF21 (to TS10239) 560 627
Black Antifouling Paint 317 (to TS10239) 560 627
Blackfriar Anti Mould Solution 195/9 295 298
Blackfriar Wood Preserver 170
Blackfriars Gold Star Clear 121 191 227
Blackfriars Green Wood Preserver 73
Blackfriars New Wood Preserver 123 226
Blade 336
Blattenex 20 418
Blended Coal Tar Creosote 69
Blueline Copper SBA100 547
Blueline SBA900 Series 596 742 786 884
Blueline SPC Tin Free SBA700 Series 625 881
Blueline Tropical SBA300 566 669
BMS Biowash 295 298
BN Algae Remover 277
BN Mosskiller 277
Bolt Ant and Crawling Insect Killer 407 457
Bolt Crawling Insect Killer 350 425
Bolt Dry Flying Insect Killer 419
Bolt Flying Insect Killer 446
Bolt Natural Flying Insect Killer 419
Boot Top 628
Boot Top Plus 661 895
Boot Top Plus (Gull White) 503 629
Boots Ant and Crawling Insect Killer 465 476
Boots Ant Killer Powder 392
Boots Ant Trap 466
Boots Dry Fly and Wasp Killer 461 473
Boots Dry Fly and Wasp Killer 3 461 473
Boots Electric Mosquito Killer 331
Boots Flea Spray 465 476
Boots Fly and Wasp Killer 446
Boots Moth Killer 360
Boots Pump Action Fly and Wasp Killer 419
Boots Slow Release Fly and Wasp Killer 360
Boots Slow Release Fly Killer 360
Boots Small Space Fly and Moth Killer 360
Boots Small Space Moth and Fly Killer 360
Boots UK Flying Insect Killer 331

Bop   *366 411 463 465 470 476*
Bop Flying and Crawling Insect Killer (MC)   *353 462 469*
Boracol 10 Rh   *131 282 294 39*
Boracol 10RH Surface Biocide   *273 293*
Boracol 20   *129*
Boracol 20 Rh   *131 39*
Boracol B40   *129*
Boracol B8.5 RH Mouldicide/Wood Preservative   *131 282 294 39*
Boric Acid Concentrate   *339*
Boric Acid Powder   *339*
Brimpex ULV 1500   *465 476*
Britannic K504 Long Life Anti-Fouling   *547*
Britannic M351 Anti-Fouling Extra   *547*
Britannic M451 Antifouling Super   *547*
Broads Antifouling   *547*
Broads Black Antifouling   *516 619 826*
Broads Freshwater Antifouling   *511 695*
Broads Red Antifouling   *495 822*
Broads Sweetwater Antifouling   *503 629*
Brolac Fungicidal Solution   *310*
Bronze   *530*
Brown Timber Preservative 2   *223*
Brown Wood Preservative   *223*
Brunol ATP   *152 171*
Brunol ATP New   *188 222*
Brunol OPA   *175 9*
Brunol PBO   *190 229*
Brunol PC   *174*
Brunol PP   *14 176*
Brunol PY   *174*
Brunol Special P   *190 229*
Brunol SPI   *14 176*
Brunol STP   *188 222*
Brunosol Concentrate   *204 309*
BS 144 Creosote   *69*
Builders Mate Creosote   *69*
Busan 1009   *159 4*
Buzz-Off   *331*
Buzz-Off 2   *331*
C-Worthy   *613 834*
Camco Ant Powder   *327*
Camco Insect Spray   *329 447*
Camrex A/F 1   *547*
Camrex A/F 2   *547*
Camrex A/F 3   *601 760*
Camrex A/F 4   *547*
Camrex A/F 5   *601 760*
Camrex C-Clean EXA   *594 737 781 844*
Camrex C-Clean HB EXA   *594 737 781 844*
Camrex C-Clean HB LA   *594 737 781 844*
Camrex C-Clean LA   *594 737 781 844*
Canovel – Vac Attack Fragrant Insecticidal Carpet Freshener   *475*

Canovel Pet Bedding and Household Spray   *386 428*
Carbo Creosote   *69*
Carbo Wood Preservative   *250 73*
Cedarwood Protector   *111*
Cedarwood Special   *111*
Cedasol 2304 RTU   *156 241*
Cedasol 2320   *156 241*
Cedasol Ready To Use (2306)   *198 267*
Cedasol Ready To Use (2310)   *178 20*
Celbor   *216 59*
Celbor M   *43*
Celbrite M   *108 38*
Celbrite MT   *110 161 35 54*
Celbrite NAP 100   *212*
Celbrite TC   *159 4*
Celbronze B   *108 38*
Celcure A Concentrate   *205 28 87*
Celcure A Fluid 10   *205 28 87*
Celcure A Fluid 6   *205 28 87*
Celcure A Paste   *205 28 87*
Celcure AO   *27 66 79*
Celcure B   *208 64 90*
Celcure CB Paste   *132 68 82*
Celcure CB Salts   *206 46 65 89*
Celcure CB90   *207 53 88*
Celcure CCA Type C   *27 66 79*
Celcure O   *208 64 90*
Celgard CF   *206 46 65 89*
Celgard FP   *129*
Celkil 90   *15 5*
Celpruf BZ   *256 60*
Celpruf BZ WR   *256 60*
Celpruf BZP   *196 257 58*
Celpruf BZP WR   *196 257 58*
Celpruf CP Special   *126 239*
Celpruf JP   *167 245*
Celpruf JP WR   *167 245*
Celpruf PK   *150 168 242*
Celpruf PK WR   *150 168 242*
Celpruf Primer   *235*
Celpruf Primer TN   *233*
Celpruf Primer ZV   *265*
Celpruf TN   *233*
Celpruf TN WR   *233*
Celpruf TNM   *192 234*
Celpruf TNM WR   *192 234*
Celpruf TP   *248*
Celpruf TP WR   *248*
Celpruf TPM   *194 249*
Celpruf TMP WR   *194 249*
Celpruf TZC   *102 220*
Celpruf ZO   *265*
Celpruf ZO WR   *265*
Celpruf ZOP   *198 267*

Celpruf ZOP WR   *198 267*
Cementone Multiplus   *103 225*
Cementone Woodworm Killer   *95*
Challenger Antifouling   *499 551*
Cheetah Antifouling   *626 892*
Chel Creosote   *92*
Chel Dry Rot Killer for Masonry and Brickwork   *33 7*
Chel Wood Preserver   *170*
Chel Wood Preserver/Woodworm Dry Rot Killer   *170*
Chel-Wood Preserver/Woodworm Dry Rot Killer   *195 263*
Chel-Wood Preserver/Woodworm Dry Rot Killer BP   *190 229*
Chelec Creosote   *92*
Chelwash   *269 278*
Chem-Kil   *435 460*
Chemcrest   *480*
Chemflake   *692*
Chemglide   *692*
Chemlife   *528*
Chemsilk   *692*
Chirton Ant and Insect Killer   *392*
Chirton Dry Fly and Wasp Killer   *435 460*
Chirton Fly and Wasp Killer   *435 460*
Chlorinated Rubber Antifouling AF19   *581 707*
Chocolate Antifouling AF23 (to TS10240)   *547*
Chugoku AF ST   *613 834*
Chugoku AF T   *613 834*
Ciba-Geigy KO Fly and Wasp Killer   *407 457*
Citromax Citronella Insect Repellent Liquid Candle Oil   *347*
Clarkes Woodcare   *212*
Classica 3786/093 Red   *547*
Clear Timber Preservative 2   *223*
Clear Wood Preservative   *223*
Clearmold D   *289*
Cleartreat   *156 241*
Co-Op Creosote   *69 92*
Co-Op Exterior Wood Preserver   *255*
Coal Tar Creosote   *69*
Coal Tar Creosote Blend   *69*
Coaltec 50   *204 309*
Coaltec 50TD   *204 309*
Cobra   *577 693*
Cobra V   *613 834*
Cockroach Bait   *380*
Colourfast Protector   *108 38*
Colturiet TCN Antifouling (Set and N Set)   *588 719 871 882*
Commercial Copolymer Antifouling   *697 733 770*
Commercial Copolymer Extra Antifouling   *512 698 726 757*
Conc Quat   *277*

Conductive Pilt 80 RFU   *10*
Conductive Pilt RFU   *235*
Control Fluid FB   *43*
Cooper Fly Spray   *419*
Cooper's Copolymer Antifouling   *613 834*
Coopers Fly Spray BB (Ready to Use)   *332 337*
Coopers Fly Spray EB (Ready to Use)   *338 438*
Coopers Fly Spray N   *419*
Coopers Fly Spray N (Ready to Use)   *419*
Coopers Head to Tail Animal Bedding Spray   *396 429*
Coopers Stomoxin P   *392*
Coopex 25% EC   *392*
Coopex Insect Powder   *392*
Coopex Maxi Smoke Generator   *392*
Coopex Mini Smoke Generator   *392*
Coopex S25% EC   *402 440*
Coopex Smoke Generator   *392*
Coopex WP   *392*
Copper Bottom Paint   *613 834*
Copper Naphthenate Solution to BS 5056   *73*
Copper/Bronze Antifouling Powder AF26   *529*
Copperbot   *530*
Coppercoat   *547 626 892*
Copperpaint   *547*
Corry's Fly and Wasp Killer   *407 457*
Cover Plus Clear Wood Preservative   *250*
Crackdown   *355*
Craig and Rose Green Antifouling   *527 557*
Craig and Rose Red Oxide Antifouling   *527 557*
Crawling Insect and Ant Killer   *333 393*
Creepy Crawly Gun!   *333 393*
Creosote   *69 92*
Creosote (Light Brown and Dark Brown)   *92*
Creosote 131C   *92*
Creosote Blend   *69 92*
Creosote Blend MK1 (Medium Dark)   *69*
Creosote Blend Mk3 (Light Golden)   *69*
Creosote Blended Wood Preservative   *69*
Creosote BS 144   *69*
Creosote BS 144 (2)   *69*
Creosote BS 144 (3)   *69*
Creosote BS 144 Type III   *69*
Creosote Coke Oven Oil   *92*
Creosote Emulsion   *92*
Creosote MK2   *92*
Croda Green Wood Preserver   *73*
Croda Timber Protector   *8 96*
Croda Wood Preserver   *6*
Crodec Antifouling Super   *526 535 558*
Cromessol Fly-Away   *360*
Cromessol Insect Killer   *435 460*
Crown Fungicidal Wash   *295 298*
Crown Fungicide Insecticide Concentrate   *103 225*

Crown Timber Preservative 255
Crown Trade Fungicidal Solution 310
Crown Trade Stronghold Fungicidal Solution 310
Crown Woodworm Concentrate 95
Cruiser Copolymer 653 787 890
Cruiser Premium 566 626 669 892
Cruiser Premium (Light Grey) 547
Cruiser Superior 633 659 670 887
Cruiser Superior (White) 503 629
Cruiser Superior 100 Series 515 630
Crysolite Glaramara Concentrated Timber Treatment 1
Crystic Copperclad 70PA 530
Cube Concentrate Fluid for Dry Rot 204 309
Cube F and I Concentrate/Grade BL 154 228
Cube Fluid for Dry Rot 204 309
Cube Preserver – Grade B – Light Brown 223
Cube Solvent F and I Fluid/Grade BL 154 228
Culmstock Mothproofer 360
Culmstock Slow Release Fly Killer 360
Cuprinol 5 Star Complete Wood Treatment S 178 20
Cuprinol Combination Grade S 178 20
Cuprinol Cuprotect Exterior Fungicidal Wash 277
Cuprinol Cuprotect Exterior Fungicide 277
Cuprinol Cuprotect Fungicidal Patio Cleaner 277
Cuprinol Cuprotect Patio Cleaner 277
Cuprinol Decorative Wood Preserver 112 19
Cuprinol Decorative Wood Preserver Red Cedar 112 19
Cuprinol Difusol S 195 263
Cuprinol Dry Rot Killer S for Brickwork and Masonry 309
Cuprinol Hardwood Basecoat 111
Cuprinol Hardwood Basecoat Meranti 111
Cuprinol Insecticidal Emulsion Concentrate 174
Cuprinol Low Odour 5 Star Complete Wood Treatment 178 20
Cuprinol Low Odour Decorative Wood Preserver 112 19
Cuprinol Low Odour Decorative Wood Preserver Red Cedar 112 19
Cuprinol Low Odour Woodworm Killer 174
Cuprinol Magnatreat F 112 19
Cuprinol Magnatreat XQD 112 19
Cuprinol No More Mould Fungicidal Spray 282 294
Cuprinol Preservative Base 112 19
Cuprinol Preservative-Wood Hardener 111
Cuprinol Wet and Dry Rot Killer for Timber(s) 18
Cuprinol Wood Preserver 18
Cuprinol Wood Preserver Clear S 18
Cuprinol Wood Preserver Dark Oak S 18
Cuprinol Wood Preserver Green S 16
Cuprinol Woodworm Killer S 174
Cuprinol Woodworm Killer S (Aerosol) 174
Cuprinol WP CPTF 2 91
Cuprinol WP CPTF 3 83
Cuprinol WP CPTF 4 265
Cuprinol WP CPTF 5 264
Cuprisol F 18
Cuprisol FN 178 20
Cuprisol P 17 177 21
Cuprisol WR 178 20
Cuprisol XQD 18
Cuprisol XQD Special 178 20
Cupron Plus TF 613 834
Cupron TF 613 834
Cuprotect Exterior Fungicide 277
Cuprotect Fungicidal Spray 282 294
Cuprotect Interior Mould Killer 282 294
Cuprotect Patio Cleaner 277
Cut End 178 20
Cymperator 348
Cyperguard BC Wood Treatment 103 225
Cyperguard Woodworm Killer 95
Cyperkill 10 348
Cyperkill 10 WP 348
Cyperkill Plus WP 352 449
Cypermethrin 10% EC 348
Cypermethrin 10% WP 348
Cypermethrin Lacquer 348
Cypermethrin PH-10EC 348
D-Stroy 407 457
Dacrylate Fungicidal Wash Solution 295 298
Dairy Fly Spray 419
Dapaflow Antifoul 565 666 837
Dark Brown Creosote 69
Dark Creosote Emulsion 69
Dark Green Timber Preservative 2 224 77
Dark Green Wood Preservative 224 77
Days Farm Fly Spray 419
DDVP (Toxicant) Strip 360
Deadly Nightshade 269 278
Decor-8 Wood Preservative Brown 156 241
Decor-8 Wood Preservative Clear 156 241
Decor-8 Wood Preservative Green 157 238 75
Decor-8 Wood Preservative Red Cedar 156 241
Deepflo 11 Inorganic Boron Masonry Biocide 282 294
Deepkill 14 176
Deepkill F 10
Deepkill I 174
Deepwood "Clear" Insecticide Emulsion Concentrate 174
Deepwood 20 Inorganic Boron Wood Preservative 131 39

Deepwood 50 Inorganic Boron Wood Preservative Paste   *131 282 294 39*
Deepwood 8 Micro Emulsifiable Insecticide Concentrate   *174*
Deepwood Fl Dual Purpose Emulsion Concentrate   *190 229*
Deepwood Standard Emulsion Concentrate   *174*
Defest Flea Free   *363 381*
Delta Cockroach and Ant Spray   *407 457*
Delta Fly Spray   *465 476*
Delta Insect Powder   *392*
Demise   *371*
Demon 40 WP   *348*
Denso Mouldshield Biocidal Cleanser   *292*
Denso Mouldshield Surface Biocide   *292*
Deosan Fly Spray   *465 476*
Dethlac Insecticidal Lacquer   *356*
Detia Ant Bait   *466*
Detia Crawling Insect Spray   *396 429*
Detia Pyrethrum Spray   *419*
Devatern 0.5L   *95*
Devatern 1.0 L   *95*
Devatern 1.0 LP   *101 165*
Devatern EC   *95*
Devatern Wood Preserver   *103 225*
Devchlor Antifouling Paint Red   *601 760*
Devchlor Chlorinated Rubber Antifouling   *581 707*
Devcol Household Insect Spray   *329 447*
Devcol Liquid Ant Killer   *416*
Devcol Universal Pest Powder   *419*
Devcol Wasp Killer   *416*
Devoe ABC 2 Antifouling   *604 789 893*
Devoe ABC 3 Black Antifouling   *547*
Devoe ABC 3 Red Antifouling   *547*
Devran 222 All Seasons Permanent   *601 760*
Devran MCP Antifouling Red   *590 723*
Devran MCP Antifouling Red Brown   *590 723*
Di-Fly   *446*
Dianex   *443*
Dipsar   *156 241*
Dipsar G R   *105 266*
DIY Time Creosote   *92*
Do It All Fungicidal Solution   *310*
Doff Ant Killer Spray   *348*
Doff Antlak   *356*
Domexyl Paste Wood Preservative   *190 229*
Doom Ant and Crawling Insect Killer Aerosol   *374 454*
Doom Ant and Insect Powder   *375*
Doom Flea Killer   *376 456*
Doom Moth Proofer Aerosol   *376 456*
Doom Tropical Strength Fly and Wasp Killer Aerosol   *407 457*

Double Action Timber Preservative for Doors   *123 226*
Double Action Wood Preservative   *123 226*
Double Shield Antifouling   *537 559*
Dragon Brand Moth Balls   *387*
Dragon Flykiller   *407 457*
Dricon   *43*
Drione   *419*
Dry Fly Killer   *446*
Dry Pin   *129*
Dry Rot and Wet Rot Fluid   *164 302*
Dry Rot Fluid (D) EC   *256 60*
Dry Rot Treatment for Masonry   *309*
Dualprime F   *111*
Dulux Exterior Preservative Basecoat   *124 231 253*
Dulux Preservative Wood Primer   *247 251 63*
Dulux Weathershield Exterior Preservative Primer   *247 251 63*
Dulux Weathershield Fungicidal Solution   *295 298*
Dulux Weathershield Preservative Basecoat   *247 251 63*
Dulux Weathershield Timber Preservative   *258 78*
Dursban 4TC   *342*
Dursban LO   *342*
Earnshaws Fungal Wash   *269 278*
Ecology Fungicide Insecticide Aqueous (Concentrate) Wood Preservative Dual Purpose   *190 229*
Ecology Fungicide Insecticide Concentrate   *103 225*
Empire 20   *342*
Ensele 3424   *209 86*
Ensele 3426   *51 84*
Ensele 3427   *51 84*
Ensele 3430   *214*
Envar 6/14 EC   *465 476*
Envar 6/14 Oil   *465 476*
Environmental Timber Treatment   *174*
Environmental Timber Treatment Paste   *174*
Environmental Woodrot Treatment   *135 137 295 298*
Envoy K926 High Performance Anti-Fouling   *606 797*
Envoy L840 Anti-Fouling (BSC 356)   *705*
Envoy TF 100   *547*
Envoy TF 200   *482 548*
Envoy TF 300   *578 696*
Envoy TF 400   *481 500 562 654 818*
Envoy TF 500   *481 500 562 654 818*
Equatorial   *625 881*
Esmi CA: 20 Concentrated Algicide   *277*

Esmi Mouldicide for Walls, Ceilings and Cupboards *277*
ETOC Liquid Vaporiser *417*
Even TF *502 617 823*
Even TF Light Grey *506 656 824*
EXL-AF *606 797*
EXL-AF E *606 797*
Exterior Mouldicide *280 288 291*
Exterior Preservative Primer *128 260*
Exterior Ronseal Satin Wood Finish *111*
Exterior Wood Preserver S *18*
Fads Homestyle Fungicidal Solution *310*
Fads Wood Preservative Brown *156 241*
Fads Wood Preservative Clear *156 241*
Fads Wood Preservative Green *157 238 75*
Fads Wood Preservative Red Cedar *156 241*
Febflex Fungicide *295 298*
Febwood Brown *235*
Febwood Exterior Wood Preserver *235*
Febwood Green *235*
Febwood Preservative WP3 *235*
Febwood Wood Preserver *128 260*
Fen-Tan *212*
Fence Protector *204*
Fence 'n' Shed Concentrate *204*
Fendona 1.5 SC *323*
Fendona 25 SC *323*
Fendona 6SC *323*
Fendona ASC *323*
Fendona Lacquer *323*
Fendona WP *323*
Fenitrothion Dusting Powder *371*
Fenitrothion Emulsion Concentrate *371*
Fenitrothion Wettable Powder *371*
Fentex Elite *160 3 56*
Fentex NP-UF *1*
Ficam 20W *327*
Ficam D *327*
Ficam Plus *328 421*
Ficam W *327*
Ficam Wasp and Hornet Killer *327*
Ficam Wasp and Hornet Spray *327*
First Class Mosquito Killer Travel Pack *331*
Flag Brand Wood Preservative *235*
Flag Brand Wood Preservative Green *73*
Flamil Finale *396 429*
Flamil Finale Super *419*
Flea Kill from Bob Martin. Insecticidal Carpet and Furniture Deodoriser *392*
Fleegard *333 393*
Flex-o-Flake *692*
Flexarb Clear Preservative *250*
Flexarb Timber Coating *111*
Flexgard IV Waterbase Preservative *483*
Flexgard V Waterbase Preservative *483*
Flexgard VI Waterbase Preservative *485 546*
Flit Flying and Crawling Insect Killer *419*
Floret Fast Knock Down *467*
Florida Antifouling AF13 *581 707*
Fly and Ant Killer *474 477*
Fly and Wasp Killer *407 446 457*
Fly Killer *360 435 460*
Fly Spray *407 457*
Fly, Wasp and Mosquito Killer *461 473*
Flying Insect Killer *360 374 454*
Flying Insect Killer Faster Knockdown *374 454*
Flykil *360*
Fongix SE Total Treatment for Wood *179 276 303 31*
Forcecontact AFUR514 (Base) *563 663*
Forest Friends Crawling Pest Control *333 393*
Forest Friends Flying Pest Control *333 393*
Fortefog *419*
Four Seasons Creosote *69*
Fresh-A-Pet Insecticidal Rug and Carpet Freshener *392*
Freshways Slow Release Insect Killer *360*
Fumite Lindane Generator Size 10 *145 375*
Fumite Lindane Generator Size 40 *145 375*
Fumite Lindane Pellet No 3 *375*
Fumite Lindane Pellet No 4 *375*
Fungaside *284 313*
Fungicidal Algaecidal Bacteriacidal Wash *269 278*
Fungicidal Preservative *18*
Fungicidal Solution *277*
Fungicidal Solution FL.2 *269 278*
Fungicidal Wall Solution *270*
Fungicidal Wash *135 137 269 278 295 298 309 310*
Fungicidal Wash Solution *269 278*
Fungicide Insecticide *103 225*
Fungicide Insecticide Wood Preservative *156 241*
Fungicide/Insecticide Microemulsion M9 *14 176*
Fungiguard *269 278*
Fungishield Sterilising Solution Concentrate GS36 *271*
Fungishield Sterilising Solution GS37 *271*
Fungo *292*
Fungol Primer 55 *116 141 148*
Funnel Trap Insecticidal Strip *360*
Gainpaste *103 225*
Gainserv 140 *134 213*
Gainserv 680 *164*
Gainserv 681 *73*
Gainserv 682 *164*
Gainserv 683 *164*
Gainserv 684 *109 37 99*

Gainserv 685   *212 311*
Gainserv 686   *95*
Gainserv Concentrate   *134 213*
Gainserv Polymeric   *134 213*
Gallwey ABS   *134 213*
Gallwey BMC   *159 4*
Gallwey SA   *15 5*
Gammexane Smoke Generator No 22   *145 375*
Gelert Mosquito Repellent   *467*
General Purpose Antifouling Paint   *547*
Glen Wood Care Green   *73*
Glen Wood Care Wood Preservative   *32*
Globol Pyrethrum Electrical Evaporator   *331*
Globol Shake and Spray Insect Killer   *419*
Globol Small Space Fly and Moth Strip   *360*
Gloquat RP   *277 32*
Gold Label Kennel and Stable Powder   *392*
Golden Creosote   *92*
Goodboy Household Flea Spray   *465 476*
Grafo Anti-Foul SW   *566 669*
Granyte Farmcare Green Wood Preservative   *73*
Grassline ABL Anti-Fouling Type M127   *702 709*
Grassline ABL Anti-Fouling Type M190   *702 709*
Grassline ABL Anti-Fouling Type M336   *579 703 764*
Grassline ABL Anti-Fouling Type M349   *579 703 764*
Grassline ABL Anti-Fouling Type M349 HS   *579 703 764*
Grassline ABL Anti-Fouling Type M397   *482 548*
Grassline ABL Anti-Fouling Type M398   *579 703 764*
Grassline ABL Antifouling Type M129   *702 709*
Grassline ABL Antifouling Type M349   *594 737 781 844*
Grassline EL Anti-Fouling Type L975   *606 797*
Grassline TF Anti-Fouling Type M394   *626 892*
Grassline TF Anti-Fouling Type M396   *482 548*
Grassline XL Anti-Fouling Type K968B   *581 707*
Grassline XL3 Anti-Fouling Type L806 B   *606 797*
Grassline XLS Anti-Fouling Type M299   *581 707*
Grassline XLS Antifouling Type M299   *581 707*
Great Mills Ant Killer Spray   *348*
Great Mills Exterior Preservative Primer   *128 260*
Great Mills Wood Preservative Brown   *156 241*
Great Mills Wood Preservative Clear   *156 241*
Great Mills Wood Preservative Green   *157 238 75*
Great Mills Wood Preservative Red Cedar   *156 241*
Green   *182 76*
Green Plus Wood Preserver   *73*
Green Range Dipsar   *105 266*
Green Range Dual Purpose AQ   *103 225*
Green Range Fungicidal Concentrate   *309*
Green Range Fungicidal Solvent   *223*
Green Range Mayonnaise   *195 263*
Green Range Murosol 20   *295 298*
Green Range Timber Treatment Paste   *195 263*
Green Range Woodworm Killer   *95*
Green Range Woodworm Killer AQ   *95*
Green Range Wykamol Plus   *103 225*
Green Wood Preservative   *73*
Greenco Ant Killer   *416*
Greenerway Crawling Pest Control   *333 393*
Greenerway Flying Pest Control   *333 393*
Greenhills Creosote   *69*
Gummipaint A/F   *490 655 820*
Haden Mosquito and Flying Insect Killer   *331*
Hai-Hong's Antifouling Nautic HI-7695   *593 735 774 798*
Hail Flying Insect Killer   *407 457*
Hail Plus Crawling Insect Killer   *410 464 472*
Halcyon 5000 (Base)   *563 663*
Halodec   *292*
Halophane 105   *292*
Halophane 106M   *292*
Halophane Bonding Solution   *289*
Halophane No. 1 Aerosol   *292*
Halophane No. 3 Aerosol   *292*
Halophen BM1165L   *289*
Hard Racing   *547*
Hard Racing Antifouling   *516 619 826*
Hard Racing Antifouling White   *519 658 829*
Hardwood Protector   *111*
Hartz Rid Flea Carpet Freshener   *392*
Hempel's Antifouling 761GB   *547*
Hempel's Antifouling 762GB   *560 627*
Hempel's Antifouling 763GB   *628*
Hempel's Antifouling 7650   *640 713 768*
Hempel's Antifouling 7659   *640 713 768*
Hempel's Antifouling Bravo 7610   *601 760*
Hempel's Antifouling Classic 7611   *601 760*
Hempel's Antifouling Classic 7611 Red (Tin Free) 5000   *502 617 823*
Hempel's Antifouling Classic 7633   *601 760*
Hempel's Antifouling Classic 76540   *502 617 823*
Hempel's Antifouling Classic 7654A   *623 874 894*
Hempel's Antifouling Classic 7654B   *571 683 838*
Hempel's Antifouling Classic 7654C   *489 614 819*
Hempel's Antifouling Classic 7654E   *522 621 830*
Hempel's Antifouling Classic 7655   *583 711 765*
Hempel's Antifouling Classic 7677   *602 773 799*
Hempel's Antifouling Classic Tin Free 7654   *499 551*

HSE PRODUCT NAME INDEX

Hempel's Antifouling Clear 0777 *725 771*
Hempel's Antifouling Combic 71990 *502 617 823*
Hempel's Antifouling Combic 71992 *502 617 823*
Hempel's Antifouling Combic 7199A *623 874 894*
Hempel's Antifouling Combic 7199B *571 683 838*
Hempel's Antifouling Combic 7199C *489 614 819*
Hempel's Antifouling Combic 7199E *522 621 830*
Hempel's Antifouling Combic 7699 *591 729 766*
Hempel's Antifouling Combic 7699B *595 741 785 873 885*
Hempel's Antifouling Combic Tin Free 71990 *502 617 823*
Hempel's Antifouling Forte 7620 *583 711 765*
Hempel's Antifouling Forte Tin Free 7625 *499 551*
Hempel's Antifouling Hi Build 7600 *568 672 712 762*
Hempel's Antifouling Mille 7670 *645 736 775 800*
Hempel's Antifouling Mille 7678 *561 644 730 761*
Hempel's Antifouling Mille Copper 7671 *532 728 759*
Hempel's Antifouling Mille Dynamic *502 617 823*
Hempel's Antifouling Mille Dynamic 717 GB *504 632 665 859*
Hempel's Antifouling Mille MS7672 *593 735 774 798*
Hempel's Antifouling Nautic 71900 *502 617 823*
Hempel's Antifouling Nautic 71902 *502 617 823*
Hempel's Antifouling Nautic 7190A *623 874 894*
Hempel's Antifouling Nautic 7190B *571 683 838*
Hempel's Antifouling Nautic 7190C *489 614 819*
Hempel's Antifouling Nautic 7190E *522 621 830*
Hempel's Antifouling Nautic 7673 *593 735 774 798*
Hempel's Antifouling Nautic 7674 *593 735 774 798*
Hempel's Antifouling Nautic 7680 *734 777 804*
Hempel's Antifouling Nautic 7690B *595 741 785 873 885*
Hempel's Antifouling Nautic 7691B *595 741 785 873 885*
Hempel's Antifouling Nautic HI 7691 *591 729 766*
Hempel's Antifouling Nautic HI 7690 *591 729 766*
Hempel's Antifouling Nautic HI 7695 *593 735 774 798*

Hempel's Antifouling Nautic Tin Free 7190 *502 617 823*
Hempel's Antifouling Nautic Tin Free 7190E *522 621 830*
Hempel's Antifouling Nordic 7133 *547*
Hempel's Antifouling Oceanic 7640 *772 805*
Hempel's Antifouling Olympic 7155 *584 714 780 843*
Hempel's Antifouling Olympic HI-7661 *502 617 823*
Hempel's Antifouling Olympic Tin Free 7154 *499 551*
Hempel's Antifouling Paint 161P (Red and Chocolate to TS10240) *547*
Hempel's Antifouling Rennot 7150 *563 663*
Hempel's Antifouling Rennot 7177 *563 663*
Hempel's Antifouling Tin Free 740GB *493 550*
Hempel's Antifouling Tin Free 741GB *486 549*
Hempel's Antifouling Tin Free 742GB *563 663*
Hempel's Antifouling Tin Free 743GB *520 554*
Hempel's Antifouling Tin Free 745GB *499 551*
Hempel's Antifouling Tin Free 750GB *499 551*
Hempel's Antifouling Tin Free 751GB *520 554*
Hempel's Antifouling Tin Free 752GB *493 550*
Hempel's Antifouling Tin Free 7661 *499 551*
Hempel's Antifouling Tin Free 7662 *520 554*
Hempel's Antifouling Tropic 7644 *601 760*
Hempel's Bravo 7610A *571 683 838*
Hempel's Copper Bottom Paint 7116 *613 834*
Hempel's Excelsior CR Antifouling 7643 *705*
Hempel's Hard Racing 76380 *506 656 824*
Hempel's Hard Racing 7638A *636 684 852*
Hempel's Hard Racing 76480 *502 617 823*
Hempel's Hard Racing 7648A *571 683 838*
Hempel's Hard Racing Copper Free 7649 *772 805*
Hempel's Mille Alu 71601 *506 656 824*
Hempel's Mille Alu 71602 *636 684 852*
Hempel's Mille Dynamic 71600 *506 656 824*
Hempel's Mille Dynamic 7160A *636 684 852*
Hempel's Mille Dynamic 71700 *502 617 823*
Hempel's Mille Dynamic 7170A *571 683 838*
Hempel's Tin Free Antifouling 744GB *547*
Hempel's Tin Free Antifouling 7626 *520 554*
Hempel's Tin Free Antifouling 7660 *547*
Hempel's Tin Free Hard Racing 7648 *503 629*
Hempel's Antifouling Bravo Tin Free 7610 *502 617 823*
Heritage Woodworm Killer *174*
Hickson Antiblu 3737 *162 55*
Hickson Antiblu 3738 *162 55*
Hickson Antiblu 3739 *159 4*
Hickson Antiborer 3767 *95*
Hickson Antiborer 3768 *174*
Hickson NP-1 *107 12*

Hickson Timbercare 2511 Brown   69
Hickson Timbercare WRQD   156 241
Hickson Woodex   126 239
High Performance Antifouling Paint   601 760
Homebase Antkiller Spray   348
Homebase Creosote   69 92
Homebase Fungicidal Wash   277
Homebase Mould Cleaner   277
Homecharm Wood Preservative Brown   156 241
Homecharm Wood Preservative Clear   156 241
Homecharm Wood Preservative Green   157 238 75
Homecharm Wood Preservative Red Cedar   156 241
Homework All Purpose Wood Preserver   255
Homework Exterior Wood Preservative   255
Hospital Flying Insect Spray   435 460
Household Flea Spray   396 407 429 457
Household Flea Spray from Bob Martin   392
Howes Olympic Algaecide   277
Hyperion Mould Inhibiting Solution   295 298
Icon 2.5 EC   382
Imersol 2410   198 267
Imersol WRQD Ready to Use (2523)   198 267
Imperator Fog S 18/6   402 440
Imperator Fog Super S 30/15   402 440
Imperator ULV S 6/3   402 440
Imperator ULV Super S 10/4   402 440
Impra-Color   111 119 183
Impra-Holzschutzgrund (Primer)   120 189 221
Impra-Sanol   174
Industrial Hygiene Controllable Cassette Insect Killer   360
Industrial Wood Preservative AA 155   126 239
Injecta CCA-C   205 28 87
Insecticidal Aerosol   396 429
Insecticide Aerosol Single Dose for Aircraft Disinsection   475
Insecticide Fungicide Wood Preservative   190 229
Insectrol   333 393
Insectrol Professional   333 393
Insektigun   333 393
Instasective   339
Inter 100   625 881
Interclene AQ HZA700 Series (Base)   692
Interclene Extra BAA100 Series   547
Interclene Extra BCA500 Series   601 760
Interclene Premium BCA300 Series   547
Interclene Super BCA400 Series   601 760
Interclene Super BCA400 Series (BCA400 Red)   547
Interclene Underwater Premium BCA468 Red   547
Interior Mould Remover Wipes   277

International Intertox Blue Peter   111
International TBT Free Copolymer Antifouling BQA100 Series   547
International TBT Free Copolymer Antifouling BQA200 Series   625 881
International Tin Free SPC BNA100 Series   566 669
Intersleek BXA 810/820   692
Intersleek BXA560 Series (Base)   692
Intersleek BXA580 Series (Base)   692
Intersleek FCS HKA560 Series (Base)   692
Intersleek FCS HKA580 Series (Base)   692
Intersmooth Hisol 2000 BFA270 Series   596 742 786 884
Intersmooth Hisol 9000 BFA970 Series   596 742 786 884
Intersmooth Hisol BFA250/BFA900 Series   596 742 786 884
Intersmooth Hisol BFA948 Orange   647 743 788 889
Intersmooth Hisol BFC270/950/970 Series   521 592 727 758
Intersmooth Hisol SPC Antifouling BFA949 Red   643 732 769
Intersmooth Hisol Tin Free BGA620 Series   566 669
Intersmooth SPC Antifouling BF0250 Series   591 729 766
Intersmooth SPC BFA040/BFA050 Series   643 732 769
Intersmooth SPC BFA090/BFA190 Series   591 729 766
Intersmooth Tin Free BGA530 Series   566 669
Interspeed 2000   503 507 629 660 879
Interspeed 2000 (White)   503 629
Interspeed Antifouling BW0900 Series   499 551
Interspeed Extra (BWA500 Red)   625 881
Interspeed Extra BLA200 Series   601 760
Interspeed Extra BWA500 Red   566 669
Interspeed Extra BWA750 Red   499 551
Interspeed Extra Red BJA450 Series   601 760
Interspeed Extra Strong   566 669
Interspeed Premium BWA900 Red   566 669
Interspeed Super BJA600 Series   547
Interspeed Super BWA900 Red   499 551 566 669
Interspeed Super BWA909 Black   566 669
Interspeed System 2 BRA140 Series   601 760
Interspeed System 2 BRA140/BRA240 Series   625 881
Interspeed System 2 BRA142 Brown   566 669
Interspeed System 2 BRA143 Brown   499 551
Interspeed System 2 BRA240 Series   547
Interspeed System 2 BRO142/240 Series   547
Interswift BKA000/700 Series   596 742 786 884

Interswift BKC000/700 Series   *569 673 731 763*
Interswift Tin Free BQA400 Series   *566 669*
Interswift Tin-Free SPC BTA540 Series   *566 669*
Interviron BQA200 Series   *625 881*
Interviron BQA400 Series   *566 669*
Interviron BQA450 Series   *499 551*
Interviron BQA750 Series (Base)   *499 551*
Interviron Super BNA400 Series   *566 669*
Interviron Super BQA400 Series   *566 669*
Interviron Super Tin-Free Polishing Antifouling BQO400 Series   *499 551*
Interviron Super Tin-Free Polishing Antifouling BQO420 Series   *520 554*
Iodofenphos Granular Bait   *380*
Iscosan Fungicidal Solution   *269 278*
Janisolve Fly Killer   *435 460*
Jertox Moth Balls   *387*
Jertox Moth Crystals   *318*
Jewson Creosote   *69 92*
Jeyes Crawling Insect Spray   *474 477*
Jeyes Expel Ant and Crawling Insect Killer   *465 476*
Jeyes Expel Ant Killer Powder   *392*
Jeyes Expel Ant Trap   *466*
Jeyes Expel Flea Killer   *465 476*
Jeyes Expel Fly and Wasp Killer   *446*
Jeyes Expel Moth Killer   *360*
Jeyes Expel Plug-In Flying Insect Killer   *467*
Jeyes Expel Pump Action Fly and Wasp Killer   *419*
Jeyes Expel Slow Release Fly and Wasp Killer   *360*
Jeyes Flying Insect Spray   *419*
Jeyes Kontrol Ant and Crawling Insect Killer   *465 476*
Jeyes Kontrol Ant Killer Powder   *392*
Jeyes Kontrol Flea Killer   *465 476*
Jeyes Kontrol Fly and Wasp Killer   *446*
Jeyes Kontrol Moth Killer   *360*
Jeyes Kontrol Plug-In Flying Insect Killer   *467*
Jeyes Kontrol Slow Release Fly and Wasp Killer   *360*
Jeyes Slow Release Fly Killer Controllable Cassette   *360*
Jeyes Small Space Fly and Moth Strip   *360*
Johnson Wax Raid Fly and Wasp Killer   *446*
Johnson's Carpet Flea Guard   *392*
Johnson's Flea Guard   *392*
Johnson's Household Flea Spray   *396 429*
Johnstone's Fungicidal Wash   *269 278*
KO2 Selkil   *407 457*
Kalon Creosote   *92*
Kamone Colour 60-1   *641 722 783 855*
Kamone Colour 60-2   *635 677 796 853*
Kamone FRP-60 1   *720*

Kamone FRP-60 2   *675 795*
Kamone FRP-DC 1   *639 708*
Kamone FRP-DC 2   *634 676 794*
Karate Dairy Fly Spray   *419*
Kartini Antifouling AF24   *547*
Kathon 886F   *15 5*
Kemira CCA Type BS   *205 28 87*
Kemwood CCA Type BS   *205 28 87*
Kibes Sterilising Solution Concentrate GS 36   *271*
Killgerm Boric Acid Powder   *339*
Killgerm Carbaryl 5% Dust   *340*
Killgerm Dimethoate Extra   *369 453*
Killgerm Fenitrothion 40 WP   *371*
Killgerm Fenitrothion 50 EC   *371*
Killgerm Fenitrothion-Pyrethroid Concentrate   *374 454*
Killgerm Masonry Sterilant   *236 24 275 312*
Killgerm Precor RTU   *386 428*
Killgerm Precor ULV I   *386 428*
Killgerm Precor ULV III   *349 384*
Killgerm Propoxur 20 EC   *418*
Killgerm Py-Kill W   *446*
Killgerm Pyrethrum Spray   *419*
Killgerm Terminate   *342*
Killgerm ULV 400   *419*
Killgerm ULV 500   *465 476*
Killgerm Wood Protector   *103 225*
Killgerm Woodworm Killer   *95*
Killoff Aerosol Insect Killer   *465 476*
Killoff Ant Killer Powder   *392*
Kingfisher Fungicidal Wall Solution   *295 298*
Kingfisher Timber Paste   *136 138 185*
Kingfisher Wood Preservative   *223 224 77*
Kingston Dual Purpose Fluid   *156 241*
Kleeneze Anti-Mould Spray   *272 286*
Kleeneze Fly Killer   *446*
Kleenoff Crawling Insect Spray   *474 477*
Kleenoff Flying Insect Spray   *419*
Kleenoff Slow Release Fly Killer Controllable Cassette   *360*
Kleenoff Small Space Fly and Moth Strip   *360*
Knox Out 2 FM   *356*
Konk 1 (B) Flying Insect Killer   *419*
Konk I   *419*
Kontrol Kitchen Size Fly Killer   *360*
Kudos   *370 394*
Kwik Insecticide   *474 477*
Kybosh   *396 429*
Langlow Clear Wood Preserver   *103 225*
Langlow Creosote   *69*
Langlow Timbershield   *22*
Langlow Wood Preservative   *111*
Langlow Wood Preservative Green   *73*
Langlow Wood Preserver Formulation A   *111*

Langlow Wood Preserver Formulation B *223*
Lanosol P *392*
Lanosol RTU *392*
Lanstar Creosote *92*
Laporte CCA AWPA Type C *27 66 79*
Laporte CCA Oxide Type 1 *27 66 79*
Laporte CCA Oxide Type 2 *27 66 79*
Laporte CCA Type 1 *205 28 87*
Laporte CCA Type 2 *205 28 87*
Laporte Cut End Preservative *209 86*
Laporte Mould-Ex *15 5*
Laporte Permatreat *156 241*
Laporte Permatreat Cut End Preservative *156 241*
Laporte Permatreat with Water Repellant *156 241*
Laporte Permawood CCA *205 28 87*
LAR-VAC 100 *150 168 242*
Larsen Brown Wood Preservative *156 241*
Larsen Clear Wood Preservative *153 173 243*
Larsen Concentrated Algicide *295 298*
Larsen Creosote *92*
Larsen Green Wood Preservative *73*
Larsen Joinery Grade *172 246*
Larsen Joinery Grade 2 *223*
Larsen Wood Preservative Clear 2 and Brown 2 *190 229*
Larsen Woodworm Killer 2 *190 229*
Larvac 300 *195 263*
Larvex-100 *341 391*
Larvex-15 *341 391*
Laser Insect Killer *365 408 452*
Laura Ashley Home Fungicidal Wash *310*
Le Marin *501 573 680 862*
Leader *575 688*
Leyland Creosote *69 92*
Leyland Sterilisation Wash *277*
Leyland Timbrene Environmental Formula *255*
Leyland Timbrene Green Environmental Formula *73*
Leyland Timbrene Supreme Environmental Formula *121 191 227*
Leyland Timbrene Woodworm Killer *174*
Leyland Universal Preservative Base *255*
Lichenite *289*
Light Alloy *691*
Lily Antifouling IV *542 587 716 833*
Lily Antifouling Super Tropical *541 582 706*
Linotol Fungicidal Wash *295 298*
Lister Teak Dressing *111*
Littac *323*
Lloyds Supersave Slow Release Fly and Wasp Killer *360*
Lloyds Supersave Small Space Fly and Moth Strip *360*

Long Life TF *502 617 823*
Low Odour Woodworm Killer *174*
LPL Biocidal Wash *277*
Lynx Antifouling *491*
Lynx Metal Free Antifouling *511 695*
Lyvia Mosquito Killer *331*
M-Tec Biocide *295 298*
MacPherson Anti-Mould Solution *135 137 295 298*
MacPherson Antimould Solution (RFU) *295 298*
Mafu Ant and Insect Powder *392*
Mafu Ant Killer *413*
Magicote Fungicidal Concentrate *310*
Magicote Masonry Fungicidal Solution *310*
Mar-Cide *204 309*
Mar-Kil S *190 229*
Mar-Kil W *190 229*
Marclear Antifouling *613 834*
Marine Gold DX White-1 *671 710 751 865*
Marine Gold DX White-2 *648 747 812 858*
Marine Gold DX-1 *638 689 721 740 782 854*
Marine Gold DX-2 *649 750 856*
Marley Patio Cleaner *269 278*
Masterstroke Wood Preserver *127 252*
Maxforce Bait Station *377*
Maxforce Bait Stations *377*
Maxforce Gel *377*
Maxforce Pharaoh's Ant Killer *377*
McDougall Rose Hi-Life Creosote *92*
Mebon Longlife Anti-Fouling (T) 6-2-09 *601 760*
Mebon Longlife Anti-Fouling 6-2-16 *601 760*
Mebon Tropical Anti-Fouling Paint 6-2-01 *525 556*
Mect *159 4*
Medium Brown Creosote *69*
Mekseal *208 64 90*
Meksol *205 28 87*
Mekure *205 28 87*
Mekure T1 Oxide *27 66 79*
Mekure T2 *205 28 87*
Mekure T2 Oxide *27 66 79*
Meridian MP40 Antifouling *481 500 562 654 818*
Metalife Fungicidal Wash *135 137 295 298*
Micrapor Environmental Flea Spray *392*
Microguard FI Concentrate *181 57*
Microguard Mouldicidal Wood Preserver *279 287 34 45*
Microguard Permethrin Concentrate *174*
Microguard Woodworm Fluid *174*
Micromite *371*
Micron *626 892*
Micron (Gull White) *499 551*
Micron 25 Plus *647 743 788 889*
Micron 400 Series *547*
Micron 500 Series *566 669*

Micron 600 Series   *624 877*
Micron CSC   *566 626 669 892*
Micron CSC 100 Series   *547*
Micron CSC 200 Series   *514 553*
Micron CSC 300 Series   *524 555*
Micron CSC Dover White   *499 551*
Micron Plus Antifouling   *566 669*
Microstel   *269 278*
Microtech Dual Purpose AQ   *11 97*
Microtech Woodworm Killer AQ   *95*
Middletons Creosote   *69*
Miricoat AF Coating   *530*
Mitrol PQ 8   *163*
Morrisons Creosote   *69*
Mosgo   *212 311*
Mosgo P   *212 311*
Mosiguard Shield   *392*
Moskil Mosquito Repellent   *467*
Mosqui—Go Electric   *331*
Mosquito Killer Travel Pack   *331*
Mosquito Repellent   *331*
Moss-Cure   *277*
Mosscheck   *289*
Moth and Fly Killer   *360*
Moth Repellent   *318*
Mothaks   *320 388*
Motox   *373 409 455*
Mould Cleaner   *271*
Mould Cure   *280 288 291*
Mould Inhibitor   *106 289*
Mould Killer   *277*
Mouldcheck Barrier   *289*
Mouldcheck Barrier 2   *289*
Mouldcheck Spray   *292*
Mouldcheck Sterilizer   *292*
Mouldrid   *135 137 295 298*
MPX   *503 507 629 660 879*
MPX (White)   *503 629*
MRS Clear Mould Fungicidal Wash   *295 298*
Multispray   *346 400 424*
Murphy Kil-Ant Powder   *327*
Murphy Kil-Ant Ready To Use   *348*
Muscatrol   *392*
Mycodet   *292*
Mycofen Barrier   *292*
Mycofen Concentrate   *277*
Mycospray   *277*
Mystox BTL   *152 171*
Mystox BTV   *152 171*
Natural Ant Gun   *416*
Natural Wasp Gun   *416*
Natural Wasp Killer   *416*
Neopybuthrin Premium   *402 440*
Neopybuthrin S200Q   *402 440*
Neopybuthrin S9E   *402 440*

Neopybuthrin Super   *402 440*
Ness Dual Purpose   *103 225*
Ness Dual Purpose AQ   *103 225*
Ness Fungicidal Concentrate   *309*
Ness Fungicidal Solvent   *223*
Ness Mayonnaise Wood Treatment   *149 166*
Ness Woodworm Killer   *95*
Ness Woodworm Killer AQ   *95*
Net Clean Beta   *491*
Net Clean Gamma   *534*
New Cut 'n' Spray   *43*
New Formula Cedarwood   *111*
New Seige II   *348*
New Super Raid Insecticide   *410 464 472*
New Tetracide   *343 351 423*
New Vapona Ant Killer   *352 449*
New Vapona Fly and Wasp Killer   *412 471*
New Vapona Fly Killer Dry Fomulation   *461 473*
Nipacide DP30   *292*
Nippon Ant and Crawling Insect Killer   *407 457*
Nippon Ant Killer Powder   *392*
Nippon Fly Killer Pads   *333 393*
Nippon Fly Killer Spray   *407 457*
Nippon Flying Insect Killer Tablets   *331*
Nippon Killaquer Crawling Insect Killer   *407 457*
Nippon Ready For Use Ant and Crawling Insect Killer   *392*
Nippon Ready For Use Fly Killer Spray   *474 477*
Nisa Fly and Wasp Killer   *407 457*
Nitromors Creosote   *69 92*
Nitromors Mould Remover   *269 278*
Nitromors Timbrene Clear   *115 140*
Nitromors Timbrene Clear Environmental Formula   *123 226*
Nitromors Timbrene Environmental Formula Wood Preservative   *255*
Nitromors Timbrene Supreme   *117 143 184*
Nitromors Timbrene Supreme Environmental Formula   *121 191 227*
Nitromors Timbrene Woodworm Killer   *174*
Noa-Noa Rame   *571 613 683 834 838*
Nomad Residex P   *392*
Norden Durahart   *652 767*
Norden Kobberstoff Green   *601 760*
Norden Kobberstoff Red   *547*
Norden Non-Stop Black   *642 724*
Norden Non-Stop Blue   *642 724*
Norden Non-Stop Red   *642 724*
Norden Non-Stop White   *642 724*
Norimp 2000 Black   *547*
North Atlantic/Launching Antifouling AF2 Red   *613 834*
North Atlantic/Launching Antifouling F2/801 Black   *613 834*
Nu Wave A/F Flat Bottom   *576 690 749 840*

Nu Wave A/F Vertical Bottom   576 690 749 840
Nubex Emulsion Concentrate (Low Odour)   95
Nubex Emulsion Concentrate C (Low Odour)   95
Nubex Emulsion Concentrate CB (Low Odour)   103 225
Nubex Emulsion Grade L   145
Nubex Emulsion LB (Low Odour)   154 228
Nubex Fungicide FS15   212 311
Nubex Fungicide QT Concentrate   237 284 313 42
Nubex Standard Emulsion Concentrate LT   156 241
Nubex WDR   135 137 295 298
Nubex Woodworm All Purpose CB   103 225
Nubex Woodworm All Purpose LT   156 241
Nuodex 87   135 137 295 298
Nut Brown Creosote   92
Nuvan Staykil   363 381
Nuvanol N 500 FW   380
Nuvanol N 500 SC   380
Oakmere Creosote Type 2   92
Odex Fly Spray   333 393 446
Oko Pet Care Insecticidal Carpet Freshener   475
One Shot Space Spray   392
One-Shot Aircraft Aerosol Insect Control   475
OS Color Wood Stain and Preservative   122 197 261
OS Color WR   118 122 144 193 197 240 261
Outright Household Flea Spray   392
Ovoline 275 Golden Creosote   69
Palace Base Coat Wood Preservative   223
Palace Fungicidal Wash   310
Palace Microfine Fungicide Concentrate   223
Palace Microfine Fungicide Insecticide (Concentrate)   103 225
Palace Microfine Insecticide Concentrate   95
Palace Mould Remover   295 298
Palace Timbertreat Ecology   195 263
Palace Woodworm Killer   145
Palavac Industrial Timber Preservative   104 259
Pale Green Timber Preservative 2   224 77
Pale Green Wood Preservative   224 77
Panacide M   292
Panacide M21   292
Panaclean 736   292
Pandrol Timbershield Rods   129
Paneltone   1
Paramos   277
Patente Laxe   499 551
Patriot Flying and Crawling Insect Killer   419
Payless Creosote   92
Payless Fungicidal Wash   277
Payless Wood Preservative Brown   156 241
Payless Wood Preservative Clear   156 241
Payless Wood Preservative Green   157 238 75

Payless Wood Preservative Red Cedar   156 241
Payless Wood Preserver   111
Payless Wood Preserver—Clear   103 225
Payless Wood Preserver—Red Cedar   223
Payless Wood Preserver—Universal   190 229
PC Insect Killer   333 393
PC Pest Control Spray   333 393
PC-D   309
PC-D Concentrate   309
PC-F/A   235
PC-F/B   223
PC-H/3   95
PC-H/4   103 225
PC-I   145
PC-K   216 59
PC-XJ/2   223
PCX 12/P Concentrate   95
PCX-12/P   95
PCX-122   103 225
PCX-122 Concentrate   103 225
Pedigree Exelpet Bedding Spray   474 477
Penetone Fly Killer   435 460
Penguin Non-Stop   565 666 837
Penguin Non-Stop White   631 667 851
Penguin Racing   613 834
Peripel   392
Peripel 55   392
Perma AQ Dual Purpose   190 229
Permacide Masonry Fungicide   269 278
Permadex Masonry Fungicide Concentrate   295 298
Permadip 9 Concentrate   129
Permagard FWS   269 278
Permalene Satin Wood Stain   111
Permapaste PB   190 229
Permarock Fungicidal Wash   269 278
Permasect 0.5 Dust   392
Permasect 10 WP   392
Permasect Powder   392
Permatreat Paste   190 229
Permethrin   392
Permethrin Dusting Powder   392
Permethrin F and I Concentrate   190 229
Permethrin PH—25WP   392
Permethrin Wettable Powder   392
Permethrin WW Conc   174
Permoglaze Micatex Fungicidal Treatment   135 137 295 298
Permolite Low Odour Concentrate P   174
Perycut Cockroach Mat   333 393
Pesguard NS 6/14 EC   465 476
Pesguard NS 6/14 Oil   465 476
Pesguard OBA F 7305 C   465 476
Pesguard OBA F 7305 D   465 476
Pesguard OBA F 7305 E   465 476

Pesguard OBA F 7305 F   *465 476*
Pesguard OBA F 7305 G   *465 476*
Pesguard OBA F 7305 B   *465 476*
Pesguard PS 102   *474 477*
Pesguard PS 102A   *474 477*
Pesguard PS 102B   *474 477*
Pesguard PS 102C   *474 477*
Pesguard PS 102D   *474 477*
Pesguard WBA F-2656   *465 476*
Pesguard WBA F-2692   *465 476*
Pesguard WBA F2714   *478 479*
Pestkill Household Spray   *465 476*
Pestkill Insect Spray   *407 457*
Pestrin   *392*
Pet Fresh Insecticidal Carpet Freshener   *475*
Pharorid   *383*
Pharorid S   *443*
Phernal Brand Moth Balls   *387*
Phillips Pestkil Karpet Killer   *475*
Pif Paf Crawling Insect Killer   *333 393*
Pif Paf Crawling Insect Killer Aerosol   *333 393 404 441 458*
Pif Paf Crawling Insect Powder   *392*
Pif Paf Fly Spray   *419*
Pif Paf Flying Insect Killer   *333 393*
Pif Paf Flying Insect Killer Aerosol   *333 393*
Pif Paf Insecticide   *364 403 439*
Pif Paf Mosquito Mats   *331*
Pilot Antifouling   *547*
Pilt 80 RFU   *10*
Pilt NF4 RFU   *10*
PLA Products Dry Rot Killer   *309*
PLA Products Fungicidal Wash   *309*
PLA Products Woodworm Killer   *152 171*
PLA Wood Preserver   *255*
Plantsafe Autumn Gold   *204*
POB Mould Treatment   *269 278*
POB Wood Preservative Brown   *156 241*
POB Wood Preservative Clear   *156 241*
POB Wood Preservative Green   *157 238 75*
POB Wood Preservative Red Cedar   *156 241*
Polycell Mould Cleaner   *295 298*
Polyphase Emulsifiable Concentrate   *10*
Polyphase Solvent Based Concentrate   *10*
Polyphase Solvent Based Ready For Use   *10*
Polyphase Water Based Concentrate   *10*
Polyphase Water Based Ready For Use   *10*
Precor Plus Premise Spray   *385 395*
Precor ULV II   *385 395*
Premier Pro-Tec (S) Environmental Formula Wood Preservative   *255*
Premiere Fly and Wasp Killer   *446*
Premium Grade Wood Treatment   *196 257 58*
Premium Wood Treatment   *195 263 306 317*
Preservative For Wood Black   *150 168 242*

Preservative For Wood Green   *73*
Prevent   *419*
Preventol OF   *6*
Pro AM Timbertreat Brown   *164*
Prop-n-Drive   *503 629*
Propeller T.F.   *503 629*
Protect A Fence   *214*
Protek 9 Star Wood Protection   *129*
Protek Blue   *159 4*
Protek Double 9 Star Wood Protection   *129*
Protek Shedstar   *129*
Protek Wood Protection (FT Grade 20:1)   *159 4*
Protek Wood Protection (FT Grade 9:1)   *159 4*
Protek Wood Protection Fencegrade   *129*
Protek Wood Protection Fencegrade 9:1   *129*
Protek Wood Protection Shed Grade   *129*
Protek Woodstar   *129*
Protim 200   *151 169 254*
Protim 200C   *149 166*
Protim 210   *156 241*
Protim 210 C   *156 241*
Protim 210 CWR   *156 241*
Protim 210 WR   *156 241*
Protim 215 PP   *235*
Protim 220   *198 267*
Protim 220 CWR   *198 267*
Protim 220 WR   *198 267*
Protim 23 WR   *156 241*
Protim 230   *192 234*
Protim 230 WR   *192 234*
Protim 240   *194 249*
Protim 240 C   *194 249*
Protim 240 CWR   *194 249*
Protim 240 WR   *194 249*
Protim 250   *14 176*
Protim 250 WR   *14 176*
Protim 260 F   *223*
Protim 340   *187 203*
Protim 340 WR   *187 203*
Protim 80   *150 168 242*
Protim 80 C   *150 168 242*
Protim 80 CWR   *150 168 242*
Protim 80 Oil Brown   *150 168 242*
Protim 80 WR   *150 168 242*
Protim 800   *178 20*
Protim 800 C   *178 20*
Protim 800 C Oil Brown   *178 20*
Protim 800 CWR   *178 20*
Protim 800 WR   *178 20*
Protim 800P   *17 177 21*
Protim 80C Oil Brown   *150 168 242*
Protim 90   *156 241*
Protim AQ   *174*
Protim Aquachem-Insecticidal Emulsion P   *174*
Protim B10   *129*

Protim Brown   150 168 242
Protim Brown 800   178 20
Protim CBC   103 225
Protim CCA Oxide—Type II   27 66 79
Protim CCA Oxide 50   27 66 79
Protim CCA Oxide 58   27 66 79
Protim CCA Oxide 72   27 66 79
Protim CCA Salts Type 2   205 28 87
Protim CDB   190 229
Protim Cedar   126 239
Protim Curative 800   178 20
Protim Curative F   223
Protim Curative GP   190 229
Protim Curative Z   198 267
Protim Curative ZF   265
Protim Exterior Brown   164
Protim FDR 230   233
Protim FDR 240   248
Protim FDR 250   10
Protim FDR 800   18
Protim FDR-H   167 245
Protim FDR210   235
Protim Fentex Europa 1 RFU   2 44
Protim Fentex Europa I   2 44
Protim Fentex Green Concentrate   134 213
Protim Fentex Green RFU   134 213
Protim Fentex M   134 213
Protim Fentex WR   212
Protim GC   164
Protim Grade Basic   150 168 242
Protim Grade Basic C   150 168 242
Protim Grade Basic CWR   150 168 242
Protim Grade Basic WR   150 168 242
Protim Green E   182 76
Protim Green WR   73
Protim Injection Fluid   223
Protim Insecticidal Emulsion 8   174
Protim Insecticidal Emulsion C   95
Protim Insecticidal Emulsion P   174
Protim Joinery Lining   167 245
Protim Joinery Lining 280   6
Protim JP   167 245
Protim JP 210   235
Protim JP 220   265
Protim JP 240   248
Protim JP 250   10
Protim JP 800   18
Protim Kleen II   134 213
Protim Panelguard   212
Protim Paste   156 241
Protim Paste 220   198 267
Protim Paste 220 F   265
Protim Paste 800   178 20
Protim Paste 800 F   18
Protim Paste P   174

Protim Plug Compound   212 311
Protim R Clear   167 245
Protim R Coloured   235
Protim R Coloured 333   235
Protim R Coloured Plus   126 239
Protim Solignum Softwood Basestain CS   126 239
Protim Stainguard   159 4
Protim TWR   156 241
Protim Wall Solution II   295 298
Protim Wall Solution II Concentrate   295 298
Protim WB12   152 171
Protim Woodworm Killer   145
Protim Woodworm Killer C   95
Protim Woodworm Killer P   174
Protim WR 220   198 267
Protim WR 260   190 229
Protim WR 800   178 20
Protrol   378
Protrol Plus Crack and Crevice Aerosol   405 442
Protrol Plus ULV   379 427
Provence Fly and Wasp Killer   446
Puma Antifouling   547
Purge Insect Destroyer   407 457
PY Kill 25 Plus   465 476
Pybuthrin 2/16   419
Pybuthrin 33   419
Pybuthrin 33 BB   332 337
Pybuthrin Fly Killer   333 393
Pybuthrin Fly Spray   333 393
Pybuthrin ULV   419
Pynamin Forte Mat 120   467
Pynosect 10   407 457
Pynosect 2   435 460
Pynosect 4   435 460
Pynosect 6   392
Pynosect Extra Fog   407 457
Pynosect PCO   392
Pyrakill Flying Insect Spray   446
Pyrasol C RTU   348
Pyrasol CP   348
Pyrematic Flying Insect Killer   419
Pyrolith 3505 Ready To Use   216 59
Quantum Hygiene Fly Killer   435 460
Rabamarine A/F No 1000 (A-Sol/B-Sol)   609 808 849
Rabamarine A/F No 1000SP (A-Sol/B-Sol)   609 808 849
Rabamarine A/F No 100S   607 806 848
Rabamarine A/F No 2500 HS   576 690 749 840
Rabamarine A/F No 2500M HS   576 690 749 840
Raffaello Alloy   506 656 824
Raffaello Plus Tin Free   499 551
Raid Ant and Crawling Insect Killer   352 449

HSE PRODUCT NAME INDEX

Raid Ant and Roach Killer   *407 457*
Raid Ant Bait   *342*
Raid Ant Bait (NZ)   *339*
Raid Ant Bait C   *327*
Raid Cockroach Killer Formula 2   *350 425*
Raid Dry Natural Flying Insect Killer   *419*
Raid Fly and Wasp Killer   *461 473*
Raid Flying Insect Killer   *446*
Raid House and Plant   *465 476*
Raid Mothproofer   *392*
Raid Residual Crawling Insect Killer   *352 449*
Raid Wasp Nest Destroyer   *344 448*
Raid Yardguard   *412 471*
Rainstopper Deadly Nightshade Biowash   *310*
Ranch Preservative   *123 226*
Ravax AF   *613 834*
Ravax Anti-Fouling   *613 834*
Ravax Anti-Fouling HB   *613 834*
Ravax Anti-Fouling ND   *613 834*
Ready Kill   *348*
Red Antifouling AF22 (To TS10240)   *547*
Red Can Fly Killer   *446*
Reldan 50 EC   *345*
Remecology Fungicide Insecticide R9   *190 229*
Remecology Insecticide R8   *174*
Remecology Spirit Based Fungicide R5   *10*
Remecology Spirit Based Insecticide R6   *174*
Remecology Spirit Based K7   *190 229*
Remecology Timber Preservative Paste   *14 176*
Remtox AQ Fungicide R7   *10*
Remtox AQ Fungicide/Insecticide R3   *14 176*
Remtox Borocol 10 RH Masonry Biocide   *273 293*
Remtox Borocol 20 Wood Preservative   *129*
Remtox Boron Rods   *129*
Remtox Dry Rot F.W.S.   *270*
Remtox Dry Rot Paint   *10 270*
Remtox Dual Purpose Paste K9   *14 176*
Remtox Fungicidal Wall Solution RS   *270*
Remtox Fungicide Microemulsion M7   *10*
Remtox Fungicide Paste K6   *10*
Remtox Fungicide/Insecticide Microemulsion M9   *14 176*
Remtox FWS (Low Odour)   *309*
Remtox Insecticide Microemulsion M8   *174*
Remtox Insecticide Paste R4   *174*
Remtox Microactive Fungicide W7   *10*
Remtox Microactive FWS W6   *270*
Remtox Microactive Insecticide W8   *174*
Remtox Remwash Extra   *277*
Remtox Spirit Based F/I K7   *14 176*
Rentex   *210 211 26*
Rentokil Ant and Insect Powder Professional   *327*

Rentokil Dry Rot and Wet Rot Treatment   *255 315*
Rentokil Dry Rot Fluid (E)   *285 43*
Rentokil Dry Rot Fluid (E) For Bonded Warehouses   *285 43*
Rentokil Dry Rot Paste (D)   *256 60*
Rentokil Dual Purpose Fluid   *195 263*
Rentokil Houseplant Insect Killer   *431 434*
Rentokil Iodofenphos Gel   *380*
Rentokil Mould Cure Spray   *280 288 291*
Rentokil Wasp Nest Killer   *340*
Rentokil Wood Preservative   *195 263*
Rentokil Woodworm Killer   *174*
Rentokil Woodworm Treatment   *174*
Residex   *419*
Residex P   *392*
Residroid   *407 457*
Residual Powerkill   *392*
Resigen   *402 440*
Resistone PC   *22 271*
Reslin 25P   *402 440*
Reslin 25S   *402 440*
Reslin 25SE   *402 440*
Reslin Premium   *402 440*
Reslin Super   *402 440*
RLT Bactdet   *292*
RLT Clearmold Spray   *289*
RLT Clearmould Spray   *289*
RLT Halophen   *289*
RLT Halophen DS   *289*
Roachbuster   *339*
Rodewod 10 OL   *29*
Rodewod 50 SL   *29*
Ronseal Fencelife   *61*
Ronseal Low Odour Wood Preserver   *255*
Ronseal Trade High Build Preservative Woodstain   *111*
Ronseal Trade Low Build Preservative Woodstain   *123 226*
Ronseal Trade Preservative Woodstain Basecoat and Colour Harmoniser   *123 226*
Ronseal Wood Preservative Tablets   *129*
Roofguard Smoke Generator   *145 375*
Roxem C   *274 304 307 321 397 420*
Roxem D   *283 301 305 308 330 389 398 430*
Russco Strips   *360*
Ruwa Bronze Bottom Paint   *530*
Ruwa Vinyl Anti-Fouling TL   *547*
S C Johnson Raid   *465 476*
S C Johnson Raid Ant and Cockroach Killer   *352 449*
Sactif Flying Insect Killer   *407 457*
Sadolin Base No 561-2611   *14 176*
Sadolin Bioclean 979-9020   *310*
Sadolin New Base   *123 226*

465

Sadolin Sadovac 35 *13 146*
Sadolin Sadovac No 561-2392 *235*
Safe Kill RTU *336*
Safeguard Antiflame 4050 WD Wood Preservative *129*
Safeguard BP Dual Purpose Wood Preservative *190 229*
Safeguard BP O/S Wood Preservative *190 229*
Safeguard Deepwood 1 Insecticide Emulsion Concentrate *174*
Safeguard Deepwood Fungicide *223*
Safeguard Deepwood I Insecticide Emulsion Concentrate *174*
Safeguard Deepwood II Dual Purpose Emulsion Concentrate *190 229*
Safeguard Deepwood III Timber Treatment *190 229*
Safeguard Deepwood IV Woodworm Killer *174*
Safeguard Deepwood Paste *190 229*
Safeguard Deepwood Surface Biocide Concentrate *295 298*
Safeguard Fungicidal Micro Emulsifiable Concentrate *10 270*
Safeguard Fungicidal Wall Solution *309*
Safeguard Mould and Moss Killer *295 298*
Safeguard Woodworm Killer *174*
Safetray SL *29*
Sainsbury's Crawling Insect and Ant Killer *333 393*
Sainsbury's Fly and Wasp Killer *407 457*
Sainsbury's Fly, Wasp and Mosquito Killer *461 473*
Sandtex Fungicide *310*
Sanmex Fly and Wasp Killer *407 457*
Sanmex Supakil Insecticide *407 457*
Saturin E10 *190 229*
Saturin E30 *309*
Saturin E5 *190 229*
SC Johnson Flying Insect Killer *419*
SC Johnson Raid Fly and Wasp Killer *446*
Scanvac Cedar, No 561-2508 *235*
Scomet *530*
Scorpio Fly Spray for Flying Insects *407 457*
Seajet 033 *571 683 838*
Seashield (Base) *563 663*
Seatender 10 *613 834*
Seatender 12 *487 572 679 861*
Seatender 15 *484 523 615 835*
Seatender 7 *613 834*
Sectacide 50 EC *371*
Secto Ant and Crawling Insect Lacquer *356*
Secto Ant and Crawling Insect Spray *474 477*
Secto Ant Bait *327*
Secto Extra Strength Insect Killer *392*

Secto Flea Free Insecticidal Rug and Carpet Freshener *392*
Secto Fly Killer *407 457*
Secto Fly Killer Living Room Size *360*
Secto Household Flea Spray *396 429*
Secto Kil-a-line *356*
Secto Mini-Space Insect Killers *360*
Secto Moth Killer *319 361*
Secto Slow Release Fly Killer Kitchen Size *360*
Sergeant's Dust Mite Patrol Injector and Spray *465 476*
Sergeant's Dust Mite Patrol Spray *465 476*
Sergeant's Insect Patrol *465 476*
Sergeants Car Patrol *475*
Sergeants Carpet Patrol *475*
Sergeants Dust Mite Patrol *475*
Sergeants Mite Patrol *475*
Sergeants Rug Patrol *475*
Sevin D *340*
Shearwater Racing Antifouling *547*
Shelltox Cockroach and Crawling Insect Killer 2 *352 449*
Shelltox Cockroach and Crawling Insect Killer 3 *352 449*
Shelltox Cockroach and Crawling Insect Killer 4 *352 449*
Shelltox Crawling Insect Killer Liquid Spray *354 468*
Shelltox Extra Flykiller *367 451*
Shelltox Flying Insect Killer 2 *465 476*
Shelltox Flying Insect Killer 3 *465 476*
Shelltox Flykiller *367 451*
Shelltox Insect Killer *407 457*
Shelltox Insect Killer 2 *407 457*
Shelltox Mat 1 *331*
Shelltox Mat 2 *467*
Shelltox Super Flying Insect Killer *465 476*
Sherley's Rug-de-Bug *392*
Siege *348*
Siege II *348*
Sigma Anti-Fouling Broken White DL-2253 *628*
Sigma Antifouling CR *542 587 716 833*
Sigma Antifouling IV *542 587 716 833*
Sigma Pilot Antifouling LL *543 589 718 870 880*
Sigma Pilot Antifouling LL/TF *545 622 869 891*
Sigma Pilot Antifouling TA *600 753 883*
Sigma Pilot Ecol Antifouling *545 622 869 891*
Sigmaplane Ecol AF First Coat *540 574 682 863*
Sigmaplane Ecol Antifouling *498 544 571 618 683 836 838*
Sigmaplane HA Antifouling *599 752 872 886*
Sigmaplane HB *599 752 872 886*
Sigmaplane TA Antifouling *600 753 883*
Sigmaplane Tin Free AF SO 0671 *539 567 668*
Sigmaplane Tin Free AF SO 0672 *497 538 552*

Sigmathrift Antifouling 600 753 883
Signpost Creosote 92
Sinesto B 215 25
Siphotrol Premise Spray 385 395
Skeetal Flowable Concentrate 325 326
Skeetal Flowable Concentrate (Aerial) 325
Slipstream Antifouling 563 663
Slow Release Fly Killer Controllable Cassette 360
Smite 345
Snowcem Algicide 277
Solignum Anti Fungi Concentrate 309
Solignum Colourless 195 263
Solignum Dark Brown 69
Solignum Dry Rot Killer 295 298
Solignum Dry Rot Killer Concentrate 295 298
Solignum Fencing Fluid 92
Solignum Fungicide 295 298
Solignum Gold 92
Solignum Green 182 76
Solignum Medium Brown 69
Solignum Remedial Concentrate 190 229
Solignum Remedial Fluid PB 190 229
Solignum Remedial Mayonnaise 195 263
Solignum Remedial PB 190 229
Solignum Universal 198 267
Solignum Wood Preservative Paste 195 263
Solignum Woodworm Killer 174
Solignum Woodworm Killer Concentrate 174
Solignum Woodworm Killer Concentrate P 174
Solignum Woodworm Killer Trade 174
Solvent Black Preserver 235
Sorex Fly Spray RTU 465 476
Southdown Creosote 92
Sov AQ Micro F 10
Sov AQ Micro F/I 14 176
Sov AQ Micro I 174
Sovac F 10
Sovac F/I 14 176
Sovac F/I WR 14 176
Sovac FWR 10
Sovac I 174
Sovaq Micro F 10
Sovaq Micro F/I 14 176
Sovaq Micro I 174
Sovereign AQ F/I 14 176
Sovereign AQ/FT 223
Sovereign AQF 10
Sovereign Aqueous Fungicide/Insecticide 2 190 229
Sovereign Aqueous Insecticide 2 174
Sovereign Fungicidal Wall Solution 309
Sovereign Insecticide/Fungicide 2 190 229
Sovereign Timber Preservative 223

Sovereign Timber Preservative Dark Green 224 77
Sovereign Timber Preservative Pale Green 224 77
Sovereign Timber Rod 129
SP.153 Sterilising Detergent Wash 292
SP.154 Fungicidal and Bactericidal Treatment 292
Speedclean Antifouling 613 834
Spencer Fungicidal Treatment 269 278
Spencer Wood Preservative 250
Spira "No Bite" Mosquito Killer 331
Spira No Bite Outdoor Mosquito Coils 322
SPL Wood Preservative Brown 156 241
SPL Wood Preservative Clear 156 241
SPL Wood Preservative Green 157 238 75
SPL Wood Preservative Red Cedar 156 241
Spraydex Ant & Insect Killer 333 393
Spraydex Houseplant Spray 333 393
Spraydex Insect Killer 333 393
Square Deal Deep Protection Wood Preserver 255
Square Deal Fungicidal Solution 277
Square Deal Fungicidal Wash 277
Square Deal Wood Preserver 255
St Michael Flyspray 446
Stapro Insecticide 392
Sterilising Fluid 108 281 290 38
Stop Insect Killer 407 457
Stradz Mosquito Killer 331
Sumithion 20% MC 371
Sumithrin 10 Sec 475
Sumithrin 10 Sec Carpet Treatment 475
Supabug Crawling Insect Killer 334 399 422
Supaswat Insect Killer 365 408 452 465 476
Super Andy Man Wood Protection 129
Super Andyman Creosote 69
Super Fly Spray 465 476
Super Quality Antifouling 603 778 846
Super Raid II 465 476
Super Raid Insect Killer Formula 2 446
Super Shelltox Crawling Insect Killer 352 449
Super Shelltox Flying Insect Killer 465 476
Super Tropical Antifouling 547
Super Tropical Extra Antifouling 547
Superdrug Fly and Wasp Killer Pump Spray 419
Superdrug Fly Killer 407 457
Superdrug Slow Release Fly Killer 360
Superdrug Small Space Fly/Moth Strip 360
Supergrade Wood Preserver 149 166
Supergrade Wood Preserver Black 149 166
Supergrade Wood Preserver Brown 149 166
Supergrade Wood Preserver Clear 149 166
Supergrade Wood Preserver Green 73
Superspeed 613 834

Supertropical Antifouling AF5   *581 707*
Superyacht 800 Antifouling   *650 754 888*
Superyacht Antifouling   *647 743 788 889*
Swak Natural   *419*
Swat   *435 460*
Swat A   *435 460*
Takata LLL Antifouling   *590 723*
Takata LLL Antifouling Hi-Solid   *590 723*
Takata LLL Antifouling LS   *590 723*
Takata LLL Antifouling LS Hi-Solid   *590 723*
Takata LLL Antifouling No 2001   *590 723*
Takata Seaqueen   *651 755 896*
Tanalith (3419) CBC   *47 67 81*
Tanalith 3302   *27 66 79*
Tanalith 3313   *27 66 79*
Tanalith 3357   *205 28 87*
Tanalith 3422   *50 80*
Tanalith 3485   *219 49 72*
Tanalith 3487   *48 71*
Tanalith C3310   *27 66 79*
Tanalith CBC Paste 3402   *207 53 88*
Tanalith CL (3354)   *205 28 87*
Tanalith CP 3353   *205 28 87*
Tanalith Oxide C3309   *27 66 79*
Tanalith Oxide C3314   *27 66 79*
Tanamix 3743   *15 5*
Target Flying Insect Killer   *474 477*
Teak Oil   *111*
Teamac Killa   *585 715 815*
Teamac Killa Copper   *610 814*
Teamac Killa Copper Plus   *488 536 564 611 664 813 816 832*
Teamac Standard Copper   *547*
Teamac Super Tropical   *612 817 850*
Teamac Wood Preservative   *250*
Teamac Woodtec Green   *73*
Tecca CCB1   *201 52 85*
Tecca P2   *205 28 87*
Teepol Products Vapona Fly Killer   *360*
Teknar HP-D   *325*
Tesco Dark Brown Creosote   *92*
Tesco Light Brown Creosote   *92*
Tetra Concentrated Mould Cleaner   *269 278*
Texas Ant Gun   *348*
Texas Creosote   *69 92*
Texas Exterior Wood Preserver   *128 235 260*
Texas Wood Preservative Brown   *156 241*
Texas Wood Preservative Green   *157 238 75*
Texas Wood Preservative Red Cedar   *156 241*
TFA 10   *487 572 679 861*
TFA 10 H   *487 572 679 861*
TFA 10 HG   *484 523 615 835*
TFA 10 LA   *489 614 819*
TFA 10G   *484 523 615 835*
TFA-10 LA Light/Dark   *571 683 838*

TFA-10 Light/Dark   *571 683 838*
TFA-20   *571 683 838*
TFA-30   *566 669*
TFA-30 White   *633 670*
Thompson's Interior and Exterior Fungicidal Spray   *295 298*
Thompson's Interior Mould Killer   *295 298*
Tiger Cruising   *613 834*
Tiger Tin Free Antifouling   *563 663*
Tiger White   *503 629*
Tigerline Antifouling   *503 629*
Timber Preservative   *255*
Timber Preservative Clear TFP7   *125 232 262*
Timber Preservative Green   *73*
Timber Preservative Solvent Based Fungicidal Insecticide 10   *156 241*
Timber Preservative Solvent Based Fungicidal Insecticide 30   *155 230 244*
Timber Treat   *156 241*
Timbercare Microporous Exterior Preservative   *128 260*
Timbercare WRQD Ready To Use   *194 249*
Timbercol Preservative (MBT) Concentrate   *158*
Timbercol Preservative Concentrate   *159 200 4*
Timberdip   *22*
Timberglow   *22*
Timberglow Concentrate   *130 23*
Timberglow Extra   *130 23*
Timberguard 1 Plus 7 Concentrate   *22*
Timberlife   *111*
Timberlife Extra   *111*
Timbermate Clear Wood Preservative   *103 225*
Timbermate Green   *73*
Timberplus   *170*
Timbertex PI   *129*
Timbertex PI/2   *204*
Timbertone S7 Preservative   *158*
Timbertreat Green Preservative   *73*
Timbertreat Multi Purpose Preservative   *36 98*
Timbertreat Wood Preservative   *32*
Timbertreat Wood Preservative—Clear   *36 98*
Timbertreat Woodworm and Dry Rot Killer   *36 98*
Timbertreat Woodworm and Dry Rot Killer—Water Based   *36 98*
Timbor   *129*
Timbor Paste   *129*
Timbrene Green Wood Preserver   *73*
Timbrene Supreme   *121 191 227*
Timbrene Wood Preserver   *255*
Timbrene Woodworm Killer   *174*
Timbrol Creosote Blend   *69*
Tinocide Insecticidal Lacquer   *348*
Titan FGA Antifouling   *518 620 828*
Titan FGA Antifouling White   *517 657 827*
Titan Tin Free   *493 550*

HSE PRODUCT NAME INDEX

Titan Tin Free Antifouling *547*
Torkill Fungicidal Solution 'W' *295 298*
Tox Exterminating Fly and Wasp Killer *374 454*
Trade Ronseal Low Odour Wood Preserver *255*
Travis Perkins Creosote *92*
Trawler *547*
Trawler Tin-Free *511 695*
TRC Water Repellent Wood Preserver *250*
Trend Wood Preservative Brown *156 241*
Trend Wood Preservative Clear *156 241*
Trend Wood Preservative Green *157 238 75*
Trend Wood Preservative Red Cedar *156 241*
Trilanco Fly Killer *435 460*
Trilanco Fly Spray *419*
Trimethrin 20S *174*
Trimethrin 2AQ *174*
Trimethrin 3AQ *174*
Trimethrin 30S *174*
Trimethrin 6 *195 263*
Trimethrin AQ Plus *190 229*
Trimethrin OS Plus *190 229*
Tripaste PP *14 176*
Trisol 21 *309*
Tritec *174*
Tritec Plus *190 229*
Triton Tripaste *195 263*
Tropical Antifouling (New Version) *700 776 802*
Tropical Antifouling (Old Version) *699 756 793 801*
Tropical Antifouling AF4 *581 707*
Tropical Super Service Antifouling Paint *547*
TS 10239 Antifouling *560 627*
TS 10240 Antifouling ADA160 Series *547*
Turbair Beetle Killer *372 401 433*
Tymasil *390*
Ultrabond Wood Preservative *111*
Ultrabond Woodworm Killer *123 226*
Unisil S Silicone Waterproofing Solution *295 298*
Unitas Antifouling Paint Black *560 627*
Unitas Antifouling Paint Chocolate *547*
Unitas Antifouling Paint Red *547*
Unitas Antifouling Paint White *628*
Unitas Fungicidal Wash – Exterior (Solvent Based) *295 298*
Unitas Fungicidal Wash – Interior (Water Based) *295 298*
Universal *178 20*
Universal Fluid (Grade PB) *190 229*
Universal Wood Preservative *100 114*
Universal Wood Preserver *190 229*
Vacsele *194 249*
Vacsele 2611 *198 267*
Vacsele P2312 *198 267*
Vacsele Ready To Use (2605) *198 267*
Vacsol (2:1 Conc) 2711/2712 *198 267*
Vacsol 2234 J Conc *235*
Vacsol 2622/2623 WR *192 234*
Vacsol 2625/2626 WR 2:1 Concentrate *192 234*
Vacsol 2652/2653 JWR *233*
Vacsol 2709/2710 *198 267*
Vacsol 2713/2714 *192 234*
Vacsol 2716/2717 2:1 Concentrate *192 234*
Vacsol 2746/2747 J *233*
Vacsol 2:1 Concentrate *198 267*
Vacsol J (2:1 Conc) 2744/2745 *265*
Vacsol J 2742/2743 *265*
Vacsol J 2:1 Concentrate *265*
Vacsol J Ready To Use *248 265*
Vacsol J RTU *235*
Vacsol JWR (2:1 Conc) 2642/2643 *265*
Vacsol JWR 2640/2641 *265*
Vacsol JWR 2:1 Concentrate *265*
Vacsol JWR Concentrate *235*
Vacsol JWR Ready To Use *248 265*
Vacsol JWR RTU *235*
Vacsol MWR Concentrate 2203 *156 241*
Vacsol MWR Ready For Use *194 249*
Vacsol MWR Ready To Use 2204 *156 241*
Vacsol P (2310) *198 267*
Vacsol P 2304 RTU *156 241*
Vacsol P Ready To Use (2334) *235*
Vacsol P Ready To Use (2335) *265*
Vacsol Ready To Use *198 267*
Vacsol WR (2:1 Conc) 2614/2615 *198 267*
Vacsol WR 2612/2613 *198 267*
Vacsol WR 2:1 Concentrate *198 267*
Vacsol WR Concentrate 2115 *156 241*
Vacsol WR Ready To Use *194 198 249 267*
Vacsol WR Ready To Use 2116 *156 241*
Vallance Fungicidal Wash *295 298*
Valspar Creosote *92*
Valspar Shed and Fence Preservative *204*
Valspar Timberguard Shed and Fence Preservative *204*
Vapona Ant and Crawling Insect Killer *333 393*
Vapona Ant and Crawling Insect Killer Aerosol *352 449*
Vapona Ant and Crawling Insect Powder *392*
Vapona Ant and Crawling Insect Spray *354 468*
Vapona Ant Trap *466*
Vapona Antpen *348*
Vapona Carpet and Household Flea Powder *392*
Vapona Fly and Wasp Killer *333 393*
Vapona Fly and Wasp Killer Spray *367 451*
Vapona Fly Killer *360*
Vapona Flypen *348*
Vapona Green Arrow Fly and Wasp Killer *419*
Vapona House and Plant Fly Aerosol *419*
Vapona House and Plant Fly Spray *465 476*
Vapona Moth Killer *360*

Vapona Moth Proofer *392*
Vapona Mothaks *320 388*
Vapona Plug-In Flying Insect Killer *467*
Vapona Professional Cockroach Killer *360*
Vapona Small Space Fly Killer *360*
Vapona Wasp and Fly Killer *367 451*
Vapona Wasp and Fly Killer Spray *465 476*
VC 17-Victory Antifouling *503 629*
VC 17M EP-Antifouling *530*
VC 17M Tropicana *496 531*
VC Aqua 12 *633 670*
VC Offshore *503 629*
VC Offshore 602 *499 551*
VC Offshore Extra 100 Series *624 877*
VC Offshore Extra Strength *499 551*
VC Offshore SP Antifouling *624 877*
VC Prop-o-Drev *503 629*
VC17M *530*
Vet-Kem Acclaim *432 445*
Vet-Kem Pump Spray *406 444*
Vijurrax Spray *357 362*
Vinilstop 9926 Red *547*
Vinyl AF *547*
Vinyl Antifouling 2000 *613 834*
Vinyl Antifouling AF12 *581 707*
Vitapet Flearid Household Spray *465 476*
Vitax Kontrol Ant Killer Powder *392*
Vulcanite CA:20 Concentrated Algicide *277*
Vulcanite Mouldicide For Walls, Ceilings and Cupboards *277*
Vulcanite Timbertreat Green Preservative *73*
Vulcanite Timbertreat Multi Purpose Preservative *36 98*
Vulcanite Timbertreat Wood Preservative *32*
Vulcanite Timbertreat Wood Preservative Clear *36 98*
Vulcanite Timbertreat Woodworm and Dry Rot Killer Water Based *36 98*
W.S.P. Wood Treatment *204*
Wahl Envoyage Mosquito Killer *331*
Wasp Nest Destroyer *340 435 460*
Wasp Nest Killer Professional *327*
Waspend *415 436 459*
Waspex *380*
Water Based Wood Preserver (Concentrate) *11 97*
Water Based Woodworm Killer *95*
Water Repellent Pink Primer *149 166*
Water-Based Wood Preserver *11 97*
Waterways *547*
Waterways Antifouling *566 669*
Wax Polish *145*
Weathershield Exterior Preservative Basecoat *124 231 253*

Weathershield Exterior Timber Preservative *258 78*
Weathershield Fungicidal Solution *135 137 295 298*
Weathershield Fungicidal Wash *135 137 295 298*
Weathershield Preservative Primer *124 231 253*
Wellcome Crawling Insect Killer *402 440*
Wellcome Environmental Health Fly Spray *333 393*
Wellcome Fly Killer *333 393*
Wellcome Multishot Aircraft Aerosol *475*
Wellcome One-Shot Aircraft Aerosol *475*
Wellcome Pif Paf Crawling Insect Killer *333 393*
Wellcome Pif Paf Flying Insect Killer *333 393*
Whiskas Bedding Pest Control Spray *474 477*
Whiskas Exelpet Bedding Spray *474 477*
Wickes All Purpose Wood Treatment *178 20*
Wickes Creosote *69*
Wickes Exterior Wood Preserver *18*
Wickes Fungicidal Wash *271*
Wickes Shed and Fence Treatment *204*
Wickes Wood Preservative Brown *156 241*
Wickes Wood Preservative Clear *156 241*
Wickes Wood Preservative Green *157 238 75*
Wickes Wood Preservative Red Cedar *156 241*
Wickes Wood Preserver *204*
Wickes Woodworm Killer *174*
Wilco Wood Preservative Brown *156 241*
Wilco Wood Preservative Clear *156 241*
Wilco Wood Preservative Green *157 238 75*
Wilco Wood Preservative Red Cedar *156 241*
Wilko Ant Killer Lacquer *356*
Wilko Ant Killer Spray *348*
Wilko Clear Wood Preserver *150 168 242*
Wilko Creosote *69*
Wilko Exterior System Preservative Primer *128 260*
Wilko Exterior Wood Preserver *235*
Wilko Red Cedar Wood Preserver *126 239*
Willo-Zawb *475*
Willodorm *475*
Wintox *359 368 450*
Wocosen 12 OL *187 203*
Wocosen S *202*
Wolmanit CB *201 52 85*
Wolmanit CB-A *207 53 88*
Wolmanit CB-P *201 52 85*
Wolvac 55 *116 141 148*
Wood Preservative *111 250*
Wood Preservative—AA155/01 *235*
Wood Preservative AA 155/00 *126 239*
Wood Preservative AA 155/01 *235*
Wood Preservative AA 155/03 *126 239*
Wood Preservative AA155 *126 239*

Wood Preservative Clear   *195 263*
Wood Preservative Type 3   *69*
Wood Preserver   *223 265*
Wood Preserver Green   *73*
Woodex Intra   *126 239*
Woodman Creosote and Light Brown
   Creosote   *92*
Woodtreat   *149 166*
Woodtreat 25   *14 176*
Woodtreat BP   *195 263*
Woodworm Fluid   *145*
Woodworm Fluid (B) EC   *174*
Woodworm Fluid (B) Emulsion Concentrate   *174*
Woodworm Fluid (B) For Bonded
   Warehouses   *174*
Woodworm Fluid B   *174*
Woodworm Fluid FB   *43*
Woodworm Fluid For Bonded Warehouses and
   Distilleries   *145*
Woodworm Fluid FP   *174*
Woodworm Fluid Minimum Odour   *145*
Woodworm Fluid Z Emulsion Concentrate   *174*
Woodworm Furniture Polish   *174*

Woodworm Killer   *174 190 229 95*
Woodworm Killer (Grade P)   *174*
Woodworm Killer and Rot Treatment   *179 276 303 31*
Woodworm Paste   *145*
Woodworm Paste B   *174*
Woodworm Roof Void Paste   *174*
Woodworm Treatment Spray   *174*
Woolworth Mosquito Killer   *331*
Wudfil Wet Rot Treatment   *235*
Xyladecor Matt U 404   *113 30*
Xyladecor Matt U-4010   *117 143 184*
Xylamon Brown U 101 C   *179 31*
Xylamon Brown U1011   *142 186*
Xylamon Curative U 152 G/H   *179 31*
Xylamon Primer Dipping Stain U415   *113 30*
Xylamon Primer Dipping Stain U411   *115 140*
Z-Stop Anti-Wasp Strip   *392*
Z.144 Fungicidal Wash   *269 278*
Zap Pest Control   *333 393*
Zodiac and Household Flea Spray   *432 445*
Zodiac and Pump Spray   *406 444*

6
HSE ACTIVE INGREDIENT INDEX

1,4-Dichlorobenzene *318*
1,4-Dichlorobenzene and Dichlorvos *319*
1,4-Dichlorobenzene and Naphthalene *320*
2,3,5,6-Tetrachloro-4-(Methyl Sulphonyl) Pyridine *480*
2,3,5,6-Tetrachloro-4-(Methyl Sulphonyl) Pyridine and 2-Methylthio-4-Tertiary-Butylamino-6-Cyclopropylamino-S-Triazine and Cuprous Oxide and Cuprous Thiocyanate and Zinc Oxide *481*
2,3,5,6-Tetrachloro-4-(Methyl Sulphonyl) Pyridine and Cuprous Oxide *482*
2,4,5,6-Tetrachloro Isophthalonitrile *483*
2,4,5,6-Tetrachloro Isophthalonitrile and 4,5-Dichloro-2-N-Octyl-4-Isothiazolin-3-One and Cuprous Oxide and Zinc Oxide *484*
2,4,5,6-Tetrachloro Isophthalonitrile and Copper Sulphate *485*
2,4,5,6-Tetrachloro Isophthalonitrile and Cuprous Oxide *486*
2,4,5,6-Tetrachloro Isophthalonitrile and Cuprous Oxide and Dichlorophenyl Dimethylurea and Zinc Oxide *487*
2,4,5,6-Tetrachloro Isophthalonitrile and Cuprous Oxide and Zinc Naphthenate *488*
2,4,5,6-Tetrachloro Isophthalonitrile and Cuprous Oxide and Zinc Oxide *489*
2,4,5,6-Tetrachloro Isophthalonitrile and Cuprous Thiocyanate and Zinc Oxide *490*
2-(Thiocyanomethylthio) Benzothiazole *1*
2-(Thiocyanomethylthio) Benzothiazole *491*
2-(Thiocyanomethylthio) Benzothiazole and 2-Methylthio-4-Tertiary-Butylamino-6-Cyclopropylamino-S-Triazine and Dichlorophenyl Dimethylurea and Zinc Oxide and Zinc Pyrithione *492*
2-(Thiocyanomethylthio) Benzothiazole and Boric Acid *2*
2-(Thiocyanomethylthio) Benzothiazole and Boric Acid and Methylene Bis (Thiocyanate) *3*
2-(Thiocyanomethylthio) Benzothiazole and Cuprous Oxide *493*
2-(Thiocyanomethylthio) Benzothiazole and Methylene Bis (Thiocyanate) *4*
2-Methyl-4-Isothiazolin-3-One and 5-Chloro-2-Methyl-4-Isothiazolin-3-One *5*
2-Methylthio-4-Tertiary-Butylamino-6-Cyclopropylamino-S-Triazine and Cuprous Oxide and Zinc Oxide *494*
2-Methylthio-4-Tertiary-Butylamino-6-Cyclopropylamino-S-Triazine and Copper Metal *496*
2-Methylthio-4-Tertiary-Butylamino-6-Cyclopropylamino-S-Triazine and Copper Resinate and Cuprous Oxide *497*
2-Methylthio-4-Tertiary-Butylamino-6-Cyclopropylamino-S-Triazine and Copper Resinate and Cuprous Oxide and Zinc Oxide *498*
2-Methylthio-4-Tertiary-Butylamino-6-Cyclopropylamino-S-Triazine and Cuprous Oxide *499*
2-Methylthio-4-Tertiary-Butylamino-6-Cyclopropylamino-S-Triazine and Cuprous Oxide and Cuprous Thiocyanate and Zinc Oxide and 2,3,5,6-Tetrachloro-4-(Methyl Sulphonyl) Pyridine *500*
2-Methylthio-4-Tertiary-Butylamino-6-Cyclopropylamino-S-Triazine and Cuprous Oxide and Dichlorophenyl Dimethylurea and Zinc Oxide *501*
2-Methylthio-4-Tertiary-Butylamino-6-Cyclopropylamino-S-Triazine and Cuprous Oxide and Zinc Oxide *502*
2-Methylthio-4-Tertiary-Butylamino-6-Cyclopropylamino-S-Triazine and Cuprous Thiocyanate *503*
2-Methylthio-4-Tertiary-Butylamino-6-Cyclopropylamino-S-Triazine and Cuprous Thiocyanate and Dichlofluanid and Zinc Oxide *504*
2-Methylthio-4-Tertiary-Butylamino-6-Cyclopropylmino-S-Triazine and Cuprous Thiocyanate and Dichlorophenyl Dimethylurea and Zinc Oxide *505*
2-Methylthio-4-Tertiary-Butylamino-6-Cyclopropylamino-S-Triazine and Cuprous Thiocyanate and Zinc Oxide *506*
2-Methylthio-4-Tertiary-Butylamino-6-Cyclopropylamino-S-Triazine and Cuprous Thiocyanate and Zineb *507*
2-Methylthio-4-Tertiary-Butylamino-6-Cyclopropylamino-S-Triazine and Dichlorophenyl Dimethylurea and Zinc Oxide and Zinc Pyrithione *508*
2-Methylthio-4-Tertiary-Butylamino-6-Cyclopropylamino-S-Triazine and Dichlorophenyl Dimethylurea and Zinc Oxide and Zinc Pyrithione and 2-(Thiocyanomethylthio) Benzothiazole *509*
2-Methylthio-4-Tertiary-Butylamino-6-Cyclopropylamino-S-Triazine and Dichlorophenyl Dimethylurea and Zinc Pyrithione *510*
2-Methylthio-4-Tertiary-Butylamino-6-Cyclopropylamino-S-Triazine and Thiram *511*
2-Methylthio-4-Tertiary-Butylamino-6-Cyclopropylamino-S-Triazine and Thiram and Tributyltin Methacrylate and Tributyltin Oxide *512*

HSE ACTIVE INGREDIENT INDEX

2-Methylthio-4-Tertiary-Butylamino-6-Cyclopropylamino-S-Triazine and Tributyltin Methacrylate and Zinc Oxide *513*
2-Methylthio-4-Tertiary and Zinc Oxide *495*
2-Methylthio-4-Tertiary-Butylamino-6-Cyclopropylamino-S-Triazine and Cuprous Oxide *514*
2-Methylthio-4-Tertiary-Butylamino-6-Cyclopropylamino-S-Triazine and Cuprous Thiocyanate *515*
2-Methylthio-4-Tertiary-Butylamino-6-Cyclopropylamino-S-Triazine and Cuprous Oxide and Zinc Oxide *516*
2-Methylthio-4-Tertiary-Butylamino-6-Cyclopropylamino-S-Triazine and Cuprous Thiocyanate and Zinc Oxide *517*
2-Methylthio-4-Tertiary-Butylamino-6-Cyclopropylamino-S-Triazine and Cuprous Oxide and Zinc Oxide *518*
2-Methylthio-4-Tertiary-Butylamino-6-Cyclopropylamino-S-Triazine and Cuprous Thiocyanate and Zinc Oxide *519*
2-Phenylphenol *268*
2-Phenylphenol *6*
2-Phenylphenol and Benzalkonium Chloride *269*
2-Phenylphenol and Benzalkonium Chloride *7*
2-Phenylphenol and Cypermethrin *8*
2-Phenylphenol and Permethrin *9*
3-Iodo-2-Propynyl-N-Butyl Carbamate *10*
3-Iodo-2-Propynyl-N-Butyl Carbamate *270*
3-Iodo-2-Propynyl-N-Butyl Carbamate and Cypermethrin *11*
3-Iodo-2-Propynyl-N-Butyl Carbamate and Dialkyldimethyl Ammonium Chloride *12*
3-Iodo-2-Propynyl-N-Butyl Carbamate and Gamma-HCH *13*
3-Iodo-2-Propynyl-N-Butyl Carbamate and Permethrin *14*
4,5-Dichloro-2-N-Octyl-4-Isothiazolin-3-One and Cuprous Oxide *520*
4,5-Dichloro-2-N-Octyl-4-Isothiazolin-3-One and Cuprous Oxide and Tributyltin Methacrylate and Tributyltin Oxide *521*
4,5-Dichloro-2-N-Octyl-4-Isothiazolin-3-One and Cuprous Oxide and Zinc Oxide *522*
4,5-Dichloro-2-N-Octyl-4-Isothiazolin-3-One and Cuprous Oxide and Zinc Oxide and 2,4,5,6-Tetrachloro Isophthalonitrile *523*
4,5-Dichloro-2-N-Octyl-4-Isothiazolin-3-One and Cuprous Oxide *524*
4-Chloro-Meta-Cresol and Cuprous Oxide *525*
5-Chloro-2-Methyl-4-Isothiazolin-3-One and 2-Methyl-4-Isothiazolin-3-One *15*
Acypetacs Copper *16*

Acypetacs Copper and Acypetacs Zinc and Permethrin *17*
Acypetacs Zinc *18*
Acypetacs Zinc and Dichlofluanid *19*
Acypetacs Zinc and Permethrin *20*
Acypetacs Zinc and Permethrin and Acypetacs Copper *21*
Alkylaryltrimethyl Ammonium Chloride *22*
Alkylaryltrimethyl Ammonium Chloride *271*
Alkylaryltrimethyl Ammonium Chloride and Boric Acid *272*
Alkylaryltrimethyl Ammonium Chloride and Disodium Octaborate *23*
Alkylaryltrimethyl Ammonium Chloride and Disodium Octaborate *273*
Alkylaryltrimethyl Ammonium Chloride and Permethrin and Pyrethrins *274*
Alkylaryltrimethyl Ammonium Chloride and Permethrin and Pyrethrins *321*
Alkylaryltrimethyl Ammonium Chloride and Tributyltin Oxide *24*
Alkylaryltrimethyl Ammonium Chloride and Tributyltin Oxide *275*
Alkyltrimethyl Ammonium Chloride and Sodium Tetraborate *25*
Allethrin *322*
Alphacypermethrin *323*
Ammonium Bifluoride and Sodium Dichromate and Sodium Fluoride *26*
Arsenic Pentoxide and Chromium Trioxide and Copper Oxide *27*
Arsenic Pentoxide and Copper Sulphate and Sodium Dichromate *28*
Arsenic Trioxide and Copper Naphthenate and Cuprous Oxide *526*
Arsenic Trioxide and Cuprous Oxide *527*
Azaconazole *29*
Azaconazole and Dichlofluanid *30*
Azaconazole and Permethrin *276*
Azaconazole and Permethrin *31*
Azamethiphos *324*
Bacillus Thuringiensis Var Israelensis *325*
Bacillus Thuringiensis Var Israelensis (Serotype H14) Fermenter Product *600 326*
Bendiocarb *327*
Bendiocarb and Pyrethrins *328*
Bendiocarb and Tetramethrin *329*
Benzalkonium Chloride *277*
Benzalkonium Chloride *32*
Benzalkonium Chloride and 2-Phenylphenol *278*
Benzalkonium Chloride and 2-Phenylphenol *33*
Benzalkonium Chloride and Boric Acid *279*
Benzalkonium Chloride and Boric Acid *34*

HSE ACTIVE INGREDIENT INDEX

Benzalkonium Chloride and Boric Acid and Dialkyldimethyl Ammonium Chloride and Methylene Bis (Thiocyanate)   *35*
Benzalkonium Chloride and Carbendazim and Dialkyldimethyl Ammonium Chloride   *280*
Benzalkonium Chloride and Cypermethrin   *36*
Benzalkonium Chloride and Cypermethrin and Dialkyldimethyl Ammonium Chloride   *37*
Benzalkonium Chloride and Dialkyldimethyl Ammonium Chloride   *281*
Benzalkonium Chloride and Dialkyldimethyl Ammonium Chloride   *38*
Benzalkonium Chloride and Disodium Octaborate   *282*
Benzalkonium Chloride and Disodium Octaborate   *39*
Benzalkonium Chloride and Gamma-HCH   *40*
Benzalkonium Chloride and Naphthalene and Permethrin and Pyrethrins   *283*
Benzalkonium Chloride and Naphthalene and Permethrin and Pyrethrins   *330*
Benzalkonium Chloride and Permethrin   *41*
Benzalkonium Chloride and Tributyltin Oxide   *284*
Benzalkonium Chloride and Tributyltin Oxide   *42*
Bioallethrin   *331*
Bioallethrin and Bioresmethrin   *332*
Bioallethrin and Permethrin   *333*
Bioallethrin and Permethrin and Pyrethrins   *334*
Bioallethrin and S-Bioallethrin   *335*
Bioresmethrin   *336*
Bioresmethrin and Bioallethrin   *337*
Bioresmethrin and S-Bioallethrin   *338*
Boric Acid   *285*
Boric Acid   *339*
Boric Acid   *43*
Boric Acid and 2-(Thiocyanomethylthio) Benzothiazole   *44*
Boric Acid and Alkylaryltrimethyl Ammonium Chloride   *286*
Boric Acid and Benzalkonium Chloride   *287*
Boric Acid and Benzalkonium Chloride   *45*
Boric Acid and Chromium Acetate and Copper Sulphate and Sodium Dichromate   *46*
Boric Acid and Chromium Trioxide and Copper Oxide   *47*
Boric Acid and Copper Carbonate Hydroxide   *48*
Boric Acid and Copper Carbonate Hydroxide and Tebuconazole   *49*
Boric Acid and Copper Oxide   *50*
Boric Acid and Copper Sulphate   *51*
Boric Acid and Copper Sulphate and Potassium Dichromate   *52*
Boric Acid and Copper Sulphate and Sodium Dichromate   *53*

Boric Acid and Dialkyldimethyl Ammonium Chloride and Methylene Bis (Thiocyanate) and Benzalkonium Chloride   *54*
Boric Acid and Methylene Bis (Thiocyanate)   *55*
Boric Acid and Methylene Bis (Thiocyanate) and 2-(Thiocyanomethylthic) Benzothiazole   *56*
Boric Acid and Permethrin   *57*
Boric Acid and Permethrin and Zinc Octoate   *58*
Boric Acid and Sodium Tetraborate   *59*
Boric Acid and Zinc Octoate   *60*
Carbaryl   *340*
Carbendazim   *61*
Carbendazim and Dialkyldimethyl Ammonium Chloride and Benzalkonium Chloride   *288*
Carbendazim and Furmecyclox   *62*
Carbendazim and Tributyltin Oxide and Zinc Naphthenate   *63*
Ceto-Stearyl Diethoxylate and Oleyl Monoethoxylate   *341*
Chlorpyrifos   *342*
Chlorpyrifos and Cypermethrin and Pyrethrins   *343*
Chlorpyrifos and Tetramethrin   *344*
Chlorpyrifos-Methyl   *345*
Chlorpyrifos-Methyl and Permethrin and Pyrethrins   *346*
Chromium Acetate and Copper Sulphate and Sodium Dichromate   *64*
Chromium Acetate and Copper Sulphate and Sodium Dichromate and Boric Acid   *65*
Chromium Trioxide and Copper Oxide and Arsenic Pentoxide   *66*
Chromium Trioxide and Copper Oxide and Boric Acid   *67*
Chromium Trioxide and Copper Oxide and Disodium Octaborate   *68*
Cis 1-(3-Chloroallyl)-3,5,7-Triaza-1-Azonia Adamantane Chloride   *528*
Citronella Oil   *347*
Coal Tar Creosote   *69*
Coal Tar Creosote and Creosote   *70*
Copper Bronze Powder   *529*
Copper Carbonate Hydroxide and Boric Acid   *71*
Copper Carbonate Hydroxide and Tebuconazole and Boric Acid   *72*
Copper Metal   *530*
Copper Metal and 2-Methylthio-4-Tertiary-Butylamino-6-Cyclopropylamino-S-Triazine   *531*
Copper Metal and Tributyltin Methacrylate and Tributyltin Oxide   *532*
Copper Metal and Zinc Oxide   *533*
Copper Naphthenate   *534*
Copper Naphthenate   *73*

HSE ACTIVE INGREDIENT INDEX

Copper Naphthenate and Cuprous Oxide and Arsenic Trioxide 535
Copper Naphthenate and Cuprous Oxide and Dichlofluanid and Zinc Naphthenate and Zinc Oxide 536
Copper Naphthenate and Disodium Octaborate 74
Copper Naphthenate and Gamma-HCH and Tributyltin Oxide 75
Copper Naphthenate and Permethrin 76
Copper Naphthenate and Tri(Hexylene Glycol) Biborate 77
Copper Naphthenate and Zinc Octoate 78
Copper Oxide and Arsenic Pentoxide and Chromium Trioxide 79
Copper Oxide and Boric Acid 80
Copper Oxide and Boric Acid and Chromium Trioxide 81
Copper Oxide and Disodium Octaborate and Chromium Trioxide 82
Copper Resinate and Cuprous Oxide 537
Copper Resinate and Cuprous Oxide and 2-Methylthio-4-Tertiary-Butylamino-6-Cyclopropylamino-S-Triazine 538
Copper Resinate and Cuprous Oxide and Dichlorophenyl Dimethylurea 539
Copper Resinate and Cuprous Oxide and Dichlorophenyl Dimethylurea and Zinc Oxide 540
Copper Resinate and Cuprous Oxide and Tributyltin Fluoride 541
Copper Resinate and Cuprous Oxide and Tributyltin Fluoride and Zinc Oxide 542
Copper Resinate and Cuprous Oxide and Tributyltin Fluoride and Zinc Oxide and Zineb 543
Copper Resinate and Cuprous Oxide and Zinc Oxide and 2-Methylthio-4-Tertiary-Butylamino-6-Cyclopropylamino-S-Triazine 544
Copper Resinate and Cuprous Oxide and Zinc Oxide and Zineb 545
Copper Salt Of Synthetic Carboxylic Acids (C8-C12) 83
Copper Sulphate and 2,4,5,6-Tetrachloro Isophthalonitrile 546
Copper Sulphate and Boric Acid 84
Copper Sulphate and Potassium Dichromate and Boric Acid 85
Copper Sulphate and Sodium Dichromate 86
Copper Sulphate and Sodium Dichromate and Arsenic Pentoxide 87
Copper Sulphate and Sodium Dichromate and Boric Acid 88
Copper Sulphate and Sodium Dichromate and Boric Acid and Chromium Acetate 89
Copper Sulphate and Sodium Dichromate and Chromium Acetate 90
Copper Versatate 91
Creosote 92
Creosote and Coal Tar Creosote 93
Creosote and TC Oil 94
Cuprous Oxide 547
Cuprous Oxide and 2,3,5,6-Tetrachloro-4-(Methyl Sulphonyl) Pyridine 548
Cuprous Oxide and 2,4,5,6-Tetrachloro Isophthalonitrile 549
Cuprous Oxide and 2-(Thiocyanomethylthio) Benzothiazole 550
Cuprous Oxide and 2-Methylthio-4-Tertiary-Butylamino-6-Cyclopropylamino-S-Triazine 551
Cuprous Oxide and 2-Methylthio-4-Tertiary-Butylamino-6-Cyclopropylamino-S-Triazine and Copper Resinate 552
Cuprous Oxide and 2-Methylthio-4-Tertiary-Butylamino-6-Cyclopropylamino-S-Triazine 553
Cuprous Oxide and 4,5-Dichloro-2-N-Octyl-4-Isothiazolin-3-One 554
Cuprous Oxide and 4,5-Dichloro-2-N-Octyl-4-Isothiazolin-3-one 555
Cuprous Oxide and 4-Chloro-Meta-Cresol 556
Cuprous Oxide and Arsenic Trioxide 557
Cuprous Oxide and Arsenic Trioxide and Copper Naphthenate 558
Cuprous Oxide and Copper Resinate 559
Cuprous Oxide and Cuprous Sulphide 560
Cuprous Oxide and Cuprous Thiocyanate and Tributyltin Methacrylate and Tributyltin Oxide 561
Cuprous Oxide and Cuprous Thiocyanate and Zinc Oxide and 2,3,5,6-Tetrachloro-4-(Methyl Sulphonyl) Pyridine and 2-Methylthio-4-Tertiary-Butylamino-6-Cyclopropylamino-S-Triazine 562
Cuprous Oxide and Dichlofluanid 563
Cuprous Oxide and Dichlofluanid and Zinc Naphthenate and Zinc Oxide and Copper Naphthenate 564
Cuprous Oxide and Dichlofluanid and Zinc Oxide 565
Cuprous Oxide and Dichlorophenyl Dimethylurea 566
Cuprous Oxide and Dichlorophenyl Dimethylurea and Copper Resinate 567
Cuprous Oxide and Dichlorophenyl Dimethylurea and Tributyltin Fluoride and Tributyltin Oxide 568

Cuprous Oxide and Dichlorophenyl Dimethylurea and Tributyltin Methacrylate and Tributyltin Oxide *569*
Cuprous Oxide and Dichlorophenyl Dimethylurea and Tributyltin Methacrylate and Tributyltin Oxide and Zinc Oxide *570*
Cuprous Oxide and Dichlorophenyl Dimethylurea and Zinc Oxide *571*
Cuprous Oxide and Dichlorophenyl Dimethylurea and Zinc Oxide and 2,4,5,6-Tetrachloro Isophthalonitrile *572*
Cuprous Oxide and Dichlorophenyl Dimethylurea and Zinc Oxide and 2-Methylthio-4-Tertiary-Butylamino-6-Cyclopropylamino-S-Triazine *573*
Cuprous Oxide and Dichlorophenyl Dimethylurea and Zinc Oxide and Copper Resinate *574*
Cuprous Oxide and Folpet *575*
Cuprous Oxide and Maneb and Tributyltin Methacrylate and Zinc Oxide *576*
Cuprous Oxide and Oxytetracycline Hydrochloride *577*
Cuprous Oxide and Thiram *578*
Cuprous Oxide and Tributyltin Acrylate and Tributyltin Oxide *579*
Cuprous Oxide and Tributyltin Acrylate and Zinc Oxide *580*
Cuprous Oxide and Tributyltin Fluoride *581*
Cuprous Oxide and Tributyltin Fluoride and Copper Resinate *582*
Cuprous Oxide and Tributyltin Fluoride and Tributyltin Oxide *583*
Cuprous Oxide and Tributyltin Fluoride and Tributyltin Oxide and Zinc Oxide *584*
Cuprous Oxide and Tributyltin Fluoride and Zinc Naphthenate *585*
Cuprous Oxide and Tributyltin Fluoride and Zinc Oxide *586*
Cuprous Oxide and Tributyltin Fluoride and Zinc Oxide and Copper Resinate *587*
Cuprous Oxide and Tributyltin Fluoride and Zinc Oxide and Zineb *588*
Cuprous Oxide and Tributyltin Fluoride and Zinc Oxide and Zineb and Copper Resinate *589*
Cuprous Oxide and Tributyltin Methacrylate *590*
Cuprous Oxide and Tributyltin Methacrylate and Tributyltin Oxide *591*
Cuprous Oxide and Tributyltin Methacrylate and Tributyltin Oxide and 4,5-Dichloro-2-N-Octyl-4-Isothiazolin-3-One *592*
Cuprous Oxide and Tributyltin Methacrylate and Tributyltin Oxide and Triphenyltin Fluoride *593*
Cuprous Oxide and Tributyltin Methacrylate and Tributyltin Oxide and Zinc Oxide *594*
Cuprous Oxide and Tributyltin Methacrylate and Tributyltin Oxide and Zinc Oxide and Zineb *595*
Cuprous Oxide and Tributyltin Methacrylate and Tributyltin Oxide and Zineb *596*
Cuprous Oxide and Tributyltin Methacrylate and Zinc Metal *597*
Cuprous Oxide and Tributyltin Methacrylate and Zinc Metal and Zinc Oxide *598*
Cuprous Oxide and Tributyltin Methacrylate and Zinc Oxide and Zineb *599*
Cuprous Oxide and Tributyltin Methacrylate and Zineb *600*
Cuprous Oxide and Tributyltin Oxide *601*
Cuprous Oxide and Tributyltin Oxide and Triphenyltin Fluoride *602*
Cuprous Oxide and Tributyltin Oxide and Zinc Oxide *603*
Cuprous Oxide and Tributyltin Oxide and Ziram *604*
Cuprous Oxide and Tributyltin Polysiloxane and Zinc Oxide *605*
Cuprous Oxide and Triphenyltin Fluoride *606*
Cuprous Oxide and Triphenyltin Fluoride and Zinc Oxide *607*
Cuprous Oxide and Triphenyltin Hydroxide *608*
Cuprous Oxide and Triphenyltin Hydroxide and Zinc Oxide *609*
Cuprous Oxide and Zinc Naphthenate *610*
Cuprous Oxide and Zinc Naphthenate and 2,4,5,6-Tetrachloro Isophthalonitrile *611*
Cuprous Oxide and Zinc Naphthenate and Zinc Oxide *612*
Cuprous Oxide and Zinc Oxide *613*
Cuprous Oxide and Zinc Oxide and 2,4,5,6-Tetrachloro Isophthalonitrile *614*
Cuprous Oxide and Zinc Oxide and 2,4,5,6-Tetrachloro Isophthalonitrile and 4,5-Dichloro-2-N-Octyl-4-Isothiazolin-3-One *615*
Cuprous Oxide and Zinc Oxide and 2-Methylthio-4-Tertiary-Butylamino-6-Cyclopropylamino-S-Triazine *616*
Cuprous Oxide and Zinc Oxide and 2-Methylthio-4-Tertiary-Butylamino-6-Cyclopropylamino-S-Triazine *617*
Cuprous Oxide and Zinc Oxide and 2-Methylthio-4-Tertiary-Butylamino-6-Cyclopropylamino-S-Triazine and Copper Resinate *618*
Cuprous Oxide and Zinc Oxide and 2-Methylthio-4-Tertiary-Butylamino-6-Cyclopropylamino-S-Triazine *619*
Cuprous Oxide and Zinc Oxide and 2-Methylthio-4-Tertiary-Butylamino-6-Cyclopropylamino-S-Triazine *620*

HSE ACTIVE INGREDIENT INDEX

Cuprous Oxide and Zinc Oxide and 4,5-Dichloro-2-N-Octyl-4-Isothiazolin-3-One *621*
Cuprous Oxide and Zinc Oxide and Zineb and Copper Resinate *622*
Cuprous Oxide and Zinc Oxide and Ziram *623*
Cuprous Oxide and Zinc Pyrithione *624*
Cuprous Oxide and Zineb *625*
Cuprous Oxide and Ziram *626*
Cuprous Sulphide and Cuprous Oxide *627*
Cuprous Thiocyanate *628*
Cuprous Thiocyanate and 2-Methylthio-4-Tertiary-Butylamino-6-Cyclopropylamino-S-Triazine *629*
Cuprous Thiocyanate and 2-Methylthio-4-Tertiary-Butylamino-6-Cyclopropylamino-S-Triazine *630*
Cuprous Thiocyanate and Dichlofluanid and Zinc Oxide *631*
Cuprous Thiocyanate and Dichlofluanid and Zinc Oxide and 2-Methylthio-4-Tertiary-Butylamino-6-Cyclopropylamino-S-Triazine *632*
Cuprous Thiocyanate and Dichlorophenyl Dimethylurea *633*
Cuprous Thiocyanate and Dichlorophenyl Dimethylurea and Triphenyltin Chloride *634*
Cuprous Thiocyanate and Dichlorophenyl Dimethylurea and Triphenyltin Chloride and Zinc Oxide *635*
Cuprous Thiocyanate and Dichlorophenyl Dimethylurea and Zinc Oxide *636*
Cuprous Thiocyanate and Dichlorophenyl Dimethylurea and Zinc Oxide and 2-Methylthio-4-Tertiary-Butylamino-6-Cyclopropylamino-S-Triazine *637*
Cuprous Thiocyanate and Maneb and Tributyltin Meso-Dibromosuccinate and Tributyltin Methacrylate and Tributyltin Oxide and Zinc Oxide *638*
Cuprous Thiocyanate and Tributyltin Fluoride *639*
Cuprous Thiocyanate and Tributyltin Fluoride and Tributyltin Oxide *640*
Cuprous Thiocyanate and Tributyltin Meso-Dibromosuccinate and Tributyltin Oxide and Zinc Oxide *641*
Cuprous Thiocyanate and Tributyltin Methacrylate *642*
Cuprous Thiocyanate and Tributyltin Methacrylate and Tributyltin Oxide *643*
Cuprous Thiocyanate and Tributyltin Methacrylate and Tributyltin Oxide and Cuprous Oxide *644*
Cuprous Thiocyanate and Tributyltin Methacrylate and Tributyltin Oxide and Triphenyltin Fluoride *645*
Cuprous Thiocyanate and Tributyltin Methacrylate and Tributyltin Oxide and Zinc Oxide *646*
Cuprous Thiocyanate and Tributyltin Methacrylate and Tributyltin Oxide and Zineb *647*
Cuprous Thiocyanate and Tributyltin Methacrylate and Zinc Metal and Zinc Oxide *648*
Cuprous Thiocyanate and Tributyltin Methacrylate and Zinc Oxide *649*
Cuprous Thiocyanate and Tributyltin Methacrylate and Zineb *650*
Cuprous Thiocyanate and Tributyltin Methacrylate and Ziram *651*
Cuprous Thiocyanate and Tributyltin Oxide *652*
Cuprous Thiocyanate and Tributyltin Oxide and Zineb *653*
Cuprous Thiocyanate and Zinc Oxide and 2,3,5,6-Tetrachloro-4-(Methyl Sulphonyl) Pyridine and 2-Methylthio-4-Tertiary-Butylamino-6-Cyclopropylamino-S-Triazine and Cuprous Oxide *654*
Cuprous Thiocyanate and Zinc Oxide and 2,4,5,6-Tetrachloro Isophthalonitrile *655*
Cuprous Thiocyanate and Zinc Oxide and 2-Methylthio-4-Tertiary-Butylamino-6-Cyclopropylamino-S-Triazine *656*
Cuprous Thiocyanate and Zinc Oxide and 2-Methylthio-4-Tertiary-Butylamino-6-Cyclopropylamino-S-Triazine *657*
Cuprous Thiocyanate and Zinc Oxide and 2-Methylthio-4-Tertiary-Butylamino-6-Cyclopropylamino-S-Triazine *658*
Cuprous Thiocyanate and Zineb *659*
Cuprous Thiocyanate and Zineb and 2-Methylthio-4-Tertiary-Butylamino-6-Cyclopropylamino-S-Triazine *660*
Cuprous Thiocyanate and Ziram *661*
Cypermethrin *348*
Cypermethrin *95*
Cypermethrin and 2-Phenylphenol *96*
Cypermethrin and 3-Iodo-2-Propynyl-N-Butyl Carbamate *97*
Cypermethrin and Benzalkonium Chloride *98*
Cypermethrin and Dialkyldimethyl Ammonium Chloride and Benzalkonium Chloride *99*
Cypermethrin and Dichlofluanid *100*
Cypermethrin and Methoprene *349*
Cypermethrin and Pentachlorophenol *101*
Cypermethrin and Pyrethrins *350*
Cypermethrin and Pyrethrins and Chlorpyrifos *351*
Cypermethrin and Tebuconazole *102*
Cypermethrin and Tetramethrin *352*
Cypermethrin and Tetramethrin and d-Allethrin *353*

Cypermethrin and Tri(Hexylene Glycol)
  Biborate   *103*
Cypermethrin and Zinc Octoate   *104*
Cypermethrin and Zinc Versatate   *105*
Cypermethrin and d-Allethrin   *354*
Deltamethrin   *355*
Dialkyldimethyl Ammonium Chloride   *106*
Dialkyldimethyl Ammonium Chloride   *289*
Dialkyldimethyl Ammonium Chloride and
  3-Iodo-2-Propynyl-N-Butyl Carbamate   *107*
Dialkyldimethyl Ammonium Chloride and
  Benzalkonium Chloride   *108*
Dialkyldimethyl Ammonium Chloride and
  Benzalkonium Chloride   *290*
Dialkyldimethyl Ammonium Chloride and
  Benzalkonium Chloride and
  Carbendazim   *291*
Dialkyldimethyl Ammonium Chloride and
  Benzalkonium Chloride and
  Cypermethrin   *109*
Dialkyldimethyl Ammonium Chloride and
  Methylene Bis (Thiocyanate) and Benzalkonium
  Chloride and Boric Acid   *110*
Diazinon   *356*
Diazinon and Dichlorvos   *357*
Diazinon and Pyrethrins   *358*
Dichlofluanid   *111*
Dichlofluanid   *662*
Dichlofluanid and Acypetacs Zinc   *112*
Dichlofluanid and Azaconazole   *113*
Dichlofluanid and Cuprous Oxide   *663*
Dichlofluanid and Cypermethrin   *114*
Dichlofluanid and Furmecyclox   *115*
Dichlofluanid and Furmecyclox and Gamma-
  HCH   *116*
Dichlofluanid and Furmecyclox and
  Permethrin   *117*
Dichlofluanid and Furmecyclox and Permethrin
  and Tributyltin Oxide   *118*
Dichlofluanid and Permethrin   *119*
Dichlofluanid and Permethrin and
  Tebuconazole   *120*
Dichlofluanid and Permethrin and Tri(Hexylene
  Glycol) Biborate   *121*
Dichlofluanid and Permethrin and Zinc
  Octoate   *122*
Dichlofluanid and Tri(Hexylene Glycol)
  Biborate   *123*
Dichlofluanid and Tri(Hexylene Glycol) Biborate
  and Zinc Naphthenate   *124*
Dichlofluanid and Tri(Hexylene Glycol) Biborate
  and Zinc Octoate   *125*
Dichlofluanid and Tributyltin Oxide   *126*
Dichlofluanid and Zinc Naphthenate   *127*

Dichlofluanid and Zinc Naphthenate and Zinc
  Oxide and Copper Naphthenate and Cuprous
  Oxide   *664*
Dichlofluanid and Zinc Octoate   *128*
Dichlofluanid and Zinc Oxide and 2-Methylthio-4-
  Tertiary-Butylamino-6-Cyclopropylamino-
  S-Triazine and Cuprous Thiocyanate   *665*
Dichlofluanid and Zinc Oxide and Cuprous
  Oxide   *666*
Dichlofluanid and Zinc Oxide and Cuprous
  Thiocyanate   *667*
Dichlorophen   *292*
Dichlorophen and Dichlorvos and
  Tetramethrin   *359*
Dichlorophenyl Dimethylurea and Copper
  Resinate and Cuprous Oxide   *668*
Dichlorophenyl Dimethylurea and Cuprous
  Oxide   *669*
Dichlorophenyl Dimethylurea and Cuprous
  Thiocyanate   *670*
Dichlorophenyl Dimethylurea and Tributyltin
  Fluoride and Tributyltin Methacrylate and Zinc
  Oxide   *671*
Dichlorophenyl Dimethylurea and Tributyltin
  Fluoride and Tributyltin Oxide and Cuprous
  Oxide   *672*
Dichlorophenyl Dimethylurea and Tributyltin
  Methacrylate and Tributyltin Oxide and Cuprous
  Oxide   *673*
Dichlorophenyl Dimethylurea and Tributyltin
  Methacrylate and Tributyltin Oxide and Zinc
  Oxide and Cuprous Oxide   *674*
Dichlorophenyl Dimethylurea and Triphenyltin
  Chloride   *675*
Dichlorophenyl Dimethylurea and Triphenyltin
  Chloride and Cuprous Thiocyanate   *676*
Dichlorophenyl Dimethylurea and Triphenyltin
  Chloride and Zinc Oxide and Cuprous
  Thiocyanate   *677*
Dichlorophenyl Dimethylurea and Zinc
  Oxide   *678*
Dichlorophenyl Dimethylurea and Zinc Oxide and
  2,4,5,6-Tetrachloro Isophthalonitrile and
  Cuprous Oxide   *679*
Dichlorophenyl Dimethylurea and Zinc Oxide and
  2-Methylthio-4-Tertiary-Butylamino-6-
  Cyclopropylamino-S-Triazine and Cuprous
  Oxide   *680*
Dichlorophenyl Dimethylurea and Zinc Oxide and
  2-Methylthio-4-Tertiary-Butylamino-6-
  Cyclopropylamino-S-Triazine and Cuprous
  Thiocyanate   *681*
Dichlorophenyl Dimethylurea and Zinc Oxide and
  Copper Resinate and Cuprous Oxide   *682*

Dichlorophenyl Dimethylurea and Zinc Oxide and Cuprous Oxide 683
Dichlorophenyl Dimethylurea and Zinc Oxide and Cuprous Thiocyanate 684
Dichlorophenyl Dimethylurea and Zinc Oxide and Zinc Pyrithione and 2-(Thiocyanomethylthio) Benzothiazole and 2-Methylthio-4-Tertiary-Butylamino-6-Cyclopropylamino-S-Triazine 685
Dichlorophenyl Dimethylurea and Zinc Oxide and Zinc Pyrithione and 2-Methylthio-4-Tertiary-Butylamino-6-Cyclopropylamino-S-Triazine 686
Dichlorophenyl Dimethylurea and Zinc Pyrithione and 2-Methylthio-4-Tertiary-Butylamino-6-Cyclopropylamino-S-Triazine 687
Dichlorvos 360
Dichlorvos and 1,4-Dichlorobenzene 361
Dichlorvos and Diazinon 362
Dichlorvos and Iodofenphos 363
Dichlorvos and Permethrin and S-Bioallethrin 364
Dichlorvos and Permethrin and Tetramethrin 365
Dichlorvos and Permethrin and Tetramethrin and d-Allethrin 366
Dichlorvos and Tetramethrin 367
Dichlorvos and Tetramethrin and Dichlorophen 368
Dimethoate and Tetramethrin 369
Disodium Octaborate 129
Disodium Octaborate and Alkylaryltrimethyl Ammonium Chloride 130
Disodium Octaborate and Alkylaryltrimethyl Ammonium Chloride 293
Disodium Octaborate and Benzalkonium Chloride 131
Disodium Octaborate and Benzalkonium Chloride 294
Disodium Octaborate and Chromium Trioxide and Copper Oxide 132
Disodium Octaborate and Copper Naphthenate 133
Disodium Octaborate and Sodium Pentachlorophenoxide 134
Dodecylamine Lactate and Dodecylamine Salicylate 135
Dodecylamine Lactate and Dodecylamine Salicylate 295
Dodecylamine Lactate and Dodecylamine Salicylate and Permethrin 136
Dodecylamine Lactate and Dodecylamine Salicylate and Tributyltin Oxide 296
Dodecylamine Lactate and Dodecylamine Salicylate and Zinc Octoate 297
Dodecylamine Salicylate and Dodecylamine Lactate 137
Dodecylamine Salicylate and Dodecylamine Lactate 298
Dodecylamine Salicylate and Permethrin and Dodecylamine Lactate 138
Dodecylamine Salicylate and Tributyltin Oxide and Dodecylamine Lactate 299
Dodecylamine Salicylate and Zinc Octoate and Dodecylamine Lactate 300
Encapsulated Permethrin and Pyrethrin 370
Fenitrothion 371
Fenitrothion and Permethrin and Resmethrin 372
Fenitrothion and Permethrin and Tetramethrin 373
Fenitrothion and Tetramethrin 374
Folpet and Cuprous Oxide 688
Furmecyclox and Carbendazim 139
Furmecyclox and Dichlofluanid 140
Furmecyclox and Gamma-HCH and Dichlofluanid 141
Furmecyclox and Permethrin 142
Furmecyclox and Permethrin and Dichlofluanid 143
Furmecylclox and Permethrin and Tributyltin Oxide and Dichlofluanid 144
Gamma-HCH 145
Gamma-HCH 375
Gamma-HCH and 3-Iodo-2-Propynyl-N-Butyl Carbamate 146
Gamma-HCH and Benzalkonium Chloride 147
Gamma-HCH and Dichlofluanid and Furmecyclox 148
Gamma-HCH and Pentachlorophenol 149
Gamma-HCH and Pentachlorophenol and Tributyltin Oxide 150
Gamma-HCH and Pentachlorophenol and Zinc Naphthenate 151
Gamma-HCH and Pentachlorophenyl Laurate 152
Gamma-HCH and Pentachlorophenyl Laurate and Tributyltin Oxide 153
Gamma-HCH and Tetramethrin 376
Gamma-HCH and Tri(Hexylene Glycol) Biborate 154
Gamma-HCH and Tri(Hexylene Glycol) Biborate and Tributyltin Oxide 155
Gamma-HCH and Tributyltin Oxide 156
Gamma-HCH and Tributyltin Oxide and Copper Naphthenate 157
Hydramethylnon 377
Hydroprene 378
Hydroprene and Pyrethrins 379
Iodofenphos 380

HSE ACTIVE INGREDIENT INDEX

Iodofenphos and Dichlorvos 381
Lambda-Cyhalothrin 382
Maneb and Tributyltin Meso-Dibromosuccinate and Tributyltin Methacrylate and Tributyltin Oxide and Zinc Oxide and Cuprous Thiocyanate 689
Maneb and Tributyltin Methacrylate and Zinc Oxide and Cuprous Oxide 690
Methoprene 383
Methoprene and Cypermethrin 384
Methoprene and Permethrin 385
Methoprene and Pyrethrins 386
Methylene Bis (Thiocyanate) 158
Methylene Bis (Thiocyanate) 691
Methylene Bis (Thiocyanate) and 2-(Thiocyanomethylthio) Benzothiazole 159
Methylene Bis (Thiocyanate) and 2-(Thiocyanomethylthio) Benzothiazole and Boric Acid 160
Methylene Bis (Thiocyanate) and Benzalkonium Chloride and Boric Acid and Dialkyldimethyl Ammonium Chloride 161
Methylene Bis (Thiocyanate) and Boric Acid 162
Naphthalene 387
Naphthalene and 1,4-Dichlorobenzene 388
Naphthalene and Permethrin and Pyrethrins and Benzalkonium Chloride 301
Naphthalene and Permethrin and Pyrethrins and Benzalkonium Chloride 389
Natamycin 390
No Recognised Active Ingredient 692
Oleyl Monoethoxylate and Ceto-Stearyl Diethoxylate 391
Oxine-Copper 163
Oxytetracycline Hydrochloride and Cuprous Oxide 693
Pentachlorophenol 164
Pentachlorophenol 302
Pentachlorophenol and Cypermethrin 165
Pentachlorophenol and Gamma-HCH 166
Pentachlorophenol and Tributyltin Oxide 167
Pentachlorophenol and Tributyltin Oxide and Gamma-HCH 168
Pentachlorophenol and Zinc Naphthenate and Gamma-HCH 169
Pentachlorophenyl Laurate 170
Pentachlorophenyl Laurate and Gamma-HCH 171
Pentachlorophenyl Laurate and Tributyltin Oxide 172
Pentachlorophenyl Laurate and Tributyltin Oxide and Gamma-HCH 173
Permethrin 174
Permethrin 392
Permethrin and 2-Phenylphenol 175

Permethrin and 3-Iodo-2-Propynyl-N-Butyl Carbamate 176
Permethrin and Acypetacs Copper and Acypetacs Zinc 177
Permethrin and Acypetacs Zinc 178
Permethrin and Azaconazole 179
Permethrin and Azaconazole 303
Permethrin and Benzalkonium Chloride 180
Permethrin and Bioallethrin 393
Permethrin and Boric Acid 181
Permethrin and Copper Naphthenate 182
Permethrin and Dichlofluanid 183
Permethrin and Dichlofluanid and Furmecyclox 184
Permethrin and Dodecylamine Lactate and Dodecylamine Salicylate 185
Permethrin and Encapsulated Permethrin 394
Permethrin and Furmecyclox 186
Permethrin and Methoprene 395
Permethrin and Propiconazole 187
Permethrin and Pyrethrins 396
Permethrin and Pyrethrins and Alkylaryltrimethyl Ammonium Chloride 304
Permethrin and Pyrethrins and Alkylaryltrimethyl Ammonium Chloride 397
Permethrin and Pyrethrins and Benzalkonium Chloride and Naphthalene 305
Permethrin and Pyrethrins and Benzalkonium Chloride and Naphthalene 398
Permethrin and Pyrethrins and Bioallethrin 399
Permethrin and Pyrethrins and Chlorpyrifos-Methyl 400
Permethrin and Resmethrin and Fenitrothion 401
Permethrin and S-Bioallethrin 402
Permethrin and S-Bioallethrin and Dichlorvos 403
Permethrin and S-Bioallethrin and Tetramethrin 404
Permethrin and S-Hydroprene 405
Permethrin and S-Methoprene 406
Permethrin and Tebuconazole 188
Permethrin and Tebuconazole and Dichlofluanid 189
Permethrin and Tetramethrin 407
Permethrin and Tetramethrin and Dichlorvos 408
Permethrin and Tetramethrin and Fenitrothion 409
Permethrin and Tetramethrin and d-Allethrin 410
Permethrin and Tetramethrin and d-Allethrin and Dichlorvos 411
Permethrin and Tri(Hexylene Glycol) Biborate 190
Permethrin and Tri(Hexylene Glycol) Biborate and Dichlofluanid 191

HSE ACTIVE INGREDIENT INDEX

Permethrin and Tributyltin Naphthenate *192*
Permethrin and Tributyltin Oxide and Dichlofluanid and Furmecyclox *193*
Permethrin and Tributyltin Phosphate *194*
Permethrin and Zinc Octoate *195*
Permethrin and Zinc Octoate *306*
Permethrin and Zinc Octoate and Boric Acid *196*
Permethrin and Zinc Octoate and Dichlofluanid *197*
Permethrin and Zinc Versatate *198*
Permethrin and d-Allethrin *412*
Phoxim *413*
Pirimiphos-Methyl *199*
Pirimiphos-Methyl *414*
Pirimiphos-Methyl and Resmethrin and Tetramethrin *415*
Potassium 2-Phenylphenoxide *200*
Potassium Dichromate and Boric Acid and Copper Sulphate *201*
Potassium Salts of Fatty Acids *416*
Prallethrin *417*
Propiconazole *202*
Propiconazole and Permethrin *203*
Propoxur *418*
Pyrethrins *419*
Pyrethrins and Alkylaryltrimethyl Ammonium Chloride and Permethrin *307*
Pyrethrins and Alkylaryltrimethyl Ammonium Chloride and Permethrin *420*
Pyrethrins and Bendiocarb *421*
Pyrethrins and Bioallethrin and Permethrin *422*
Pyrethrins and Chlorpyrifos and Cypermethrin *423*
Pyrethrins and Chlorpyrifos-Methyl and Permethrin *424*
Pyrethrins and Cypermethrin *425*
Pyrethrins and Diazinon *426*
Pyrethrins and Hydroprene *427*
Pyrethrins and Methoprene *428*
Pyrethrins and Permethrin *429*
Pyrethrins and Permethrin and Naphthalene and Benzalkonium Chloride *308*
Pyrethrins and Permethrin and Naphthalene and Benzalkonium Chloride *430*
Pyrethrins and Resmethrin *431*
Pyrethrins and S-Methoprene *432*
Resmethrin and Fenitrothion and Permethrin *433*
Resmethrin and Pyrethrins *434*
Resmethrin and Tetramethrin *435*
Resmethrin and Tetramethrin and Pirimiphos-Methyl *436*
S-Bioallethrin and Bioallethrin *437*
S-Bioallethrin and Bioresmethrin *438*
S-Bioallethrin and Dichlorvos and Permethrin *439*
S-Bioallethrin and Permethrin *440*
S-Bioallethrin and Tetramethrin and Permethrin *441*
S-Hydroprene and Permethrin *442*
S-Methoprene *443*
S-Methoprene and Permethrin *444*
S-Methoprene and Pyrethrins *445*
Sodium 2-Phenylphenoxide *204*
Sodium 2-Phenylphenoxide *309*
Sodium Dichromate and Arsenic Pentoxide and Copper Sulphate *205*
Sodium Dichromate and Boric Acid and Chromium Acetate and Copper Sulphate *206*
Sodium Dichromate and Boric Acid and Copper Sulphate *207*
Sodium Dichromate and Chromium Acetate and Copper Sulphate *208*
Sodium Dichromate and Copper Sulphate *209*
Sodium Dichromate and Sodium Fluoride and Ammonium Bifluoride *210*
Sodium Fluoride and Ammonium Bifluoride and Sodium Dichromate *211*
Sodium Hypochlorite *310*
Sodium Hypochlorite *694*
Sodium Pentachlorophenoxide *212*
Sodium Pentachlorophenoxide *311*
Sodium Pentachlorophenoxide and Disodium Octaborate *213*
Sodium Tetraborate *214*
Sodium Tetraborate and Alkyltrimethyl Ammonium Chloride *215*
Sodim Tetraborate and Boric Acid *216*
TC Oil and Creosote *217*
Tebuconazole *218*
Tebuconazole and Boric Acid and Copper Carbonate Hydroxide *219*
Tebuconazole and Cypermethrin *220*
Tebuconazole and Dichlofluanid and Permethrin *221*
Tebuconazole and Permethrin *222*
Tetramethrin *446*
Tetramethrin and Bendiocarb *447*
Tetramethrin and Chlorpyrifos *448*
Tetramethrin and Cypermethrin *449*
Tetramethrin and Dichlorophen and Dichlorvos *450*
Tetramethrin and Dichlorvos *451*
Tetramethrin and Dichlorvos and Permethrin *452*
Tetramethrin and Dimethoate *453*
Tetramethrin and Fenitrothion *454*
Tetramethrin and Fenitrothion and Permethrin *455*
Tetramethrin and Gamma-HCH *456*

Tetramethrin and Permethrin *457*
Tetramethrin and Permethrin and
  S-Bioallethrin *458*
Tetramethrin and Pirimiphos-Methyl and
  Resmethrin *459*
Tetramethrin and Resmethrin *460*
Tetramethrin and d-Allethrin *461*
Tetramethrin and d-Allethrin and
  Cypermethrin *462*
Tetramethrin and d-Allethrin and Dichlorvos and
  Permethrin *463*
Tetramethrin and d-Allethrin and Permethrin *464*
Tetramethrin and d-Phenothrin *465*
Thiram and 2-Methylthio-4-Tertiary-Butylamino-6-
  Cyclopropylamino-S-Triazine *695*
Thiram and Cuprous Oxide *696*
Thiram and Tributyltin Methacrylate and Tributyltin
  Oxide *697*
Thiram and Tributyltin Methacrylate and Tributyltin
  Oxide and 2-Methylthio-4-Tertiary-Butylamino-6-
  Cyclopropylamino-S-Triazine *698*
Thiram and Tributyltin Oleate and Tributyltin
  Tetrachlorophthalate and Triphenyltin
  Fluoride *699*
Thiram and Tributyltin Oxide and Triphenyltin
  Fluoride *700*
Tri(Hexylene Glycol) Biborate *223*
Tri(Hexylene Glycol) Biborate and Copper
  Naphthenate *224*
Tri(Hexylene Glycol) Biborate and
  Cypermethrin *225*
Tri(Hexylene Glycol) Biborate and
  Dichlofluanid *226*
Tri(Hexylene Glycol) Biborate and Dichlofluanid
  and Permethrin *227*
Tri(Hexylene Glycol) Biborate and Gamma-
  HCH *228*
Tri(Hexylene Glycol) Biborate and
  Permethrin *229*
Tri(Hexylene Glycol) Biborate and Tributyltin
  Oxide and Gamma-HCH *230*
Tri(Hexylene Glycol) Biborate and Zinc
  Naphthenate and Dichlofluanid *231*
Tri(Hexylene Glycol) Biborate and Zinc Octoate
  and Dichlofluanid *232*
Tributyltin Acrylate *701*
Tributyltin Acrylate and Tributyltin Fluoride *702*
Tributyltin Acrylate and Tributyltin Oxide and
  Cuprous Oxide *703*
Tributyltin Acrylate and Zinc Oxide and Cuprous
  Oxide *704*
Tributyltin Fluoride *705*
Tributyltin Fluoride and Copper Resinate and
  Cuprous Oxide *706*
Tributyltin Fluoride and Cuprous Oxide *707*

Tributyltin Fluoride and Cuprous
  Thiocyanate *708*
Tributyltin Fluoride and Tributyltin Acrylate *709*
Tributyltin Fluoride and Tributyltin Methacrylate
  and Zinc Oxide and Dichlorophenyl
  Dimethylurea *710*
Tributyltin Fluoride and Tributyltin Oxide and
  Cuprous Oxide *711*
Tributyltin Fluoride and Tributyltin Oxide and
  Cuprous Oxide and Dichlorophenyl
  Dimethylurea *712*
Tributyltin Fluoride and Tributyltin Oxide and
  Cuprous Thiocyanate *713*
Tributyltin Fluoride and Tributyltin Oxide and Zinc
  Oxide and Cuprous Oxide *714*
Tributyltin Fluoride and Zinc Naphthenate and
  Cuprous Oxide *715*
Tributyltin Fluoride and Zinc Oxide and Copper
  Resinate and Cuprous Oxide *716*
Tributyltin Fluoride and Zinc Oxide and Cuprous
  Oxide *717*
Tributyltin Fluoride and Zinc Oxide and Zineb and
  Copper Resinate and Cuprous Oxide *718*
Tributyltin Fluoride and Zinc Oxide and Zineb and
  Cuprous Oxide *719*
Tributyltin Meso-Dibromosuccinate *720*
Tributyltin Meso-Dibromosuccinate and Tributyltin
  Methacrylate and Tributyltin Oxide and Zinc
  Oxide and Cuprous Thiocyanate and
  Maneb *721*
Tributyltin Meso-Dibromosuccinate and Tributyltin
  Oxide and Zinc Oxide and Cuprous
  Thiocyanate *722*
Tributyltin Methacrylate and Cuprous Oxide *723*
Tributyltin Methacrylate and Cuprous
  Thiocyanate *724*
Tributyltin Methacrylate and Tributyltin
  Oxide *725*
Tributyltin Methacrylate and Tributyltin Oxide and
  2-Methylthio-4-Tertiary-Butylamino-6-
  Cyclopropylamino-S-Triazine and Thiram *726*
Tributyltin Methacrylate and Tributyltin Oxide and
  4,5-Dichloro-2-N-Octyl-4-Isothiazolin-3-One
  and Cuprous Oxide *727*
Tributyltin Methacrylate and Tributyltin Oxide and
  Copper Metal *728*
Tributyltin Methacrylate and Tributyltin Oxide and
  Cuprous Oxide *729*
Tributyltin Methacrylate and Tributyltin Oxide and
  Cuprous Oxide and Cuprous Thiocyanate *730*
Tributyltin Methacrylate and Tributyltin Oxide and
  Cuprous Oxide and Dichlorophenyl
  Dimethylurea *731*
Tributyltin Methacrylate and Tributyltin Oxide and
  Cuprous Thiocyanate *732*

HSE ACTIVE INGREDIENT INDEX

Tributyltin Methacrylate and Tributyltin Oxide and Thiram *733*
Tributyltin Methacrylate and Tributyltin Oxide and Triphenyltin Fluoride *734*
Tributyltin Methacrylate and Tributyltin Oxide and Triphenyltin Fluoride and Cuprous Oxide *735*
Triphenyltin Methacrylate and Tributyltin Oxide and Triphenyltin Fluoride and Cuprous Thiocyanate *736*
Tributyltin Methacrylate and Tributyltin Oxide and Zinc Oxide and Cuprous Oxide *737*
Tributyltin Methacrylate and Tributyltin Oxide and Zinc Oxide and Cuprous Oxide and Dichlorophenyl Dimethylurea *738*
Tributyltin Methacrylate and Tributyltin Oxide and Zinc Oxide and Cuprous Thiocyanate *739*
Tributyltin Methacrylate and Tributyltin Oxide and Zinc Oxide and Cuprous Thiocyanate and Maneb and Tributyltin Meso-Dibromosuccinate *740*
Tributyltin Methacrylate and Tributyltin Oxide and Zinc Oxide and Zineb and Cuprous Oxide *741*
Tributyltin Methacrylate and Tributyltin Oxide and Zineb and Cuprous Oxide *742*
Tributyltin Methacrylate and Tributyltin Oxide and Zineb and Cuprous Thiocyanate *743*
Tributyltin Methacrylate and Triphenyltin Fluoride *744*
Tributyltin Methacrylate and Zinc Metal and Cuprous Oxide *745*
Tributyltin Methacrylate and Zinc Metal and Zinc Oxide and Cuprous Oxide *746*
Tributyltin Methacrylate and Zinc Metal and Zinc Oxide and Cuprous Thiocyanate *747*
Tributyltin Methacrylate and Zinc Oxide and 2-Methylthio-4-Tertiary-Butylamino-6-Cyclopropylamino-S-Triazine *748*
Tributyltin Methacrylate and Zinc Oxide and Cuprous Oxide and Maneb *749*
Tributyltin Methacrylate and Zinc Oxide and Cuprous Thiocyanate *750*
Tributyltin Methacrylate and Zinc Oxide and Dichlorophenyl Dimethylurea and Tributyltin Fluoride *751*
Tributyltin Methacrylate and Zinc Oxide and Zineb and Cuprous Oxide *752*
Tributyltin Methacrylate and Zineb and Cuprous Oxide *753*
Tributyltin Methacrylate and Zineb and Cuprous Thiocyanate *754*
Tributyltin Methacrylate and Ziram and Cuprous Thiocyanate *755*
Tributyltin Naphthenate *233*
Tributyltin Naphthenate and Permethrin *234*

Tributyltin Oleate and Tributyltin Tetrachlorophthalate and Triphenyltin Fluoride and Thiram *756*
Tributyltin Oxide *235*
Tributyltin Oxide and 2-Methylthio-4-Tertiary-Butylamino-6-Cyclopropylamino-S-Triazine and Thiram and Tributyltin Methacrylate *757*
Tributyltin Oxide and 4,5-Dichloro-2-N-Octyl-4-Isothiazolin-3-One and Cuprous Oxide and Tributyltin Methacrylate *758*
Tributyltin Oxide and Alkylarylmethyl Ammonium Chloride *236*
Tributyltin Oxide and Alkylarylmethyl Ammonium Chloride *312*
Tributyltin Oxide and Benzalkonium Chloride *237*
Tributyltin Oxide and Benzalkonium Chloride *313*
Tributyltin Oxide and Copper Metal and Tributyltin Methacrylate *759*
Tributyltin Oxide and Copper Naphthenate and Gamma-HCH *238*
Tributyltin Oxide and Cuprous Oxide *760*
Tributyltin Oxide and Cuprous Oxide and Cuprous Thiocyanate and Tributyltin Methacrylate *761*
Tributyltin Oxide and Cuprous Oxide and Dichlorophenyl Dimethylurea and Tributyltin Fluoride *762*
Tributyltin Oxide and Cuprous Oxide and Dichlorophenyl Dimethylurea and Tributyltin Methacrylate *763*
Tributyltin Oxide and Cuprous Oxide and Tributyltin Acrylate *764*
Tributyltin Oxide and Cuprous Oxide and Tributyltin Fluoride *765*
Tributyltin Oxide and Cuprous Oxide and Tributyltin Methacrylate *766*
Tributyltin Oxide and Cuprous Thiocyanate *767*
Tributyltin Oxide and Cuprous Thiocyanate and Tributyltin Fluoride *768*
Tributyltin Oxide and Cuprous Thiocyanate and Tributyltin Methacrylate *769*
Tributyltin Oxide and Dichlofluanid *239*
Tributyltin Oxide and Dichlofluanid and Furmecyclox and Permethrin *240*
Tributyltin Oxide and Dodecylamine Lactate and Dodecylamine Salicylate *314*
Tributyltin Oxide and Gamma-HCH *241*
Tributyltin Oxide and Gamma-HCH and Pentachlorophenol *242*
Tributyltin Oxide and Gamma-HCH and Pentachlorophenyl Laurate *243*
Tributyltin Oxide and Gamma-HCH and Tri(Hexylene Glycol) Biborate *244*
Tributyltin Oxide and Pentachlorophenol *245*

Tributyltin Oxide and Pentachlorophenyl Laurate 246
Tributyltin Oxide and Thiram and Tributyltin Methacrylate 770
Tributyltin Oxide and Tributyltin Methacrylate 771
Tributyltin Oxide and Triphenyltin Fluoride 772
Tributyltin Oxide and Triphenyltin Fluoride and Cuprous Oxide 773
Tributyltin Oxide and Triphenyltin Fluoride and Cuprous Oxide and Tributyltin Methacrylate 774
Tributyltin Oxide and Triphenyltin Fluoride and Cuprous Thiocyanate and Tributyltin Methacrylate 775
Tributyltin Oxide and Triphenyltin Fluoride and Thiram 776
Tributyltin Oxide and Triphenyltin Fluoride and Tributyltin Methacrylate 777
Tributyltin Oxide and Zinc Naphthenate and Carbendazim 247
Tributyltin Oxide and Zinc Oxide and Cuprous Oxide 778
Tributyltin Oxide and Zinc Oxide and Cuprous Oxide and Dichlorophenyl Dimethylurea and Tributyltin Methacrylate 779
Tributyltin Oxide and Zinc Oxide and Cuprous Oxide and Tributyltin Fluoride 780
Tributyltin Oxide and Zinc Oxide and Cuprous Oxide and Tributyltin Methacrylate 781
Tributyltin Oxide and Zinc Oxide and Cuprous Thiocyanate and Maneb and Tributyltin Meso-Dibromosuccinate and Tributyltin Methacrylate 782
Tributyltin Oxide and Zinc Oxide and Cuprous Thiocyanate and Tributyltin Meso-Dibromosuccinate 783
Tributyltin Oxide and Zinc Oxide and Cuprous Thiocyanate and Tributyltin Methacrylate 784
Tributyltin Oxide and Zinc Oxide and Zineb and Cuprous Oxide and Tributyltin Methacrylate 785
Tributyltin Oxide and Zineb and Cuprous Oxide and Tributyltin Methacrylate 786
Tributyltin Oxide and Zineb and Cuprous Thiocyanate 787
Tributyltin Oxide and Zineb and Cuprous Thiocyanate and Tributyltin Methacrylate 788
Tributyltin Oxide and Ziram and Cuprous Oxide 789
Tributyltin Phosphate 248
Tributyltin Phosphate and Permethrin 249
Tributyltin Polysiloxane and Zinc Metal 790
Tributyltin Polysiloxane and Zinc Oxide 791
Tributyltin Polysiloxane and Zinc Oxide and Cuprous Oxide 792
Tributyltin Tetrachlorophthalate and Triphenyltin Fluoride and Thiram and Tributyltin Oleate 793
Trichlorphon 466
Triphenyltin Chloride and Cuprous Thiocyanate and Dichlorophenyl Dimethylurea 794
Triphenyltin Chloride and Dichlorophenyl Dimethylurea 795
Triphenyltin Chloride and Zinc Oxide and Cuprous Thiocyanate and Dichlorophenyl Dimethylurea 796
Triphenyltin Fluoride and Cuprous Oxide 797
Triphenyltin Fluoride and Cuprous Oxide and Tributyltin Methacrylate and Tributyltin Oxide 798
Triphenyltin Fluoride and Cuprous Oxide and Tributyltin Oxide 799
Triphenyltin Fluoride and Cuprous Thiocyanate and Tributyltin Methacrylate and Tributyltin Oxide 800
Triphenyltin Fluoride and Thiram and Tributyltin Oleate and Tributyltin Tetrachlorophthalate 801
Triphenyltin Fluoride and Thiram and Tributyltin Oxide 802
Triphenyltin Fluoride and Tributyltin Methacrylate 803
Triphenyltin Fluoride and Tributyltin Methacrylate and Tributyltin Oxide 804
Triphenyltin Fluoride and Tributyltin Oxide 805
Triphenyltin Fluoride and Zinc Oxide and Cuprous Oxide 806
Triphenyltin Hydroxide and Cuprous Oxide 807
Triphenyltin Hydroxide and Zinc Oxide and Cuprous Oxide 808
Zinc Metal and Cuprous Oxide and Tributyltin Methacrylate 809
Zinc Metal and Tributyltin Polysiloxane 810
Zinc Metal and Zinc Oxide and Cuprous Oxide and Tributyltin Methacrylate 811
Zinc Metal and Zinc Oxide and Cuprous Thiocyanate and Tributyltin Methacrylate 812
Zinc Naphthenate 250
Zinc Naphthenate and 2,4,5,6-Tetrachloro Isophthalonitrile and Cuprous Oxide 813
Zinc Naphthenate and Carbendazim and Tributyltin Oxide 251
Zinc Naphthenate and Cuprous Oxide 814
Zinc Naphthenate and Cuprous Oxide and Tributyltin Fluoride 815
Zinc Naphthenate and Dichlofluanid 252
Zinc Naphthenate and Dichlofluanid and Tri(Hexylene Glycol) Biborate 253
Zinc Naphthenate and Gamma-HCH and Pentachlorophenol 254

HSE ACTIVE INGREDIENT INDEX

Zinc Naphthenate and Zinc Oxide and Copper Naphthenate and Cuprous Oxide and Dichlofluanid *816*
Zinc Naphthenate and Zinc Oxide and Cuprous Oxide *817*
Zinc Octoate *255*
Zinc Octoate *315*
Zinc Octoate and Boric Acid *256*
Zinc Octoate and Boric Acid and Permethrin *257*
Zinc Octoate and Copper Naphthenate *258*
Zinc Octoate and Cypermethrin *259*
Zinc Octoate and Dichlofluanid *260*
Zinc Octoate and Dichlofluanid and Permethrin *261*
Zinc Octoate and Dichlofluanid and Tri(Hexylene Glycol) Biborate *262*
Zinc Octoate and Dodecylamine Lactate and Dodecylamine Salicylate *316*
Zinc Octoate and Permethrin *263*
Zinc Octoate and Permethrin *317*
Zinc Oxide and 2,3,5,6-Tetrachloro-4-(Methyl Sulphonyl) Pyridine and 2-Methylthio-4-Tertiary-Butylamino-6-Cyclopropylamino-S-Triazine and Cuprous Oxide and Cuprous Thiocyanate *818*
Zinc Oxide and 2,4,5,6-Tetrachloro Isophthalonitrile and Cuprous Oxide *819*
Zinc Oxide and 2,4,5,6-Tetrachloro Isophthalonitrile and Cuprous Thiocyanate *820*
Zinc Oxide and 2-Methylthio-4-Tertiary-Butylamino-6-Cyclopropylamino-S-Triazine and Cuprous Oxide *821*
Zinc Oxide and 2-Methylthio-4-Tertiary *822*
Zinc Oxide and 2-Methylthio-4-Tertiary-Butylamino-6-Cyclopropylamino-S-Triazine and Cuprous Oxide *823*
Zinc Oxide and 2-Methylthio-4-Tertiary-Butylamino-6-Cyclopropylamino-S-Triazine and Cuprous Thiocyanate *824*
Zinc Oxide and 2-Methylthio-4-Tertiary-Butylamino-6-Cyclopropylamino-S-Triazine and Tributyltin Methacrylate *825*
Zinc Oxide and 2-Methylthio-4-Tertiary-Butylamino-6-Cyclopropylamino-S-Triazine and Cuprous Oxide *826*
Zinc Oxide and 2-Methylthio-4-Tertiary-Butylamino-6-Cyclopropylamino-S-Triazine and Cuprous Thiocyanate *827*
Zinc Oxide and 2-Methylthio-4-Tertiary-Butylamino-6-Cyclopropylamino-S-Triazine and Cuprous Oxide *828*
Zinc Oxide and 2-Methylthio-4-Tertiary-Butylamino-6-Cyclopropylamino-S-Triazine and Cuprous Thiocyanate *829*
Zinc Oxide and 4,5-Dichloro-2-N-Octyl -4-Isothiazolin-3-One and Cuprous Oxide *830*
Zinc Oxide and Copper Metal *831*
Zinc Oxide and Copper Naphthenate and Cuprous Oxide and Dichlofluanid and Zinc Naphthenate *832*
Zinc Oxide and Copper Resinate and Cuprous Oxide and Tributyltin Fluoride *833*
Zinc Oxide and Cuprous Oxide *834*
Zinc Oxide and Cuprous Oxide and 4,5-Dichloro-2-N-Octyl-4-Isothiazolin-3-One and 2,4,5,6-Tetrachloro Isophthalonitrile *835*
Zinc Oxide and Cuprous Oxide and Copper Resinate and 2-Methylthio-4-Tertiary-Butylamino-6-Cyclopropylamino-S-Triazine *836*
Zinc Oxide and Cuprous Oxide and Dichlofluanid *837*
Zinc Oxide and Cuprous Oxide and Dichlorophenyl Dimethylurea *838*
Zinc Oxide and Cuprous Oxide and Dichlorophenyl Dimethylurea and Tributyltin Methacrylate and Tributyltin Oxide *839*
Zinc Oxide and Cuprous Oxide and Maneb and Tributyltin Methacrylate *840*
Zinc Oxide and Cuprous Oxide and Tributyltin Acrylate *841*
Zinc Oxide and Cuprous Oxide and Tributyltin Fluoride *842*
Zinc Oxide and Cuprous Oxide and Tributyltin Fluoride and Tributyltin Oxide *843*
Zinc Oxide and Cuprous Oxide and Tributyltin Methacrylate and Tributyltin Oxide *844*
Zinc Oxide and Cuprous Oxide and Tributyltin Methacrylate and Zinc Metal *845*
Zinc Oxide and Cuprous Oxide and Tributyltin Oxide *846*
Zinc Oxide and Cuprous Oxide and Tributyltin Polysiloxane *847*
Zinc Oxide and Cuprous Oxide and Triphenyltin Fluoride *848*
Zinc Oxide and Cuprous Oxide and Triphenyltin Hydroxide *849*
Zinc Oxide and Cuprous Oxide and Zinc Naphthenate *850*
Zinc Oxide and Cuprous Thiocyanate and Dichlofluanid *851*
Zinc Oxide and Cuprous Thiocyanate and Dichlorophenyl Dimethylurea *852*
Zinc Oxide and Cuprous Thiocyanate and Dichlorophenyl Dimethylurea and Triphenyltin Chloride *853*

HSE ACTIVE INGREDIENT INDEX

Zinc Oxide and Cuprous Thiocyanate and Maneb and Tributyltin Meso-Dibromosuccinate and Tributyltin Methacrylate and Tributyltin Oxide *854*

Zinc Oxide and Cuprous Thiocyanate and Tributyltin Meso-Dibromosuccinate and Tributyltin Oxide *855*

Zinc Oxide and Cuprous Thiocyanate and Tributyltin Methacrylate *856*

Zinc Oxide and Cuprous Thiocyanate and Tributyltin Methacrylate and Tributyltin Oxide *857*

Zinc Oxide and Cuprous Thiocyanate and Tributyltin Methacrylate and Zinc Metal *858*

Zinc Oxide and Dichlofluanid and Cuprous Thiocyanate and 2-Methylthio-4-Tertiary-Butylamino-6-Cyclopropylamino-S-Triazine *859*

Zinc Oxide and Dichlorophenyl Dimethylurea *860*

Zinc Oxide and Dichlorophenyl Dimethylurea and Cuprous Oxide and 2,4,5,6-Tetrachloro Isophthalonitrile *861*

Zinc Oxide and Dichlorophenyl Dimethylurea and Cuprous Oxide and 2-Methylthio-4-Tertiary-Butylamino-6-Cyclopropylamino-S-Triazine *862*

Zinc Oxide and Dichlorophenyl Dimethylurea and Cuprous Oxide and Copper Resinate *863*

Zinc Oxide and Dichlorophenyl Dimethylurea and Cuprous Thiocyanate and 2-Methylthio-4-Tertiary-Butylamino-6-Cyclopropylamino-S-Triazine *864*

Zinc Oxide and Dichlorophenyl Dimethylurea and Tributyltin Fluoride and Tributyltin Methacrylate *865*

Zinc Oxide and Tributyltin Polysiloxane *866*

Zinc Oxide and Zinc Pyrithione and 2-(Thiocyanomethylthio) Benzothiazole and 2-Methylthio-4-Tetiary-Butylamino-6-Cyclopropylamino-S-Triazine and Dichlorophenyl Dimethylurea *867*

Zinc Oxide and Zinc Pyrithione and 2-Methylthio-4-Tertiary-Butylamino-6-Cyclopropylamino-S-Triazine and Dichlorophenyl Dimethylurea *868*

Zinc Oxide and Zineb and Copper Resinate and Cuprous Oxide *869*

Zinc Oxide and Zineb and Copper Resinate and Cuprous Oxide and Tributyltin Fluoride *870*

Zinc Oxide and Zineb and Cuprous Oxide and Tributyltin Fluoride *871*

Zinc Oxide and Zineb and Cuprous Oxide and Tributyltin Methacrylate *872*

Zinc Oxide and Zineb and Cuprous Oxide and Tributyltin Methacrylate and Tributyltin Oxide *873*

Zinc Oxide and Ziram and Cuprous Oxide *874*

Zinc Pyrithione and 2-(Thiocyanomethylthio) Benzothiazole and 2-Methylthio-4-Tertiary-Butylamino-6-Cyclopropylamino-S-Triazine and Dichlorophenyl Dimethylurea and Zinc Oxide *875*

Zinc Pyrithione and 2-Methylthio-4-Tertiary-Butylamino-6-Cyclopropylamino-S-Triazine and Dichlorophenyl Dimethylurea *876*

Zinc Pyrithione and Cuprous Oxide *877*

Zinc Pyrithione and Zinc Oxide and Dichlorophenyl Dimethylurea and 2-Methylthio-4-Tertiary-Butylamino-6-Cyclopropylamino-S-Triazine *878*

Zinc Salt Of Synthetic (C8-C12) Carboxylic Acids *264*

Zinc Versatate *265*

Zinc Versatate and Cypermethrin *266*

Zinc Versatate and Permethrin *267*

Zineb and 2-Methylthio-4-Tertiary-Butylamino-6-Cyclopropylamino-S-Triazine and Cuprous Thiocyanate *879*

Zineb and Copper Resinate and Cuprous Oxide and Tributyltin Fluoride and Zinc Oxide *880*

Zineb and Cuprous Oxide *881*

Zineb and Cuprous Oxide and Tributyltin Fluoride and Zinc Oxide *882*

Zineb and Cuprous Oxide and Tributyltin Methacrylate *883*

Zineb and Cuprous Oxide and Tributyltin Methacrylate and Tributyltin Oxide *884*

Zineb and Cuprous Oxide and Tributyltin Methacrylate and Tributyltin Oxide and Zinc Oxide *885*

Zineb and Cuprous Oxide and Tributyltin Methacrylate and Zinc Oxide *886*

Zineb and Cuprous Thiocyanate *887*

Zineb and Cuprous Thiocyanate and Tributyltin Methacrylate *888*

Zineb and Cuprous Thiocyanate and Tributyltin Methacrylate and Tributyltin Oxide *889*

Zineb and Cuprous Thiocyanate and Tributyltin Oxide *890*

Zineb and Zinc Oxide and Cuprous Oxide and Copper Resinate *891*

Ziram and Cuprous Oxide *892*

Ziram and Cuprous Oxide and Tributyltin Oxide *893*

Ziram and Cuprous Oxide and Zinc Oxide *894*

Ziram and Cuprous Thiocyanate *895*

Ziram and Cuprous Thiocyanate and Tributyltin Methacrylate *896*

d-Allethrin *467*
d-Allethrin and Cypermethrin *468*
d-Allethrin and Cypermethrin and Tetramethrin *469*
d-Allethrin and Dichlorvos and Permethrin and Tetramethrin *470*
d-Allethrin and Permethrin *471*
d-Allethrin and Permethrin and Tetramethrin *472*
d-Allethrin and Tetramethrin *473*
d-Allethrin and d-Phenothrin *474*
d-Phenothrin *475*
d-Phenothrin and Tetramethrin *476*
d-Phenothrin and d-Allethrin *477*
d-Phenothrin and d-Tetramethrin *478*
d-Tetramethrin and d-Phenothrin *479*

Printed in the United Kingdom for HMSO
Dd300145 C35 2/95 G559 10170

THE PESTICIDES REGISTER

For all those involved with pesticides, whether as manufacturers, suppliers, users or advisers, **The Pesticides Register** is the only official monthly listing of UK approvals and other official announcements on pesticides and covers new developments in pesticides law. It includes:

— Details of new approvals, both full and provisional, making key information widely available for the first time; reproduces key points in full, e.g. statutory conditions of sale, supply, storage, use and advertisement;

— A separate section for off-label approvals. In particular this will benefit growers who are required by law to read the approvals before using the products.

— Official notices listing revocations, suspensions and amendments to the conditions of approval.

— Announcements on UK pesticides policy.

— News of routine reviews of older pesticides.

— Details on where to obtain information on pesticides.

Approval details of the following:

Agricultural, horticultural, forestry and home garden products;

Wood preservatives, public hygiene and household insecticides;

Surface biocides and antifouling products.

Commodity substances and adjuvants

Published by **HMSO** for the **Ministry Of Agriculture, Fisheries And Food** and the **Health And Safety Executive**. Please order from HMSO bookshops listed on the back cover, or by using the form overleaf.

1995 Rates:
Annual Subscription £58 (Twelve Issues)
Single Copy £4.95

Order Form

Post and packing free

☐ Please set up an annual subscription to The Pesticides Register @ £58
OR
☐ Please send me a single copy of The Pesticides Register (latest issue) @ £4.95

☐ I enclose a cheque for £ _____ made payable to HMSO

☐ Please debit my

Access/Visa/Amex/Connect Account ☐☐☐☐☐☐☐☐☐☐☐☐☐☐☐☐

Signature _____ Expiry date _____

☐ Please charge to my HMSO Account No. _____

PLEASE COMPLETE IN BLOCK CAPITALS

Name _____

Address _____

_____

_____ Postcode _____

When completed, send to:

    HMSO Books
    PO Box 276
    London
    SW8 5DT

9